What SARS-CoV-2 Variants Have Taught Us: Evolutionary Challenges of RNA Viruses

What SARS-CoV-2 Variants Have Taught Us: Evolutionary Challenges of RNA Viruses

Editors

Ahmed El-Shamy
Mohamed M. Ibrahim

MDPI

Basel • Beijing • Wuhan • Barcelona • Belgrade • Novi Sad • Cluj • Manchester

Editors
Ahmed El-Shamy
California Northstate University
Elk Grove, CA, USA

Mohamed M. Ibrahim
Misr University for Science & Technology
6th of October, Egypt

Editorial Office
MDPI
St. Alban-Anlage 66
4052 Basel, Switzerland

This is a reprint of articles from the Special Issue published online in the open access journal *Viruses* (ISSN 1999-4915) (available at: https://www.mdpi.com/journal/viruses/special_issues/sars_variants).

For citation purposes, cite each article independently as indicated on the article page online and as indicated below:

Lastname, A.A.; Lastname, B.B. Article Title. *Journal Name* **Year**, *Volume Number*, Page Range.

ISBN 978-3-0365-9831-4 (Hbk)
ISBN 978-3-0365-9832-1 (PDF)
doi.org/10.3390/books978-3-0365-9832-1

Cover image courtesy of Ahmed El-Shamy

Contents

About the Editors ix

Shymaa E. Bilasy, Tutik Sri Wahyuni, Mohamed Ibrahim and Ahmed El-Shamy
What SARS-CoV-2 Variants Have Taught Us: Evolutionary Challenges of RNA Viruses
Reprinted from: *Viruses* **2024**, *16*, 139, doi:10.3390/v16010139 1

Andrés Carrazco-Montalvo, Andrés Herrera-Yela, Damaris Alarcón-Vallejo, Diana Gutiérrez-Pallo, Isaac Armendáriz-Castillo, Derly Andrade-Molina, et al.
Omicron Sub-Lineages (BA.1.1.529 + BA.*) Current Status in Ecuador
Reprinted from: *Viruses* **2022**, *14*, 1177, doi:10.3390/v14061177 5

Hancong Sun, Jinghan Xu, Guanying Zhang, Jin Han, Meng Hao, Zhengshan Chen, et al.
Developing Pseudovirus-Based Neutralization Assay against Omicron-Included SARS-CoV-2 Variants
Reprinted from: *Viruses* **2022**, *14*, 1332, doi:10.3390/v14061332 13

Emily J. Aston, Michael G. Wallach, Aarthi Narayanan, Sofia Egaña-Labrin and Rodrigo A. Gallardo
Hyperimmunized Chickens Produce Neutralizing Antibodies against SARS-CoV-2
Reprinted from: *Viruses* **2022**, *14*, 1510, doi:10.3390/v14071510 23

Katja Verena Goller, Juliane Moritz, Janine Ziemann, Christian Kohler, Karsten Becker, Nils-Olaf Hübner and the CoMV-Gen Study Group
Differences in Clinical Presentations of Omicron Infections with the Lineages BA.2 and BA.5 in Mecklenburg-Western Pomerania, Germany, between April and July 2022
Reprinted from: *Viruses* **2022**, *14*, 2033, doi:10.3390/v14092033 33

Josilene R. Pinheiro, Esther C. dos Reis, Jéssica P. Farias, Mayanna M. C. Fogaça, Patrícia de S. da Silva, Itana Vivian R. Santana, et al.
Impact of Early Pandemic SARS-CoV-2 Lineages Replacement with the Variant of Concern P.1 (Gamma) in Western Bahia, Brazil
Reprinted from: *Viruses* **2022**, *14*, 2314, doi:10.3390/v14102314 43

Emilia Caputo and Luigi Mandrich
SARS-CoV-2: Searching for the Missing Variants
Reprinted from: *Viruses* **2022**, *14*, 2364, doi:10.3390/v14112364 55

Daniele Focosi, Massimo Franchini and Arturo Casadevall
On the Need to Determine the Contribution of Anti-Nucleocapsid Antibodies as Potential Contributors to COVID-19 Convalescent Plasma Efficacy
Reprinted from: *Viruses* **2022**, *14*, 2378, doi:10.3390/v14112378 63

Akira Aoki, Hirokazu Adachi, Yoko Mori, Miyabi Ito, Katsuhiko Sato, Masayoshi Kinoshita, et al.
Rapid Identification of SARS-CoV-2 Omicron BA.5 Spike Mutation F486V in Clinical Specimens Using a High-Resolution Melting-Based Assay
Reprinted from: *Viruses* **2022**, *14*, 2401, doi:10.3390/v14112401 69

Dhruv Chauhan, Nikhil Chakravarty, Arjit Vijey Jeyachandran, Akshaya Jayakarunakaran, Sanjeev Sinha, Rakesh Mishra, et al.
In Silico Genome Analysis Reveals the Evolution and Potential Impact of SARS-CoV-2 Omicron Structural Changes on Host Immune Evasion and Antiviral Therapeutics
Reprinted from: *Viruses* **2022**, *14*, 2461, doi:10.3390/v14112461 77

Anzhalika Sidarovich, Nadine Krüger, Cheila Rocha, Luise Graichen, Amy Kempf, Inga Nehlmeier, et al.
Host Cell Entry and Neutralization Sensitivity of SARS-CoV-2 Lineages B.1.620 and R.1
Reprinted from: *Viruses* **2022**, *14*, 2475, doi:10.3390/v14112475 . . . . **91**

Varsha Ravi, Aparna Swaminathan, Sunita Yadav, Hemant Arya and Rajesh Pandey
SARS-CoV-2 Variants of Concern and Variations within Their Genome Architecture: Does Nucleotide Distribution and Mutation Rate Alter the Functionality and Evolution of the Virus?
Reprinted from: *Viruses* **2022**, *14*, 2499, doi:10.3390/v14112499 . . . . **103**

Shuk-Ching Wong, Albert Ka-Wing Au, Janice Yee-Chi Lo, Pak-Leung Ho, Ivan Fan-Ngai Hung, Kelvin Kai-Wang To, et al.
Evolution and Control of COVID-19 Epidemic in Hong Kong
Reprinted from: *Viruses* **2022**, *14*, 2519, doi:10.3390/v14112519 . . . . **121**

Deling Shi, Changkai Bu, Peng He, Yuefan Song, Jonathan S. Dordick, Robert J. Linhardt, et al.
Structural Characteristics of Heparin Binding to SARS-CoV-2 Spike Protein RBD of Omicron Sub-Lineages BA.2.12.1, BA.4 and BA.5
Reprinted from: *Viruses* **2022**, *14*, 2696, doi:10.3390/v14122696 . . . . **151**

Louis Nevejan, Sien Ombelet, Lies Laenen, Els Keyaerts, Thomas Demuyser, Lucie Seyler, et al.
Severity of COVID-19 among Hospitalized Patients: Omicron Remains a Severe Threat for Immunocompromised Hosts
Reprinted from: *Viruses* **2022**, *14*, 2736, doi:10.3390/v14122736 . . . . **165**

Petra Šimičić and Snježana Židovec-Lepej
A Glimpse on the Evolution of RNA Viruses: Implications and Lessons from SARS-CoV-2
Reprinted from: *Viruses* **2023**, *15*, 1, doi:10.3390/v15010001 . . . . **177**

Gregory Mathez, Trestan Pillonel, Claire Bertelli and Valeria Cagno
Alpha and Omicron SARS-CoV-2 Adaptation in an Upper Respiratory Tract Model
Reprinted from: *Viruses* **2023**, *15*, 13, doi:10.3390/v15010013 . . . . **199**

Reed L. Berkowitz and David A. Ostrov
The Elusive Coreceptors for the SARS-CoV-2 Spike Protein
Reprinted from: *Viruses* **2023**, *15*, 67, doi:10.3390/v15010067 . . . . **207**

Jacob Warger and Silvana Gaudieri
On the Evolutionary Trajectory of SARS-CoV-2: Host Immunity as a Driver of Adaptation in RNA Viruses
Reprinted from: *Viruses* **2023**, *15*, 70, doi:10.3390/v15010070 . . . . **221**

Srijan Chatterjee, Manojit Bhattacharya, Sagnik Nag, Kuldeep Dhama and Chiranjib Chakraborty
A Detailed Overview of SARS-CoV-2 Omicron: Its Sub-Variants, Mutations and Pathophysiology, Clinical Characteristics, Immunological Landscape, Immune Escape, and Therapies
Reprinted from: *Viruses* **2023**, *15*, 167, doi:10.3390/v15010167 . . . . **237**

Arup Acharjee, Arka Ray, Akanksha Salkar, Surbhi Bihani, Chaitanya Tuckley, Jayanthi Shastri, et al.
Humoral Immune Response Profile of COVID-19 Reveals Severity and Variant-Specific Epitopes: Lessons from SARS-CoV-2 Peptide Microarray
Reprinted from: *Viruses* **2023**, *15*, 248, doi:10.3390/v15010248 . . . . **265**

Jacques Fantini, Fodil Azzaz, Henri Chahinian and Nouara Yahi
Electrostatic Surface Potential as a Key Parameter in Virus Transmission and Evolution: How to Manage Future Virus Pandemics in the Post-COVID-19 Era
Reprinted from: *Viruses* **2023**, *15*, 284, doi:10.3390/v15020284 . . . . **283**

Carolina Torres, Mercedes Nabaes Jodar, Dolores Acuña, Romina Micaela Zambrana Montaño, Andrés Carlos Alberto Culasso, Ariel Fernando Amadio, et al.
Omicron Waves in Argentina: Dynamics of SARS-CoV-2 Lineages BA.1, BA.2 and the Emerging BA.2.12.1 and BA.4/BA.5
Reprinted from: *Viruses* **2023**, *15*, 312, doi:10.3390/v15020312 . . . . **303**

Sanaa M. Kamal, Moheyeldeen Mohamed Naghib, Moataz Daadour, Mansour N. Alsuliman, Ziad G. Alanazi, Abdulaziz Abdullah Basalem, et al.
The Outcome of BNT162b2, ChAdOx1-Sand mRNA-1273 Vaccines and Two Boosters: A Prospective Longitudinal Real-World Study
Reprinted from: *Viruses* **2023**, *15*, 326, doi:10.3390/v15020326 . . . . **317**

Yamina L. Carattini, Anthony Griswold, Sion Williams, Ranjini Valiathan, Yi Zhou, Bhavarth Shukla, et al.
Combined Use of RT-qPCR and NGS for Identification and Surveillance of SARS-CoV-2 Variants of Concern in Residual Clinical Laboratory Samples in Miami-Dade County, Florida
Reprinted from: *Viruses* **2023**, *15*, 593, doi:10.3390/v15030593 . . . . **337**

About the Editors

Ahmed El-Shamy

I am assistant Dean of Research, the Founding Director of Master of Pharmaceutical Sciences (MPS) Program and associate professor of Virology at College of Graduate Studies, California Northstate University. Under my directorship, three MPS classes (a total of 68 students) were graduated with 100% graduation rate, over 60% and 40% of the MPS graduates are currently MD candidates and pharmaceutical industry employees, respectively. In 2021, I completed the Program for Leading Innovations in Health Care and Education in Harvard Medical School. As an instructor and educator, I strove to create an environment that was conducive to learning, I developed and coordinated multiple didactic courses in multidisciplinary areas of biomedical sciences, including Biostatistics, Translational Medicine, Emerging Viruses, antivirals and Graduate Seminar. My research focuses on the interaction between viruses and host cells at the molecular and cellular level; with special focus on viral hepatitis and RNA viruses such as SARS-CoV-2. In 2009, I received a PhD in Molecular Virology from Kobe University, Japan and I received my post-doctoral training at the Division of Liver Diseases at Icahn School of Medicine at Mount Sinai, New York City. I published over a 40 peer reviewed publications with 13 publications as the first or corresponding author. Three of my first author publications have been published in the highest impact liver journals (two in Hepatology (IF: 17.42) and one in Journal of Hepatology (IF: 30.1)). Furthermore, I have established a global network with world-leading virologists. In 2014, I received Japan Society for the Promotion of Science Award; and in 2013, I received The Encouragement State Prize in Medical Sciences from the Academy of Scientific Research and Technology, Egypt. My PhD study was selected as the Medical School Excellent Paper for the 2008–2009 Academic Year from Kobe University, Japan (El-Shamy et al., Hepatology. 2008; 48:38–47).

Mohamed M. Ibrahim

Prof. Dr. Mohamed M. Ibrahim is a full professor at College of Biotechnology, MUST since 2016. His scientific interest includes application of biotechnology in molecular genetic diagnosis of diseases and studying multidrug resistant bacteria as well as biodegradation and bioremediation of organic pollutants. He has been nominated as a Vice President for Graduate Studies and Research (from Feb., 2018–July, 2021). After promoting as an assistant professor, he has been also assigned as a vice dean for graduate studies and research, College of Biotechnology, MUST (from July, 2011–Aug., 2021). Immediately after he obtained his Ph.D. from Department of Genetics, Faculty of Agriculture, Ain Shams University in 2005. Professor Mohamed got his B.Sc. and M. Sc. from Department of Genetics, Faculty of Agriculture, Ain Shams University in 1995 and 2000, respectively.

viruses

MDPI

Editorial

What SARS-CoV-2 Variants Have Taught Us: Evolutionary Challenges of RNA Viruses

Shymaa E. Bilasy [1], Tutik Sri Wahyuni [2], Mohamed Ibrahim [3] and Ahmed El-Shamy [4,*]

1 College of Dental Medicine, California Northstate University, Elk Grove, CA 95757, USA; shymaa.bilasy@cnsu.edu
2 Department of Pharmaceutical Science, Faculty of Pharmacy, Natural Product Medicine Research and Development, Institute Tropical Disease, Airlangga University, Surabaya 60286, Indonesia; tutik-s-w@ff.unair.ac.id
3 College of Biotechnology, Misr University for Science and Technology, Giza, Egypt; mohamed.ibrahim@must.edu.eg
4 Master of Pharmaceutical Sciences Program, College of Graduate Studies, California Northstate University, Elk Grove, CA 95757, USA
* Correspondence: ahmed.elshamy@cnsu.edu

Since its discovery in 2019, SARS-CoV-2 still makes the headline news. SAR-CoV-2 is an emerging RNA virus that spread from a seafood market in Wuhan, China, and caused the COVID-19 pandemic. SARS-CoV-2 is a positive sense single-stranded RNA virus of about 30 kb in length that is derived from the *Coronaviridae*. Up to September 2023, more than 770 million COVID-19 cases and 6.9 million deaths were reported to the World Health Organization (WHO). Around 37 variants of SARS-CoV-2 were reported and they were categorized as variants of concern (VOC) or variants of interest (VOI) or variants being monitored (VBM) [1]. The WHO labeled SARS-CoV-2 Omicron (B.1.1.529) as a variant of concern in November 2021 [2]. This variant was first reported in South Africa, but it quickly became the dominant circulating SARS-CoV-2 worldwide [3,4]. The Omicron viruses continued to genetically evolve with more sub-lineages added to its phylogenetic tree. The genetic divergence of Omicron has been associated with changes in the viral transmissibility, virulence, and ability to evade protective immune response conferred by natural infection or vaccination [5].

To shed light on the evolutionary behavior of RNA viruses and how it shapes their epidemiolocal fitness and pathological features, *Viruses* developed a Special Issue entitled "What SARS-CoV-2 Variants Have Taught Us: Evolutionary Challenges of RNA Viruses". This issue included twenty-four research topics comprising different aspects of SARS-CoV-2 infection. Viral lineages in different geographical locations were discussed by several authors. Omicron and its sub-variants were responsible for COVID-19 infections in several countries. Given the large number of mutations in Omicron compared to its previous predecessors, the existence of missing SARS-CoV-2 variants was addressed in the *Viruses* Special Issue. Interestingly, phylogenetic analyses suggested the presence of intermediate variants between SARS-CoV-2 Omicron and Delta variants, which might have not been documented. In addition, developing rapid and accurate diagnostic methodology was another point explored in this Special Issue. Targeted reverse-transcriptase quantitative polymerase chain reaction (RT-qPCR) was highlighted in this Special Issue to accurately identify new variants. Compared to NGS, the targeted RT-qPCR-based method is more cost-effective and flexible and can provide near real-time changes in variant prevalence.

In general, RNA viruses have high genetic variability due to fast, low-fidelity replication. Mutations, recombination, and reassortment are the main mechanisms responsible for genetic change and evolution [6]. Those changes can eventually affect viral fitness. In this *Viruses* Special Issue, the genome architecture of different SARS-CoV-2 variants was analyzed with the aim of discovering mutations correlated with viral pathogenicity. A

Citation: Bilasy, S.E.; Wahyuni, T.S.; Ibrahim, M.; El-Shamy, A. What SARS-CoV-2 Variants Have Taught Us: Evolutionary Challenges of RNA Viruses. *Viruses* **2024**, *16*, 139. https://doi.org/10.3390/v16010139

Received: 4 December 2023
Accepted: 12 January 2024
Published: 18 January 2024

higher number of mutations were associated with mortality cases in both Delta and th Omicron variants. Further, compared to MERS-CoV and SARS-CoV-1, the SARS-CoV- genome was biased towards a lower GC dinucleotide content, which may explain it moderate virulence.

Several articles addressed the immune response to SARS-CoV-2 infection, which revealed several immunogenic regions. Immunogenic regions may be correlated to the disease severity and/or potentially used as serological markers.

Numerous mutations in Omicron and its sub-lineages altered the transmission dynamics and pathophysiology. Sun et al. reported neutralization assays using diluted plasma samples from COVID-19 convalescent patients using a SARS-CoV-2 pseudovirus. This model contained the SARS-CoV-2 spike protein on an HIV-1 backbone and a luciferase reporter gene [7]. Although pseudoviruses cannot reflect the behavior of the full virus, they are considered a versatile research tool that can be conveniently performed at a lower biosafety level and can easily introduce mutations to reflect the newly discovered variants. Among the Omicron viruses, several variants and subvariants are classified as "variants being monitored" according to the updated guidelines developed by the WHO [1]. For some time, the BA.2.86 and XBB.1.1, a recombinant form of two BA.2 sub-lineages, were dominating the community spread. Recently the EG.5 (Eris) accounts for 25% of the new cases. The EG.5 variant descends from the XBB.1.9.2 subvariant. However, it contains an additional FLip mutation, Phe456Leu, and its subvariant, EG.5.1, has another spike protein mutation (Q52H) [8]. Therefore, pseudoviruses can be instrumental in exploring the significance of those mutations.

In conclusion, as an emerging RNA virus, SARS-CoV-2 will continue to evolve, and the threat of a viral outbreak will always be present. This virus has displayed a remarkable ability to mutate and generate new variants, and it has been proven to be a highly adaptable and evolving pathogen, leading to public health concerns. This Special Issue of *Viruses* sheds light on the evolutionary nature of SARS-CoV-2 and its potential implications for public health. Recognizing the ever-evolving nature of SARS-CoV-2 underscores the importance of continued surveillance, vaccination strategies, and adaptable public health responses to address this dynamic threat.

Conflicts of Interest: The authors declare no conflicts of interest.

List of Contributions:

1. Carrazco-Montalvo, A.; Herrera-Yela, A.; Alarcón-Vallejo, D.; Gutiérrez-Pallo, D.; Armendáriz-Castillo, I.; Andrade-Molina, D.; Muñoz-Mawyin, K.; Fernández-Cadena, J.; Morey-León, G. USFQ-COVID-19 Consortium; CRN Influenza y OVR—INSPI; Patiño, L. Omicron Sub-Lineages (BA.1.1.529 + BA.*) Current Status in Ecuador. *Viruses* **2022**, *14*, 1177. https://doi.org/10.3390/v14061177.
2. Sun, H.; Xu, J.; Zhang, G.; Han, J.; Hao, M.; Chen, Z.; Fang, T.; Chi, X.; Yu, C. Developing Pseudovirus-Based Neutralization Assay against Omicron-Included SARS-CoV-2 Variants. *Viruses* **2022**, *14*, 1332. https://doi.org/10.3390/v14061332.
3. Aston, E.; Wallach, M.; Narayanan, A.; Egaña-Labrin, S.; Gallardo, R. Hyperimmunized Chickens Produce Neutralizing Antibodies against SARS-CoV-2. *Viruses* **2022**, *14*, 1510. https://doi.org/10.3390/v14071510.
4. Goller, K.; Moritz, J.; Ziemann, J.; Kohler, C.; Becker, K.; Hübner, N. The CoMV-Gen Study Group Differences in Clinical Presentations of Omicron Infections with the Lineages BA.2 and BA.5 in Mecklenburg-Western Pomerania, Germany, between April and July 2022. *Viruses* **2022**, *14*, 2033. https://doi.org/10.3390/v14092033.
5. Pinheiro, J.; dos Reis, E.; Farias, J.; Fogaça, M.; da Silva, P.; Santana, I.; Rocha, A.; Vidal, P.; Simões, R.; Luiz, W.; et al. Impact of Early Pandemic SARS-CoV-2 Lineages Replacement with the Variant of Concern P.1 (Gamma) in Western Bahia, Brazil. *Viruses* **2022**, *14*, 2314. https://doi.org/10.3390/v14102314.
6. Caputo, E.; Mandrich, L. SARS-CoV-2: Searching for the Missing Variants. *Viruses* **2022**, *14*, 2364. https://doi.org/10.3390/v14112364.

7. Focosi, D.; Franchini, M.; Casadevall, A. On the Need to Determine the Contribution of Anti-Nucleocapsid Antibodies as Potential Contributors to COVID-19 Convalescent Plasma Efficacy. *Viruses* **2022**, *14*, 2378. https://doi.org/10.3390/v14112378.
8. Aoki, A.; Adachi, H.; Mori, Y.; Ito, M.; Sato, K.; Kinoshita, M.; Kuriki, M.; Okuda, K.; Sakakibara, T.; Okamoto, Y.; et al. Rapid Identification of SARS-CoV-2 Omicron BA.5 Spike Mutation F486V in Clinical Specimens Using a High-Resolution Melting-Based Assay. *Viruses* **2022**, *14*, 2401. https://doi.org/10.3390/v14112401.
9. Chauhan, D.; Chakravarty, N.; Jeyachandran, A.; Jayakarunakaran, A.; Sinha, S.; Mishra, R.; Arumugaswami, V.; Ramaiah, A. In Silico Genome Analysis Reveals the Evolution and Potential Impact of SARS-CoV-2 Omicron Structural Changes on Host Immune Evasion and Antiviral Therapeutics. *Viruses* **2022**, *14*, 2461. https://doi.org/10.3390/v14112461.
10. Sidarovich, A.; Krüger, N.; Rocha, C.; Graichen, L.; Kempf, A.; Nehlmeier, I.; Lier, M.; Cossmann, A.; Stankov, M.; Schulz, S.; et al. Host Cell Entry and Neutralization Sensitivity of SARS-CoV-2 Lineages B.1.620 and R.1. *Viruses* **2022**, *14*, 2475. https://doi.org/10.3390/v14112475.
11. Ravi, V.; Swaminathan, A.; Yadav, S.; Arya, H.; Pandey, R. SARS-CoV-2 Variants of Concern and Variations within Their Genome Architecture: Does Nucleotide Distribution and Mutation Rate Alter the Functionality and Evolution of the Virus? *Viruses* **2022**, *14*, 2499. https://doi.org/10.3390/v14112499.
12. Wong, S.; Au, A.; Lo, J.; Ho, P.; Hung, I.; To, K.; Yuen, K.; Cheng, V. Evolution and Control of COVID-19 Epidemic in Hong Kong. *Viruses* **2022**, *14*, 2519. https://doi.org/10.3390/v14112519.
13. Shi, D.; Bu, C.; He, P.; Song, Y.; Dordick, J.; Linhardt, R.; Chi, L.; Zhang, F. Structural Characteristics of Heparin Binding to SARS-CoV-2 Spike Protein RBD of Omicron Sub-Lineages BA.2.12.1, BA.4 and BA.5. *Viruses* **2022**, *14*, 2696. https://doi.org/10.3390/v14122696.
14. Nevejan, L.; Ombelet, S.; Laenen, L.; Keyaerts, E.; Demuyser, T.; Seyler, L.; Soetens, O.; Van Nedervelde, E.; Naesens, R.; Geysels, D.; et al. Severity of COVID-19 among Hospitalized Patients: Omicron Remains a Severe Threat for Immunocompromised Hosts. *Viruses* **2022**, *14*, 2736. https://doi.org/10.3390/v14122736.
15. Šimičić, P.; Židovec-Lepej, S. A Glimpse on the Evolution of RNA Viruses: Implications and Lessons from SARS-CoV-2. *Viruses* **2023**, *15*, 1. https://doi.org/10.3390/v15010001.
16. Mathez, G.; Pillonel, T.; Bertelli, C.; Cagno, V. Alpha and Omicron SARS-CoV-2 Adaptation in an Upper Respiratory Tract Model. *Viruses* **2023**, *15*, 13. https://doi.org/10.3390/v15010013.
17. Berkowitz, R.; Ostrov, D. The Elusive Coreceptors for the SARS-CoV-2 Spike Protein. *Viruses* **2023**, *15*, 67. https://doi.org/10.3390/v15010067.
18. Warger, J.; Gaudieri, S. On the Evolutionary Trajectory of SARS-CoV-2: Host Immunity as a Driver of Adaptation in RNA Viruses. *Viruses* **2023**, *15*, 70. https://doi.org/10.3390/v15010070.
19. Chatterjee, S.; Bhattacharya, M.; Nag, S.; Dhama, K.; Chakraborty, C. A Detailed Overview of SARS-CoV-2 Omicron: Its Sub-Variants, Mutations and Pathophysiology, Clinical Characteristics, Immunological Landscape, Immune Escape, and Therapies. *Viruses* **2023**, *15*, 167. https://doi.org/10.3390/v15010167.
20. Acharjee, A.; Ray, A.; Salkar, A.; Bihani, S.; Tuckley, C.; Shastri, J.; Agrawal, S.; Duttagupta, S.; Srivastava, S. Humoral Immune Response Profile of COVID-19 Reveals Severity and Variant-Specific Epitopes: Lessons from SARS-CoV-2 Peptide Microarray. *Viruses* **2023**, *15*, 248. https://doi.org/10.3390/v15010248.
21. Fantini, J.; Azzaz, F.; Chahinian, H.; Yahi, N. Electrostatic Surface Potential as a Key Parameter in Virus Transmission and Evolution: How to Manage Future Virus Pandemics in the Post-COVID-19 Era. *Viruses* **2023**, *15*, 284. https://doi.org/10.3390/v15020284.
22. Torres, C.; Nabaes Jodar, M.; Acuña, D.; Montaño, R.; Culasso, A.; Amadio, A.; Aulicino, P.; Ceballos, S.; Cacciabue, M.; Debat, H.; et al. Omicron Waves in Argentina: Dynamics of SARS-CoV-2 Lineages BA.1, BA.2 and the Emerging BA.2.12.1 and BA.4/BA.5. *Viruses* **2023**, *15*, 312. https://doi.org/10.3390/v15020312.
23. Kamal, S.; Naghib, M.; Daadour, M.; Alsuliman, M.; Alanazi, Z.; Basalem, A.; Alaskar, A.; Saed, K. The Outcome of BNT162b2, ChAdOx1-Sand mRNA-1273 Vaccines and Two Boosters: A Prospective Longitudinal Real-World Study. *Viruses* **2023**, *15*, 326. https://doi.org/10.3390/v15020326.
24. Carattini, Y.; Griswold, A.; Williams, S.; Valiathan, R.; Zhou, Y.; Shukla, B.; Abbo, L.; Parra, K.; Jorda, M.; Nimer, S.; et al. Combined Use of RT-qPCR and NGS for Identification and Surveillance of SARS-CoV-2 Variants of Concern in Residual Clinical Laboratory Samples in Miami-Dade County, Florida. *Viruses* **2023**, *15*, 593. https://doi.org/10.3390/v15030593.

References

1. Statement on the Update of WHO's Working Definitions and Tracking System for SARS-CoV-2 Variants of Concern and Variants of Interest. Available online: https://www.who.int/news/item/16-03-2023-statement-on-the-update-of-who-s-working-definitions-and-tracking-system-for-sars-cov-2-variants-of-concern-and-variants-of-interest (accessed on 1 December 2023).
2. SARS-CoV-2 Variant Classifications and Definitions. Available online: https://www.cdc.gov/coronavirus/2019-ncov/variants/variant-classifications.html (accessed on 1 December 2023).
3. Classification of Omicron (B.1.1.529): SARS-CoV-2 Variant of Concern. Available online: https://www.who.int/news/item/26-11-2021-classification-of-omicron-(b.1.1.529)-sars-cov-2-variant-of-concern (accessed on 1 December 2023).
4. Vitiello, A.; Ferrara, F.; Auti, A.M.; Di Domenico, M.; Boccellino, M. Advances in the Omicron variant development. *J. Intern. Med.* **2022**, *292*, 81–90. [CrossRef] [PubMed]
5. Carabelli, A.M.; Peacock, T.P.; Thorne, L.G.; Harvey, W.T.; Hughes, J.; Consortium, C.-G.U.; Peacock, S.J.; Barclay, W.S.; de Silva, T.I.; Towers, G.J.; et al. SARS-CoV-2 variant biology: Immune escape, transmission and fitness. *Nat. Rev. Microbiol.* **2023**, *21*, 162–177. [CrossRef] [PubMed]
6. Simicic, P.; Zidovec-Lepej, S. A Glimpse on the Evolution of RNA Viruses: Implications and Lessons from SARS-CoV-2. *Viruses* **2022**, *15*, 1. [CrossRef] [PubMed]
7. Sun, H.; Xu, J.; Zhang, G.; Han, J.; Hao, M.; Chen, Z.; Fang, T.; Chi, X.; Yu, C. Developing Pseudovirus-Based Neutralization Assay against Omicron-Included SARS-CoV-2 Variants. *Viruses* **2022**, *14*, 1332. [CrossRef] [PubMed]
8. Dyer, O. COVID-19: Infections climb globally as EG.5 variant gains ground. *BMJ* **2023**, *382*, 1900. [PubMed]

viruses

MDPI

Communication

Omicron Sub-Lineages (BA.1.1.529 + BA.*) Current Status in Ecuador

Andrés Carrazco-Montalvo [1,*], Andrés Herrera-Yela [1], Damaris Alarcón-Vallejo [1], Diana Gutiérrez-Pallo [1], Isaac Armendáriz-Castillo [2,3], Derly Andrade-Molina [4], Karen Muñoz-Mawyin [4], Juan Carlos Fernández-Cadena [5], Gabriel Morey-León [6], USFQ-COVID-19 Consortium [†], CRN Influenza y OVR—INSPI [‡] and Leandro Patiño [7,*]

1 Instituto Nacional de Investigación en Salud Pública "Leopoldo Izquieta Pérez", Centro de Referencia Nacional de Genómica, Secuenciación y Bioinformática, Quito 170136, Ecuador; mherrera@inspi.gob.ec (A.H.-Y.); dalarcon@inspi.gob.ec (D.A.-V.); dgutierrez@inspi.gob.ec (D.G.-P.)
2 Instituto Nacional de Investigación en Salud Pública "Leopoldo Izquieta Pérez", Coordinación Zonal 9, Quito 170136, Ecuador; isaac.arcas@gmail.com
3 Facultad de Ingenierías y Ciencias Aplicadas, Universidad Internacional SEK, Quito 170302, Ecuador
4 Omics Sciences Laboratory, Faculty of Medical Sciences, Universidad Espíritu Santo, Samborondón 092301, Ecuador; dmandrademolina@uees.edu.ec (D.A.-M.); kemunoz@espol.edu.ec (K.M.-M.)
5 INTERLAB, Guayaquil 090512, Ecuador; jfernandez@interlabsa.com
6 Facultad de Ciencias Medicas, Universidad de Guayaquil, Guayaquil 090514, Ecuador; gabriel.moreyl@ug.edu.ec
7 Instituto Nacional de Investigación en Salud Pública "Leopoldo Izquieta Pérez"—Dirección Técnica de Investigación, Desarrollo e Innovación, Guayaquil 170403, Ecuador
* Correspondence: acarrazco@inspi.gob.ec (A.C.-M.); lpatino@inspi.gob.ec (L.P.)
† USFQ-COVID-19 Consortium are listed in acknowledgments.
‡ CRN Influenza y OVR—INSPI are listed in acknowledgments.

Citation: Carrazco-Montalvo, A.; Herrera-Yela, A.; Alarcón-Vallejo, D.; Gutiérrez-Pallo, D.; Armendáriz-Castillo, I.; Andrade-Molina, D.; Muñoz-Mawyin, K.; Fernández-Cadena, J.C.; Morey-León, G.; USFQ-COVID-19 Consortium; et al. Omicron Sub-Lineages (BA.1.1.529 + BA.*) Current Status in Ecuador. *Viruses* **2022**, *14*, 1177. https://doi.org/10.3390/v14061177

Academic Editors: Ahmed Elshamy and Mohamed Ibrahim

Received: 19 April 2022
Accepted: 25 May 2022
Published: 28 May 2022

Publisher's Note: MDPI stays neutral with regard to jurisdictional claims in published maps and institutional affiliations.

Abstract: The Omicron variant of SARS-CoV-2 is the latest pandemic lineage causing COVID-19. Despite having a vaccination rate ≥85%, Ecuador recorded a high incidence of Omicron from December 2021 to March 2022. Since Omicron emerged, it has evolved into multiple sub-lineages with distinct prevalence in different regions. In this work, we use all Omicron sequences from Ecuador available at GISAID until March 2022 and the software Nextclade and Pangolin to identify which lineages circulate in this country. We detected 12 different sub-lineages (BA.1, BA.1.1, BA.1.1.1, BA.1.1.14, BA.1.1.2, BA.1.14, BA.1.15, BA.1.16, BA.1.17, BA.1.6, BA.2, BA.2.3), which have been reported in Africa, America, Europe, and Asia, suggesting multiple introduction events. Sub-lineages BA.1 and BA.1.1 were the most prevalent. Genomic surveillance must continue to evaluate the dynamics of current sub-lineages, the early introduction of new ones and vaccine efficacy against evolving SARS-CoV-2.

Keywords: COVID-19; Omicron; sub-lineages; Ecuador

1. Introduction

The novel coronavirus identified as severe acute respiratory syndrome coronavirus 2 (SARS-CoV-2) is the infectious agent causing the COVID-19 pandemic [1]; by April 2022, a total of 483,556,595 positive cases and 6,132,461 deaths worldwide had been reported [2]. The mutation rate of the virus is still in debate, but after two years, it has evolved into different variants, some of which have higher transmission rates [3]. Some model studies suggest that the effective (instantaneous) reproduction number of Omicron could be from 3.19 (95% CI: 2.82–3.61) [4,5] to 3.3 (95% CI: 2.0–7.8) [6], which is greater than that of the Delta variant. Moreover, each emerging variant could diverge into sub-lineages that need to be surveilled locally in order to evaluate any change in their transmission dynamics.

Omicron, unlike previous variants, harbors a wide variety of mutations within its genome [7]. Fifteen mutations have been reported only in the Receptor Binding Domain (RBD) region, which enable Omicron to be more transmissible by allowing the virus to bind more easily to the human angiotensin-converting enzyme 2 (ACE2) protein as compared with the original strain [8]. Other relevant mutations include R203K and G204R, which are linked to viral replication [4]. According to WHO, by April 2022 [9], five major sub-lineages originating from BA.1.1.529 had been detected globally: BA.1, BA.2, BA.3, BA.4, BA.5 with different frequencies in different regions.

Ecuador is among the countries with a vaccine coverage ≥85%; however, following the introduction of the Omicron Variant (lineage BA.1.1.529 in December 2021) [10] the Ministry of Public Health reported the highest frequency of positive cases since the beginning of the pandemic, accumulating more than 300,000 cases from January to March 2022 [11]. By the end of March, sub-lineage BA.2 was also detected.

At least four institutions from the public and private sector are involved locally in the genomic surveillance of the virus and are submitting sequences to the public repository GISAID with a two-week frequency [12]. This allows us to understand the epidemiological trends and incidence of the COVID-19 disease and the effects of the detection and spread of new variants [13]. Given the emergence of sub-lineages, this work aims to identify which are circulating and prevailing in Ecuador. Our analyses focused on the presence and trend of each Omicron sub-lineage (BA.1.1.529 + BA.*) reported since epidemiological week 49 in 2021, when we detected the first case in Ecuador [10].

2. Materials and Methods

2.1. Sequence Production and Data Collection

The institutions involved in SARS-CoV-2 sequencing in Ecuador use different platforms including MinION (Oxford Nanopore), MiSeq and MiniSeq (Illumina). We downloaded 1245 Omicron sequences submitted to GISAID from Ecuador up until March 2022; 703 obtained from "Instituto Nacional de Investigación en Salud Publica (INSPI)" using MinION (ONT) and MiSeq (Illumina), 456 obtained from "Universidad San Francisco de Quito-USFQ" using MinION and 86 sequences obtained from "Universidad de Especialidades Espiritu Santo" which used MiniSeq (Illumina). Starting from the date on which the first case of Omicron was detected in Ecuador, a comparison of sequences was performed by epidemiological week for Omicron vs. Delta.

2.2. Lineage Assignment and Phylogenetics

Sequences were classified by epidemiological week and then submitted to Nextclade [14] and Pangolin COVID-19 [15] for clade and lineage assignment. Lineage nomenclature was assigned according to the PANGO (Phylogenetic Assignment of Named Global Outbreak Lineages) software updated in March 2022. It relies on establishing a numerical value to descendants that meet certain conditions that belong to lineages A or B, with a maximum of three sublevels, whereby new lineages will be assigned with a letter [15]. The criteria used for lineage assignment involved minimum lineage size, genome quality, genetic specificity, and epidemiological significance, which vary over time and depend on the degree of adaptation [16]. Consequently, each lineage is assigned a unique alphanumeric code that includes partial information regarding the phylogenetic history of that lineage based on a common ancestor [15].

A stacked bar chart of lineages by epidemiological week was made using R [17] and GraphPad Prism software [18]. A phylogenetic tree was built in Nextclade using the nearest neighbor method and visualized by Nextrain Auspice [14].

3. Results

The first case of Omicron was detected in Ecuador in epidemiological week (EW) 49 of 2021, it co-circulated with Delta until EW 52 of 2021 and EW 05 of 2022. From EW 06 Omicron became the only variant detected in all the samples sequenced, Figure 1.

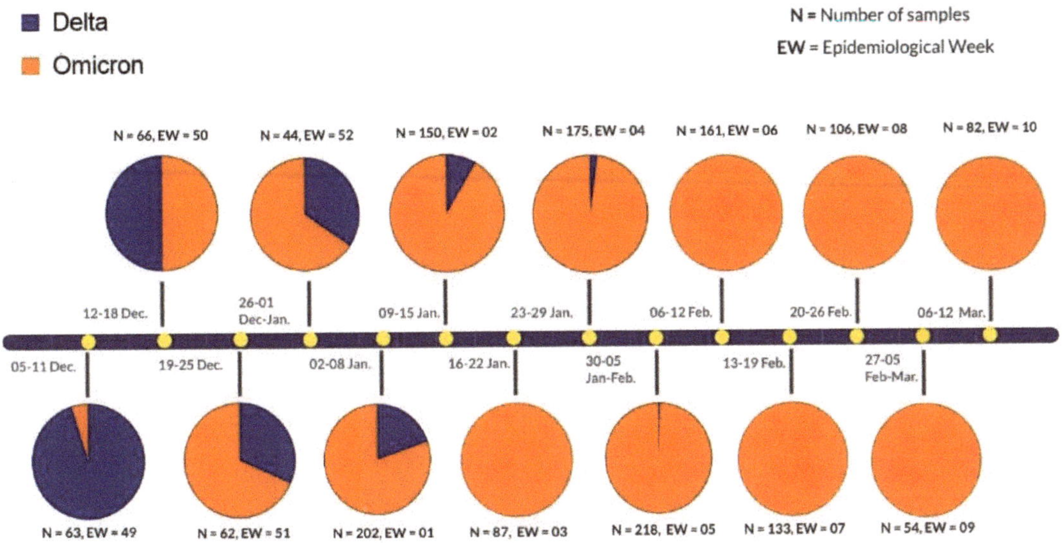

Figure 1. Variants of concern in Ecuador since Omicron's first detection.

We detected 12 sub-lineages of the Omicron variant (BA.1.1.529 + BA.*) circulating in Ecuador, with different worldwide origins and mutation numbers (Table 1). All sub-lineages display the highest number of mutations in the spike protein. BA.2 and BA.2.3 sub-lineages differ from the others in the number of mutations in five out of nine genes (ORF1a, ORF1b, S, M, N). The complete set of mutations is detailed in Table S1 [15].

Table 1. List of Omicron sub-lineages circulating in Ecuador. For each Omicron sub-lineage, the place/s of origin and mutation numbers that characterize them are recorded.

Sub-Lineage	Clade Names	Origin	Genes								
			ORF1a	ORF1b	S	ORF3a	E	M	ORF6	ORF8	N
BA.1	21K (Omicron)	South Africa	8	2	33	0	1	3	0	1	4
BA.1.1	21K (Omicron)	South Africa	8	2	32	0	1	3	0	1	4
BA.1.1.1	21K (Omicron)	Europe	8	2	31	0	1	3	0	1	4
BA.1.1.14	21K (Omicron)	Europe	8	2	30	0	1	3	0	1	4
BA.1.1.2	21K (Omicron)	Japan	9	2	34	0	1	3	0	1	4
BA.1.14	21K (Omicron)	Brazil	8	2	21	0	1	3	0	1	4
BA.1.15	21K (Omicron)	USA	8	2	30	1	1	3	0	1	5
BA.1.16	21K (Omicron)	UK	8	2	29	0	1	3	0	1	4
BA.1.17	21K (Omicron)	Europe	9	2	30	0	1	3	0	1	4
BA.1.6	21K (Omicron)	Canada and Sint Maarten	8	3	30	0	1	3	0	1	4
BA.2	21L (Omicron)	India and South Africa	9	4	29	1	1	2	1	1	5
BA.2.3	21L (Omicron)	Philippines	10	4	29	2	1	2	1	1	5

Figure 2a shows the number of cases for each sub-lineage from EW 49 of 2021 to EW 10 of 2022. BA.1 was the first sub-lineage detected followed by BA.1.1 and BA.1.15 which were detected on EW 50 of 2021. BA.1.16 and BA.1.17 appeared on EW 51 of 2021, all remaining sub-lineages have been detected since EW 2 of 2022. BA.1 and BA.1.1 were found across all epidemiological weeks, while the growth and diversity of the other sub-lineages were proportional with the increase of cases; however, both decreased together when positive cases dropped (Table S2). Among all the sequences analyzed, 62.33% corresponded to the BA.1.1 sub-lineage, 24.82% to BA.1, 6.18% to BA.1.14, 4.33% to BA.1.15, 1.12% to BA.1.17, the remaining sub-lineages BA.1.6, BA.1.16, BA.2, BA.2.3, BA.1.1.1; BA.1.1.2, BA.1.1.14, were found with a frequency ≤1%.

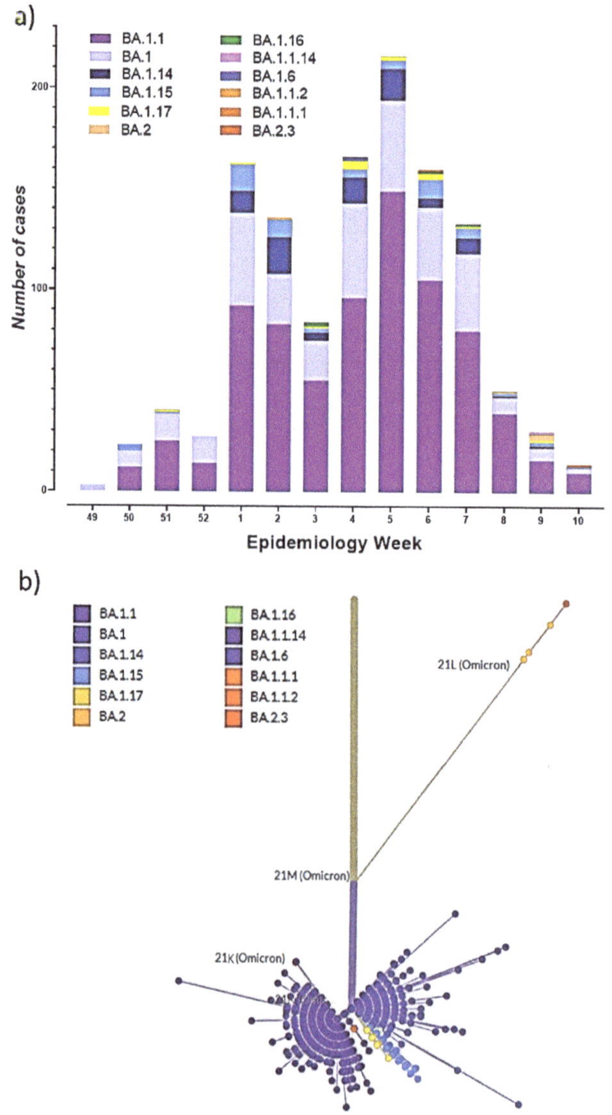

Figure 2. Omicron sub-lineages (BA.1.1.529 + BA.*) in Ecuador. (**a**) Sub-lineages by epidemiological week, (**b**) phylogenetic tree of all lineages.

The phylogenetic analysis of the sub-lineages detected is shown in Figure 2b. The sub-lineages clustered into two clades: 21K and 21L, which diverged from 21M. The highest diversity was in the 21K clade, forming two groups with BA.1 and BA.1.1. The 21K clade only showed two sub-lineages corresponding to BA.2 and BA.2.3.

Among the most predominant sub-lineages registered in Ecuador, BA.1 appeared in 162 countries showing higher prevalence in the UK at 43.0%, USA 22.0%, Denmark 5.0%, Germany 4.0%, and Brazil 3.0%; BA.1.1 appeared in 154 countries showing higher prevalence in the USA at 48.0%, UK 22.0%, Germany 6.0%, Canada 4.0%, and France 2.0%; BA.1.14 appeared in 82 countries with higher prevalence in Brazil at 19.0%, Belgium 17.0%, UK 13.0%, Denmark 10.0% and Germany 7.0%; and BA.1.15 appeared in 124 countries with most prevalence in the USA at 69.0%, the UK 14.0%, Canada 4.0%, Germany 2.0% and Mexico 1.0% [15].

The BA.2 and BA.2.3 sub-lineages were recently reported in Ecuador; these currently predominate in several countries, so we analyzed their worldwide prevalence: BA.2 appeared in 119 countries with higher prevalence in the UK at 40.0%, Denmark 16.0%, Germany 13.0%, the USA 7.0% and France 4.0% [15]; and BA.2.3 has been registered in 81 countries with higher prevalence in the UK at 31.0%, the USA 23.0%, Canada 10.0%, Germany 4.0%, and South Korea 4.0% [15].

4. Discussion

Following Omicron's first identification, an increasing number of sub-lineages are being reported globally. Of all the worldwide Omicron sequences available in GISAD, at least 36 sub-lineages (all in the major BA.1–BA.5 as reported by WHO) had been identified by April 2022 [9,19]. In Ecuador, up to March 2022, we had detected 12 Omicron sub-lineages (BA.1.1.529 + BA.*) also reported in Africa, America, Europe, and Asia, suggesting multiple introduction events. The detected sub-lineages harbor from 40 to 54 mutations (Table 1), of which approximately 30 contribute to amino acid changes in the SARS-CoV-2 spike protein [20].

During the Omicron wave in Ecuador, the major number of cases occurred between epidemiological weeks one and seven of 2022 (Figure 2a). However, in EW04 and EW05, the Delta variant was still circulating in less populated provinces in the country. At the end of December 2021, sub-lineages BA.1 and BA.1.1 were the most predominant, which agreed with the global trends [21]. In Europe, Denmark had reported the prevalence of BA.1, BA.1.1, and BA.2 sub-lineages in the same period [22]. In Italy, the predominant sub-lineages were BA.1.15, BA.1.1, and BA.1.17 [19]. A study conducted in Hong Kong using 542 Omicron sequences, showed the BA.2.2, BA.1 and BA.1.1 sub-lineages predominated [23]. According to PAHO reports, BA.1 and BA.1.1 were also identified in more than 97% of the cases registered in the Americas [24]. Different Omicron sub-lineages (BA.1.1.529 + BA.*) have also been reported in other Latin American countries; a study by the University of Feevale in Brazil has reported the circulation of seven sub-lineages in that country, with BA.1, BA.1.1 and lately BA.2 being the predominant ones [25]. A similar pattern occurred in Chile, where the Ministry of Health reported the circulation of 10 sub-lineages, with BA.1.1 and BA.2 displacing BA.1 [26].

Our analysis shows the divergence of the 21K and 21L clades from 21M, which is related to three mutations in 21L not present in 21K: they comprise Nsp3 (G489S, analogous to the A488S mutation found in Delta VOC), Nsp4 (L438F, analogous to the L438P mutation found in Lambda VOC) and Nsp6 which displaced the C-terminus in the 21K clade [27]. Differences in the number of N mutations were also identified, the BA.2 and BA.2.3 sub-lineages showed an R203K/G204R mutation in the nucleocapsid protein. This mutation has also been reported by Wu, H., et al. (2021), being associated with the appearance of new variants [28].

In this study, five samples were identified within Clade 21L (BA.2 and BA.2.3 lineages) in EW 10; due to the high transmissibility and predominance of BA.2 and BA.2.3 reported in other countries, it is important to survey them in Ecuador during the following months. According to the United Kingdom Health Security Agency, the BA.2 sub-lineage has shown a higher growth rate than BA.1; in England, BA.2 showed 75% more spreading when compared to BA.1 [29]. Similar behavior was observed in Denmark, where BA.2 had an accelerated growth compared to BA.1, becoming the dominant sub-lineage [30]. The behavior of BA.2 may be related to mutations found in the spike protein of the virus [30]. According to the detail of the mutations presented in the results section of this study, one of the main differences between BA.1 and BA.2 is the absence of the 143/145 deletion in BA.2, as mentioned by Colson et al.; this difference produces a flattening of the surface of the N terminal domain (NTD) that could facilitate the initial interaction of the virus with the lipid rafts and would explain its greater transmissibility [31]. Another difference is that BA.1 and BA.3 sub-lineages have a deletion 69/70 in the spike protein, which is not found in BA.2 [31].

BA.2 sub-lineage was detected in Ecuador in EW 10, coinciding with a drop in case numbers, probably due to a high incidence of the first Omicron sub-lineage (BA.1.1.529 + BA.*) combined with the high vaccination coverage. The COVID-19 positivity from EW 19 was $\leq$5% [32]. Despite a low positivity rate detected in the last weeks, Public Health Authorities need to keep all the genomic surveillance strategies active to evaluate any change in the predominance of the Omicron sub-lineage (BA.1.1.529 + BA.*) currently circulating in the country. Furthermore, early detection of new lineages reported worldwide will help to improve current public health policies and influence decisions such as not requiring people to wear a mask in outdoor areas of conveyances. Furthermore, these reports will provide information on the efficacy of the vaccines as well as the future reinforcement doses required to protect the most vulnerable population from the emergence of new variants and sub-lineages.

The current pandemic has highlighted the constant danger of the emergence or re-emergence of viruses with pandemic potential. SARS-CoV-2 has demonstrated a high capacity for evolution and adaptability; it has mutated at an accelerated rate, leading to new variants [33]. Therefore, the response to the possible emergence of new viruses involves scientific and technological strengthening of research to consolidate a surveillance system with a preventive approach for the containment of future outbreaks [34].

Supplementary Materials: The following supporting information can be downloaded at: https://www.mdpi.com/article/10.3390/v14061177/s1, Table S1: Detail of mutations by sub-lineage, Table S2: Number of cases by epidemiological week.

Author Contributions: Conceptualization, L.P.; A.C.-M.; methodology, L.P.; A.C.-M.; A.H.-Y.; D.A.-V.; D.G.-P.; I.A.-C.; D.A.-M.; K.M.-M.; J.C.F.-C.; G.M.-L.; USFQ-COVID-19 Consortium; software, L.P.; A.C.-M.; A.H.-Y.; D.A.-V.; D.G.-P.; I.A.-C.; D.A.-M.; K.M.-M.; J.C.F.-C.; G.M.-L.; USFQ-COVID-19 Consortium; validation L.P.; A.C.-M.; A.H.-Y.; D.A.-V.; D.G.-P.; I.A.-C.; D.A.-M.; K.M.-M.; J.C.F.-C.; G.M.-L.; USFQ-COVID-19 Consortium; formal analysis, L.P.; A.C.-M.; A.H.-Y.; D.A.-V.; D.G.-P.; I.A.-C.; D.A.-M.; K.M.-M.; J.C.F.-C.; G.M.-L.; USFQ-COVID-19 Consortium; investigation, L.P.; A.C.-M.; A.H.-Y.; D.A.-V.; D.G.-P.; I.A.-C.; D.A.-M.; K.M.-M.; J.C.F.-C.; G.M.-L.; USFQ-COVID-19 Consortium; resources, L.P.; A.C.-M.; A.H.-Y.; D.A.-V.; D.G.-P.; I.A.-C.; D.A.-M.; K.M.-M.; J.C.F.-C.; G.M.-L.; USFQ-COVID-19 Consortium; data curation, L.P.; A.C.-M.; A.H.-Y.; D.A.-V.; D.G.-P.; writing—original draft preparation, L.P.; A.C.-M.; A.H.-Y.; D.A.-V.; D.G.-P.; writing—review and editing, L.P.; A.C.-M.; A.H.-Y.; D.A.-V.; D.G.-P.; I.A.-C.; D.A.-M.; K.M.-M.; J.C.F.-C.; G.M.-L.; USFQ-COVID-19 Consortium; CRN Influenza y OVR—INSPI; visualization, L.P.; A.C.-M.; A.H.-Y.; D.A.-V.; D.G.-P.; I.A.-C.; D.A.-M.; K.M.-M.; J.C.F.-C.; G.M.-L.; USFQ-COVID-19 Consortium; CRN Influenza y OVR—INSPI; supervision, L.P.; A.C.-M.; A.H.-Y.; D.A.-V.; D.G.-P.; I.A.-C.; D.A.-M.; K.M.-M.; J.C.F.-C.; G.M.-L.; USFQ-COVID-19 Consortium; project administration, L.P.; A.C.-M.; A.H.-Y.; D.A.-V.; D.G.-P.; I.A.-C.; D.A.-M.; K.M.-M.; J.C.F.-C.; G.M.-L.; USFQ-COVID-19 Consortium. All authors have read and agreed to the published version of the manuscript.

Funding: This research received no external funding.

Institutional Review Board Statement: Not applicable.

Informed Consent Statement: Not applicable.

Data Availability Statement: Not applicable.

Acknowledgments: We thank the technical staff of different institutions contributing to sequencing in Ecuador including the Centro de Referencia Nacional de Influenza y Otros virus Respiratorios and the Dirección Técnica de Investigación, Desarrollo e Innovación, both from INSPI, technical staff from Universidad de Especialidades Espíritu Santo, INTERLAB, Universidad de Guayaquil and USFQ-COVID-19 Consortium. The members of USFQ-COVID-19 are: Mateo Carvajal, Erika Muñoz, Rommel Guevara, Sully Márquez, Belén Prado-Vivar, Michelle Grunauer, Gabriel Trueba, Patricio Rojas-Silva, Verónica Barragan, and Paúl Cárdenas. The members of CRN Influenza y OVR are: Alfredo Bruno, Jimmy Garcés, Maritza Olmedo, Michelle Páez, and Rubén Armas-Gonzalez.

Conflicts of Interest: The authors declare no conflict of interest.

References

1. Wu, F.; Zhao, S.; Yu, B.; Chen, Y.; Wang, W.; Song, Z.; Hu, Y.; Tao, Z.; Tian, J.; Pei, Y.; et al. A new coronavirus associated with human respiratory disease in China. *Nature* **2020**, *579*, 265–269. [CrossRef] [PubMed]
2. World Health Organization. WHO Coronavirus (COVID-19) Dashboard. Available online: https://covid19.who.int/ (accessed on 7 April 2022).
3. Hu, B.; Guo, H.; Zhou, P.; Shi, Z. Characteristics of SARS-CoV-2 and COVID-19. *Nat. Rev. Microbiol.* **2021**, *19*, 141–154. [CrossRef] [PubMed]
4. Khandia, R.; Singhal, S.; Alqahtani, T.; Kamal, M.A.; El-Shall, N.; Nainu, F.; Desingu, P.; Dhama, K. Emergence of SARS-CoV-2 Omicron (B.1.1.529) variant, salient features, high global health concerns and strategies to counter it amid ongoing COVID-19 pandemic. *Environ. Res.* **2022**, *209*, 112816. [CrossRef] [PubMed]
5. Ito, K.; Piantham, C.; Nishiura, H. Relative instantaneous reproduction number of Omicron SARS-CoV-2 variant with respect to the Delta variant in Denmark. *J. Med. Virol.* **2022**, *94*, 2265–2268. [CrossRef]
6. Nishiura, H.; Ito, K.; Anzai, A.; Kobayashi, T.; Piantham, C.; Rodríguez-Morales, A.J. Relative reproduction number of SARS-CoV-2 Omicron (B.1.1.529) compared with Delta variant in South Africa. *J. Clin. Med.* **2021**, *11*, 30. [CrossRef]
7. Jung, C.; Kmiec, D.; Koepke, L.; Zech, F.; Jacob, T.; Sparrer, K.; Kirchhoff, F. Omicron: What makes the latest SARS-CoV-2 variant of concern so concerning? *J. Virol.* **2022**, *96*, e02077-21. [CrossRef]
8. Lupala, C.; Ye, Y.; Chen, H.; Su, X.D.; Liu, H. Mutations on RBD of SARS-CoV-2 Omicron variant result in stronger binding to human ACE2 receptor. *Biochem. Biophys. Res. Commun.* **2022**, *590*, 34–41. [CrossRef]
9. Emergence of Sub-Lineage and Recombination Events: Genetic Evolution of SARS-CoV-2. Available online: https://www.paho.org/en/documents/emergence-sub-lineage-and-recombination-events-genetic-evolution-sars-cov-2-13-april-2022 (accessed on 13 April 2022).
10. Carrazco, A.; Armendáriz, I.; Tello, C.; Morales, D.; Armas, R.; Guizado, D.; León, A.; Ramos, D.; Fuentes, B.; USFQ-Consortium; et al. First detection of SARS-CoV-2 variant B.1.1.529 (Omicron) in Ecuador. *New Microbes New Infect.* **2021**, *45*, 100951. [CrossRef]
11. Ministerio de Salud Pública. Available online: https://www.salud.gob.ec/informes-de-situacion-sitrep-e-infografias-covid-19-desde-26-07-2021/ (accessed on 7 April 2022).
12. Khare, S.; Gurry, C.; Freitas, L.; Schultz, M.; Bach, G.; Diallo, A.; Akite, N.; Ho, J.; Lee, R.; Yeo, W.; et al. GISAID's Role in Pandemic Response. *China CDC Wkly.* **2021**, *3*, 1049–1051. [CrossRef]
13. Resende, P.; Delatorre, E.; Gräf, T.; Mir, D.; Couto, F.; Reis, L.; Dias, A.; Da Fonseca, A.; Ogrzewalska, M.; Caetano, B.; et al. Evolutionary Dynamics and Dissemination Pattern of the the Early Pandemic Phase in Brazil. *Front. Microbiol.* **2021**, *11*, 615280. [CrossRef]
14. Aksamentov, I.; Roemer, C.; Hodcroft, E.; Neher, R. Nextclade: Clade assignment, mutation calling and quality control for viral genomes. *J. Open Source Softw.* **2021**, *6*, 3773. [CrossRef]
15. Rambaut, A.; Holmes, E.; O'Toole, Á.; Verity, H.; McCrone, J.; Ruis, C.; du Plessis, L.; Pybus, O. A dynamic nomenclature proposal for SARS-CoV-2 lineages to assist genomic epidemiology. *Nat. Microbiol.* **2020**, *5*, 1403–1407. [CrossRef] [PubMed]
16. González, F.; Shaw, M.; Phan, T.; Kulkarni, U.; Paraskevis, D.; Luciani, F.; Kimura, H.; Sironi, M. One year into the pandemic: Short-term evolution of SARS-CoV-2 and emergence of new lineages. *Infect. Genet. Evol.* **2021**, *92*, 104869.
17. R Core Team. *R: A Language and Environment for Statistical Computing*; R Foundation for Statistical Computing: Vienna, Austria, 2021; Available online: https://www.R-project.org/ (accessed on 14 April 2022).
18. GraphPad Software. Available online: www.graphpad.com (accessed on 13 April 2022).
19. The Maravi Post. Expansion of L452R-Positive SARS-CoV-2 Omicron Variant, Northern Lombardy, Italy. Available online: https://www.maravipost.com/expansion-of-l452r-positive-sars-cov-2-omicron-variant-northern-lombardy-italy/ (accessed on 14 April 2022).

20. Mallapaty, S. Where did Omicron come from? Three key theories. *Nature* **2022**, *602*, 26–28. [CrossRef]
21. Thomas, L. Omicron BA.1.1 and BA.2 Subvariants with New Mutations in New Zealand and Hong Kong. 2022. Available online: https://www.news-medical.net/news/20220320/Omicron-BA11-and-BA2-subvariants-with-new-mutations-in-New-Zealand-and-Hong-Kong.aspx (accessed on 13 May 2022).
22. Genomic Overview of SARS-CoV-2 in Denmark. Available online: https://www.covid19genomics.dk/statistics (accessed on 8 April 2022).
23. Chen, L.; Abdullah, M.; Chan, W.; Chan, B.; Jonathan, D.; Chu, W.; Lu, L.; Zhang, X.; Zhao, Y.; Chuang, W.; et al. Contribution of low population immunity to the severe Omicron BA.2 outbreak in Hong Kong. 2022; manuscript in preparation. Available online: https://www.researchsquare.com/article/rs-1512533/v1(accessed on 18 April 2022).
24. UNAM. Programa de Vigilancia Genómica de SARS-CoV-2 Realizado por el CoViGen-Mex. Available online: http://mexcov2.ibt.unam.mx:8080/COVID-TRACKER/open/reports/23%20de%20Marzo%20del%202022.pdf (accessed on 13 May 2022).
25. Rapid Establishment of Omicron Variant in Rio Grande do Sul State, Brazil. Available online: https://virological.org/t/rapid-establishment-of-omicron-variant-in-rio-grande-do-sul-state-brazil/783 (accessed on 13 April 2022).
26. Government of Chile. Epidemiology Department. Available online: https://www.minsal.cl/wp-content/uploads/2022/04/Informe_Variantes-N%C2%B028.pdf (accessed on 8 April 2022).
27. Wu, C.R.; Yin, W.C.; Jiang, Y.; Xu, H.E. Structure genomics of SARS-CoV-2 and its Omicron variant: Drug design templates for COVID-19. *Acta Pharmacol. Sin.* **2022**. [CrossRef]
28. Wu, H.; Xing, N.; Meng, K.; Fu, B.; Xue, W.; Dong, P.; Tang, W.; Xiao, Y.; Liu, G.; Luo, H.; et al. Nucleocapsid mutations R203K/G204R increase the infectivity, fitness, and virulence of SARS-CoV-2. *Cell Host Microbe* **2021**, *29*, 1788–1801.e6. [CrossRef]
29. UK Health Security Agency. SARS-CoV-2 Variants of Concern and Variants under Investigation in England. Technical Briefing 39. Available online: https://assets.publishing.service.gov.uk/government/uploads/system/uploads/attachment_data/file/1063424/Tech-Briefing-39-25March2022_FINAL.pdf (accessed on 13 May 2022).
30. Callaway, F. Why does the Omicron sub-variant spread faster than the original? *Nature* **2022**, *602*, 556–557. [CrossRef]
31. Colson, P.; Delerce, J.; Beye, M.; Levasseur, A.; Boschi, C.; Houhamdi, L.; Fournier, P.E. First cases of infection with the 21L/BA. 2 Omicron variant in Marseille, France. *J. Med. Virol.* **2022**, *94*, 3421–3430. [CrossRef]
32. Ministerio de Salud Pública. Datos Epidemiológicos COVID-19. Available online: https://app.powerbi.com/view?r=eyJrIjoiNjkzNjUyNGMtZjc1NC00OWU0LWEzMWQtOTFkY2ZkYjZkMWIwIiwidCI6IjcwNjIyMGRiLTliMjktNGU5MS1hODI1LTI1NmIwNmQyNjlmMyJ9&pageName=ReportSection33e2bb803a8b183d9100 (accessed on 17 May 2022).
33. Trilla, A. One world, one health: The novel coronavirus COVID-19 epidemic. *Med. Clin. (Engl. Ed.)* **2020**, *154*, 175–177. [CrossRef]
34. Singh, J.; Pandit, P.; McArthur, A.G.; Banerjee, A.; Mossman, K. Evolutionary trajectory of SARS-CoV-2 and emerging variants. *Virol. J.* **2021**, *18*, 166. [CrossRef] [PubMed]

viruses

MDPI

Article

Developing Pseudovirus-Based Neutralization Assay against Omicron-Included SARS-CoV-2 Variants

Hancong Sun, Jinghan Xu, Guanying Zhang, Jin Han, Meng Hao, Zhengshan Chen, Ting Fang, Xiangyang Chi * and Changming Yu *

Institute of Biotechnology, Academy of Military Medical Sciences, Beijing 100071, China; sun_hancong@163.com (H.S.); pjxjh333@163.com (J.X.); zhangguanying_123@outlook.com (G.Z.); harncy@126.com (J.H.); haorm_23@126.com (M.H.); czs0076@163.com (Z.C.); 15510742074@163.com (T.F.)
* Correspondence: xiangyangchi@163.com (X.C.); yuchangming@126.com (C.Y.)

Abstract: The global spread of SARS-CoV-2 and its variants poses a serious threat to human health worldwide. Recently, the emergence of Omicron has presented a new challenge to the prevention and control of the COVID-19 pandemic. A convenient and reliable in vitro neutralization assay is an important method for validating the efficiency of antibodies, vaccines, and other potential drugs. Here, we established an effective assay based on a pseudovirus carrying a full-length spike (S) protein of SARS-CoV-2 variants in the HIV-1 backbone, with a luciferase reporter gene inserted into the non-replicate pseudovirus genome. The key parameters for packaging the pseudovirus were optimized, including the ratio of the S protein expression plasmids to the HIV backbone plasmids and the collection time for the Alpha, Beta, Gamma, Kappa, and Omicron pseudovirus particles. The pseudovirus neutralization assay was validated using several approved or developed monoclonal antibodies, underscoring that Omicron can escape some neutralizing antibodies, such as REGN10987 and REGN10933, while S309 and ADG-2 still function with reduced neutralization capability. The neutralizing capacity of convalescent plasma from COVID-19 convalescent patients in Wuhan was tested against these pseudoviruses, revealing the immune evasion of Omicron. Our work established a practical pseudovirus-based neutralization assay for SARS-CoV-2 variants, which can be conducted safely under biosafety level-2 (BSL-2) conditions, and this assay will be a promising tool for studying and characterizing vaccines and therapeutic candidates against Omicron-included SARS-CoV-2 variants.

Keywords: SARS-CoV-2 variants; Omicron; pseudovirus; neutralization assay; convalescent plasma

Citation: Sun, H.; Xu, J.; Zhang, G.; Han, J.; Hao, M.; Chen, Z.; Fang, T.; Chi, X.; Yu, C. Developing Pseudovirus-Based Neutralization Assay against Omicron-Included SARS-CoV-2 Variants. *Viruses* **2022**, *14*, 1332. https://doi.org/10.3390/v14061332

Academic Editors: Ahmed Elshamy and Mohamed Ibrahim

Received: 5 April 2022
Accepted: 16 June 2022
Published: 18 June 2022

1. Introduction

During the COVID-19 pandemic, the emergence of several highly transmissible SARS-CoV-2 variants, especially Omicron (B.1.1.529), has attracted enough attention worldwide. Some mutations in SARS-CoV-2 spike (S) protein can alter the antigenic properties with various distinct mechanisms, leading to their potential to be more transmissible, virulent, pathogenic, or evade immunity induced by previous infection or vaccination [1–8]. N501Y, which is present in lineages Alpha (B.1.17), Beta (B.1.351), Gamma (P.1), and Omicron, is responsible for higher affinity to ACE2 and increased infectivity due to a large phenolic group that makes two additional contacts with ACE2 residues [6,9,10]. Moreover, variants carrying E484K such as Beta and Gamma have been shown to contribute to the escape of some neutralizing antibodies and resistance to convalescent sera and postvaccination sera [10,11]. The replacement of a glutamate residue with a lysine residue causes a change in the biophysical properties of an epitope residue, diminishing some antibodies binding directly [12]. Moreover, lineages Beta and Gamma possess alternative amino acid substitutions K417N/T, which facilitate immune escape [13,14]. Regarding the Kappa (B.1.617.1) variant, the presence of the substitutions L452R and E484Q has been shown to affect antibody recognition, cause immune escape, and improve infectivity [15,16]. Currently, the Omicron variant contains a total of 59 mutations in its genome, with as many as 37 mutations occurring in the spike protein, including S371L, K417N, N440K, G446S, S477N, T478K, E484A, Q493R, G496S, Q498R, N501Y, and Y505H. In addition to some mutations that contribute to enhancing its transmissibility significantly [7,9,17], there are also some mutations that may affect partial therapeutic antibodies either through altering the conformation of mixed protein/carbohydrate epitope involving N343-N-linked glycan [18], such as S371L, or through changing its surface charge distribution, such as N440K, T478K, and E484A [7,19]. In Cao's recent study, Omicron could lead to significant humoral immune evasion and potential antigenic shifting with more than 85% of the tested human-neutralizing antibodies being escaped [20].

Due to the high risk of SARS-CoV-2 infection, the cultivation of an authentic virus requires a laboratory with a high level of biosafety, at least a biosafety level 3 (BSL-3) laboratory equipped with a negative pressure system, which limits the throughput and accessibility of authentic virus neutralization assays. An alternative method is packaging convenient and reliable replication-defective pseudovirus expressing the S protein that can be used under BSL-2 conditions. To date, the development of SARS-CoV-2 pseudoviruses using human immunodeficiency virus (HIV)-based lentiviral particles [21,22], murine leukemia virus (MLV)-based retroviral particles [21–23], or vesicular stomatitis virus (VSV)-based systems [21,24–26] has become a powerful tool for evaluating the efficacy of therapeutic drugs and vaccines, and the results from such pseudovirus neutralization assays correlate well with the results of measurements using authentic viruses [22,27].

HIV-1 has been frequently utilized as a vector virus for the creation of pseudotyped viruses harboring foreign virus surface protein [28,29]. Its genome contains genes encoding three major viral structural proteins, namely, gag, pol, and env [30]; two regulatory proteins, namely, tat and rev; and four accessory proteins that help complete viral packaging, namely, vpr, vif, vpu, and nef [31]. In this study, we used an HIV-based lentiviral system to produce pseudoviruses displaying the spike protein of SARS-CoV-2 variants on their surface, in which the firefly luciferase gene was inserted into the pNL4-3 nef gene [32], leading to frame shifts in env and vpr. The SARS-CoV-2 pseudoviruses constructed from this system are replication-defective viruses, yet they are competent for a single round of infection to host cells in a similar way as authentic viruses. The pseudovirus-based neutralization assay generated in this work is safe, convenient, and useful for the evaluation of vaccines and therapeutic candidates against SARS-CoV-2 variants.

2. Materials and Methods

2.1. Plasmids and Cells

The full length of the S gene of Alpha (GISAID accession ID: EPI_ISL_708969), Beta (GISAID accession ID: EPI_ISL_712081), Gamma (GISAID accession ID: EPI_ISL_792680), Kappa (GISAID accession ID: EPI_ISL_1360306) or Omicron/BA.1 (GISAID accession ID: EPI_ISL_12422410) was codon-optimized, synthesized, and cloned into the pUC57 vector by General Biosystems Inc. (Anhui, China). The S genes of SARS-CoV-2 variants were then amplified using the primers 5′- TATCGATCCGGAGGTACCATGG-3′ and 5′-TTATCAGTGATGGTGATG-3′ and cloned into the expression vector pCAGGS with the NEBuilder® HiFi DNA Assembly Cloning Kit (NEB), generating the pCAGGS-Alpha-S, pCAGGS-Beta-S, pCAGGS-Gamma-S, pCAGGS-Kappa-S, and pCAGGS-Omicron-S plasmids. The constructed recombinant plasmids bearing the S protein of SARS-CoV-2 variants were confirmed by DNA sequencing. The pDC316-WT-S containing a full-length S gene from Wuhan-Hu-1 and HIV backbone vector pNL4-3.Luc.R-E- were stored in our laboratory. ACE2-293T cells, which are HEK293T cells overexpressing ACE2 receptor, were produced and kept in our laboratory.

2.2. Analysis of SARS-CoV-2 Variant S Protein Expression

A total of 10^6 HEK293T cells in 4 mL growth medium were seeded in each well of a 6-well plate 16 h before transfection. The pCAGGS-Alpha-S, pCAGGS-Beta-S, pCAGGS-Gamma-S, pCAGGS-Kappa-S, pCAGGS-Omicron-S, and the reference pDC316-WT-S plasmids were individually transfected into HEK293T cells. HEK293T cells transfected with an empty pCAGGS vector were used as the negative control. After 48 h of incubation, the cells were fixed with 100% methanol for 30 min at −20 °C and then blocked in PBS containing 2% FBS for 1 h at room temperature. The cells transfected with the pCAGGS-Omicron-S plasmid were then incubated with primary antibody (Sino Biological, 40591-MM41), which is specific to Omicron S protein, whereas the other cells were incubated with primary antibody (Sino Biological, 40591-MM43), which is specific to S proteins of SARS-CoV-2 variants except Omicron, at 5 μg/mL for 1 h at 37 °C, followed by a further incubation at 37 °C for 1 h with Alexa Fluor 488 goat anti-mouse secondary antibody (Abcam, ab150117) at 2 μg/mL. For nuclear staining, cells were treated with DAPI for 10 min at room temperature. Stained sections were analyzed with BioTek Cytation1 Cell Imaging Multi-Mode Plate Readers.

2.3. Production and Titration of SARS-CoV-2 Pseudotyped Variants

HEK293T cells were inoculated in cell dishes and grown overnight at 37 °C with 5% CO_2 until the confluency for adherent cells reached 70–90%. The recombinant spike protein expression plasmids (pCAGGS-Alpha-S, pCAGGS-Beta-S, pCAGGS-Gamma-S, pCAGGS-Kappa-S, or pCAGGS-Omicron-S) were cotransfected with the HIV backbone vector pNL4-3.Luc.R-E- at different ratios into HEK293T cells with the Turbofect transfection reagent (Thermo Scientific, Waltham, MA, USA), respectively. The supernatants containing SARS-CoV-2 variant pseudoviruses were harvested 21–64 h after transfection and filtered through a 0.45 μm filter. The supernatants were then aliquoted into 2-mL cryotubes and stored at −80 °C.

The titer of the SARS-CoV-2 variant pseudovirus was measured by quantification of the luciferase activity. Supernatants containing SARS-CoV-2 variant pseudoviruses at the volume of 50 μL were used to infect 2×10^4 ACE2-293T cells in 100 μL DMEM with 10% FBS in each well of 96-well plates. The well without the addition of the pseudovirus served as the cell control. After a 48-h incubation in a 5% CO_2 environment at 37 °C, the culture supernatant was removed gently to leave 100 μL in each well, and then 100 μL of luciferase substrate (Perkin Elmer, Waltham, MA, USA) was added to each well. Two minutes after incubation at room temperature, 150 μL lysate was transferred to white solid 96-well plates (Costar, Washington, DC, USA) for the detection of luminescence using a TECAN Spark multifunctional microplate detector.

2.4. Neutralization Assay

For the neutralization assay, 50 μL pseudoviruses (~4×10^5 RLU) were incubated with serial dilutions of plasma samples (dilutions of 1:10, 30, 90, 270, 810, 2430, 7290, and 21,870) from COVID-19 convalescent patients and healthy individuals or monoclonal antibodies for 1 h at 37 °C, and then 2×10^4 ACE2-293T cells were added to each well. Cells without viruses, plasma, or antibodies were used as blank controls, and cells with viruses but without plasma or antibodies were used as virus controls. Luciferase activities were measured 48 h after infection, and the percent neutralization was calculated as 100% − (sample signals − blank control signals)/(virus control signals − blank control signals) × 100%. A three-parameter logistical analysis was performed on the full dilution series using Prism 8 (GraphPad Software, San Diego, CA, USA). All data are presented as the means ± standard deviations (SDs).

3. Results

3.1. Construction of the Recombinant Plasmids Expressing SARS-CoV-2 Variants Spike Proteins

The full-length S protein genes of the Alpha, Beta, Gamma, Kappa, and Omicron variants (Figure 1A) were synthesized and individually cloned into the pCAGGS vector, generating the pCAGGS-Alpha-S, pCAGGS-Beta-S, pCAGGS-Gamma-S, pCAGGS-Kappa-S, and pCAGGS-Omicron-S recombinant plasmids, respectively. These recombinant plasmids and the pDC316-WT-S were transfected into HEK293T cells, and the expression of S proteins on the HEK293T cell surface was evaluated using immunofluorescence (Figure 1B). The results showed that the S proteins of the WT, Alpha, Beta, Gamma, Kappa, and Omicron variants were expressed on the surface of HEK293T cells, whereas HEK293T cells transfected with the empty pCAGGS vector did not express any S protein. These data suggest that the recombinant plasmids bearing the S protein of WT and its variants can be used to package pseudotyped viruses.

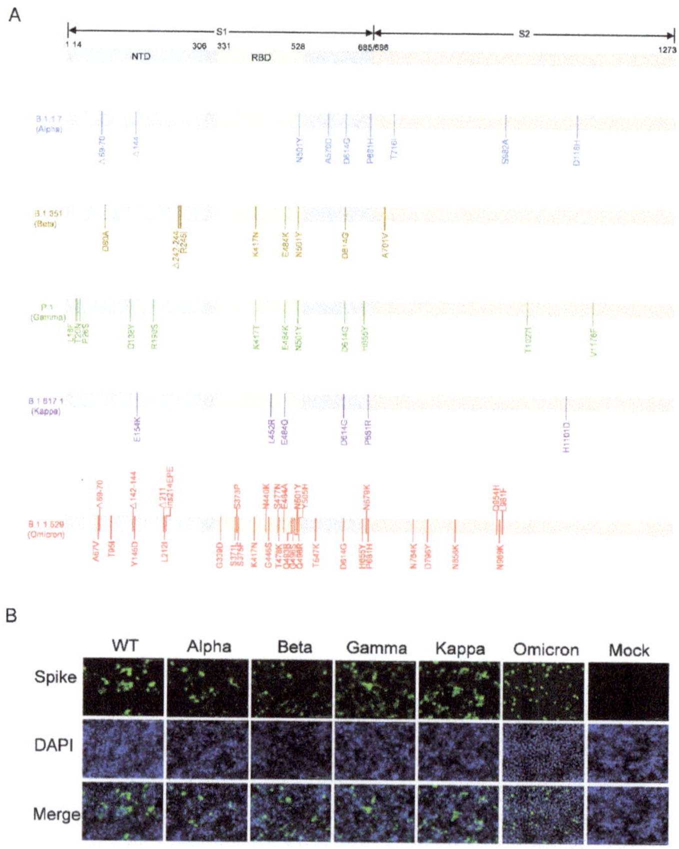

Figure 1. Detection of SARS-CoV-2 variants S protein expression in HEK293T cells. (**A**) Schematic overview of spike protein of SARS-CoV-2 variants, including Alpha (B.1.17), Beta (B.1.351), Gamma (P.1), Kappa (B.1.617.1), and Omicron (B.1.1.529). Amino acid mutations in comparison to the Wuhan-Hu-1 sequence are indicated. RBD, receptor binding domain; NTD, N-terminal domain. (**B**) Detection of SARS-CoV-2 S protein expression in HEK293T cells by immunofluorescence. The recombinant plasmids containing full-length S genes of SARS-CoV-2 variants were individually transfected into HEK293T cells. Cells transfected with an empty pCAGGS vector with the same procedure were used as the negative control. The cells were fixed after 48 h of incubation and labeled with the corresponding antibodies. Nuclei were stained with DAPI.

3.2. Optimization of SARS-CoV-2 Variants Pseudovirus Production

To generate the SARS-CoV-2 variant pseudoviruses, we used an HIV backbone vector-based pseudovirus packaging system (Figure 2A). The HIV backbone vector, pNL4-3.Luc.R-E-, was derived from pNL4-3 vector, in which the nef gene was replaced by the firefly luciferase gene, resulting in frame shifts in env and vpr. The recombinant spike protein expression plasmids (pCAGGS-Alpha-S, pCAGGS-Beta-S, pCAGGS-Gamma-S, pCAGGS-Kappa-S, or pCAGGS-Omicron-S) were cotransfected with the pNL4-3.Luc.R-E- plasmids into HEK293T cells, respectively, at ratios of 1:30, 1:60, 1:90, 1:120, 1:150, and 1:180. As shown in Figure 2B, the supernatants containing SARS-CoV-2 variant pseudoviruses were harvested 21–64 h post-transfection. ACE2-293T cells were infected with SARS-CoV-2 variant pseudoviruses for 48 h, and then viral titers were determined by measuring the relative luminescence units (RLU). The RLU values reached the peak level at 52 h post-infection for the pseudotyped Alpha, Beta, and Kappa variants at ratios of 1:60, 1:120, and 1:90, whereas the RLU level was highest for the Gamma variant at 64 h post-infection at a ratio of 1:60. For the Omicron pseudotyped virus, the highest viral titer was observed at 59 h after transfection at a ratio of 1:150 with approximately 1.6×10^6 RLU.

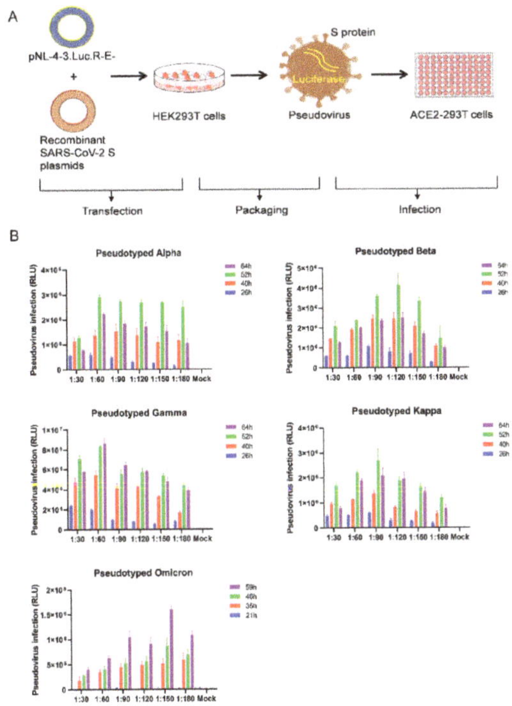

Figure 2. Optimization of SARS-CoV-2 variants pseudovirus production. (**A**) Schematic representation of the SARS-CoV-2 variants pseudovirus production and neutralization assay. The HIV backbone vector pNL4-3.Luc.R-E- plasmids were cotransfected with pCAGGS-Alpha-S, pCAGGS-Beta-S, pCAGGS-Gamma-S, pCAGGS-Kappa-S, or pCAGGS-Omicron-S, respectively into HEK293T cells to package the pseudotyped lentiviral particles. The supernatants containing SARS-CoV-2 variants pseudovirus with S protein were collected and then ACE2-293T cells were used to measure the pseudoviral titer. (**B**) Effect of the ratio of the recombinant S protein expression plasmids to the HIV backbone plasmids and the collection time for pseudovirus particles on the production of pseudovirus. Cells without pseudovirus infection were used as background. The data were expressed as mean relative luciferase units (RLU) ± standard deviation (SD) of 3 parallel wells in 96-well culture plates.

3.3. *Validation of the Neutralization Sensitivity of SARS-CoV-2 Pseudotyped Variants*

To examine the neutralization sensitivity of the Omicron variant and the other SARS-CoV-2 variants, we evaluated the neutralizing activities of several antibodies that have obtained emergency use authorization (REGN10987 [33], REGN10933 [33], S309 [34], BRII-198 [35] and LY-CoV1404 [36]) or are being studied in clinical trials presently (ADG-2 [37]) against these SARS-CoV-2 pseudotyped variants (Figure 3A). Consistent with the previously reported results [20,38,39], Omicron/BA.1 can escape some neutralizing antibodies, such as REGN10987 and REGN10933, while S309 and ADG-2 function with reduced neutralization capability. Luckily, LY-CoV1404 and BRII-198 were able to potently neutralize Alpha, Beta, Gamma, Kappa, and Omicron/BA.1 pseudoviruses. Moreover, we assessed the neutralizing capacities of the plasma from ten COVID-19 convalescent patients [40], who recovered from infection of the Wuhan-Hu-1 strain, using the pseudovirus neutralization assays we established. As shown in Figure 3B, although all the plasma showed inhibition against pseudotyped wild-type (WT) virus, Alpha, Gamma, and Kappa variants with different average IC_{50} titers (462.9, 1818, 290.6, and 203.9), all of them we tested showed weak inhibition against Beta. When it turned to Omicron, only 1/10 (ConV-9) convalescent plasma showed detectable activity, and the reduction level was much higher than that of Beta. This result is consistent with the work of Cameroni et al. [41], indicating that Omicron was highly resistant to the plasma from the convalescent patients in Wuhan. Most control plasma samples from healthy individuals showed no inhibitory activity against all tested pseudoviruses, only few of them had extremely weak detectable responses, but the inhibition rates have not reached 50% in initial dilution.

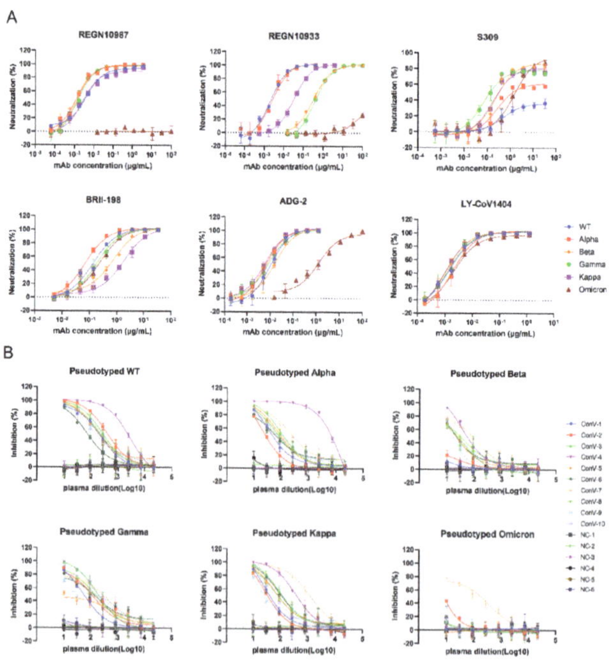

Figure 3. Validation of the neutralization sensitivity of SARS-CoV-2 pseudotyped variants. (**A**) Neutralizing curves of monoclonal antibodies against pseudotyped SARS-CoV-2 variants. Data are representative of at least two independent experiments. Mean ± SD was shown. (**B**) The inhibition activity of ten COVID-19 convalescent plasma samples against pseudotyped SARS-CoV-2 variants. Six plasma samples from healthy individuals were tested as negative controls (NC). The initial dilutions for both positive and negative samples were 1:10, followed by a 3-fold serial dilution. Samples were tested in triplicates and the experiments were repeated at least twice. Data from one of at least two independent experiments are presented in Mean ± SD.

4. Discussion

SARS-CoV-2 variants are highly pathogenic coronaviruses, and experiments with these viruses need to be performed in the biosafety level 3 laboratory with appropriate qualifications, which has become a bottleneck in drug and vaccine development. Pseudovirus neutralization assay based on retroviruses or HIV backbones is an alternative approach to authentic virus neutralization assay and can be used safely and conveniently to a greater extent for drug screening and vaccine evaluation [22,42–44].

Many factors affect the packaging efficiency of pseudoviruses, including the expression level of the viral envelope protein, the ratio of packaging plasmid combinations, the efficiency of plasmid transfection, the growth state of HEK293T cells before transfection, and the collection time of pseudovirus supernatants. In our study, the ratio of the spike protein expression plasmids to the HIV backbone plasmids considerably influenced the production of pseudovirus, and this ratio differed between wild-type SARS-CoV-2 and its variants. It is necessary to determine the optimal ratio of packaging plasmid combinations before large-scale pseudovirus packaging. Moreover, the optimal ratio of the spike protein expression plasmids to the HIV backbone plasmids in our study (1:60 to 1:120) were much lower than that in the previous studies (1:1 to 1:9) [21,45]. In addition, we found that a relatively high viral titer can be obtained from 50 h to 64 h after transfection, indicating that optimizing the collection time of pseudovirus supernatants is also crucial for improving the pseudovirus yield.

We used several approved monoclonal antibodies to validate the pseudovirus neutralization assay we established, including REGN10987, REGN10933, S309, BRII-198, LY-CoV1404, and ADG-2; the results were consistent with the previous research [20,38,39], which suggests this pseudovirus-based neutralization assay is a reliable alternative for the rapid detection of neutralizing antibodies against SARS-CoV-2 and its variants when BSL-3 facilities are not available. We also showed that convalescent plasma from COVID-19 patients blocked the entry of pseudotyped SARS-CoV-2 wild-type, Alpha, Beta, Gamma, and Kappa into ACE2-293T cells, while all convalescent plasma samples except ConV-9 failed to prevent the Omicron infection, underscoring the immune evasion of Omicron. However, one limitation of this assay was that we did not detect other receptors in addition to ACE2 receptor, the key determinant of SARS-CoV-2 attachment to target cells [46,47]. Recent studies have reported that SARS-CoV-2 also has AXL, KREMEN1, and ASGR1 in addition to ACE2 as its candidate receptors [48,49]. The role of other SARS-CoV-2 receptors in pseudovirus neutralization assay could not be ignored.

Overall, the spike proteins of SARS-CoV-2 variants were expressed on the surface of HEK293T cells to produce pseudovirus particles using a lentivirus vector-based pseudovirus system in optimized packaging conditions. This pseudovirus system can be used in animals to evaluate the in-vivo efficacy of vaccines or antibodies [50,51]. Taken together, this convenient and reliable pseudovirus system can be widely used for developing SARS-CoV-2 vaccines and therapeutic drugs and for studying SARS-CoV-2 infection.

Author Contributions: Conceptualization, H.S., C.Y. and X.C.; investigation, H.S., J.X., G.Z. and J.H.; resources, M.H. and T.F.; data curation, Z.C.; writing—original draft preparation, H.S., X.C. and C.Y.; writing—review and editing, all authors. All authors have read and agreed to the published version of the manuscript.

Funding: This research was supported by grants from the National Natural Science Foundation of China (projects 81803429) and the Young Elite Scientists Sponsorship Program by CAST.

Institutional Review Board Statement: The study was conducted in accordance with the Declaration of Helsinki, and approved by the Research Ethics Committee of Wuhan Infectious Disease Hospital, Hubei Province, China (approval number: KY-2020-11.01).

Informed Consent Statement: The Research Ethics Committee waived the requirement informed consent before the study started because of the urgent need to collect epidemiological and clinical data. We analyzed all the data anonymously.

Data Availability Statement: Not applicable.

Acknowledgments: We thank GISAID and associated laboratories and researchers for the shared sequence information.

Conflicts of Interest: The authors declare no conflict of interest. The funders had no role in the design of the study; in the collection, analyses, or interpretation of data; in the writing of the manuscript, or in the decision to publish the results.

References

1. Garcia-Beltran, W.F.; Lam, E.C.; St Denis, K.; Nitido, A.D.; Garcia, Z.H.; Hauser, B.M.; Feldman, J.; Pavlovic, M.N.; Gregory, D.J.; Poznansky, M.C.; et al. Multiple SARS-CoV-2 variants escape neutralization by vaccine-induced humoral immunity. *Cell* **2021**, *184*, 2372–2383.e9. [CrossRef] [PubMed]
2. Faria, N.R.; Mellan, T.A.; Whittaker, C.; Claro, I.M.; Candido, D.D.S.; Mishra, S.; Crispim, M.A.E.; Sales, F.C.S.; Hawryluk, I.; McCrone, J.T.; et al. Genomics and epidemiology of the P.1 SARS-CoV-2 lineage in Manaus, Brazil. *Science* **2021**, *372*, 815–821. [CrossRef] [PubMed]
3. Wibmer, C.K.; Ayres, F.; Hermanus, T.; Madzivhandila, M.; Kgagudi, P.; Oosthuysen, B.; Lambson, B.E.; de Oliveira, T.; Vermeulen, M.; van der Berg, K.; et al. SARS-CoV-2 501Y.V2 escapes neutralization by South African COVID-19 donor plasma. *Nat. Med.* **2021**, *27*, 622–625. [CrossRef] [PubMed]
4. Kuzmina, A.; Khalaila, Y.; Voloshin, O.; Keren-Naus, A.; Boehm-Cohen, L.; Raviv, Y.; Shemer-Avni, Y.; Rosenberg, E.; Taube, R. SARS-CoV-2 spike variants exhibit differential infectivity and neutralization resistance to convalescent or post-vaccination sera. *Cell Host Microbe* **2021**, *29*, 522–528.e2. [CrossRef]
5. Shen, X.; Tang, H.; McDanal, C.; Wagh, K.; Fischer, W.; Theiler, J.; Yoon, H.; Li, D.; Haynes, B.F.; Sanders, K.O.; et al. SARS-CoV-2 variant B.1.1.7 is susceptible to neutralizing antibodies elicited by ancestral spike vaccines. *Cell Host Microbe* **2021**, *29*, 529–539.e3. [CrossRef]
6. Koehler, M.; Ray, A.; Moreira, R.A.; Juniku, B.; Poma, A.B.; Alsteens, D. Molecular insights into receptor binding energetics and neutralization of SARS-CoV-2 variants. *Nat. Commun.* **2021**, *12*, 6977. [CrossRef]
7. Cui, Z.; Liu, P.; Wang, N.; Wang, L.; Fan, K.; Zhu, Q.; Wang, K.; Chen, R.; Feng, R.; Jia, Z. Structural and functional characterizations of infectivity and immune evasion of SARS-CoV-2 Omicron. *Cell* **2022**, *185*, 860–871.e13. [CrossRef]
8. Hoffmann, M.; Krüger, N.; Schulz, S.; Cossmann, A.; Rocha, C.; Kempf, A.; Nehlmeier, I.; Graichen, L.; Moldenhauer, A.-S.; Winkler, M.S. The Omicron variant is highly resistant against antibody-mediated neutralization: Implications for control of the COVID-19 pandemic. *Cell* **2022**, *185*, 447–456.e11. [CrossRef]
9. Starr, T.N.; Greaney, A.J.; Hilton, S.K.; Ellis, D.; Crawford, K.H.D.; Dingens, A.S.; Navarro, M.J.; Bowen, J.E.; Tortorici, M.A.; Walls, A.C.; et al. Deep Mutational Scanning of SARS-CoV-2 Receptor Binding Domain Reveals Constraints on Folding and ACE2 Binding. *Cell* **2020**, *182*, 1295–1310.e20. [CrossRef]
10. Xie, X.; Liu, Y.; Liu, J.; Zhang, X.; Zou, J.; Fontes-Garfias, C.R.; Xia, H.; Swanson, K.A.; Cutler, M.; Cooper, D.; et al. Neutralization of SARS-CoV-2 spike 69/70 deletion, E484K and N501Y variants by BNT162b2 vaccine-elicited sera. *Nat. Med.* **2021**, *27*, 620–621. [CrossRef]
11. Liu, Z.; VanBlargan, L.A.; Bloyet, L.M.; Rothlauf, P.W.; Chen, R.E.; Stumpf, S.; Zhao, H.; Errico, J.M.; Theel, E.S.; Liebeskind, M.J.; et al. Identification of SARS-CoV-2 spike mutations that attenuate monoclonal and serum antibody neutralization. *Cell Host Microbe* **2021**, *29*, 477–488.e4. [CrossRef] [PubMed]
12. Harvey, W.T.; Carabelli, A.M.; Jackson, B.; Gupta, R.K.; Thomson, E.C.; Harrison, E.M.; Ludden, C.; Reeve, R.; Rambaut, A.; Peacock, S.J. SARS-CoV-2 variants, spike mutations and immune escape. *Nat. Rev. Microbiol.* **2021**, *19*, 409–424. [CrossRef] [PubMed]
13. Khan, A.; Zia, T.; Suleman, M.; Khan, T.; Ali, S.S.; Abbasi, A.A.; Mohammad, A.; Wei, D.Q. Higher infectivity of the SARS-CoV-2 new variants is associated with K417N/T, E484K, and N501Y mutants: An insight from structural data. *J. Cell. Physiol.* **2021**, *236*, 7045–7057. [CrossRef] [PubMed]
14. Wang, R.; Zhang, Q.; Ge, J.; Ren, W.; Zhang, R.; Lan, J.; Ju, B.; Su, B.; Yu, F.; Chen, P.; et al. Analysis of SARS-CoV-2 variant mutations reveals neutralization escape mechanisms and the ability to use ACE2 receptors from additional species. *Immunity* **2021**, *54*, 1611–1621.e5. [CrossRef] [PubMed]
15. Deng, X.; Garcia-Knight, M.A.; Khalid, M.M.; Servellita, V.; Wang, C.; Morris, M.K.; Sotomayor-González, A.; Glasner, D.R.; Reyes, K.R.; Gliwa, A.S.; et al. Transmission, infectivity, and neutralization of a spike L452R SARS-CoV-2 variant. *Cell* **2021**, *184*, 3426–3437.e8. [CrossRef] [PubMed]
16. McCallum, M.; Walls, A.C.; Sprouse, K.R.; Bowen, J.E.; Rosen, L.E.; Dang, H.V.; De Marco, A.; Franko, N.; Tilles, S.W.; Logue, J.; et al. Molecular basis of immune evasion by the Delta and Kappa SARS-CoV-2 variants. *Science* **2021**, *374*, 1621–1626. [CrossRef]
17. Mannar, D.; Saville, J.W.; Zhu, X.; Srivastava, S.S.; Berezuk, A.M.; Tuttle, K.S.; Marquez, A.C.; Sekirov, I.; Subramaniam, S. SARS-CoV-2 Omicron variant: Antibody evasion and cryo-EM structure of spike protein-ACE2 complex. *Science* **2022**, *375*, 760–764. [CrossRef]
18. Sztain, T.; Ahn, S.H.; Bogetti, A.T.; Casalino, L.; Goldsmith, J.A.; Seitz, E.; McCool, R.S.; Kearns, F.L.; Acosta-Reyes, F.; Maji, S.; et al. A glycan gate controls opening of the SARS-CoV-2 spike protein. *Nat. Chem.* **2021**, *13*, 963–968. [CrossRef]
19. Jawad, B.; Adhikari, P.; Podgornik, R.; Ching, W.Y. Binding Interactions between Receptor-Binding Domain of Spike Protein and Human Angiotensin Converting Enzyme-2 in Omicron Variant. *J. Phys. Chem. Lett.* **2022**, *13*, 3915–3921. [CrossRef]
20. Cao, Y.; Wang, J.; Jian, F.; Xiao, T.; Song, W.; Yisimayi, A.; Huang, W.; Li, Q.; Wang, P.; An, R. Omicron escapes the majority of existing SARS-CoV-2 neutralizing antibodies. *Nature* **2022**, *602*, 657–663. [CrossRef]

21. Johnson, M.C.; Lyddon, T.D.; Suarez, R.; Salcedo, B.; LePique, M.; Graham, M.; Ricana, C.; Robinson, C.; Ritter, D.G. Optimized pseudotyping conditions for the SARS-COV-2 spike glycoprotein. *J. Virol.* **2020**, *94*, e01062-20. [CrossRef] [PubMed]
22. Giroglou, T.; Cinatl, J., Jr.; Rabenau, H.; Drosten, C.; Schwalbe, H.; Doerr, H.W.; Von Laer, D. Retroviral vectors pseudotyped with severe acute respiratory syndrome coronavirus S protein. *J. Virol.* **2004**, *78*, 9007–9015. [CrossRef] [PubMed]
23. Zheng, Y.; Larragoite, E.T.; Williams, E.S.; Lama, J.; Cisneros, I.; Delgado, J.C.; Slev, P.; Rychert, J.; Innis, E.A.; Coiras, M. Neutralization assay with SARS-CoV-1 and SARS-CoV-2 spike pseudotyped murine leukemia virions. *Virol. J.* **2021**, *18*, 1. [CrossRef] [PubMed]
24. Salazar-García, M.; Acosta-Contreras, S.; Rodríguez-Martínez, G.; Cruz-Rangel, A.; Flores-Alanis, A.; Patiño-López, G.; Luna-Pineda, V.M. Pseudotyped vesicular stomatitis virus-severe acute respiratory syndrome-coronavirus-2 spike for the study of variants, vaccines, and therapeutics against coronavirus disease 2019. *Front. Microbiol.* **2021**, *12*, 817200. [CrossRef] [PubMed]
25. Ou, X.; Liu, Y.; Lei, X.; Li, P.; Mi, D.; Ren, L.; Guo, L.; Guo, R.; Chen, T.; Hu, J. Characterization of spike glycoprotein of SARS-CoV-2 on virus entry and its immune cross-reactivity with SARS-CoV. *Nat. Commun.* **2020**, *11*, 1620. [CrossRef] [PubMed]
26. Zhang, L.; Li, Q.; Liang, Z.; Li, T.; Liu, S.; Cui, Q.; Nie, J.; Wu, Q.; Qu, X.; Huang, W. The significant immune escape of pseudotyped SARS-CoV-2 Variant Omicron. *Emerg. Microbes Infect.* **2022**, *11*, 1–5. [CrossRef]
27. Xiong, H.-L.; Wu, Y.-T.; Cao, J.-L.; Yang, R.; Liu, Y.-X.; Ma, J.; Qiao, X.-Y.; Yao, X.-Y.; Zhang, B.-H.; Zhang, Y.-L. Robust neutralization assay based on SARS-CoV-2 S-protein-bearing vesicular stomatitis virus (VSV) pseudovirus and ACE2-overexpressing BHK21 cells. *Emerg. Microbes Infect.* **2020**, *9*, 2105–2113. [CrossRef]
28. Zhao, G.; Du, L.; Ma, C.; Li, Y.; Li, L.; Poon, V.K.; Wang, L.; Yu, F.; Zheng, B.-J.; Jiang, S. A safe and convenient pseudovirus-based inhibition assay to detect neutralizing antibodies and screen for viral entry inhibitors against the novel human coronavirus MERS-CoV. *Virol. J.* **2013**, *10*, 266. [CrossRef]
29. Du, L.; Zhao, G.; Zhang, X.; Liu, Z.; Yu, H.; Zheng, B.-J.; Zhou, Y.; Jiang, S. Development of a safe and convenient neutralization assay for rapid screening of influenza HA-specific neutralizing monoclonal antibodies. *Biochem. Biophys. Res. Commun.* **2010**, *397*, 580–585. [CrossRef]
30. Han, C.; Johnson, J.; Dong, R.; Kandula, R.; Kort, A.; Wong, M.; Yang, T.; Breheny, P.J.; Brown, G.D.; Haim, H. Key Positions of HIV-1 Env and Signatures of Vaccine Efficacy Show Gradual Reduction of Population Founder Effects at the Clade and Regional Levels. *mBio* **2020**, *11*, e00126-20. [CrossRef]
31. Strebel, K. HIV accessory proteins versus host restriction factors. *Curr. Opin. Virol.* **2013**, *3*, 692–699. [CrossRef] [PubMed]
32. Lassen, K.G.; Hebbeler, A.M.; Bhattacharyya, D.; Lobritz, M.A.; Greene, W.C. A flexible model of HIV-1 latency permitting evaluation of many primary CD4 T-cell reservoirs. *PLoS ONE* **2012**, *7*, e30176. [CrossRef] [PubMed]
33. Hansen, J.; Baum, A.; Pascal, K.E.; Russo, V.; Giordano, S.; Wloga, E.; Fulton, B.O.; Yan, Y.; Koon, K.; Patel, K.; et al. Studies in humanized mice and convalescent humans yield a SARS-CoV-2 antibody cocktail. *Science* **2020**, *369*, 1010–1014. [CrossRef] [PubMed]
34. Pinto, D.; Park, Y.-J.; Beltramello, M.; Walls, A.C.; Tortorici, M.A.; Bianchi, S.; Jaconi, S.; Culap, K.; Zatta, F.; De Marco, A. Cross-neutralization of SARS-CoV-2 by a human monoclonal SARS-CoV antibody. *Nature* **2020**, *583*, 290–295. [CrossRef]
35. Ju, B.; Zhang, Q.; Ge, J.; Wang, R.; Sun, J.; Ge, X.; Yu, J.; Shan, S.; Zhou, B.; Song, S.; et al. Human neutralizing antibodies elicited by SARS-CoV-2 infection. *Nature* **2020**, *584*, 115–119. [CrossRef]
36. Westendorf, K.; Žentelis, S.; Wang, L.; Foster, D.; Vaillancourt, P.; Wiggin, M.; Lovett, E.; van der Lee, R.; Hendle, J.; Pustilnik, A.; et al. LY-CoV1404 (bebtelovimab) potently neutralizes SARS-CoV-2 variants. *Cell Rep.* **2022**, *39*, 110812. [CrossRef]
37. Rappazzo, C.G.; Tse, L.V.; Kaku, C.I.; Wrapp, D.; Sakharkar, M.; Huang, D.; Deveau, L.M.; Yockachonis, T.J.; Herbert, A.S.; Battles, M.B.; et al. Broad and potent activity against SARS-like viruses by an engineered human monoclonal antibody. *Science* **2021**, *371*, 823–829. [CrossRef]
38. Iketani, S.; Liu, L.; Guo, Y.; Liu, L.; Chan, J.F.-W.; Huang, Y.; Wang, M.; Luo, Y.; Yu, J.; Chu, H. Antibody evasion properties of SARS-CoV-2 Omicron sublineages. *Nature* **2022**, *604*, 553–556. [CrossRef]
39. Liu, L.; Iketani, S.; Guo, Y.; Chan, J.F.-W.; Wang, M.; Liu, L.; Luo, Y.; Chu, H.; Huang, Y.; Nair, M.S. Striking antibody evasion manifested by the Omicron variant of SARS-CoV-2. *Nature* **2022**, *602*, 676–681. [CrossRef]
40. Chi, X.; Yan, R.; Zhang, J.; Zhang, G.; Zhang, Y.; Hao, M.; Zhang, Z.; Fan, P.; Dong, Y.; Yang, Y. A neutralizing human antibody binds to the N-terminal domain of the Spike protein of SARS-CoV-2. *Science* **2020**, *369*, 650–655. [CrossRef]
41. Cameroni, E.; Bowen, J.E.; Rosen, L.E.; Saliba, C.; Zepeda, S.K.; Culap, K.; Pinto, D.; VanBlargan, L.A.; De Marco, A.; di Iulio, J. Broadly neutralizing antibodies overcome SARS-CoV-2 Omicron antigenic shift. *Nature* **2022**, *602*, 664–670. [CrossRef] [PubMed]
42. Catanese, M.T.; Dorner, M. Advances in experimental systems to study hepatitis C virus in vitro and in vivo. *Virology* **2015**, *479*, 221–233. [CrossRef] [PubMed]
43. Yang, Y.; Du, L.; Liu, C.; Wang, L.; Ma, C.; Tang, J.; Baric, R.S.; Jiang, S.; Li, F. Receptor usage and cell entry of bat coronavirus HKU4 provide insight into bat-to-human transmission of MERS coronavirus. *Proc. Natl. Acad. Sci. USA* **2014**, *111*, 12516–12521. [CrossRef] [PubMed]
44. Nie, J.; Li, Q.; Wu, J.; Zhao, C.; Hao, H.; Liu, H.; Zhang, L.; Nie, L.; Qin, H.; Wang, M. Establishment and validation of a pseudovirus neutralization assay for SARS-CoV-2. *Emerg. Microbes Infect.* **2020**, *9*, 680–686. [CrossRef]
45. Crawford, K.H.; Eguia, R.; Dingens, A.S.; Loes, A.N.; Malone, K.D.; Wolf, C.R.; Chu, H.Y.; Tortorici, M.A.; Veesler, D.; Murphy, M. Protocol and reagents for pseudotyping lentiviral particles with SARS-CoV-2 spike protein for neutralization assays. *Viruses* **2020**, *12*, 513. [CrossRef]
46. Hoffmann, M.; Kleine-Weber, H.; Schroeder, S.; Krüger, N.; Herrler, T.; Erichsen, S.; Schiergens, T.S.; Herrler, G.; Wu, N.-H.; Nitsche, A. SARS-CoV-2 cell entry depends on ACE2 and TMPRSS2 and is blocked by a clinically proven protease inhibitor. *Cell* **2020**, *181*, 271–280.e8. [CrossRef]

47. Walls, A.C.; Park, Y.J.; Tortorici, M.A.; Wall, A.; McGuire, A.T.; Veesler, D. Structure, Function, and Antigenicity of the SARS-CoV-2 Spike Glycoprotein. *Cell* **2020**, *181*, 281–292.e6. [CrossRef]
48. Wang, S.; Qiu, Z.; Hou, Y.; Deng, X.; Xu, W.; Zheng, T.; Wu, P.; Xie, S.; Bian, W.; Zhang, C.; et al. AXL is a candidate receptor for SARS-CoV-2 that promotes infection of pulmonary and bronchial epithelial cells. *Cell Res.* **2021**, *31*, 126–140. [CrossRef]
49. Gu, Y.; Cao, J.; Zhang, X.; Gao, H.; Wang, Y.; Wang, J.; He, J.; Jiang, X.; Zhang, J.; Shen, G.; et al. Receptome profiling identifies KREMEN1 and ASGR1 as alternative functional receptors of SARS-CoV-2. *Cell Res.* **2022**, *32*, 24–37. [CrossRef]
50. Zhang, L.; Li, Q.; Liu, Q.; Huang, W.; Nie, J.; Wang, Y. A bioluminescent imaging mouse model for Marburg virus based on a pseudovirus system. *Hum. Vaccines Immunother.* **2017**, *13*, 1811–1817. [CrossRef]
51. Tseng, S.H.; Lam, B.; Kung, Y.J.; Lin, J.; Liu, L.; Tsai, Y.C.; Ferrall, L.; Roden, R.B.S.; Wu, T.C.; Hung, C.F. A novel pseudovirus-based mouse model of SARS-CoV-2 infection to test COVID-19 interventions. *J. Biomed. Sci.* **2021**, *28*, 34. [CrossRef] [PubMed]

viruses

MDPI

Article

Hyperimmunized Chickens Produce Neutralizing Antibodies against SARS-CoV-2

Emily J. Aston [1], Michael G. Wallach [2], Aarthi Narayanan [3], Sofia Egaña-Labrin [4] and Rodrigo A. Gallardo [4,*]

1 Department of Animal Science, College of Agricultural and Environmental Sciences, University of California-Davis, Davis, CA 95616, USA; aston.emily@gmail.com
2 School of Life Sciences, Faculty of Science, University of Technology Sydney, Sydney, NSW 2007, Australia; michael.wallach@uts.edu.au
3 National Center for Biodefense and Infectious Diseases, George Mason University, Fairfax, VA 22030, USA; anaraya1@gmu.edu
4 Department of Population Health and Reproduction, School of Veterinary Medicine, University of California-Davis, Davis, CA 95616, USA; seganal@ucdavis.edu
* Correspondence: ragallardo@ucdavis.edu

Abstract: The novel severe acute respiratory syndrome (SARS) coronavirus, SARS-CoV-2, is responsible for the global COVID-19 pandemic. Effective interventions are urgently needed to mitigate the effects of COVID-19 and likely require multiple strategies. Egg-extracted antibody therapies are a low-cost and scalable strategy to protect at-risk individuals from SARS-CoV-2 infection. Commercial laying hens were hyperimmunized against the SARS-CoV-2 S1 protein using three different S1 recombinant proteins and three different doses. Sera and egg yolk were collected at three and six weeks after the second immunization for enzyme-linked immunosorbent assay and plaque-reduction neutralization assay to determine antigen-specific antibody titers and neutralizing antibody titers, respectively. In this study we demonstrate that hens hyperimmunized against the SARS-CoV-2 recombinant S1 and receptor binding domain (RBD) proteins produced neutralizing antibodies against SARS-CoV-2. We further demonstrate that antibody production was dependent on the dose and type of antigen administered. Our data suggests that antibodies purified from the egg yolk of hyperimmunized hens can be used as immunoprophylaxis in humans at risk of exposure to SARS-CoV-2.

Keywords: COVID-19; SARS-CoV-2; chicken; passive immunization; antibodies; neutralizing antibodies; egg

Citation: Aston, E.J.; Wallach, M.G.; Narayanan, A.; Egaña-Labrin, S.; Gallardo, R.A. Hyperimmunized Chickens Produce Neutralizing Antibodies against SARS-CoV-2. *Viruses* **2022**, *14*, 1510. https://doi.org/10.3390/v14071510

Academic Editors: Lianpan Dai and Ahmed Elshamy

Received: 14 March 2022
Accepted: 7 July 2022
Published: 9 July 2022

1. Introduction

The novel severe acute respiratory syndrome (SARS) coronavirus, SARS-CoV-2, is the etiologic agent of COVID-19, a newly emerged viral respiratory disease in humans. First identified in late 2019, the disease reached pandemic status and continues to be a significant public health threat worldwide. The COVID-19 landscape is rapidly evolving, and effective strategies to mitigate its spread are urgently needed. Notably, the first emergency use authorizations for a vaccine for the prevention of COVID-19 were issued on 11 December 2020 [1]. Multiple vaccines and treatments have since been developed and approved for emergency use. Although vaccines appear reasonably effective at reducing the severity of illness associated with COVID-19, the recent emergence of new variants threatens the efficacy of these vaccines. Therefore, ongoing research to develop new vaccines, antiviral drugs, and antibody therapies is critical to reducing the devastating impact of the pandemic. Successful COVID-19 intervention likely requires multiple approaches, and antibody therapies are an attractive strategy to protect at-risk individuals from SARS-CoV-2 infection.

The genetic material of coronaviruses is composed of a single-stranded, positive-sense RNA genome [2]. This genome is the largest of the RNA viruses and ranges from 27 to

30 kilobase pairs [2]. SARS-CoV-2 is a betacoronavirus belonging to the family *Coronaviridae* [2]. Coronaviruses cause respiratory and gastrointestinal disease in a wide range of animal species, including mammals and birds [2]. Alphacoronaviruses and betacoronaviruses are known to infect humans and other mammals, and gammacoronaviruses and deltacoronaviruses infect birds, although some can infect mammals [2]. The most notable and extensively studied gammacoronavirus in chickens is infectious bronchitis virus (IBV), against which most commercial hens are immunized [2].

Coronavirus entry into cells is mediated by the spike protein, most specifically its S1 portion, a peplomer-like structure anchored to the virus membrane in the form of a trimer. In addition, the S1 subunit of this protein determines virus variability and elicits neutralizing antibodies [2–4]. On the globular head of each SARS-CoV-2 S1 protein is a receptor-binding domain (RBD), which specifically recognizes the human angiotensin-converting enzyme 2 (ACE2) [5]. The RBD is the most antigenic region of the spike protein in coronaviruses and is thus an attractive site for therapies that target cell entry of the virus.

One such therapy was approved by the U.S. Food and Drug Administration on 23 August 2020 for an emergency use authorization for investigational convalescent plasma to treat hospitalized COVID-19 patients [6]. The principle behind this therapy is that recovering COVID-19-infected individuals develop SARS-CoV-2-specific antibodies that can be recovered from plasma and administered to ill patients to neutralize the virus if applied systemically. This passive transfer of antibodies is not limited only to SARS-CoV-2-infected patients. Researchers are also applying this concept of virus neutralization to generate SARS-CoV-2-specific antibodies in animals to be used for passive immunization against the virus in humans [7]. Harvesting antibodies from eggs laid by hens that have been immunized against the spike protein of SARS-CoV-2 is an attractive model to produce protective antibodies due to the scalability, convenience, and low cost [7].

Chickens produce immunoglobulin Y (IgY), which is a homologue of mammalian IgG. An average egg yolk yields 50–100 mg of IgY, of which 2–10% comprise specific antibodies [8,9]. When hyperimmunizing hens, the amount of antigen-specific IgY produced will depend on the age of the hen, adjuvant, route of application, as well as the dose, antigenicity, and molecular weight of antigen administered to each hen [10–13]. A laying hen lays an average of 300 eggs per year, which corresponds to approximately 15–30 g of IgY. Thus, a considerable amount of polyclonal antibody can be non-invasively recovered from the eggs laid by immunized chickens. Other advantages of using IgY in human applications are that it is well tolerated and can be administered orally [14,15]. Furthermore, IgY neither binds to human rheumatoid factors nor activates the human complement system [14]; therefore, reducing the risk of inflammatory reactions as a secondary effect of using antibodies produced in different species. These characteristics make chicken IgY a promising source of new therapies for human viral diseases such as COVID-19 in addition to vaccination strategies [16].

Here, we demonstrate that hens hyperimmunized against the SARS-CoV-2 recombinant receptor binding domain and/or S1 protein produced neutralizing antibodies against SARS-CoV-2. By hyperimminunizing and testing for antibodies against SARS-CoV-2 via enzyme-linked immunosorbent assay (ELISA) and plaque-reduction neutralization assay (PRNA), we further demonstrate that antibody production was dependent on the dose and type of antigen administered. Our data suggests that antibodies purified from the egg yolk of hyperimmunized chickens can be used as immunoprophylaxis in humans at risk of exposure to SARS-CoV-2.

2. Materials and Methods

2.1. Constructs and Virus

The constructs used for vaccines A and B included the SARS-CoV-2 S1-derived RBD protein #0570, which included amino acids 328-533 and a His tag (vaccine A) and SARS-CoV S1-derived RBD amino acids 319-542 and a His tag (vaccine B), using a bacterial based in vitro expression system [17]. The reference virus used to obtain these constructs was the

Wuhan strain. The construct used for vaccine C consisted of the SARS-CoV-2 S1 protein, including amino acids 16-685 and an Fc-tag, which was expressed in HEK293 cells (Cat # S1N-C5255, AcroBiosystems, Newark, DE, USA); these proteins were glycosylated.

Virus neutralization experiments were performed using the 2019-nCoV/USA-WA1/2020 strain sourced from an infected patient in Washington state.

2.2. Experimental Design

Two hundred 85-week-old laying hens were transported from a table egg layer farm in the California Central Valley to the UC Davis Teaching and Research Animal Care Services facility, where they were placed on pine shavings in climate-controlled BSL-2 rooms. After one week of acclimatization, the hens received two immunizations administered twelve days apart. Vaccines were prepared as an oil–water emulsion with an equal volume of Freund's incomplete adjuvant (Thermo Scientific, IL, USA) using an Ultra-Turrax T25 High-Speed Homogenizer (IKA, Staufen, Germany) at 25,000 rpm for 10 min. The experimental groups were as follows: vaccine A twice (A/A), vaccine B twice (B/B), vaccine C twice (C/C), vaccine C followed by vaccine B (C/B), and adjuvant only (negative control). Each treatment group was divided into the following subgroups receiving a 2.5, 5, or 50 µg dose. All vaccines were administered by intramuscular route in the pectoral muscle in 0.5 mL volumes. Blood was collected from the ulnar vein upon arrival, 21 days following the second immunization, and at the end of the experiment (6 weeks post second dose). Blood was centrifuged and serum was stored at 4 °C for ELISA. One month following the second immunization, eggs were collected every week and stored at 4 °C for IgY extraction. In the sixth week after the last immunization, the hens were humanely euthanized by CO_2 inhalation, immediately followed by cervical dislocation. A schematic of the experimental design is in Figure 1.

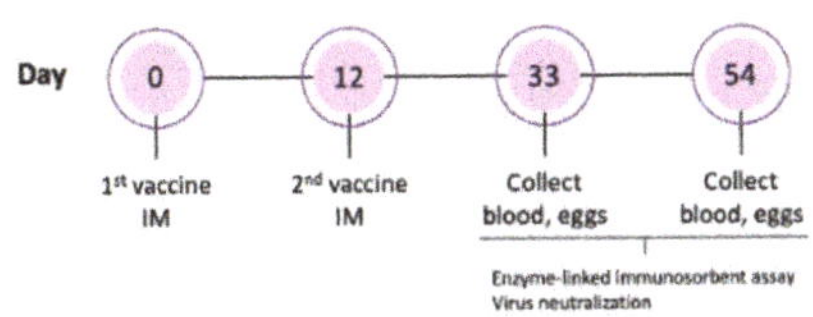

Group	Vaccine type	Dose	# birds
1	B	2.5 µg	17
2	B	5 µg	17
3	B	50 µg	17
4	A	2.5 µg	17
5	A	5 µg	17
6	A	50 µg	18
7	C	2.5 µg	18
8	C	5 µg	18
9	C	50 µg	18
10	CB	2.5 µg	18
11	CB	5 µg	18
12	CB	50 µg	18
13	No Vaccine	-	8

Figure 1. Schematic of the experimental design.

2.3. IgY Extraction from Egg Yolk

IgY extraction was performed as described previously [18], with minor modifications. Nine eggs per group were cracked open. The egg white was discarded, and the egg yolk was washed with phosphate-buffered saline (PBS). The egg yolk was punctured, and the vitelline membrane was discarded. The yolk was transferred to a 50 mL sterile centrifuge tube and Milli-Q® water (Millipore purification system, Bedford, MA, USA) was added for a yolk dilution of 1:5. The tube was vortexed vigorously and stored overnight at 4 °C. The following day, the tubes were centrifuged at 1000× *g* for 30 min at room temperature. Since the supernatant was still yellow, a second dilution/centrifugation step was performed. Ten mL of the supernatant was collected and mixed with 40 mL of Milli-Q® water for a final yolk dilution of 1:10, and centrifugation was performed at 2000× *g* at 15 °C for 30 min. The supernatant, now clear, was obtained and filtered through a Millex -HV 0.45 μm filter (Millipore, Bedford, MA, USA) and stored at 4 °C. To measure the IgY extraction efficiency, protein concentration of the yolk extract from 9 egg yolks was individually measured and compared with protein concentration from 10 individual serum samples obtained at necropsy from vaccine C group at 50 μg using the Take 3 plate on an Epoch Microplate Spectrophotometer (BioTek Instruments, Inc., Winooski, VT, USA).

2.4. IgY Antibody Titers

Serum and yolk IgY titers were measured by enzyme-linked immunosorbent assay (ELISA). Wells were coated with 0.5 μg of antigens A, B, or C in coating buffer and incubated at 4 °C overnight. All wash steps were performed with PBS-Tween 20 (0.05% Tween 20). Plates were blocked with 5% milk in PBS-Tween 20 for 1 h at room temperature. Serum samples were diluted 1:3200 in Sample Diluent, provided by the manufacturer, and purified yolk samples were diluted 1:400 in Milli-Q® water. The rest of the ELISA was performed using the reagents provided by the commercial IDEXX infectious bronchitis Ab Test (IDEXX, Westbrook, ME, USA), according to the manufacturer's instructions. Data were presented as sample to positive (S/P) ratio, which was calculated from the optical density (O/D) values using the following equation: (mean of test sample—mean of negative control)/(mean of positive control—mean of negative control). Three positive controls (A, B, and C) were prepared as a reference standard for each antigen. Each positive control was derived from sera from a hen immunized twice with 50 μg of one of the corresponding vaccines (A, B, or C) and selected for high IgY titer. Positive control sera were diluted in two-fold steps from 1:25 to 1:12,800 with the manufacturer-provided dilution buffer and incubated in wells coated with either 0.5 or 1 μg of corresponding antigen (A, B, or C). The final serum dilution and antigen coating concentration were determined by plotting the absorbance against the serum dilution and selecting the dilutions associated with the highest absorbance before saturation was observed. One negative control was used for all antigens. An S/P ratio allows titers to be compared across multiple plates, but S/P ratios should not be compared between antigens or between serum and egg yolk-derived IgY because a different positive control was used for each antigen and sample type.

2.5. Plaque Reduction Neutralization Assay

Antibodies derived from pooled sera and egg yolk from hens receiving vaccine A/A, B/B, C/C, and B/C at 50 μg dose were assessed for their ability to neutralize SARS-CoV-2 using a plaque reduction neutralization assay (PRNA). Briefly, the antibodies were diluted 1:10 in culture medium and incubated in two-fold serial dilutions with 100 plaque forming units (PFUs) of the Washington Strain of SARS-CoV-2 (USA-WA1/2020). SARS-CoV-2 was diluted in supplemented DMEM at a 1:1 ratio. The virus was then added to the antibody samples and allowed to incubate for 1 h at 37 °C and 5% CO_2. After incubation, viral plaque assay was conducted to quantify viral titers. Twelve-well plates were previously seeded with Vero cells (ATCC CCL-81) at a density of 2×10^5 cells per well. Media was aspirated from plates and virus-antibody samples were transferred to wells, one sample per well. Plates were incubated for 1 h at 37 °C and 5% CO_2. After infection, a 1:1 overlay consisting of 0.6% agarose and 2× Eagle's Minimum Essential Medium without phenol red, supplemented with 10% fetal bovine serum (FBS), non-essential amino acids (Gibco, 11140-050), 1 mM sodium pyruvate, 2 mM L-glutamine, and 1% P/S was added to each well. Plates were incubated at 37 °C for 48 h. Cells were fixed with 10% formaldehyde for 1 h at room temperature. Formaldehyde was aspirated and the agarose overlay was removed. Cells were stained with crystal violet (1% CV *w*/*v* in a 20% ethanol solution). Viral titer of SARS-CoV-2 was determined by counting the number of plaques and represented as plaque-forming units (PFU) and as relative percentage of neutralization with respect to the negative control.

2.6. Statistical Analysis

The data were analyzed using GraphPad Prism v.8.4.3 software (GraphPad Software, Inc., La Jolla, CA, USA; www.graphpad.com (accessed on 20 January 2022); RRID:SCR_002798). A Kruskal–Wallis test with Dunn's post-test was performed to compare treatment groups for each vaccine combination. Virus neutralization data by groups was compared by unpaired t tests. Significant differences were determined at $p < 0.05$.

3. Results

3.1. Antigen-Specific IgY Antibodies Were Detected in the Serum of Hyperimmunized Hens

Commercial laying hens were immunized against the SARS-CoV-2 S1 protein, and blood was collected 21 days following the second vaccination for antigen-specific ELISAs. Birds receiving 50 μg doses from all vaccines demonstrated strong antibody responses, which were higher ($p < 0.05$) than negative control titers (Figure 2). At the lower doses (2.5 and 5 μg), only birds receiving vaccine C demonstrated positive antibody responses, which were significantly higher than non-immunized control titers at 2.5 μg. Notably, antibody titers among birds administered vaccine C were not significantly different between doses, whereas antibody titers among birds administered vaccines A and B were significantly higher in the 50 μg groups compared with their respective 2.5 and 5 μg groups (Figure 2A–C).

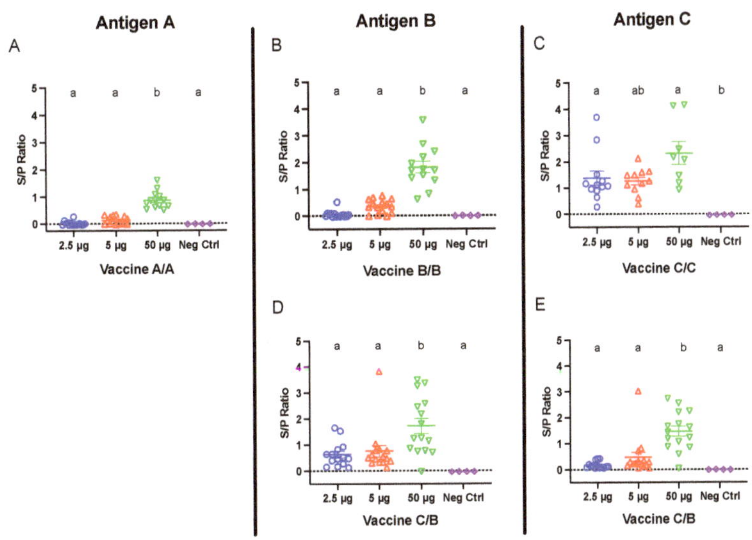

Figure 2. IgY titers in serum collected 21 days following the second vaccination for groups receiving (**A**) vaccine A/A, (**B**) vaccine B/B, (**C**) vaccine C/C, and (**D**,**E**) vaccine C/B combinations. Hens were inoculated twelve days apart. Antigen A-, B-, and C-coated plates were used for panels (**A**), (**B**,**D**), and (**C**,**E**), respectively. Error bars indicate the mean and standard error. Means without a common letter (a, b) are significantly different ($p < 0.05$).

To test the immunogenicity of combining two vaccines, one group of birds received primary vaccine C followed by a boost with vaccine B. ELISAs were performed 21 days following the second immunization to measure both antigen C- and B-specific antibody responses. The birds receiving 50 μg of vaccine C/B developed strong antigen B- and C-specific IgY titers (Figure 2D,E), which were significantly greater than their respective negative control titers and titers among birds receiving 2.5 and 5 μg doses.

3.2. Antigen A-, B-, and C-Specific IgY antibodies Were Detected in the Yolk of Hyperimmunized Hens

Beginning one month following the second immunization, eggs were collected weekly for ELISAs to detect antigen A-, B-, and C-specific IgY titers extracted from egg yolks. Yolk extracts were diluted 1:400, and specific IgY for each antigen was measured by ELISA using antigen A-, B-, or C-covered plates (Figure 3). Antigen-specific IgY titers extracted from yolks collected 6 weeks after the second immunization were significant for all tested antigens in birds receiving 50 μg doses, regardless of which vaccine they were administered. Among the birds administered vaccine A/A, egg yolk-derived antibodies were detectable at the lower immunization doses, but the titers were only significant for the 2.5 μg dose. Notably, the vaccine A/A 50 μg titers were significantly greater than the 2.5 and 5 μg titers. Egg yolk-derived IgY antibodies from birds receiving the lower doses for vaccine B/B were not significantly different from negative control titers. In contrast, egg yolk-derived IgY of hens from the vaccine C/C groups showed significant antibody titers at all doses (Figure 3A–C).

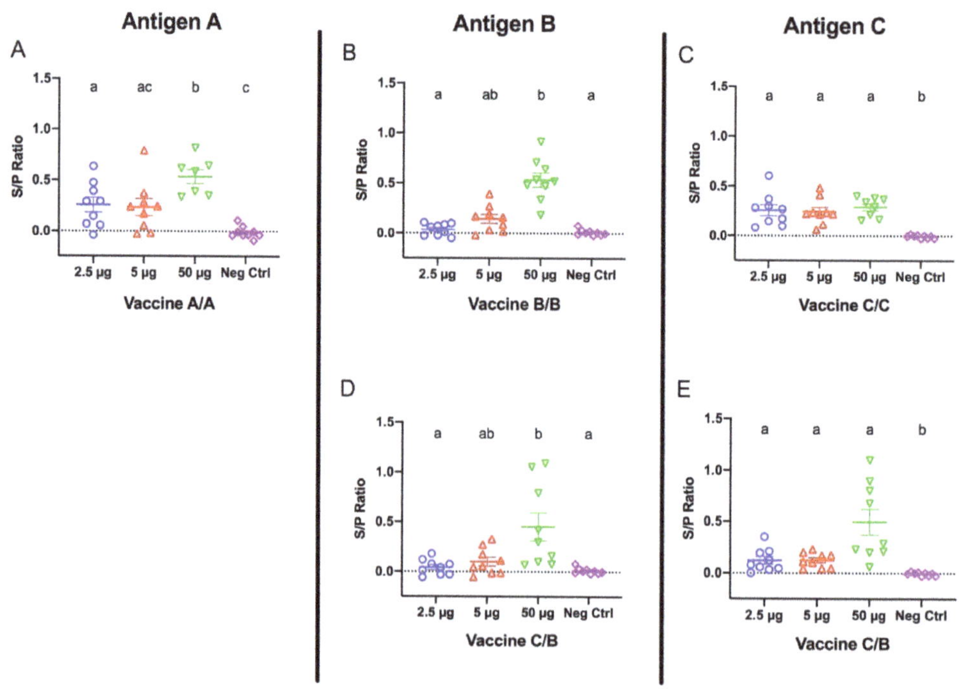

Figure 3. IgY titers from egg yolk collected six weeks following the second vaccination for groups receiving (**A**) vaccine A/A, (**B**) vaccine B/B, (**C**) vaccine C/C, and (**D**,**E**) vaccine C/B combinations. Hens were inoculated twelve days apart. Antigen A-, B-, and C-coated plates were used for panels (**A**), (**B**,**D**) and (**C**,**E**), respectively. Error bars indicate the mean and standard error. Means without a common letter (a–c) are significantly different ($p < 0.05$).

Egg yolk extractions were also analyzed for C- and B-specific antibody responses in hens receiving a combination of vaccine C and B as first and second dose, respectively. B-specific antibody titers in hens administered vaccine C/B were significant ($p < 0.05$) only in hens receiving the 50 μg doses (Figure 3D), which was consistent with the data reported for the hens receiving vaccine B/B. In contrast, C-specific antibody titers in hens receiving vaccine C/B were significant at 2.5, 5, and 50 μg doses, which corresponded to observations noted in hens in the vaccine C/C group (Figure 3E).

Egg yolk extracts showed a mean protein concentration of 2.42 mg/mL with a standard deviation of 0.48, whereas serum protein concentration had a mean of 56.71 mg/mL with a 2.2 standard deviation.

3.3. *Hyperimmunized Hens Produced Neutralizing Antibodies against SARS-CoV-2*

Hens immunized against the SARS-CoV-2 S1 protein were tested for neutralizing antibodies in sera at three weeks after the second immunization and in egg yolk extractions at six weeks following the second immunization. Negative controls included IgY from serum and egg yolk of hens that were either infectious bronchitis virus (IBV)-vaccinated or IBV-naïve. The IBV-naïve birds were tested regularly to confirm the absence of IBV antibodies by ELISA.

Sera and egg yolk IgY extractions from both IBV-vaccinated and IBV-naïve controls did not neutralize the virus. Sera from hens receiving vaccine A/A and B/B neutralized the virus up to 1:20, whereas sera from hens receiving vaccines C/C and C/B neutralized the virus up to 1:40 and 1:80, respectively (Figure 4A). IgY extractions from egg yolk neutralized the virus up to 1:10 among only the hens in the vaccine C/C and C/B groups, and no neutralization was observed among egg yolk samples from hens receiving vaccines A/A and B/B (Figure 4B).

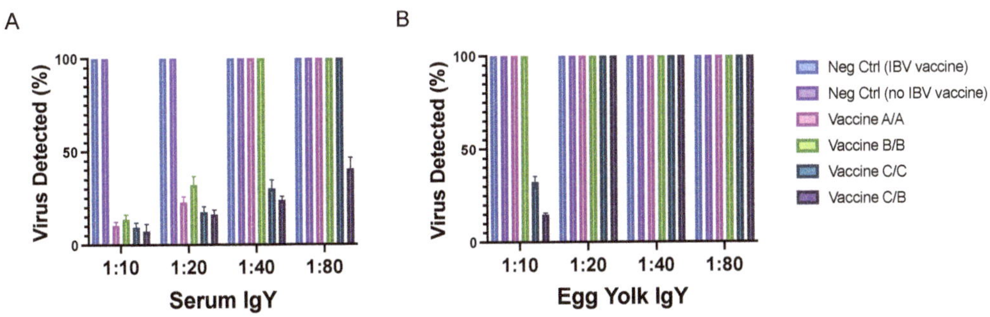

Figure 4. A plaque reduction neutralization assay (PRNA) was performed to detect SARS-CoV-2-specific neutralizing antibodies in (**A**) sera and (**B**) egg yolk IgY extractions from hens receiving 50 µg of vaccine A/A, B/B, C/C, and C/B. Controls included negative control hens that were either vaccinated or not vaccinated for IBV. Sera was collected at three weeks after the second vaccination, and egg yolk was harvested beginning at six weeks after the second vaccination. Viral titer of SARS-CoV-2 was determined by counting the number of plaques and represented as relative percentage of neutralization with respect to the negative control. The PRNA was performed in triplicate.

4. Discussion

In this study we demonstrate that laying hens hyperimmunized against three different SARS-CoV-2 recombinant spike proteins produced specific antibodies in sera. We further confirmed that the dose and type of antigen administered influenced the levels of antibody titers. Doses were selected based on previous experience and considering a low, medium, and high protein content in each vaccine. In addition, we showed that all vaccines induced neutralizing antibodies against SARS-CoV-2, but that vaccines against the glycosylated S1 protein (C/C) and combination of glycosylated S1 and non-glycosylated RBD of S1 (C/B) led to better neutralization. Combination of C and B antigens was used due to the results obtained in the ELISA from the collected sera showing better responses than for the A antigen. In addition, glycosylated antigen was used as a primer due to its superior response in comparison with the other tested antigens. Specifically, neutralization was observed in all groups, but it was at lower serum concentrations among the hens receiving the full-length glycosylated S1 I protein at least once, whereas virus neutralization by IgY extracted from egg yolk occurred only in the vaccine C/C and C/B groups.

Compared with the negative control group, the full length and tagged S1 protein elicited a higher level ($p < 0.05$) of IgY titers in serum and egg yolk at all doses, whereas immunization with the RBD fragments (A or B; non-tagged) induced antibody titers higher than the negative control ($p < 0.05$) only when immunized at the highest dose of 50 μg. This observation could be explained by the structural difference between antigens. In contrast to antigens A and B, in which the S1 fragments contained only the receptor-binding domain, antigen C comprised the entire extracellular domain of the S1 protein. The increased size and number of potential epitopes may be responsible for the enhanced antigenicity of vaccine C. Another possible explanation could be that antigen C was produced using a eukaryotic HEK293 expression system. In contrast to the prokaryotic expression system used for manufacturing vaccines A and B, HEK293 cells are capable of the post-translational modifications necessary for the correct folding of the protein [19]. Thus, modifications such as phosphorylation or glycosylation of the recombinant protein could potentially enhance immunogenicity [19]. Our data suggest that antibodies generated from vaccine C could therefore be administered at a fraction of the dose necessary for seroconversion to occur. This is crucial when considering potential applications of IgY in prevention strategies against COVID-19.

We also demonstrated that a combination approach using vaccine C followed by vaccine B generated good IgY titers specific for antigen C, which were comparable to the titers observed with vaccine C/C. Similarly, vaccine C/B IgY titers specific for antigen B corresponded to vaccine B/B titers. Importantly, the vaccine C/B combination resulted in neutralizing antibody titers that were at least as good as the titers induced by vaccine C/C. Since the production of proteins using the eukaryotic HEK293 expression system is more costly than using a prokaryotic system, a vaccine combination approach may prove to be a more economical, but equally effective, immunization strategy.

Since the hens had been vaccinated for IBV, a chicken gammacoronavirus, it was important to determine whether cross-neutralization could be observed between samples from IBV-vaccinated hens and SARS-CoV-2. Virus neutralization was not observed by IBV-vaccinated hens, which suggests that IBV antibodies do not cross-react with SARS-CoV-2. Although unsurprising given the enormous genetic variability of coronaviruses, this observation underscores the importance of ensuring that the spike protein used for vaccination is genetically similar to the spike protein of the circulating virus.

As expected, SARS-CoV-2-specific IgY titers were also detected in egg yolk. As shown in our results, IgY titers in yolk were lower than in serum. These outcomes might be related to the efficiency of our yolk antibody extraction method, which is based only on ultrapure water and centrifugation. The finding of a reduced egg yolk extract protein concentration compared to serum protein concentration further supports this claim. This extraction inefficiency might also explain the lower neutralization activity demonstrated by the yolk IgY in the plaque neutralization test. The egg yolk is a complex structure composed of proteins, lipids, vitamins, and minerals. The main components of the egg yolk are lipids in ~65% of dry matter, and the lipid-to-protein ratio is about 2:1 [20]. In addition, lipids in the yolk are exclusively associated with lipoprotein assemblies, i.e., triglycerides, phospholipids, and cholesterol [20]. This intricate lipid association with proteins can complicate protein extraction from the yolk if the lipids are not diluted appropriately, which might have been the case with our IgY extraction method. Anecdotally, we have worked with a private company upscaling the hyperimmunization and egg production systems. They have used a polyethylene glycol precipitation to extract IgY from yolks with better results (Michelle Hawkins, CAMAS, personal communication). The polyethylene glycol method has been tested by several researchers [12,21] and seems to be a better method to address the issue of fats present in the egg yolk. This methodology is crucial for pure antibody extractions and formulation of preventative products, such as local use sprays containing IgY.

A recent publication demonstrated that IgY extracted from egg yolk of hens immunized with the RBD of SARS-CoV-2 could prevent entry and replication of SARS-CoV-2 in vitro [22]. These results strengthen the case for administering egg yolk-derived IgY to reduce SARS-CoV-2 infection. In this study the neutralizing antibody titer was relatively low and for commercial success will require further optimization to increase the titer. As an example, the egg yolk extraction method could be optimized to improve IgY extraction efficacy. The most likely application of such an approach involves a spray to coat the nose and throat of at-risk humans to reduce the risk of SARS-CoV-2 infection.

For some of the measured responses, differences were minor; this might be explained by the size limitations of our BSL2 facilities. While a sample size of 30 hens per group would have given a statistical power of 80% and 5% significance, we were only able to house 17 to 18 birds per group, limiting our statistical power.

The IgY neutralizing effects measured in this study were performed using the USA-WA1/2020. Like many other RNA viruses including the coronaviruses, SARS-CoV-2 is prone to frequent mutations [23,24]. Thus far, thousands of mutations have been observed at the nucleotide level in the SARS-CoV-2 genome, indicating that SARS-CoV-2 has already become heterogeneous [23]. The emergence of more prevalent new variants has already resulted in reduced efficacy of current vaccines, so it is critical to closely monitor the evolution of SARS-CoV-2 to ensure that prophylactic antibody administration targeted toward the SARS-CoV-2 RBD retains sufficient neutralizing activity.

Author Contributions: Conceptualization: M.G.W. Study design: R.A.G. Execution of experiment and sample collection and processing: E.J.A., S.E.-L. and R.A.G. Design and performance of virus neutralization assay: A.N. Data analysis: E.J.A. and R.A.G. Writing: E.J.A. All authors have read and agreed to the published version of the manuscript.

Funding: Funding for the experiments were obtained from a collaborative research agreement between the University of California, School of Veterinary Medicine and SPARK at Stanford, Stanford University, School of Medicine. Agreement number: 62366195.

Institutional Review Board Statement: Animal care and experimental procedures were performed in compliance with all federal and institutional animal use guidelines. The University of California is an Association for Assessment and Accreditation of Laboratory Animal Care (AAALAC) accredited institution. The Institution for Animal Care and Use Committee (IACUC) protocol number was 20,851. This study is reported in accordance to ARRIVE guidelines. All experimental procedures were approved by the University of California, Office of Research.

Informed Consent Statement: Not applicable.

Acknowledgments: We thank Daria Mochly-Rosen for her contributions to the project. In addition, we are grateful to Maria Timofeveya and Patrick Montine for assisting with the bird experiment and sample processing. Also, the California poultry industry California poultry federation (CPF) and Pacific egg and poultry association (PEPA) for their support. Finally, JS-West for the donation of the birds used in this study.

Conflicts of Interest: The authors declare no conflict of interest.

References

1. U.S. Food & Drug Administration. FDA Takes Key Action in Fight against COVID-19 by Issuing Emergency Use Authorization for First COVID-19 Vaccine. Available online: https://www.fda.gov/news-events/press-announcements/fda-takes-key-action-fight-against-covid-19-issuing-emergency-use-authorization-first-covid-19 (accessed on 5 December 2021).
2. Jackwood, M.W.; de Wit, S. Infectious Bronchitis. In *Diseases of Poultry*, 14th ed.; John Wiley & Sons, Inc.: Hoboken, NJ, USA, 2020; pp. 167–188.
3. Cavanagh, D.; Davis, P.J. Coronavirus IBV: Removal of spike glycopolypeptide S1 by urea abolishes infectivity and haemagglutination but not attachment to cells. *J. Gen. Virol.* **1986**, *67*, 1443–1448. [CrossRef] [PubMed]
4. Niesters, H.G.M.; Kusters, J.G.; Lenstra, J.A.; Spaan, W.J.M.; Horzined, M.C.; van der Zeijst, B.A.M. The neutralization epitopes on the spike protein of infectious bronchitis virus and their antigenic variation. *Adv. Exp. Med. Biol.* **1987**, *218*, 483–492. [CrossRef] [PubMed]
5. Shang, J.; Wan, Y.; Luo, C.; Ye, G.; Geng, Q.; Auerbach, A.; Li, F. Cell entry mechanisms of SARS-CoV-2. *Proc. Natl. Acad. Sci. USA* **2020**, *117*, 11727–11734. [CrossRef] [PubMed]

6. U.S. Food & Drug Administration. FDA Issues Emergency Use Authorization for Convalescent Plasma as Potential Promising COVID-19 Treatment, Another Achievement in Administration's Fight against Pandemic. Available online: https://www.fda.gov/news-events/press-announcements/fda-issues-emergency-use-authorization-convalescent-plasma-potential-promising-covid-19-treatment (accessed on 5 December 2021).
7. Jensenius, J.C.; Anderson, I.; Hau, J.; Crone, M.; Koch, C. Eggs: Conveniently packaged antibodies. Methods for purification of yolk IgG. *J. Immunol. Methods* **1981**, *46*, 63–68. [CrossRef]
8. Júnior, Á.F.; Santiago, F.M.; Silva, M.V.; Ferreira, F.B.; Júnior, A.G.M.; Mota, C.M.; Faria, M.S.; Filho, H.H.S.; Silva, D.A.O.; Cunha-Júnior, J.P.; et al. Production, characterization and applications for *Toxoplasma gondii*-specific polyclonal chicken egg yolk immunoglobulins. *PLoS ONE* **2012**, *7*, e40391. [CrossRef]
9. Sudjarwo, S.A.; Eraiko, K.; Sudjarwo, G.W.; Koerniasari. The potency of chicken egg yolk immunoglobulin (IgY) specific as immunotherapy to *Mycobacterium tuberculosis* infection. *J. Adv. Pharm. Technol. Res.* **2017**, *8*, 91–96. [CrossRef] [PubMed]
10. Polson, A.; von Wechmar, M.B.; Fazakerley, G. Antibodies to proteins from yolk of immunized hens. *Immunol. Commun.* **1980**, *9*, 495–514. [CrossRef] [PubMed]
11. Schade, R.; Calzado, E.G.; Sarmiento, R.; Chacana, P.A.; Porankiewicz-Asplund, J.; Terzolo, H.R. Chicken egg yolk antibodies (IgY-technology): A review of progress in production and use in research and human and veterinary medicine. *Altern. Lab. Anim.* **2005**, *33*, 129–154. [CrossRef] [PubMed]
12. Pauly, D.; Chacana, P.A.; Calzado, E.G.; Brembs, B.; Schade, R. IgY technology: Extraction of chicken antibodies from egg yolk by polyethylene glycol (PEG) precipitation. *J. Vis. Exp.* **2011**, *51*, e3084. [CrossRef] [PubMed]
13. Grando, T.H.; Baldissera, M.D.; de Sá, M.F.; do Carmo, G.M.; Porto, B.C.Z.; Aguirre, G.S.V.; Azevedo, M.I.; de Jesus, F.P.K.; Santurio, J.M.; Sagrillo, M.R.; et al. Avian antibodies (IgY) against *Trypanosoma cruzi*: Purification and characterization studies. *J. Immunol. Methods* **2017**, *449*, 56–61. [CrossRef] [PubMed]
14. Abbas, A.T.; El-Kafrawy, S.A.; Sohrab, S.S.; Azhar, E.I.A. IgY antibodies for the immunoprophylaxis and therapy of respiratory infections. *Hum. Vaccin. Immunother.* **2019**, *15*, 264–275. [CrossRef] [PubMed]
15. Pereira, E.P.V.; van Tilburg, M.F.; Florean, E.O.P.T.; Guedes, M.I.F. Egg yolk antibodies (IgY) and their applications in human and veterinary health: A review. *Int. Immunopharmacol.* **2019**, *73*, 293–303. [CrossRef] [PubMed]
16. Fu, C.Y.; Huang, H.; Wang, X.M.; Liu, Y.G.; Wang, Z.G.; Cui, S.J.; Gao, H.L.; Li, Z.; Li, J.P.; Kong, Z.G. Preparation and evaluation of anti-SARS coronavirus IgY from yolks of immunized SPF chickens. *J. Virol. Methods* **2006**, *133*, 112–115. [CrossRef] [PubMed]
17. Kim, D.M.; Swartz, J.R. Efficient production of a bioactive, multiple disulfide-bonded protein using modified extracts of *Escherichia coli*. *Biotechnol. Bioeng.* **2004**, *85*, 122–129. [CrossRef] [PubMed]
18. Bird, C.R.; Thorpe, R. Purification of immunoglobulin Y (IgY) from chicken eggs. In *The Protein Protocols Handbook*; Humana Press: Totowa, NJ, USA, 2009; pp. 1779–1781.
19. Jefferis, R. Posttranslational Modifications and the Immunogenicity of Biotherapeutics. *J. Immunol. Res.* **2016**, *2016*, 535872. [CrossRef] [PubMed]
20. Anton, M. Composition and structure of hen egg yolk. In *Bioactive Egg Compounds*; Springer: Berlin/Heidelberg, Germany, 2007; pp. 1–6.
21. Shafiq, M.K. Isolation of egg yolk immunoglobulins [IgY] by chloroform polyethylene glycol technique and assaying of antibodies against avian infectious bronchitis. *Vet. Med. J.* **1997**, *45*, 273–278.
22. Wei, J.; Lu, Y.; Rui, Y.; Zhu, X.; He, S.; Wu, S.; Xu, Q. A Chicken IgY Can Efficiently Inhibit the Entry and Replication of SARS-CoV-2 by Targeting the ACE2 Binding Domain In Vitro. 2021. Available online: https://www.biorxiv.org/content/10.1101/2021.02.16.430255v1.full (accessed on 5 December 2021).
23. Islam, M.F.; Hoque, M.N.; Rahman, M.S.; Rubayat Ul Alam, A.S.M.; Akther, M.; Akter Puspo, J.; Akter, S.; Sultana, M.; Crandall, K.A.; Hossain, M.A. Genome-wide analysis of SARS-CoV-2 virus strains circulating worldwide implicates heterogeneity. *Sci. Rep.* **2020**, *10*, 14004. [CrossRef] [PubMed]
24. Wang, R.; Hozumi, Y.; Yin, C.; Wei, G.W. Decoding SARS-CoV-2 transmission and evolution and ramifications for COVID-19 diagnosis, vaccines, and medicine. *J. Chem. Inf. Model.* **2020**, *60*, 5853–5865. [CrossRef] [PubMed]

viruses

MDPI

Communication

Differences in Clinical Presentations of Omicron Infections with the Lineages BA.2 and BA.5 in Mecklenburg-Western Pomerania, Germany, between April and July 2022

Katja Verena Goller [1,*], Juliane Moritz [1], Janine Ziemann [1], Christian Kohler [2], Karsten Becker [2,†], Nils-Olaf Hübner [1,3,*,†] and the CoMV-Gen Study Group [‡]

1 Central Unit for Infection Control and Prevention, University Medicine Greifswald, 17475 Greifswald, Germany
2 Friedrich-Loeffler-Institute of Medical Microbiology, University Medicine Greifswald, 17475 Greifswald, Germany
3 Institute for Hygiene and Environmental Medicine, University Medicine Greifswald, 17475 Greifswald, Germany
* Correspondence: katja.goller@med.uni-greifswald.de (K.V.G.); nils.huebner@med.uni-greifswald.de (N.-O.H.)
† These authors contributed equally to this work.
‡ The CoMV-Gen Study Group are listed in the acknowledgements; comv-gen@med.uni-greifswald.de.

Abstract: Knowledge on differences in the severity and symptoms of infections with the SARS-CoV-2 Omicron variants BA.2 (Pango lineage B.1.529.2) and BA.5 (Pango lineage B.1.529.5) is still scarce. We investigated epidemiological data available from the public health authorities in Mecklenburg-Western Pomerania, Northeast Germany, between April and July 2022 retrospectively. Comparative analyses revealed significant differences between recorded symptoms of BA.2 and BA.5 infected individuals and found strong correlations of associations between symptoms. In particular, the symptoms 'chills or sweating', 'freeze' and 'runny nose' were more frequently reported in BA.2 infections. In contrast, 'other clinical symptoms' appeared more frequently in Omicron infections with BA.5. However, the results obtained in this study provide no evidence that BA.5 has a higher pathogenicity or causes a more severe course of infection than BA.2. To our knowledge, this is the first report on clinical differences between the current Omicron variants BA.2 and BA.5 using public health data. Our study highlights the value of timely investigations of data collected by public health authorities to gather detailed information on the clinical presentation of different SARS-CoV-2 subvariants at an early stage.

Keywords: SARS-CoV-2; Omicron; BA.2; BA.5; symptoms; Germany

Citation: Goller, K.V.; Moritz, J.; Ziemann, J.; Kohler, C.; Becker, K.; Hübner, N.-O.; the CoMV-Gen Study Group. Differences in Clinical Presentations of Omicron Infections with the Lineages BA.2 and BA.5 in Mecklenburg-Western Pomerania, Germany, between April and July 2022. *Viruses* **2022**, *14*, 2033. https://doi.org/10.3390/v14092033

Academic Editors: Ahmed El-Shamy and Mohamed Ibrahim

Received: 8 August 2022
Accepted: 8 September 2022
Published: 13 September 2022

1. Introduction

Starting in December 2021 and ongoing until July 2022, infections with the Severe Acute Respiratory Syndrome Virus 2 (SARS-CoV-2) in Germany have mainly been caused by the Omicron variant (Phylogenetic Assignment of Named Global Outbreak (Pango) lineage: B.1.1.529) and its subvariants BA.1, BA.2, BA.3, BA.4 and BA.5, including their respective sub-lineages [1]. In December 2021, the Omicron subvariant BA.2 was reported for the first time in Germany [1,2]. Since then, it has rapidly replaced the Omicron subvariant BA.1 that had been dominating until then and was the dominant variant by the end of February 2022 [1]. Concurrently, by the end of February 2022, the Omicron variant BA.5, first detected in April 2022 in South-Africa where it was responsible for a fifth wave between April and June 2022 [3,4], appeared in Germany [1]. The BA.5 variant increased rapidly and became the dominating variant in Germany by calendar week 23 (June 2022) [1].

Little is known about differences in the severity of disease between BA.2 and BA.5 infections. However, a study in animals recently suggested that BA.4 and BA.5 may be more pathogenic than BA.2 [5], and there is also evidence that the Omicron variants BA.4 and BA.5 exhibit a higher transmissibility than BA.2 [6].

Interestingly, to the best of our knowledge, differences in clinical presentations between the currently circulating Omicron variants BA.2 and BA.5 have not yet been reported on the basis of public health data. We therefore conducted a comparative analysis between BA.2 and BA.5 lineages using official reporting data from the public health authorities from the German federal state Mecklenburg-Western Pomerania.

2. Materials and Methods

In order to investigate differences in symptoms caused by infections with the Omicron lineages of BA.2 and BA.5, including their respective sub-lineages, data available from the public health authorities from the federal state Mecklenburg-Western Pomerania, Germany, were used. For the German federal state of Mecklenburg-Western Pomerania, the CoMV-Gen Study Group (www.comv-gen.de (accessed on 8 August 2022)) is entrusted by the state government with monitoring the circulating SARS-CoV-2 -variants in collaboration with local diagnostic laboratories and public health authorities. On behalf of the state government, the CoMV-Gen project's mission is conducting the SARS-CoV-2 variant surveillance in Mecklenburg-Western Pomerania, including gathering genetic information from target PCRs as well as from whole genome sequencing analyses in collaboration with local diagnostic laboratories, and connecting these with epidemiological data available from the public health authorities. All confirmed positive cases recorded between 25 April and 13 July 2022 (covering twelve calendar weeks) in the public health surveillance system (Surveillance Outbreak Response Management and Analysis System (SORMAS) [7]) by the public health authorities from Mecklenburg-Western Pomerania were analyzed retrospectively. Data on symptoms were mainly collected by structured self-reporting questionnaires, and to a lesser extent, from interviews with the cases. Data from individuals that were notified as being infected by Omicron were collected. All cases reported as being infected by Omicron BA.2 or BA.5 lineages up to 13 July 2022 were considered for analyses. According to the standardized and cross-sectional German Corona Consensus Dataset (GECCO) [8] for research, symptoms relevant to COVID-19 were filtered out and frequencies were compared. In order to investigate the infection status across the years of age of the positive cases, five age groups in 20-year intervals were defined: 0–19, 20–39, 40–59, 60–79 and 80 and more years of age.

Ethical approval was granted by the ethics committee of the University of Greifswald, Germany (BB 125/21). Only anonymized aggregated data are shown in this manuscript and only symptomatic cases were used to compare symptoms when the exact lineages were assigned and reported in the system either as BA.2 or BA.5.

Statistical analyses were conducted by using SPSS 28.0 (IBM SPSS Statistics for Windows, Version 28.0. Armonk, NY, USA: IBM Corp.). For comparative analyses, Pearson's chi-square tests were conducted, and the 95% confidence intervals (CI95) were determined A p-value of 0.05 was considered statistically significant. In order to analyze correlations between symptoms for both variants, non-parametric Spearman Rho correlation coefficients were determined for symptoms that occurred with a minimum overall frequency of 10% of cases.

3. Results

In total, 1027 cases with confirmed Omicron infections were reported in SORMAS during the study period (Figure 1). Of those, 607 cases were deposited as Omicron-positive in SORMAS but were not assigned to a specific sub-lineage, while 405 cases were characterized as BA.2 (n = 246) or BA.5 (n = 159), respectively. Fifteen cases belonged to other Omicron lineages including BA.1 (six cases) and BA.4 (nine cases). These 15 cases, as well as the unassigned cases, were excluded from the analyses.

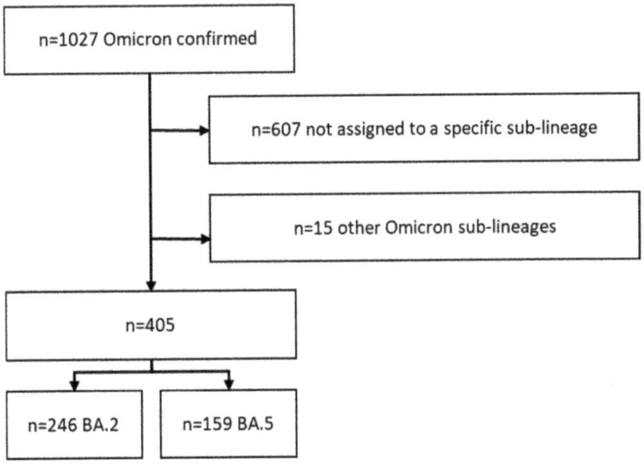

Figure 1. Sample size of the study population. In total, 405 assigned BA.2 and BA.5 cases were included in the analyses.

Information on infection status per age group, sex, symptomatic and vaccination status are summarized in Table 1. Statistical analyses revealed no significant differences between the age groups, sex and vaccination status of BA.2 and BA.5 cases (Table 1). Within the BA.2 cases 124 individuals (50.1%), and of the BA.5 cases 69 individuals (43.3%) reported one or more symptoms (Pearson's chi square statistics: $p = 0.561$).

Table 1. Epidemiological characteristics and analyses on information from recorded BA.2 and BA.5 cases.

Characteristics		BA.2		BA.5		*p* Value [1]
		n	[%]	*n*	[%]	
Sex	Female	126	51.2	96	60.4	0.135
	Male	118	48.0	61	38.4	
	Unknown	2	0.8	2	1.3	
Age groups	0–19	25	10.2	13	8.2	0.759
	20–39	76	30.9	48	30.2	
	40–59	94	38.2	70	44.0	
	60–79	43	17.5	24	15.1	
	80+	7	2.8	3	1.9	
	Unknown	1	0.4	1	0.6	
Symptoms	Yes	124	50.1	69	43.4	0.186
	No	122	49.9	90	56.6	
Vaccination [2]	Yes	75	57.7	60	49.2	0.124
	No	42	32.3	54	44.3	
	Unknown	13	10	8	6.6	

[1] *p* values are shown from Pearson's chi square statistics. [2] No records from 174 cases.

A summary of the statistical analyses including CI95 intervals for each symptom and results from Pearson's chi-square statistics are given in Table 2. No records for 'acute respiratory distress syndrome', 'respiratory insufficiency/assisted ventilation' and 'oxygen saturation < 94%' were given.

Table 2. Differences between reported symptoms from symptomatic BA.2 and BA.5 cases (CI, confidence interval (CI 95% [lower–upper]); Positive, positively reported; n.r., not reported.).

Symptom	Status	BA.2			BA.5			*p*-Value [1]
		n	%	CI 95 [%]	*n*	%	CI 95 [%]	
Breathing difficulties/dyspnoea	Positive n.r.	18 106	14.5 85.5	[9.2–21.5] [78.5–90.8]	4 65	5.8 94.2	[2.0–13.2] [86.8–98.0]	0.097
Chills or sweating	Positive n.r.	22 102	17.7 82.3	[11.8–25.2] [74.8–88.2]	3 66	4.3 95.7	[1.2–11.1] [88.9–98.8]	**0.012**
Cough	Positive n.r.	105 19	84.7 15.3	[77.6–90.2] [9.8–22.4]	52 17	75.4 24.6	[64.3–84.3] [15.7–35.7]	0.125
Diarrhea	Positive n.r.	12 112	9.7 90.3	[5.4–15.8] [84.2–94.6]	9 60	13 87	[6.7–22.5] [77.5–93.3]	0.631
Fever	Positive n.r.	43 81	34.7 65.3	[26.7–43.3] [56.7–73.3]	34 35	49.3 50.7	[37.7–60.9] [39.1–62.3]	0.065
Freeze	Positive n.r.	29 95	23.4 76.6	[16.6–31.4] [68.6–83.4]	6 63	8.7 91.3	[3.7–17.0] [83.0–96.3]	**0.011**
Headache	Positive n.r.	68 56	54.8 45.2	[46.1–63.4] [36.6–53.9]	37 32	53.6 46.4	[41.9–65.0] [35.0–58.1]	0.881
Increased heart rate/tachycardia	Positive n.r.	12 112	9.7 90.3	[5.4–15.8] [84.2–94.6]	3 66	4.3 95.7	[1.2–11.1] [88.9–98.8]	0.264
Loss of smell	Positive n.r.	15 109	12.1 87.9	[7.2–18.7] [81.3–92.8]	15 54	21.7 78.3	[13.3–32.5] [67.5–86.7]	0.097
Loss of taste	Positive n.r.	25 99	20.2 79.8	[13.8–27.9] [72.1–86.2]	17 52	24.6 75.4	[15.7–35.7] [64.3–84.3]	0.585
Muscle or body aches	Positive n.r.	40 84	32.3 67.7	[24.5–40.8] [59.2–75.5]	28 41	40.6 59.4	[29.6–52.4] [47.6–70.4]	0.273
Nausea	Positive n.r.	8 116	6.5 93.5	[3.1–11.8] [88.2–96.9]	6 63	8.7 91.3	[3.7–17.0] [83.0–96.3]	0.774
Other symptoms [2]	Positive n.r.	39 85	31.5 68.5	[23.8–40.0] [60.0–76.2]	40 29	58.0 42.0	[46.2–69.1] [30.9–53.8]	**<0.001**
Pneumonia	Positive n.r.	2 122	1.6 98.4	[0.3–5.1] [94.9–99.7]	0 69	0 100	[–] [–]	0.289
Rapid breathing/tachypnea	Positive n.r.	2 122	1.6 98.4	[0.3–5.1] [94.9–99.7]	2 67	2.9 97.1	[0.6–9.0] [91.0–99.4]	0.618
Runny nose	Positive n.r.	110 14	88.7 11.3	[82.3–93.4] [6.6–17.7]	52 17	75.4 24.6	[64.3–84.3] [15.7–35.7]	**0.023**
Severe feeling of sickness	Positive n.r.	40 84	32.3 67.7	[24.5–40.8] [59.2–75.5]	30 39	43.5 56.5	[32.2–55.2] [44.8–67.8]	0.159
Sore throat/pharyngitis	Positive n.r.	77 47	62.1 37.9	[53.4–70.3] [29.7–46.6]	37 32	53.6 46.4	[41.9–65.0] [35.0–58.1]	0.538

[1] In bold: significant *p* values are shown from Pearson's chi square statistics. [2] 'Other symptoms' are explained in the text.

The absolute and relative frequency of the symptoms of cases of our study population either infected with BA.2 or BA.5 are additionally shown in Figure 2. Altogether, 21 specific symptoms (according to the GECCO-criteria [8]) are of relevance for COVID-19 infections and the content of the questionnaire. The most prominent symptoms reported in BA.2 as well as BA.5 cases were respiratory symptoms such as 'runny nose', 'cough' and 'pharyngitis'. On the other hand, severe symptoms such as 'acute respiratory distress syndrome', 'oxygen saturation below 94%' and 'respiratory insufficiency that needs assisted ventilation' were not reported for both variants.

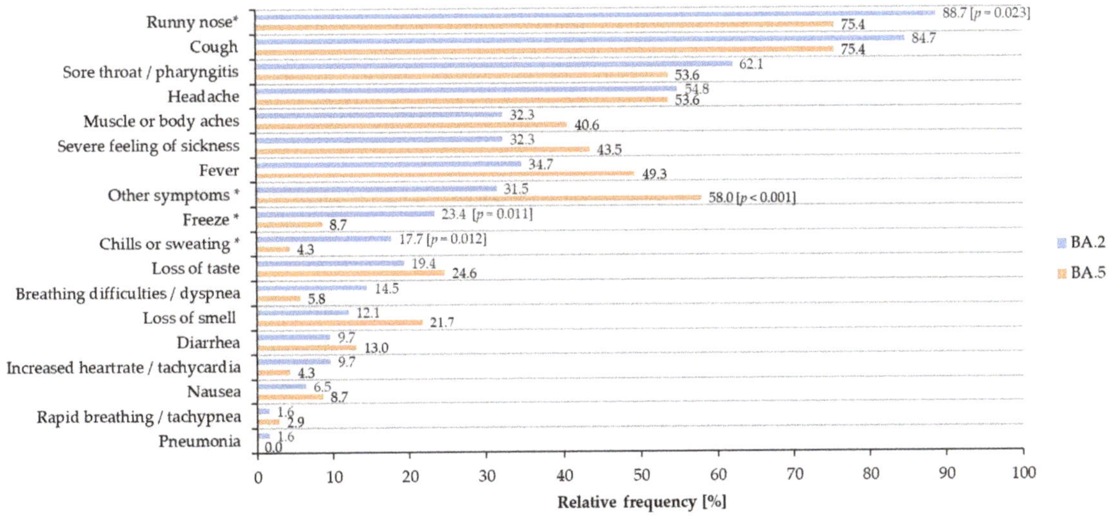

Figure 2. Relative frequency of the most prominent 18 symptoms of infected symptomatic BA.2 and BA.5 cases. * p values are given in squared brackets for significant results.

However, the Omicron BA.2 and BA.5 lineages also showed marked differences in the frequency of various symptoms, as shown in Table 2 and Figure 2. Significant differences in symptoms 'chills or sweating' and 'freeze' between BA.2 and BA.5 infections were observed. These symptoms were more frequently notified from individuals infected with BA.2 ('chills or sweating': $p = 0.012$; 17.7%, CI95: 11.8–25.2%; 'freeze': $p = 0.011$; 23.4%; CI95: 16.6–31.4%) than from BA.5 cases ('chills or sweating': 4.3%, CI95: 1.2–11.1%; 'freeze': 8.7%; CI95: 3.7–17.0%), with 88.7% of BA.2 cases more frequently reporting 'runny nose' ($p = 0.023$; CI95: 82.3–93.4%) than BA.5 cases (75.4%; CI95: 64.3–84.3%). In contrast, BA.5 cases significantly more often reported 'other symptoms' ($p < 0.001$; 58.0%, CI95: 46.2–69.1%) than BA.2 cases (31.5%, CI95: 23.8–40.0%). 'Other symptoms' included a wide range of symptoms such as 'dullness', 'exhaustion, 'fatigue', 'dizziness' among others that are all not items in the GECCO catalogue. This might indicate a more elaborated set of symptoms reported from BA.5 cases.

Correlations between symptoms for both variants are shown in Table 3.

As shown in Table 3, 'loss of smell' was strongly correlated with 'loss of taste' (Spearman Rho = 0.571, $p < 0.01$); 'freeze' was strongly correlated with 'chills or sweating' (Spearman Rho = 0.579, $p < 0.01$), as well as with 'muscle or body aches' (Spearman Rho = 0.328, $p < 0.01$); and 'headache' was correlated to 'runny nose' associated with 'cough' (Spearman Rho = 0.334, $p < 0.01$).

Table 3. Non-parametric correlation between different symptoms of both BA.2 and BA.5 positive cases. Spearman-Rho correlation coefficients > 0.3. are given in bold.

	Symptoms												
	Chills or Sweating	**Cough**	**Diarrhea**	**Breathing Difficulties/Dyspnea**	**Fever**	**Headache**	**Muscle or Body Aches**	**Rapid Breathing/Tachypnea**	**Runny Nose**	**Sore Throat/Pharyngitis**	**Loss of Taste**	**Loss of Smell**	**Freeze**
1-14 Chills or sweating	1.000												
1-14 Cough	0.026	1.000											
1-14 Diarrhea	0.063	0.082	1.000										
1-14 Breathing difficulties/dyspnea	0.299 [2]	0.046	0.136	1.000									
1-14 Fever	0.095	0.064	0.191 [2]	0.107	1.000								
1-14 Headache	0.198 [2]	0.096	0.019	0.230 [2]	0.278 [2]	1.000							
2-14 Muscle or body aches	0.232 [2]	0.047	0.125	0.179 [1]	0.174 [1]	**0.414** [2]	1.000						
1-14 Rapid breathing/tachypnea	0.161 [1]	0.07	0.066	0.291 [2]	0.179 [1]	0.133	0.121	1.000					
1-14 Runny nose	0.169 [1]	**0.334** [2]	0.108	0.113	−0.105	0.053	−0.002	0.064	1.000				
1-14 Sore throat/pharyngitis	0.07	0.169 [1]	0.122	0.1	0.205 [2]	0.190 [2]	0.217 [2]	−0.027	0.181 [1]	1.000			
1-14 Loss of taste	0.096	0.059	0.219 [2]	0.127	0.109	0.079	0.242 [2]	0.188 [2]	0.094	0.107	1.000		
1-14 Loss of smell	0.133	0.022	0.263 [2]	0.251 [2]	0.147 [1]	0.163 [1]	0.192 [2]	0.239 [2]	0.149 [1]	0.124	**0.571** [2]	1.000	
1-14 Freeze	**0.579** [2]	0.122	0.181 [1]	0.254 [2]	0.111	0.269 [2]	**0.328** [2]	0.12	0.206 [2]	0.009	0.045	0.021	1.000

[1] Significant correlations $p < 0.05$. [2] Significant correlations $p < 0.01$.

4. Discussion

In this study, we conducted a comparative analysis of the clinical presentations between the Omicron BA.2 and BA.5 lineages, including their respective sub-lineages, using official reporting data from the public health authorities of the Germany federal state Mecklenburg-Western Pomerania. To the best of our knowledge, this is the first time that the differences in the clinical presentations between these two current Omicron lineages have been reported using public health data.

Both sub-lineages showed a comparable severity of infection with approximately 50% of cases being symptomatic. Overall, the symptoms 'runny nose', 'cough', 'sore throat' and 'headache' were most often reported by symptomatic cases in both variants, while 'pneumonia' and other symptoms of a severe course of the disease were rare or not reported. However, the BA.2 and BA.5 lineages showed that differences in the frequency of certain symptoms—e.g., 'other symptoms' that are not part of the GECCO list of relevance for COVID-19 infections—were reported more frequently from BA.5 than from BA.2 cases. In contrast, the symptoms 'chills or sweating', 'freeze' and 'runny nose' were more frequently reported in BA.2 cases. This fits well with the literature, with 'headache', 'runny nose', 'sore throat', 'sneezing', 'persistent cough' and 'fever' as the most prominent symptoms reported for COVID-19 infections [9].

Our analysis shows some strong correlations between symptoms. Some of these correlations can be explained by the fact that they are part of the respiratory symptom cluster [10], and some are clearly distinguishable but closely related, such as 'loss of smell' and 'loss of taste', respectively. Others can be explained by the fact that they are difficult for patients to distinguish, e.g., 'freeze' and 'chills or sweating', and some can be interpreted as a distinctive clinical course, e.g., 'chills or sweating' and 'muscle or body aches'. Other studies most prominently reported SARS-CoV-2 symptoms such as 'loss of smell' and 'loss of taste', mainly when associated with 'fever', and 'cough' in association with 'shortness of breath' and 'chest pain' [10]. However, different SARS-CoV-2 variants are known to exhibit different symptoms and influence the course of infections [2,11,12]. For BA.2, the European Centre for Disease Prevention and Control (ECDC) reports reduced evidence and for BA.5 no evidence for the impact on severity so far [13]. However, different variants and sub-variants have genetically determined differences that have been shown to correlate with different clinical features and outcomes [5,11].

Our study has several limitations that should be considered for the interpretation: First, the estimated differences in symptoms between the BA.2 and BA.5 lineages could be biased due to a low incidence situation at the beginning of a new wave. Second, the database provided by SORMAS is mainly based on self-reports of the patients who were either interviewed by the public health authorities staff or filled in standardized questionnaires on their own. In particular, the recording of symptoms depends on the timing of the query, the patient's physical condition and their willingness to provide information. As the public health departments and the use of SORMAS is decentralized, the way in which data were collected (either by interview or questionnaire) may have some differences between counties and over time. Although response is mandatory, it is quite possible that, for example, severe cases who were too ill to answer the questions were not fully recorded, as hospital record data can, but which is not systematically used. Since the questioning is one-time after the notification date, it is also conceivable that, depending on the day of the response of the patient, not all symptoms were recorded; later, potential hospitalization during the course of infection may, thus, not have been reported. Another possible systematic error is the partial overlap of symptoms in the symptom categories. For example, the symptom 'freeze' resembles the symptom 'chills' and, thus, could be reported either in both categories or only in one. This could also explain the strong correlation between both symptoms. Third, only cases about which data on the Omicron lineage were present were included in the study. Therefore, only a small number (less than 5%) of confirmed cases could be included. This could aggravate the fact that, to date, many people use commercial self-testing systems when acquiring respiratory or other typical symptoms for COVID-19 infections and may, in some cases, not report their infection status to the public health authorities. Additionally, sub-lineages in the BA.2 and BA.5 groups were pooled in our analysis, despite the fact that the genetic heterogenicity amongst these sub-lineages steadily increased over the study period.

However, our study has some strengths: The data comprised official reporting data for a geographically distinct region and a manageable time frame. The Omicron lineages were well comparable in terms of age and sex structure, vaccination and frequency of symptomatic cases. Thus, it is not to be expected that external, undetected influences have significantly distorted the results.

In summary, our results provide no evidence that BA.5 has a higher pathogenicity than BA.2. However, approximately 50% of cases developed symptoms which could lead to sick leave. If the incidence is high, this could represent a significant social burden, especially in critical infrastructure and healthcare areas. Differences in symptoms between variants should continue to be closely monitored in order to identify symptomatic patterns early.

5. Conclusions

To our knowledge, we firstly report profound differences in symptoms between the Omicron lineages BA.2 and BA.5 based on public health data. Our results show some significant differences between the lineages but provide no evidence that BA.5 leads to a more severe course of disease than BA.2. Our study highlights the opportunities and challenges of analyzing routine data collected by the public health authorities in Germany to gather detailed information on variant behavior at an early stage.

The rapid identification of new variants or subvariants of SARS-CoV-2 is important for surveillance, in order to closely monitor the occurrence of infection and to enable initiating protective countermeasures. Timely investigations of public health data connected to confirmed circulating SARS-CoV-2 variants, including reported symptoms, enables the assessment of the current situation and the chance to react quickly to manage the pandemic.

Author Contributions: Conceptualization, N.-O.H., K.V.G. and K.B.; formal analysis, K.V.G., J.M. and N.-O.H.; resources, J.Z., C.K. and the CoMV-Gen Study Group; data curation, K.V.G., J.M. and J.Z.; writing—original draft preparation, K.V.G.; writing—review and editing, K.B., C.K., J.M., J.Z. and the CoMV-Gen Study Group; project administration, N.-O.H. and K.B. All authors have read and agreed to the published version of the manuscript.

Funding: This study was conducted in the framework of the CoMV-Gen project, and the study, as well as the APC, was funded by the Ministry of Social Affairs, Health and Sports of the Federal State Government of Mecklenburg-Western Pomerania, Germany. Number: CoMV-Gen.

Institutional Review Board Statement: Ethical approval was granted by the ethics committee of the University of Greifswald, Germany (BB 125/21; date of approval:10 August 2021).

Informed Consent Statement: Patient consent was waived since only anonymized aggregated data are shown.

Data Availability Statement: Not applicable.

Acknowledgments: We thank the public health authorities of Mecklenburg-Western Pomerania, Germany for providing us with data from the German surveillance system SORMAS and the collaborating diagnostic laboratories within the CoMV-Gen study group. The members of the CoMV-Gen Study Group are: Nils-Olaf Hübner (Central Unit for Infection Control and Prevention, University Medicine Greifswald, Greifswald, Germany; Institute for Hygiene and Environmental Medicine, University Medicine Greifswald, Germany); Karsten Becker (Friedrich Loeffler-Institute of Medical Microbiology, University Medicine Greifswald, Germany); Katja Verena Goller (Central Unit for Infection Control and Prevention, University Medicine Greifswald, Greifswald, Germany); Juliane Moritz (Central Unit for Infection Control and Prevention, University Medicine Greifswald, Greifswald, Germany); Janine Ziemann (Central Unit for Infection Control and Prevention, University Medicine Greifswald, Greifswald, Germany; KOMPASS e.V.—MRE-Netzwerk Mecklenburg-Vorpommern, Greifswald, Germany); Christian Kohler (Friedrich Loeffler-Institute of Medical Microbiology, University Medicine Greifswald, Germany); Jürgen Bohnert (Friedrich Loeffler-

Institute of Medical Microbiology, University Medicine Greifswald, Germany); Lena Ulm (Friedrich Loeffler-Institute of Medical Microbiology, University Medicine Greifswald, Germany); Kathrin Zimmermann (Friedrich Loeffler-Institute of Medical Microbiology, University Medicine Greifswald, Germany); Lars Kaderali (Institute of Bioinformatics, University Medicine Greifswald, Greifswald, Germany); Ana Tzvetkova (Institute of Bioinformatics, University Medicine Greifswald, Greifswald, Germany); Martin Beer (Institute of Diagnostic Virology, Friedrich-Loeffler-Institut—Federal Research Institute for Animal Health, Greifswald-Insel Riems, Germany); Anne Pohlmann (Institute of Diagnostic Virology, Friedrich-Loeffler-Institut—Federal Research Institute for Animal Health, Greifswald-Insel Riems, Germany); Jacqueline King (Institute of Diagnostic Virology, Friedrich-Loeffler-Institut—Federal Research Institute for Animal Health, Greifswald-Insel Riems, Germany); Lina Stacker (Institute of Diagnostic Virology, Friedrich-Loeffler-Institut—Federal Research Institute for Animal Health, Greifswald-Insel Riems, Germany); Martina Littmann (LAGuS—Landesamt für Gesundheit und Soziales Mecklenburg-Vorpommern, Rostock, Germany); Tilo Sasse (LAGuS—Landesamt für Gesundheit und Soziales Mecklenburg-Vorpommern, Rostock, Germany); Kathrin Badstübner-Meeske (LAGuS—Landesamt für Gesundheit und Soziales Mecklenburg-Vorpommern, Rostock, Germany); Mandy Wietzke (LAGuS—Landesamt für Gesundheit und Soziales Mecklenburg-Vorpommern, Rostock, Germany); Elena Demihovska (LAGuS—Landesamt für Gesundheit und Soziales Mecklenburg-Vorpommern, Rostock, Germany); Ute Siering (District of Ludwigslust und Parchim, Ludwigslust, Germany); Robert Stach (District of Nordwestmecklenburg, Wismar, Germany); Gabriele Neumann (District of Nordwestmecklenburg, Wismar, Germany); Gerit Hübner (Municipality of Schwerin, Schwerin, Germany); Jasmin von Gadow (Municipality of Schwerin, Schwerin, Germany); Cornelia Ruhnau (District of Mecklenburgische Seenplatte, Neubrandenburg, Germany); Franz-Josef Stein (District of Mecklenburgische Seenplatte, Neubrandenburg, Germany); Kristin von der Oelsnitz (District of Rostock, Güstrow, Germany); Sylvia Krause (District of Rostock, Güstrow, Germany); Markus Schwarz (Municipality of Rostock, Rostock, Germany); Andi Friese (Municipality of Rostock, Rostock, Germany); Marlies Kühn (District of Vorpommern-Greifswald, Greifswald, Germany); Stefanie Hahn (District of Vorpommern-Greifswald, Greifswald, Germany); Marcel Neumeier (District of Vorpommern-Greifswald, Greifswald, Germany); Jörg Heusler (District of Vorpommern-Rügen, Stralsund, Germany); Kirstin Bengard (District of Vorpommern-Rügen, Stralsund, Germany); Egon Werle (Dietrich- Bonhoeffer-Klinikum Neubrandenburg (DBKNB) GmbH, Neubrandenburg, Germany); Eleonora Wasmund (Dietrich- Bonhoeffer-Klinikum Neubrandenburg (DBKNB) GmbH, Neubrandenburg, Germany); Lutz Briedigkeit (Helios Kliniken Schwerin, Schwerin, Germany); Kristian Meinck (IMD Labor Greifswald MVZ Labor Greifswald GmbH, Greifswald, Germany); Veronika Balau (IMD Labor Greifswald MVZ Labor Greifswald GmbH, Greifswald, Germany); Anja Kruggel (IMD Labor Greifswald MVZ Labor Greifswald GmbH, Greifswald, Germany); Mandy Wudtke (IMD Labor Greifswald MVZ Labor Greifswald GmbH, Greifswald, Germany); Ingo Neumann (Labor MVZ Westmecklenburg, Schwerin, Germany); Kirsten Gerst (LALLF—Landesamt für Landwirtschaft, Lebensmittelsicherheit und Fischerei Mecklenburg-Vorpommern, Rostock, Germany); Nele Matthes (LALLF—Landesamt für Landwirtschaft, Lebensmittelsicherheit und Fischerei Mecklenburg-Vorpommern, Rostock, Germany); Anja Balz (Medizinisches Labor Rostock, Rostock, Germany); Antje Lange (Medizinisches Labor Rostock, Rostock, Germany); Franziska Arnold (MVZ Martinsried GmbH, Martinsried, Germany); Oliver Wachter (MVZ Martinsried GmbH, Martinsried, Germany); Sedat Karadeniz (MVZ Martinsried GmbH, Martinsried, Germany); Christoph Kopp (MVZ Martinsried GmbH, Martinsried, Germany); Patricia Stunic (MVZ Martinsried GmbH, Martinsried, Germany); Bernhard Miller (MVZ Labor PD Volkmann und Kollegen Karlsruhe GbR, Karlsruhe, Germany); Fabian Ripp (MVZ Labor PD Volkmann und Kollegen Karlsruhe GbR, Karlsruhe, Germany); Barbara Hartmann (MVZ Labor PD Volkmann und Kollegen Karlsruhe GbR, Karlsruhe, Germany); Alisa Mitin (MVZ Labor PD Volkmann und Kollegen Karlsruhe GbR, Karlsruhe, Germany); Stephan Schaefer (MVZ Labor Limbach Vorpommern-Rügen GmbH, Stralsund, Germany); Yvonne Gerdes (MVZ Labor Limbach Vorpommern-Rügen GmbH, Stralsund, Germany); Philipp Warnke (University Medicine Rostock—Institut für Medizinische Mikrobiologie, Virologie und Hygiene, Rostock, Germany); and Moritz Jansson (University Medicine Rostock—Institut für Medizinische Mikrobiologie, Virologie und Hygiene, Rostock, Germany).

Conflicts of Interest: The authors declare no conflict of interest.

References

1. Robert Koch Institut (RKI). Wöchentlicher Lagebericht des RKI zur Coronavirus-Krankheit-2019 (COVID-19). Available online: http://www.rki.de/DE/Content/InfAZ/N/Neuartiges_Coronavirus/Daten/VOC_VOI_Tabelle.html (accessed on 7 July 2022).

2. Sievers, C.; Zacher, B.; Ullrich, A.; Huska, M.; Fuchs, S.; Buda, S.; Haas, W.; Diercke, M.; an der Heiden, M.; Kröger, S. SARS-CoV-2 Omicron variants BA.1 and BA.2 both show similarly reduced disease severity of COVID-19 compared to Delta, Germany, 2021 to 2022. *Eurosurveillance* **2022**, *27*, 2200396. [CrossRef] [PubMed]
3. Tegally, H.; Moir, M.; Everatt, J.; Giovanetti, M.; Scheepers, C.; Wilkinson, E.; Subramoney, K.; Makatini, Z.; Moyo, S.; Amoako, D.G.; et al. Emergence of SARS-CoV-2 Omicron lineages BA.4 and BA.5 in South Africa. *Nat. Med.* **2022**, 1–6. [CrossRef] [PubMed]
4. Viana, R.; Moyo, S.; Amoako, D.G.; Tegally, H.; Scheepers, C.; Althaus, C.L.; Anyaneji, U.J.; Bester, P.A.; Boni, M.F.; Chand, M.; et al. Rapid epidemic expansion of the SARS-CoV-2 Omicron variant in southern Africa. *Nature* **2022**, *603*, 679–686. [CrossRef] [PubMed]
5. Kimura, I.; Yamasoba, D.; Tamura, T.; Nao, N.; Oda, Y.; Mitoma, S.; Ito, J.; Nasser, H.; Zahradnik, J.; Uriu, K.; et al. Virological characteristics of the novel SARS-CoV-2 Omicron variants including BA.2.12.1, BA.4 and BA.5. *bioRxiv* **2022**. [CrossRef]
6. Cao, Y.; Yisimayi, A.; Jian, F.; Song, W.; Xiao, T.; Wang, L.; Du, S.; Wang, J.; Li, Q.; Chen, X.; et al. BA.2.12.1, BA.4 and BA.5 escape antibodies elicited by Omicron infection. *Nature* **2022**, *608*, 593–602. [CrossRef] [PubMed]
7. Tom-Aba, D.; Silenou, B.C.; Doerrbecker, J.; Fourie, C.; Leitner, C.; Wahnschaffe, M.; Strysewske, M.; Arinze, C.C.; Krause, G. The Surveillance Outbreak Response Management and Analysis System (SORMAS): Digital Health Global Goods Maturity Assessment. *JMIR Public Health Surveill.* **2020**, *6*, e15860. [CrossRef] [PubMed]
8. Sass, J.; Bartschke, A.; Lehne, M.; Essenwanger, A.; Rinaldi, E.; Rudolph, S.; Heitmann, K.U.; Vehreschild, J.J.; von Kalle, C.; Thun, S. The German Corona Consensus Dataset (GECCO): A standardized dataset for COVID-19 research in university medicine and beyond. *BMC Med. Inform. Decis. Mak.* **2020**, *20*, 341. [CrossRef] [PubMed]
9. The ZOE Health Study. Omicron Symptoms: What Are They and How Long Do They Last? Available online: www.joinzoe.com/learn/omicron-symptoms (accessed on 8 July 2022).
10. Dixon, B.E.; Wools-Kaloustian, K.; Fadel, W.F.; Duszynski, T.J.; Yiannoutsos, C.; Halverson, P.K.; Menachemi, N. Symptoms and symptom clusters associated with SARS-CoV-2 infection in community-based populations: Results from a statewide epidemiological study. *PLoS ONE* **2021**, *16*, e0241875. [CrossRef] [PubMed]
11. Menni, C.; Valdes, A.M.; Polidori, L.; Antonelli, M.; Penamakuri, S.; Nogal, A.; Louca, P.; May, A.; Figueiredo, J.C.; Hu, C.; et al. Symptom prevalence, duration, and risk of hospital admission in individuals infected with SARS-CoV-2 during periods of omicron and delta variant dominance: A prospective observational study from the ZOE COVID Study. *Lancet* **2022**, *399*, 1618–1624. [CrossRef]
12. Meo, S.A.; Meo, A.S.; Al-Jassir, F.F.; Klonoff, D.C. Omicron SARS-CoV-2 new variant: Global prevalence and biological and clinical characteristics. *Eur. Rev. Med. Pharmacol. Sci.* **2021**, *25*, 8012–8018. [CrossRef]
13. European Centre for Disease Prevention and Control. Available online: https://www.ecdc.europa.eu/en/COVID-19/variants-concern (accessed on 18 July 2022).

viruses MDPI

Article

Impact of Early Pandemic SARS-CoV-2 Lineages Replacement with the Variant of Concern P.1 (Gamma) in Western Bahia, Brazil

Josilene R. Pinheiro [1,2], Esther C. dos Reis [1], Jéssica P. Farias [1], Mayanna M. C. Fogaça [1], Patrícia de S. da Silva [1,2], Itana Vivian R. Santana [1], Ana Luiza S. Rocha [1], Paloma O. Vidal [1], Rafael da C. Simões [1], Wilson B. Luiz [2], Alexander Birbrair [3,4,5], Renato S. de Aguiar [6,7], Renan P. de Souza [6], Vasco A. de C. Azevedo [6], Gepoliano Chaves [8], Aline Belmok [9], Ricardo Durães-Carvalho [10,11], Fernando L. Melo [9], Bergmann M. Ribeiro [9] and Jaime Henrique Amorim [1,2,*]

1 Center of Biological Sciences and Health, Federal University of Western Bahia, Barreiras 47805, BA, Brazil
2 Department of Biological Sciences, State University of Santa Cruz, Ilhéus 45662, BA, Brazil
3 Department of Dermatology, School of Medicine and Public Health, University of Wisconsin-Madison, Madison, WI 53706, USA
4 Department of Pathology, Federal University of Minas Gerais, Belo Horizonte 31270, MG, Brazil
5 Department of Radiology, Columbia University Medical Center, New York, NY 10032, USA
6 Department of Genetics, Ecology and Evolution, Federal University of Minas Gerais, Belo Horizonte 31270, MG, Brazil
7 D'Or Institute of Research, Rio de Janeiro 22281, RJ, Brazil
8 Department of Pediatrics, University of Chicago, Chicago, IL 60637, USA
9 Laboratory of Baculoviruses, University of Brasilia, Brasilia 70910, DF, Brazil
10 Department of Microbiology, Immunology and Parasitology, São Paulo School of Medicine, Federal University of São Paulo (UNIFESP), São Paulo 04023, SP, Brazil
11 Post-Graduate Program in Structural and Functional Biology, UNIFESP, São Paulo 04023, SP, Brazil
* Correspondence: jaime.amorim@ufob.edu.br; Tel.: +5577-3614-3218

Citation: Pinheiro, J.R.; dos Reis, E.C.; Farias, J.P.; Fogaça, M.M.C.; da Silva, P.d.S.; Santana, I.V.R.; Rocha, A.L.S.; Vidal, P.O.; Simões, R.d.C.; Luiz, W.B.; et al. Impact of Early Pandemic SARS-CoV-2 Lineages Replacement with the Variant of Concern P.1 (Gamma) in Western Bahia, Brazil. *Viruses* **2022**, *14*, 2314. https://doi.org/10.3390/v14102314

Academic Editors: Ahmed El-Shamy and Mohamed Ibrahim

Received: 27 September 2022
Accepted: 21 October 2022
Published: 21 October 2022

Abstract: Background: The correct understanding of the epidemiological dynamics of COVID-19, caused by the SARS-CoV-2, is essential for formulating public policies of disease containment. Methods: In this study, we constructed a picture of the epidemiological dynamics of COVID-19 in a Brazilian population of almost 17000 patients in 15 months. We specifically studied the fluctuations of COVID-19 cases and deaths due to COVID-19 over time according to host gender, age, viral load, and genetic variants. Results: As the main results, we observed that the numbers of COVID-19 cases and deaths due to COVID-19 fluctuated over time and that men were the most affected by deaths, as well as those of 60 or more years old. We also observed that individuals between 30- and 44-years old were the most affected by COVID-19 cases. In addition, the viral loads in the patients' nasopharynx were higher in the early symptomatic period. We found that early pandemic SARS-CoV-2 lineages were replaced by the variant of concern (VOC) P.1 (Gamma) in the second half of the study period, which led to a significant increase in the number of deaths. Conclusions: The results presented in this study are helpful for future formulations of efficient public policies of COVID-19 containment.

Keywords: COVID-19; impact; variant of concern

1. Introduction

The *Severe Acute Respiratory Syndrome Coronavirus 2* (SARS-CoV-2) belongs to the Betacoronavirus genus in the Coronaviridae family [1]. It is the causative agent of the coronavirus disease 2019 (COVID-19) [2,3], the world's major public health problem in the last three years, which affected hundreds of million people and caused more than 6 million deaths [4]. Due to the high epidemiological impact of the disease, a great effort was carried out by the scientific community around the world to rapidly study the virus and the disease. Relevant scientific progress was rapidly achieved regarding the knowledge of pathophysiology, transmission, diagnosis, and treatment [5–10]. In addition, several vaccine formulations were developed and are being shown as essential in the control of severe forms of COVID-19 worldwide [11,12]. However, the consecutive emergence of new genetic variants of SARS-CoV-2 have brought new questions and challenges to pathophysiology, transmission, diagnosis, treatment, and the use of vaccines [13].

The correct understanding of the epidemiological dynamics of COVID-19 is essential for formulating public policies of disease containment. For example, knowing when and why a new wave of COVID-19 is expected to happen can be used for preparedness in terms of the use of non-pharmacological measures, vaccination, reinforcement in the number of health professionals, reinforcements in laboratory tests availability, reinforcements in the numbers of hospital beds, and the cancellation of events, etc. [14–16]. Understanding the dynamics of COVID-19 cases and deaths over time, and according to gender and age, is also essential to predict situations of epidemiological risk and for preparedness regarding specific groups [17,18]. In addition, the knowledge about viral genetic variants circulating in a given area [19], their epidemiological impact and their probable origins, are also essential for successful public health policies of containment of importing or exporting viruses.

In this study, we aimed to have a picture of the epidemiological behavior of COVID-19 in a population study of almost 17,000 patients in 15 months. We specifically aimed to: (i) see the fluctuations of COVID-19 cases and deaths during the period of study; (ii) describe and understand the epidemiological behavior of COVID-19 according to patients gender, age, viral load, and viral genetic variants; (iii) to test possible associations between these variables; and (iv) to phylogenetically reconstruct the evolutionary relationships of SARS-CoV-2 circulating lineages sampled from the study. The results presented in this study are helpful for the future formulation of public policies of COVID-19 containment.

2. Materials and Methods

2.1. Data Collection

This is a retrospective study of the cases of COVID-19 registered in the cities of Western Bahia (west region of Bahia state, Brazil) from May 2020 to July 2021. All patient data and samples were provided by the Laboratory of Infectious Agents and Vectors from Western Bahia Federal University, located in Barreiras city, Bahia, Brazil. Information such as the patient's name, age, gender, sample identification number, collection date, RT-qPCR result for SARS-CoV-2 detection with the value of the cycle threshold (Ct), the date of onset of symptoms, the municipality of residence, and the patient's care unit, were used in this study. These data were tabulated and used in statistical analyses. All the research complied with all relevant ethical and biosafety guidelines. Ethics approval was obtained from the institutional ethics committee of the Federal University of Western Bahia (CAAE 40779420.6.0000.8060). All procedures and possible risks were explained to volunteers. Informed consent was obtained from all participants. The research was performed in accordance with relevant guidelines/regulations. The sample is composed of data from 16,908 laboratory tests, including positive and negative results of SARS-CoV-2 detection.

2.2. RNA Extraction and RT-qPCR

The nucleic acid extractions of nasopharyngeal samples were carried out using the Total RNA Purification Kit (Cellco Biotec, Sao Carlos, SP, Brazil), following the manufacturer's protocol. We also carried out viral RNA extraction using the Extracta Kit—RNA e DNA Viral (MVXA-P016FAST) (Loccus, Sao Paulo, SP, Brazil), using an Extracta32 instrument (Loccus, Sao Paulo, SP, Brazil), following the manufacturer's instructions.

Reverse transcription, followed by quantitative polymerase chain reaction (RT-qPCR) assays, were carried out as previously described by us [20]. Thermocycling was carried out in a QuantStudio 5 instrument (Applied Biosystems, Waltham, MA, USA) with a hold stage composed of a first step of 5 min at 50 °C, followed by a second step of 20 s at 95 °C. The PCR stage was composed of a first step of 15 s at 95 °C followed by a second step of

1 min at 55 °C, repeated 45 times. We also used Allplex 2019-nCOV RT-qPCR kit (Seegene, Song-pa-gu, Seoul, Republic of Korea), following the manufacturer's instructions.

2.3. Viral Genotyping by RT-qPCR

Viral variants were characterized using RhAmp technology (Integrated DNA Technologies IDT, Coralville, IA, USA) and TaqMan SARS-CoV-2 Mutation Panel (Thermo Fisher, Waltham, MA, USA) with specific primers and probes targeting the VOCs defining mutations: K417T (A22812C), E484K (G23012A), and N501Y (A23063T). RhAmp and TaqMan detailed protocols were previously reported [21,22].

2.4. SARS-CoV-2 Genome SEQUENCING

One hundred and twenty positive samples collected from May 2020 to July 2021 were sequenced using Next Generation Sequencing (NGS) on the Oxford Nanopore's MinIon platform. Viral RNA was extracted as described above. The RNA samples were submitted to reverse transcription with random primers using LunaScript®(New England Biolabs, Inc., Ipswich, MA, USA) or SuperScript®IV First-Strand Synthesis System (ThermoFisher Scientific, Waltham, MA, USA), as previously described (nCoV-2019 sequencing protocol v3 (LoCost) (protocols.io) [23]. The cDNA obtained was used as a template for the amplification of the entire genome of SARS-CoV-2 with the following primer scheme: a 400bp amplicon scheme from ARTIC nCoV-2019 sequencing protocol (v3) (nCoV-2019 sequencing protocol v3 (LoCost) (protocols.io) was used, as previously described [23]. End-prep reactions were performed with NEBNext®Ultra™ II End Repair/dA-Tailing Module, and amplicons were barcoded using the ONT Native Barcoding Expansion kit (EXP-NBD104). The barcoded samples were then combined, purified with AMPure XP Beads, and loaded onto Oxford Nanopore MinION SpotON Flow Cells R9.4.1 (Oxford Nanopore Technologies), following the manufacturer's instructions. The sequencing was carried out using the fast accuracy base-calling in the MinKNOW software. ARTIC Network's RAMPART (https://artic.network/ncov-2019, accessed on 2 December 2021) was used to monitor the sequencing run in real-time to estimate the depth of coverage (20×) across the entire genome for each barcode (https://artic.network/rampart, accessed on 2 December 2021). The analysis and consensus generation were performed according to the pipeline proposed by ARTIC Network using the Medaka protocol (https://artic.network/ncov-2019/ncov2019-bioinformatics-sop.html, accessed on 2 December 2021). All consensus genomes were deposited in the Global Initiative on Sharing Avian Influenza Data-EpiCoV (GISAID-EpiCoV) database (see Supplementary Material S1, for details)

2.5. Phylogenetic Analysis

New SARS-CoV-2 whole-genome sequences obtained here were submitted to lineages assigner Pangolin web application, available online: https://pangolin.cog-uk.io/, accessed on 1 September 2022. Initially, phylogenetic reconstructions were performed using datasets containing sequences from the study (n = 112) and 1004 representative Nextstrain's subsampling SARS-CoV-2 genomic sequences from South America countries and their territories (n = 16). Such sequences were retrieved from the beginning of the pandemic until August 2022 (https://nextstrain.org/, accessed on 1 September 2022), representing multiple SARS-CoV-2 circulating strains. In addition, high-coverage complete SARS-CoV-2 Gamma genome sequences (n = 14846) from all Brazilian States and the Federal District (n = 27), deposited up to 31 July 2022 in the GISAID-EpiCoV, were also downloaded. Data sets were filtered out by the Sequence Cleaner, a biopython-based script (https://biopython.org/wiki/Sequence_Cleaner, accessed on 1 September 2022), which comprised a set of unambiguous sequences $\geq$ 29,000 bp with 0% of Ns and degenerated nucleotides. Sequences that did not fit these criteria were automatically excluded. The outcomes were aligned with the SARS-CoV-2 reference coding-sequence (NC_045512.2) by MAFFT v.7 [24] and edited by the UGENE v.44.0 [25].

Aiming to investigate the relative amount of unresolved to fully resolved trees, the ph logenetic signal approach was explored through likelihood mapping analysis of 10,000 ra dom quartets using TREE-PUZZLE v.5.2 [26]. Then, the maximum likelihood (ML) metho was implemented by using two different command line algorithms: FastTree v.2.1.7 [27 and IQ-TREE v.2 [26]. FastTree was executed by using the GTR substitution model + CAT with 20 gamma (G) distribution parameters and a mix of Nearest-Neighbor Interchanges (NNI) and Sub-Tree-Prune-Regraft (SPR). ML from IQ-TREE was inferred using the substitution model GTR + F + I + G4, executed and optimized by the Maximum Parsimony and Neighbor-Joining trees, and hill-climbing algorithms, respectively [28]. The reliability of the nodes was analyzed by the Shimodaira–Hasegawa (SH-like) test, which uses bootstrap resampling and corrects critical values for multiple comparisons [29], and SH-aLTR/aBayes/ultrafast bootstrap support values, both with 1000 replicates. Phylogenetic trees were generated by Interactive Tree of Life [30].

2.6. Statistical Analyses

To compare the means of two groups we used *t*-student test. To compare more than two groups we used analysis of variance (ANOVA) followed by Bonferroni multiple comparison test. In addition, to verify relations between variables we carried out linear regression analyses. In all cases, statistical significance was set as $p \leq 0.05$. In some cases, we carried out descriptive statistics.

3. Results

3.1. Fluctuations of Numbers of COVID-19 Cases and Deaths from May 2020 to July 2021

The number of COVID-19 cases per day fluctuated from May 2020 to July 2021, with statistically significant peaks in August 2020, and in April and May 2021 (Figure 1A and Supplementary Table S1). The numbers of deaths per day due to COVID-19 also fluctuated in the study period (Figure 1B and Supplementary Table S2). First, relevant increases in the numbers of deaths per day were observed on July and October 2020. Then, significant increases were observed in April–June 2021. Collectively, these results indicate that the study population was relevantly affected by COVID-19, with different waves of cases and deaths over the time of study.

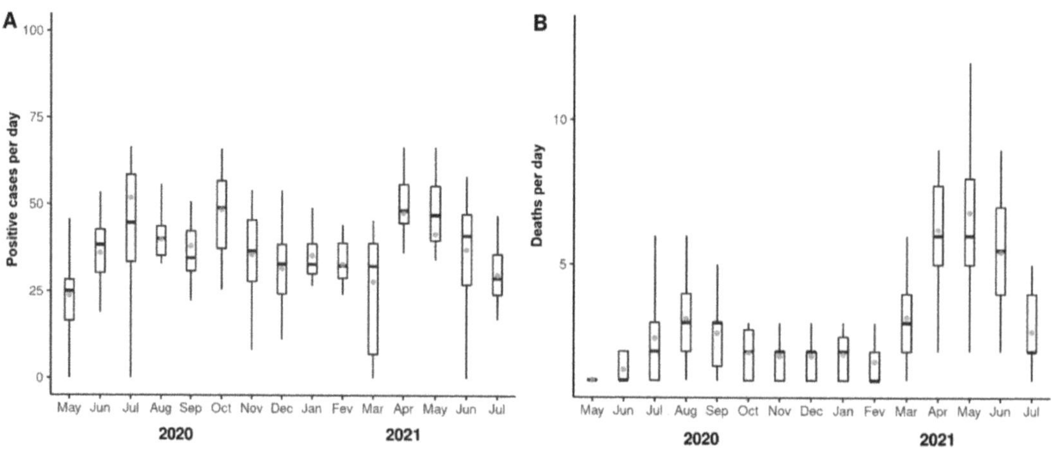

Figure 1. Fluctuations of COVID-19 cases and deaths due to COVID-19 during the period of study. (**A**) numbers of cases of COVID-19 per day during the period of study. (**B**) numbers of deaths due to COVID-19 per day during the period of study. Dots represent means. Horizontal bars represent medians.

3.2. Impact of COVID-19 According to Gender or Age

The numbers and rates of COVID-19 cases and deaths due to COVID-19 per day were also computed. As shown in Figure 2A, there was not a significant difference in the rates of cases according to gender. However, as shown in Figure 2B, the numbers of men dying of COVID-19 per day were significantly higher than those for women ($p = 0.0042$) during the study period. In addition, the proportions of cases and deaths in age groups 0–11, 12–18, 19–29, 30–44, 45–59, and 60 and above years old were observed. As shown in Figure 2C, the 30–44 years old age group was the most affected by COVID-19 cases during the study period (see Supplementary Table S3 for absolute numbers and descriptive statistics). This group was followed by those 45–59 and 60 and above years old in the proportions of COVID-19 cases.

In contrast, the group 60 and above years old, was the most affected by deaths due to COVID-19 in most of the period of study (Figure 2D and Supplementary Table S4 for absolute numbers and descriptive statistics). Such a leadership in the numbers of deaths was not observed only in the first month of the study and in May–July 2021, when the sum of numbers of deaths in the groups of 30–44 and 45–59 years old was higher than in the oldest group. It is essential to highlight that even in the three last months of the study, the oldest group was the most affected by deaths when compared pair-to-pair with the other age groups. Collectively, these results indicate that: (i) men were the most affected by deaths due to COVID-19; (ii) the age group of 30–44 years old was the most affected by COVID-19 cases; and (iii) the age group of 60 and above years old was the most affected by deaths due to COVID-19.

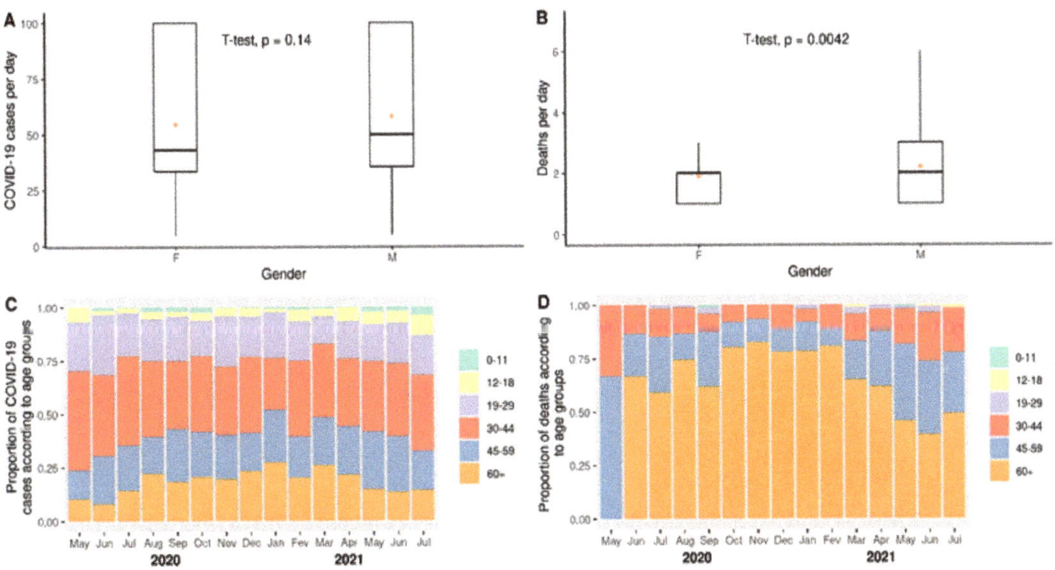

Figure 2. Impact of COVID-19 according to gender or age. Comparisons of numbers of cases per day (**A**) and numbers of deaths per day (**B**) considering the whole period of study were carried out based on Student's t-test. Significance was set as $p \leq 0.05$. In addition, proportions of cases (**C**) and deaths (**D**) according to age groups were computed. Dots represent means. Horizontal bars represent medians.

3.3. Viral Loads According to Cycle Threshold Values

The SARS-CoV-2 viral loads were inferred according to cycle thresholds (CT) found after RT-qPCR of the patient's nasopharynx swabs. As shown in Figure 3A and Supplementary Table S5, we observed the fluctuation of CT values of SARS-CoV-2-positive samples during the study period. It was possible to observe increases and decreases in the mean CT values along the months. Interestingly, a tendency to decrease was observed in the last three months. The CT values were not significantly different according to gender, as shown in Figure 3B ($p > 0.05$). In addition, when CT values were compared along months according to age groups (Figure 3C), it was possible to observe that the age group of 60 and above years old presented lower CT values in June and July 2020 ($p < 0.05$) (see Supplementary Material S2 for statistical details). However, such a decrease was not seen in the other months of the study period. Moreover, patients with ages ranging from 0 to 18 years old presented significantly diminished or increased CT values in one or two months. However, we did not find a general tendency regarding viral loads according to age group. In contrast, the CT values were shown to be significantly lower in the beginning of the symptom period ($p < 0.05$), as shown in Figure 3D (see Supplementary Material S3, for statistical details). Collectively, these results indicate that the study population presented different viral loads in nasopharynx over time, as shown by CT values. However, the CT values did not differ relevantly according to age groups. In contrast, they were shown to be significantly reduced in samples collected at the early stage of COVID-19 symptoms, which indicates higher viral loads in this period.

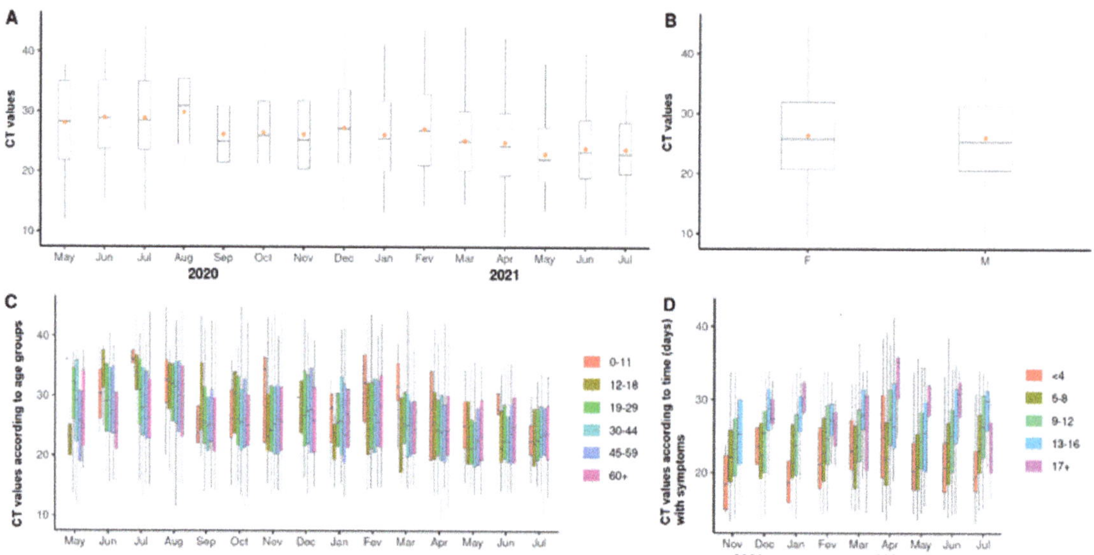

Figure 3. Viral loads according to cycle threshold (CT) values. (**A**) fluctuation of viral loads found in samples collected from the study population along the period of study. (**B**) comparison of viral loads according to gender. (**C**) viral loads according to age group for each month of the study. (**D**) viral loads according to time (days) with symptoms, from November 2020 to July 2021. Dots represent means. Horizontal bars represent medians.

3.4. *Substitution of SARS-CoV-2 Lineages and Its Impact on Local Health*

As shown in Figure 4A, from May to November 2020, only early pandemic lineages (EPLs) of SARS-CoV-2 such as B1.1, B1.1.28, B1.1.33, and N9 were found. In December 2020, the P.2 (Zeta) variant of interest (VOI) lineage was found together with EPLs. In January and February 2021, the P.2 VOI was still detected in predominance over EPLs. However, the P.1 (Gamma) variant of concern (VOC) lineage was detected in low proportions in these months. In March 2021, the Gamma VOC dominated the scenario over Zeta and prevented the B.1.1.7 (Alpha) VOC lineage fixation. These results show that the Gamma VOC was introduced in the study area/population and dominated the scene. Moreover, from April to July 2021 the Gamma VOC completely dominated the scenario and was the only SARS-CoV-2 lineage detected in nasopharynx samples of the study population.

As shown in Figure 4B, the increase in the numbers of cases caused an increase in the numbers of deaths due to COVID-19. It is important to note that the peaks of COVID-19 cases coincided with those of deaths in this study. Thus, we carried out a linear regression analysis to see if the increase in the numbers of cases determined the numbers of deaths. To see if the increase in the numbers of cases was determined by the increase in the proportions of the Gamma VOC detection, we also conducted a regression analysis (Figure 4C), which revealed no association between these variables. However, when we analyzed the association of proportions of Gamma VOC detection and numbers of deaths, we saw that the increase in the proportions of Gamma VOC detection determined the increase in numbers of deaths (Figure 4D). These results collectively indicate that the substitution of early pandemic SARS-CoV-2 lineages by the Gamma VOC caused a significant impact on the health of the population studied, with a significant increase in the numbers of deaths due to COVID-19.

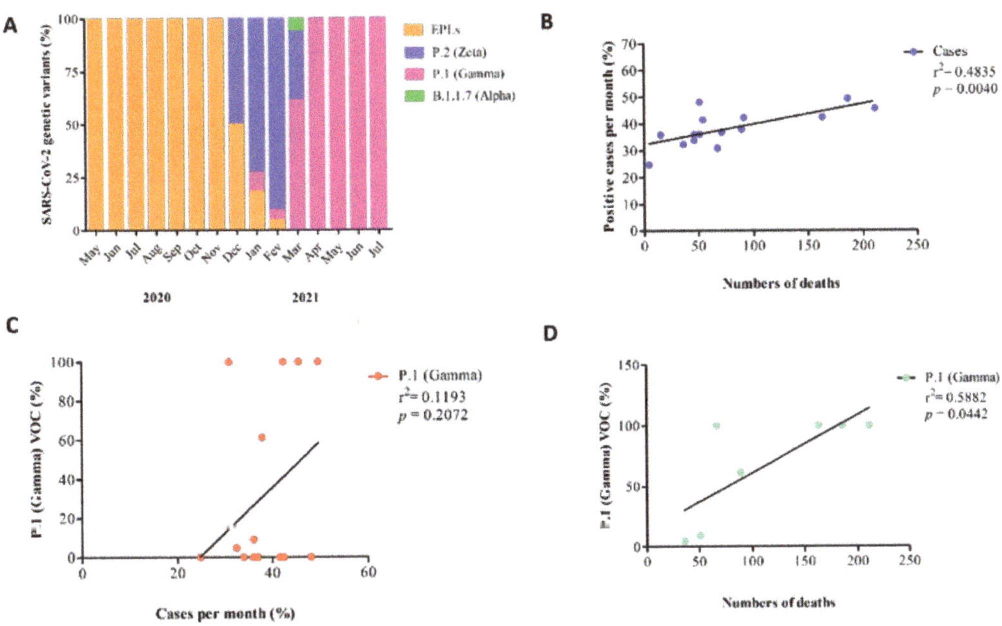

Figure 4. Substitution of SARS-CoV-2 lineages and its impact on local health. (**A**) Proportions of SARS-CoV-2 lineages found during the period of study. Viruses were classified based on genome sequencing and a specific RT-qPCR strategy capable of detecting specific mutations, as described in Section 2. EPLs, early pandemic lineages of SARS-CoV-2. (**B**) association between proportions of numbers of cases per month and numbers of deaths per month, as confirmed by linear regression analysis. (**C**) lack of association between proportions of viruses of the Gamma lineage found per month and proportions of cases per month (confirmed by linear regression analysis). (**D**) association between proportions of viruses of the Gamma lineage found per month and numbers of death per month (confirmed by linear regression analysis). Statistical significance was set as $p \leq 0.05$.

3.5. Viral Phylogeny

To phylogenetically describe circulating viruses in Western Bahia during the study period we carried out maximum-likelihood analyses. As shown in Figure 5A, SARS-CoV-2 genomes of viruses that circulated in our study area during the study period were grouped separately from those found in other countries of South America. On the other hand, they grouped with viruses found in Brazil (Figure 5B). In addition, EPLs of SARS-CoV-2 were shown to have preceded VOC in the study as also shown in Figure 5B. This result confirms that EPLs of SARS-CoV-2 were replaced by VOCs, mainly the Gamma lineage, as shown in Figure 4A. Collectively, these results show that viruses found in the study area during the study period are phylogenetically related to Brazilian isolates and that EPLs of SARS-CoV-2 were replaced by the Gamma lineage.

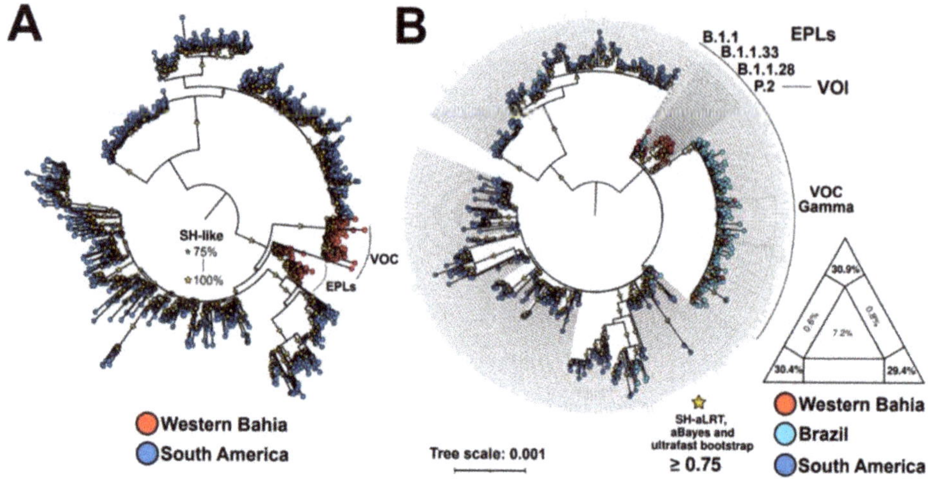

Figure 5. Maximum-likelihood midpoint rooted phylogenetic tree based on 1117 (A) and 603 (**B**) representative genome sequences of SARS-CoV-2. Nextstrain's subsampled SARS-CoV-2 genomic data from South America (including Brazil) since pandemic started up to August 2022 containing unfiltered (**A**) and filtered (**B**) sequences from the study. The SARS-CoV-2 genomes from this study are identified by the red circles. Tips are colored according to sampling locations. Yellow stars assume Shimodaira–Hasegawa (SH-like) test (**A**) and SH-aLTR/aBayes/ultrafast bootstrap support (**B**) based on 1000 replicates. Only values equal or greater than 75% are shown. Likelihood mapping of the final sequences alignment showing low phylogenetic noise, as required for reliable phylogeny inference (**B**). Abbreviations: VOC and VOI, Variant(s) of Concern and Variant(s) of Interest, respectively. EPLs, Early Pandemic Lineages. In letter B, "Brazil" represents whole-genome of SARS-CoV-2 Gamma variant from all Brazilian States and the Federal District. Branch lengths are drawn in scale of nucleotide substitutions per site according to the bar scale. Colors and symbols used in the panels are defined according to the legend to the left and right of the figure.

4. Discussion

In this study, we aimed to have a picture of the epidemiological dynamics of COVID-19 in a population of almost 17,000 patients in a period of 15 months. We studied: (i) the fluctuations of COVID-19 cases and deaths due to COVID-19 during the period of study; (ii) the epidemiological behavior of COVID-19 according to patient gender, age, viral load and viral genetic variants; (iii) associations between these variables; and (iv) viral phylogeny.

We observed that the numbers of COVID-19 cases and deaths due to COVID-19 fluctuated over time. This was an expected result, once that fluctuations in the numbers of cases and deaths were also observed worldwide, as indicated by data from the World Health Organization [31]. So far, during the pandemic, several factors have had an impact on whether the numbers of COVID-19 cases and deaths are increasing or declining in specific locations. These factors include human behavior, infection prevention policies, viral genetic mutations, the number of people who are vulnerable because they have not developed some immunity, whether from natural infection or through vaccination, and the effectiveness of vaccines over time.

In our case, the national, state, and city government authorities adopted different policies over time. The first case of COVID-19 in Bahia was reported by the state health authority on 6 March 2020, nine days after the first case in Brazil [32]. This first case involved a history of travel to Europe. In addition, the first case in Western Bahia was reported on 21 March 2020, involving a history of travel to São Paulo city [32]. It is important to highlight that the Brazilian carnival took place in February 2020, just before the first cases in our study area. The carnival is a very popular festival, which promotes intense traveling movements and direct contact with people [33]. In addition, international travelers are very frequent during the festival. Surprisingly, none or negligible infection prevention policies regarding COVID-19 were adopted by all levels of government authorities at that time, even with the WHO declaring that COVID-19 constitutes a Public Health Emergency of International Concern (PHEIC) on 30 January 2020 [31]. Thus, the carnival seems to have contributed to bringing SARS-CoV-2 to Brazil, to the Bahia state, and to Western Bahia.

Following the introduction of COVID-19 in our study area, we observed a sharpened increase in the numbers of cases and deaths from June to August 2020. Although government and health authorities had launched prevention policies [34], the June celebrations, which are very popular in Northeast Brazilian states such as Bahia, seem to have impacted human behavior. Despite decrees imposing social and physical distancing, the lack of experience with the pandemic at that time seems to have impelled people to commemorate at private celebrations. After this event, the numbers of cases per day remained elevated in comparison with the first month of study, a probable result of the spread and multiplication of SARS-CoV-2 in the population. Such a situation was dramatically changed with the replacement of the early pandemic lineages (EPLs) of SARS-CoV-2 found in 2020 with the Gamma variant of concern (VOC) in 2021.

Although the vaccination has been initiated on January 2021 at the study area, the specific groups of elderly and health professionals were vaccinated first. This was imposed by the low availability of vaccines. The introduction of the Gamma lineage in the study area in early 2021 resulted in a total domination of the scenario by the new virus lineage. In addition to the replacement of the early pandemic lineages (EPLs) of SARS-CoV-2 in the study area, an increased mortality took place, most probably related to the increased pathogenicity of the Gamma lineage in comparison to the EPLs. In this context, it is important to highlight that the proportions of adults with ages ranging from 30 to 59 dying due to COVID-19 were increased at this period. This group was vaccinated with a delay in comparison to the elderly.

In fact, all lineages of SARS-CoV-2 detected during the study produced more deaths in men than in women. It is well known that women have a better immune response than men. Generally, adult females mount stronger innate and adaptive immune responses than males. This results in the faster clearance of pathogens and greater vaccine efficacy in females than in males [35]. In addition, men tend to expose themselves more to risk [36]. These two factors seem to explain the results. On the other hand, the higher incidence of COVID-19 in people with age ranging from 30 to 44 does not have an easy explanation. We suppose that people with this age range are more economically and professionally active. They may have moved and contacted more people, exposing themselves more to infection.

Another interesting observation in our study was related to higher viral loads found in samples collected from patients in the early symptomatic period. This result by itself is in accordance with previous studies [37,38]. However, differences in viral loads were not observed according to the replacement of viral lineages. Interestingly, diminished viral loads are related to a worst outcome [38], but the more pathogenic Gamma lineage, which dominated the scenario in the five last months of study, was not detected with higher CT values (diminished viral load). It is important to highlight that a possible explanation for these results can be related not only to viral clearance, but also to the descending of infection from the nasopharynx to the lower respiratory tract, especially in severe cases.

Collectively, results presented in this study indicate that the numbers of COVID-19 cases and deaths due to COVID-19 fluctuated over time and that men were the most affected by deaths, as well as those of 60 or more years old. We also observed that individuals between 30 and 44 years old were the most affected by COVID-19 cases. In addition, the viral loads in the patient's nasopharynx were higher in the early symptomatic period. Relevantly, we found that early pandemic SARS-CoV-2 lineages were replaced by the variant of concern (VOC) P.1 (Gamma) in the second half of the study period, which led to a significant increase in the number of deaths. Although the low number of samples subjected to genomic sequencing may generate limitations regarding the time of detection of viral lineages replacements, the main conclusions are supported by robust statistical analyses. In addition, genomic surveillance was complemented by genotyping using RT-qPCR. Thus, the results presented in this study are helpful for future formulations of efficient public policies of COVID-19 containment.

Supplementary Materials: The following supporting information can be downloaded at: https://www.mdpi.com/article/10.3390/v14102314/s1. Material S1: Metadata; Material S2: Pairwise comparisons using *t* tests with pooled SD; Material S3: Pairwise comparisons using *t* tests with pooled SD; Table S1: Matrix with *p* values of comparisons (Bonferroni multiple comparison test after ANOVA) between each pair of months, regarding numbers of cases per day. Identification of months are shown in rows and columns. First, the number of the year is given and is followed by the number of the month, beginning at 1 (January) and ending at 12 (December); Table S2: Matrix with *p* values of comparisons (Bonferroni multiple comparison test after ANOVA) between each pair of months, regarding numbers of deaths per day. Identification of months are shown in rows and columns. First, the number of the year is given and is followed by the number of the month, beginning at 1 (January) and ending at 12 (December); Table S3: Absolute numbers of COVID-19 cases according to age group and month of study; Table S4: Absolute numbers of deaths due to COVID-19 according to age group and month of study; Table S5: Matrix with *p* values of comparisons (Bonferroni multiple comparison test after ANOVA) between each pair of months, regarding CT values. Identification of months are shown in rows and columns. First, the number of the year is given and is followed by the number of the month, beginning at 1 (January) and ending at 12 (December).

Author Contributions: J.R.P., E.C.d.R., J.P.F., P.d.S.d.S., M.M.C.F., I.V.R.S. and A.L.S.R. carried out samples collection, sample analyses and organization of databanks. In addition, they analyzed the data and interpreted the data generated; P.O.V., R.d.C.S., W.B.L. and A.B. (Alexander Birbrair) analyzed the data and interpreted the data generated. R.S.d.A., R.P.d.S. and V.A.d.C.A. carried out genotyping of viruses using RT-qPCR, analyzed the data and interpreted the data generated. J.R.P., G.C., A.B. (Aline Belmok), F.L.M. and B.M.R. carried out genome sequencing and assembled viral genomes. R.D.-C. carried out phylogenetic analyses, analyzed the data and interpreted the data generated. J.R.P., E.C.d.R., R.D.-C. and J.H.A. prepared figures and wrote the manuscript. J.H.A. conceived the study. All authors have read and agreed to the published version of the manuscript.

Funding: This research was funded by Financiadora de Estudos e Projetos (FINEP), project 27968—FINEP/RTR/PRPq/REDE/COVID-19—Sub: 09; FINEP 1227/21 Fase II—Corona-ômica BR MCTI Rede nacional de genomas, exoma e transcriptoma de COVID-19. This work was partially supported by Coordenação de Aperfeiçoamento de Pessoal de Nível Superior (CAPES) (Edital n° 9/2020 Prevenção e Combate a Surtos, Endemias, Epidemias e Pandemias" Project number: 88887.504570/2020-00); Rede Corona-ômica BR MCTI/FINEP (Financiadora de Estudos e Projetos, FINEP Project number: 01.20.0029.000462/20). This work was partially supported by the Fundação de Amparo à Pesquisa do Estado de São Paulo (FAPESP), Brazil, grants 2019/01255-9 and 2021/03684-4 (Young Investigator Program).

Institutional Review Board Statement: Ethics approval was obtained from institutional ethics committee of the Federal University of Western Bahia (CAAE 40779420.6.0000.8060).

Informed Consent Statement: All procedures and possible risks were explained to volunteers. Informed consent was obtained from all participants. Research was performed in accordance with relevant guidelines/regulations.

Data Availability Statement: We are submitting supporting data together with this manuscript. If necessary, additional data will be provided under request.

Acknowledgments: We are thankful to Conselho Nacional de Desenvolvimento Científico e Tecnológico (CNPq) by providing PhD research fellowships to Josilene Ramos Pinheiro. We are thankful for the support of the Regional Health Center of Western Bahia. We are also thankful for the support of the Epidemiological Surveillance Service of Barreiras city. We are also thankful to technicians of the Laboratory of Infectious Agents and Vectors. The authors also thank the National Laboratory for Scientific Computing (LNCC/MCTI, Brazil) for providing HPC resources of the Santos Dumont supercomputer (ID #45691, project "virusevolution").

Conflicts of Interest: The authors declare no conflict of interest.

References

1. Gorbalenya, A.E.; Baker, S.C.; Baric, R.S.; de Groot, R.J.; Drosten, C.; Gulyaeva, A.A.; Haagmans, B.L.; Lauber, C.; Leontovich, A.M.; Neuman, B.; et al. The species Severe acute respiratory syndrome-related coronavirus: Classifying 2019-nCoV and naming it SARS-CoV-2. *Nat. Microbiol.* **2020**, *5*, 536–544.
2. Zhou, P.; Yang, X.-L.; Wang, X.-G.; Hu, B.; Zhang, L.; Zhang, W.; Si, H.-R.; Zhu, Y.; Li, B.; Huang, C.-L.; et al. A pneumonia outbreak associated with a new coronavirus of probable bat origin. *Nature* **2020**, *579*, 270–273. [CrossRef] [PubMed]
3. Wu, F.; Zhao, S.; Yu, B.; Chen, Y.-M.; Wang, W.; Song, Z.-G.; Hu, Y.; Tao, Z.-W.; Tian, J.-H.; Pei, Y.-Y.; et al. A new coronavirus associated with human respiratory disease in China. *Nature* **2020**, *579*, 265–269. [CrossRef] [PubMed]
4. WHO Coronavirus (COVID-19) Dashboard | WHO Coronavirus Disease (COVID-19) Dashboard. Available online: https://covid19.who.int/ (accessed on 26 September 2022).
5. Wiersinga, W.J.; Rhodes, A.; Cheng, A.C.; Peacock, S.J.; Prescott, H.C. Pathophysiology, Transmission, Diagnosis, and Treatment of Coronavirus Disease 2019 (COVID-19): A Review. *JAMA* **2020**, *324*, 782–793. [CrossRef] [PubMed]
6. Farias, J.P.; da Silva, P.S.; Fogaça, M.M.C.; Santana, I.V.; Luiz, W.B.; Birbrair, A.; Amorim, J.H. The COVID-19 humoral immunological status induced by CoronaVac and AstraZeneca vaccines significantly benefits from a booster shot with the Pfizer vaccine. *J. Virol.* **2022**, *96*, e0017722. [CrossRef]
7. Oreshkova, N.; Molenaar, R.J.; Vreman, S.; Harders, F.; Munnink, B.B.O.; van der Honing, R.W.; Gerhards, N.; Tolsma, P.; Bouwstra, R.; Sikkema, R.S.; et al. SARS-CoV-2 infection in farmed minks, the Netherlands, April and May 2020. *Eurosurveillance* **2020**, *25*, 2001005. [CrossRef]
8. Triggle, C.R.; Bansal, D.; Ding, H.; Islam, M.M.; Farag, E.A.B.A.; Hadi, H.A.; Sultan, A.A. A Comprehensive Review of Viral Char-acteristics, Transmission, Pathophysiology, Immune Response, and Management of SARS-CoV-2 and COVID-19 as a Basis for Controlling the Pandemic. *Front. Immunol.* **2021**, *12*, 1–23.
9. Marcin, F.D.; Martin, S.W.; Tomas, S.; Sara, C.; Shankar, M.; Lachmann, G.; Monneret, G.; Venet, F.; Bauer, M.; Brunkhorst, F.M.; et al. The COVID-19 puzzle: Deciphering pathophysiology and phenotypes of a new disease entity | Elsevier Enhanced Reader. *Lancet Respir. Med.* **2021**, *9*, 622–642.
10. Trougakos, I.P.; Stamatelopoulos, K.; Terpos, E.; Tsitsilonis, O.E.; Aivalioti, E.; Paraskevis, D.; Kastritis, E.; Pavlakis, G.N.; Dimopoulos, M.A. Insights to SARS-CoV-2 life cycle, pathophysiology, and rationalized treatments that target COVID-19 clinical complications. *J. Biomed. Sci.* **2021**, *28*, 1–18. [CrossRef]
11. COVID-19 Vaccine Effectiveness | CDC. Available online: https://www.cdc.gov/coronavirus/2019-ncov/vaccines/effectiveness/index.html (accessed on 26 June 2022).
12. Vaccine Efficacy, Effectiveness and Protection. Available online: https://www.who.int/news-room/feature-stories/detail/vaccine-efficacy-effectiveness-and-protection (accessed on 26 June 2022).
13. Koelle, K.; Martin, M.A.; Antia, R.; Lopman, B.; Dean, N.E. The changing epidemiology of SARS-CoV-2. *Science* **2022**, *375*, 1116–1121. [CrossRef]
14. Da Silva, S.J.R.; Pena, L. Collapse of the public health system and the emergence of new variants during the second wave of the COVID-19 pandemic in Brazil. *One Health* **2021**, *13*, 100287. [CrossRef] [PubMed]
15. Xavier, D.R.; Lima e Silva, E.; Lara, F.A.; e Silva, G.R.R.; Oliveira, M.F.; Gurgel, H.; Barcelos, C. Involvement of political and so-cio-economic factors in the spatial and temporal dynamics of COVID-19 outcomes in Brazil: A population-based study. *Lancet Reg. Health—Am.* **2022**, *10*, 100221. [PubMed]
16. Brusselaers, N.; Steadson, D.; Bjorklund, K.; Breland, S.; Stilhoff, S.J.; Ewing, A.; Bergmann, S.; Steineck, G. Evaluation of science advice during the COVID-19 pandemic in Sweden. *Humanit. Soc. Sci. Commun.* **2022**, *9*, 1–17.
17. Riley, S.; Ainslie, K.E.C.; Eales, O.; Walters, C.E.; Wang, H.; Atchison, C.; Fronterre, C.; Diggle, P.J.; Ashby, D.; Donnelly, C.A.; et al. Resurgence of SARS-CoV-2: Detection by community viral surveillance. *Science* **2021**, *372*, 990–995. [CrossRef] [PubMed]
18. Wikle, N.B.; Tran, T.N.-A.; Gentilesco, B.; Leighow, S.M.; Albert, E.; Strong, E.R.; Brinda, K.; Inam, H.; Yang, F.; Hossain, S.; et al. SARS-CoV-2 epidemic after social and economic reopening in three U.S. states reveals shifts in age structure and clinical characteristics. *Sci. Adv.* **2022**, *8*, eabf9868. [CrossRef] [PubMed]
19. McCormick, K.D.; Jacobs, J.L.; Mellors, J.W. The emerging plasticity of SARS-CoV-2. *Science* **2021**, *371*, 1306–1308. [CrossRef]
20. Rocha, A.L.S.; Pinheiro, J.R.; Nakamura, T.C.; da Silva, J.D.S.; Rocha, B.G.S.; Klein, R.C.; Birbrair, A.; Amorim, J.H. Fomites and the environment did not have an important role in COVID-19 transmission in a Brazilian mid-sized city. *Sci. Rep.* **2021**, *11*, 1–8. [CrossRef]

21. A novel RT-qPCR Assay for Detection of SARS-CoV-2 Variants based on RhAmp Technology (IDT Technologies). Available online: https://www.protocols.io/view/a-novel-rt-qpcr-assay-for-detection-of-sars-cov-2-x54v9jwbmg3e/v1 (accessed on 2 May 2022).
22. Silva, J.P.; de Lima, A.B.; Alvim, L.B.; Malta, F.S.; Mendonça, C.P.; Fonseca, P.L.C.; Moreira, F.R.R.; Queiroz, D.C.; Ferreira, J.G.G.; Ferreira, A.C.S.; et al. Delta Variant of SARS-CoV-2 Replacement in Brazil: A National Epidemiologic Surveillance Program. *Viruses* **2022**, *14*, 847. [CrossRef]
23. Nascimento, V.A.D.; Corado, A.D.L.G.; Nascimento, F.O.D.; da Costa, Á.K.A.; Duarte, D.C.G.; Luz, S.L.B.; Goncalves, L.M.F.; de Jesus, M.S.; da Costa, C.F.; Delatorre, E.; et al. Genomic and phylogenetic characterisation of an imported case of sars-cov-2 in amazonas state, brazil. *Mem. Inst. Oswaldo Cruz* **2020**, *115*, 1–6. [CrossRef]
24. Katoh, K.; Misawa, K.; Kuma, K.I.; Miyata, T. MAFFT: A novel method for rapid multiple sequence alignment based on fast Fourier transform. *Nucleic Acids Res.* **2002**, *30*, 3059–3066. [CrossRef]
25. Okonechnikov, K.; Golosova, O.; Fursov, M.; UGENE Team. Unipro UGENE: A unified bioinfor-matics toolkit. *Bioinformatics* **2012**, *28*, 1166–1167. [CrossRef]
26. Minh, B.Q.; Schmidt, H.A.; Chernomor, O.; Schrempf, D.; Woodhams, M.D.; Von Haeseler, A.; Lanfear, R. IQ-TREE 2: New Models and Efficient Methods for Phylogenetic Inference in the Genomic Era. *Mol. Biol. Evol.* **2020**, *37*, 1530–1534. [CrossRef] [PubMed]
27. Price, M.N.; Dehal, P.S.; Arkin, A.P. FastTree 2—Approximately Maximum-Likelihood Trees for Large Alignments. *PLoS ONE* **2010**, *5*, e9490. [CrossRef] [PubMed]
28. Gascuel, O. BIONJ: An improved version of the NJ algorithm based on a simple model of sequence data. *Mol. Biol. Evol.* **1997**, *14*, 685–695. [CrossRef] [PubMed]
29. Shimodaira, H. An Approximately Unbiased Test of Phylogenetic Tree Selection. *Syst. Biol.* **2002**, *51*, 492–508. [CrossRef] [PubMed]
30. Letunic, I.; Bork, P. Interactive Tree of Life (iTOL) v5: An online tool for phylogenetic tree display and annotation. *Nucleic Acids Res.* **2021**, *49*, W293–W296. [CrossRef] [PubMed]
31. World Health Organization (WHO). Coronavirus COVID-19 Dashboard. Available online: https://covid19.who.int (accessed on 8 December 2021).
32. Boletins Epidemiológicos—COVID-19 Sesab. Available online: http://www.saude.ba.gov.br/temasdesaude/coronavirus/boletins-epidemiologicos-covid-19/ (accessed on 14 September 2022).
33. Página Inicial—Português (Brasil). Available online: https://www.gov.br/turismo/pt-br (accessed on 14 September 2022).
34. Planos Estaduais e Comunicados (Covid-19) | Sesab. Available online: http://www.saude.ba.gov.br/temasdesaude/coronavirus/planos-estaduais-e-comunicados-covid-19/ (accessed on 14 September 2022).
35. Klein, S.L.; Flanagan, K.L. Sex differences in immune responses. *Nat. Rev. Immunol.* **2016**, *16*, 626–638. [CrossRef]
36. Apalkova, Y.; Butovskaya, M.L.; Bronnikova, N.; Burkova, V.; Shackelford, T.K.; Fink, B. Assessment of Male Physical Risk-Taking Behavior in a Sample of Russian Men and Women. *Evol. Psychol. Sci.* **2018**, *4*, 314–321. [CrossRef]
37. He, X.; Lau, E.H.Y.; Wu, P.; Deng, X.; Wang, J.; Hao, X.; Lau, Y.C.; Wong, J.Y.; Guan, Y.; Tan, X.; et al. Temporal dynamics in viral shedding and transmissibility of COVID-19. *Nat. Med.* **2020**, *26*, 672–675. [CrossRef]
38. Argyropoulos, K.V.; Serrano, A.; Hu, J.; Black, M.; Feng, X.; Shen, G.; Call, M.; Kim, M.J.; Lytle, A.; Belovarac, B.; et al. Association of Initial Viral Load in Severe Acute Respiratory Syndrome Coronavirus 2 (SARS-CoV-2) Patients with Outcome and Symptoms. *Am. J. Pathol.* **2020**, *190*, 1881–1887. [CrossRef]

viruses

MDPI

Communication

SARS-CoV-2: Searching for the Missing Variants

Emilia Caputo [1] and Luigi Mandrich [2,*]

1 Institute of Genetics and Biophysics-IGB-CNR, "A. Buzzati-Traverso", Via Pietro Castellino 111, 80131 Naples, Italy
2 Research Institute on Terrestrial Ecosystems-IRET-CNR, Via Pietro Castellino 111, 80131 Naples, Italy
* Correspondence: luigi.mandrich@cnr.it

Abstract: Structural and phylogenetic analysis of the spike glycoprotein highlighted that the last variants, annotated as omicron, have about 30 mutations compared to the initial version reported in China, while the delta variant, supposed to be the omicron ancestor, shows only 7 mutations. Moreover, the five omicron variants were isolated between November 2021 and January 2022, and the last variant BA.2.75, unofficially named centaurus, was isolated in May 2022. It appears that, since the isolation of the delta variant (October 2020) to the omicron BA.1 (November 2021), there was an interval of one year, whereas the five omicron variants were isolated in three months, and after a successive four months period, the BA.2.75 variant was isolated. So, what is the temporal and phylogenetic correlation among all these variants? The analysis of common mutations among delta and the omicron variants revealed: (i) a phylogenetic correlation among these variants; (ii) the existence of BA.1 and BA.2 omicron variants a few months before being isolated; (iii) at least three possible intermediate variants during the evolution of omicron; (iv) the evolution of the BA.2.12.1, BA.4 and BA.5 variants from omicron BA.2; (v) the centaurus variant evolution from omicron BA.2.12.1.

Keywords: SARS-CoV-2; SARS-CoV-2 variants; phylogenetic analysis

Citation: Caputo, E.; Mandrich, L. SARS-CoV-2: Searching for the Missing Variants. *Viruses* **2022**, *14*, 2364. https://doi.org/10.3390/v14112364

Academic Editors: Ahmed El-Shamy and Mohamed Ibrahim

Received: 5 October 2022
Accepted: 25 October 2022
Published: 26 October 2022

1. Introduction

The last three years have been characterized by a worldwide pandemic due to a new type of coronavirus, called SARS-CoV-2, and reported for the first time in China in December 2019 [1]. In September 2022, WHO confirmed about 610 million cases and 6.5 million deaths by SARS-CoV-2 (https://www.who.int/publications/m/item/weekly-epidemiological-update-on-COVID-19, 28 September 2022).

SARS-CoV-2 is derived from the *Coronaviridae* viruses group [2]. It has a genome constituted of a positive sense single-stranded RNA of about 30 kb length [3], its mechanisms of infection and replication have been elucidated and a critical rule has been assigned to the membrane spike glycoprotein (S) [3,4]. Since spike protein is the most abundant and characteristic protein of SARS-CoV-2, it has been used as an antigen for the vaccine's production against the virus [4]. Furthermore, the mutations isolated on the spike protein have been used to classify and monitor the SARS-CoV-2 variants [5,6]. In fact, about 30 SARS-CoV-2 variants have been identified and classified as either Variants of Concern (VOCs) or Variants of Interest (VOIs). The VOCs show a greater virulence, transmissibility and severity of the symptoms compared to VOIs, as well as a reduced effectiveness of the vaccines [5,7]. The VOIs, instead, are mainly characterized by alterations into the receptor binding affinity [7]. According to this classification, the variants alpha, beta, gamma, delta and omicron are defined as VOCs, whereas lambda and mu are defined as VOIs [7]. Furthermore, each VOC is able to accumulate significant mutations, resulting in a rapid replacement of previous variants [7].

The SARS-CoV-2 variants have accumulated mutations, affecting the infection and diffusion of viral mechanisms, in particular, an increase in diffusion and milder symptoms have been observed [8,9]. It is important to note that SARS-CoV-2 variants are also characterized by other genomic mutations that are not present on spike protein; this is the case for the omicron BA.4 and BA.5 variants that differ from the initial version of the virus for about 50 mutations: 30 of these are on spike protein, and both variants have the same mutations, whereas they differ for other genomic mutations, which has led to having two variants with different genotypic and phenotypic characteristics [10].

Here, we analyzed at structural level the six omicron variants that have indications about their origin and phylogenesis, because they have been identified in a restricted interval time, from November 2021 to January 2022 [10], but they preset a high number of mutations in respect to the last isolated mu variant, in January 2021 [11,12].

2. Materials and Methods

The sequences of SARS-CoV-2 variants delta, omicron BA.1, BA.2, BA.2.12.1, BA.4, BA.5 and BA.2.75 were from the "expasy viralzone" web site (https://viralzone.expasy.org/9556, 17 September 2022), and were used to make the multiple sequence alignment to generate a phylogenetic three (Multiple Sequence Alignment by Clustal Omega program at https://www.ebi.ac.uk/Tools/msa/clustalo/; 17 September 2022).

3. Results

In a recent study, the SARS-CoV-2 variants were analyzed, at structural level, in order to generate a phylogenetic three, indicating a common ancestor between the delta and the five SAR-CoV-2 omicron variants [12], although the spike protein mutations on the delta variant were 7 compared to the more than 30 ones identified on the omicrons [12]. Moreover, the mu variant, reported in January 2021, was the last variant isolated before omicron BA.1 in November 2021. Successively, until January 2022, the other five omicron variants were reported, suggesting either a delay in the identification of new variants or the possibility that they have not been identified at all.

In some cases, there have been countries with high numbers of infections where the virus has evolved rapidly [13] and perhaps for politic, economic and/or technological reasons, few sequences of the viral genome have been performed; in other cases, especially in the countries where the vaccination against SARS-CoV-2 has reached high levels of coverage in the population, it was thought that it had defeated the virus and therefore the monitoring of any new variants was reduced, instead the new variants accumulating many mutations escape the antibody response, following by natural infection or vaccination [14,15].

Here, we have analyzed the sequence alignment and the common mutations among the omicron variants to reconstruct their evolutionary lineage and to identify any intermediate variants that have never been reported.

3.1. The Omicrons Origin and Evolution

The omicron variants are characterized by a high number of mutations compared to the initial version of the virus. Since the SARS-CoV-2 variants, isolated before omicron, are characterized by a low number of mutations, the omicron origin seems to be uncertain showing common mutations described in other VOCs, such as alpha, beta, gamma and delta variants [10,16], but from an evolutionary point of view, the delta variant seems to be closer to omicron than the others [12]. Thus, we cannot exclude that the omicron origin could be derived by events of genomic recombination in two VOCs, contemporarily infecting patients. Furthermore, an antigenic shift has been observed for the omicron variant that is a step change from the viral antigenicity, leading to viral escape from vaccine-acquired immunity or infection from previous variants, which is consistent with the observed increased transmissibility [17].

Starting from this information, we generated a new multiple sequences alignment among the wild type version of the spike protein and the variants delta, omicron BA.1, BA.2, BA.2.12.1, BA.4, BA.5 and the last BA.2.75, unofficially indicated as centaurus (see

Supplemental Figure S1). We identified all the spike common mutations among the variants and those present only in one or more variants, as schematically represented in Figure 1.

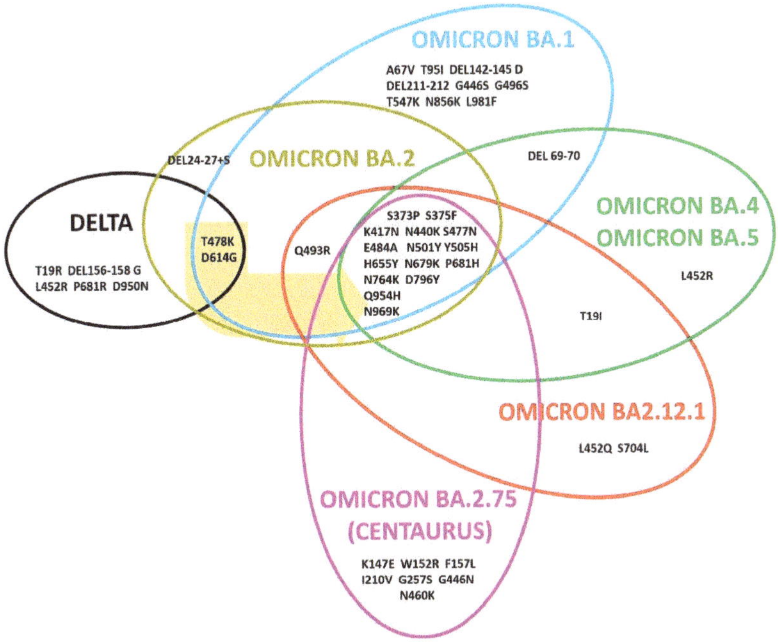

Figure 1. Schematic representation of the omicron variants evolution starting from delta. The common mutations are reported at the overlap among the variants. The mutations present in all the omicron variants are S373P, K417N, N440K, S477N, T478K, E484A, N501Y, Y505H, D614G, H655Y, N679K, P681H, N764K, D796Y, Q954H and N969K, and T478K and D614G derived from delta, indicated by the yellow arrow. The unique mutations of each variant are reported outside the overlap among the variants.

Despite the omicron variants evolved from delta, they show only two common mutations: T478K and D614G. The position 681 is mutated in delta and omicron variants, but in delta, the substitution was P681R whereas in omicron, it was P681H. The other mutations present only in delta are T19R, the deletion 156-158 with substitution in glycine, L452R and D950N.

Interestingly, the alignment revealed the common mutations in all the omicron variants, which are S373P, S375F, K417N, N440K, S477N, T478K, E484A, N501Y, Y505H, D614G, H655Y, N679K, P681H, N764K, D796Y, Q954H and N969K (Figure 1, Supplemental Figure S1).

Sequence analysis suggests that omicron BA.1 and BA.2 are evolved independently from a common ancestor, derived from delta and also having the mutations T478K and D614G (Figure 1, Supplemental Figure S1); in fact, omicron BA.1 presents nine unique and characteristic mutations, which are A67V, T95I, deletion 142-145 with substitution in aspartic, deletion 211-212, G446S, G496S, T547K, N856K and L981F, whereas omicron BA.2 shows the deletion 24-27 as a specific mutation with a substitution in serine (Figure 1, Supplemental Figure S1).

Omicron BA.2.12.1 evolved from BA.2 because it has two other specific mutations in common: S375F and Q493R, and the unique and specific mutation S704L (Figure 1, Supplemental Figure S1). From omicron BA.2.12.1, BA.4/BA.5 and BA.2.75 are separately evolved; the spike protein of the variants BA.4 and BA.5 shows the same mutations and both differ from the spike protein of BA.2.12.1 because it did not have the S704L and Q493R mutations, while presenting the deletion 69-70 in common with omicron BA.1 (Figure 1, Supplemental Figure S1).

The last isolated variant, BA.2.75, shows the 17 common mutations with the other omicrons and the unique and specific mutations K147E, W152R, F157L, I210V, G257S, G446N and N460K, indicating that BA.2.75 evolved from BA.2.12.1 (Figure 1, Supplemental Figure S1).

3.2. The Omicron BA.4 and BA.5

As reported, BA.4 and BA.5 variants have the same mutations on the spike proteins and they are similar to the BA.2.12.1 one, but they are classified as two different variants [10]. This is possible because they differ in about 50 mutations on the whole genome from the initial version of SARS-CoV-2, and 30 of them are on spike (Figure 1, Supplemental Figure S1), while about 20 are in other part of the genome.

In particular, they show a similar pattern at 5′ genome region, corresponding to the genes encoding ORF1ab and the envelope protein E. In addition, while BA.4 shows specific mutations at ORF7b, N protein and nonstructural protein 1 (NSP1), both associated to the genomic RNA, BA.5 presents specific differences at the 3′ genome region, in correspondence with ORF6 and membrane protein M encoding genes [10].

3.3. The Omicron BA.2.275

The last SARS-CoV-2 variant identified is omicron BA.2.75, unofficially known as centaurus. It was isolated in May 2022 in India [18], and it is characterized by 34 mutations on the spike protein, 17 of them are in common with the other omicron variants. Further, it shows the reversion of R493Q compared to the ancestral variant, and seven unique and specific mutations: K147E, W152R, F157L, I210V, G257S, G446N and N460K (Figure 1, Supplemental Figure S1) [18,19], which may be related to immune escape and resistance to antibody therapies, indicating a typical antigenic shift [20]. There are indications that the BA.2.75 variant was isolated for the first time in India in January 2022, and spread outside India only in May 2022. The symptoms associated with this variant are fever, cough, sputum, diarrhoea and fatigue [20], similar to those of a seasonal flu, as previously predicted [10].

4. Discussion

The analysis of the alignment of the wild type version, delta and the omicron variants highlight important information about the origin and the evolution of these variants. The timing of their evolution is different from their isolation date. In fact, they are not evolved in sequence from BA.1 up to BA.2.75 but through a number of intermediate variants, which have never been isolated. In particular, by analyzing all the mutations of these variants, we obtained a more correct evolutionary lineage that includes at least three intermediates, identified through the common mutations among them (Figure 2). We supposed that there was a joint starting from the wild type version of spike protein, represented by Intermediate 1 (Figure 2), where the evolutionary branches of delta and omicron variants have separated, and this missing intermediate only has the two common mutations among them, which are T478K and D614G. From Intermediate 1 evolved Intermediate 2, which shows the 17 mutations present in all the omicron variants. In Table 1, the mutations, that we assigned to the missing SARS-CoV-2 variants in the omicron lineage are listed, indicated as Intermediate 1, 2 and 3 (Table 1).

The variants BA.1 and BA.2 evolved separately from Intermediate 2 (Figure 2); BA.1 harbors 35 mutations in its spike protein in respect to the initial version isolated in SARS-CoV-2, 10 of them are specific and not present in the other omicron. Specifically, they are A67V, DEL69-70, T95I, DEL142-145D, DEL211-212, G446S, G496S, T547K, N856K and L981F (Figure 1); whereas BA.2 lacks 13 mutations present in BA.1 but shows 8 unique mutations not found in BA.1 that are T19I, DEL24-27S, G142D, V213G, T376A, D405N, R408S and Q498R [21]. Based on the presence of a high number of different mutations between them, it is reasonable to assume that their evolution process began and ended a few months before their isolation.

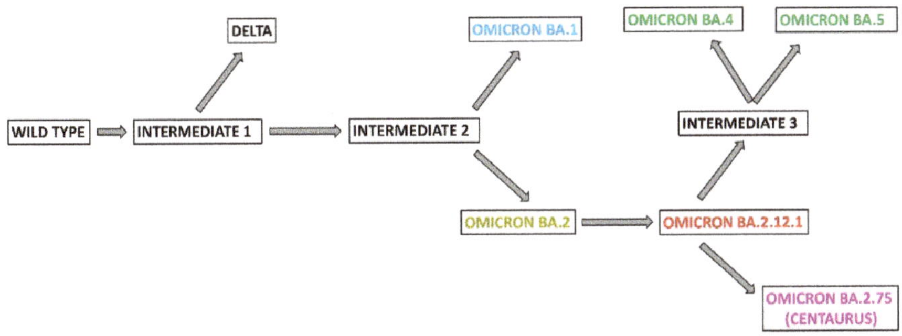

Figure 2. Phylogenetic lineage of the omicron variants starting from wild type version and the delta variant. Intermediates 1, 2 and 3 are three missing variants. Intermediate 1 has common mutations between delta and Intermediate 2. Intermediate 2 has the common mutations of all the omicron variants. Intermediate 3 has the common mutation between omicron BA.4 and BA.5.

Table 1. List of mutations that are supposed to be present in the three missing intermediates related to the omicron lineage of evolution.

Intermediate 1	Intermediate 2	Intermediate 3
T478K D614G	S373P S375F K417N N440K S477N T478K E484A N501Y Y505H D614G H655Y N679K P681H N764K D796Y Q954H N969K	T19I DEL24-27S DEL69-70 G142D V213G G339D S371F S373P S375F T376A D405N R408S K417N N440K L452R S477N T478K E484A F486V Q498R N501Y Y505H D614G H655Y N679K P681H N764K D796Y Q954H N969K

The variant BA.2.12.1 evolved from BA.2. In fact, beyond the common mutations in spike protein, BA.2.12.1 has also the substitutions T19I, L452Q and S704L [22] (Figure 2).

Following the scheme of Figure 2, BA.4, BA.5 and BA.2.75 evolved from the BA.2.12.1 variant. The SARS-CoV-2 variant classification also takes into account other mutations present in the genome that change the characteristics of the virus. In fact, the spike proteins of BA.4 and BA.5 are identical but they differ for mutations in E and M genes [10]. Considering that the deletion 69-70 is present also in BA.1 (Figure 1) and in alpha variant, it has been suggested that BA.4 and BA.5 diverged via recombinant event with other variants [22]. Another study suggested that BA.4 and BA.5 evolved by a recombination with mouse SARS-CoV-2 virus, and in this case, mouse would have been the host for the evolution [23]. Regarding our analysis, it suggests that BA.4 and BA.5 evolved from BA.2.12.1 (Figure 2). In this case, we identified a third intermediate having all the mutations only on spike protein present in BA.4 and BA.5, and not on the other regions of the viral genome. Then, subsequent mutations on the genome led to the two variants BA.4 and BA.5; in Table 1 are listed the mutations characterizing the Intermediate 3.

The last omicron variant to be isolated was BA.2.75; it evolved from BA.2.12.1 (Figure 2). Sequence analysis highlighted that BA.2.75 compared to BA.2.12.1 shows seven new specific mutations: K147E, W152R, F157L, I210V, G275S, G446N and N460K, and reverted to the mutation Q493R (Figure 1) [18]. Epidemiological data reveal that omicron BA.2.75 is present in more than 30 countries worldwide and it does not appear to be dangerous, keeping the hospitalization of patients unchanged [24]. Probably, it is necessary to study in more detail the characteristics of this new variant and to use the new vaccines designed against both the BA.4/5 and BA.2.75 variants, to reduce the risk of a new critical healthcare wave of infection.

5. Conclusions

During the pandemic, about 30 variants were isolated, but in many cases, it was not possible to follow their evolution in time. After three years of the pandemic, we observed a new evolutionary lineage represented by the omicron variants, which have completely replaced the pre-existing variants. The timeline of the omicron evolution variants is different from the date of their isolation, in fact, a retroactive analysis revealed that omicron BA.1 was present in Europe at least 10 days before its isolation in South Africa [25]. Moreover, five different omicron variants were reported in three months, which differ among them from 8 to 18 mutations (Figure 1), suggesting that many intermediate variants were lost.

This study referred to a small group of evolutionarily correlated variants and strongly supported the existence of at least three important intermediates, which never have been isolated. To date, if we consider that about 30 SARS-CoV-2 variants have been isolated, we can state that there are many other missing variants.

Supplementary Materials: The following supporting information can be downloaded at: https://www.mdpi.com/article/10.3390/v14112364/s1, Figure S1: Sequences alignment. SARS-CoV-2 wild type spike protein and its variants, delta and omicron, were aligned by using the Clustal Omega program (https://www.ebi.ac.uk/Tools/msa/clustalo/, 17 September 2022). For the alignment, the default parameters were used (Dealign Input Sequences: no (false); Number of Combined Iterations: 0; Max Guide Tree Iterations: −1 (off); Max HMM Iterations: −1 (off); Use mBed-like clustering during subsequent iterations: yes (true); mBed-like Clustering Guide-tree: yes (true)).

Author Contributions: E.C. and L.M. contributed to the study conception, design, writing, review and editing of the manuscript. All authors have read and agreed to the published version of the manuscript.

Funding: This work was supported by IGB-CNR and IRET-CNR for overheads.

Institutional Review Board Statement: Not applicable.

Informed Consent Statement: Not applicable.

Data Availability Statement: Not applicable.

Acknowledgments: We are grateful to Bernard Loeffler and Francesca Varrone for their excellent editing assistance.

Conflicts of Interest: The authors declare no conflict of interest.

References

1. Hu, B.; Guo, H.; Zhou, P.; Shi, Z.L. Characteristics of SARS-CoV-2 and COVID-19. *Nat. Rev. Microbiol.* **2021**, *19*, 141–154. [CrossRef]
2. Chan, J.F.W.; Kok, K.H.; Zhu, Z.; Chu, H.; Wang-TO, K.K.; Yuan, S.; Yuen, K.-Y. Genomic characterization of the 2019 novel human-pathogenic coronavirus isolated from a patient with atypical pneumonia after visiting Wuhan. *Emerg. Microbes Infect.* **2020**, *9*, 221–236. [CrossRef]
3. Wang, H.; Li, X.; Li, T.; Zhang, S.; Wang, L.; Wu, X.; Liu, J. The genetic sequence, origin, and diagnosis of SARS-CoV-2. *Eur. J. Clin. Microbiol. Infect. Dis.* **2020**, *39*, 1629–1635. [CrossRef]
4. Huang, Y.; Yang, C.; Xu, X.-F.; Xu, W.; Liu, S.W. Structural and functional properties of SARS-CoV-2 spike protein: Potential antivirus drug development for COVID-19. *Acta Pharmacol. Sin.* **2020**, *41*, 1141–1149. [CrossRef]
5. Mistry, P.; Barmania, F.; Mellet, J.; Peta, K.; Strydom, A.; Viljoen, I.M.; James, W.; Gordon, S.; Pepper, M.S. SARS-CoV-2 Variants, Vaccines, and Host Immunity. *Front. Immunol.* **2022**, *12*, 809244. [CrossRef]
6. Harvey, W.T.; Carabelli, A.M.; Jackson, B.; Gupta, R.K.; Thomson, E.C.; Harrison, E.M.; Ludden, C.; Reeve, R.; Rambaut, A.; COVID-19 Genomics UK (COG-UK) Consortium; et al. SARS-CoV-2 variants, spike mutations and immune escape. *Nat. Rev. Microbiol.* **2021**, *19*, 409–424. [CrossRef]
7. Flores-Vega, V.R.; Monroy-Molina, J.V.; Jimenez-Hernandez, L.E.; Torres, A.G.; Santos-Preciado, J.I.; Rosales-Reyes, R. SARS-CoV-2: Evolution and emergence of new viral variants. *Viruses* **2022**, *14*, 653. [CrossRef] [PubMed]
8. Korber, B.; Fischer, W.M.; Gnanakaran, S.; Yoon, H.; Theiler, J.; Abfalterer, W.; Hengartner, N.; Giorgi, E.E.; Bhattacharya, T.; Foley, B.; et al. Tracking Changes in SARS-CoV-2 Spike: Evidence that D614G Increases Infectivity of the COVID-19 Virus. *Cell* **2021**, *182*, 812–827.e19. [CrossRef]
9. Rochman, N.D.; Wolf, Y.I.; Faure, G.; Mutz, P.; Zhang, F.; Koonin, E.V. Ongoing global and regional adaptive evolution of SARS-CoV-2. *Proc. Natl. Acad. Sci. USA* **2021**, *118*, e2104241118. [CrossRef] [PubMed]
10. Tegally, H.; Moir, M.; Everatt, J.; Giovanetti, M.; Scheepers, C.; Wilkinson, E.; Subramoney, K.; Makatini, Z.; Moyo, S.; Amoako, D.G.; et al. Emergence of SARS-CoV-2 Omicron lineages BA.4 and BA.5 in South Africa. *Nat. Med.* **2022**, *28*, 1785–1790. [CrossRef] [PubMed]

11. Hernandez-Ortiz, J.; Cardona, A.; Ciuoderis, K.; Averhoff, F.; Maya, M.-A.; Cloherty, G.; Osorio, J.E. Assessment of SARS-CoV-2 Mu Variant Emergence and Spread in Colombia. *JAMA Netw. Open* **2022**, *5*, e224754. [CrossRef] [PubMed]
12. Caputo, E.; Mandrich, L. Structural and Phylogenetic Analysis of SARS-CoV-2 Spike Glycoprotein from the Most Widespread Variants. *Life* **2022**, *12*, 1245. [CrossRef]
13. Wang, R.; Chen, J.; Gao, K.; Wei, G.-W. Vaccine-escape and fast-growing mutations in the United Kingdom, the United States, Singapore, Spain, India, and other COVID-19-devastated countries. *Genomics* **2021**, *113*, 2158–2170. [CrossRef] [PubMed]
14. Garcia-Beltran, W.F.; Lam, E.C.; St Denis, K.; Nitido, A.D.; Garcia, Z.H.; Hauser, B.M.; Feldman, J.; Pavlovic, M.N.; Gregory, D.J.; Poznansky, M.C.; et al. Multiple SARS-CoV-2 variants escape neutralization by vaccine-induced humoral immunity. *Cell* **2021**, *184*, 2372–2383.e9. [CrossRef]
15. Tuekprakhon, A.; Nutalai, R.; Dijokaite-Guraliuc, A.; Zhou, D.; Ginn, H.M.; Selvaraj, M.; Liu, C.; Mentzer, A.J.; Supasa, P.; Duyvesteyn, H.M.E.; et al. Antibody escape of SARS-CoV-2 Omicron BA.4 and BA.5 from vaccine and BA.1 serum. *Cell* **2022**, *185*, 2422–2433.e13. [CrossRef] [PubMed]
16. Rahimi, F.; Talebi Bezmin Abadi, A. Is Omicron the last SARS-CoV-2 Variant of Concern? *Arch. Med. Res.* **2022**, *53*, 336–338. [CrossRef]
17. Talenti, A.; Hodcroft, E.B.; Robertson, D.L. The evolution and biology of SARS-CoV-2 variants. *Cold Spring Harb. Perspect. Med.* **2022**, *12*, a041390. [CrossRef]
18. Zappa, M.; Verdecchia, P.; Angeli, F. Knowing the new Omicron BA.2.75 variant ('Centaurus'): A simulation study. *Eur. J. Intern. Med.* **2022**, *105*, 107–108. [CrossRef]
19. Gruell, H.; Vanshylla, K.; Tober-Lau, P.; Hillus, D.; Sander, L.E.; Kurth, F.; Klein, K. Neutralisation sensitivity of the SARS-CoV-2 omicron BA.2.75 sublineage. *Lancet Infect. Dis.* **2022**, *22*, 1422–1423. [CrossRef]
20. Hirotsu, Y.; Omata, M. Detection of the Omicron BA.2.75 subvariant in Japan. *J. Infect.* **2022**, S0163-4453(22)00523-0. [CrossRef]
21. Shrestha, L.B.; Foster, C.; Rawlinson, W.; Tedla, N.; Bull, R.A. Evolution of the SARS-CoV-2 omicron variants BA.1 to BA.5: Implications for immune escape and transmission. *Rev. Med. Virol.* **2022**, *32*, e2381. [CrossRef] [PubMed]
22. Gruell, H.; Vanshylla, K.; Korenkov, M.; Tober-Lau, P.; Zehner, M.; Münn, F.; Janicki, H.; Augustin, M.; Schommers, P.; Sander, L.E.; et al. SARS-CoV-2 Omicron sublineages exhibit distinct antibody escape patterns. *Cell Host Microbe* **2022**, *30*, 1231–1241.e6. [CrossRef] [PubMed]
23. Sun, Y.; Lin, W.; Dong, W.; Xu, J. Origin and evolutionary analysis of the SARS-CoV-2 Omicron variant. *J. Biosaf. Biosecur.* **2022**, *4*, 33–37. [CrossRef] [PubMed]
24. Callaway, E. Will 'Centaurus' be the next global coronavirus variant? Indian cases offer clues. *Nature* **2022**, *608*, 462–463. [CrossRef]
25. Poudel, S.; Ishak, A.; Perez-Fernandez, J.; Garcia, E.; Leon-Figueroa, D.A.; Romanì, L.; Bonilla-Aldana, K.; Rodriguez-Morales, A.J. Highly mutated SARS-CoV-2 Omicron variant sparks significant concern among global experts—What is known so far? *Trav. Med. Infect. Dis.* **2022**, *45*, 102234. [CrossRef] [PubMed]

viruses

MDPI

Opinion

On the Need to Determine the Contribution of Anti-Nucleocapsid Antibodies as Potential Contributors to COVID-19 Convalescent Plasma Efficacy

Daniele Focosi [1,*], Massimo Franchini [2] and Arturo Casadevall [3]

1 North-Western Tuscany Blood Bank, Pisa University Hospital, 56124 Pisa, Italy
2 Division of Hematology and Transfusion Medicine, Carlo Poma Hospital, 46100 Mantua, Italy
3 Department of Molecular Microbiology and Immunology, Johns Hopkins Bloomberg School of Public Health, Baltimore, MD 21287, USA
* Correspondence: daniele.focosi@gmail.com

Abstract: Historically the therapeutic potential of polyclonal passive immunotherapies in viral diseases has been related to antiviral neutralizing antibodies, but there is also considerable evidence that non-neutralizing antibodies can translate into clinical benefit as well. In the setting of SARS-CoV-2 infection, we review here in vitro and in vivo evidence supporting a contributing role for anti-nucleocapsid antibodies. Retrospective investigation of anti-nucleocapsid antibody levels in randomized clinical trials of COVID-19 convalescent plasma is warranted to better understand whether there is an association with efficacy or lack thereof.

Keywords: COVID-19; SARS-CoV-2; nucleocapsid; protein localization; antibody therapy

Citation: Focosi, D.; Franchini, M.; Casadevall, A. On the Need to Determine the Contribution of Anti-Nucleocapsid Antibodies as Potential Contributors to COVID-19 Convalescent Plasma Efficacy. *Viruses* **2022**, *14*, 2378. https://doi.org/10.3390/v14112378

Academic Editors: Ahmed El-Shamy and Mohamed Ibrahim

Received: 12 October 2022
Accepted: 26 October 2022
Published: 27 October 2022

Many patients with infectious diseases cannot tolerate the side effect of small chemical antimicrobials, making antibody a safer prophylactic and therapeutic alternative. This is case for immunocompromised individuals with COVID-19 where the ritonavir component of the Paxlovid antiviral formulation can interfere with other drugs needed to treat their underlying condition. When matching viruses and therapeutic antibodies, most attention has focused so far on neutralizing antibodies (nAb), i.e., antibodies of different isotypes (IgG_1, IgG_3, IgM, and IgA) that reduce viral infection of replication-competent cells in in vitro viral neutralization tests (VNT) [1]. With regards to SARS-CoV-2, the causative agent of COVID-19, this in vitro neutralization reflects largely the presence of antibodies to the region of the Spike protein that interacts with the ACE2 receptor on human cells. Hence, for COVID-19, nAbs consist mostly of antibodies to the Spike protein. B-cells making these antibodies recovered from convalescent donors have been used to make the therapeutic monoclonal antibodies. Furthermore, the titer of nAbs has been the major correlate of protection after Spike-based vaccination used to assess vaccine efficacy [2]. For COVID19 convalescent plasma (CCP), the therapeutic potential has been generally correlated with the nAb titer. This assumption has been validated on both mechanistic studies that established its antiviral activity [3], and clinical studies that show dose-response relationships between the nAb titer and efficacy [4–6]. However, the undoubted importance of nAbs does not exclude the possibility that more antiviral antibodies could be associated with additional clinical benefits. This principle was illustrated in studies with vesicular stomatitis virus where in vitro neutralization titer correlated with avidity and neutralization rate constant, but in vivo efficacy was independent of in vitro neutralizing activity (PMID: 9197261).

In contrast to the SARS-CoV-2 Spike protein, the structural nucleocapsid (N) phosphoprotein is highly conserved among human coronaviruses, where it is essential for linking the viral genome to the viral membrane [7]. N is classically considered an internal protein of SARS-CoV-2, and, as such, to be only useful for eliciting T-cell mediated immune responses [8,9]. Orthologous N protein from influenza A [10], measles [11], respiratory syncytial [12], lymphocytic choriomeningitis [13], and human immunodeficiency viruses [14] are expressed on the surface of infected cells, where they can be the target for antibody dependent cellular cytotoxicity mechanism (ADCC). Not surprisingly, about 10^4–10^5 SARS-CoV-2 N proteins occur on the surface of a range of different SARS-CoV-2-infected cell types, either natural (Vero, Caco-2, Calu-3) or humanized (BHK-21_hACE2 or CHOK1_hACE2, and HEK293-FT_hACE2) [15]. SARS-CoV-2 N is likely placed on the cell surface and secreted through a non-canonical pathway that bypasses insertion into the endoplasmic reticulum [16]. N released by SARS-CoV-2 infected cells or N-expressing transfected cells binds to heparan sulfate, which promotes Spike-ACE2 interaction [17] and heparin on neighboring cells, which may contribute to coagulopathy, and also neutralizes the biological activity of many different chemokines, blocking chemotaxis of immune effector cells [15]. Freely circulating N protein can also activate the complement cascade via the alternative pathway, thus potentially contributing to the inflammatory changes that are associated with severe COVID-19 [18,19]. Hence, there is considerable direct and circumstantial evidence that the N protein contributes to the pathogenesis of COVID-19, and if that is the case, it is reasonable to posit that N-binding antibodies can contribute to host defense through ADCC, and by interfering with its deleterious effects on immune function.

Antibodies to the N protein are elicited after immunization with experimental N-based vaccines [8,9,20,21] and after natural SARS-CoV-2 infection. In the pre-vaccine era, the occurrence of antibodies to N following natural infection was almost universal and levels remained detectable for more than 6 months [22–24]. The finding that each 1-log increase in SARS-CoV-2 viral copies at diagnosis was associated with 90% higher odds of seroconversion for N antibodies [25] suggests that high levels of circulating N, which are also associated with the severity of pulmonary illness and clinically important patient outcomes [26], are required to elicit this response.

Whether antibodies to the N protein exert any protective role has been the subject of sporadic investigations. mAbs targeting SARS-CoV-2 N protein can inhibit free N-induced MASP-2 activation in vitro [18], and mAbs to the related alphacoronavirus mouse hepatitis virus (MHV) N protein exert anti-viral activity in vitro in the presence of complement and in vivo [27,28]. A very interesting animal model showed that C57BL/6 mice prime-boosted with an adenovirus serotype five vector expressing N developed anti-N antibodies 2 weeks later, which were unable to neutralize live authentic SARS-CoV-2. However, when their sera was transfused to *naïve* K18-hACE2 mice, followed by intranasal challenge with 10^3 PFU SARS-CoV-2 USA-WA1/2020, the animals experienced lung viral loads 14-fold lower than in controls at day 4 [29]. Very few clinical studies have investigated the correlation between antibody levels to N protein in CCP and clinical outcome. In a retrospective observational study in 96 hospitalized patients, Cain et al. found no statistically significant difference in neither mortality nor time from transfusion to death between patients receiving CCP with low vs. high antibody levels to N protein. Unfortunately no multivariate analysis was conducted to account for antibody levels to spike protein in the plasma [30]. When it comes to randomized controlled trials (RCT), the gold standard of evidence-based medicine, only a single RCT assessed antibody to N protein in CCP. A reanalysis of the Penn2CCP RCT data showed that the clinical benefit of CCP was related to a shift towards reduced inflammatory S responses and enhanced N humoral responses [31]. Furthermore, CCP induced immunomodulatory changes to recipient humoral profiles (including more anti-inflammatory S-specific Fc glycans) persisted for at least 2 months, marked by the selective evolution of anti-inflammatory Fc-glycan profiles and persistently expanded N-specific humoral immunity following CCP therapy [32].

Many different ingredients in CCP can lead to clinical benefit [33] and there is now suggestive evidence for adding antibodies to N protein to the list of potential antiviral ingredients. Mechanisms of action other than neutralization are needed to explain the potential clinical benefit in vivo of antibodies to N protein. In this regard, antibodies to the SARS-CoV-2 N protein, once bound to the surface of N-expressing cells, activate Fc receptors (FcR)-expressing cells [15]. In murine influenzavirus models, IgG to N protein specifically promoted virus clearance by using a mechanism involving both FcRs and $CD8^+$ T lymphocytes [34]. Accordingly, antibodies to SARS-CoV-2 N protein exert relevant antibody-dependent NK cell activation (ADNKA) after infection, driving high levels of pro-inflammatory cytokine production for more than 6 months [35].

Of interest and in contrast to the highly variable Spike protein, N protein has been mostly conserved so far in SARS-CoV-2 evolution, with antibodies to N showing cross-reactivity across sublineages [36]. Hence, if any, the therapeutic benefit from old CCP units could be preserved against the most recent SARS-CoV-2 variants of concern. However, there is great cross-reactivity for antibodies to N proteins among coronaviruses, and N protein epitopes are shared between SARS-CoV-2 and alphacoronaviruses. Hence, one cannot be certain whether any correlation between antibody to N protein and protection was a result from previous endemic coronavirus infection(s), which are highly prevalent in the human populations worldwide, or from the recent SARS-CoV-2 infection. Nevertheless, the amino acid sequences of the entire N protein of common coronaviruses are sufficiently dissimilar to that of SARS-CoV-2, with only the conserved residues at the N-terminal domain of NP showing a high degree of similarity. Consequently, usage of an N-terminally truncated nucleocapsid protein (ΔN-NP) could provide better specificity for discriminating among coronaviruses [37], with epitope mapping unveiling the 155–171 epitope [38] and 255–346 [39] as highly immunogenic and specific. High titer responses against N of alpha-coronaviruses have been detected during early COVID-19 stages, raising the possibility that SARS-CoV-2 infection boosted pre-existing immunity [40], without clear correlations with disease severity [41]. That said, it is likely that not all anti-N antibodies are equally beneficial: e.g., occurrence of antibodies to a 21-residue epitope from N (termed Ep9) [42], lack of antibodies against the seasonal betacoronavirus OC43 N [43], or occurrence of IgG to -alphacoronaviruses (NL-63 and 229E) N protein [40] have all been associated with severe COVID-19. Notably, in individuals with severe COVID-19 (such as those admitted to ICU), N-specific antibody titers prevail over anti-Spike titer [40,44,45].

There is unfortunately an unpredictable conundrum between the increased potency of CCP from vaccinated donors (so-called VaxPlasma or "hybrid plasma" or VaxCCP) and anti-N antibodies. The superiority of VaxCCP over CCP is due to the 10–100-fold higher anti-Spike nAb titers seen in VaxCCP compared to CCP, and their heterologous nature neutralizing most, if not all, SARS-CoV-2 variants of concern. Nevertheless, it has been shown that after vaccine breakthrough infection the occurrence of antibody to N protein, compared to unvaccinated subjects, is dramatically decreased from 93% to just 50% at day 54 post-infection [22,25]. The few individuals that mount antibody responses to N protein after vaccine breakthrough infection make lower titers [46], and substantial seroreversion of N total immunoglobulin has also been found shortly after vaccine breakthrough infections [47]. As such, VaxCCP has much lower content of anti-N, which will be an unlikely contributor to clinical benefit in the future usages.

The available evidence provides a compelling case for analyzing the antibody content to N protein on stored samples from the dozens of RCTs completed studying CCP efficacy against COVID-19. There are thousands of such CCP remnant samples available that could be studied for antibodies to N and correlated to clinical outcome data. Although the results of such a study will likely have no immediate implications for the current pandemic, it has the potential to inform future pandemics from related coronaviruses or different viruses expressing orthologous protein N's. At the very least, we should miss this opportunity to further dissect the contribution of humoral immunity to N protein in COVID-19. In particular, under the upcoming "mutations wave" dominated by Spike R346X- and K444X-harboring Omicron sublineages, CCP will remain a fundamental weapon against COVID19 in immunocompromised patients [48], who are at higher risk for more severe presentation and are not protected by vaccine boosts.

Funding: This research received no external funding.

Institutional Review Board Statement: Not applicable.

Informed Consent Statement: Not applicable.

Data Availability Statement: This manuscript generated no new data.

Conflicts of Interest: D.F. and M.F. were co-investigators in the TSUNAMI RCT of CCP. A.C. was a co-investigator in the CSSC-004 RCT of CCP.

References

1. Focosi, D.; Mazzetti, P.; Pistello, M.; Maggi, F. Viral infection neutralization tests: A focus on SARS-CoV-2 with implications for convalescent plasma therapy. *Rev. Med Virol.* **2020**, *31*, e2170. [CrossRef] [PubMed]
2. Khoury, D.S.; Cromer, D.; Reynaldi, A.; Schlub, T.E.; Wheatley, A.K.; Juno, J.A.; Subbarao, K.; Kent, S.J.; Triccas, J.A.; Davenport, M.P. Neutralizing antibody levels are highly predictive of immune protection from symptomatic SARS-CoV-2 infection. *Nat. Med.* **2021**, *27*, 1205–1211. [CrossRef] [PubMed]
3. Natarajan, H.; Crowley, A.R.; Butler, S.E.; Xu, S.; Weiner, J.A.; Bloch, E.M.; Littlefield, K.; Wieland-Alter, W.; Connor, R.I.; Wright, P.F.; et al. Markers of Polyfunctional SARS-CoV-2 Antibodies in Convalescent Plasma. *mBio* **2021**, *12*. [CrossRef] [PubMed]
4. Joyner, M.J.; Carter, R.E.; Senefeld, J.W.; Klassen, S.A.; Mills, J.R.; Johnson, P.W.; Theel, E.S.; Wiggins, C.C.; Bruno, K.A.; Klompas, A.M.; et al. Convalescent Plasma Antibody Levels and the Risk of Death from COVID-19. *N. Engl. J. Med.* **2021**, *384*, 1015–1027. [CrossRef]
5. Libster, R.; Pérez Marc, G.; Wappner, D.; Coviello, S.; Bianchi, A.; Braem, V.; Esteban, I.; Caballero, M.T.; Wood, C.; Berrueta, M.; et al. Early High-Titer Plasma Therapy to Prevent Severe COVID-19 in Older Adults. *N. Engl. J. Med.* **2021**, *384*, 610–618. [CrossRef]
6. Maor, Y.; Cohen, D.; Paran, N.; Israely, T.; Ezra, V.; Axelrod, O.; Shinar, E.; Izak, M.; Rahav, G.; Rahimi-Levene, N.; et al. Compassionate use of convalescent plasma for treatment of moderate and severe pneumonia in COVID-19 patients and association with IgG antibody levels in donated plasma. *EClinicalMedicine* **2020**, *26*, 100525. [CrossRef]
7. Dinesh, D.C.; Chalupska, D.; Silhan, J.; Koutna, E.; Nencka, R.; Veverka, V.; Boura, E. Structural basis of RNA recognition by the SARS-CoV-2 nucleocapsid phosphoprotein. *PLoS Pathog.* **2020**, *16*, e1009100. [CrossRef]
8. Dangi, T.; Class, J.; Palacio, N.; Richner, J.M.; Penaloza MacMaster, P. Combining spike- and nucleocapsid-based vaccines improves distal control of SARS-CoV-2. *Cell Rep.* **2021**, *36*, 109664. [CrossRef]
9. Matchett, W.E.; Joag, V.; Stolley, J.M.; Shepherd, F.K.; Quarnstrom, C.F.; Mickelson, C.K.; Wijeyesinghe, S.; Soerens, A.G.; Becker, S.; Thiede, J.M.; et al. Cutting Edge: Nucleocapsid Vaccine Elicits Spike-Independent SARS-CoV-2 Protective Immunity. *J. Immunol.* **2021**, *207*, 376–379. [CrossRef]
10. Yewdell, J.W.; Frank, E.; Gerhard, W. Expression of influenza A virus internal antigens on the surface of infected P815 cells. *J. Immunol.* **1981**, *126*, 1814–1819.
11. Marie, J.C.; Saltel, F.; Escola, J.M.; Jurdic, P.; Wild, T.F.; Horvat, B. Cell surface delivery of the measles virus nucleoprotein: A viral strategy to induce immunosuppression. *J. Virol.* **2004**, *78*, 11952–11961. [CrossRef] [PubMed]
12. Céspedes, P.F.; Bueno, S.M.; Ramírez, B.A.; Gomez, R.S.; Riquelme, S.A.; Palavecino, C.E.; Mackern-Oberti, J.P.; Mora, J.E.; Depoil, D.; Sacristán, C.; et al. Surface expression of the hRSV nucleoprotein impairs immunological synapse formation with T cells. *Proc. Natl. Acad. Sci. USA* **2014**, *111*, E3214–E3223. [CrossRef] [PubMed]
13. Straub, T.; Schweier, O.; Bruns, M.; Nimmerjahn, F.; Waisman, A.; Pircher, H. Nucleoprotein-specific nonneutralizing antibodies speed up LCMV elimination independently of complement and FcγR. *Eur. J. Immunol.* **2013**, *43*, 2338–2348. [CrossRef] [PubMed]
14. Ikuta, K.; Morita, C.; Miyake, S.; Ito, T.; Okabayashi, M.; Sano, K.; Nakai, M.; Hirai, K.; Kato, S. Expression of human immunodeficiency virus type 1 (HIV-1) gag antigens on the surface of a cell line persistently infected with HIV-1 that highly expresses HIV-1 antigens. *Virology* **1989**, *170*, 408–417. [CrossRef]
15. Lopez-Munoz, A.; Kosik, I.; Holly, J.; Yewdell, J. Cell Surface SARS-CoV-2 Nucleocapsid Protein Modulates Innate and Adaptive Immunity. *Sci. Adv.* **2021**, *8*, eabp9770. [CrossRef]
16. Rabouille, C. Pathways of Unconventional Protein Secretion. *Trends Cell Biol.* **2017**, *27*, 230–240. [CrossRef]
17. Clausen, T.M.; Sandoval, D.R.; Spliid, C.B.; Pihl, J.; Perrett, H.R.; Painter, C.D.; Narayanan, A.; Majowicz, S.A.; Kwong, E.M.; McVicar, R.N.; et al. SARS-CoV-2 Infection Depends on Cellular Heparan Sulfate and ACE2. *Cell* **2020**, *183*, 1043–1057.e15. [CrossRef]
18. Gao, T.; Hu, M.; Zhang, X.; Li, H.; Zhu, L.; Liu, H.; Dong, Q.; Zhang, Z.; Wang, Z.; Hu, Y.; et al. Highly pathogenic coronavirus N protein aggravates lung injury by MASP-2-mediated complement over-activation. *medRxiv* **2020**. [CrossRef]
19. Kang, S.; Yang, M.; He, S.; Wang, Y.; Chen, X.; Chen, Y.Q.; Hong, Z.; Liu, J.; Jiang, G.; Chen, Q.; et al. A SARS-CoV-2 antibody curbs viral nucleocapsid protein-induced complement hyperactivation. *Nat. Commun.* **2021**, *12*, 2697. [CrossRef]
20. Chiuppesi, F.; Zaia, J.A.; Frankel, P.H.; Stan, R.; Drake, J.; Williams, B.; Acosta, A.M.; Francis, K.; Taplitz, R.A.; Dickter, J.K.; et al. Safety and immunogenicity of a synthetic multiantigen modified vaccinia virus Ankara-based COVID-19 vaccine (COH04S1): An open-label and randomised, phase 1 trial. *Lancet Microbe* **2022**, 3, e252–e264. [CrossRef]

21. Hajnik, R.L.; Plante, J.A.; Liang, Y.; Alameh, M.G.; Tang, J.; Bonam, S.R.; Zhong, C.; Adam, A.; Scharton, D.; Rafael, G.H.; et al. Dual spike and nucleocapsid mRNA vaccination confer protection against SARS-CoV-2 Omicron and Delta variants in preclinical models. *Sci. Transl. Med.* **2022**, *14*, eabq1945. [CrossRef] [PubMed]
22. Navaratnam, A.M.D.; Shrotri, M.; Nguyen, V.; Braithwaite, I.; Beale, S.; Byrne, T.E.; Fong, W.L.E.; Fragaszy, E.; Geismar, C.; Hoskins, S.; et al. Nucleocapsid and spike antibody responses post virologically confirmed SARS-CoV-2 infection: An observational analysis in the Virus Watch community cohort. *medRxiv* **2022**. [CrossRef]
23. Krutikov, M.; Palmer, T.; Tut, G.; Fuller, C.; Azmi, B.; Giddings, R.; Shrotri, M.; Kaur, N.; Sylla, P.; Lancaster, T.; et al. Prevalence and duration of detectable SARS-CoV-2 nucleocapsid antibodies in staff and residents of long-term care facilities over the first year of the pandemic (VIVALDI study): Prospective cohort study in England. *Lancet Healthy Longev.* **2021**. [CrossRef]
24. Brochot, E.; Demey, B.; Touzé, A.; Belouzard, S.; Dubuisson, J.; Schmit, J.-L.; Duverlie, G.; Francois, C.; Castelain, S.; Helle, F. Anti-spike, Anti-nucleocapsid and Neutralizing Antibodies in SARS-CoV-2 Inpatients and Asymptomatic Individuals. *Front. Microbiol.* **2020**, *11*, 584251. [CrossRef] [PubMed]
25. Follmann, D.; Janes, H.E.; Buhule, O.D.; Zhou, H.; Girard, B.; Marks, K.; Kotloff, K.; Desjardins, M.; Corey, L.; Neuzil, K.M.; et al. Antinucleocapsid Antibodies After SARS-CoV-2 Infection in the Blinded Phase of the Randomized, Placebo-Controlled mRNA-1273 COVID-19 Vaccine Efficacy Clinical Trial. *Ann. Intern. Med.* **2022**, *175*, 1258–1265. [CrossRef] [PubMed]
26. The Association of Baseline Plasma SARS-CoV-2 Nucleocapsid Antigen Level and Outcomes in Patients Hospitalized With COVID-19. *Ann. Intern. Med.* **2022**. [CrossRef]
27. Nakanaga, K.; Yamanouchi, K.; Fujiwara, K. Protective effect of monoclonal antibodies on lethal mouse hepatitis virus infection in mice. *J. Virol.* **1986**, *59*, 168–171. [CrossRef]
28. Lecomte, J.; Cainelli-Gebara, V.; Mercier, G.; Mansour, S.; Talbot, P.J.; Lussier, G.; Oth, D. Protection from mouse hepatitis virus type 3-induced acute disease by an anti-nucleoprotein monoclonal antibody. Brief report. *Arch. Virol.* **1987**, *97*, 123–130. [CrossRef]
29. Dangi, T.; Sanchez, S.; Park, M.; Class, J.; Richner, M.C.; Richner, J.M.; Penaloza-MacMaster, P. Nucleocapsid-specific humoral responses improve the control of SARS-CoV-2. *bioRxiv* **2022**. [CrossRef]
30. Cain, W.V.; Sill, A.M.; Solipuram, V.; Weiss, J.J.; Miller, C.B.; Jelsma, P.F. Efficacy of COVID-19 Convalescent Plasma Based on Antibody Concentration. *Adv. Hematol.* **2022**, *2022*, 7992927. [CrossRef]
31. Herman, J.D.; Wang, C.; Loos, C.; Yoon, H.; Rivera, J.; Eugenia Dieterle, M.; Haslwanter, D.; Jangra, R.K.; Bortz, R.H.; Bar, K.J.; et al. Functional convalescent plasma antibodies and pre-infusion titers shape the early severe COVID-19 immune response. *Nat. Commun.* **2021**, *12*, 1–11. [CrossRef] [PubMed]
32. Herman, J.D.; Wang, C.; Burke, J.S.; Zur, Y.; Compere, H.; Kang, J.; Macvicar, R.; Shin, S.; Frank, I.; Siegel, D.; et al. A role for Nucleocapsid-specific antibody function in COVID-19 Convalescent plasma therapy. *medRxiv* **2022**. [CrossRef]
33. Focosi, D.; Franchini, M.; Pirofski, L.-a.; Burnouf, T.; Fairweather, D.; Joyner, M.J.; Casadevall, A. COVID-19 Convalescent Plasma Is More than Neutralizing Antibodies: A Narrative Review of Potential Beneficial and Detrimental Co-Factors. *Viruses* **2021**, *13*, 1594. [CrossRef] [PubMed]
34. LaMere, M.W.; Lam, H.T.; Moquin, A.; Haynes, L.; Lund, F.E.; Randall, T.D.; Kaminski, D.A. Contributions of antinucleoprotein IgG to heterosubtypic immunity against influenza virus. *J. Immunol.* **2011**, *186*, 4331–4339. [CrossRef]
35. Fielding, C.; Sabberwal, P.; Williamson, J.; Greenwood, E.; Crozier, T.; Zelek, W.; Seow, J.; Graham, C.; Huettner, I.; Edgeworth, J.; et al. ADNKA overcomes SARS-CoV2-mediated NK cell inhibition through non-spike antibodies. *bioRxiv* **2021**. [CrossRef]
36. Rak, A.; Donina, S.; Zabrodskaya, Y.; Rudenko, L.; Isakova-Sivak, I. Cross-Reactivity of SARS-CoV-2 Nucleocapsid-Binding Antibodies and Its Implication for COVID-19 Serology Tests. *Viruses* **2022**, *14*, 2041. [CrossRef]
37. Yamaoka, Y.; Jeremiah, S.S.; Miyakawa, K.; Saji, R.; Nishii, M.; Takeuchi, I.; Ryo, A. Whole Nucleocapsid Protein of Severe Acute Respiratory Syndrome Coronavirus 2 May Cause False-Positive Results in Serological Assays. *Clin. Infect. Dis. Off. Publ. Infect. Dis. Soc. Am.* **2021**, *72*, 1291–1292. [CrossRef]
38. Frigerio, R.; Musicò, A.; Strada, A.; Mussida, A.; Gagni, P.; Bergamaschi, G.; Chiari, M.; Barzon, L.; Gori, A.; Cretich, M. Epitope Mapping on Microarrays Highlights a Sequence on the N Protein with Strong Immune Response in SARS-CoV-2 Patients. *Methods Mol. Biol.* **2022**, *2578*, 209–217. [CrossRef]
39. Lu, R.-M.; Ko, S.-H.; Chen, W.-Y.; Chang, Y.-L.; Lin, H.-T.; Wu, H.-C. Monoclonal Antibodies against Nucleocapsid Protein of SARS-CoV-2 Variants for Detection of COVID-19. *Int. J. Mol. Sci.* **2021**, *22*, 12412. [CrossRef]
40. Nückel, J.; Planatscher, E.; Mohr, A.W.; Deichl, K.; Mijočević, H.; Feuerherd, M.; Wolff, L.; Erber, J.; Schneider, J.; Quante, M.; et al. Association between IgG responses against the nucleocapsid proteins of alphacoronaviruses and COVID-19 severity. *Front. Immunol.* **2022**, *13*, 889836. [CrossRef]
41. Sermet-Gaudelus, I.; Temmam, S.; Huon, C.; Behillil, S.; Gajdos, V.; Bigot, T.; Lurier, T.; Chrétien, D.; Backovic, M.; Delaunay-Moisan, A.; et al. Prior infection by seasonal coronaviruses, as assessed by serology, does not prevent SARS-CoV-2 infection and disease in children, France, April to June 2020. *Eurosurveillance* **2021**, *26*, 2001782. [CrossRef] [PubMed]
42. Sen, S.R.; Sanders, E.C.; Gabriel, K.N.; Miller, B.M.; Isoda, H.M.; Salcedo, G.S.; Garrido, J.E.; Dyer, R.P.; Nakajima, R.; Jain, A.; et al. Predicting COVID-19 Severity with a Specific Nucleocapsid Antibody plus Disease Risk Factor Score. *mSphere* **2021**, *6*. [CrossRef]
43. Dugas, M.; Grote-Westrick, T.; Merle, U.; Fontenay, M.; Kremer, A.E.; Vollenberg, R.; Lorentzen, E.; Tiwari-Heckler, S.; Duchemin, J.; Ellouze, S.; et al. Lack of antibodies against seasonal coronavirus OC43 nucleocapsid protein identifies patients at risk of critical COVID-19. *medRxiv* **2020**. [CrossRef]

44. Röltgen, K.; Powell, A.E.; Wirz, O.F.; Stevens, B.A.; Hogan, C.A.; Najeeb, J.; Hunter, M.; Wang, H.; Sahoo, M.K.; Huang, C.; et al. Defining the features and duration of antibody responses to SARS-CoV-2 infection associated with disease severity and outcome. *Sci. Immunol.* **2020**, *5*. [CrossRef]
45. Atyeo, C.; Fischinger, S.; Zohar, T.; Slein, M.D.; Burke, J.; Loos, C.; McCulloch, D.J.; Newman, K.L.; Wolf, C.; Yu, J.; et al. Distinct Early Serological Signatures Track with SARS-CoV-2 Survival. *Immunity* **2020**, *53*, 524–532.e4. [CrossRef]
46. Delgado, J.F.; Vidal-Pla, M.; Moya, M.C.; Espasa, M.; Casabella, A.; Seda, M.; Calvet, J.; Gratacós, J.; Serrano, R.M.; Peña, P. SARS-CoV-2 Spike Protein Vaccine-Induced Immune Imprinting Reduces Nucleocapsid Protein Antibody Response in SARS-CoV-2 Infection. *J. Immunol. Res.* **2022**, *2022*, 8287087. [CrossRef]
47. Loesche, M.; Karlson, E.W.; Talabi, O.; Zhou, G.; Boutin, N.; Atchley, R.; Loevinsohn, G.; Chang, J.B.P.; Hasdianda, M.A.; Okenla, A.; et al. Longitudinal SARS-CoV-2 Nucleocapsid Antibody Kinetics, Seroreversion, and Implications for Seroepidemiologic Studies. *Emerg. Infect. Dis.* **2022**, *28*, 1859–1862. [CrossRef]
48. Senefeld, J.W.; Franchini, M.; Mengoli, C.; Cruciani, M.; Zani, M.; Gorman, E.K.; Focosi, D.; Casadevall, A.; Joyner, M.J. COVID-19 convalescent plasma for the treatment of immunocompromised patients: A systematic review. *medRxiv* **2022**. [CrossRef]

viruses

MDPI

Article

Rapid Identification of SARS-CoV-2 Omicron BA.5 Spike Mutation F486V in Clinical Specimens Using a High-Resolution Melting-Based Assay

Akira Aoki [1], Hirokazu Adachi [2], Yoko Mori [1], Miyabi Ito [2], Katsuhiko Sato [2], Masayoshi Kinoshita [2], Masahiro Kuriki [2,3], Kenji Okuda [2,4], Toru Sakakibara [2,5], Yoshinori Okamoto [1] and Hideto Jinno [1,*]

1 Faculty of Pharmacy, Meijo University, 150 Yagotoyama, Tempaku-ku, Nagoya 468-8503, Japan
2 Aichi Prefectural Institute of Public Health, 7-6 Nagare, Tsuji-machi, Kita-ku, Nagoya 462-8576, Japan
3 Kiyosu Health Center, 129 Haruhi, Furikata, Kiyosu 452-0961, Japan
4 Handa Health Center, 1-45-4 Deguchi-cho, Handa 475-0903, Japan
5 Nishio Health Center, 12 Shimoda, Yorizumi-cho, Nishio 445-0073, Japan
* Correspondence: jinno@meijo-u.ac.jp

Abstract: Severe acute respiratory syndrome coronavirus 2 (SARS-CoV-2) Omicron subvariant BA.5 emerged as of February 2022 and replaced the earlier Omicron subvariants BA.1 and BA.2. COVID-19 genomic surveillance should be continued as new variants seem to subsequently appear, including post-BA.5 subvariants. A rapid assay is needed to differentiate between the currently dominant BA.5 variant and other variants. This study successfully developed a high-resolution melting (HRM)-based assay for BA.4/5-characteristic spike mutation F486V detection and demonstrated that our assay could discriminate between BA.1, BA.2, and BA.5 subvariants in clinical specimens. The mutational spectra at two regions (G446/L452 and F486) for the variant-selective HRM analysis was the focus of our assay. The mutational spectra used as the basis to identify each Omicron subvariant were as follows: BA.1 (G446S/L452/F486), BA.2 (G446/L452/F486), and BA.4/5 (G446/L452R/F486V). Upon mutation-coding RNA fragment analysis, the wild-type fragments melting curves were distinct from those of the mutant fragments. Based on the analysis of 120 clinical samples (40 each of subvariants BA.1, BA.2, and BA.5), this method's sensitivity and specificity were determined to be more than 95% and 100%, respectively. These results clearly demonstrate that this HRM-based assay is a simple screening method for monitoring Omicron subvariant evolution.

Keywords: severe acute respiratory syndrome coronavirus 2; high-resolution melting; Omicron subvariant; BA.5; F486V mutation

Citation: Aoki, A.; Adachi, H.; Mori, Y.; Ito, M.; Sato, K.; Kinoshita, M.; Kuriki, M.; Okuda, K.; Sakakibara, T.; Okamoto, Y.; et al. Rapid Identification of SARS-CoV-2 Omicron BA.5 Spike Mutation F486V in Clinical Specimens Using a High-Resolution Melting-Based Assay. *Viruses* **2022**, *14*, 2401. https://doi.org/10.3390/v14112401

Academic Editors: Ahmed El-Shamy and Mohamed Ibrahim

Received: 27 September 2022
Accepted: 28 October 2022
Published: 29 October 2022

1. Introduction

The coronavirus disease 2019 (COVID-19) outbreak was caused by the severe acute respiratory syndrome coronavirus 2 (SARS-CoV-2) that emerged in China in late 2019 [1–3]. The number of global COVID-19 cases surpassed 600 million according to the World Health Organization (WHO) data (Weekly epidemiological update on COVID-19, 7 September 2022), including over 6.4 million deaths. A number of SARS-CoV-2 variants emerged worldwide, some of which could potentially increase public health risk. The WHO has classified these as variants of concern (VOCs) and variants of interest (VOIs) since 2020. The COVID-19 genomic surveillance, tracking VOCs and VOIs, is essential in protecting against the spread of high-risk variants and monitoring new variants.

The novel VOC, the Omicron variant, emerged from South Africa in November 2021 [4–6]. The original Omicron variant BA.1 quickly spread to many countries and exceeded the other variants, including the Delta variant. The BA.2 subvariant of Omicron, which was more transmissible than BA.1 [7], rapidly increased since early 2022. The new Omicron subvariants BA.4 and BA.5 were first detected in South Africa in January and February

2022, respectively. The two subvariants subsequently became mainstream in the country. Although BA.4 did not spread further around the world, BA.5 quickly spread across the world after a few months from BA.2 becoming the dominant variant. Furthermore, close monitoring of the Omicron variants' evolution should continue since additional Omicron subvariants seem to emerge subsequently [8].

Next-generation sequencing (NGS) is the gold standard for determining the whole viral genome. NGS data leads to assigning a lineage or clade to each SARS-CoV-2 strain. A screening assay, which identifies a SARS-CoV-2 point mutation, can contribute to the rapid identification of the individual variants within a day as NGS analysis requires a few days to acquire the data. A few previous studies have shown that screening assays can differentiate the Omicron subvariants BA.1 and BA.2 based on their characteristic mutations [9–12]. The BA.5 subvariant has two characteristic spike mutations, L452R and F486V (BA.4 sequence identical to BA.5 sequence in spike protein) [13,14]. Wilhelm et al. [15] reported that Omicron subvariants BA.4 and BA.5 in wastewater could be detected by commercially available kits for L452R and E484A/F486V. However, there are no or limited studies that demonstrate Omicron subvariant BA.5 identification in clinical specimens by a screening assay.

High-resolution melting (HRM) analysis is a post-PCR genotyping technique based on an amplicons' melting behavior. HRM-based assay is a less-costly and quick-build method as the assay does not need a specific probe. Our HRM-based assay was previously demonstrated to discriminate between the Omicron BA.1 and BA.2 subvariants targeting three mutations in the spike receptor-binding domain (RBD), R408, G446/L452, and T477/T478 [16]. In this study, we aimed to (i) develop an HRM-based assay to detect the F486V mutation and (ii) validate the HRM-based assay for L452/G446 and F486 sites in 120 clinical samples (40 each of the BA.1, BA.2, and BA.5 subvariants). This HRM-based assay would contribute to rapid discernment between BA.1 (G446S, L452, and F486), BA.2 (G446, L452, and F486), and BA.4/5 (G446, L452R, and F486V) (Table 1).

Table 1. Receptor-binding domain (RBD) amino acid substitutions in Omicron BA.1, BA.2, BA.4, and BA.5 subvariants.

Pangolin	RBD Amino Acid Substitutions
BA.1	G339D, S371L, S373P, S375F, K417N, N440K, **G446S**, S477N, T478K, E484A, Q493R, G496S, Q498R, N501Y, Y505H
BA.2	G339D, S371F, S373P, S375F, T376A, D405N, R408S, K417N, N440K, S477N, T478K, E484A, Q493R, Q498R, N501Y, Y505H
BA.4/5	G339D, S371F, S373P, S375F, T376A, D405N, R408S, K417N, N440K, **L452R**, **F486V**, S477N, T478K, E484A, Q498R, N501Y, Y505H

Substitutions detected by this high-resolution melting assay are marked in bold.

2. Materials and Methods

2.1. Ethics Statement

The Meijo University Research Ethics Committee (Approval number: 2020-17-2) and Aichi Prefectural Institute of Public Health (Approval number: 20E-4) approved this project, and was carried out according to the Infectious Diseases Control Law of Japan.

2.2. Preparation of Standard RNA Fragments: In Vitro T7 Transcription

The SARS-CoV-2 sequence was obtained from the NCBI (GenBank ID: MN908947), the GISAID database (www.gisaid.org/; accessed on 9 August 2022), and the Pango nomenclature system (https://cov-lineages.org/lineages.html; accessed on 9 August 2022). Eurofins Genomics K.K provided the three RBD DNA fragments (wild-type, F486V mutant, BA.4/5 variant mutant; 600–1000 bp in length) with a 5′ T7 upstream promoter sequence. (Tokyo, Japan). In vitro T7 transcription was performed as previously described [16,17]. The synthesized single-stranded RNA fragments were used as reverse transcriptase (RT)-PCR amplification templates.

2.3. RT-PCR Amplification: First PCR

A one-step RT-PCR kit (One Step PrimeScript III RT-qPCR Mix, with UNG; TaKaRa Bio Inc., Kusatsu, Japan) was used to perform RT-PCR in a single closed tube in accordance with the manufacturer's instructions. The primer pairs used for RT-PCR amplification are listed in Table 2. Each DNA fragment was observed as a single, correctly sized band (290 bp). RT-PCR amplification was performed as previously described [16,17]. The reaction mixture was diluted 10,000-fold with water and used as a template for the second PCR and HRM analyses after amplification.

Table 2. Primers sequences in reverse transcriptase (RT)-PCR amplification and high-resolution melting (HRM) analysis.

Primer	Sequence
	RT-PCR amplification
Fwd primer	5′-TTACAGGCTGCGTTATAG-3′
Rev primer	5′-ACAAACAGTTGCTGGTGCAT-3′
	HRM analysis
G446-L452 Fwd primer	5′-GGCTGCGTTATAGCTTGGAATTCTAACAATCTT-3′
G446-L452 Rev primer	5′-TCAAAAGGTTTGAGATTAGACTTCC-3′
F486 Fwd primer	5′-ACACCTTGTAATGGTGTTGAAGGT-3′
F486 Rev primer	5′-AGTGGGTTGGAAACCATATGATTGTAA-3′

Fwd: forward; Rev: reverse.

2.4. HRM Analysis: Second PCR

An HRM reagent (MeltDoctor HRM Master Mix; Thermo Fisher Scientific, Waltham, MA, USA) was used to perform HRM according to the manufacturer's instructions. The primer pairs for HRM analysis are listed in Table 2. Each DNA fragment was observed as a single, correctly sized band (G466-L452, 104 bp; F486, 69 bp). The G446-L452 forward primer design was based on the N440 coding sequence, the F486 forward primer design was based on the T478 and E484 coding sequences, and the F486 reverse primer design was based on the Q493 coding sequence to avoid the potential influences of the N440, T478, E484, and Q493 mutations as shown in Figure 1. A real-time PCR system (LightCycler 96 System; Roche Diagnostics, Basel, Switzerland) was used to perform all reactions in duplicate. PCR amplification and HRM curves were performed as previously described [16,17].

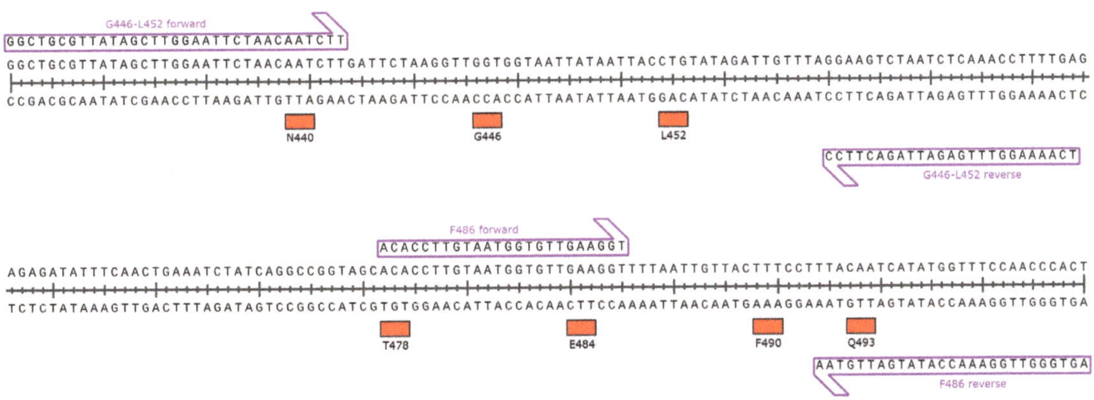

Figure 1. Primer annealing site schematic map for high-resolution melting analyses.

2.5. Clinical Samples

From December 2021 to August 2022, 120 nasopharyngeal swabs or saliva samples were collected from suspected COVID-19 and those detected with COVID-19 by the Aichi Prefectural Institute of Public Health. The Ct values in the clinical quantitative PCR test for these samples ranged from 14–35. RNA purification and whole-genome sequencing were performed as previously described [16].

3. Results

3.1. Development of HRM Analysis for F486V Mutation Detection

First, the HRM-based assay for detecting F486V was developed, a characteristic mutation of BA.4/5. There are several SARS-CoV-2 variants possessing mutations near the F486 site, such as T478K, E484K, E484A, and Q493R. The risk of a false positive or false negative will likely increase in these variants using this assay. The primer sets were thus designed to avoid these mutations (Figure 1). The normalized melting curves and peaks for F486, F486V, and BA.4/5 RBD plots are shown in Figure 2. The F486V RBD plot was different from the F486 RBD plot. Additionally, the BA.4/5 RBD plot was in agreement with the F486V plot. These results suggested that this HRM-based assay can discriminate between the Omicron subvariant BA.4/5 (as the F486V variant) and other subvariants, including BA.1 and BA.2 (as the F486 wild-type).

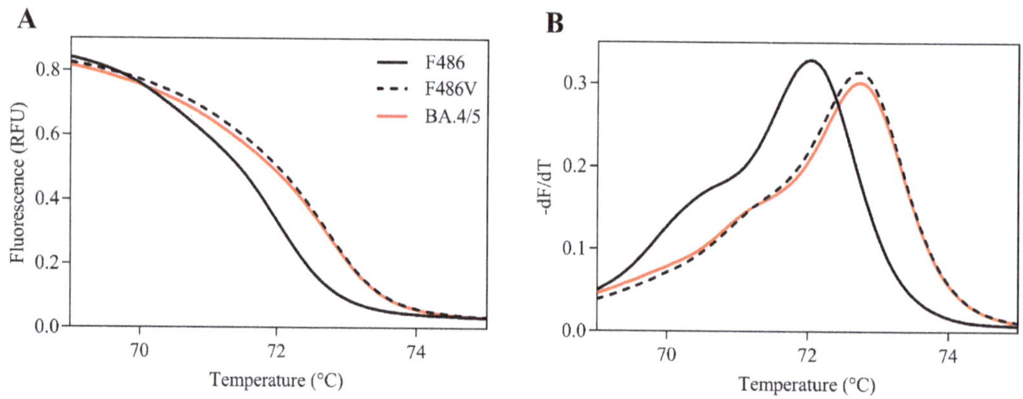

Figure 2. Positive control RNAs normalized melting curves and melting peaks for the F486 site. Normalized melting curve plots (**A**) and melting peak plots (**B**) for the F486 site were acquired using F486 receptor-binding domain standard fragments (RBD; solid black line), F486V RBD (dashed black line), and BA.4/5 RBD (solid red line).

3.2. Validation of HRM Analysis for the G446/L452 and F486 Sites in Clinical Samples

Second, HRM-based assay was validated for G446/L452 and F486 sites in clinical samples infected with BA.1, BA.2, or BA.5. A total of 120 clinical samples (n = 40, BA.1; n = 40, BA.2; n = 40, BA.5) were randomly selected after whole-genome sequencing by NGS (Table 3). RBD sequences in all samples were identical to the reference sequence (BA.1, G446S/L452/F486; BA.2, G446/L452/F486; and BA.5, G446/L452R/F486V) confirmed by NGS data. HRM analysis together with positive control RNAs was used to analyze clinical samples. The Gene Scanning Software automatically classified samples into wild-type or mutant strains based on HRM curves after HRM analysis. The melting peak plots of representative samples at two RBD regions are shown in Figure 3. At the G446/L452 site, all BA.1 and BA.2 samples were correctly classified as G446S mutant and wild-type, respectively. The two BA.5 sample plots disagreed with the L452R RBD plot. All samples were correctly classified as each variant at the F486 site. Therefore, the sensitivity and specificity of the current HRM-based assay were calculated based on these results. The sensitivity and specificity of the assay for the two regions were more than 95% and 100%, respectively (Table 4).

Table 3. Pangolin of 120 clinical samples by whole-genome sequencing.

BA.1			BA.2			BA.5		
Pangolin	**No. of Samples**	**Ct Mean Value (Range)**	**Pangolin**	**No. of Samples**	**Ct Mean Value (Range)**	**Pangolin**	**No. of Samples**	**Ct Mean Value (Range)**
BA.1	11	25.2 (22.2–29.9)	BA.2	10	23.0 (18.0–27.45)	BA.5	3	23.2 (16.9–27.9)
BA.1.1	5	24.2 (23.0–26.8)	BA.2.3	14	25.7 (20.1–34.9)	BA.5.1	2	22.2 (22.1–22.3)
BA.1.1.2	8	24.6 (20.0–29.9)	BA.2.3.1	7	23.9 (18.8–28.4)	BA.5.2	12	25.0 (19.2–29.1)
B.1.1.529	16	27.4 (21.1–32.6)	BA.2.3.13	1	20.8	BA.5.2.1	14	23.8 (14.3–29.1)
			BA.2.10	4	24.7 (22.9–28.7)	BA.5.3.1	1	22.1
			BA.2.10.2	1	23.6	BE.1.1	2	21.3 (17.9–24.8)
			BA.2.24	2	21.4 (18.4–24.5)	BF.5	6	24.1 (22.3–24.9)
			BA.2.29	1	28.8			

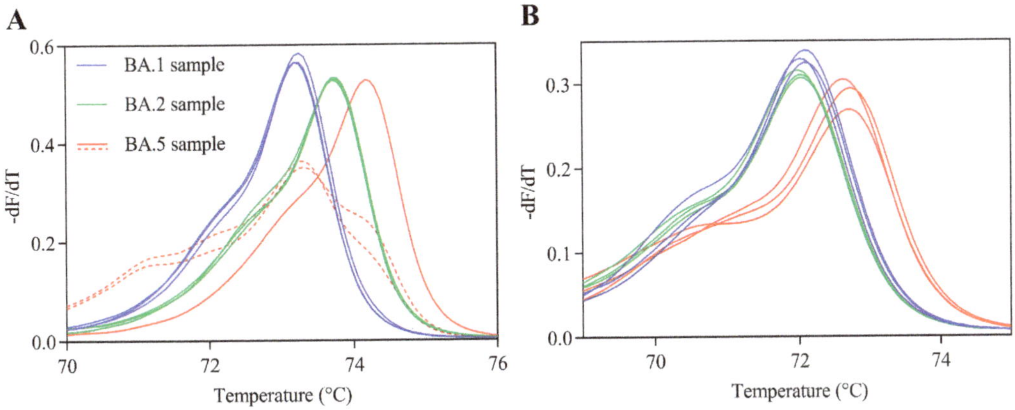

Figure 3. Melting peaks of clinical samples for two receptor-binding domain regions. Melting peak plots for the G446/L452 (**A**) and F486 (**B**) sites were acquired using clinical samples with three BA.1 samples (blue line), three BA.2 samples (green line), and three BA.5 samples (red line). Solid lines indicate true positive and true negative samples. Dashed lines indicate false positive samples.

Table 4. High-resolution melting assay sensitivity and specificity compared with next-generation sequencing.

	G446S Detection	L452R Detection	F486V Detection
Sensitivity [a]	100% (40/40)	95.0% (38/40)	100% (40/40)
Specificity [b]	100% (80/80)	100% (80/80)	100% (80/80)

[a] Sensitivity: True positive number/(True positive number + False Positive number) × 100. [b] Specificity: False negative number/(False negative number + True negative number) × 100.

4. Discussion

This is the first study conducted on developing a screening assay that can distinguish between the Omicron subvariant BA.5 and other subvariants in clinical samples. The assay costs less than USD 5 per sample, and the results can be obtained within 4 h. This HRM-based assay can be a powerful tool as a high-throughput screening test to identify the SARS-CoV-2 variants in many clinical samples.

The Omicron BA.5 subvariant seems to lead to lower mortality and hospitalization rates than earlier variants; however, the COVID-19 pandemic will not end this year [8]. A new high-risk variant will likely happen subsequently as the Omicron BA.5 subvariant is not considered the final SARS-CoV-2 variant [8]. Therefore, COVID-19 genomic surveillance should continue monitoring the next strain's emergence. The BA.2.75 variant called Centaurus (non-WHO labeling) is one of the Omicron subvariants first detected in

India [18–21]. BA.2.75's potential infectivity and immune evasion are unclear, indicating that the variant should be closely monitored. The assay for N460S mutation detection was still developed as our HRM-based assay can be constructed quickly, which is a BA.2.75 subvariant characteristic mutation. We will continue developing HRM-based assays to detect new variants with potentially high public health risks.

This study needs to be interpreted in the context of its limitations. First, this assay cannot differentiate between BA.4 and BA.5 subvariants since both variants possess identical spike protein sequences (Table 1). The assay will be developed to identify the BA.4 specific mutations as necessary, such as OFR1a-DEL141/143, ORF7b-D61L, and N-P151S [22]. Second, this assay was validated using a limited sample number. Our HRM-based assay should analyze a larger sample size. Third, the current assay's detection limit should be determined using low-copy virus samples with Ct values of more than 35.

5. Conclusions

We developed an HRM-based assay that detects the SARS-CoV-2 spike mutation F486V, known as the Omicron subvariants BA.4/5-characteristic mutation. Based on the analysis of clinical samples with BA.1, BA.2, and BA.5, our assay showed enough sensitivity and specificity to distinguish among the three subvariants. Further studies are needed to validate this HRM-based assay using diverse samples with Ct values of more than 35. However, the current HRM-based assay can construct a cost-effective screening test without any specific probe.

Author Contributions: All the authors participated in the interpretation of the results. A.A. designed and conducted the HRM analyses and wrote the original manuscript draft. H.A. and M.I. conducted RNA purification, NGS analysis, and edited the manuscript. Y.M. performed sequence analyses. Y.O. planned the study and edited the manuscript. K.S., M.K. (Masayoshi Kinoshita), M.K. (Masahiro Kuriki), K.O., T.S. and H.J. planned the study and supervised the project. All authors have read and agreed to the published version of the manuscript.

Funding: This work was supported in part by the DAIKO FOUNDATION and Meijo University Research Project for Countermeasures Against COVID-19.

Institutional Review Board Statement: The Research Ethics Committee of Meijo University (approval No. 2020-17-4) and the Aichi Prefectural Institute of Public Health (approval No. 20E-4) approved this project.

Informed Consent Statement: Patient consent was waived because all clinical samples were obtained through the administrative actions process for COVID-19 diagnosis according to the Infectious Diseases Control Law in Japan.

Data Availability Statement: Data are unsuitable for public deposition due to ethical restrictions and the Infectious Diseases Control Law in Japan. The data presented in this study are available on request from the corresponding author. The data are not publicly available due to privacy restrictions.

Acknowledgments: The authors would like to thank the staff of the Laboratory of Virology, Department of Microbiology and Medical Zoology, Aichi Prefectural Institute of Public Health, who performed COVID-19 PCR testing, RNA purification from clinical samples, and NGS analysis.

Conflicts of Interest: The authors declare no competing interest.

References

1. Coronaviridae Study Group of the International Committee on Taxonomy of Viruses. The species severe acute respiratory syndrome-related coronavirus: Classifying 2019-nCoV and naming it SARS-CoV-2. *Nat. Microbiol.* **2020**, *5*, 536–544. [CrossRef] [PubMed]
2. Wu, F.; Zhao, S.; Yu, B.; Chen, Y.M.; Wang, W.; Song, Z.G.; Hu, Y.; Tao, Z.W.; Tian, J.H.; Pei, Y.Y.; et al. Author Correction: A new coronavirus associated with human respiratory disease in China. *Nature* **2020**, *580*, E7. [CrossRef] [PubMed]
3. Zhou, P.; Yang, X.L.; Wang, X.G.; Hu, B.; Zhang, L.; Zhang, W.; Si, H.R.; Zhu, Y.; Li, B.; Huang, C.L.; et al. A pneumonia outbreak associated with a new coronavirus of probable bat origin. *Nature* **2020**, *579*, 270–273. [CrossRef] [PubMed]
4. Karim, S.S.A.; Karim, Q.A. Omicron SARS-CoV-2 variant: A new chapter in the COVID-19 pandemic. *Lancet* **2021**, *398*, 2126–2128. [CrossRef]

5. Petersen, E.; Ntoumi, F.; Hui, D.S.; Abubakar, A.; Kramer, L.D.; Obiero, C.; Tambyah, P.A.; Blumberg, L.; Yapi, R.; Al-Abri, S.; et al. Emergence of new SARS-CoV-2 Variant of Concern Omicron (B.1.1.529)—Highlights Africa's research capabilities, but exposes major knowledge gaps, inequities of vaccine distribution, inadequacies in global COVID-19 response and control efforts. *Int. J. Infect. Dis.* **2022**, *114*, 268–272. [CrossRef]
6. Viana, R.; Moyo, S.; Amoako, D.G.; Tegally, H.; Scheepers, C.; Althaus, C.L.; Anyaneji, U.J.; Bester, P.A.; Boni, M.F.; Chand, M.; et al. Rapid epidemic expansion of the SARS-CoV-2 Omicron variant in southern Africa. *Nature* **2022**, *603*, 679–686. [CrossRef]
7. Yamasoba, D.; Kimura, I.; Nasser, H.; Morioka, Y.; Nao, N.; Ito, J.; Uriu, K.; Tsuda, M.; Zahradnik, J.; Shirakawa, K.; et al. Virological characteristics of the SARS-CoV-2 Omicron BA.2 spike. *Cell* **2022**, *185*, 2103–2115.e19. [CrossRef]
8. Callaway, E. What Omicron's BA.4 and BA.5 variants mean for the pandemic. *Nature* **2022**, *606*, 848–849. [CrossRef]
9. Ayadi, W.; Taktak, A.; Gargouri, S.; Smaoui, F.; Chtourou, A.; Skouri-Gargouri, H.; Derbel, R.; Sassi, A.H.; Gargouri, A.; Hammami, A.; et al. Development of a simple genotyping method based on indel mutations to rapidly screen SARS-CoV-2 circulating variants: Delta, Omicron BA.1 and BA.2. *J. Virol. Methods* **2022**, *307*, 114570. [CrossRef]
10. Hirotsu, Y.; Maejima, M.; Shibusawa, M.; Natori, Y.; Nagakubo, Y.; Hosaka, K.; Sueki, H.; Mochizuki, H.; Tsutsui, T.; Kakizaki, Y.; et al. Classification of Omicron BA.1, BA.1.1, and BA.2 sublineages by TaqMan assay consistent with whole genome analysis data. *Int. J. Infect. Dis.* **2022**, *122*, 486–491. [CrossRef]
11. Koshikawa, T.; Miyoshi, H. High-resolution melting analysis to discriminate between the SARS-CoV-2 Omicron variants BA.1 and BA.2. *Biochem. Biophys. Rep.* **2022**, *31*, 101306. [CrossRef] [PubMed]
12. Lin, H.; Liang, Y.; Zou, L.; Li, B.; Zhao, J.; Wang, H.; Sun, J.; Deng, X.; Tang, S. Combination of isothermal recombinase-aided amplification and CRISPR-Cas12a-mediated assay for rapid detection of major severe acute respiratory syndrome coronavirus 2 variants of concern. *Front. Microbiol.* **2022**, *13*, 945133. [CrossRef] [PubMed]
13. Cao, Y.; Yisimayi, A.; Jian, F.; Song, W.; Xiao, T.; Wang, L.; Du, S.; Wang, J.; Li, Q.; Chen, X.; et al. BA.2.12.1, BA.4 and BA.5 escape antibodies elicited by Omicron infection. *Nature* **2022**, *608*, 593–602. [CrossRef]
14. Tegally, H.; Moir, M.; Everatt, J.; Giovanetti, M.; Scheepers, C.; Wilkinson, E.; Subramoney, K.; Makatini, Z.; Moyo, S.; Amoako, D.G.; et al. Emergence of SARS-CoV-2 Omicron lineages BA.4 and BA.5 in South Africa. *Nat. Med.* **2022**, *28*, 1785–1790. [CrossRef] [PubMed]
15. Wilhelm, A.; Agrawal, S.; Schoth, J.; Meinert-Berning, C.; Bastian, D.; Orschler, L.; Ciesek, S.; Teichgräber, B.; Wintgens, T.; Lackner, S.; et al. Early detection of SARS-CoV-2 Omicron BA.4 and BA.5 in German wastewater. *Viruses* **2022**, *14*, 1876. [CrossRef] [PubMed]
16. Aoki, A.; Adachi, H.; Mori, Y.; Ito, M.; Sato, K.; Okuda, K.; Sakakibara, T.; Okamoto, Y.; Jinno, H. Discrimination of SARS-CoV-2 Omicron sublineages BA.1 and BA.2 using a high-resolution melting-based assay: A pilot study. *Microbiol. Spectr.* **2022**, *10*, e01367-01322. [CrossRef] [PubMed]
17. Aoki, A.; Adachi, H.; Mori, Y.; Ito, M.; Sato, K.; Okuda, K.; Sakakibara, T.; Okamoto, Y.; Jinno, H. A rapid screening assay for L452R and T478K spike mutations in SARS-CoV-2 Delta variant using high-resolution melting analysis. *J. Toxicol. Sci.* **2021**, *46*, 471–476. [CrossRef] [PubMed]
18. Zappa, M.; Verdecchia, P.; Angeli, F. Knowing the new Omicron BA.2.75 variant ('Centaurus'): A simulation study. *Eur. J. Intern. Med.* **2022**, *105*, 107–108. [CrossRef]
19. Sheward, D.J.; Kim, C.; Fischbach, J.; Muschiol, S.; Ehling, R.A.; Björkström, N.K.; Hedestam, G.B.K.; Reddy, S.T.; Albert, J.; Peacock, T.P.; et al. Evasion of neutralising antibodies by omicron sublineage BA.2.75. *Lancet Infect. Dis.* **2022**, *22*, 1421–1422. [CrossRef]
20. Scarpa, F.; Sanna, D.; Azzena, I.; Giovanetti, M.; Benvenuto, D.; Angeletti, S.; Ceccarelli, G.; Pascarella, S.; Casu, M.; Fiori, P.L.; et al. On the SARS-CoV-2 BA.2.75 variant: A genetic and structural point of view. *J. Med. Virol.* 2022, *in press*. [CrossRef]
21. Mohapatra, R.K.; Kandi, V.; Mishra, S.; Sarangi, A.K.; Pradhan, M.K.; Mohapatra, P.K.; Behera, A.; Dhama, K. Emerging novel sub-lineage BA.2.75: The next dominant omicron variant? *Int. J. Surg.* **2022**, *104*, 106835. [CrossRef]
22. Desingu, P.A.; Nagarajan, K. The emergence of Omicron lineages BA.4 and BA.5, and the global spreading trend. *J. Med. Virol.* **2022**, *94*, 5077–5079. [CrossRef] [PubMed]

viruses

MDPI

Article

In Silico Genome Analysis Reveals the Evolution and Potential Impact of SARS-CoV-2 Omicron Structural Changes on Host Immune Evasion and Antiviral Therapeutics

Dhruv Chauhan [1], Nikhil Chakravarty [2], Arjit Vijey Jeyachandran [3], Akshaya Jayakarunakaran [4], Sanjeev Sinha [5], Rakesh Mishra [1], Vaithilingaraja Arumugaswami [3,6,*] and Arunachalam Ramaiah [1,*]

1 Tata Institute for Genetics and Society, Centre at inStem, Bangalore 560065, India
2 Department of Epidemiology, University of California, Los Angeles, CA 90095, USA
3 Department of Molecular and Medical Pharmacology, University of California, Los Angeles, CA 90095, USA
4 Department of Molecular, Cell and Developmental Biology, University of California, Los Angeles, CA 90095, USA
5 All India Institute of Medical Sciences, New Delhi 110029, India
6 Eli and Edythe Broad Center of Regenerative Medicine and Stem Cell Research, University of California, Los Angeles, CA 90095, USA
* Correspondence: varumugaswami@mednet.ucla.edu (V.A.); arunachalam@tigs.res.in (A.R.)

Citation: Chauhan, D.; Chakravarty, N.; Jeyachandran, A.V.; Jayakarunakaran, A.; Sinha, S.; Mishra, R.; Arumugaswami, V.; Ramaiah, A. In Silico Genome Analysis Reveals the Evolution and Potential Impact of SARS-CoV-2 Omicron Structural Changes on Host Immune Evasion and Antiviral Therapeutics. *Viruses* **2022**, *14*, 2461. https://doi.org/10.3390/v14112461

Academic Editors: Ahmed El-Shamy and Mohamed Ibrahim

Received: 20 September 2022
Accepted: 4 November 2022
Published: 6 November 2022

Abstract: New variants of SARS-CoV-2 continue to evolve. The novel SARS-CoV-2 variant of concern (VOC) B.1.1.529 (Omicron) was particularly menacing due to the presence of numerous consequential mutations. In this study, we reviewed about 12 million SARS-CoV-2 genomic and associated metadata using extensive bioinformatic approaches to understand how evolutionary and mutational changes affect Omicron variant properties. Subsampled global data based analysis of molecular clock in the phylogenetic tree showed 29.56 substitutions per year as the evolutionary rate of five VOCs. We observed extensive mutational changes in the spike structural protein of the Omicron variant. A total of 20% of 7230 amino acid and structural changes exclusive to Omicron's spike protein were detected in the receptor binding domain (RBD), suggesting differential selection pressures exerted during evolution. Analyzing key drug targets revealed mutation-derived differential binding affinities between Delta and Omicron variants. Nine single-RBD substitutions were detected within the binding site of approved therapeutic monoclonal antibodies. T-cell epitope prediction revealed eight immunologically important functional hotspots in three conserved non-structural proteins. A universal vaccine based on these regions may likely protect against all these SARS-CoV-2 variants. We observed key structural changes in the spike protein, which decreased binding affinities, indicating that these changes may help the virus escape host cellular immunity. These findings emphasize the need for continuous genomic surveillance of SARS-CoV-2 to better understand how novel mutations may impact viral spread and disease outcome.

Keywords: SARS-CoV-2; Omicron; COVID-19; evolution; receptor binding domain; T-cell epitope; spike protein; drug

1. Introduction

Severe acute respiratory syndrome coronavirus 2 (SARS-CoV-2) is a positive-sense, single-stranded RNA virus with approximately 30 kilobases of genome coding for 29 proteins from 15 open reading frames. SARS-CoV-2, which shows zoonotic origination from bats, was first detected in Wuhan, China in 2019 [1–3]. The causative virus of coronavirus disease 2019 (COVID-19), SARS-CoV-2 has infected over 596 million people worldwide, prompting a massive human health crisis [4]. As the pandemic progresses, new variants of SARS-CoV-2 continue to emerge with amplified virulence or transmission rates. Host immunological responses, viral adaptation to human hosts, lack of global vaccine coverage, and failure to follow the preventive procedures (e.g., wearing a well-fitting mask, keeping hands clean, and maintaining physical distance) have contributed to the continuous emergence of novel SARS-CoV-2 variants. To date, Alpha, Beta, Gamma, Delta, and Omicron variants of concern (VOCs) have been identified. Like other RNA viruses, SARS-CoV-2 has been continuously evolving through genetic mutations or viral recombination and structural changes, including insertions and deletions in different viral genes. These genetic changes have impacted physical and biological properties of SARS-CoV-2, resulting in increased viral spread, reinfection rates, and clinical severity, as reported among recent VOCs.

Lineages of SARS-CoV-2, such as B.1.1.7 (Alpha variant) in the United Kingdom, B.1.351 (Beta variant) in South Africa, P.1 (Gamma variant) in Brazil, and B.1.617.2 (Delta variant) in India, were purported based on mutations at various genes [5–12]. The Delta variant is responsible for more infections and spreads faster than the previously reported Alpha and Beta VOCs, thus attaining much international attention [13]. On 26 November 2021, the World Health Organization (WHO) designated the novel variant B.1.1.529 (Omicron variant), initially reported from South Africa, as a VOC based on the evidence that several mutations, particularly in the Spike protein, may have an impact on viral behavior such as transmissibility, infectivity, clinical severity, risk of reinfection, and the potential impact on diagnostics, prevention, and treatments [13]. Initial reports illustrated that Omicron has increased potential for faster viral spread than Delta and reinfection of fully vaccinated, boosted, and previously infected individuals [14]. The Omicron variant has also shown increased potential to evade pre-existing antibodies, T-cell immunity, and overall human immune action [14]. Thus, this variant has become more prevalent worldwide, overtaking Delta and other VOCs and quickly spreading to all continents [15]. With the Omicron variant being the leading cause of mortality and morbidity in the current phase of the COVID-19 pandemic, we analyzed the genome of VOCs, including the Omicron variant, with the aim of gaining insights into geographical distributions, evolutionary relationships, and the potential impact of key mutations identified in Omicron variants in altering therapeutics and immunity.

2. Materials and Methods

2.1. SARS-CoV-2 Data Analyses

The metadata and genomic and amino acid sequences of SARS-CoV-2 were retrieved from the *Global Initiative on Sharing All Influenza Database* (GISAID) (https://www.gisaid.org/, accessed on 10 October 2022). The VOC metadata was used to analyze the distribution of SARS-CoV-2 infection among different age groups and genders globally and those specific to the Omicron variant. All existing SARS-CoV-2 sequences submitted (6,104,697 Omicron sequences and 5,897,516 sequences for other VOCs) in the GISAID were utilized for this study. The number of VOC sequences reported in GISAID across the world was shown on a world map using packages geojsonio v0.10.0 and ggplot2 v3.3.6 in R v3.6.3. A density plot was generated for comparing the probability distributions of five VOC sequences based on age using ggplot2 v3.3.6 in R v3.6.3. The mutation metadata were organized into groups with respect to VOCs, with non-VOC data filtered out for further analysis. VOC mutations were mapped to the four structural proteins: envelope (E), membrane (M), nucleocapsid (N) and spike (S) and non-structural proteins nsp1–16, ns3, ns6, ns7a, ns7b, and ns8. We compared the structural protein mutational landscape of the five VOCs (Alpha, Beta, Gamma, Delta, and Omicron) to identify unique and common mutations. Mutations in different functional domains of spike were further analyzed. To illustrate the mutations in the context of 3D protein structure, the protein's tertiary monomeric structure for the reference Wuhan-strain spike protein was modelled based on the template structure (PDB ID: 6VXX) in the SWISS-MODEL server [16]. The PyMol (https://pymol.org/2/, accessed on 2 January 2022) program was used to visualize the protein 3D structures. For the phylogenetic analysis, the subset of globally circulating VOCs' sequences, created by Nextstrain (accessed on 22 October 2022), was used (Table S1) [17]. The molecular clock tree option was used to visualize the phylogenetic tree.

2.2. SARS-CoV-2 Target Protein and Drug Interaction Analysis

A total of eight representative inhibitors—namely parecoxib, chlortetracycline, ivermectin B1a, ivermectin B1b, sulfasalazine, remdesivir, atovaquone, and nirmatrelvir (a component of Paxlovid (Pfizer))—were selected as described in the previous reports [18,19] for a molecular docking study. To investigate the impact of novel mutations in Omicron proteins on the potency of the above-mentioned inhibitors, we first modelled the spike structural protein and three non-structural proteins nsp3, nsp5, and nsp12 using known protein structures (PDB ID: 7CWN, 6Y2E, 6LU7, and 6M71, respectively) as a template in the SWISS-MODEL server. For additional comparison, we modelled the structures of these four proteins from a reference EPI_ISL_402124 and a representative Delta variant (EPI_ISL_1470937, EPI_ISL_4577393). Subsequently, these predicted structures were used for molecular docking with their respective drugs in AutoDock Vina v1.1.2 [20]. The drug-binding cavities in the protein structures were determined based on a previous study [21]. The interacting residues of the target protein and the nature of their interactions were identified using LigPlot+ v2.2.4 [22]. A lower binding affinity score indicates higher affinity of the drug to the given protein. Hence, we only considered the best docking model for each protein exhibiting the high binding affinity score to the drug, as they are more likely to bind and inhibit the virus.

2.3. Prediction of T-Cell Epitopes (TCEs) and Their Binding Affinity to Predominant HLA Alleles

To identify the potential impact of Omicron strain mutations in human host immunity, amino acid sequences for the nsp3, nsp5, nsp12, and spike proteins of the reference strain and two variants, i.e., Delta and Omicron, were used to predict the 15-mer CD4 TCEs using the *Immune Epitope Database* and Analysis Resource (IEDB) MHC-II binding prediction tool under the IEDB-recommended prediction method (v2.22) [23]. In order to identify the most immunodominant CD4 epitopes among the predicted potential 15 mer peptides from these four proteins, the binding affinity was predicted using the six most prevalent Human Leukocyte Antigen (HLA; MHC Class-II) alleles (http://allelefrequencies.net/, accessed on 8 January 2022) reported from global populations [24]. The epitope binding prediction results are given in units of IC 50 nM, where a lower number indicates high binding affinity. Thus, we considered the peptides having high binding affinity score (score IC $\leq$ 10 nM) as having the potential to bind to T-cell receptors and stimulate an effective adaptive T-cell immune response [2]. We also identified the mean binding affinity score (IC $\leq$ 10 nM) of the predicted peptides with HLA alleles of the global population to evaluate the generality of those predicted epitopes.

3. Results and Discussion

3.1. Geographical Distribution of Omicron Variant

The metadata analysis of 12,002,213 genome sequences (including Alpha: 1,192,900; Beta: 43,874; Gamma: 129,871; Delta: 4,530,871; Omicron: 6,104,697) revealed that a greater number of sequences were reported from Europe, followed by North America (Figure 1A–C). This sequence data also show that the number of Omicron sequences deposited in GISAID elevated the total number of the remaining four VOC sequences. The comparison of the distributions of five SARS-CoV-2 VOC sequences revealed that people within the age group of 26–35 years were predominantly infected with the Omicron variant, which is consistent with the overall pattern observed across SARS-CoV-2 VOC infections (Figure 1D). While 60% of all VOC sequences have no gender information, the demographic data from the remaining 4,796,586 (21.1% female; 18.9% male) sequences revealed that there is a propensity for females to be susceptible to COVID-19) (Figure 1E). This genomic epidemiological data shows a correlation between the gender and age composition of the under-studied COVID-19 patients. This further illustrates the necessity of thorough genomic surveillance with more completed demographic data to assess the emerging and circulating SARS-CoV-2 variants and prevalence of COVID-19 worldwide.

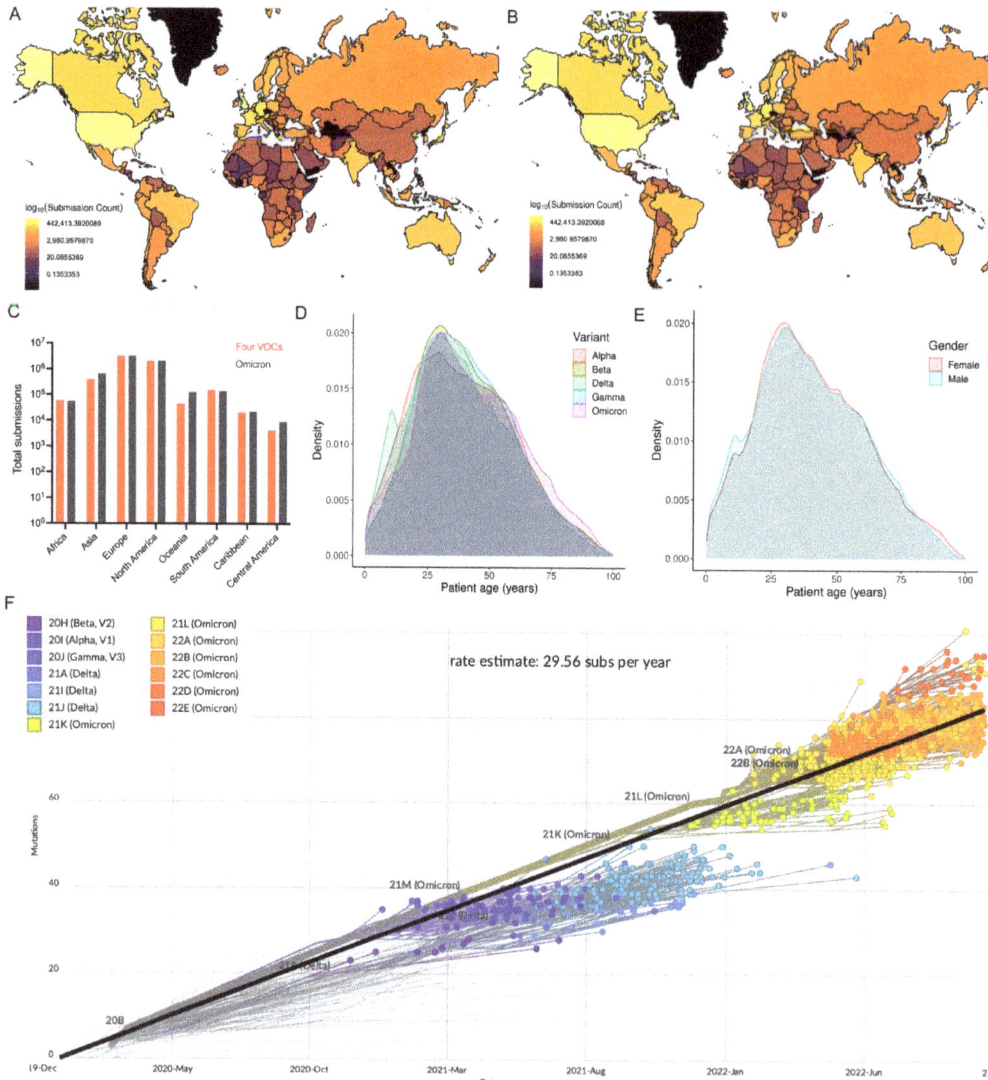

Figure 1. Global distribution of SARS-CoV-2 VOCs as reported in GISAID. (**A**) Number of Omicron and (**B**) all VOC genome sequences reported in GISAID. (**C**) Number of four VOCs and Omicron genome sequences (continent-wise) are presented. (**D**) Density plot shows the probability density distribution of five VOC genomic sequences based on age of COVID-19 cases. Only 38% (4.55 million) of the VOC sequences described the patient's age (0–100 years) were considered. (**E**) Density plot displays the distribution of the consolidated number of all five VOC genomic sequences based on gender. (**F**) Phylogenetic tree based on a subset of globally circulating VOCs sequences created by Nextstrain (accessed on 22 October 2022).

To study phylogenetic relationships, the subset data of globally circulating five VOCs sequences were analyzed in Nextstrain (Figure 1F). The molecular clock phylogeny revealed that (i) nearly all five VOCs emerged independently during late 2020 to late 2021, (ii) genome sequences of the same VOCs were more closely related, and (iii) the Omicron variant is the most commonly circulating variant currently found to have more than ~40 mutational differences relative to a reference strain, confirming that evolutionary pressures shaped the novel Omicron variant to distinguish itself from other variants. The evolutionary rate of five VOCs has been estimated to be 29.56 substitutions per year. Notably, pairwise comparison of a genome sequence of the Omicron variant (EPI_ISL_7547731) with other variants (Alpha EPI_ISL_7856427; Beta EPI_ISL_7814263; Gamma EPI_ISL_7846411; and Delta EPI_ISL_7861981) showed 95.6–96.1% genetic similarity. The Omicron variant shares 96% genome similarity with the reference strain. Interestingly, the reference strain shares a similar percent homology (though not identical changes) with bat coronavirus (bat/Yunnan/RaTG13/2013) [2], indicating the impact of genetic changes and structural variations in driving the evolution of SARS-CoV-2 variants.

3.2. Potential Effect of Omicron Mutations on Host Immune Response

The comparison of the whole mutational profiles of proteome of five SARS-CoV-2 VOCs identified 33,075 signature mutations and structural changes exclusive to the Omicron variant (Figure 2A), in which 10,280 amino acid substitutions and structural changes were uniquely identified in the Omicron structural proteins (Figure 2B; Tables S2 and S3). These were not detected in the other four VOCs. If more than one amino acid or structural changes were identified in one position, then each change was considered individually for the analysis. We also identified that 74,359 mutational/structural changes commonly occurred in Omicron and Alpha variants, which is a larger number than common changes shared between Omicron and Beta (18,135) or Gamma (28,805) but a lesser number than those shared with Delta (105,454) (Figure 2A). All five VOCs shared a total of 10,471 common changes, suggesting these more evolutionarily conserved sites can be targeted for making novel disease control strategies. Among the 10,471 common changes, 1741 were identified in the spike protein (Figure 2A,B; Table S4), with three high-prevalence mutations (D614G, 99.50%; T478K, 83.28%; G142D, 73.15%) occurring in more than 70% of all VOC sequences studied (Figure 2C), whereas 7230 of 33,075 Omicron-specific signature changes were localized in the spike protein (Figure 2D; Tables S2 and S5). Remarkably, 1459 of 7230 changes identified in the spike were exclusively detected in the receptor binding domain (RBD) of the Omicron variant (Table S2). While all RBD mutations appear to be important to the behavior of SARS-CoV-2, identifying the precise pathobiological effects of these unique Omicron signature mutations, especially high-prevalence changes on ACE2 binding affinity, will facilitate our understanding of the variant's rapid transmissibility, infectivity, and disease severity. A recent affinity and kinetics study reported that S477N and N501Y amino acid changes in other VOCs enhance transmission mainly by increasing binding to the ACE2 receptor, while the K417N mutation aids immune escape [25], suggesting that these changes in Omicron may also be involved in the same function. Nine single amino acid substitutions in the RBD (S373P, frequency 46.91%; K417N, 41.9%; N440K, 40.47%; G446S, 14.95%; S477N, 46.72%; T478K, 83.28%; G496S, 18.08%; Q498R, 45.62%; N501Y, 56.85%) (Table S4) were detected commonly within the binding site of the monoclonal antibodies, providing a potential reason for the observed loss of antibody binding or neutralization. However, further detailed studies are required to verify the impact of amino acid changes in the spike protein, particularly on RBD and available monoclonal antibody therapeutics [13].

A comparison of three other structural proteins—M, E, and N—among the five variants revealed that 194, 66, and 677 changes are commonly shared, respectively. However, in the Omicron variant, there were 1011 (M), 204 (E), and 1835 (N) changes exclusively identified in other three structural proteins (Tables S3 and S5). Similarly, analysis of the mutations in the non-structural proteins of Omicron showed 83,965 amino acid substitutions, 6877 deletions, and 7310 insertions in nsp1–16 polyproteins, whereas 8438 amino acid substitutions, 1902 insertion, and 585 deletions were found in the rest of the proteins ns3, ns6, ns7a, ns7b, and ns8 (Table S6). Overall, our analysis showed that, among the Omicron's structural and non-structural proteins, the spike and polyproteins have undergone major genetic changes [26,27]. Moreover, further studies are required to verify the impact of these amino acid and structural changes in the currently circulating novel Omicron lineages, such as BA.4, BA.5, and BQ.1.

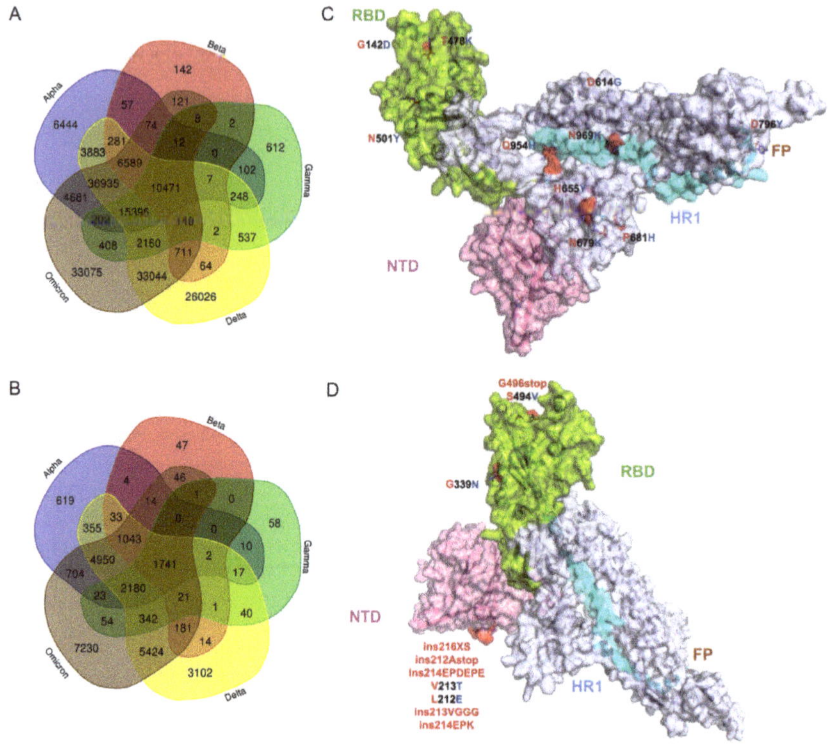

Figure 2. The common and exclusive mutations as well as structural changes in the proteome of Omicron variants. (**A**) Comparison of mutational profile data between 5 VOCs. In total, 12,002,213 sequences retrieved from GISAID's EpiCoV database were used for this analysis. (**B**) Comparison of spike mutational profile data between 5 VOCs. (**C**) Illustration of the top 10 high-prevalence common spike glycoprotein mutations (D614G, frequency 99.50%; T478K, 83.28%; G142D, 73.15%; P681H, 60.31%; N501Y, 56.86%; H655Y, 51.69%; N679K, 50.54%; D796Y, 49.77%; N969K, 49.65%; Q954H, 49.54%) shared by 5 VOCs. (**D**) Illustration of the top 10 high-prevalence mutation and structural changes (G339N, 0.145%; ins216XS, 0.060%; ins212Astop; 0.373%; ins214EPDEPE, 0.021%; V213T, 0.020%; L212E, 0.020%; ins213VGGG, 0.016%; ins214EPK, 0.014%; G496stop, 0.010%; S494V, 0.010%) that are exclusive to Omicron variants on the monomer's tertiary structure. Amino acid substitutions and structural changes in Omicron variants relative to the reference strain are represented in red stick models. Four domains in spike are highlighted as follows: (i) green for receptor binding domain (RBD), (ii) purple for N-terminal domain (NTD), (iii) brown for fusion peptide (FP) domain, and (iv) cyan for the heptad repeats-1 (HR1) domain. Grey is for inter-domain regions.

Further, we analyzed the influence of Omicron's spike mutations on neuropilin 1 (NRP-1) host receptor binding. It is known that three amino acid changes in the furin cleavage site (H655Y, N679K, P681H) of the spike protein could aid viral transmission. For instance, the P681H mutation in the Alpha variant has been found to be involved in enhancing spike cleavage, resulting in increased viral transmission. Our analysis showed that these three mutations are identified in the Omicron variant, suggesting a possible increase in viral transmission. We also interpret that the proline-681 to histidine change in Omicron spike, resulting in a hydrophobic side chain substituted by a charged side chain, alters the interaction between the neuropilin 1 (NRP-1)-B1 domain with spike CendR peptide (furin cleavage site), similar to that observed in the Alpha variant [28]. This mutation shortens the distance between H681 residue and the interacting surface residues of NRP1 by 2 Å (reference with P681 = 16 Å; variant with H681 = 14 Å) [28], suggesting a strong interaction between the Omicron spike protein and NRP1, which may result in enhanced penetration of the virus into the central nervous system (CNS). Altogether, these amino acid changes in the spike protein reveal that infection by the Omicron variant is possibly enhanced by invasion of the CNS and potentially increased transmission compared to the early SARS-CoV-2 variant.

3.3. Molecular Docking Analysis of Omicron Proteins with Known Antiviral Drugs

COVID-19 patient management can be improved by complementing current approved treatments with repurposing existing drugs as an imperative option. While the spike protein is a primary target for vaccine and neutralizing antibody-based therapeutic development, recent studies have demonstrated alternative promising antiviral targets, including nsp3 (papain-like protease), nsp5 (main protease), and nsp12 (RNA-dependent RNA polymerase, RdRp) [29–31]. Thus, we performed molecular docking analysis for eight antiviral drugs against these four target proteins (Spike, nsp3, nsp5, and nsp12) to identify if mutations in the Omicron variant alter the binding efficiency of the drugs. We also compared the overall patterns with respective proteins of reference and the Delta variant. The binding efficiency of inhibitors were characterized based on a scoring function output of AutoDock Vina v1.1.2 [20]. The ligands with the lowest binding affinity (lowest docking score) were considered potential inhibitors of SARS-CoV-2. The mutations in Omicron proteins that change drug binding affinity are illustrated in Figure 3, and the interacting residues and scoring outputs of the drugs are itemized in Table 1 and Figure S1.

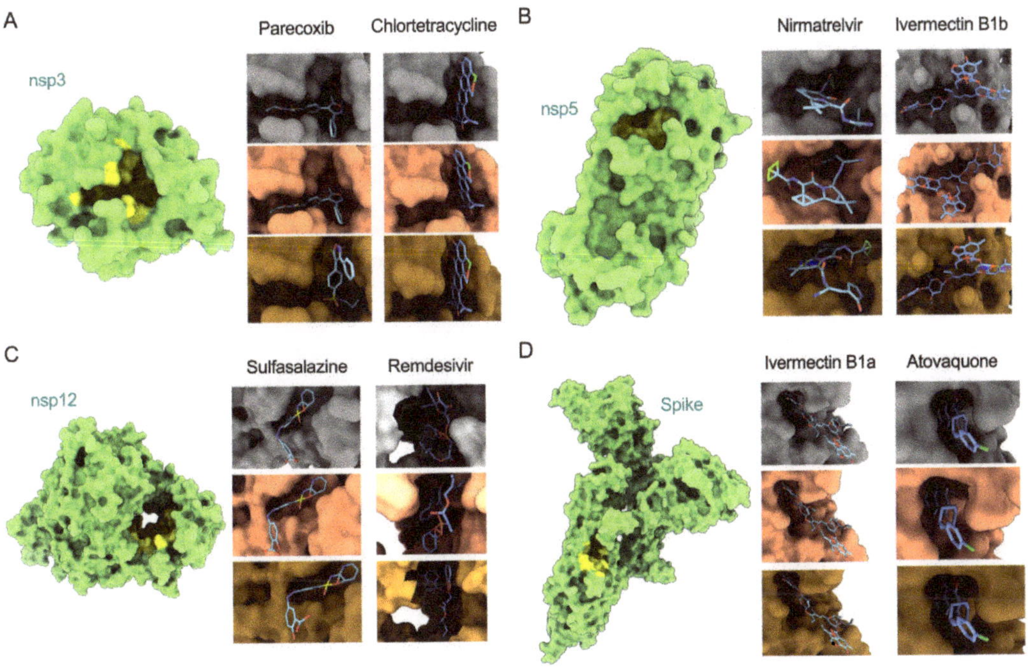

Figure 3. Surface model of binding of inhibitors to the active sites of nsp3, nsp5, nsp12, and spike (models with green color, left). Illustration of docking models (reference = grey; Delta = salmon; Omicron = goldenrod) for binding energy conformations are provided for (**A**) parecoxib and chlortetracycline bound to nsp3, (**B**) nirmatrelvir and ivermectin B1b bound to nsp5, (**C**) sulfasalazine and remdesivir bound to nsp12, and (**D**) ivermectin B1a and atovaquone bound to spike.

Table 1. Molecular docking analysis of eight antiviral compounds (parecoxib, chlortetracycline, ivermectin B1a, ivermectin B1b, sulfasalazine, remdesivir, atovaquone, and nirmatrelvir) against their respective SARS-CoV-2 target proteins (nsp3, nsp5, nsp12, and spike). The 3D structure of four target proteins of the reference strain and the two variants Delta and Omicron were modeled before docking with inhibitors. The hydrogen bond-forming residues from the pool of interacting residues of each target protein are highlighted in bold. While we display all interacting residues from the reference strain, the interacting residues that were different from the reference are displayed for two variants for better readability. Refer to Figure S1 for visualization of inhibitors and target interactions.

Protein	Drug	Variant	Binding Affinity (kcal/mol)	Residues Involved in Interactions
nsp3	Parecoxib	Reference	−8.4	Ala242, Gly250, Gly251, Gly252, **Val253**, Ala333, Gly334, Ile335, Phe336, Ala358, **Phe360**, Leu364, Val539
		Delta	−8.4	Val359
		Omicron	−8.2	Asp226, Ile227, Ala256, Pro329, **Leu330**, Val359, Asp361
	Chlortetracycline	Reference	−8.5	Asp226, Ile227, Val228, Val253, Ala256, Pro329, **Leu330**, Ala333, Gly334, Val359, Phe360, Asp361, Leu364
		Delta	−8.5	No unique residues identified
		Omicron	−8.5	No unique residues identified
nsp5	Nirmatrelvir	Reference	−7.7	Phe140, Leu141, Asn142, Cys145, **His163**, Met165, Glu166, His172, Arg188, Gln189, Thr190, Gln192
		Delta	−7.6	His41, Met49, Leu167, Pro168, Asp187
		Omicron	−7.8	His41, Met49, Asp187
	Ivermectin B1b	Reference	−10.2	Thr26, **His41**, Ser46, Met49, Asn119, Gly143, Cys145, Met165, **Glu166**, Pro168, Arg188, Gln189, Thr190, Gln192
		Delta	−9.8	Thr24, Thr25, Asn142, Ala191
		Omicron	−10.2	No unique residues identified
nsp12	Sulfasalazine	Reference	−9.6	**Arg583**, Gly584, Gly597, **Ser592**, **His599**, **Asn600**, Met601, Lys603, **Thr604**
		Delta	−9.6	No unique residues identified
		Omicron	−9.3	No unique residues identified
	Remdesivir	Reference	−6.4	Arg553, Asp618, Tyr619, Pro620, **Lys621**, Cys622, Asp623, Asn691, Ser759, Asp760, Asp761, Glu811, Ser814
		Delta	−7.4	Tyr455, Arg624, Ser682, Thr687
		Omicron	−8	Tyr455, Arg533, Lys545, Arg555, Thr556, Trp617, Arg624, Ser682
Spike	Ivermectin B1a	Reference	−10	Lys811, Pro812, Ser813, Arg815, Asp820, Phe823, Gly832, Phe833, Ile834, Lys854, **Val860**, Thr866, Glu868
		Delta	−10	No unique residues identified
		Omicron	−10	No unique residues identified
	Atovaquone	Reference	−7.3	Thr732, Leu828, Ala831, Gly832, Phe833, Ile834, Val860, Pro862
		Delta	−7.3	No unique residues identified
		Omicron	−7.3	No unique residues identified

Our analysis showed that remdesivir exhibited lower binding affinity with the Delta variant's nsp12 (−7.4 kcal/mol delta; −6.4 kcal/mol reference strain) but exhibited higher binding affinity with Omicron's nsp12 (binding energy: −8.0 kcal/mol) by interacting with fifteen residues, in which two (Cys622, Asn691) interact via hydrogen bonds (Figures 3 and S1; Table 1). This suggests that the active site of the Omicron RdRp could be considered further to combat the COVID-19 pandemic. Similarly, nirmatrelvir showed a high binding affinity with Omicron's nsp3 (−7.8 kcal/mol) by interacting with eleven residues, of which one (Glu166) formed a hydrogen bond. Interestingly, this drug exhibited similar binding affinity with the Delta variant and reference strain, suggesting that this drug may be an ideal candidate to treat multiple variants. However, upon widespread use of these drugs, there may be a possibility for emergence of drug-resistant viral strains.

The binding affinities of all other six drugs (chlortetracycline, parecoxib, ivermectin B1b, sulfasalazine, ivermectin B1a, and atovaquone) were similar against each of the assessed viral proteins from the two variants and reference strain, suggesting nearly-conserved features of interacting residues from these proteins and the uniformity of these drugs in inhibiting SARS-CoV-2 strains. Only atovaquone uniformly interacted with the least number of residues (eight) of the spike protein (binding energy: −7.3 kcal/mol) across all three variants. However, no residues in the binding cavities interacted via hydrogen bonds. Notably, ivermectin B1a (−10 kcal/mol), ivermectin B1b (−10.2 to −9.8 kcal/mol), and sulfasalazine (−9.6 to −9.3 kcal/mol) exhibited high binding affinity with spike, nsp5, and nsp12 proteins, respectively. Altogether, the docking analysis revealed that: (i) the potential active cavities of the proteins studied are nearly conserved in different variants of SARS-CoV-2 and exhibit similar binding affinities, and (ii) remdesivir and nirmatrelvir may serve as active therapeutic drugs targeting the highly conserved nsp12 and nsp5 proteins, respectively, in the Omicron variant.

3.4. Impact of Key Amino Acid and Structural Changes in T-Cell Epitopes in Modulating the Host Cellular Immune Response

Immunologically targeting SARS-CoV-2 proteins, especially the full-length spike protein and its RBD, using a vaccine can result in the generation of viral variants with immune-evasive mutations. The presentation of viral epitopes by class II MHC (HLA) isoforms plays a vital role in stimulating CD4^{+} Th1 and Th2 cell-mediated immune responses. As the binding affinity of HLA factors with SARS-CoV-2 CD4 epitopes determine host immune responses and the outcome of COVID-19, we analyzed the impact of Omicron's mutations in modulating the host cellular immune response in silico. With respect to this immunity hypothesis, we predicted the peptides from the Omicron spike, nsp3, nsp5, and nsp12 proteins to identify the impact of these mutations on epitope binding affinity with the HLA alleles. Binding affinity of the peptides from these four proteins were compared with those of the Delta variant and a reference strain based on the predominant HLA alleles reported in the global population [2,18]. Complete analyses of peptides from four proteins of a reference strain and two variants showed that 20.5–32% of the predicted 15-mer peptides exhibited high binding affinity (HBA) to at least one of the global HLA types analyzed. Among these, most of the HBA T-cell epitopes (TCEs) were identified from nsp12 (32%) (Figure 4A). Because defining common epitopes from different variants is crucial for designing a universally potent subunit vaccine, we conducted a comparative assessment of HBA TCEs of the four proteins. Our analysis revealed that almost all HBA TCEs (63–529) of three non-structural proteins were commonly shared by the three strains assessed (Figures 4B and S2). While these HBA TCEs of nsp3, nsp5, and nsp12 proteins are highly conserved among the reference and two VOCs, the spike displayed 2–68 HBA TCEs that are exclusive to reference and two VOCs (Figure 4B). Remarkably, the Omicron spike protein was predicted to contain a maximum of 68 HBA TCEs resulting from extensive novel mutations. The HBA T-cell epitopes identified in the functionally important regions of spike and non-structural proteins may critically modulate host immune responses to circulating SARS-CoV-2 variants. While many of the weak binding affinity regions in these proteins may favor viral evasion of host antiviral immunity, the HBA TCEs possibly induce protective immune response.

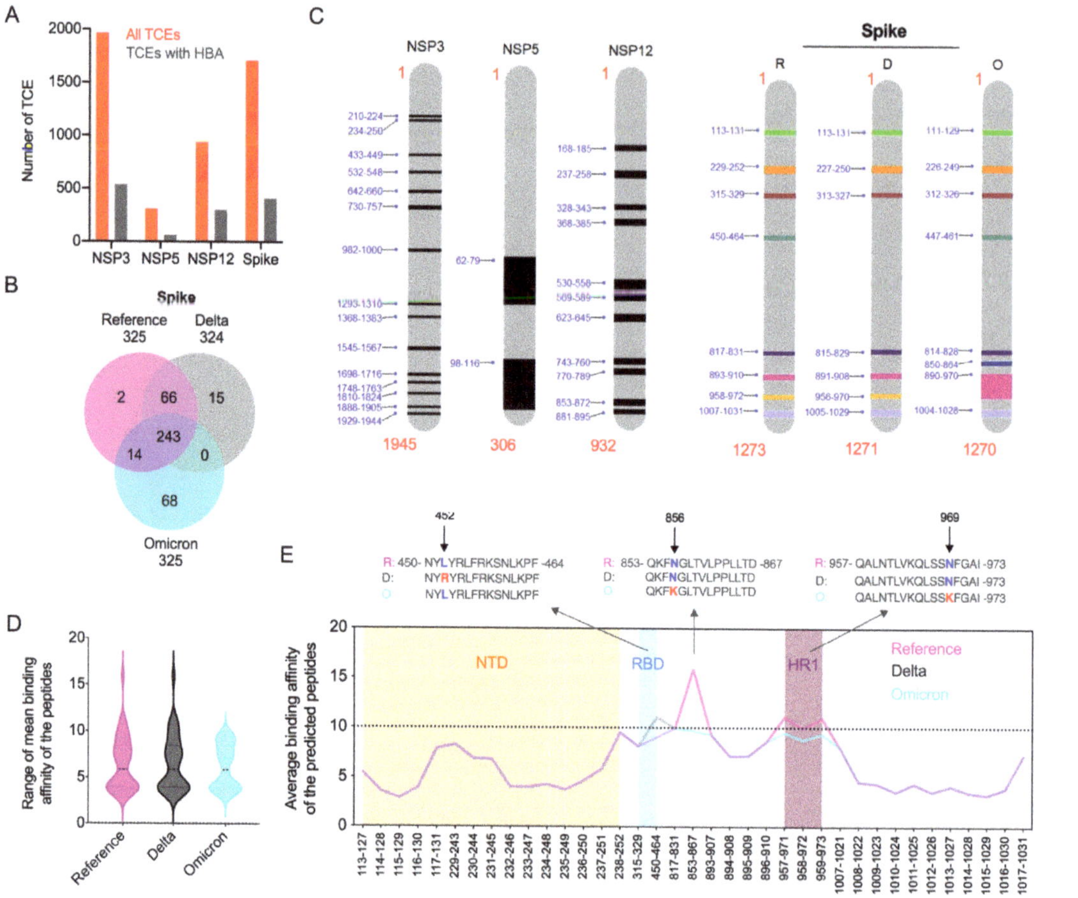

Figure 4. Prediction of high binding affinity CD4 TCEs of SARS-CoV-2 variants. (**A**) Overall distribution of HBA T-cell epitopes predicted from all nsp3, nsp5, nsp12, and spike proteins of three SARS-CoV-2 variants: reference, Delta, and Omicron. For each protein, the total number of predicted unique peptides (red) and the number of these peptides identified with HBA (grey) to at-least one prevalent global HLA allele is shown. (**B**) Comparison of high binding affinity TCEs from reference, Delta, and Omicron spike proteins. (**C**) The average binding affinity was calculated from the binding affinity score of each predicted fifteen amino acids peptides from four proteins with six prevalent HLA types studied. The functional hotspots in conserved nsp3, nsp5, and nsp12 proteins and spike are highlighted. Functional hotspots in spike predicted from reference (R) and two variants (D, Delta; O, Omicron) are shown. (**D**) The range and density of the median average binding affinity scores of the peptides predicted from the spike of the reference and two VOCs. For the comparison, the corresponding mean HBA TCEs in the reference and two variants (HBA score IC <= 10 nM or more in some peptides) are included. (**E**) The expanded version of panel D. The mean binding affinity score of each predicted peptide from the spike of three variants are plotted. Peptide coordinates provided in the *X*-axis are based on their locus in the reference protein. Mean HBA TCEs identified in NTD domain (orange), RBD domain (blue), and HR1 domain (plum) are highlighted. The key mutations that alter the binding affinity of the TCEs of Omicron and Delta variants are shown in the alignment of the predicted peptides.

Since our preliminary epitope analyses identified the majority of HBA TCEs in the conserved nsp3, nsp5, and nsp12 proteins, we considered that these proteins may be more important in cell-mediated immune responses to SARS-CoV-2 infection than previously appreciated. Thus, we further analyzed the epitopes of these proteins along with the spike protein to understand the immunologically significant regions or functional hotspots within these proteins. Thus, to identify the functional hotspots in the four proteins that could potentially interact with predominant HLA alleles, the mean binding affinity was calculated for each of the predicted peptides of these proteins. We considered only CD4 TCEs with a high-confidence mean binding affinity score of IC $\leq$ 10 nM for further analysis. The data indicated 2–15 hotspots across the non-structural proteins from all three variants, in which the maximum number of hotspots were identified in nsp3 (Figure 4C). We identified 7–8 functional hotspots in the spike protein, in which the N-terminal domain (NTD; position aa 13-305) covered a maximum of 15 HBA TCEs (sliding window approach), while the receptor binding domain (RBD; aa 319-541) and heptad repeat region 1 (HR1; aa 912–984) secured the lowest number (aa 2-3) of HBA TCEs. Seven total HBA TCEs covering the Omicron spike domain spanning region from aa 890-970 (includes partial HR1) were identified to be a longer and immunologically important hotspot. Altogether, these findings indicate that added novel mutations in TCEs of the Omicron spike protein may have biological significance in SARS-CoV-2 immune evasion from T-cell response elicited by either natural infection or vaccination. While peptide regions in non-structural proteins may actively stimulate host immunity, much of the spike protein may favor host immune evasion.

Key amino acid changes in viral proteins have significant impact on epitope binding affinity, which changes viral behavior. With this hypothesis, we compared the epitopes with high mean HBA in four proteins across two variants and a reference strain. Since no amino acid changes in the mean HBA epitopes of non-structural proteins were observed, we focused on the highly mutated spike protein for further analysis. The range and density of the median average binding affinity scores of the peptides predicted from the spike protein of the three SARS-CoV-2 viruses showed that all mean HBA TCEs of Omicron exhibited high binding affinity (Figure 4D). For comparison, the corresponding mean HBA TCEs in the other strains (HBA score IC <= 10 nM or more in some peptides) were included. We identified two key mutations in the Omicron spike protein, namely hydrophilic/amphipathic changes N856K and N969K, leading to improved HBA for the TCEs carrying lysine (Figure 4E; Table S7), indicating that these changes may actively stimulate host immune response. While the former mutation was present in one HBA TCE (853–867), the latter was present in three HBA epitopes (957–971, 958–972, 959–973). The spike protein of the Delta strain had hydrophobic to hydrophilic amino acid change (L452R), which led to low binding affinity. In comparison, no amino acid changes were observed in the Omicron variant at this position relative to the reference strain, implying that these changes appear to favor the ability of the Delta variant to evade host immunity.

We also identified key structural changes in the Omicron spike protein, leading to decreased binding affinity (higher score), implying that these changes may facilitate viral evasion of host immunity. For instance, hydrophilic amino acid changes R19T in the Omicron spike protein (5-LVLLPLVSSQCVNLTTRTQL-24) relative to that of Delta, covering six 15 mer epitopes, led to reduced HBA binding affinity for the TCEs carrying threonine (average of mean HBA changes from 23.8 to 30.7). Similarly, two mutations in the Omicron spike protein, i.e., hydrophilic changes N348D and Q352H (Delta 936-LSSTASALGKLQ**N**VVN**Q**NA QALNTLV-961; Omicron 935-LSSTASALGKLQ**D**VVN**H**NAQALNTLV-960), also led to decreased HBA (average of mean HBA from 35.2 to 49.5) for the TCEs carrying aspartic acid and histidine. The combined amino acid changes G142D in the Omicron and Delta spike proteins compared to reference and three-amino-acid deletion 143-YYH-145 in the Omicron spike protein relative to Delta/reference (Delta 136-CNDPFLDVYYHKNNKSWMES-155; Omicron 134-CNDPFLD—HKNNKSWMESEFR-153) decreased the binding affinity enormously (38.3/46.1 to 69.9), indicating that these structural changes may help virus to escape from the host cellular immunity.

4. Conclusions

Our analysis based on about 12 million SARS-CoV-2 genomic sequences belonging to five VOCs showed that structural changes, including point mutations and indels occurring among different proteins, probably contribute significantly to SARS-CoV-2 genetic diversification and the generation of new variants. However, we discovered some immunologically important point mutations that are probably driven by selection by the human immune system, while the remaining mutations likely occurred due to adaptation to the host organism. The eight drugs we studied interacted with nearly-conserved residues in the binding cavities of four proteins from different SARS-CoV-2 variants, suggesting the potential of inhibitory action against the Omicron variant. Our exploration identified the eight functional hotspots in the three conserved non-structural proteins nsp3, nsp5, and nsp12, which were predicted to be functionally more important in immune responses to SARS-CoV-2 infections than previously appreciated, as the universal vaccine based on these conserved regions can effectively induce cellular immune responses to all the SARS-CoV-2 variants. Key structural changes in the Omicron spike protein at the epitope level may contribute to the wide range of immune evasion and high level of transmissibility of this novel SARS-CoV-2 variant among humans. Our analysis suggests that development of a novel vaccine may need to consider including epitopes from non-structural proteins and the 37 immunodominant CD4 epitopes of spike protein that we identified to induce an efficient T-cell response (Table S7). Further wet-lab experiments are required to confirm the immunological implications and the potential contribution of immunodominant epitopes in T-cell functional assays as well as their role in SARS-CoV-2 clearance from infected hosts. Together, our findings provide insights into the impact of novel mutations in the Omicron variant on drug interactions and CD4 epitope binding affinity to predominant HLA alleles. Our study also affirms the requirement for continuous global genomic surveillance of circulating SARS-CoV-2 variants, such as BA.4, BA.5, and BQ.1, to better understand the spread of emerging novel variants. As the virus evolves to become seasonally endemic, it is fundamentally important to continually evaluate the SARS-CoV-2 genomic surveillance and COVID-19-control strategies.

Supplementary Materials: The following supporting information can be downloaded at: https://www.mdpi.com/article/10.3390/v14112461/s1, Figure S1: Interactions between inhibitors and target proteins of reference, Delta, and Omicron variants (column 1–3, respectively). The name of the target protein's residues taking part in hydrophobic interactions are represented in black, whereas those taking part in hydrogen bonds are represented in green. Hydrophobic interactions are represented as red, continuous lines on a semicircle, whereas the hydrogen bonds are represented as a broken green line extending between the interacting atoms of residues and the inhibitor. (A) The interactions of nsp3 with parecoxib or chlortetracycline and nsp5 with nirmatrelvir or ivermectin B1b are shown. (B) The interactions of nsp12 with sulfasalazine or remdesivir and spike with ivermectin B1a or atovaquone are shown; Figure S2: Comparison of high binding affinity TCEs from reference, Delta, and Omicron nsp3, nsp5, and nsp12; Table S1: Virus nomenclature and metadata details of the sub-dataset of globally circulating SARS-CoV-2 belonging to five VOCs; Table S2: Common and unique mutations of spike protein among the VOC, with emphasis on Omicron variant. The nature of the mutations in each VOC are identified in comparison with a reference strain sequence (EPI_ISL_402124). This table summarizes common and unique mutations to their respective domains (NTD, N-terminal domain; RBD, receptor binding domain; HR1 and HR-2, heptad repeats-1 and -2; FP, fusion peptide; TM, transmembrane; CT, cytoplasmic tail) in the tertiary monomeric structure of the spike protein. Uncharacterized amino acids are represented as "X". Insertions (prefix "ins"), deletions (suffix "del"), and point mutations are represented in this table by a single-letter abbreviation of amino acids followed by their position in the reference protein; Table S3: Common and unique mutations of envelope (E), membrane (M), and nucleocapsid (N) protein among the VOCs, with an emphasis to Omicron variant. The nature of the mutations in each VOC are identified in comparison with a reference strain sequence (EPI_ISL_402124). This table summarizes common and unique mutations to their respective domains in the tertiary monomeric structure of E, M, and N. Uncharacterized amino acids are represented as "X". Insertions (prefix "ins"), deletions (suffix "del"), and point mutations are represented in the table by a single-letter abbreviation of amino acids followed by their position in the reference protein; Table S4: Mutations, insertions, and deletions that are commonly present in the structural proteins (spike, envelope, membrane, and nucleocapsid) of the five VOCs provided along with the frequency (%) of each change in 12,002,213 VOC sequences; Table S5: Mutations, insertions, and deletions that are exclusively present in the structural proteins (spike, envelope, membrane, and nucleocapsid) of the Omicron variant provided along with the frequency (%) of each change in 6,104,697 Omicron sequences; Table S6: Mutations found in various proteins (spike, envelope, membrane, nucleocapsid, nsp1–16, ns3, ns6, ns7a, ns7b, and ns8) of the Omicron variant. The mutations are identified in comparison with a reference strain sequence (EPI_ISL_402124). This table characterizes the mutations as insertions, deletions, or point mutations. All the mutations of structural proteins (spike, envelope, membrane, and nucleocapsid) are represented by a single-letter abbreviation of amino acids followed by their position in the reference. All the mutations of non-structural proteins (nsp1–16, ns3, ns6, ns7a, ns7b, and ns8) are represented by a prefix of the protein name followed by the single-letter abbreviation of amino acids and their position in the respective amino acid sequence of the reference; Table S7: Mean high binding affinity score of TCEs predicted from the spike of all three viral variants. The starting and ending positions of the epitopes in the spike, mean HBA score, and amino acid sequences of the epitopes are provided.

Author Contributions: A.R. and V.A. conceptualized the study; D.C. and A.R. performed bioinformatics analysis of the genome data; D.C., N.C., A.V.J., A.J., S.S., R.M., V.A. and A.R. involved in data interpretation; D.C., A.R. and V.A. wrote the manuscript with input from all authors. All authors have read and agreed to the published version of the manuscript.

Funding: D.C., R.M. and A.R. are supported by the Tata Institute for Genetics and Society. This study is partly supported by National Institute of Health (NIH) awards 1R01EY032149-01, 1R01DK132735-01, and 5R01AI163216-02 to V.A.

Institutional Review Board Statement: Not applicable.

Informed Consent Statement: Not applicable.

Acknowledgments: We gratefully acknowledge the authors as well as the originating and submitting Laboratories for their sequence and metadata shared through GISAID and GenBank on which this in silico research analysis is based.

Conflicts of Interest: The authors have declared that no competing interests exist.

References

1. Huang, C.; Wang, Y.; Li, X.; Ren, L.; Zhao, J.; Hu, Y.; Zhang, L.; Fan, G.; Xu, J.; Gu, X.; et al. Clinical Features of Patients Infected with 2019 Novel Coronavirus in Wuhan, China. *Lancet* **2020**, *395*, 497–506. [CrossRef]
2. Ramaiah, A.; Arumugaswami, V. Insights into Cross-Species Evolution of Novel Human Coronavirus SARS-CoV-2 and Defining Immune Determinants for Vaccine Development. *bioRxiv* **2020**. 2020.01.29.925867. [CrossRef]
3. Kannan, S.R.; Spratt, A.N.; Sharma, K.; Chand, H.S.; Byrareddy, S.N.; Singh, K. Omicron SARS-CoV-2 Variant: Unique Features and Their Impact on Pre-Existing Antibodies. *J. Autoimmun.* **2022**, *126*, 102779. [CrossRef] [PubMed]
4. WHO Coronavirus (COVID-19) Dashboard. Available online: https://covid19.who.int/?adgroupsurvey=\{adgroupsurvey\}&gclid=Cj0KCQjw7KqZBhCBARIsAI-fTKJFN2-Suk6nKcfohSYp0SPeXye_qAQ5Wv-ou27miS3x3RhSkhh9SnMaAlTpEALw_wcB (accessed on 25 August 2022).
5. Tegally, H.; Wilkinson, E.; Giovanetti, M.; Iranzadeh, A.; Fonseca, V.; Giandhari, J.; Doolabh, D.; Pillay, S.; San, E.J.; Msomi, N.; et al. Detection of a SARS-CoV-2 variant of concern in South Africa. *Nature* **2021**, *592*, 438–443. [CrossRef] [PubMed]
6. Cherian, S.; Potdar, V.; Jadhav, S.; Yadav, P.; Gupta, N.; Das, M.; Rakshit, P.; Singh, S.; Abraham, P.; Panda, S.; et al. SARS-CoV-2 Spike Mutations, L452R, T478K, E484Q and P681R, in the Second Wave of COVID-19 in Maharashtra, India. *Microorganisms* **2021**, *9*, 1542. [CrossRef] [PubMed]
7. Edara, V.-V.; Pinsky, B.A.; Suthar, M.S.; Lai, L.; Davis-Garner, M.E.; Floyd, K.; Flowers, M.W.; Wrammert, J.; Hussaini, L.; Ciric, C.R.; et al. Infection and Vaccine-Induced Neutralizing Antibody Responses to the SARS-CoV-2 B.1.617.1 Variants. *N. Engl. J. Med.* **2021**, *385*, 664–666. [CrossRef]
8. Hoffmann, M.; Hofmann-Winkler, H.; Krüger, N.; Kempf, A.; Nehlmeier, I.; Graichen, L.; Arora, P.; Sidarovich, A.; Moldenhauer, A.-S.; Winkler, M.S.; et al. SARS-CoV-2 Variant B.1.617 Is Resistant to Bamlanivimab and Evades Antibodies Induced by Infection and Vaccination. *Cell Rep.* **2021**, *36*, 109415. [CrossRef]
9. Leung, K.; Shum, M.H.; Leung, G.M.; Lam, T.T.; Wu, J.T. Early Transmissibility Assessment of the N501Y Mutant Strains of SARS-CoV-2 in the United Kingdom, October to November 2020. *Eurosurveillance* **2021**, *26*, 2002106. [CrossRef]

10. Mlcochova, P.; Kemp, S.A.; Dhar, M.S.; Papa, G.; Meng, B.; Ferreira, I.A.T.M.; Datir, R.; Collier, D.A.; Albecka, A.; Singh, S.; et al. SARS-CoV-2 B.1.617.2 Delta Variant Replication and Immune Evasion. *Nature* **2021**, *599*, 114–119. [CrossRef]
11. Yadav, P.D.; Sapkal, G.N.; Abraham, P.; Ella, R.; Deshpande, G.; Patil, D.Y.; Nyayanit, D.A.; Gupta, N.; Sahay, R.R.; Shete, A.M.; et al. Neutralization of Variant under Investigation B.1.617 with Sera of BBV152 Vaccinees. *Clin. Infect. Dis.* **2021**, *74*, 366–368. [CrossRef]
12. Faria, N.R.; Claro, I.M.; Candido, D.; Franco, L.A.M.; Andrade, P.S.; Coletti, T.M.; Silva, C.A.M.; Sales, F.C.; Manuli, E.R.; Aguiar, R.S.; et al. Genomic Characterisation of an Emergent SARS-CoV-2 Lineage in Manaus: Preliminary Findings. Available online: https://virological.org/t/genomic-characterisation-of-an-emergent-sars-cov-2-lineage-in-manaus-preliminary-findings/586 (accessed on 5 January 2022).
13. CDC. COVID-19. Available online: https://public4.pagefreezer.com/browse/CDC%20Covid%20Pages/15-07-2022T12:20/https://www.cdc.gov/coronavirus/2019-ncov/science/science-briefs/scientific-brief-omicron-variant.html (accessed on 20 December 2021).
14. Espenhain, L.; Funk, T.; Overvad, M.; Edslev, S.M.; Fonager, J.; Ingham, A.C.; Rasmussen, M.; Madsen, S.L.; Espersen, C.H.; Sieber, R.N.; et al. Epidemiological Characterisation of the First 785 SARS-CoV-2 Omicron Variant Cases in Denmark, December 2021. *Eurosurveillance* **2021**, *26*, 2101146. [CrossRef] [PubMed]
15. Wang, L.; Cheng, G. Sequence Analysis of the Emerging SARS-CoV-2 Variant Omicron in South Africa. *J. Med. Virol.* **2021**, *94*, 1728–1733. [CrossRef] [PubMed]
16. Waterhouse, A.; Bertoni, M.; Bienert, S.; Studer, G.; Tauriello, G.; Gumienny, R.; Heer, F.T.; de Beer, T.A.; Rempfer, C.; Bordoli, L.; et al. SWISS-MODEL: Homology Modelling of Protein Structures and Complexes. *Nucleic Acids Res.* **2018**, *46*, W296–W303. [CrossRef]
17. Hadfield, J.; Megill, C.; Bell, S.M.; Huddleston, J.; Potter, B.; Callender, C.; Sagulenko, P.; Bedford, T.; Neher, R.A. Nextstrain: Real-time tracking of pathogen evolution. *Bioinformatics* **2018**, *34*, 4121–4123. [CrossRef] [PubMed]
18. Jamalipour Soufi, G.; Iravani, S. Potential Inhibitors of SARS-CoV-2: Recent Advances. *J. Drug Target.* **2020**, *29*, 349–364. [CrossRef] [PubMed]
19. Mengist, H.M.; Dilnessa, T.; Jin, T. Structural Basis of Potential Inhibitors Targeting SARS-CoV-2 Main Protease. *Front. Chem.* **2021**, *9*, 622898. [CrossRef]
20. Trott, O.; Olson, A.J. AutoDock Vina: Improving the Speed and Accuracy of Docking with a New Scoring Function, Efficient Optimization, and Multithreading. *J. Comput. Chem.* **2009**, *31*, 455–461. [CrossRef]
21. Marak, B.N.; Dowarah, J.; Khiangte, L.; Singh, V.P. Step toward Repurposing Drug Discovery for COVID-19 Therapeutics through in Silico Approach. *Drug Dev. Res.* **2020**, *82*, 374–392. [CrossRef]
22. Laskowski, R.A.; Swindells, M.B. LigPlot+: Multiple Ligand—Protein Interaction Diagrams for Drug Discovery. *J. Chem. Inf. Model.* **2011**, *51*, 2778–2786. [CrossRef]
23. Jensen, K.K.; Andreatta, M.; Marcatili, P.; Buus, S.; Greenbaum, J.A.; Yan, Z.; Sette, A.; Peters, B.; Nielsen, M. Improved Methods for Predicting Peptide Binding Affinity to MHC Class II Molecules. *Immunology* **2018**, *154*, 394–406. [CrossRef]
24. Ramaiah, A.; Koralur, M.C.; Dasch, G.A. Complexity of Type-Specific 56 KDa Antigen CD4 T-Cell Epitopes of Orientia Tsutsugamushi Strains Causing Scrub Typhus in India. *PLoS ONE* **2018**, *13*, e0196240. [CrossRef] [PubMed]
25. Barton, M.I.; MacGowan, S.A.; Kutuzov, M.A.; Dushek, O.; Barton, G.J.; van der Merwe, P.A. Effects of Common Mutations in the SARS-CoV-2 Spike RBD and Its Ligand, the Human ACE2 Receptor on Binding Affinity and Kinetics. *eLife* **2021**, *10*, e70658. [CrossRef] [PubMed]
26. Gao, S.-J.; Guo, H.; Luo, G. Omicron variant (B.1.1.529) of SARS-CoV-2, a global urgent public health altert! *J. Med. Virol.* **2022**, *94*, 1255–1256. [CrossRef] [PubMed]
27. Zhou, W.; He, P.; Li, J.; Liu, H.; Shi, M.; Yu, J.; Wei, H. Steep decline in binding capability of SARS-CoV-2 omicron variant (B.1.1.529) RBD to the antibodies in early COVID-19 convalescent sera and inactivated vaccine sera. *Viruses* **2022**, *14*, 335. [CrossRef]
28. Chakravarty, N.; Senthilnathan, T.; Paiola, S.; Gyani, P.; Cario, C.; Urena, E.; Jeysankar, A.; Jeysankar, P.; Ignatius Irudayam, J.; Subramanian, N.; et al. Neurological Pathophysiology of SARSCoV2 and Pandemic Potential RNA Viruses: A Comparative Analysis. *FEBS Lett.* **2021**, *595*, 2854–2871. [CrossRef]
29. Mariano, G.; Farthing, R.J.; Lale-Farjat, S.L.; Bergeron, J.R. Bergeron, Structural Characterization of SARSCoV2: Where We Are, and Where We Need to Be. *Front. Mol. Biosci.* **2020**, *7*, 605236. [CrossRef]
30. Pachetti, M.; Marini, B.; Benedetti, F.; Giudici, F.; Mauro, E.; Storici, P.; Masciovecchio, C.; Angeletti, S.; Ciccozzi, M.; Gallo, R.C.; et al. Emerging SARSCoV2 Mutation Hot Spots Include a Novel RNA-dependent-RNA Polymerase Variant. *J. Transl. Med.* **2020**, *18*, 179. [CrossRef]
31. Ullrich, S.; Nitsche, C. The SARSCoV2 Main Protease as Drug Target. *Bioorganic Med. Chem. Lett.* **2020**, *30*, 127377. [CrossRef]

viruses

MDPI

Brief Report

Host Cell Entry and Neutralization Sensitivity of SARS-CoV-2 Lineages B.1.620 and R.1

Anzhalika Sidarovich [1,2], Nadine Krüger [1], Cheila Rocha [1,2], Luise Graichen [1,2], Amy Kempf [1,2], Inga Nehlmeier [1], Martin Lier [3], Anne Cossmann [4], Metodi V. Stankov [4,5], Sebastian R. Schulz [6], Georg M. N. Behrens [4,5,7], Hans-Martin Jäck [6], Stefan Pöhlmann [1,2,*] and Markus Hoffmann [1,2,*]

1 Infection Biology Unit, German Primate Center, 37077 Göttingen, Germany
2 Faculty of Biology and Psychology, Georg-August-University Göttingen, 37073 Göttingen, Germany
3 Department of Anesthesiology, University of Göttingen Medical Center, Georg-August University of Göttingen, Robert-Koch-Straße 40, 37075 Göttingen, Germany
4 Department for Rheumatology and Immunology, Hannover Medical School, 30625 Hannover, Germany
5 German Centre for Infection Research (DZIF), Partner Site Hannover-Braunschweig, 30625 Hannover, Germany
6 Division of Molecular Immunology, Department of Internal Medicine 3, Friedrich-Alexander University of Erlangen-Nürnberg, 91054 Erlangen, Germany
7 Centre for Individualized Infection Medicine (CiiM), Feodor-Lynen-Straße 7, 30625 Hannover, Germany
* Correspondence: spoehlmann@dpz.eu (S.P.); mhoffman@dpz.eu (M.H.)

Abstract: The spike (S) protein of severe acute respiratory syndrome coronavirus 2 (SARS-CoV-2) facilitates viral entry into host cells and is the key target for neutralizing antibodies. The SARS-CoV-2 lineage B.1.620 carries fifteen mutations in the S protein and is spread in Africa, the US and Europe, while lineage R.1 harbors four mutations in S and infections were observed in several countries, particularly Japan and the US. However, the impact of the mutations in B.1.620 and R.1 S proteins on antibody-mediated neutralization and host cell entry are largely unknown. Here, we report that these mutations are compatible with robust ACE2 binding and entry into cell lines, and they markedly reduce neutralization by vaccine-induced antibodies. Our results reveal evasion of neutralizing antibodies by B.1.620 and R.1, which might have contributed to the spread of these lineages.

Keywords: SARS-CoV-2; spike protein; B.1.620; R.1; cell entry; neutralization; antibody evasion; ACE2 binding

Citation: Sidarovich, A.; Krüger, N.; Rocha, C.; Graichen, L.; Kempf, A.; Nehlmeier, I.; Lier, M.; Cossmann, A.; Stankov, M.V.; Schulz, S.R.; et al. Host Cell Entry and Neutralization Sensitivity of SARS-CoV-2 Lineages B.1.620 and R.1. *Viruses* **2022**, *14*, 2475. https://doi.org/10.3390/v14112475

Academic Editors: Ahmed El-Shamy and Mohamed Ibrahim

Received: 13 October 2022
Accepted: 4 November 2022
Published: 9 November 2022

1. Introduction

The severe acute respiratory syndrome coronavirus 2 (SARS-CoV-2) is responsible for the coronavirus disease 2019 (COVID-19) pandemic. Vaccines protect against severe COVID-19, and vaccine-induced neutralizing antibodies are believed to be important for protection [1–3]. Furthermore, recombinant, monoclonal neutralizing antibodies are used for COVID-19 treatment [4,5]. The viral spike (S) protein employs the cellular receptor ACE2 [6,7] and an S protein-activating cellular protease (TMPRSS2 or cathepsin L) for host cell entry. Importantly, the S protein interface with ACE2 is a key target for neutralizing antibodies [8]. Mutations in the S proteins of emerging SARS-CoV-2 lineages can allow evasion of neutralizing antibodies and may alter virus–host cell interactions during viral entry, thereby potentially modulating viral transmissibility. However, the S proteins of several SARS-CoV-2 lineages remain to be analyzed for their capacity to mediate viral entry and their neutralization sensitivity. Here, we analyzed the S proteins of lineages B.1.620 and R.1.

2. Materials and Methods

2.1. Cell Culture

HEK-293T (human, female, kidney; ACC-635, DSMZ; RRID: CVCL_0063), Vero (African green monkey kidney, female, kidney; CRL-1586, ATCC; RRID: CVCL_0574, kindly provided by Andrea Maisner) and Huh-7 cells (human, male, liver; JCRB Cat# JCRB0403; RRID: CVCL_0336, kindly provided by Thomas Pietschmann) were maintained in Dulbecco's modified Eagle medium (DMEM, PAN-Biotech, Aidenbach, Germany). Calu-3 (human, male, lung; HTB-55, ATCC; RRID: CVCL_0609, kindly provided by Stephan Ludwig) and Caco-2 cells (human, male, colon; HTB-37, ATCC, RRID: CVCL_0025) were maintained in minimum essential medium (Thermo Fisher Scientific, Waltham, MA, USA). All media were supplemented with 10% fetal bovine serum (Biochrom, Berlin, Germany) and 100 U/mL penicillin and 0.1 mg/mL streptomycin (PAA Laboratories GmbH, Cölbe, Germany). Furthermore, Calu-3 and Caco-2 cells received 1× non-essential amino acid solution (from 100× stock, PAA Laboratories GmbH) and 1 mM sodium pyruvate (Thermo Fisher Scientific). All cell lines were incubated at 37 °C in a humidified atmosphere containing 5% CO_2. Cell lines were validated by STR-typing, amplification and sequencing of a fragment of the cytochrome c oxidase gene, and/or microscopic examination with respect to their growth characteristics. In addition, cell lines were regularly tested for mycoplasma contamination. Transfection of cells was carried out by calcium-phosphate precipitation.

2.2. Plasmids

Plasmids encoding DsRed, VSV-G (vesicular stomatitis virus glycoprotein), SARS-CoV-2 S B.1 (codon optimized, contains C-terminal truncation of 18 amino acid), SARS-CoV-2 S B.1.617.2, and soluble human ACE2 (angiotensin-converting enzyme 2) have been previously described [9–12]. Spike (S) mutations of SARS-CoV-2 lineage B.1.620 (GISAID Accession ID: EPI_ISL_1540680) and R.1 (GISAID Accession ID: EPI_ISL_3183767) were introduced into the expression plasmid for the S protein of SARS-CoV-2 B.1 by hybrid PCR using overlapping primers. PCR products purified from an agarose gel (NucleoSpin Gel and PCR Clean-up, Macherey-Nagel, Düren, Germany) were mixed and subjected to PCR with primers corresponding to the 3′ and 5′ ends full-length S protein sequence. Generated open reading frames were ligated with linearized pCG1 plasmid (kindly provided by Roberto Cattaneo, Mayo Clinic College of Medicine, Rochester, MN, USA). All S protein sequences were verified by sequencing (Microsynth SeqLab, Göttingen, Germany).

2.3. Production of Pseudotype Particles

Production of rhabdoviral pseudotypes bearing SARS-CoV-2 spike protein has been previously described [13]. In brief, 293T cells were transfected with expression plasmid for SARS-CoV-2 S protein, VSV-G or control plasmid by calcium-phosphate precipitation. At 24 h posttransfection, cells were inoculated with VSV*ΔG-FLuc [14], a replication-deficient vesicular stomatitis virus that lacks the genetic information for VSV-G and instead codes for two reporter proteins, enhanced green fluorescent protein (eGFP) and firefly luciferase (FLuc) (kindly provided by Gert Zimmer) at a multiplicity of infection of 3. Following 1 h incubation, the inoculum was removed, and cells were washed with phosphate-buffered saline (PBS). Subsequently, cells received culture medium containing anti-VSV-G antibody (culture supernatant from I1-hybridoma cells; ATCC no. CRL-2700; except for cells expressing VSV-G, which received only medium) to neutralize residual input virus. After 16–18 h, the culture supernatant was harvested, separated from cellular debris by centrifugation for 10 min at 4000× g at room temperature, and the clarified supernatants were stored at −80 °C.

2.4. Analysis of Spike Protein-Mediated Cell Entry

For cell entry study, target cells were seeded in 96-well plates. At 20 h post seeding, the cells were inoculated with equal volumes of pseudotype particles. At 18 h postinoculation, pseudotype entry efficiency was quantified by measuring the activity of virus-encoded luciferase. For this, cells were lysed using PBS containing 0.5% triton X-100 (Carl Roth, Karlsruhe, Germany) for 30 min at RT. Afterwards, cell lysates were transferred into white 96-well plates and mixed with luciferase substrate (Beetle-Juice, PJK, Kleinblittersdorf, Germany) before luminescence was measured using a Hidex Sense Plate luminometer (Hidex, Turku, Finland).

2.5. Production of Soluble ACE2

The production of soluble human ACE2 equipped with the Fc-portion of human immunoglobulin G at the C-terminus (solACE2-Fc) has been described in detail previously [15]. Briefly, 293T cells were seeded and transfected with expression plasmid for soluble hACE2. After overnight incubation, the medium was replaced, and the cells further incubated for 38 h before the supernatant was collected and centrifuged. The clarified supernatant was concentrated (100×) using a Vivaspin protein concentrator column (molecular weight cut-off of 30 kDa; Sartorius, Göttingen, Germany). The concentrated soluble ACE2 was stored at −80 °C.

2.6. Analysis of ACE2 Binding by Flow Cytometry

In order to test the binding of the different S proteins to ACE2, 293T cells were seeded in 6-well plates and transfected with expression plasmid for the respective SARS-CoV-2 S protein by calcium-phosphate precipitation. Cells transfected with empty plasmid served as a negative control. At 24 h posttransfection, the medium was replaced. At 48 h posttransfection, the culture medium was removed, cells were resuspended in PBS, transferred into 1.5 mL reaction tubes and pelleted by centrifugation. All centrifugation steps were carried out at room temperature at 600× *g* for 5 min. Subsequently, the supernatant was aspirated and the cells were washed with PBS containing 1% bovine serum albumin (BSA, PBS-B) and pelleted by centrifugation. Next, cell pellets were resuspended in 250 µL PBS-B containing soluble hACE2-Fc (1:200) and rotated at 4 °C for 60 min using a Rotospin rotator disk (IKA). Then, cells were pelleted, washed and resuspended in 250 µL PBS-B containing anti-human AlexaFlour-488-conjugated antibody (1:200; Thermo Fisher Scientific) and rotated again for 60 min at 4 °C. Finally, the cells were washed with PBS-B, resuspended in 100 µL PBS-B and subjected to flow cytometric analysis using an ID7000 Spectral Cell Analyzer (Sony Biotechnology, San Jose, CA, USA). Median channel fluorescence data were further analyzed using the ID7000 software.

2.7. Collection of Serum and Plasma Samples

Healthcare professionals vaccinated with either two doses of the mRNA vaccine BNT162b2 (BNT) or a first dose of the vectored vaccine AZD1222 (AZ) followed by a second dose of BNT were recruited as part of prospective studies investigating seroconversion within the healthcare system (e.g., CoCo (COVID-19 Contact) study, https://www.cocostudie.de/, accessed on 1 October 2022). Specific details on the samples can be found in Table S1. Serum samples were heat-inactivated at 56 °C for 30 min prior to neutralization experiments.

2.8. Neutralization Assay

For neutralization assay, S protein bearing pseudotype particles were pre-incubated at 37 °C for 30 min in the presence of different concentrations of monoclonal antibody (Casirivimab, Imdevimab, Bamlanivimab, Etesevimab, Sotrovimab or an unrelated human control IgG) (concentration spectrum: from 10 to 10–5 µg/mL). Alternatively, pseudotype particles were pre-incubated in the presence of different concentrations of plasma or serum from vaccinated individuals (diluted from 1:25 to 1:6400). Following incubation, mixtures were inoculated onto Vero cells. Pseudotype particles incubated with medium served as controls. Transduction efficiency was determined at 16–18 h postinoculation as described above.

2.9. Data Analysis

Data analysis was carried out using Microsoft Excel (as part of Microsoft Office Professional Plus, version 2016, Microsoft Corporation) and GraphPad Prism version 8.3.0 (GraphPad Software). Statistical significance was assessed using either one-way ANOVA with Dunnett's post hoc test (data on S protein particle incorporation and cleavage, ACE2 binding, and S protein-driven cell entry) or the Friedman test with Dunn's comparisons test (neutralization data). Only p-values of 0.05 or lower were considered statistically significant ($p > 0.05$, not significant (ns); $p \leq 0.05$, *; $p \leq 0.01$, **; $p \leq 0.001$, ***). In order to calculate the serum/plasma dilutions that result in half-maximal inhibition of S protein-driven cell entry (neutralizing titer 50, NT50), a non-linear regression model was used.

3. Results

3.1. The Spike Proteins of SARS-CoV-2 Lineages B.1.620 and R.1 Differ Greatly with Respect to the Number of Mutations

The SARS-CoV-2 lineage B.1.620 was first observed in western and central Africa (earliest sequences in the GISAID (Global Initiative on Sharing All Influenza Data) database were reported from Senegal, Cameroon and the Central African Republic) and dispersed into neighboring countries, Asia, Europe, and North and Central America in early to mid-2021 (Figure S1A). It carries a unique combination of mutations in the S protein [16] (Figure 1A), some of which have also been observed in the variants of concern (VOC) B.1.1.7 (Alpha), B.1.351 (Beta), P.1 (Gamma) and B.1.1.529 (Omicron). The N-terminal domain (NTD) of the S protein, which contains an antigenic supersite [17–20], is heavily mutated in the B.1.620 lineage, and the mutations may reduce binding of neutralizing antibodies. Furthermore, mutations S477N and E484K, which are located in the receptor binding domain (RBD) (Figure 1A), might modulate ACE2 interactions [21,22] and reduce neutralization sensitivity to RBD-specific antibodies [23–26]. Finally the B.1.620 spike protein contains mutation D614G, which is associated with increased transmissibility [27,28], and mutation P681H, which is located at the N-terminus of the S1/S2 cleavage site but does not increase spike protein cleavage by furin [29,30]. In the first half of 2021, the SARS-CoV-2 lineage R.1 (sublineage of B.1.1.316) spread to at least 30 countries, with the majority of cases observed in Japan and the US [31,32] (Figure S1A). In Japan, its prevalence reached 40% [33], but after a short period of expansion, infections declined and R.1 was replaced by the B.1.1.7 and B.1.617.2 lineages [31,32,34]. In contrast to the B.1.620 S protein, its counterpart in R.1 is not heavily mutated. It harbors the E484K and D614G mutations described above, as well as one mutation in the NTD (W152L), which is believed to be associated with evasion of neutralizing antibodies [35,36], and one mutation in the in the S2 subunit (G769V) (Figure 1A).

3.2. The Spike Proteins of SARS-CoV-2 Lineages B.1.620 and R.1 Differ Regarding Cleavability, Particle Incorporation and ACE2 Binding Compared to the Spike Protein of the SARS-CoV-2 B.1 Lineage

We investigated host cell entry of B.1.620 and R.1 and its inhibition using rhabdoviral reporter particles pseudotyped with the respective S proteins, which are an adequate and well-established model for SARS-CoV-2 entry into cells and is inhibition by neutralizing antibodies [37]. The S proteins of B.1 (identical to the S protein of the Wuhan-Hu-1 isolate, except for the presence of mutation D614G), which circulated early in the pandemic, and B.1.617.2 (Delta variant) served as controls. Immunoblot analyses revealed that the S proteins of B.1.620 and R.1 were robustly incorporated into particles (Figure 1B), although incorporation of R.1 S protein was reduced as compared to the other S proteins studied. All S proteins were cleaved at the S1/S2 site, as expected. Cleavage efficiency of B.1.6120 and particularly R.1 S proteins was reduced, while cleavage of B.1.617.2 S protein was augmented (although this effect was not statistically significant) relative to B.1. spike (Figure 1B), in keeping with the published data [11]. Further, the S2 band of B.1.620 migrated slightly faster as compared to the other S2 bands (Figure 1B) and the underlying reasons are at present unclear. Binding of S protein expressing cells to ACE2 fused to the Fc portion of human immunoglobulin G revealed strong ACE2 binding to R.1 S protein, while binding to B.1.620 and B.1.617.2 S proteins was reduced as compared to B.1 S protein (Figure 1C). Next, we analyzed host cell entry mediated by the B.1.620 and R.1 S proteins. For this, we employed the human cell lines 293T (kidney), Huh-7 (liver), Caco-2 (colon) and Calu-3 (lung) and the African green monkey cell line Vero (kidney) as targets. The B.1.620 and R.1 S proteins mediated entry into 293T, Huh-7, Caco-2 and Calu-3 cells with similar efficiency as the B.1 S protein, while entry into Vero cells was significantly less efficient (Figure 1D and Figure S1B). Further, Calu-3 cell entry driven by the B.1.617.2 S protein was enhanced relative to B.1 S protein, in agreement with published data [11,38,39] (Figure 1D).

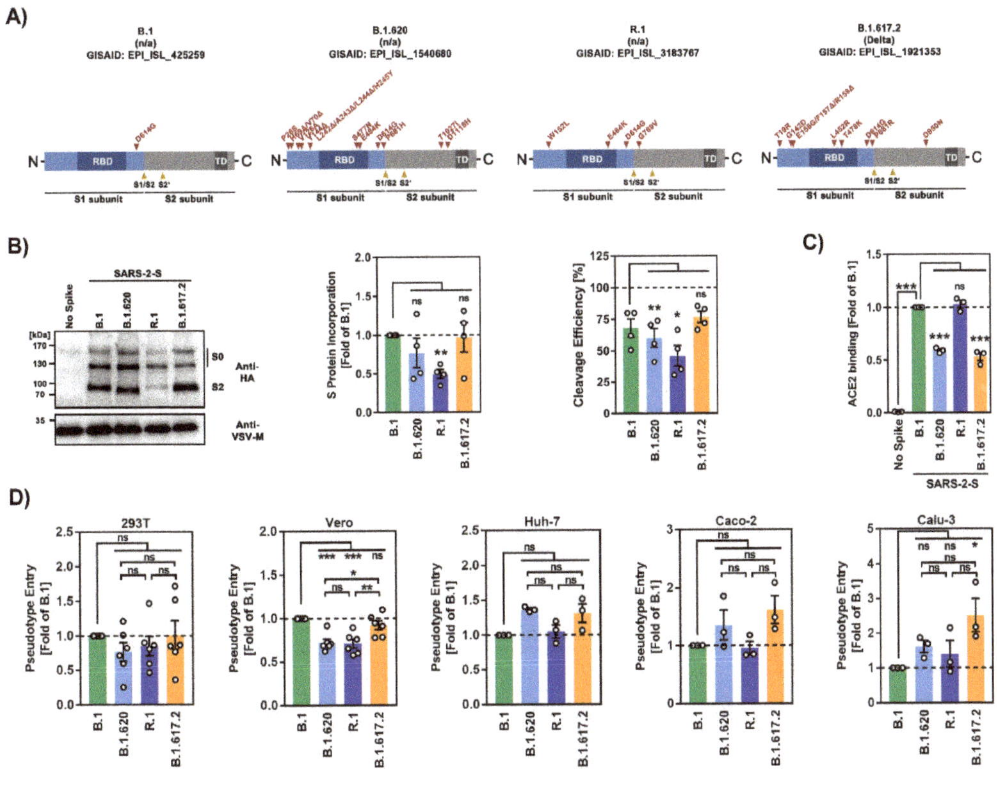

Figure 1. Spike proteins of SARS-CoV-2 lineages B.1.620 and R.1 differ regarding cleavability, particle incorporation and ACE2 binding compared to the spike protein of the SARS-CoV-2 B.1 lineage. (**A**) Schematic overview of the S protein domain organization of SARS-CoV-2 lineages B.1, B.1.620, R.1 and B.1.617.2. The location of mutations compared to the S protein of the original virus (Wuhan-Hu-01 isolate) is shown. RBD, receptor-binding domain; TD, transmembrane domain; S1/S2 and S2′, cleavage sites for host cell proteases. (**B**) Particle incorporation of SARS-CoV-2 S proteins. The incorporation of S proteins into VSV (vesicular stomatitis virus) pseudotypes was analyzed by immunoblot using an antibody against a C-terminal hemagglutinin (HA) tag (left panel). Bands corresponding to uncleaved precursor SARS-CoV-2 S protein (S0) and S2 subunit are labeled. Detection of VSV-M was used as loading control. A representative blot is shown, and similar results were obtained in four independent experiments. Total (mean) levels of SARS-CoV-2 S protein in particles were quantified with respect to the corresponding VSV-M signals and subsequently normalized (B.1 = 1, middle panel). Further, cleavage efficiency for each S protein was quantified (right panel). For this, total S protein signals were set as 100% and the relative percentage of S0 and S2 signals was determined. The

mean from four independent experiments is shown. Error bars indicate the standard error of the mean (SEM). Statistical significance of differences between WT and variant S proteins was analyzed by one-way analysis of variance (ANOVA) with Dunnett's post hoc test ($p > 0.05$, not significant (ns); $p \leq 0.05$, *; $p \leq 0.01$, **). **(C)** Strong ACE2 binding of R.1 S protein. Transfected 293T cells expressing the indicated S proteins were incubated with soluble ACE2 containing a C-terminal Fc-tag. Subsequently, the cells were stained with anti-human AlexaFlour-488-conjugated secondary antibody and subjected to flow cytometric analysis. Cells transfected with empty plasmid served as negative control and ACE2 binding was normalized against B.1 (=1). The mean data of three biological replicates is shown, error bars represent the SEM. The statistical significance of differences between WT and variant S proteins was analyzed by one-way ANOVA with Dunnett's post hoc test ($p > 0.05$, ns; $p \leq 0.001$, ***). **(D)** B.1.620 and R.1 S proteins drive efficient entry into human cell lines. Particles pseudotyped with the indicated S proteins were inoculated onto four different human cell lines (293T, Huh-7, Caco-2, Calu-3) and one African green monkey cell line (Vero). Transduction efficiency was quantified by measuring virus-encoded luciferase activity in cell lysates at 16–18 h post transduction. Presented are the mean data from three to six biological replicates (each conducted with technical quadruplicates) for which transduction was normalized against B.1 (=1). Error bars indicate the SEM. Statistical significance of differences between was analyzed by one-way ANOVA with Dunnett's post hoc test ($p > 0.05$, ns; $p \leq 0.05$, *; $p \leq 0.01$, **; $p \leq 0.001$, ***; please see also Figure S1B).

3.3. The Spike Proteins of SARS-CoV-2 Lineages B.1.620 and R.1 Display Reduced Sensitivity to Neutralization by Antibodies Induced upon Vaccination and Clinically-Used Monoclonal Antibodies

We next studied susceptibility of the B.1.620 and R.1 S proteins to antibody-mediated neutralization, employing plasma and/or serum from individuals who had either received two immunizations with the mRNA vaccine BNT162b2 (BNT), or a first dose of the vectored vaccine AZD1222 (AZ) followed by a second dose of BNT (Table S1), which represented widely used vaccination regimens in Germany at the time when B.1.620 and R.1 circulated. Neutralization of particles bearing the B.1.620 and R.1 S proteins was 3.1- and 2.1-fold, respectively, less efficient than that of particles bearing the B.1 S protein, and was comparable to that measured for particles bearing the B.1.617.2 S protein (Figure 2A). Finally, we analyzed inhibition of the B.1.620 and R.1 S proteins by monoclonal antibodies used for COVID-19 therapy (Figure 2B). Four out of five antibodies inhibited all S proteins tested efficiently and to roughly comparable levels. In contrast, all S proteins with the exception of the B.1 S protein were largely fully resistant against Bamlanivimab (Figure 2B), which is line with previous reports on B.1.617.2 [40,41] or the presence of mutation E484K in the case of B.1.620 and R.1 [42].

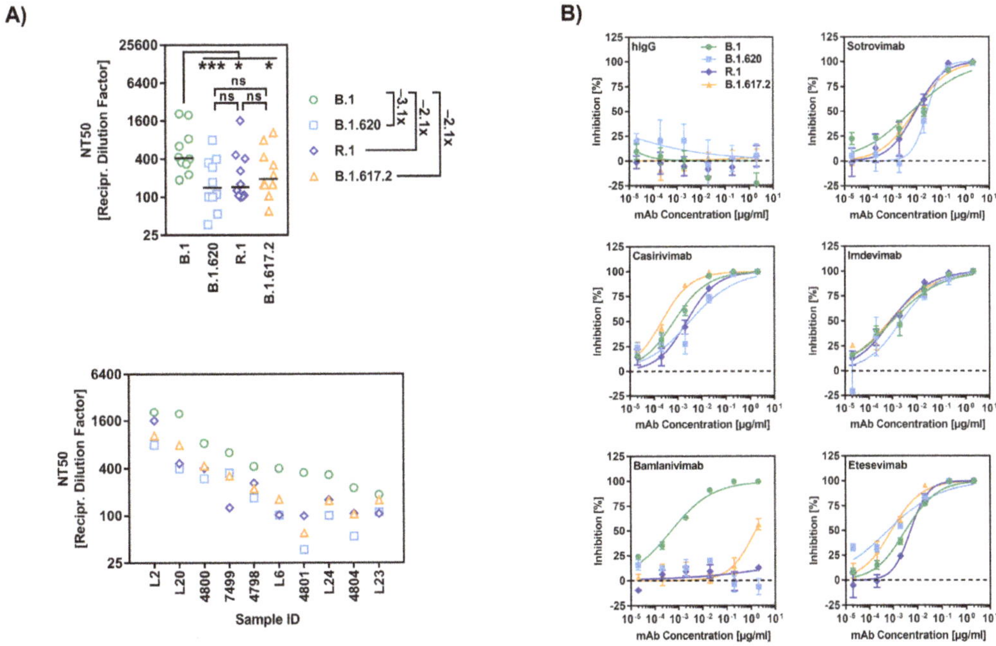

Figure 2. SARS-CoV-2 lineages B.1.620 and R.1 evade antibody-mediated neutralization. (**A**,**B**) The S proteins of SARS-CoV-2 B.1.620 and R.1 evade neutralization by antibodies induced by vaccination or employed for COVID-19 therapy. S protein-bearing particles were incubated at 37 °C for 30 min in the presence of the indicated plasma samples from BNT/BNT or AZ/BNT vaccinated individuals (panel (**A**)) or therapeutic monoclonal antibodies (panel (**B**)) before being inoculated onto Vero cells. Transduction efficiency was quantified as stated for Figure 1D and used to calculate the plasma dilution factor that leads to a 50% reduction in transduction (NT50, panel (**A**)). Data for ten serum samples from vaccinated donors are presented. Black lines indicate the median and numbers on the right represent the fold change in NT50 compared to B.1. Statistical significance of differences between individual groups was analyzed by Friedman test with Dunn's multiple comparisons test (panel (**A**); $p > 0.05$, ns; $p \leq 0.05$, *; $p \leq 0.001$, ***; please see also Figure S1C).

4. Discussion

We observed that the S proteins of B.1.620 and R.1 drive robust cell entry into various cell lines and evade antibody-mediated neutralization with similar efficiency as the B.1.617.2 S protein. Some of our observations are noteworthy:

Cleavage of the B.1.620 and particularly R.1 S proteins was less efficient, while cleavage of the B.1.617.2 S protein was slightly more efficient (not statistically significant) than the B.1. S protein. The latter phenotype might result from the presence of the P681R mutation in the B.1.617.2 S protein, which is located within the S1/S2 site and increases cleavability, transmissibility and pathogenicity [38,39]. Why the S2 band of B.1.620 S protein migrated faster during gel electrophoresis is at present unclear. However, the faster migration might reflect cleavage at a site different from the canonical S1/S2 site or altered posttranslational modifications.

The finding that binding of B.1.620 S protein to soluble ACE2 was less efficient as compared to B.1 S protein is somewhat surprising as it has been previously reported that RBD mutation S477N strengthens ACE2 binding [21], whereas RBD mutation E484K slightly reduces ACE2 interaction [43]. However, in the context of the Omicron S protein, it has also been shown that the combination of several RBD mutations that reduce ACE2 interaction with some RBD mutations that strengthen ACE2 interaction can result in an overall increase of ACE2 binding [44], and the opposite trend might be true for the combination of RBD mutations S477N and E484K. The R.1 S protein bound to ACE2 with higher efficiency as compared to the B.1 S protein, despite harboring RBD mutation E484K that is associated with a subtle reduction in ACE2 interaction efficiency [43]. Here, one can speculate that the reduced cleavage phenotype of the R.1 S protein compared to the B.1 S protein might restrict the conformational flexibility of the R.1 S protein and thus may favor a conformation required for efficient ACE2 binding. While we did not specifically test this hypothesis, it should be noted that Zhang and colleagues made a similar observation when they compared ACE2 binding of the S protein of an early SARS-CoV-2 isolate (without D614G mutation) and a mutant version thereof that contained an altered S1/S2 cleavage site and therefore was not cleaved by furin [45]. In addition, we note that ACE2 binding to B.1.617.2 S protein was less efficient than the B.1 S protein, although in a previous study we detected comparable binding [11]; subtle differences in experimental conditions might be responsible. Finally, it should be stated that staining with the neutralizing antibody Imdevimab and subsequent FACS analysis revealed robust expression of all S proteins at the cell surface (Figure S2), indicating that differences in ACE2 binding were not due to differences in S protein surface expression.

Regarding host cell entry, B.1.620 and R.1 S proteins did not mediate increased entry into any of the cell lines tested as compared to B.1 spike. In contrast, the B.1.617.2 S protein facilitated entry into Calu-3 lung cells with higher efficiency than the B.1 S protein, in keeping with our published data [11], and this phenotype was most likely due to mutation P681R. Thus, P681R increases S protein cleavage at the S1/S2 site [38,39], which is a prerequisite for S protein activation by TMPRSS2 and Calu-3 cell entry [46].

The observation that the S proteins of B.1.620 and R.1 were resistant against the therapeutic antibody Bamlanivimab is not surprising given that mutation E484K, which is present in the RBDs of both S proteins, is located in the Bamlanivimab epitope and confers Bamlanivimab resistance [47]. Further, the evasion of vaccination-induced neutralizing antibodies by B.1.620 does not come as a surprise, considering that this lineage harbors several mutations in the NTD and RBD, some of which are known to reduce antibody-mediated neutralization. In contrast, the observation that R.1 S protein evaded antibody-mediated neutralization with similar efficiency as B.1.617.2 S protein was surprising since R.1 S harbors only two additional mutations in the S1 subunit of the S protein, W152L and E484K, compared to B.1 S. The role of E484K in evasion of neutralizing antibodies is well established [26]. However, the robust evasion of neutralization by antibodies induced by BNT/BNT or AZ/BNT vaccination suggests a substantial contribution of W152L, which has also been suggested by other studies [35,36], although functional data are so far missing. Collection of serum/plasma from BNT/BNT-vaccinated individuals was carried out within one month after the second vaccination, while samples from AZ/BNT-vaccinated individuals were taken within two to four months after the second vaccination. While the discrepancy in sampling time between BNT/BNT- and AZ/BNT-vaccinated groups constitutes a limitation of this study, we did not observe differences in the extent of immune evasion by B.1.617.2, B.1.620 and R.1 for the two vaccination groups.

Collectively, our results, which await confirmation with authentic virus, suggest that B.1.620 and R.1 evade neutralizing antibodies with similar efficiency as B.1.617.2, and should thus be able to spread in an immunologically non-naïve target population.

Supplementary Materials: The following supporting information can be downloaded at: https://www.mdpi.com/article/10.3390/v14112475/s1, Figure S1: Epidemiology, cell entry and neutralization of SARS-CoV-2 lineages B.1.620 and R.1; Figure S2: Surface expression of SARS-CoV-2 lineage B.1.620 and R.1 spike proteins; Table S1: Information on vaccinee sera.

Author Contributions: Conceptualization, S.P. and M.H.; formal analysis, A.S., N.K., C.R., S.P. and M.H.; investigation, A.S., N.K., C.R., L.G., A.K. and I.N.; resources, M.L., A.C., M.V.S., S.R.S., G.M.N.B. and H.-M.J.; writing—original draft preparation, A.S., S.P. and M.H.; writing—review and editing, all authors; funding acquisition, N.K., G.M.N.B., H.-M.J. and S.P. All authors have read and agreed to the published version of the manuscript.

Funding: This research was funded by the German Federal Ministry of Education and Research (Bundesministerium für Bildung und Forschung), grant numbers 1KI2074A (NK), 01KI2006D (SP), 01KI20328A (SP), 01KX2021 (SP), 01KI2043 (H-MJ), NaFoUniMedCOVID19-COVIM: 01KX2021 (H-MJ), the Ministry for Science and Culture of Lower Saxony (Niedersächsisches Ministerium für Wissenschaft und Kultur), grant numbers 14-76103-184 (SP) and MWK HZI COVID-19 (SP), the Bavarian State Ministry for Science and the Arts (Bayerisches Staatsministerium für Wissenschaft und Kunst), grant number RTG1660 (H-MJ), the German Research Foundation (Deutsche Forschungsgemeinschaft), grant numbers PO 716/11-1 (SP), PO 716/14-1 (SP) and TRR130 (H-MJ), the German Center for Infection Research, grant number 80018019238 (GMNB), and the European Regional Development Fund, grant number Defeat Corona: ZW7-8515131 (GMNB).

Institutional Review Board Statement: Vaccinee sera were collected after approval given by the ethics committee of the University Medicine Göttingen (reference number: 8/9/20) and the Institutional Review Board of Hannover Medical School (8973_BO_K_2020, amendment 10 December 2020).

Informed Consent Statement: Written informed consent was obtained from each participant for the use of vaccinee sera for research.

Data Availability Statement: The data presented in this study are available on request from the corresponding authors.

Acknowledgments: We thank Roberto Cattaneo, Stephan Ludwig, Andrea Maisner, Thomas Pietschmann and Gert Zimmer for providing reagents. Further, we gratefully acknowledge the originating laboratories responsible for obtaining the specimens, as well as the submitting laboratories where the genome data were generated and shared via GISAID, on which this research is based.

Conflicts of Interest: A.K., I.N., S.P. and M.H. conduct contract research (testing of vaccinee sera for neutralizing activity against SARS-CoV-2) for Valneva unrelated to this work. G.M.N.B. served as advisor for Moderna and S.P. served as advisor for BioNTech, unrelated to this work. All other authors declare no competing interests.

References

1. Favresse, J.; Gillot, C.; Di Chiaro, L.; Eucher, C.; Elsen, M.; Van Eeckhoudt, S.; David, C.; Morimont, L.; Dogné, J.-M.; Douxfils, J. Neutralizing Antibodies in COVID-19 Patients and Vaccine Recipients after Two Doses of BNT162b2. *Viruses* **2021**, *13*, 1364. [CrossRef] [PubMed]
2. Baden, L.R.; El Sahly, H.M.; Essink, B.; Kotloff, K.; Frey, S.; Novak, R.; Diemert, D.; Spector, S.A.; Rouphael, N.; Creech, C.B.; et al. Efficacy and Safety of the mRNA-1273 SARS-CoV-2 Vaccine. *N. Engl. J. Med.* **2020**, *384*, 403–416. [CrossRef] [PubMed]
3. Polack, F.P.; Thomas, S.J.; Kitchin, N.; Absalon, J.; Gurtman, A.; Lockhart, S.; Perez, J.L.; Pérez Marc, G.; Moreira, E.D.; Zerbini, C.; et al. Safety and Efficacy of the BNT162b2 mRNA COVID-19 Vaccine. *N. Engl. J. Med.* **2020**, *383*, 2603–2615. [CrossRef] [PubMed]
4. Baum, A.; Ajithdoss, D.; Copin, R.; Zhou, A.; Lanza, K.; Negron, N.; Ni, M.; Wei, Y.; Mohammadi, K.; Musser, B.; et al. REGN-COV2 antibodies prevent and treat SARS-CoV-2 infection in rhesus macaques and hamsters. *Science* **2020**, *370*, 1110–1115. [CrossRef]
5. Chen, P.; Nirula, A.; Heller, B.; Gottlieb, R.L.; Boscia, J.; Morris, J.; Huhn, G.; Cardona, J.; Mocherla, B.; Stosor, V.; et al. SARS-CoV-2 Neutralizing Antibody LY-CoV555 in Outpatients with COVID-19. *N. Engl. J. Med.* **2020**, *384*, 229–237. [CrossRef]
6. Hoffmann, M.; Kleine-Weber, H.; Schroeder, S.; Kruger, N.; Herrler, T.; Erichsen, S.; Schiergens, T.S.; Herrler, G.; Wu, N.H.; Nitsche, A.; et al. SARS-CoV-2 Cell Entry Depends on ACE2 and TMPRSS2 and Is Blocked by a Clinically Proven Protease Inhibitor. *Cell* **2020**, *181*, 271–280.e278. [CrossRef]
7. Zhou, T.; Tsybovsky, Y.; Gorman, J.; Rapp, M.; Cerutti, G.; Chuang, G.-Y.; Katsamba, P.S.; Sampson, J.M.; Schön, A.; Bimela, J.; et al. Cryo-EM Structures of SARS-CoV-2 Spike without and with ACE2 Reveal a pH-Dependent Switch to Mediate Endosomal Positioning of Receptor-Binding Domains. *Cell Host Microbe* **2020**, *28*, 867–879.e865. [CrossRef]
8. Yang, Y.; Du, L. SARS-CoV-2 spike protein: A key target for eliciting persistent neutralizing antibodies. *Signal Transduct. Target. Ther.* **2021**, *6*, 95. [CrossRef]
9. Brinkmann, C.; Hoffmann, M.; Lubke, A.; Nehlmeier, I.; Kramer-Kuhl, A.; Winkler, M.; Pohlmann, S. The glycoprotein of vesicular stomatitis virus promotes release of virus-like particles from tetherin-positive cells. *PLoS ONE* **2017**, *12*, e0189073. [CrossRef]
10. Hoffmann, M.; Arora, P.; Gross, R.; Seidel, A.; Hornich, B.F.; Hahn, A.S.; Kruger, N.; Graichen, L.; Hofmann-Winkler, H.; Kempf, A.; et al. SARS-CoV-2 variants B.1.351 and P.1 escape from neutralizing antibodies. *Cell* **2021**, *184*, 2384–2393.e2312. [CrossRef]
11. Arora, P.; Sidarovich, A.; Kruger, N.; Kempf, A.; Nehlmeier, I.; Graichen, L.; Moldenhauer, A.S.; Winkler, M.S.; Schulz, S.; Jack, H.M.; et al. B.1.617.2 enters and fuses lung cells with increased efficiency and evades antibodies induced by infection and vaccination. *Cell Rep.* **2021**, *37*, 109825. [CrossRef] [PubMed]

12. Hoffmann, M.; Kruger, N.; Schulz, S.; Cossmann, A.; Rocha, C.; Kempf, A.; Nehlmeier, I.; Graichen, L.; Moldenhauer, A.S.; Winkler, M.S.; et al. The Omicron variant is highly resistant against antibody-mediated neutralization: Implications for control of the COVID-19 pandemic. *Cell* **2022**, *185*, 447–456.e411. [CrossRef] [PubMed]
13. Kleine-Weber, H.; Elzayat, M.T.; Wang, L.; Graham, B.S.; Muller, M.A.; Drosten, C.; Pohlmann, S.; Hoffmann, M. Mutations in the Spike Protein of Middle East Respiratory Syndrome Coronavirus Transmitted in Korea Increase Resistance to Antibody-Mediated Neutralization. *J. Virol.* **2019**, *93*, e01381-18. [CrossRef] [PubMed]
14. Berger Rentsch, M.; Zimmer, G. A vesicular stomatitis virus replicon-based bioassay for the rapid and sensitive determination of multi-species type I interferon. *PLoS ONE* **2011**, *6*, e25858. [CrossRef] [PubMed]
15. Hoffmann, M.; Zhang, L.; Kruger, N.; Graichen, L.; Kleine-Weber, H.; Hofmann-Winkler, H.; Kempf, A.; Nessler, S.; Riggert, J.; Winkler, M.S.; et al. SARS-CoV-2 mutations acquired in mink reduce antibody-mediated neutralization. *Cell Rep.* **2021**, *35*, 109017. [CrossRef] [PubMed]
16. Dudas, G.; Hong, S.L.; Potter, B.I.; Calvignac-Spencer, S.; Niatou-Singa, F.S.; Tombolomako, T.B.; Fuh-Neba, T.; Vickos, U.; Ulrich, M.; Leendertz, F.H.; et al. Emergence and spread of SARS-CoV-2 lineage B.1.620 with variant of concern-like mutations and deletions. *Nat. Commun.* **2021**, *12*, 5769. [CrossRef]
17. Chi, X.; Yan, R.; Zhang, J.; Zhang, G.; Zhang, Y.; Hao, M.; Zhang, Z.; Fan, P.; Dong, Y.; Yang, Y.; et al. A neutralizing human antibody binds to the N-terminal domain of the Spike protein of SARS-CoV-2. *Science* **2020**, *369*, 650–655. [CrossRef]
18. Liu, L.; Wang, P.; Nair, M.S.; Yu, J.; Rapp, M.; Wang, Q.; Luo, Y.; Chan, J.F.W.; Sahi, V.; Figueroa, A.; et al. Potent neutralizing antibodies against multiple epitopes on SARS-CoV-2 spike. *Nature* **2020**, *584*, 450–456. [CrossRef]
19. McCallum, M.; De Marco, A.; Lempp, F.A.; Tortorici, M.A.; Pinto, D.; Walls, A.C.; Beltramello, M.; Chen, A.; Liu, Z.; Zatta, F.; et al. N-terminal domain antigenic mapping reveals a site of vulnerability for SARS-CoV-2. *Cell* **2021**, *184*, 2332–2347.e2316. [CrossRef]
20. Suryadevara, N.; Shrihari, S.; Gilchuk, P.; VanBlargan, L.A.; Binshtein, E.; Zost, S.J.; Nargi, R.S.; Sutton, R.E.; Winkler, E.S.; Chen, E.C.; et al. Neutralizing and protective human monoclonal antibodies recognizing the N-terminal domain of the SARS-CoV-2 spike protein. *Cell* **2021**, *184*, 2316–2331.e2315. [CrossRef]
21. Singh, A.; Steinkellner, G.; Köchl, K.; Gruber, K.; Gruber, C.C. Serine 477 plays a crucial role in the interaction of the SARS-CoV-2 spike protein with the human receptor ACE2. *Sci. Rep.* **2021**, *11*, 4320. [CrossRef] [PubMed]
22. Mejdani, M.; Haddadi, K.; Pham, C.; Mahadevan, R. SARS-CoV-2 receptor-binding mutations and antibody contact sites. *Antib. Ther.* **2021**, *4*, 149–158. [CrossRef] [PubMed]
23. Jangra, S.; Ye, C.; Rathnasinghe, R.; Stadlbauer, D.; Alshammary, H.; Amoako, A.A.; Awawda, M.H.; Beach, K.F.; Bermúdez-González, M.C.; Chernet, R.L.; et al. SARS-CoV-2 spike E484K mutation reduces antibody neutralisation. *Lancet Microbe* **2021**, *2*, e283–e284. [CrossRef]
24. Jangra, S.; Ye, C.; Rathnasinghe, R.; Stadlbauer, D.; group, P.V.I.s.; Krammer, F.; Simon, V.; Martinez-Sobrido, L.; García-Sastre, A.; Schotsaert, M. The E484K mutation in the SARS-CoV-2 spike protein reduces but does not abolish neutralizing activity of human convalescent and post-vaccination sera. *medRxiv* **2021**, arXiv:2021.2001.2026.21250543. [CrossRef]
25. Lassaunière, R.; Polacek, C.; Fonager, J.; Bennedbæk, M.; Boding, L.; Rasmussen, M.; Fomsgaard, A. Neutralisation of the SARS-CoV-2 Delta sub-lineage AY.4.2 and B.1.617.2 + E484K by BNT162b2 mRNA vaccine-elicited sera. *medRxiv* **2021**, arXiv:2021.2011.2008.21266075. [CrossRef]
26. Chen, R.E.; Zhang, X.; Case, J.B.; Winkler, E.S.; Liu, Y.; VanBlargan, L.A.; Liu, J.; Errico, J.M.; Xie, X.; Suryadevara, N.; et al. Resistance of SARS-CoV-2 variants to neutralization by monoclonal and serum-derived polyclonal antibodies. *Nat. Med.* **2021**, *27*, 717–726. [CrossRef]
27. Plante, J.A.; Liu, Y.; Liu, J.; Xia, H.; Johnson, B.A.; Lokugamage, K.G.; Zhang, X.; Muruato, A.E.; Zou, J.; Fontes-Garfias, C.R.; et al. Spike mutation D614G alters SARS-CoV-2 fitness. *Nature* **2021**, *592*, 116–121. [CrossRef]
28. Zhou, B.; Thao, T.T.N.; Hoffmann, D.; Taddeo, A.; Ebert, N.; Labroussaa, F.; Pohlmann, A.; King, J.; Steiner, S.; Kelly, J.N.; et al. SARS-CoV-2 spike D614G change enhances replication and transmission. *Nature* **2021**, *592*, 122–127. [CrossRef]
29. Lubinski, B.; Fernandes, M.H.V.; Frazier, L.; Tang, T.; Daniel, S.; Diel, D.G.; Jaimes, J.A.; Whittaker, G.R. Functional evaluation of the P681H mutation on the proteolytic activation of the SARS-CoV-2 variant B.1.1.7 (Alpha) spike. *iScience* **2022**, *25*, 103589. [CrossRef]
30. Arora, P.; Sidarovich, A.; Graichen, L.; Hornich, B.; Hahn, A.; Hoffmann, M.; Pohlmann, S. Functional analysis of polymorphisms at the S1/S2 site of SARS-CoV-2 spike protein. *PLoS ONE* **2022**, *17*, e0265453. [CrossRef]
31. Hirotsu, Y.; Omata, M. Detection of R.1 lineage severe acute respiratory syndrome coronavirus 2 (SARS-CoV-2) with spike protein W152L/E484K/G769V mutations in Japan. *PLoS Pathog.* **2021**, *17*, e1009619. [CrossRef] [PubMed]
32. Hirotsu, Y.; Omata, M. SARS-CoV-2 B.1.1.7 lineage rapidly spreads and replaces R.1 lineage in Japan: Serial and stationary observation in a community. *Infect. Genet. Evol.* **2021**, *95*, 105088. [CrossRef] [PubMed]
33. Sekizuka, T.; Itokawa, K.; Hashino, M.; Okubo, K.; Ohnishi, A.; Goto, K.; Tsukagoshi, H.; Ehara, H.; Nomoto, R.; Ohnishi, M.; et al. A discernable increase in the severe acute respiratory syndrome coronavirus 2 R.1 lineage carrying an E484K spike protein mutation in Japan. *Infect. Genet. Evol.* **2021**, *94*, 105013. [CrossRef] [PubMed]
34. Nagano, K.; Tani-Sassa, C.; Iwasaki, Y.; Takatsuki, Y.; Yuasa, S.; Takahashi, Y.; Nakajima, J.; Sonobe, K.; Ichimura, N.; Nukui, Y.; et al. SARS-CoV-2 R.1 lineage variants that prevailed in Tokyo in March 2021. *J. Med. Virol.* **2021**, *93*, 6833–6836. [CrossRef]

35. Cavanaugh, A.M.; Fortier, S.; Lewis, P.; Arora, V.; Johnson, M.; George, K.; Tobias, J.; Lunn, S.; Miller, T.; Thoroughman, D.; et al. COVID-19 Outbreak Associated with a SARS-CoV-2 R.1 Lineage Variant in a Skilled Nursing Facility After Vaccination Program—Kentucky, March 2021. *MMWR Morb. Mortal Wkly. Rep.* **2021**, *70*, 639–643. [CrossRef]
36. Ip, J.D.; Kok, K.H.; Chan, W.M.; Chu, A.W.; Wu, W.L.; Yip, C.C.; To, W.K.; Tsang, O.T.; Leung, W.S.; Chik, T.S.; et al. Intra-host non-synonymous diversity at a neutralizing antibody epitope of SARS-CoV-2 spike protein N-terminal domain. *Clin. Microbiol. Infect.* **2021**, *27*, 1350.e1351–1350.e1355. [CrossRef]
37. Schmidt, F.; Weisblum, Y.; Muecksch, F.; Hoffmann, H.H.; Michailidis, E.; Lorenzi, J.C.C.; Mendoza, P.; Rutkowska, M.; Bednarski, E.; Gaebler, C.; et al. Measuring SARS-CoV-2 neutralizing antibody activity using pseudotyped and chimeric viruses. *J. Exp. Med.* **2020**, *217*. [CrossRef]
38. Liu, Y.; Liu, J.; Johnson, B.A.; Xia, H.; Ku, Z.; Schindewolf, C.; Widen, S.G.; An, Z.; Weaver, S.C.; Menachery, V.D.; et al. Delta spike P681R mutation enhances SARS-CoV-2 fitness over Alpha variant. *Cell Rep.* **2022**, *39*, 110829. [CrossRef]
39. Saito, A.; Irie, T.; Suzuki, R.; Maemura, T.; Nasser, H.; Uriu, K.; Kosugi, Y.; Shirakawa, K.; Sadamasu, K.; Kimura, I.; et al. Enhanced fusogenicity and pathogenicity of SARS-CoV-2 Delta P681R mutation. *Nature* **2022**, *602*, 300–306. [CrossRef]
40. Arora, P.; Kempf, A.; Nehlmeier, I.; Graichen, L.; Sidarovich, A.; Winkler, M.S.; Schulz, S.; Jack, H.M.; Stankov, M.V.; Behrens, G.M.N.; et al. Delta variant (B.1.617.2) sublineages do not show increased neutralization resistance. *Cell. Mol. Immunol.* **2021**, *18*, 2557–2559. [CrossRef]
41. Mlcochova, P.; Kemp, S.A.; Dhar, M.S.; Papa, G.; Meng, B.; Ferreira, I.; Datir, R.; Collier, D.A.; Albecka, A.; Singh, S.; et al. SARS-CoV-2 B.1.617.2 Delta variant replication and immune evasion. *Nature* **2021**, *599*, 114–119. [CrossRef] [PubMed]
42. Hoffmann, M.; Sidarovich, A.; Arora, P.; Kruger, N.; Nehlmeier, I.; Kempf, A.; Graichen, L.; Winkler, M.S.; Niemeyer, D.; Goffinet, C.; et al. Evidence for an ACE2-Independent Entry Pathway That Can Protect from Neutralization by an Antibody Used for COVID-19 Therapy. *mBio* **2022**, *13*, e0036422. [CrossRef] [PubMed]
43. Yuan, M.; Huang, D.; Lee, C.D.; Wu, N.C.; Jackson, A.M.; Zhu, X.; Liu, H.; Peng, L.; van Gils, M.J.; Sanders, R.W.; et al. Structural and functional ramifications of antigenic drift in recent SARS-CoV-2 variants. *Science* **2021**, *373*, 818–823. [CrossRef] [PubMed]
44. Mannar, D.; Saville, J.W.; Zhu, X.; Srivastava, S.S.; Berezuk, A.M.; Tuttle, K.S.; Marquez, A.C.; Sekirov, I.; Subramaniam, S. SARS-CoV-2 Omicron variant: Antibody evasion and cryo-EM structure of spike protein-ACE2 complex. *Science* **2022**, *375*, 760–764. [CrossRef] [PubMed]
45. Zhang, L.; Jackson, C.B.; Mou, H.; Ojha, A.; Peng, H.; Quinlan, B.D.; Rangarajan, E.S.; Pan, A.; Vanderheiden, A.; Suthar, M.S.; et al. SARS-CoV-2 spike-protein D614G mutation increases virion spike density and infectivity. *Nat. Commun.* **2020**, *11*, 6013. [CrossRef]
46. Hoffmann, M.; Kleine-Weber, H.; Pohlmann, S. A Multibasic Cleavage Site in the Spike Protein of SARS-CoV-2 Is Essential for Infection of Human Lung Cells. *Mol. Cell* **2020**, *78*, 779–784.e775. [CrossRef]
47. Starr, T.N.; Greaney, A.J.; Dingens, A.S.; Bloom, J.D. Complete map of SARS-CoV-2 RBD mutations that escape the monoclonal antibody LY-CoV555 and its cocktail with LY-CoV016. *Cell Rep. Med.* **2021**, *2*, 100255. [CrossRef]

viruses

MDPI

Article

SARS-CoV-2 Variants of Concern and Variations within Their Genome Architecture: Does Nucleotide Distribution and Mutation Rate Alter the Functionality and Evolution of the Virus?

Varsha Ravi [1,†], Aparna Swaminathan [1,†], Sunita Yadav [1], Hemant Arya [1] and Rajesh Pandey [1,2,*]

1 INtegrative GENomics of HOst-PathogEn (INGEN-HOPE) Laboratory, Division of Immunology and Infectious Disease Biology, CSIR-Institute of Genomics and Integrative Biology (CSIR-IGIB), Mall Road, Delhi 110007, India

2 Academy of Scientific and Innovative Research (AcSIR), Ghaziabad 201002, India

* Correspondence: rajeshp@igib.in or rajesh.p@igib.res.in; Tel.: +91-9811029551

† These authors contributed equally to this work.

Abstract: SARS-CoV-2 virus pathogenicity and transmissibility are correlated with the mutations acquired over time, giving rise to variants of concern (VOCs). Mutations can significantly influence the genetic make-up of the virus. Herein, we analyzed the SARS-CoV-2 genomes and sub-genomic nucleotide composition in relation to the mutation rate. Nucleotide percentage distributions of 1397 in-house-sequenced SARS-CoV-2 genomes were enumerated, and comparative analyses (i) within the VOCs and of (ii) recovered and mortality patients were performed. Fisher's test was carried out to highlight the significant mutations, followed by RNA secondary structure prediction and protein modeling for their functional impacts. Subsequently, a uniform dinucleotide composition of AT and GC was found across study cohorts. Notably, the N gene was observed to have a high GC percentage coupled with a relatively higher mutation rate. Functional analysis demonstrated the N gene mutations, C29144T and G29332T, to induce structural changes at the RNA level. Protein secondary structure prediction with N gene missense mutations revealed a differential composition of alpha helices, beta sheets, and coils, whereas the tertiary structure displayed no significant changes. Additionally, the N gene CTD region displayed no mutations. The analysis highlighted the importance of N protein in viral evolution with CTD as a possible target for antiviral drugs.

Keywords: molecular modeling; mutation analysis; nucleotide diversity; RNA secondary structure; VOCs

Citation: Ravi, V.; Swaminathan, A.; Yadav, S.; Arya, H.; Pandey, R. SARS-CoV-2 Variants of Concern and Variations within Their Genome Architecture: Does Nucleotide Distribution and Mutation Rate Alter the Functionality and Evolution of the Virus? *Viruses* **2022**, *14*, 2499. https://doi.org/10.3390/v14112499

Academic Editors: Ahmed El-Shamy and Mohamed Ibrahim

Received: 16 September 2022
Accepted: 9 November 2022
Published: 11 November 2022

1. Introduction

The origin of new RNA viral quasispecies pathogenic to humans, particularly severe acute respiratory syndrome coronavirus 2 (SARS-CoV-2), has made the research community come together and put massive collective efforts towards a better understanding of the viral RNA genome dynamics [1]. Despite great heterogeneity displayed by the RNA viruses within their population, higher mutation rates are not always reflected in the rapid evolution. There are conditions in which viruses replicate efficiently and yet accumulate a few viable mutations through which they can attain genome stability [2]. For instance, the 1918 H1N1 virus had evolved to suppress the CpG presence in their genome, showing a positive selection for the UpA motifs. Later, it was found that the CpG motifs can potentially trigger an antiviral response through interaction with the pattern recognition receptors (PRRs). Therefore, the selective reduction in the CpG motifs could be a viral strategy to escape the immune recognition by the host. Thus, it is important to investigate, understand, and elucidate the driving force of RNA virus heterogeneity caused by the mutations that could alter the viral genetic content [3,4]. Hence, it is vital to address questions such as what configuration of A, T/U, G, and C makes the viral genome stable? Does it vary among the different variants of a virus? What is the characteristic pattern of mutations that drives the viral RNA genome plasticity? Are these findings important for other RNA virus infections for disease severity?

The corona virus disease 2019 (COVID-19) outbreak caused by SARS-CoV-2 has been one of the most rogue pandemics in modern times, which led to unexpected infections and mortality worldwide [5]. However, the number of infected cases and mortality rates vary from country to country [6]. It is important to note that the first available genome sequence of SARS-CoV-2 placed it in the *Sarbecovirus* subgenus of the *Coronaviridae* [7]. With the genome sequence availability, it enabled an immediate analysis of its ancestry, and consequently, Zhou et al. reported a probable bat origin for the COVID-19 outbreak due to the high genetic relatedness of SARS-CoV-2 to RaTG13 [8]. Later, Makarenkov et.al suggested that the SARS-CoV-2 was a chimera of RaTG13 and pangolin coronaviruses, which reasserted that the pangolins might be an intermediate host of SARS-CoV-2 [9]. Moreover, in addition to the suggested zoonotic origin of SARS-CoV-2, several scientists proposed a potential laboratory origin of COVID-19 [10]. Although the zoonotic origin hypothesis is considered more likely, it is still not clearly defined. Hence, it is significant to study the SARS-CoV-2 genome variations due to mutations, which may help us to understand the origin of the virus.

Fortunately, worldwide sequencing and tracking of SARS-CoV-2 during the current pandemic have given us the opportunity to delve deeper into the RNA genome architecture of the virus. Consequently, the emergence of SARS-CoV-2 variants of concern (VOCs) has led to the natural selection of mutations with distinct functional consequences, which has globally augmented the propagation of the virus over the course of time [11,12]. Subsequently, the Delta variant (B.1.617.2) was observed to be more contagious, with heightened reinfections and mortality rates [13]. Furthermore, the Omicron variant was often attributed to causing a milder disease manifestation [14], but it simultaneously led to increased cases of hospital admissions in children compared to the previous waves of infections [15]. Recently, a study by Saifi et.al highlighted that the mutations associated with COVID-19 mortality patients potentially reduced the drug-binding efficiency of remdesivir in comparison with the recovered of both Delta and Omicron variant-infected patients [16]. Additionally, it is important to note that despite mass immunization of populations against the virus at an unprecedented rate, several vaccine breakthrough infection cases are being reported globally. Consequently, the spike mutations L425R,Y453F [17], and N439K [18] in the receptor-binding motif have been shown to evade the human host immune response and potentially reduce antibody neutralization.

Although intensive efforts have been made to study the spike region dynamics, the mutations outside of the spike region are also likely to contribute to the viral evolution and adaptation. In this regard, a study by Thorne et. al. reported that the Alpha variant has increased sub-genomic RNA and protein levels of the N, ORF9b, and ORF6—the genes which are known to facilitate the virus escaping the human host's innate immune response [19]. Therefore, it is essential (i) to understand the seemingly independent but interrelated genome architecture of the SARS-CoV-2 VOCs; (ii) study the evolution of the RNA genomes, which contributes to differential mutation rates; and (iii) perform evaluation of disease severity caused by different variants, which can help in understanding the variants' impacts on public health and decision making [20].

To gain insights into the viral genome architecture that can differentially distinguish the SARS-CoV-2 VOCs combined with their clinical outcomes, we comprehensively analyzed the distinct nucleotide composition that provides stability to the genome and alters the transmission rate and disease severity over the course of time. Herein, we enumerated the AT and GC composition of the SARS-CoV-2 genome at the lineage level—Alpha, Delta, and Omicron; and for the clinical outcomes—Delta recovered (DR), Delta mortality (DM), Omicron recovered (OR), and Omicron mortality (OM). Subsequently, we investigated the mutation profile of these groups and integrated it with the nucleotide composition of the

SARS-CoV-2 genome as well as different genomic regions. Further, two-tailed Fisher's test was performed to highlight the statistically significant mutations, and their functional impacts were elucidated through RNA secondary structure prediction and protein modeling. The findings highlighted the importance of the SARS-CoV-2 N gene that could serve as a potential target for antiviral drugs, which can aid us in better management of COVID-19 with future VOCs on the anvil.

2. Materials and Methods

2.1. Genome Sequencing of SARS-CoV-2

Whole genome sequencing of 1397 SARS-CoV-2 samples was performed using the Oxford Nanopore Technology (ONT, Oxford, UK) and Illumina sequencing platforms (Illumina, San Diego, CA, USA).

2.1.1. Nanopore Sequencing

A total of 310 samples were sequenced using ONT library preparation protocol—PCR tiling of SARS-CoV-2 virus with rapid barcoding (version: PCTR_9125_v110_revB_24Mar2021, ONT, Oxford, UK). In brief, 50 ng of total RNA was taken to synthesize single-stranded cDNA using LunaScript RT SuperMix (New England Biolabs, Cat. No. E3010L, Ipswich, MA, USA)). The cDNA–RNA hybrid was used to amplify the SARS-CoV-2 genome with its specific primer sets (IDT product number: 10007184) and Q5® High-Fidelity 2X master mix (New England Biolabs, Cat. No. M0494S). For sequencing library preparation, the amplified products were ligated with rapid barcode sequences (SQK-RBK110.96) followed by SPRI bead purification. The purified library was then ligated with an adapter protein and loaded on the MinION Mk1B or MinION Mk1C platform.

2.1.2. Illumina Sequencing

Sequencing library preparation was performed using Illumina COVIDSeq for the 1087 samples (Cat. No. 20043675 and reference guide: 1000000126053 v04). The total RNA was utilized to synthesize cDNA, and the viral genome was further amplified using two separate PCR reactions. The pooled amplicons then underwent tagmentation to fragment and tag amplicons with adapter sequences. This was followed by post-tagmentation cleanup, a second round of PCR amplification, and ligation of index adapters. The indexed amplicons were pooled and cleaned using the purification beads. The pooled library was then quantified using Qubit dsDNA HS Assay kit (Cat. No. Q32854). A loading concentration of 11 pM was prepared by denaturing and diluting the libraries in accordance with the MiSeq System Denature and Dilute Libraries Guide (Illumina, Document no. 15039740 v10). Sequencing was performed on the MiSeq system using the MiSeq Reagent Kit v3 (150 cycles) and 2×75 bps read length.

2.2. Sequencing Data Analysis

2.2.1. Nanopore Sequencing

The ARTIC end-to-end pipeline [21] (2021) was used for the analysis of MinION raw fast5 files up to the variant calling. Raw fast5 files of the samples were base-called and demultiplexed using Guppy basecaller, which uses the base-calling algorithms of Oxford Nanopore Technologies with a Phred quality cutoff score > 7 on a GPU-Linux accelerated computing machine. Reads with a Phred quality score of less than 7 were discarded to filter the low-quality reads. The resultant demultiplexed fastq files were normalized by a read length of 1200 (approximate size of amplicons) for further downstream analysis and aligned to the SARS-CoV-2 reference (MN908947.3) using the aligner Minimap2 v2.17 [22]. Nanopolish was used to index the raw fast5 files for variant calling from the minimap output files. To create the consensus fasta, bcftools v1.8 [23] was used over the normalized minimap2 output.

2.2.2. Illumina Sequencing

Fastqc [24] was performed for all the raw fastq files generated from the Illumina sequencing in order to check the Phred quality scores of all the sequences. A Phred quality score threshold of >20 was used for filtering reads from all the samples. Subsequently, adapter trimming was performed using the Trim Galore tool [25], and alignment of the sequences with the SARS-CoV-2 genome was performed using the HISAT2 algorithm [26,27]. BEDTools was used to generate the consensus fasta using the unaligned/filtered reads, and variant calling was performed using the high-quality reads. The sequencing depth and genome coverage for all the samples are available in the Table S1.

2.3. Data Collection and Patient Categorization

A total of 1397 SARS-CoV-2 positive nasopharyngeal RNA samples included in the study were anonymized. Moreover, all the samples had ≥80% genome coverage and an average sequencing depth of >100X (Supplementary File S1: Table S1). Further, the samples were stratified at two levels based on the lineage infected and the clinical outcome of the COVID-19 patients, recovered and mortality. Random sampling was performed at the second level of classification using python packages to avoid any sample size bias.

2.4. Nucleotide Composition Analysis

The percentage distribution of A, T/U, G, and C were calculated using the aligned Fasta sequence of the study groups: (i) VOCs—Alpha, Delta, and Omicron, and (ii) Delta recovered (DR), Delta mortality (DM), Omicron recovered (OR), and Omicron mortality (OM). Furthermore, the distributions of AT/U and GC dinucleotides were also calculated for the sample cohorts. All the calculations were performed for the whole genome of SARS-CoV-2 as well as for the different genomic regions within the SARS-CoV-2 genome (Supplementary File S2: Table S2).

2.5. Mutation and Statistical Analysis

To obtain the mutation spectra of our cohort, the VCFs of Alpha, Delta, and Omicron as well as DR, DM, OR, and OM subgroups were merged separately using bcftools [23]. Subsequently, the relative frequency of each mutation was calculated (Supplementary File S3: Table S3). To highlight the statistically significant mutations, Fisher's exact test was performed for lineage-based and clinical-outcome-based groups separately using python programming. Furthermore, the Phi correlation coefficient test using R programming was carried out to examine the associations of the significant mutations with the clinical outcomes—recovered and mortality (Supplementary File S4: Table S4).

2.6. RNA Secondary Structure Prediction

To elucidate the functional consequence of mutations, RNA secondary structure prediction was performed using the RNAfold program [28], one of the core programs of the Vienna RNA package that can predict the minimum free energy (MFE) using the dynamic programming algorithm [29]. Selective statistically significant synonymous mutations were used to determine their impact on RNA structure as compared to the wild-type SARS-CoV-2 by taking the 250 nucleotides upstream and downstream of the mutation sites. The minimum free energy (MFE) was obtained for both wild type and mutant, which is an important indicator to highlight whether the mutations affect the folding stability of the respective RNA structure (Supplementary File S5: Table S5).

2.7. Molecular Modeling and Optimization

To determine the effects of mutations on the protein structure, molecular modeling was performed for selected significant mutations. We retracted the protein Fasta sequence from Uniprot (accession id: P0DTC9.1) and modeled the protein using the Phyre Server [30]. The modeled structure was further optimized using the WHAT IF web interface [31] and validated through Ramachandran plot analysis (Supplementary File S6: Table S6).

Subsequently, the modeled protein was used for secondary structure prediction using the PDBsum online server [32]. Key mutations were incorporated in the modeled protein using PyMOL software [33]. Energy minimization is a key computational step for obtaining a minimized stable structure of the modeled protein. In this study, GROMACS software [34] was used to minimize the wild-type and mutated proteins. AMBER force field [35] was used to optimize the geometry of the modeled/generated N proteins, and the steepest descent method was used for energy minimization.

3. Results

3.1. Sample Segregation

The lineage-level stratification of our cohort (n = 1397) yielded a total of three groups—Alpha (n = 34), Delta (n = 320), and Omicron (n = 1043). Subsequently, the secondary categorization based on clinical outcome level resulted in four groups—Delta recovered (DR; n = 100), Delta mortality (DM; n = 100), Omicron recovered (OR; n = 17), and Omicron mortality (OM; n = 17) (Supplementary File S1: Table S1). It is important to mention that the selection of 17 OR samples was based on random sampling for clinical outcome-driven stratification taking gender and age into consideration (Figure 1A).

3.2. Nucleotide Composition Analysis

3.2.1. Nucleotide Distribution across the Whole Genome

The nucleotide composition analysis performed both at the lineage and outcome levels revealed a uniform distribution of A, T/U, G, and C across all the VOCs groups in comparison with the Wuhan reference genome, with T/U having the highest representation amongst all the nucleotides (~32%) (Supplementary File S2: Table S2), which was consistent with the other reported studies. Further, looking into the dinucleotide distribution, we found a uniform distribution of AT/U (61.9–62%) and GC (35.88–35.96%) within the VOCs and in comparison to the reference Wuhan strain (62.03% and 37.97%, respectively), both at the lineage and outcome levels (Figure 1B,C). Notably, though the nucleotide distribution was uniform and similar, we found a small percentage difference between the AT/GC compositions, which in terms of numbers is a handful of nucleotides that may have functional role across the VOCs, genomic regions, and clinical outcomes (Supplementary File S2). Even the microscale difference in the nucleotide abundance is important, as it could possibly be present in the specific domains of the SARS-CoV-2 genome, which can affect RNA structure and the protein domains.

3.2.2. Nucleotide Distribution across Different Genomic Regions

Furthermore, delineating from the whole genome level to different sub-genomic regions of SARS-CoV-2, the analysis highlighted a distinct composition of AT/GC across the genomic regions in the respective groups (Supplementary File S2: Table S2). Intriguingly, we found the structural gene, N, to have the highest GC content across the lineage-based groups, followed by the M and ORF3a genes. On the other hand, ORF6 had the lowest GC percentage in Alpha and Delta, whereas ORF7b was the lowest in Omicron, followed by ORF10 across all the VOCs (Figure 2A,B). The same characteristics were reflected in the recovered and mortality patients of Delta and Omicron (Figure 2C). This possibly highlights the significance of other structural genes during viral evolution in addition to the widely studied genes, ORF1ab and spike.

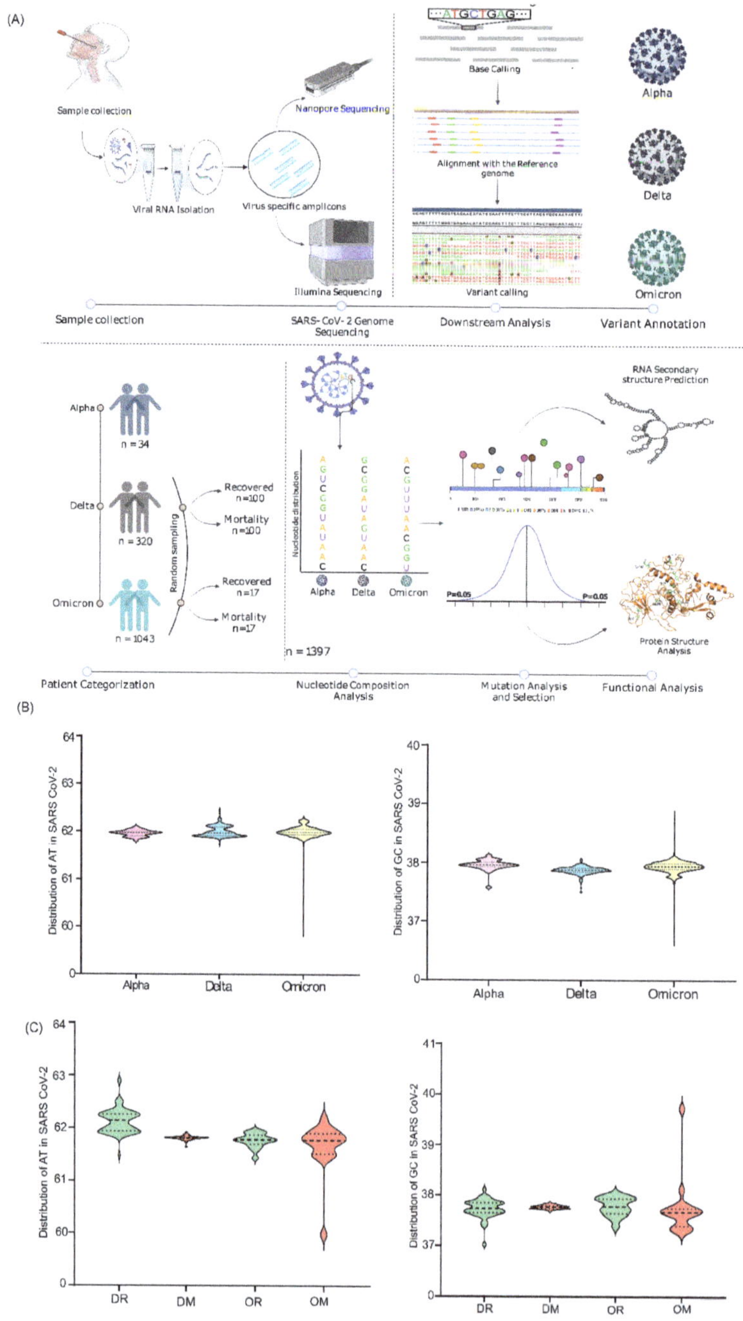

Figure 1. Study design and the Dinucleotide distribution across VOCs at the lineage and clinical outcome level. (**A**) the steps highlight the sample distribution across Alpha, Delta, and Omicron, sequencing data processing, and downstream analysis of the significant mutations. (**B**) AT/U and GC distribution in the genomes of the Alpha, Delta, and Omicron samples included in the study. (**C**) AT/U and GC distribution across the clinical outcomes with VOCs infection—DR, DM, OR, and OM.

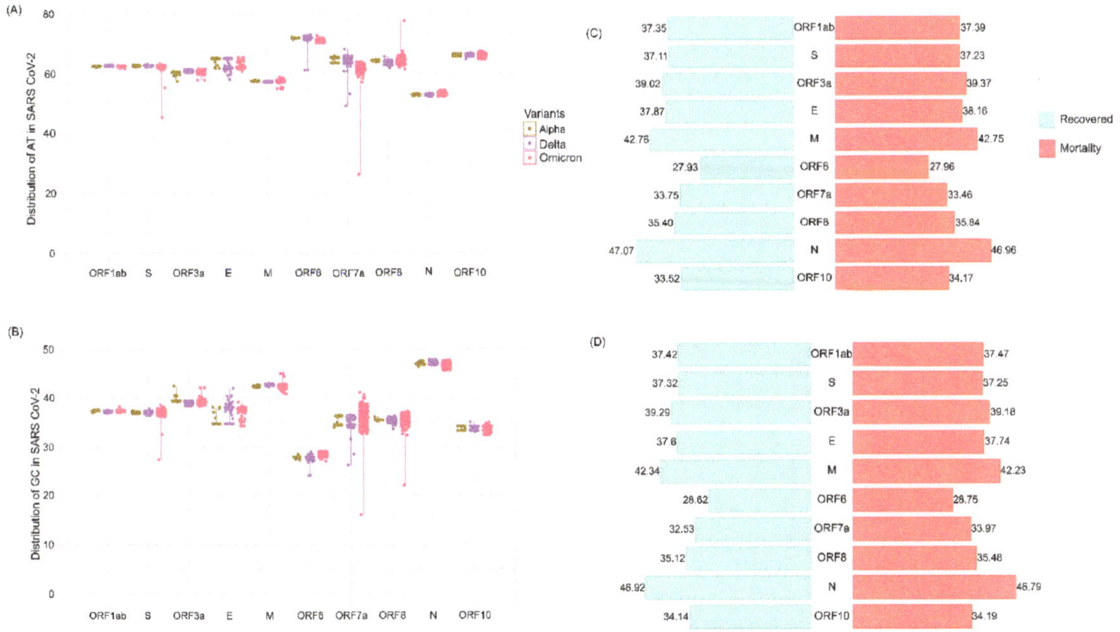

Figure 2. Dinucleotide distribution across the genomic regions of lineage- and clinical-outcome-based sub-groups. (**A**) AT/U percentage across distinct sub-genomic regions of Alpha, Delta, and Omicron. (**B**) GC percentage across different sub-genomic regions of Alpha, Delta, and Omicron. (**C**) Average GC distribution in the clinical outcome groups of Delta—Recovered and Mortality. (**D**) Average GC distribution in the clinical outcome group of Omicron—Recovered and Mortality.

3.3. Mutation Analysis Revealed High Mutation Rate in the N Gene with Highest GC Percentage

To examine the mutational diversity in our cohort, individual mutation analysis was carried out across all 1397 samples. The analysis revealed a total of 1777 mutations, of which 1718 were SNPs, 37 were deletions, and the remaining 3 were insertions (Supplementary File S3). Upon further classification of mutations as synonymous and nonsynonymous, we observed a total of 694 and 950 mutations, respectively (Figure 3A and Supplementary File S3: Table S3). To comprehend the effects of mutations vis-a-vis nucleotide distribution within genomes and the different sub-genomic regions, we classified the variations into (i) mutations that increase the GC content (GC-up) and (ii) decrease the GC content (GC-down). GC-up mutations consisted of A > G, A > C, T > C, and T > G, whereas GC-down mutations were G > A, G > T, C > A, and C > T. Here, we observed an overall higher number of GC-down mutations across the genomes, and notably, the C > T variation was present in a higher number (Figure 3B). Further examination of GC-up and GC-down mutations across the different sub-genomic regions also revealed a dominant representation of GC-down mutations compared to the GC-up mutations.

Further, to investigate the relative abundance of mutations present in different genomic regions across the lineage- and clinical-outcome-based categories, the mutations were normalized with respect to their corresponding gene length. Subsequently, these genomic regions were correlated with their average GC %age distribution. The analysis revealed two genes, N and ORF3a, to have a higher GC percentage (46.9% and 39.3%) and mutation rate (9.29% and 10.51%), respectively. Strikingly, we observed a very high mutation rate in ORF6 (10.22) even though it had the lowest GC percentage amongst all the sub-genomic regions (Figure 3C). This characteristic mutation pattern was consistent in the genomic regions of outcome-based patient categories—Delta (recovered and mortality) and Omicron (recovered and mortality) (Figure 3D). This seems to highlight that the selection pressure in the coding regions is higher at the amino acid level [36].

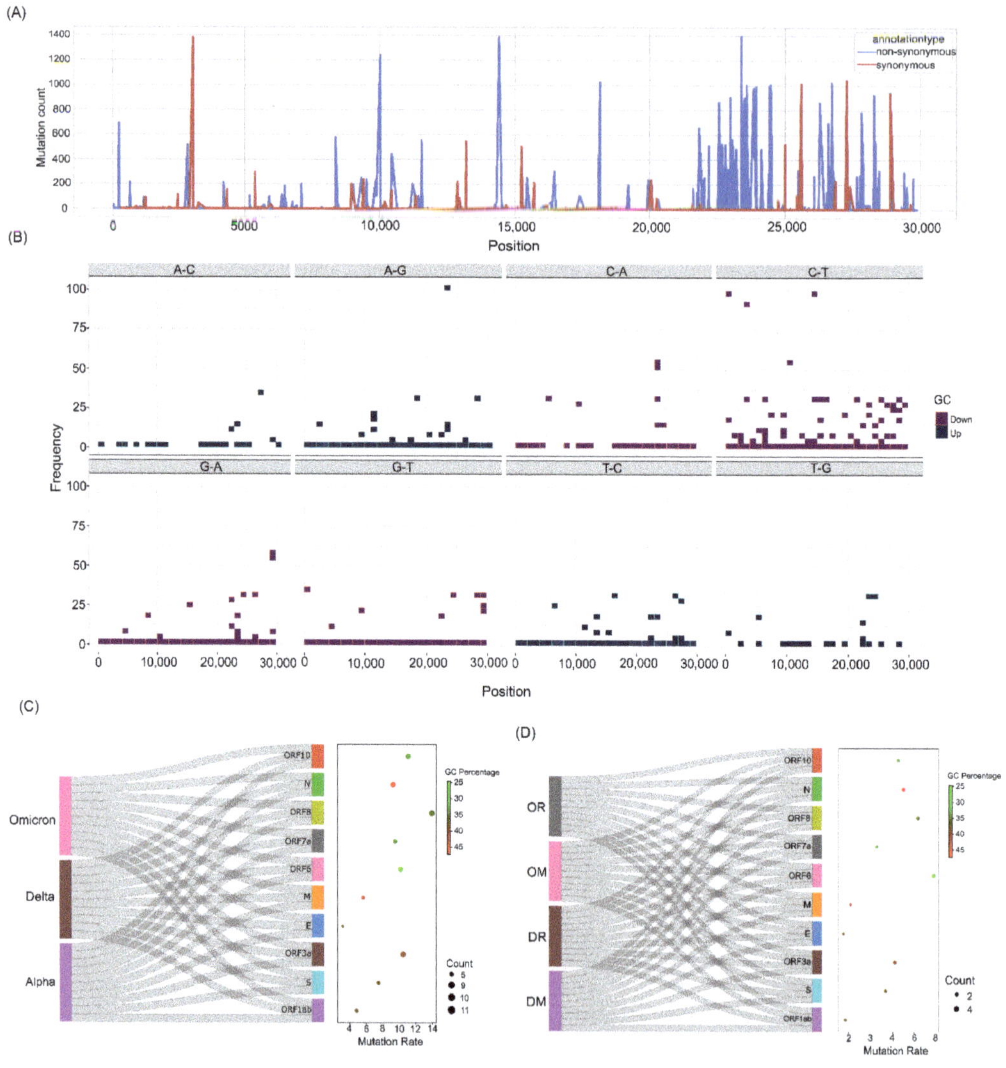

Figure 3. Mutation spectra of the samples and their correlation with GC percentage. (**A**) Synonymous and nonsynonymous mutation distribution across all the samples. (**B**) GC-up and GC-down mutations in the SARS-CoV-2 genomes. (**C**) Correlation of GC percentage and the mutation rate at the lineage level—Alpha, Delta, and Omicron. (**D**) Correlation of GC percentage and mutation rate in the clinical outcome groups—DR, DM, OR, and OM.

3.4. *Statistically Significant Mutations within the Different Categories*

To gain further insights into the mutation associations with different VOCs and clinical outcomes, Fisher's test was carried out along with Phi coefficient analysis to understand the direction of association of the mutations. Resultantly, we obtained a total of 432 mutations at the lineage level (Figure 4A), where Delta had the highest percentage of mutation distributions at 46%, followed by Alpha at 29.6% and Omicron at 24.3% (Figure 4B and Supplementary File S4: Table S4). Moreover, a higher number of mutations was predominantly seen in the ORF1ab, spike, and nucleocapsid region overall as well as within each VOC—Alpha, Delta, and Omicron. Herein, notably, we observed that three mutations, G28881A, G28882A, and G28883C, in the N gene co-occurred and were present at a high frequency in Alpha and Omicron, whereas they were missing from the Delta lineage. Though these mutations were seen in Alpha and Omicron at a considerable frequency, they were significantly associated with the Omicron lineage (p value = 7.7×10^{-225}; 6.41×10^{-202}; 1.47×10^{-204}). Interestingly, we also observed the occurrence of two different mutations at the same position 28881 in the N gene, where G28881A was significantly associated with Omicron (p value = 7.7×10^{-225}) and G28881T was significantly associated with Delta (p value = 1.53×10^{-271}).

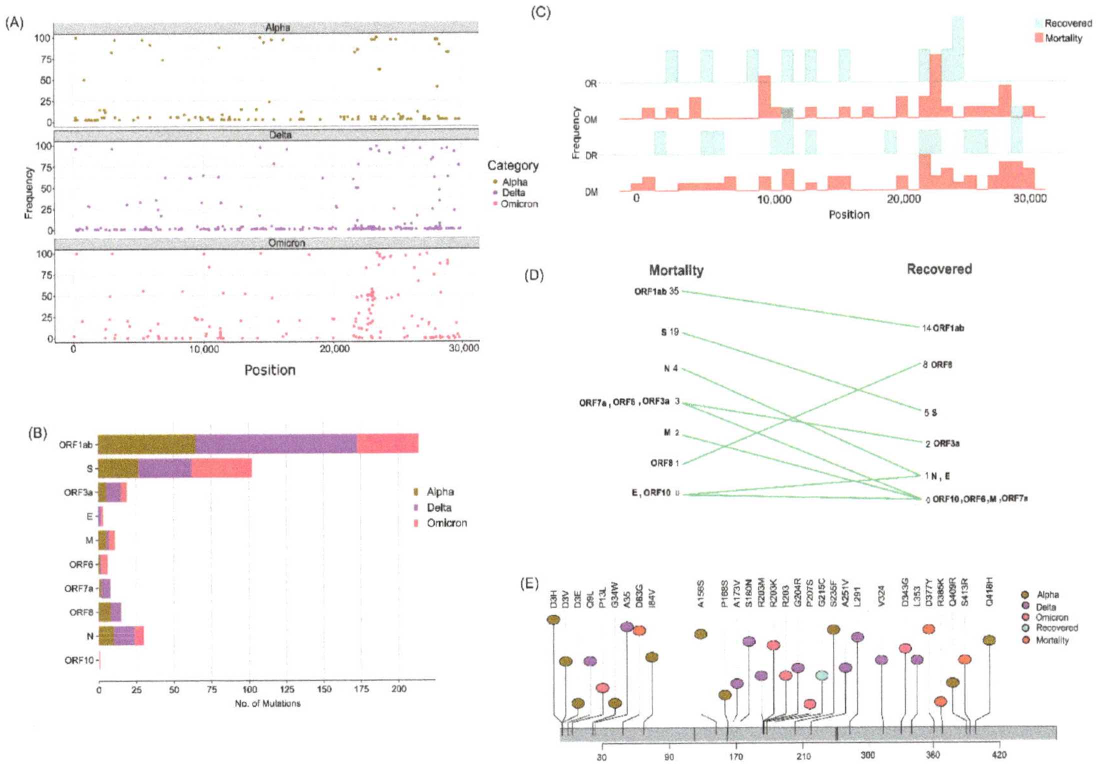

Figure 4. Significant mutations across the study cohorts in the whole genome and different sub-genomic regions. (**A**) Significant mutations associated with Alpha, Delta, and Omicron. (**B**) Mutation count across the sub-genomic regions of SARS-CoV-2. (**C**) Significant mutations associated with the clinical outcome groups of DR, DM, OR, and OM. (**D**) Mutation count in the clinical outcome groups correlated with the distinct genomic regions of SARS-CoV-2. (**E**) Lollipop plot displaying significant mutations of N gene across the study cohorts.

Although the mutation G28881A was present in Alpha at a high frequency (82.35%), it is interesting to note that it became a clade-defining mutation in the Omicron lineage with an observed mutation skip in Delta during its evolution. In contrast, the frequency of the mutation G28881T was much higher (96.25%) in Delta, whereas it was zero in Alpha and 1.24% in Omicron, thus showing a significant predominant association with the Delta lineage only. Additionally, we also noticed the co-occurrence of three other mutations—G28280C, A28281T, and T28282A—with equal high-frequency presence (91.17%) in Alpha but which became deselected in Delta and Omicron during their evolution.

Fisher's test performed at the outcome level in the Delta cohort revealed a total of 61 mutations, of which 14 mutations were significantly associated with the recovered group and 47 mutations were significantly associated with the mortality group. On the other hand, we observed 10 mutations in the Omicron recovered and 31 mutations in the Omicron mortality group (Figure 4C). Intriguingly, a higher number of mutations was associated with mortality cases in both Delta and the Omicron (Figure 4D). Additionally, the genomic regions of ORF1ab, spike, and N genes were observed to have a higher number of mutations. With the highest GC percentage and a high mutation rate in the N gene, the mutations significantly associated with the N gene were further delineated to study the functional impacts on the RNA structure and the protein. Figure 4D depicts the significant mutations in the N gene.

3.5. RNA Secondary Structure Modulated by the Specific Mutations

To elucidate the impact of mutations on the RNA secondary structure, we selected five synonymous mutations from the N gene (Figure 5A). While the mutation G28882A was significantly associated with Omicron, the rest of the four mutations were significantly associated with Delta (Figure 5A). From the analysis, we found that the RNA secondary structure is changed when C to U and G to U occur at positions 29144 and 29332, respectively. Subsequently, we observed that both G29332U mutations formed a stem instead of hairpin loop in the mutant with distinct minimum free energy (MFE) (Figure 5). Intriguingly, the mutation C29144U resulted in more negative MFE, which enhanced the RNA thermodynamic stability, while the mutation G29332U gave rise to less negative MFE, which showed a destabilizing effect on the RNA (Figure 5A). Additionally, secondary structure prediction for all the N gene synonymous mutations was performed, and MFEs were observed in the RNA secondary structures (Supplementary File S5: Table S5).

(A)

Nucleotide	Position	Start	Stop	Gene	Amino acid	Mutation type	Variant	MFE_ref	MFE_mut	Structure change
G28378T	28378	28128	28628	N	35A	Synonymous variant	Delta	−139.2	−139.2	No change
G28882A	28882	28632	29132	N	203R	Synonymous variant	Omicron	−164.9	−163.2	No change
C29144T	*29144*	*28894*	*29394*	*N*	*291L*	*Synonymous variant*	*Delta*	*−147.4*	*−148*	*Loop-stem*
C29245G	29245	28995	29495	N	324V	Synonymous variant	Delta	−149.1	149.1	No change
G29332T	*29332*	*29082*	*29582*	*N*	*353L*	*Synonymous variant*	*Delta*	*−134.5*	*−133.7*	*Loop-stem*

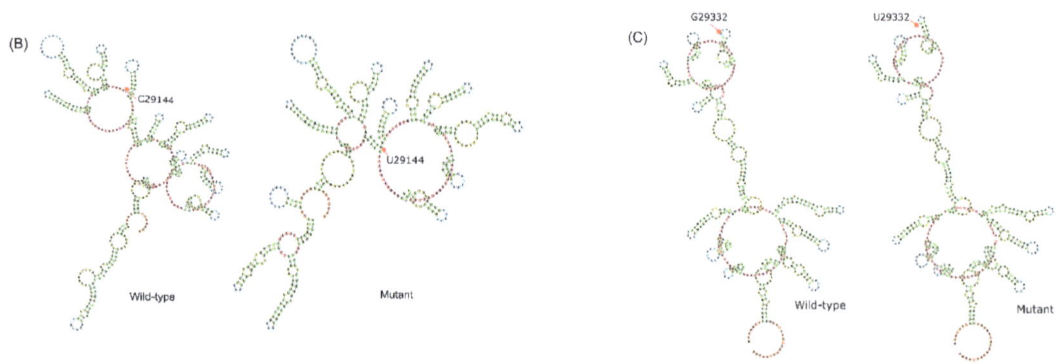

Figure 5. Synonymous mutations in the N gene and their effect on the RNA structural changes. (**A**) Table depicting the synonymous mutations of the N gene with two mutations (Orange color text) reflecting the RNA structure modulation. (**B**) RNA secondary structure of C29144T mutation, the wild type and the mutant with red arrows depicting the site of mutation. (**C**) RNA secondary structure of G29332T mutation, the wild type and the mutant with red arrows depicting the site of mutation.

3.6. Structural Modification in the N Protein

The statistically significant missense mutations from the N gene of the lineage- and the outcome-level groups (Table 1) were used for protein secondary structure prediction, using PDBsum with the default parameters. Notably, there were no significant N gene mutations in the OR group. The result showed that the wild-type N protein was made up of 16.7% helix, 7.9% beta-sheet, and 75.4% coil region. Similarly, the secondary structures of the other variants (Alpha, Delta, Omicron, DR, DM, and OM) were observed in the ranges of 16.7–18.1% (helix), 6.9–7.9% (beta-sheet), and 74–75.8% (coil region) (Figure 6A and Table 1). To understand a protein's biological function, the three-dimensional structure (arrangement of atoms of amino acids in 3D) plays an important role. Further, for tertiary structure prediction, the mutations were incorporated into the structure of the N protein (wild type), which generated six structures of different variants (Figure 6B). The structures were validated via Ramachandran plots (Supplementary File S6: Table S6) and were found to have more than 97% residues in the allowed regions. After validation, energy minimization (using GROMACS) of the modeled and the generated structures was performed and achieved the most stable conformation of the protein. The minimized N protein structures (variants) were superimposed with the wild type and were found to have RMSD in the range of 0.066–0.209 Å, which suggests that wild-type and the variants of N protein are structurally similar.

Table 1. Secondary structure changes in overall N Protein.

N Protein	Mutations Incorporated	Helix (%)	Beta Strand (%)	Coil Region (%)
Wild type	-	16.7	7.9	75.4
Alpha	D3H/V/E, G34W, I84V, A156S, P168S, S235F, Q409R, and Q148H	17.9	7.9	74.2
Delta	Q9L, D63G, A173V, S180N, R203M, P207S, G215C, A251V, D377Y, and R385K	17.4	7.9	74.7
Omicron	P13L, R203K, G204R, D343G, and S413R	18.1	7.9	74
DR	G215C	17.2	6.9	75.8
DM	D63G, D377Y, and R385K	17.2	6.9	75.8
OM	S413R	17.2	6.9	75.8

3.7. Validation of the Findings in Independent Cohorts

We carried out the same analysis using two datasets of different origins, (i) Indian but non-CSIR-IGIB, with India being a very vast country with a diverse population and geographical regions, and (ii) from the USA. A total of 1007 SARS-CoV-2 genome sequence data were retrieved from GISAID with a filter of the mentioned origins (500 sequences from India and 507 from the USA) and the timeline of December 2021 to February 2022 (Supplementary File S8: Table S8). Since our study focused on only the protein-coding genes, the same was considered to interpret the results of the USA and the Indian origin samples.

The analysis across the USA sample cohort showed the N gene to have the highest GC% (46.97%) followed by the M (42%) and ORF3a (39.32%) genes. Subsequently, the normalized mutation rate comparison analysis showed the ORF3a to have a 14.86% mutation rate, followed by the N (8.41%) and the M (6.58%) genes. A similar trend was observed in the Indian sample cohort as well, where the N gene had the highest GC content of 47.01%, followed by the M (42%) and ORF3a (39.27%) genes (Supplementary File S9: Table S9). The mutation analysis revealed a higher mutation rate in the M gene with 7.32%, followed by the N and ORF3a (6.98 and 6.76%) genes, respectively (Supplementary File S10: Table S10a–d).

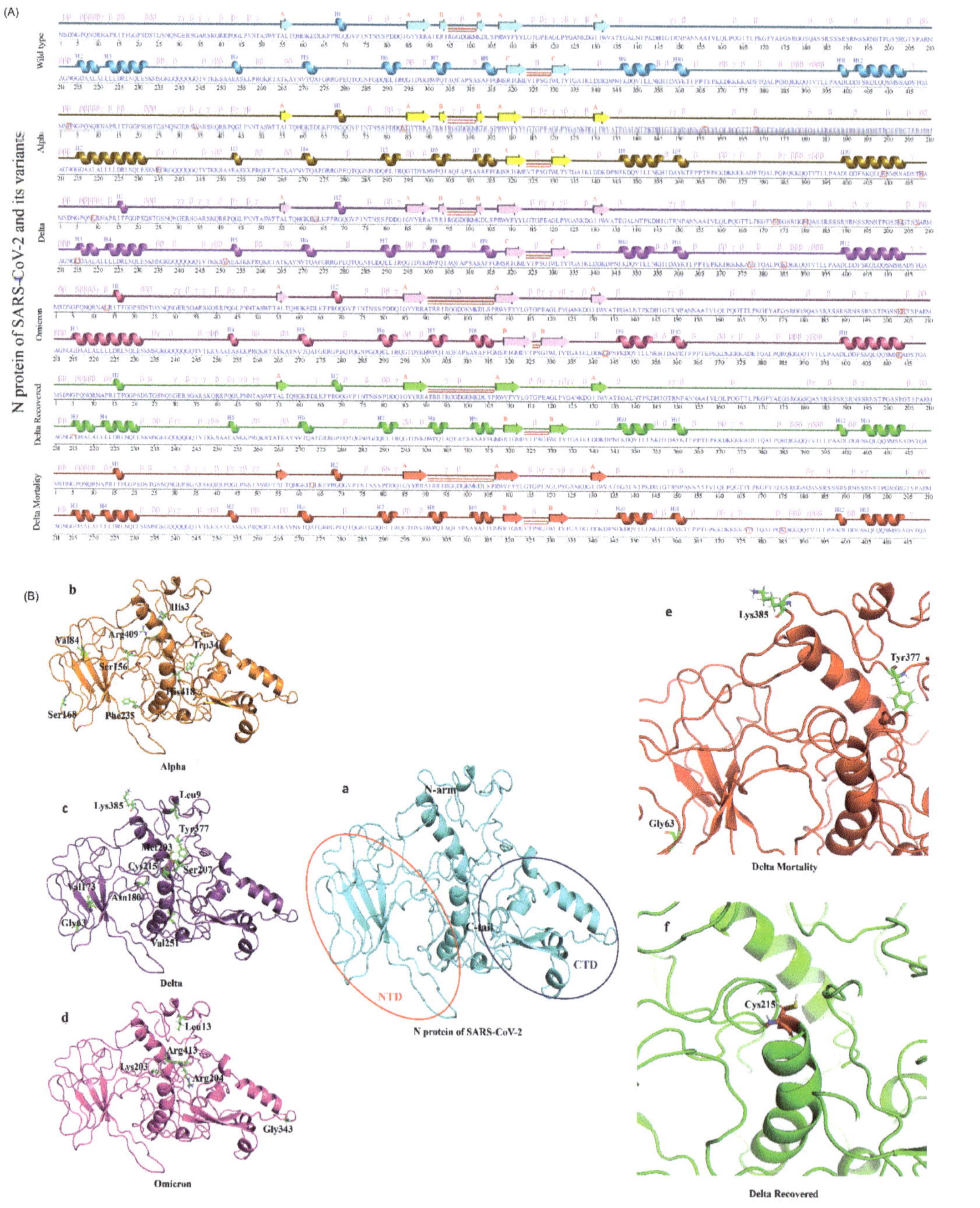

Figure 6. 2D and 3D structural changes in the N protein with respect to the variants and their clinical outcomes. (**A**) Secondary structure representation of the targeted N protein and its variants, along with the "□" represents the mutated amino acid(s). (**B**) Molecular modeling of the N protein wild type (**a**) (Red circle represents the NTD domain and the dark blue color represents the CTD domain), Alpha (**b**), Delta (**c**), Omicron (**d**), DM (**e**), and DR (**f**).

Combinedly, correlating the GC% of the respective genes and the normalized mutation rates, the N gene shows the highest GC content coupled with a high mutation rate in the SARS-CoV-2 genomes, which is consistent with the findings drawn from our in-house sample dataset of 1397 samples. Supplementary File S8 contains the metadata, Supplementary File S9 shows the average AT and GC percentage of the sample cohorts, and Supplementary File S10 shows the mutations with their normalized rates.

4. Discussion

RNA viruses exist as dynamic and diverse populations shaped by constant mutation(s) and selection for better adaptability to different micro-environments in the multitude of hosts [37]. One such RNA virus, SARS-CoV-2, caused the worldwide COVID-19 pandemic and has had devastating impacts on public health and the economy. At the same time, undoubtedly, the control measures that were taken to contain the infection, with vaccine development being the primary measure to immunize the population against the virus, progressed at an unprecedented rate as an immediate response to COVID-19. [38]. Yet, several vaccination breakthrough cases were reported and became less common with time globally, which potentially raises questions about the efficacy of the developed vaccines [39–42]. This reinforces the need to explore the multi-dimensionalities of the underlying observation, wherein it is imperative to study the pathogen characteristics leading to different disease trajectories [43–45]. With the emergence of several variants of SARS-CoV-2, which led to a big surge of infections in different populations, the central question concerns whether the viral sequences have evolved to differentially optimize genome stability through mutations that could affect the genome nucleotide composition or vice versa [1]. To gain insights into this issue, we comprehensively studied if there could be a differential nucleotide composition that provides stability to the genome to alter the transmission rate and disease severity over the course of time.

The nucleotide composition analysis across all the VOC groups at the lineage and outcome level demonstrated a higher presence of the nucleotide U, which is consistent with other reported studies [46–48]. This could be a result of natural selection because higher U content and smaller genome size can make the virus replicate more efficiently. Moreover, less host energy is required to disrupt the viral RNA secondary structures with relatively higher U content, which can make the virus more infectious, thus essentially increasing the transmission rate [49]. Consequently, we also observed that the mutations C > T and G > T were much higher in number, reflecting the SARS-CoV-2 genome bias towards a higher U content. Particularly, the C > T variation was observed in the highest number among all the study samples, possibly because C to T changes require a mere deamination of the C nucleotide. Consequently, the lower abundance of the GC dinucleotide has been widely seen in SARS-CoV-2 virus genomes and is also correlated with its moderate virulence as compared to MERS-CoV and SARS-CoV-1 [46]. Though the comparative nucleotide composition analysis demonstrated an average minimal sequence divergence across the VOC groups at the lineage and clinical outcome levels, it is imperative to explore the functional importance of the differential nucleotide numbers seen across the same groups.

Upon examining the nucleotide diversity across the distinct SARS-CoV-2 sub-genomic regions, the N gene showed the highest GC percent of around 47%, which is higher than the average GC percentage of the entire genome. The N gene is reported to be one of the important structural proteins in a virus particle that can modulate the genome transcription and the virulence [50]. Moreover, the SARS-CoV-2 N gene was observed to be highly conserved [51]. Interestingly, in our study, we found the N gene to have a higher mutation ratio after the ORF3a gene, combined with the highest GC percentage across all the groups. Furthermore, Fisher's test was performed to see the statistical significance of these mutations, which also revealed a high number of mutations in the N region of the SARS-CoV-2 genome after ORF1ab and spike—at the lineage level (Alpha, Delta, and Omicron) and the clinical outcome level (Delta—recovered and mortality; Omicron—recovered and mortality). Notably, the N gene mutations that are almost exclusive to Delta—D63G (95.6%)

and D377Y (92.05%)—were significantly associated with mortality and were also reported in breakthrough infections [52]. Strikingly, the mutation D63G was reported to be present in the recombinant strain of Delta and Omicron [53].

Elucidating the functional impact of mutations is very important in order to determine the stability of virus propagation over the course of time. In our study, we found that in the N gene, the RNA secondary structure was changed when C to T and G to T mutations occurred at positions 29144 and 29332, respectively. While the mutation C29144U stabilized the RNA structure, G29332U resulted in a comparatively unstable structure with lower negative MFE than the wild type. Interestingly, there is evidence from the previous study that the most frequent mutations to occur in the SARS-CoV-2 genomes are C to T and G to A. These mutations are related to the role of the cell-derived apolipoprotein B mRNA-editing enzyme, which leads to C-to-U deamination of a single cytidine base in the nuclear apoB transcript, introducing a translational termination. Besides that, the G29332U mutation reduced the folding stability of the RNA secondary structure, which could affect the polypeptide translation and folding. The previous studies suggested that stable RNA structures play a key role in reducing the translation speed to prevent "ribosomal traffic jams" so that the newly translated polypeptides can fold properly [54].

Furthermore, we studied the N gene missense mutations' effects on protein stability. SARS-CoV-2's N protein is a 419 AA, 45.6 kDa, positively charged, unstable hydrophobic, and poorly heat-resistant protein that is essential for virus survival [55]. In this study, the secondary and tertiary structure of the N protein (wild type and mutants) were predicated computationally, and we found that they are structurally similar. Few variations observed in the secondary structures, such as the helix (15–18 AA), were found in Delta, Omicron, DR, DM, and OM, whereas they were absent (converted into coil) in the wild type and Alpha. A beta-sheet structure (93–94 AA) was present in the wild type, Alpha, and Delta, whereas it was missing (converted into coil) in Omicron, DM, DR, and OM. Similarly, the coil region (220–222 AA and 353–356 AA) was converted into a helix in the Alpha variant. Moreover, Alpha, Delta, and Omicron had helix regions at 398–414 AA, whereas the wild type at 401–402 AA and OM, DR, and DM at 401–403 AA had coil regions between the helix structures. In three-dimensional structures, a lower deviation was observed after the superimposition of the variant's N protein structure with the wild type. Most of the mutations were found in the RNA-binding domain or NTD and intrinsically domain regions (IDRs) of the SARS-CoV-2's N protein. Interestingly, no mutations were reported in the C-terminal domain except D343G in the Omicron variant, which suggests that CTD of the N protein could be used as a potential drug-binding site. Several researchers reported that the CTD of the N protein plays an essential role in viral RNA binding, packaging of the SARS-CoV-2 viral RNA, and transcriptional regulations [56,57]. A drug repurposing approach or novel lead design technique may be used to identify/design drug molecules for the treatment of COVID-19 by targeting CTD of the N protein for the betterment of the disease. This is especially important as we foresee the continued evolution of the new SARS-CoV-2 sub-lineages with a regional and global footprint.

To understand this further, we additionally carried out recombination analysis for the 1397 whole genome sequences of in-house samples used in our study using RDP4 software (Supplementary File S7: Table S7). Resultantly, a total of two recombinant events were observed in a consensus between at least three detection methods, with 36% of the total genome sequences being detected as recombinant. For these recombination events, we noted the breakpoint positions to be in the 5′ region of the ORF1ab gene. However, it is important to note that the recombinant events observed were flagged by RDP4 as being possibly driven by other processes despite the support of at least three recombination detection methods. Furthermore, the recombinant sequences in the second recombinant event were labeled as possibly misidentified recombinant sequences by RDP4 (Supplementary File S7). It seems promising, but we did not detect any high-confidence recombination signals in the SARS-CoV-2 genomes of the VOCs—Alpha, Delta, and Omicron.

In conclusion, the results reported here show our efforts to comprehensively investigate the viral RNA dynamics vis-a-vis the mutations and their functional impacts on the virus. This can aid us in understanding the emergence and tracking of new variants along with the elucidation of different disease trajectories. Moreover, examining the immune escape mutations could possibly guide us in better designing of vaccinations and antiviral drugs to ameliorate the observed COVID-19 severity.

Supplementary Materials: The following supporting information can be downloaded at: https://www.mdpi.com/article/10.3390/v14112499/s1, Table S1: Sample-wise sequence information; Table S2: Average Percentage of AT and GC across sample cohort; Table S3: List of mutations across lineages and clinical outcomes; Table S4: Statistical significance of lineages and outcomes; Table S5: RNA Secondary Structure prediction; Table S6: Protein quality check; Table S7: Recombination analysis: Table S8: Metadata for sample cohort; Table S9: Average Percentage of AT and GC across USA and India (non-IGIB) sample cohort; Tables S10a–d: List of mutations and normalized mutation rates across USA and India cohort.

Author Contributions: V.R., A.S., S.Y., and H.A. performed analysis; A.S., V.R., and H.A. wrote the manuscript; R.P., designed, conceptualized, implemented, and coordinated the study. All authors contributed to the article and approved the final submission. All authors have read and agreed to the published version of the manuscript.

Funding: This research was funded by the Bill and Melinda Gates Foundation (BMGF), INV-033578; Foundation for Innovative New Diagnostics (FIND), GAP-0249; AIDS Healthcare Foundation (AHF), CLP-0043; and Indo-US Science and Technology Forum (IUSSTF), CLP-0033.

Institutional Review Board Statement: The studies involving human participants were reviewed and approved by CSIR-IGIB's Human Ethics Committee Clearance (Ref No: CSIR-IGIB/IHEC/2020-21/01). The patients/participants provided their written informed consent to participate in this study.

Informed Consent Statement: Informed consent was obtained from all subjects involved in the study.

Data Availability Statement: The clinical dataset collected and analyzed as part of this study is available in the Supplementary File in Table S1.

Acknowledgments: We acknowledge the role of Aradhita Baral as Research Manager and regular coordination with the funders and stakeholders towards smooth progress for achieving the deliverables. The authors also acknowledge the support of Anil Kumar and Nisha Rawat for SARS-CoV-2 sample transport and sample management.

Conflicts of Interest: The authors declare no conflict of interest, and the funders did not have a role in the planning and execution of the study.

References

1. Kockler, Z.W.; Gordenin, D.A. From RNA World to SARS-CoV-2: The Edited Story of RNA Viral Evolution. *Cells* **2021**, *10*, 1557. [CrossRef] [PubMed]
2. Steinhauer, D.A.; Holland, J.J. Rapid evolution of RNA viruses. *Annu. Rev. Microbiol.* **1987**, *41*, 409–433. [CrossRef]
3. Jimenez-Baranda, S.; Greenbaum, B.; Manches, O.; Handler, J.; Rabadán, R.; Levine, A.; Bhardwaj, N. Oligonucleotide motifs that disappear during the evolution of influenza virus in humans increase alpha interferon secretion by plasmacytoid dendritic cells. *J. Virol.* **2011**, *85*, 3893–3904. [CrossRef] [PubMed]
4. Vabret, N.; Bhardwaj, N.; Greenbaum, B.D. Sequence-Specific Sensing of Nucleic Acids. *Trends Immunol.* **2017**, *38*, 53–65. [CrossRef]
5. Dotan, A.; David, P.; Arnheim, D.; Shoenfeld, Y. The autonomic aspects of the post-COVID19 syndrome. *Autoimmun. Rev.* **2022**, *21*, 103071. [CrossRef] [PubMed]
6. Toyoshima, Y.; Nemoto, K.; Matsumoto, S.; Nakamura, Y.; Kiyotani, K. SARS-CoV-2 genomic variations associated with mortality rate of COVID-19. *J. Hum. Genet.* **2020**, *65*, 1075–1082. [CrossRef]
7. Boni, M.F.; Lemey, P.; Jiang, X.; Lam, T.T.-Y.; Perry, B.W.; Castoe, T.A.; Rambaut, A.; Robertson, D.L. Evolutionary origins of the SARS-CoV-2 sarbecovirus lineage responsible for the COVID-19 pandemic. *Nat. Microbiol.* **2020**, *5*, 1408–1417. [CrossRef] [PubMed]
8. Zhou, P.; Yang, X.-L.; Wang, X.-G.; Hu, B.; Zhang, L.; Zhang, W.; Shi, H.-R.; Zhu, Y.; Li, B.; Huang, C.-L.; et al. A pneumonia outbreak associated with a new coronavirus of probable bat origin. *Nature* **2020**, *579*, 270–273. [CrossRef]

9. Domingo, J.L. What we know and what we need to know about the origin of SARS-CoV-2. *Environ. Res.* **2021**, *200*, 111785. [CrossRef]
10. Makarenkov, V.; Mazoure, B.; Rabusseau, G.; Legendre, P. Horizontal gene transfer and recombination analysis of SARS-CoV-2 genes helps discover its close relatives and shed light on its origin. *BMC Ecol. Evo.* **2021**, *21*, 5. [CrossRef]
11. Karim, S.S.A.; Karim, Q.A. Omicron SARS-CoV-2 variant: A new chapter in the COVID-19 pandemic. *Lancet* **2021**, *398*, 2126–2128. [CrossRef]
12. Caniels, T.G.; Bontjer, I.; van der Straten, K.; Poniman, M.; Burger, J.A.; Appelman, B.; Ayesha Lavell, A.H.; Oomen, M.; Godeke, G.-J.; Valle, C.; et al. Emerging SARS-CoV-2 variants of concern evade humoral immune responses from infection and vaccination. *Sci. Adv.* **2021**, *7*, eabj5365. [CrossRef] [PubMed]
13. Tatsi, E.-B.; Filippatos, F.; Michos, A. SARS-CoV-2 variants and effectiveness of vaccines: A review of current evidence. *Epidemiol. Infect.* **2021**, *149*, e237. [CrossRef] [PubMed]
14. Abdullah, F.; Myers, J.; Basu, D.; Tintinger, G.; Ueckermann, V.; Mathebula, M.; Ramlall, R.; Spoor, S.; de Villiers, T.; Van der Walt, Z.; et al. Decreased severity of disease during the first global omicron variant covid-19 outbreak in a large hospital in tshwane, south africa. *Int. J. Infect. Dis.* **2022**, *116*, 38–42. [CrossRef]
15. New York State Department of Health. Available online: https://www.health.ny.gov/press/releases/2022/docs/pediatric_covid-19_hospitalization_report.pdf (accessed on 8 September 2022).
16. Saifi, S.; Ravi, V.; Sharma, S.; Swaminathan, A.; Chauhan, N.S.; Pandey, R. SARS-CoV-2 VOCs, Mutational diversity and clinical outcome: Are they modulating drug efficacy by altered binding strength? *Genomics* **2022**, *114*, 110466. [CrossRef]
17. Motozono, C.; Toyoda, M.; Zahradnik, J.; Saito, A.; Nasser, H.; Tan, T.S.; Ngare, N.; Kimura, I.; Uriu, K.; Kosugi, Y.; et al. SARS-CoV-2 spike L452R variant evades cellular immunity and increases infectivity. *Cell Host Microbe* **2021**, *29*, 1124–1136.e11. [CrossRef]
18. Thomson, E.C.; Rosen, L.E.; Shepherd, J.G.; Spreafico, R.; da Silva Filipe, A.; Wojcechowskyj, J.A.; Davis, C.; Piccoli, L.; Pascall, D.J.; Dillen, J.; et al. Circulating SARS-CoV-2 spike N439K variants maintain fitness while evading antibody-mediated immunity. *Cell* **2021**, *184*, 1171–1187.e20. [CrossRef]
19. Thorne, L.G.; Bouhaddou, M.; Reuschl, A.-K.; Zuliani-Alvarez, L.; Polacco, B.; Pelin, A.; Batra, J.; Whelan, M.V.X.; Hosmillo, M.; Fossati, A.; et al. Evolution of enhanced innate immune evasion by SARS-CoV-2. *Nature* **2022**, *602*, 487–495. [CrossRef]
20. Sigal, A.; Milo, R.; Jassat, W. Estimating disease severity of Omicron and Delta SARS-CoV-2 infections. *Nat. Rev. Immunol.* **2022**, *22*, 267–269. [CrossRef]
21. Artic Network. Available online: https://artic.network/ncov-2019/ncov2019-bioinformatics-sop.html (accessed on 8 September 2022).
22. Li, H. Minimap2: Pairwise alignment for nucleotide sequences. *Bioinformatics* **2018**, *34*, 3094–3100. [CrossRef]
23. Danecek, P.; Bonfield, J.K.; Liddle, J.; Marshall, J.; Ohan, V.; Pollard, M.O.; Whitwham, A.; Keane, T.; McCarthy, S.A.; Davies, R.M.; et al. Twelve years of SAMtools and BCFtools. *Gigascience* **2021**, *10*, giab008. [CrossRef] [PubMed]
24. Babraham Bioinformatics—FastQC A Quality Control tool for High Throughput Sequence Data. Available online: https://www.bioinformatics.babraham.ac.uk/projects/fastqc/ (accessed on 8 September 2022).
25. Babraham Bioinformatics—Trim Galore! Available online: https://www.bioinformatics.babraham.ac.uk/projects/trim_galore/ (accessed on 8 September 2022).
26. Kim, D.; Langmead, B.; Salzberg, S.L. HISAT: A fast spliced aligner with low memory requirements. *Nat. Methods* **2015**, *12*, 357–360. [CrossRef] [PubMed]
27. Quinlan, A.R.; Hall, I.M. BEDTools: A flexible suite of utilities for comparing genomic features. *Bioinformatics* **2010**, *26*, 841–842. [CrossRef] [PubMed]
28. Gruber, A.R.; Lorenz, R.; Bernhart, S.H.; Neuböck, R.; Hofacker, I.L. The Vienna RNA websuite. *Nucleic Acids Res.* **2008**, *36*, W70–W74. [CrossRef]
29. Zuker, M.; Stiegler, P. Optimal computer folding of large RNA sequences using thermodynamics and auxiliary information. *Nucleic Acids Res.* **1981**, *9*, 133–148. [CrossRef]
30. Kelley, L.A.; Mezulis, S.; Yates, C.M.; Wass, M.N.; Sternberg, M.J.E. The Phyre2 web portal for protein modeling, prediction and analysis. *Nat. Protoc.* **2015**, *10*, 845–858. [CrossRef]
31. Hekkelman, M.L.; Te Beek, T.A.H.; Pettifer, S.R.; Thorne, D.; Attwood, T.K.; Vriend, G. WIWS: A protein structure bioinformatics Web service collection. *Nucleic Acids Res.* **2010**, *38*, W719–W723. [CrossRef]
32. Laskowski, R.A.; Jabłońska, J.; Pravda, L.; Vařeková, R.S.; Thornton, J.M. PDBsum: Structural summaries of PDB entries. *Protein Sci.* **2018**, *27*, 129–134. [CrossRef]
33. PyMOL | pymol.org. Available online: https://pymol.org/2/ (accessed on 8 September 2022).
34. Van Der Spoel, D.; Lindahl, E.; Hess, B.; Groenhof, G.; Mark, A.E.; Berendsen, H.J.C. GROMACS: Fast, flexible, and free. *J. Comput. Chem.* **2005**, *26*, 1701–1718. [CrossRef]
35. Wang, J.; Wolf, R.M.; Caldwell, J.W.; Kollman, P.A.; Case, D.A. Development and testing of a general amber force field. *J. Comput. Chem.* **2004**, *25*, 1157–1174. [CrossRef]
36. Zhu, L.; Wang, Q.; Zhang, W.; Hu, H.; Xu, K. Evidence for selection on SARS-CoV-2 RNA translation revealed by the evolutionary dynamics of mutations in UTRs and CDSs. *RNA Biol.* **2022**, *19*, 866–876. [CrossRef] [PubMed]
37. Lauring, A.S.; Acevedo, A.; Cooper, S.B.; Andino, R. Codon usage determines the mutational robustness, evolutionary capacity, and virulence of an RNA virus. *Cell Host Microbe* **2012**, *12*, 623–632. [CrossRef] [PubMed]

38. Noh, J.Y.; Jeong, H.W.; Shin, E.-C. SARS-CoV-2 mutations, vaccines, and immunity: Implication of variants of concern. *Signal Transduct. Target. Ther.* **2021**, *6*, 203. [CrossRef] [PubMed]
39. Brown, C.M.; Vostok, J.; Johnson, H.; Burns, M.; Gharpure, R.; Sami, S.; Sabo, R.T.; Hall, N.; Foreman, A.; Schubert, P.L.; et al. Outbreak of SARS-CoV-2 Infections, Including COVID-19 Vaccine Breakthrough Infections, Associated with Large Public Gatherings—Barnstable County, Massachusetts, July 2021. *MMWR Morb. Mortal. Wkly. Rep.* **2021**, *70*, 1059–1062. [CrossRef]
40. Rovida, F.; Cassaniti, I.; Paolucci, S.; Percivalle, E.; Sarasini, A.; Piralla, A.; Giardina, F.; Sammartino, J.C.; Ferrari, A.; Bergami, F.; et al. SARS-CoV-2 vaccine breakthrough infections with the alpha variant are asymptomatic or mildly symptomatic among health care workers. *Nat. Commun.* **2021**, *12*, 6032. [CrossRef]
41. Blachere, N.E.; Hacisuleyman, E.; Darnell, R.B. Vaccine Breakthrough Infections with SARS-CoV-2 Variants. *N. Engl. J. Med.* **2021**, *385*, e7.
42. Philomina, J.B.; Jolly, B.; John, N.; Bhoyar, R.C.; Majeed, N.; Senthivel, V.; Fairoz, C.P.; Rophina, M.; Vasudevan, B.; Imran, M.; et al. Genomic survey of SARS-CoV-2 vaccine breakthrough infections in healthcare workers from Kerala. *India. J. Infect.* **2021**, *83*, 237–279. [CrossRef]
43. Li, B.; Deng, A.; Li, K.; Hu, Y.; Li, Z.; Shi, Y.; Xiong, Q.; Liu, Z.; Guo, Q.; Zou, L.; et al. Viral infection and transmission in a large, well-traced outbreak caused by the SARS-CoV-2 Delta variant. *Nat. Commun.* **2022**, *13*, 460. [CrossRef]
44. Butt, A.A.; Dargham, S.R.; Chemaitelly, H.; Al Khal, A.; Tang, P.; Hasan, M.R.; Coyle, P.V.; Thomas, A.G.; Borham, A.M.; Concepcion, E.G.; et al. Severity of Illness in Persons Infected With the SARS-CoV-2 Delta Variant vs Beta Variant in Qatar. *JAMA Intern Med.* **2022**, *182*, 197–205. [CrossRef]
45. Wolter, N.; Jassat, W.; Walaza, S.; Welch, R.; Moultrie, H.; Groome, M.; Amoako, D.G.; Everatt, G.; Bhiman, J.N.; Scheepers, C.; et al. Early assessment of the clinical severity of the SARS-CoV-2 omicron variant in South Africa: A data linkage study. *Lancet* **2022**, *399*, 437–446. [CrossRef]
46. Wang, Y.; Mao, J.-M.; Wang, G.-D.; Luo, Z.-P.; Yang, L.; Yao, Q.; Chen, K.P. Human SARS-CoV-2 has evolved to reduce CG dinucleotide in its open reading frames. *Sci. Rep.* **2020**, *10*, 12331. [CrossRef] [PubMed]
47. Panchin, A.Y.; Panchin, Y.V. Excessive G-U transversions in novel allele variants in SARS-CoV-2 genomes. *PeerJ* **2020**, *8*, e9648. [CrossRef] [PubMed]
48. Teng, X.; Li, Q.; Li, Z.; Zhang, Y.; Niu, G.; Xiao, J.; Yu, J.; Zhang, Z.; Song, Z. Compositional Variability and Mutation Spectra of Monophyletic SARS-CoV-2 Clades. *Genom. Proteom. Bioinform.* **2020**, *18*, 648–663. [CrossRef] [PubMed]
49. Wang, Y.; Chen, X.-Y.; Yang, L.; Yao, Q.; Chen, K.P. Human SARS-CoV-2 has evolved to increase U content and reduce genome size. *Int. J. Biol. Macromol.* **2022**, *204*, 356–363. [CrossRef]
50. Hiscox, J.A.; Wurm, T.; Wilson, L.; Britton, P.; Cavanagh, D.; Brooks, G. The coronavirus infectious bronchitis virus nucleoprotein localizes to the nucleolus. *J. Virol.* **2001**, *75*, 506–512. [CrossRef] [PubMed]
51. Wang, J.; Ji, J.; Ye, J.; Zhao, X.; Wen, J.; Li, W.; Hu, J.; Li, D.; Sun, M.; Zeng, H.; et al. The Structure Analysis and Antigenicity Study of the N Protein of SARS-CoV. *Genom. Proteom. Bioinform.* **2003**, *1*, 145–154. [CrossRef]
52. Shastri, J.; Parikh, S.; Aggarwal, V.; Agrawal, S.; Chatterjee, N.; Shah, R.; Devi, P.; Mehta, P.; Pandey, R. Severe SARS-CoV-2 Breakthrough Reinfection With Delta Variant After Recovery From Breakthrough Infection by Alpha Variant in a Fully Vaccinated Health Worker. *Front. Med.* **2021**, *8*, 737007. [CrossRef]
53. Bolze, A.; White, S.; Basler, T.; Dei Rossi, A.; Wayman, D.; Dai, H.; Roychoudhury, P.; Hayashibara, J.; Beatty, M.; Shah, S.; et al. Evidence for SARS-CoV-2 Delta and Omicron co-infections and recombination. *Med* **2022**, *3*, 1–12. [CrossRef]
54. Faure, G.; Ogurtsov, A.Y.; Shabalina, S.A.; Koonin, E.V. Role of mRNA structure in the control of protein folding. *Nucleic Acids Res.* **2016**, *44*, 10898–10911. [CrossRef]
55. Gao, T.; Gao, Y.; Liu, X.; Nie, Z.; Sun, H.; Lin, K.; Peng, H.; Wang, S. Identification and functional analysis of the SARS-COV-2 nucleocapsid protein. *BMC Microbiol.* **2021**, *21*, 58. [CrossRef]
56. Yang, M.; He, S.; Chen, X.; Huang, Z.; Zhou, Z.; Zhou, Z.; Chen, Q.; Chen, S.; Kang, S. Structural Insight Into the SARS-CoV-2 Nucleocapsid Protein C-Terminal Domain Reveals a Novel Recognition Mechanism for Viral Transcriptional Regulatory Sequences. *Front. Chem.* **2020**, *8*, 624765. [CrossRef] [PubMed]
57. Chen, C.-Y.; Chang, C.-K.; Chang, Y.-W.; Sue, S.-C.; Bai, H.-I.; Riang, L.; Hsiao, C.D.; Huang, T.H. Structure of the SARS coronavirus nucleocapsid protein RNA-binding dimerization domain suggests a mechanism for helical packaging of viral RNA. *J. Mol. Biol.* **2007**, *368*, 1075–1086. [CrossRef] [PubMed]

viruses

MDPI

Review

Evolution and Control of COVID-19 Epidemic in Hong Kong

Shuk-Ching Wong [1], Albert Ka-Wing Au [2], Janice Yee-Chi Lo [2], Pak-Leung Ho [3,4], Ivan Fan-Ngai Hung [5], Kelvin Kai-Wang To [3], Kwok-Yung Yuen [3] and Vincent Chi-Chung Cheng [1,6,*]

1 Infection Control Team, Queen Mary Hospital, Hong Kong West Cluster, Hong Kong SAR, China
2 Centre for Health Protection, Department of Health, Hong Kong SAR, China
3 Department of Microbiology, Li Ka Shing Faculty of Medicine, The University of Hong Kong, Hong Kong SAR, China
4 Carol Yu Center for Infection, The University of Hong Kong, Hong Kong SAR, China
5 Department of Medicine, Li Ka Shing Faculty of Medicine, The University of Hong Kong, Hong Kong SAR, China
6 Department of Microbiology, Queen Mary Hospital, Hong Kong SAR, China
* Correspondence: vcccheng@hku.hk

Abstract: Hong Kong SAR has adopted universal masking, social distancing, testing of all symptomatic and high-risk groups for isolation of confirmed cases in healthcare facilities, and quarantine of contacts as epidemiological control measures without city lockdown or border closure. These measures successfully suppressed the community transmission of pre-Omicron SARS-CoV-2 variants or lineages during the first to the fourth wave. No nosocomial SARS-CoV-2 infection was documented among healthcare workers in the first 300 days. The strategy of COVID-19 containment was adopted to provide additional time to achieve population immunity by vaccination. The near-zero COVID-19 situation for about 8 months in 2021 did not enable adequate immunization of the eligible population. A combination of factors was identified, especially population complacency associated with the low local COVID-19 activity, together with vaccine hesitancy. The importation of the highly transmissible Omicron variant kickstarted the fifth wave of COVID-19, which could no longer be controlled by our initial measures. The explosive fifth wave, which was partially contributed by vertical airborne transmission in high-rise residential buildings, resulted in over one million cases of infection. In this review, we summarize the epidemiology of COVID-19 and the infection control and public health measures against the importation and dissemination of SARS-CoV-2 until day 1000.

Keywords: COVID-19; SARS-CoV-2; variants; infection control

Citation: Wong, S.-C.; Au, A.K.-W.; Lo, J.Y.-C.; Ho, P.-L.; Hung, I.F.-N.; To, K.K.-W.; Yuen, K.-Y.; Cheng, V.C.-C. Evolution and Control of COVID-19 Epidemic in Hong Kong. *Viruses* 2022, *14*, 2519. https://doi.org/10.3390/v14112519

Academic Editor: Ahmed El-Shamy

Received: 20 October 2022
Accepted: 11 November 2022
Published: 14 November 2022

1. Introduction

Throughout the history of mankind, we have been constantly challenged by emerging infectious diseases, some of which caused devastating pandemics. Over the past 1500 years, we have experienced plague, tuberculosis, influenza, and acquired immunodeficiency syndrome [1]. The Justinian plague in the Old World and the medieval plague, caused by *Yersinia pestis* transmitted by fleas, aerosols, and contaminated food, killed half and one-third of world population, respectively [2,3]. Emerging viral infections became the predominant threat in the last century. The 1918–1919 influenza pandemic caused an estimated 50 million deaths worldwide [4]. Historical records reveal that a range of interventions was attempted to control the spread of influenza in 1918, including closure of schools and churches, banning of mass gatherings, mandated mask wearing, case isolation, and use of disinfection and hygiene measures [5]. The improvement of public health interventions and responses resulted in a marked reduction in the global number of deaths to between one and four million during the influenza pandemics in 1957 and 1968 [6,7]. With further introduction of therapeutics on top of infection control and public health measures during the influenza pandemic in 2009 [8], the estimated global number of deaths associated within the first 12 months was less than 0.3 million [9], which was apparently not particularly excessive compared with seasonal influenza viruses [10]. Moreover, the residual immunity of those above 60 years of age against the previous influenza H1 strain circulating prior to 1957 might have contributed to the relatively low mortality of the 2009 pandemic as well [11].

Coronavirus, a positive-sense, single-stranded RNA virus, is an important emerging pathogen in addition to influenza virus. Human coronavirus 229E and OC43 were consecutively identified as previously unrecognized etiological agents for the common cold in 1965 [12]. Human coronavirus NL63 and HKU-1 were subsequently discovered as respiratory pathogens in 2004 and 2005, respectively [13,14]. The occurrence of severe acute respiratory syndrome coronavirus 1 (SARS-CoV-1) in 2003 [15] and sporadic outbreaks of Middle East respiratory syndrome coronavirus since 2012 [16] in limited geographic areas had already alerted the world to the potential occurrence of devastating infectious diseases in the 21st century. In 2018, the World Health Organization (WHO) promulgated a list of priority diseases requiring urgent research and development attention [17], with the inclusion of Disease X, representing pathogens currently unknown that cause human disease and requiring cross-cutting preparedness [18,19]. Coronavirus disease 2019 (COVID-19), caused by severe acute respiratory syndrome coronavirus 2 (SARS-CoV-2), was described as the first Disease X [20]. The official announcement of the community-acquired pneumonia outbreak in Wuhan, Hubei Province, was made by the Health Commission of Hubei Province of the People's Republic of China on 31 December 2019 (day 1) [21]. Compared with the outbreak of SARS-CoV-1 in 2003, the time to discovery of the etiological agent of this novel coronavirus, SARS-CoV-2, was shorter [22]. The genome sequences of SARS-CoV-2 obtained from infected patients were closely related to two bat-derived SARS-like coronaviruses [23–26], analogous to the finding of SARS-like coronavirus in Chinese horseshoe bats in 2005, suggesting that bats may be the ultimate origin of coronaviruses posing a threat to humans [27,28].

The WHO declared a Public Health Emergency of International Concern on 30 January 2020 and proclaimed the outbreak a pandemic on 11 March 2020 (day 72) [29]. By then, COVID-19 had already been detected in 113 countries or territories on 5 continents, resulting in 118,319 confirmed cases all over the world [30]. As of 25 September 2022 (day 1000), 612 million confirmed cases and 6.5 million deaths had been reported globally [31]. During the early phase of the COVID-19 pandemic, various forms of lockdown on a global basis [32] led to far-reaching socioeconomic and health effects in different strata of life [33]. In the cosmopolitan city of Hong Kong, where the population density was 7126 persons per square kilometer, among the top four worldwide [34], the strategy was precise and dynamic infection control and public health measures without lockdown of the city or complete border closure. We summarized our experience in the control of COVID-19 during the evolution of SARS-CoV-2 variants [35], from the wild-type virus to Omicron sub-lineages (BA.1 and BA.2 to BA.4 and BA.5), in Hong Kong.

2. Global Epidemiology of COVID-19

Soon after the first reported case of COVID-19, the WHO was monitoring the global epidemiology of COVID-19 by regularly updating the situation report since 21 January 2020 [36]. In addition, data on the number of COVID-19 cases and deaths was also available in the public domains of the European Centre for Disease Prevention and Control [37], academic institutions [38], and public media [39–42]. On 11 March 2022 (two years after the start of the COVID-19 pandemic), there were 447 million confirmed cases, constituting 6% of the global population, about the combined population size of the 27 countries making up the European Union, with 6 million (0.075% of the world population) deaths [43]. As of 25 September 2022 (day 1000), 14 (6.3%) of 222 countries or areas had recorded over 10 million cumulative confirmed cases (Table 1). These 14 countries accounted for 407 million cases, which constituted 67% of the COVID-19 infections in the world.

Table 1. Global epidemiology of COVID-19 in terms of countries with a cumulative number of confirmed cases of more than 10 million as of 25 September 2022 (day 1000).

Countries	Cumulative Number of Confirmed Cases	Cumulative Number of Deaths among All Confirmed Cases	Incidence per 1000 Population [a]	Death per 1000 Population [a]	Case Fatality Rate (%)
United States of America	95,795,378	1,050,631	289.4	3.2	1.1
India	44,560,749	528,487	32.3	0.4	1.2
France	35,056,032	154,854	518.2	2.3	0.4
Brazil	34,624,427	685,750	162.9	3.2	2.0
Germany	32,905,086	149,368	395.7	1.8	0.5
Korea	24,620,128	28,213	480.2	0.6	0.1
United Kingdom	23,621,956	189,919	347.1	2.8	0.8
Italy	22,241,369	176,775	375.5	3.0	0.8
Japan	20,992,896	44,262	166.0	0.4	0.2
Russia	20,694,894	386,551	141.8	2.7	1.9
Turkey	16,852,382	101,068	201.6	1.2	0.6
Spain	13,393,196	113,845	282.6	2.4	0.9
Vietnam	11,467,619	43,146	117.8	0.5	0.4
Australia	10,191,312	14,853	399.7	0.6	0.2

[a] Population figures were retrieved from the European Centre for Disease Prevention and Control [37].

During the evolution of the COVID-19 pandemic, mutations of SARS-CoV-2 generated many variants. The WHO used letters of the Greek alphabet to define variants of concern (VOC) for better communication and discussion by the general public. VOCs are viruses associated with one or more of the following attributes at a degree of global public health significance, including (i) an increase in transmissibility or detrimental change in COVID-19 epidemiology, (ii) an increase in virulence or change in clinical disease presentation, and (iii) a decrease in effectiveness of public health and social measures or available diagnostics, vaccines, or therapeutics [35]. Currently and previously circulating VOCs are summarized in Table 2.

Table 2. Chronology of SARS-CoV-2 variants of concern during the COVID-19 pandemic [a].

Nomenclature by the World Health Organization	Earliest Documented Samples (Place and Date)	Duration	Pango Lineage	GISAID Clade	Nextstrain Clade
Alpha	United Kingdom, September 2020	18 December 2020 to 9 March 2022	B.1.1.7	GRY	20I (V1)
Beta	South Africa, May 2020	18 December 2020 to 9 March 2022	B.1.351	GH/501Y.V2	20H (V2)
Gamma	Brazil, November 2020	11 January 2021 to 9 March 2022	P.1	GR/501Y.V3	20J (V3)
Delta	India, October 2020	11 May 2021 to 7 June 2022	B.1.617.2	G/478K.V1	21A, 21I, 21J
Omicron	Multiple countries, November 2021	26 November 2021 onwards	B.1.1.529	GR/484A	21K, 21L, 21M, 22A, 22B, 22C, 22D

[a] The table was adopted and modified from the World Health Organization [35].

The emergence of SARS-CoV-2 VOCs has resulted in a progressive increase in transmissibility. The Alpha variant had a 43 to 90% higher reproduction rate than pre-existing variants [44], which caused a rapid expansion of the variant during autumn 2020 in the UK, especially among patients under 20 years of age [45]. The Delta variant was on average 63–167% more transmissible than the Alpha variant [46]. The effective reproduction number of the Omicron variant was 3.19 times greater than that of the Delta variant under the same epidemiological conditions [47]. The secondary attack rate of the Omicron variant was higher than Delta among household contacts, with a relative risk of 1.41 [48]. The evolution of SARS-CoV-2 VOCs led to a continuing rise in case numbers. The trend of COVID-19 cases in relation to the emergence of SARS-CoV-2 VOCs is illustrated in Figure 1.

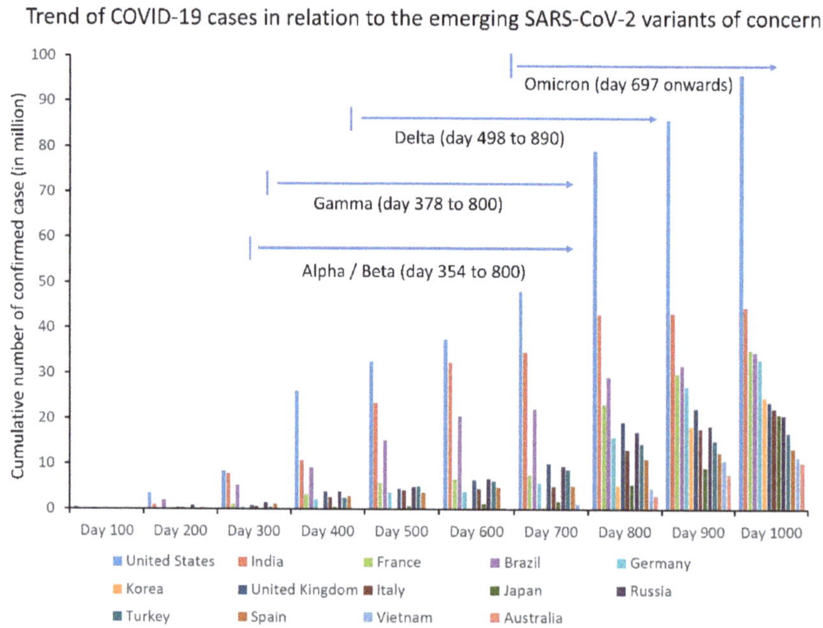

Figure 1. Cumulative number of COVID-19 cases in 14 countries reporting more than 10 million cases as of 25 September 2022 (day 1000).

3. Epidemiology of COVID-19 in Hong Kong

Since the outbreak of community-acquired pneumonia in Wuhan, Hubei Province, announced on 31 December 2019 (day 1), the Centre for Health Protection, Department of Health, Government of Hong Kong Special Administrative Region, China, had been closely monitoring the situation. A website was established for the public to access updated information, including epidemiological information of each laboratory-confirmed COVID-19 case [49]. The first COVID-19 case was a traveler from Wuhan who arrived in Hong Kong on 21 January 2020 (day 22), and was confirmed on 23 January 2020 (day 24), whereas the first locally acquired COVID-19 case was reported on 30 January 2020 (day 31). Within the first 1000 days, there were five waves of COVID-19 with a total of 1,745,505 cases recorded in Hong Kong (Table 3). Prior to the fifth wave, there were 12,636 confirmed COVID-19 cases in Hong Kong with an incidence of 1685 cases per million population. The median age was 43 years (range: 12 days to 100 years). The overall case fatality rate was 1.7% (213/12,636). The epidemiological information of 54 outbreaks in the first to the fourth waves has been reported previously [50]. During the fifth wave (primary Omicron wave), 1,200,068 COVID-19 cases were reported. The incidence of COVID-19 during the primary Omicron wave was 95 times higher than the total incidence from the first to the fourth wave in Hong Kong, with the daily number of new cases reaching a peak of >50,000 on 3 March 2022 (day 794). As of 31 May 2022 (day 883), there were 9318 deaths within the primary Omicron wave, of which 6632 (71.2%) and 2312 (24.8%) were aged ≥80 and 60–79 years, respectively. Among the deaths in the elderly population, 4934 (74.4%) and 1543 (66.7%) aged 80 and 60–79 years, respectively, were unvaccinated. A secondary wave of Omicron COVID-19 cases was observed when the BA.5 emerged in June 2022, which became the predominant virus strain in August 2022. The epidemic curves of COVID-19 in the first to the fourth wave as well as from the fifth wave onwards are illustrated in Figures 2 and 3, respectively.

Table 3. Epidemiological and virological characteristics in different waves of COVID-19 in Hong Kong.

Wave of COVID-19	Period (Duration of Each Wave, Day) [a]	Total Number of Cases [b] (Death, Case Fatality rate)	Number (%) of Cases in (Episodes of) Community Outbreaks [c]	Number (%) of Imported Cases (Remark)	Predominant Virus Strain
1st	23 January 2020 (day 24) to 14 March 2020 (day 75) (51 days)	142 (4, 2.8%)	53 (37.3%) (4)	56 (39.4%) (mainly from China)	NA
2nd	15 March 2020 (day 76) to 30 June 2020 (day 183) (108 days)	1064 (4, 0.38%)	130 (12.2%) (3)	739 (69.4%)	NA
3rd	1 July 2020 (day 184) to 31 October 2020 (day 306) (123 days)	4118 (103, 2.5%)	681 (16.5%) (23)	678 (16.5%)	B.1.1.63
4th	1 November 2020 (day 307) to 30 April 2021 (day 487) (181 days)	6451 (101, 1.6%)	1480 (22.9%) (24)	960 (14.9%)	B.1.36.27
Window phase [d]	1 May 2021 (day 488) to 30 December 2021 (day 731) (244 days)	861 [e] (1, 0.12%)	No	854 (99.2%) (without community outbreak	NA
5th (primary Omicron)	31 December 2021 (day 732) to 31 May 2022 (day 883) (152 days)	1,200,068 (9318, 0.78%) [f]	NA [g]	2292 (0.19%)	Omicron BA.2
5th (secondary Omicron)	1 June 2022 (day 884) to 25 September 2022 (day 1000) (117 days)	532,801 (585, 0.11%) [f]	NA [g]	20,519 (3.9%)	Omicron BA.5

[a] The waves of COVID-19 were defined according to an epidemiological investigation of community outbreaks or the predominant virus strain; [b] From 7 March 2022 (day 798), individuals declaring positive results of rapid antigen detection by self-testing were counted as confirmed cases, in addition to nucleic acid testing by laboratories; [c] Outbreaks involving more than 10 cases were reported; [d] No COVID-19 outbreaks were reported in the community; [e] All COVID-19 cases were diagnosed upon arrival in Hong Kong or during quarantine after arrival in Hong Kong; [f] On 23 September 2022, the Centre for Health Protection Department of Health, Hong Kong, announced an additional 153 fatal cases, which were retrospectively reported by the Hospital Authority. They were all fatal cases during the peak months from February to April 2022; [g] Due to the huge increase in cases, the epidemiological information of individual outbreak was no longer announced on the website of the Centre for Health Protection, after 6 February 2022 (day 769). NA, not applicable.

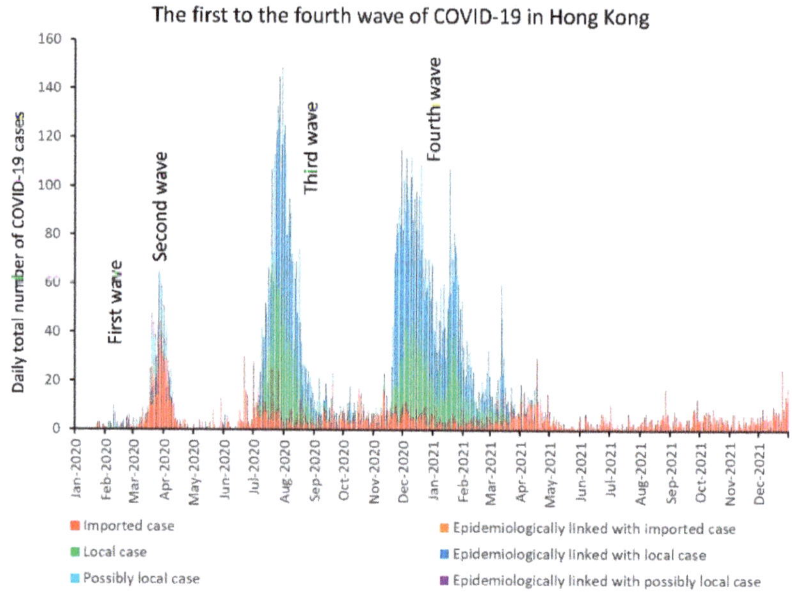

Figure 2. Epidemic curve of the first to the fourth wave of COVID-19 in Hong Kong. An imported case is defined as a case having traveled to a country or area with community transmission of COVID-19 during the incubation period (IP) (i.e., operationally defined as 7 days before symptom onset or a positive test for an asymptomatic case) or tested positive during compulsory quarantine without having any activity in the local community. A local case is defined as a case without travel history during the IP. A possibly local case is defined as those having traveled to a country or area not known to have community transmission within the IP (this category was historically used in the first two waves in early 2020 only). Imported cases, local cases, and cases epidemiologically linked with local cases constituted the main burden of COVID-19 cases.

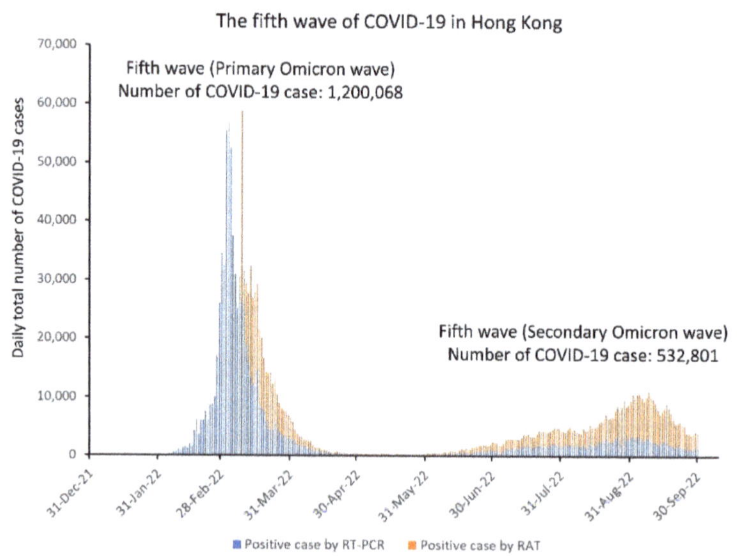

Figure 3. Epidemic curve of the fifth wave of COVID-19 in Hong Kong. RT-PCR, reverse-transcription–real-time polymerase chain reaction; RAT, rapid antigen test.

4. Evolving SARS-CoV-2 VOCs in Hong Kong

Diverse SARS-CoV-2 genomes were identified among imported infections during the first to the fourth wave [51]. The importation of different lineages, variants, and subvariants resulted in transmission chains in the local community even before declaration of the COVID-19 pandemic [52]. Except for the first and second waves, local infections were dominated by a single lineage during each wave. Based on whole-genome sequencing analysis of virus strains, 96.6% (259/268) in the third wave and 100% (73/73) in the fourth wave were B.1.1.63 and B.1.36.27 lineages, respectively. Although B.1.1.63 lineage was first detected in an imported case 2 weeks before the beginning of the third wave, B.1.36.27 lineage had circulated in Hong Kong for 2 months prior to the fourth wave [53]. During the window phase in which there was no evidence of local transmission of COVID-19 (near-zero situation), 212 virus strains from imported cases were subjected to next-generation sequencing, which revealed 42 Kappa and 70 Delta variant cases [54]. By February 2022, the Omicron variant had become the predominant virus strain, causing explosive community transmission and outbreaks in high-rise residential buildings during the fifth wave [50,55]. In addition, probable animal-to-human transmission of SARS-CoV-2 Delta variant AY.127 was documented to cause a pet shop-related COVID-19 outbreak in Hong Kong during the early phase of the fifth wave [56]. This hamster-related outbreak led to a cluster of at least 88 cases in our epidemiological investigation [57]. The evolution of the epidemic due to SARS-CoV-2 VOCs in Hong Kong is illustrated in Figure 4.

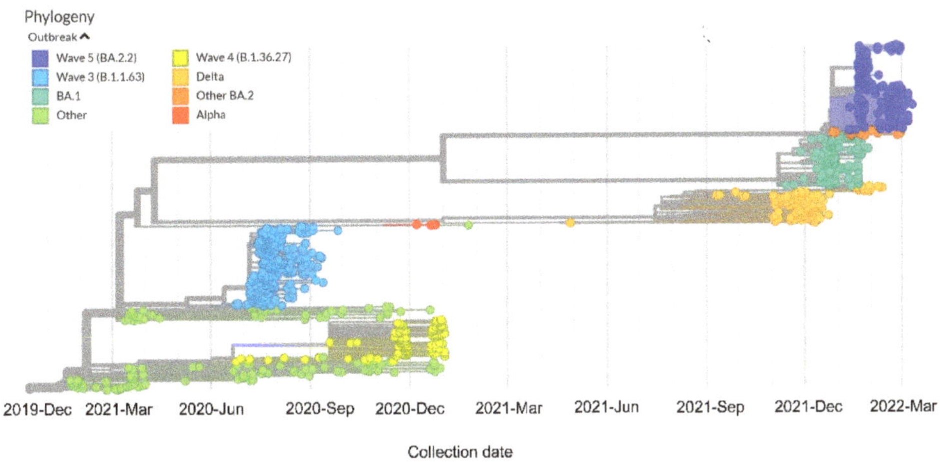

Figure 4. Time-resolved phylogenetic tree of 1168 viral genomes from January 2020 to March 2022 in Hong Kong. The tree was constructed using the Nextstrain command-line interface. The color of the dots represents the outbreak wave of a particular strain. The branch length was determined by the collection date of a sample. Wuhan-Hu-1 was used as the reference sequence during the tree construction.

5. Infection Control and Preparedness for COVID-19 in Hong Kong

5.1. Background

In Hong Kong, the SARS-CoV-1 outbreak took place in 2003. A total of 1755 persons were infected and 299 (17.0%) persons died [58]. Among the infected, 386 (22.0%) were healthcare workers, with eight fatalities, including four doctors, one nurse, and three healthcare assistants. In addition to reviewing the lessons learned from the scientific perspective [15], an independent commission of inquiry, namely, the "Select Committee to inquire into the handling of the Severe Acute Respiratory Syndrome outbreak by the Government and the Hospital Authority," was set up by the Legislative Council of the Hong Kong Special Administrative Region, China, in order to investigate the performance and accountability of the government and the Hospital Authority (the governing body of all 43 public hospitals responsible for 90% of inpatient service in Hong Kong) and their officers at the policy-making and management levels [59]. The Chief Executive of the Hong Kong Special Administrative Region also announced on 28 May 2003 the setting up of a SARS Expert Committee to (i) review the work of the government, including the Hospital Authority, in the management and the control of the outbreak; (ii) examine and review the capabilities and structure of the healthcare system in Hong Kong and the organization and operation of the Department of Health and the Hospital Authority in the prevention and management of infectious diseases such as SARS; and (iii) identify lessons to be learned and make recommendations on areas of improvements in order to better prepare our system for any future outbreaks [60,61]. Members of the committee were selected for their wide range of experience in their respective fields, which included health systems, public health, epidemiology and communicable disease control, medical expertise, and hospital management and operation.

There were 44 main recommendations by the SARS Expert Committee, as a result of which the Centre for Health Protection was established in June 2004 to protect the community from emerging or evolving public health threats [62]. Preparedness plans for emerging infectious diseases have been formulated over the years. When the outbreak of SARS-CoV-2 occurred in Hong Kong, various control measures could be implemented in a timely manner, including epidemiological investigation for the potential source of acquisition by contact tracing, regardless of symptomatic or asymptomatic cases. Risk communication to the public during the COVID-19 pandemic included daily press conferences on the latest situation and public health control measures and uploading the data and education materials to the website.

A hospital-based approach was adopted to isolate all suspected or confirmed COVID-19 patients in healthcare facilities, including airborne infection isolation rooms (AIIRs) in hospitals or community isolation and treatment facilities, during the first to the fourth wave of COVID-19 in Hong Kong. This approach of institutional containment likely minimized community transmission of COVID-19. The strategy was effective in protecting the healthcare system from being overwhelmed by uncontrolled transmission of the virus in the community. With 5% of COVID-19 patients eventually becoming critically ill, the healthcare system might have become paralyzed if the burden of infection in the community were high [63]. Prevention of nosocomial transmission of COVID-19 and protection of healthcare workers from SARS-CoV-2 remained the top priority, which dictated regular review and escalation of infection control measures during the evolution of the COVID-19 pandemic.

5.2. Infection Control for COVID-19 in the Hospitals

5.2.1. Active Surveillance

Active surveillance for patients upon admission was progressively stepped up from risk-based screening to universal admission screening, aiming at early identification and isolation of COVID-19 patients in AIIRs (Table 4). Patients fulfilling clinical and epidemiological criteria were isolated in the AIIRs and screened for SARS-CoV-2 by reverse-transcription–real-time polymerase chain reaction (RT-PCR) using nasopharyngeal specimens or saliva [64,65].

Table 4. Stepwise enhancement of active surveillance for early isolation of COVID-19 patients in airborne infection isolation rooms (AIIRs).

A	Clinical Criteria (Time of Implementation) [a]	Remark
1.	Presented with fever and acute respiratory illness or pneumonia (from day 1 to day 23)	Prepare for the importation of index patient to Hong Kong
2.	Presented with fever or acute respiratory illness or pneumonia (with effect from day 24)	In response to the importation of index patient to Hong Kong
B	**Epidemiological criteria (time of implementation) [a]**	
1.	Travel history to Wuhan, Hubei province, People's Republic of China, within 14 days before onset of symptoms, irrespective of any exposure to wet market or seafood market (from day 1 to day 16)	Prepare for the importation of index patient to Hong Kong
2.	Patient met with one of the following within 14 days prior to the onset of symptoms: (a) had visited Wuhan (regardless of whether the individual had visited wet markets or seafood markets there), (b) had visited a medical hospital in Mainland China, or (c) had had close contact with a confirmed case of the novel coronavirus while that patient was symptomatic (from day 17 to day 20)	In response to the evolving epidemic with increasing number of confirmed cases in Wuhan
3.	Patient met with one of the following within 14 days prior to the onset of symptoms: (a) had visited Hubei Province (regardless of whether the individual had visited wet markets or seafood markets there) or 2(b) or 2(c) listed above (with effect from day 21)	In response to spread of SARS-CoV-2 beyond Wuhan
4.	Universal admission screening for asymptomatic patients (with effect from 9 September 2020 (day 265) onwards)	In response to widespread transmission of SARS-CoV-2 locally

[a] Active surveillance for patients upon admission according to the clinical and epidemiological criteria was performed from 31 December 2019 (day 1).

Saliva was adopted as one of the diagnostic specimens for the detection of SARS-CoV-2 RNA in Hong Kong during the early phase of COVID-19. SARS-CoV-2 RNA was detected in self-collected saliva specimens in 91.7% (11/12) of our early COVID-19 cases [66]. Using saliva specimens, the temporal profile of viral load during the clinical course could also be demonstrated [67]. The diagnostic sensitivity of saliva is comparable to nasopharyngeal specimens using an automated point-of-care molecular assay in a local evaluation [68], as subsequently confirmed in a meta-analysis [69]. With diurnal variation in viral shedding from the upper respiratory tract, saliva was recommended to be taken preferably in the early morning to maximize the yield for screening purposes in the community [70]. Saliva was considered a promising noninvasive specimen for diagnosis and monitoring of COVID-19 patients. More importantly, it reduced the risk of exposure among healthcare workers in collecting nasopharyngeal and other respiratory specimens.

5.2.2. Training of Healthcare Workers

Provision of just-in-time education of infection control training was organized by the hospital infection control team. Staff forum and department visits were organized to update infection control knowledge, enforce hand hygiene practice using alcohol-based hand rub, and promote the use of surgical masks for both patients and healthcare workers. Furthermore, every healthcare worker who was rotated to care for COVID-19 patients in AIIRs was required to receive mandatory training by infection control nurses. The training program included the donning and doffing of personal protective equipment (PPE), the collection of respiratory specimens, proper handling and packing of clinical specimens inside the AIIR, and performing aerosol-generating procedures in accordance with the evolving infection control guidelines [71,72]. After receiving the training, each healthcare worker was required to demonstrate competence on donning and doffing of PPE, including surgical respirator, cap, face shield, gown, and gloves. The concept of directly observed donning and doffing of PPE was introduced, where the supervisor in the ward would ensure compliance with proper PPE procedures upon entering and leaving the AIIR.

Infection control nurses also performed regular and ad hoc audits to enforce compliance with infection control practices. Hand hygiene and mask compliance among healthcare workers were found to be highly satisfactory during the COVID-19 pandemic [73,74].

5.2.3. Environmental Control

Environmental control was another important aspect to reduce nosocomial transmission of COVID-19. Training of cleaning staff, especially those responsible for environmental disinfection of AIIRs, with respect to proper procedures and the use of appropriate disinfectants, was conducted by infection control nurses. A total of 377 environmental samples from AIIRs housing 21 RT-PCR-confirmed COVID-19 patients were collected during 76 episodes of room visits before daily environmental disinfection. Only 19 (5.0%) of 377 samples were RT-PCR positive from the environment [75], which was comparatively lower than isolation rooms in China (25%) and South Korea (14% to 18%) [76,77]. The use of surgical masks by COVID-19 patients, even though they were isolated in the AIIRs, might have reduced the extent of environmental contamination. Patient transfer between wards could potentially have increased the risk COVID-19 transmission within the hospital [78]. However, intra-hospital transfer of COVID-19 patients was unavoidable if the case required certain procedures, including radiological examination such as computerized tomography scanning. In addition to routine preparations made by the radiology department, the radiological examination was preferably arranged after office hours. Senior infection control nurses would provide on-site support and coordination for the entire process of patient transfer, including provision of a designated path, supervision of the donning and doffing of PPE among healthcare workers and security personnel, and monitoring the process of environmental disinfection along the path of patient transfer and in the examination room of the radiology department.

5.2.4. Air Sampling for SARS-CoV-2 RNA

To assess the risk of COVID-19 transmission in the hospital, air sampling for SARS-CoV-2 RNA was conducted in AIIRs with 12 air changes per hour during the evolution of the COVID-19 pandemic. The first imported COVID-19 case admitted into an AIIR had a moderate level of viral load of 3.3×10^6 copies/mL in pooled nasopharyngeal and throat swabs and 5.9×10^6 copies/mL in saliva. Air samples for SARS-CoV-2 RNA were collected from the AIIR using air sampler SAS Super ISO 180 model 86834 (VWR International PBI Srl, Milan, Italy). The air sampler was perpendicularly positioned at a distance of 10 cm at the level of the patient's chin, and 1000 L of air at a rate of 180 L per minute were collected for each plate containing 3 mL of viral transport medium (VTM). The patient was instructed to perform four different maneuvers (i.e., normal breathing, deep breathing, speaking "1, 2, 3" continuously, and coughing continuously) while wearing or not wearing a surgical mask with the ASTM F2100 level 1 standard. The VTM was transferred to the laboratory within 2 h for RT-PCR test. All eight air samples were negative for SARS-CoV-2 RNA [64].

Subsequently, the protocol of air sampling was modified for another six patients singly isolated in an AIIR during the first and second waves of COVID-19 by using an umbrella fitted with a transparent plastic curtain as an air shelter to cover the patient during sample collection in order to increase the proportion of exhaled air sampled and to reduce the amount of dilution by environmental air from the air ventilation system with 12 air changes per hour. Air samples of patients inside this air shelter were collected using the Sartorius MD8 air-scan sampling device (Sartorius AG, Göttingen, Germany) with sterile gelatin filters 80 mm in diameter and a pore size of 3 μm (type 17528-80-ACD) (Sartorius AG). All air samples were negative for SARS-CoV-2 RNA using the same maneuvers as described above [75]. The negative findings of air samples in the initial phase of the COVID-19 pandemic reassured our healthcare workers that the risk of airborne transmission of SARS-CoV-2 from patient to staff was negligible in AIIRs where healthcare workers donned appropriate PPE. Our finding contrasts with previous air sampling results for patients infected with the ancestral strain [79–81]. Most negative studies had <3000 L of air collection [64,75,79–81], whereas a few studies with detectable viral RNA had >5000 L of air collection [82,83]. The results of air sampling appear to be determined by the volume of air collection, which might be limited by the intrinsic characteristic of the air samplers.

Therefore, AerosolSense Sampler (Thermo Fisher Scientific Inc., Waltham, MA, USA) was applied because a larger volume of air could be collected at a flow rate of 200 L per min. In fact, air dispersal of other respiratory viruses such as parainfluenza virus, respiratory syncytial virus, adenovirus, and rhinovirus have been demonstrated in the AIIR using AerosolSense Sampler [84]. We further investigated an asymptomatic COVID-19 patient infected with a PANGO lineage B.1.525 virus, singly isolated in an AIIR, before the onset of the fifth wave. The patient had a moderate level of viral load (6.8×10^6 copies/mL in a nasopharyngeal swab sample). SARS-CoV-2 RNA was detected by air sampling (96,000 L air collected over 8 h) at a concentration of 0.005 and 0.009 copies/L, respectively, while the patient was and was not wearing a surgical mask [85]. Given the low level of air dispersal of SARS-CoV-2 RNA, the risk of inhalation of SARS-CoV-2 by healthcare workers in hospital AIIRs was considered extremely low.

During the fifth wave, general wards were temporarily converted into negative-pressure wards (NPWs) by installing mobile modular high-efficiency particulate arrestance filter units and exhaust fans in each cubicle to increase air changes per hour from 6 to 10 to accommodate COVID-19 patients. We collected air samples in the center of these NPWs for 1 to 6 h, with a total volume of 12,000 L to 72,000 L of air. Air dispersal of SARS-CoV-2 RNA was evidenced by almost 80% of air samples being positive while the patients in these NPWs were infected with the Omicron sub-lineage BA.2.2 [86]. Our serial monitoring of air sampling during the evolution of the COVID-19 pandemic demonstrates that SARS-CoV-2 variants had progressively increasing capability for airborne transmission. In addition, air dispersal of other respiratory viruses and multiple-drug-resistant bacteria were also evaluated during the COVID-19 pandemic [84,87,88]. The findings of air dispersal of SARS-CoV-2 as well as other pathogens facilitated risk assessment to guide implementation of infection prevention and control measures.

5.3. *Relieving the Burden of COVID-19 Patients in the Hospitals*

5.3.1. Proactive Screening of High-Risk Groups in Quarantine Centers

The hospital-based approach to admit all suspected and confirmed COVID-19 patients into AIIRs in hospitals posed great pressure on the bed occupancy, especially during the early phase of the COVID-19 outbreak when Hong Kong residents returned from overseas. Quarantine centers were set up by the government for quarantine of close contacts of confirmed cases and high-risk returnees, including one set up in a newly completed public housing estate, Chun Yeung Estate (CYE), which could provide 3121 residential units for the purpose of quarantine. Returning travelers from high-risk locations such as the *Diamond Princess* cruise ship and Hubei province, China, were accommodated in CYE. On 21 February 2020 (day 53), 215 SARS-CoV-2 RT-PCR-negative passengers of the *Diamond Princess* cruise ship, returning to Hong Kong from Yokohama, Japan, on two chartered flights arranged by the Hong Kong Government, were quarantined for 14 days in CYE. The quarantined persons underwent serial SARS-CoV-2 RT-PCR and serology testing on day 0 (baseline), day 4, day 8, and day 12, and were discharged from the quarantine center on day 14 if all the test results were negative for SARS-CoV-2. The clinical and virological characteristics of this cohort have been reported previously [89]. Since 4% (4/215) of quarantined persons were eventually positive for SARS-CoV-2 by RT-PCR, it was decided that 469 Hong Kong residents evacuated from Hubei province, China, on four different chartered flights on 4 and 5 March 2020 would be admitted to the quarantine center in CYE for 14 days. Seventeen (4%) of 452 persons who consented to blood testing were seropositive with the enzyme immunoassay or microneutralization test [90]. Our healthcare workers provided outreach service to screen for COVID-19 patients among these returning travelers. Only persons who had positive RT-PCR results during the quarantine period were referred to the hospital for further isolation and management.

5.3.2. Setting Up Temporary Test Centers for Suspected COVID-19 Patients

Symptomatic inbound travelers fulfilling the reporting criteria (Table 4) would be referred to hospitals for isolation and testing. The innovative idea of setting up a temporary test center at the exhibition hall of the AsiaWorld Expo (AWE) within the Hong Kong International Airport complex aimed at detecting COVID-19 infection among returning travelers aged 12 to 60 years who were clinically stable but had respiratory symptoms upon arrival. The logistics and workflow were based on the infection control principle that the hall was divided into two zones, a clean zone and a patient zone, with clear demarcation by double-fencing. The clean zone included a staff area, a gowning area, and a central command center. The patient zone included areas for patient registration, assessment, collection of clinical specimens, and waiting for diagnostic test results. Unidirectional workflow was assigned for both patients and healthcare workers. Patients who tested positive for SARS-CoV-2 were referred to the hospital for further management. The temporary test center for symptomatic travelers was under the governance of the Hospital Authority, operated by healthcare workers given just-in-time infection refresher training on site. This test center commenced service within 48 h of preparation on 20 March 2020 (day 81) [91]. During the second wave, from 20 March 2020 to 19 April 2020 (day 111), 1210 symptomatic cases that met the criteria for hospitalization under the hospital-based approach were tested at the temporary test center. Eighty-eight (7.3%) of these 1210 patients tested positive for SARS-CoV-2 [92].

5.3.3. Setting Up Community Isolation and Treatment Facilities for COVID-19 Patients

With the further surge in cases during the third wave of COVID-19 as a result of increasing community outbreaks, community isolation and treatment facilities were set up to divert clinically stable patients from hospitals. The community isolation facility was opened at Lei Yue Mun Park and Holiday Village on 24 July 2020 (day 207) [93]. Bungalows were temporarily built on two basketball courts to accommodate 120 and 250 patients in two wings at sites A and B, respectively. Patients could be transferred from hospitals to the bungalow if they were aged <50 years, were independent for activities of daily living, had no major comorbidity, were afebrile for 48 h, had no diarrhea, had normal or improving trends of hematology and biochemistry test profiles, and were not on antiviral or oxygen therapy. As for the community treatment facility, it was established at AWE and commenced operation on 1 August 2020 (day 215). Two halls at the AWE, hall 1 (500 beds; floor area 10,880 m^2 and ceiling 19 m) and hall 2 (400 beds; floor area 10,100 m^2 and ceiling 10 m), were converted into open-cubicle bays. Newly diagnosed COVID-19 patients aged 18 to 60 years were triaged for admission to the community treatment facility after medical and nursing assessment, blood tests, and chest radiography. Medically fit patients with clear chest radiography and oxygen saturation $\geq$ 96% of room air were managed in the community treatment facility until discharge [93]. This facility was built in a convention center, similar to the setup of temporary or shelter hospitals in Wuhan, China [94,95]; Singapore [96]; the United Kingdom [97,98]; and the United States of America [99]. Although the community isolation facility at Lei Yue Mun Park and Holiday Village was closed on 17 August 2020 (day 231), the community treatment facility at AWE relieved the occupancy of AIIRs by diverting COVID-19 patients from the hospitals during the third wave (1 August 2020 to 18 September 2020, serving for 49 days) and the fourth wave (25 November 2020 to 12 March 2021, serving for 108 days).

During the fifth wave (primary Omicron wave) of COVID-19 due to Omicron BA.2, there were more than 10,000 new cases per day in late February 2022. In addition to the AWE (2 January 2022 to 4 May 2022, serving 123 days), public rental buildings (~3000 residential flats), Fangcang shelters (~20,000 beds), and the Kai-Tak cruise terminal (~1000 beds) were recruited as community treatment facilities, in view of the explosive COVID-19 outbreaks in the community [50,55,100]. In anticipation of ongoing transmission of COVID-19, community isolation facilities were built in Penny's Bay (around 7000 units and 14,000 beds) [101] and Kai Tak (around 2700 units) [102], whereas the community treatment facility at AWE was re-opened for the secondary Omicron wave (22 August 2022 to 29 September 2022, serving 39 days).

5.4. COVID-19 Infection among Healthcare Workers

Infection and death among healthcare workers due to COVID-19 was reported in the early phase of the pandemic [103]. As of 8 May 2020 (day 130), a total of 152,888 infections and 14,113 deaths were reported among healthcare workers globally. Healthcare workers constituted 0.52% of COVID-19 deaths in a systematic review [104]. The WHO further estimated that between 80,000 and 180,000 healthcare workers could have died from COVID-19 in the period between January 2020 and May 2021, converging in a scenario of 115,500 deaths [105]. Although healthcare workers may acquire SARS-CoV-2 in both hospitals and the community [106], Hong Kong aimed at minimizing nosocomial infection of COVID-19 among healthcare workers [107]. Zero nosocomial COVID-19 infection among healthcare workers in the public system was achieved in the first 300 days, with a total of 78,834 COVID-19 patient days, through the implementation of a multipronged infection control strategy, including active surveillance, universal masking for patients and healthcare workers, provision of just-in-time infection control training, practicing directly observed donning and doffing of PPE, and diverting COVID-19 patients from hospitals to quarantine camps, community isolation and treatment facilities, as described above [65]. By day 300 (25 October 2020), a total of 38 healthcare workers had acquired SARS-CoV-2 in the community. The proportion of healthcare worker infection in our public healthcare system was significantly lower than that of the general population (0.04%, 38/88,960 vs. 0.07%, 5296/7403,100; $p < 0.001$), suggesting high infection control alertness among our healthcare workers during the first to third wave of COVID-19 in Hong Kong. The first nosocomial outbreak of COVID-19 was reported on 22 December 2020 (day 358) in a regional hospital, with a superspreading event due to possible airborne transmission involving 12 patients and nine healthcare workers. Whole-genome sequencing revealed that the nosocomial outbreak was attributed to SARS-CoV-2 lineage B.1.36.27 (GISAID clade GH), which was predominant in the fourth wave of COVID-19 in Hong Kong [108].

During the fifth wave, it was reported that 12,554 healthcare workers had been infected with COVID-19, overwhelming the public healthcare system as of 11 March 2022 (day 802) [109]. The proportion of healthcare worker infection was paradoxically higher than that of the general population during the corresponding period (14.1%, 12,554/88,960 vs. 8.7%, 646,800/7403,100; $p < 0.001$). The discrepancy may be related to the mandatory daily COVID-19 rapid antigen testing among healthcare workers, such that asymptomatic cases were also detected [110]. Frontline healthcare workers were more likely to report a positive COVID-19 test compared with community individuals in the United Kingdom and the United States [111]. For our infected healthcare workers, it was difficult to determine whether they had been infected in the hospital or in the community, because our staff was not working under closed-loop management like in mainland China [112]. There were no reported deaths among our healthcare workers during the COVID-19 pandemic.

6. Public Health Measures for COVID-19 in Hong Kong

6.1. Background

To quantify the public health measures, a stringency index, namely, the Oxford COVID-19 Government Response Tracker, was established for COVID-19 [113]. This index is a composite measure based on the policy responses that governments have implemented to tackle COVID-19. The data captured government policies related to closure and containment, and health and economic policy for more than 180 countries from 1 January 2020 (day 2) onwards and are accessible in the public domain [114]. The detailed methodology has been published [115]. At the time of declaration of the COVID-19 pandemic by the WHO on 11 March 2020 (day 72), the incidence of infection per 10,000 population and the number of COVID-19 deaths in Hong Kong were comparatively lower than those of the other countries in the Western Pacific, European, and American regions [107]. Hong Kong appeared to have fared better in controlling COVID-19 with public health measures [116]. The COVID-19 stringency index of Hong Kong and other countries during the evolution of the pandemic is illustrated in Figure 5.

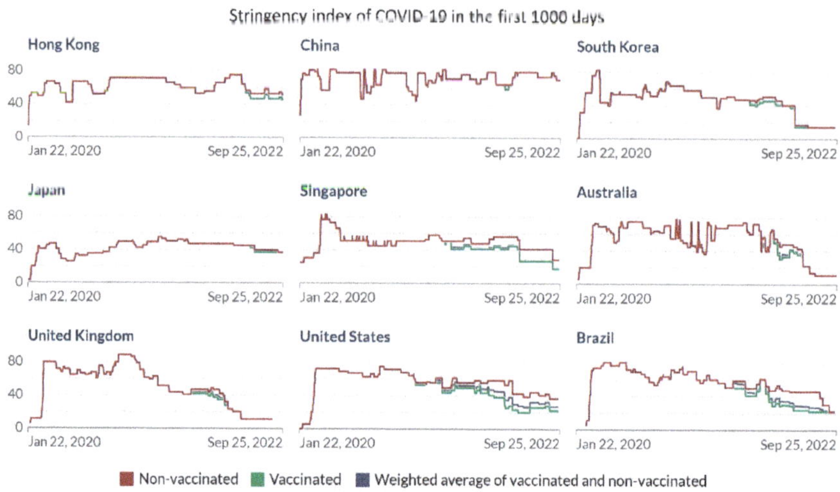

Figure 5. The evolution of the COVID-19 Stringency Index in the first 1000 days. The official announcement of the outbreak of community-acquired pneumonia in Wuhan, Hubei Province, China, was defined as day 1. The figure was generated from the website of the COVID-19 Government Response Tracker [114] and formatted for presentation.

6.2. *Universal Masking and Social Distancing for COVID-19*

Community-wide masking was practiced by the Hong Kong population at an early stage of the epidemic. The compliance of face mask usage was 96.6% (range: 95.7% to 97.2%) in all 18 administrative districts in Hong Kong [117]. This may be one of the reasons why the incidence of COVID-19 in Hong Kong (129.0 per million population) was significantly lower than that of Spain (2983.2), Italy (2250.8), Germany (1241.5), France (1151.6), the US (1102.8), the UK (831.5), Singapore (259.8), and South Korea (200.5) within the first 100 days [117]. Mandatory wearing of masks in public areas was further enacted by the law "Cap. 599I Prevention and Control of Disease (Wearing of Mask) Regulation," which went into effect on 15 July 2020 [118]. Without a total lockdown of the city or complete border closure, the aim was the implementation of precise social distancing to reduce community transmission of COVID-19. School was suspended and gradually converted to virtual class during most of the time during the first to the fourth wave of COVID-19. Work-from-home for non-essential services of civil servants was intermittently implemented. Temporary closure of community facilities such as libraries, sports centers, and leisure facilities such as cinemas, karaokes, and bars, as well as restrictions on operational hours of restaurants, was implemented during the surge of COVID-19 cases in the community [119]. Multivariate analysis of computational simulation results using the Morris Elementary Effects Method suggested that if a sufficient proportion of the population used surgical masks and followed social distancing regulations, COVID-19 could be controlled without requiring a lockdown [120].

6.3. *Quarantine Measures for Inbound Travelers*

A key component of an elimination strategy for COVID-19 was the prevention of the importation of COVID-19 cases. The quarantine measures for inbound travelers evolved to accommodate emerging SARS-CoV-2 VOCs. From the beginning, persons returning from mainland China had to undergo compulsory home quarantine for 14 days after 8 February 2020 (day 40) [121]. With the evolving pandemic, inbound travelers arriving from the specific high-risk overseas areas in the past 14 days were required to stay in a quarantine center from 1 March 2020 (day 62) [122,123] or subjected to compulsory home quarantine on 14 March 2020 (day 75) [124]. Subsequently, inbound travelers arriving from more high-risk areas were required to be quarantined for 14 days in quarantine centers after 11 May 2020 (day 133) or hotels after 25 July 2020 (day 208) [125]. On 13 November 2020 (day 319), all travelers to Hong Kong were mandated to quarantine in designated quarantine hotels for 14 days [126], and the quarantine period was further extended to 21 days on 25 December 2020 (day 361) [127]. The duration of hotel quarantine was shortened from 21 days to 14 days on 5 February 2022 (day 768) [128] and further shortened to 7 days on 1 April 2022 (day 823) for travelers vaccinated with two doses [129]. On 12 August 2022 (day 956), quarantine for inbound persons was relaxed with the implementation of the "3 + 4" model, i.e., compulsory quarantine in designated hotels for three days, followed by medical surveillance for four days, with multiple tests during the period and the monitoring period thereafter [130]. With effect from 26 September 2022 (day 1001), inbound persons were no longer required to undergo compulsory quarantine at designated quarantine hotels under the "0 + 3" model, i.e., three days of medical surveillance, followed by a four-day self-monitoring period [131]. The number of passenger arrivals at Hong Kong International Airport during the evolving quarantine policy is shown in Figure 6.

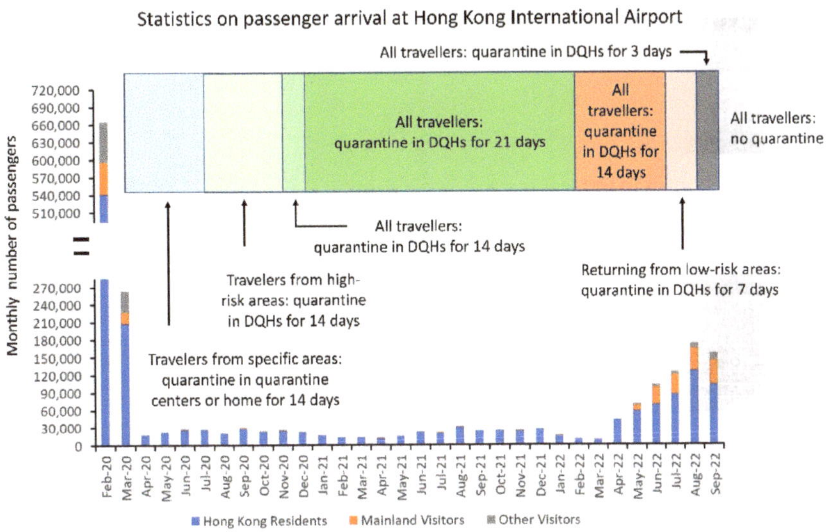

Figure 6. The number of inbound travelers, including Hong Kong residents and visitors from mainland China, arriving at Hong Kong International Airport. DQHs, designated quarantine hotels.

The designated quarantine hotels were not equipped with the standard of air ventilation systems adopted in the AIIRs of hospitals. With the increasing potential of airborne transmission of SARS-CoV-2 VOCs, recurrent episodes of cross-infection among residents in designated quarantine hotels were reported [132,133], leading to the spread of the Beta variant in the community [134]. In fact, the explosive outbreak in the fifth wave of COVID-19 in the community could also be traced to cross-infection in a designated quarantine hotel [55].

6.4. Extensive COVID-19 Testing in the Community

In addition to the progressive enhancement of COVID-19 testing in the hospitals from a risk-based approach to universal admission screening, widespread utilization of the RT-PCR test for SARS-CoV-2 was implemented in stages at the airport for inbound travelers, in government outpatient clinics for symptomatic cases, and in community testing centers for the general population with risk of exposure to COVID-19 [134]. Early recognition of asymptomatic or mildly symptomatic patients for isolation might have reduced the risk of community transmission. Compulsory testing for COVID-19 in the community was enacted by law, The Prevention and Control of Disease (Compulsory Testing for Certain Persons) Regulation, Cap. 599J, in Hong Kong [135]. Any persons who had visited public or private premises and resided in buildings where there was outbreak or transmission of COVID-19 as evidenced by epidemiological investigations or sewage surveillance were informed of the need to undergo COVID-19 testing by a compulsory testing notice. Compulsory testing operations in specified restricted geographic areas were also implemented for the control of COVID-19, as provided by restriction-testing declaration under Hong Kong legislation [136]. With the implementation of compulsory testing notices and restriction-testing declarations, mass testing of 0.81 million members of the population, along with phylogeographic analysis of positive cases, successfully controlled the community outbreak of the Beta variant in Hong Kong [134]. The number of COVID-19 tests performed in different categories during the COVID-19 pandemic is illustrated in Figure 7.

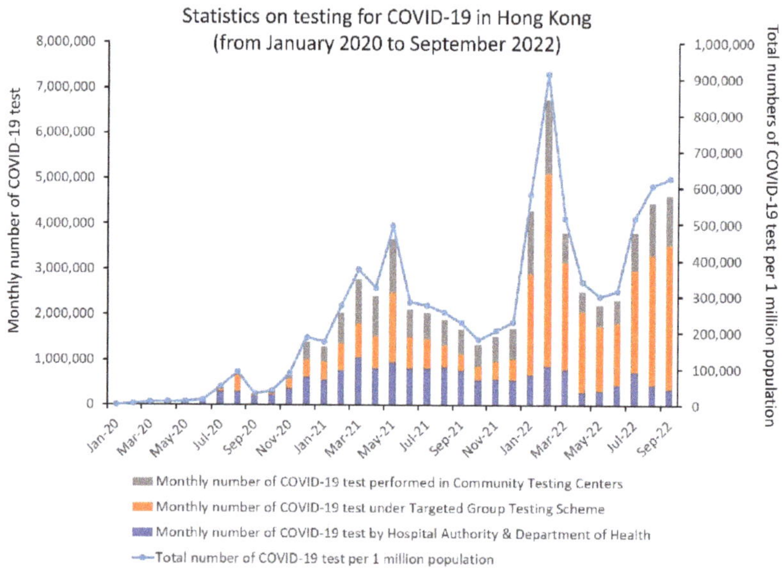

Figure 7. The number of COVID-19 tests under different categories of requests. Community testing centers provided self-paid services (from 15 November 2020, day 321) and free testing for persons under compulsory testing (from 22 November 2020, day 328). The Targeted Group Testing Scheme refers to essential staff providing services to critical infrastructure (from 14 July 2020, day 197). The Hospital Authority and Department of Health were responsible for testing hospitalized patients and out-patients during the COVID-19 pandemic.

Restriction-testing declaration for COVID-19 testing, particularly targeting residential buildings, was implemented during the fifth wave (primary and secondary Omicron waves) of COVID-19. From 31 December 2021 (day 732) to 25 September 2022 (day 1000), a total of 606,822 residents in 473 residential buildings were tested, of whom 27,726 (4.6%) tested positive for SARS-CoV-2 (Figure 8). Confirmed COVID-19 cases were required to isolate at community isolation and treatment facilities or in hospitals. The positive detection rate of SARS-CoV-2 in each high-rise residential building was as high as 30% during the peak of the fifth wave [50]. Vertical airborne transmission of SARS-CoV-2 was suspected among residents, as evidenced by the clustering of cases along vertically aligned flats with a shared drainage stack and lightwell [50].

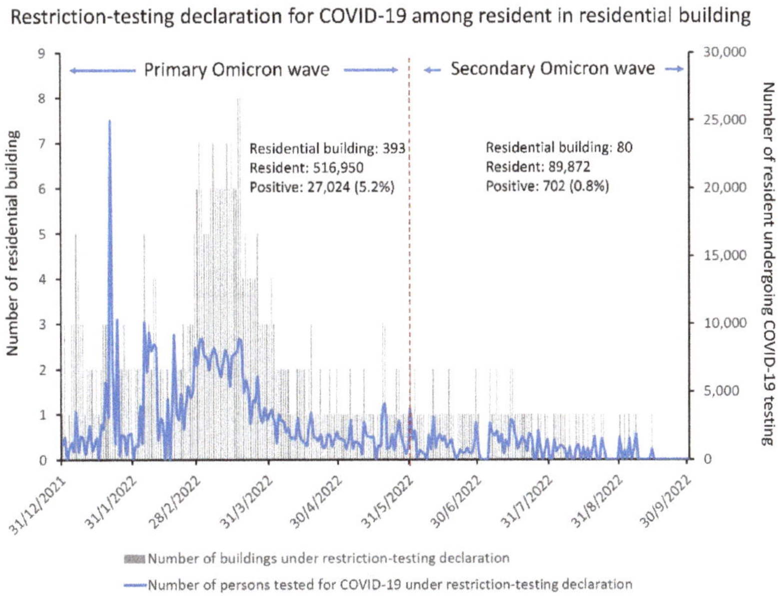

Figure 8. COVID-19 testing in the community under restriction-testing declaration for residents in residential buildings during the fifth wave (primary and secondary Omicron waves) of COVID-19 in Hong Kong.

However, no air sampling was performed to demonstrate the presence of SARS-CoV-2 RNA in high-rise residential buildings to confirm the postulation of airborne transmission. On the other hand, air dispersal of SARS-CoV-2 RNA was confirmed in the hospital setting during the handling of recurrent blockages of sewage pipes in an isolation facility designated for COVID-19 patients. During replacement of sewage pipes, infection control nurses supervised the work process and collected air samples on cutting the sewage pipes, which was considered a mechanical aerosol-generating procedure. The air sampler was placed 50 cm away from the pipe cutter. One thousand liters of air (rate 40 L/min) were collected by a gelatin filter for the first 25 min of work (air sample 1), and another 1000 L of air were collected by another gelatin filter for the next 25 min of work (air sample 2). Air dispersal of SARS-CoV-2 RNA was detected in both air sample 1 (17.5 copies/mL) and air sample 2 (16.5 copies/mL) [137].

6.5. COVID-19 Vaccination in the Community

The COVID-19 vaccination program was officially launched on 26 February 2021 (day 424) by the Government of the Hong Kong Special Administrative Region. Two formulations of COVID-19 vaccines were available: CoronaVac, an inactivated whole-cell vaccine, Sinovac Biotech (Hong Kong) Limited, and Comirnaty, a BNT162b2 mRNA vaccine, BioNTech, Fosun Pharma, in collaboration with the German drug manufacturer. Vaccination was provided for free to all eligible persons, who could be vaccinated in community vaccination centers, private clinics of medical practitioners, and public outpatient clinics and hospitals. To increase the uptake of COVID-19 vaccination, promotion to healthcare workers as well as the general population was implemented. The strategy of personal coaching was undertaken to enhance vaccination update among healthcare workers, similar to the strategy for the promotion of influenza vaccination [138]. An official website was set up to deliver updated information related to COVID-19 vaccination in multiple languages [139]. The daily number of COVID-19 vaccine doses administered was made available in the public domain of "Our World in Data" [140], and subsequently analyzed by the Centre for Health Protection for daily reporting from 5 March 2022 (day 796) [141].

Vaccine hesitancy is a major challenge in promoting the COVID-19 vaccination campaign in the locality. In a cross-sectional survey of a random sample of 1501 Hong Kong residents aged 18 years or older in April 2020, only 45.3% of the participants intended to vaccinate against SARS-CoV-2 when available [142]. Subsequently, an online survey was conducted during an early stage of a community-based COVID-19 vaccination campaign in Hong Kong. Among the 883 participants (67.5% females, 54.5% aged 18–39), 30.6% had low vaccine hesitancy, 27.4% had high vaccine hesitancy, and 27.5% had vaccine rejection [143]. Vaccine hesitancy is not an uncommon phenomenon in the general population of developed and developing countries [144–146]. Vaccine hesitancy was also observed in healthcare workers as well as medical students [147–149]. The most predominant predictors of vaccine hesitancy were a lower perceived risk of getting infected, a lower level of institutional trust, not being vaccinated against influenza, lower levels of perceived severity of COVID-19, or stronger beliefs that the vaccination would cause side effects or be unsafe, as illustrated in a systemic review on the global predictors of COVID-19 vaccine hesitancy [150]. Institutional trust is an important factor that may increase vaccine willingness and uptake. Trust in the government and civil societies tended to strengthen the positive effect of information overload and reduce the negative impact of misinformation on vaccine willingness and uptake [151]. Vaccine complacency, an unintended side effect of successful control of COVID-19, is another important factor that adversely affects the intention to receive the vaccine in Hong Kong [152]. In a cross-sectional online survey of 1205 nurses, less than two-thirds of nurses intended to take the COVID-19 vaccine when available. Stronger COVID-19 vaccination intention was associated with less complacency, together with younger age, more confidence, and more collective responsibility [153]. Similarly, lower complacency, greater anxiety, confidence in the vaccine, and collective responsibility contributed to a greater likelihood of intended vaccination in a population-based online survey in Hong Kong [154]. Other regions using strategies to suppress and almost eliminate COVID-19 including Taiwan, Macau, the Chinese mainland, Tonga, and Western Australia have also experienced vaccine complacency [155].

Therefore, an administrative measure, the Vaccine Pass, was implemented to overcome vaccine hesitancy in Hong Kong. The Vaccine Pass was incorporated in a mobile app known as LeaveHomeSafe, which could be downloaded from the App Store, Google Play, AppGallery, and APK File [156]. To protect the unvaccinated and to boost vaccination coverage in the community, the Vaccine Pass arrangement was implemented for entry into all catering business and scheduled premises from 24 February 2022 (day 422) [157,158]. The Vaccine Pass was introduced for local residents in stages (Table 5). The daily number of COVID-19 vaccines administrated in relation to the implementation of the Vaccine Pass is illustrated in Figure 9.

Up to 25 September 2022 (day 1000), 6,683,654 (90.3%) and 5,351,219 (72.3%) of the general population had received the second and third doses of the COVID-19 vaccine, respectively, in Hong Kong. Regarding the population with two doses of vaccination, the age-specific vaccination rate remained low at the extremes of age (Figure 10), which led to over 9000 deaths during the primary Omicron wave. As for vaccine efficacy, in a study covering the period between 31 December 2020 and 16 March 2022, during which 13.2 million vaccine doses had been administered, two doses of either vaccine formulation were shown to protect against severe disease and death within 28 days after confirmation of infection by a positive test. Higher effectiveness was demonstrated among the subgroup of adults aged 60 years or older who had received BNT162b2 (vaccine effectiveness 89.3%) when compared with CoronaVac (69.9%). Three doses of either vaccine offered very high levels of protection against severe or fatal outcomes (97.9%) [159]. COVID-19 vaccines have been shown to be safe in patients with cancer and chronic diseases and in people aged 60 years or older in Hong Kong [160–163].

Table 5. Implementation of the Vaccine Pass to enter specified premises in order to enhance COVID-19 vaccination coverage in Hong Kong up to 25 September 2022 (day 1000) [a].

Stage	Time Period	Requirement
First	24 February 2022 (day 422) to 29 April 2022 (day 851)	Persons $\geq$ 12 years are required to receive at least one dose of COVID-19 vaccine in order to use the Vaccine Pass to enter specified premises.
Second	30 April 2022 (day 852) to 30 May 2022 (day 882)	For persons $\geq$ 18 years: Two doses of COVID-19 vaccine. For persons aged 12 to 17 years: 1st dose, if within 6 months of 1st dose, or 2nd dose, if after 6 months from 1st dose.
Third	31 May 2022 (day 883) onwards [b]	For persons $\geq$ 18 years: 2nd dose, if within 6 months of 2nd dose, or 3rd dose, if after 6 months from 2nd dose. For persons aged 12 to 17 years: 2nd dose, if within 6 months of 2nd dose, or 3rd dose, if after 6 months from 2nd dose.

[a] Specified premises: all catering business premises (including bars or pubs), amusement game centers, bathhouses, fitness centers, places of amusement, places of public entertainment, party rooms, beauty parlors and massage establishments, clubs or nightclubs, karaoke establishments, mahjong-tin kau premises, club houses, sports premises, swimming pools, cruise ships, event premises, religious premises, barber shops or hair salons, shopping malls, department stores, supermarkets, markets, and hotels or guesthouses. [b] The government further extended the Vaccine Pass to cover children aged 5–11 years after 30 September 2022 (1st dose if within 3 months, 2nd dose if after 3 months from 1st dose).

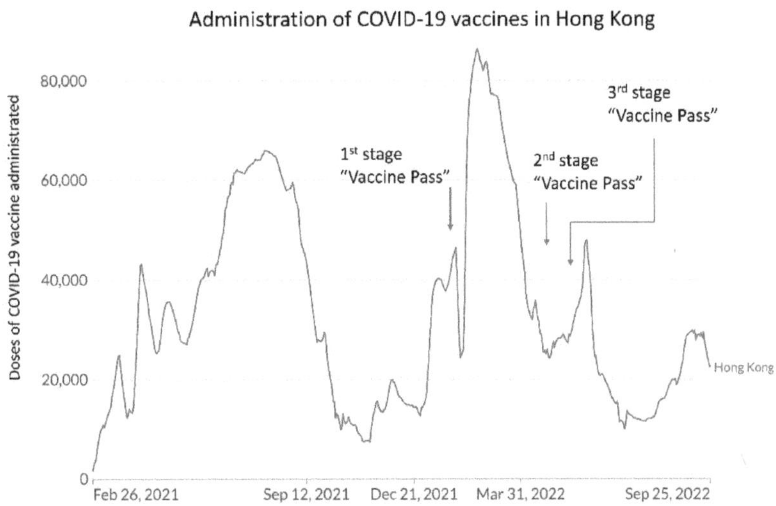

Figure 9. Daily doses of COVID-19 vaccine administrated in Hong Kong from 26 February 2021 (day 424) to 25 September 2022 (day 1000). The figure was generated from the website of "Our World in Data" [140] and formatted for presentation.

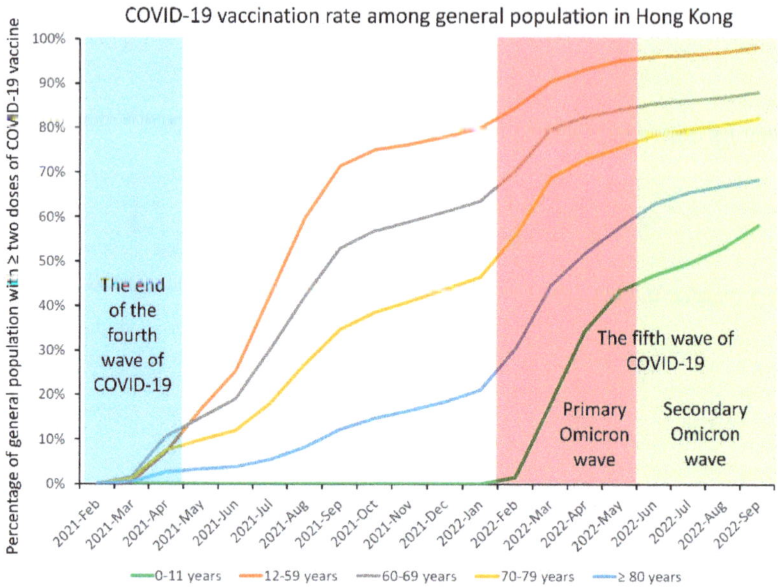

Figure 10. Monthly trend of COVID-19 vaccination rate among the general population until 25 September 2022 (day 1000) in Hong Kong. On 1 August 2022 (day 945), the Scientific Committee on Emerging and Zoonotic Diseases and the Scientific Committee on Vaccine Preventable Diseases of the Centre for Health Protection recommended using CoronaVac for those aged 6 months to under 3 years.

6.6. Other Non-Pharmaceutical Intervention in the Community

As mask-off activities such as dining and drinking in restaurants are associated with COVID-19 transmission [117], legislation was implemented on air-change requirements or utilization of air purifiers in dine-in catering premises [164]. Specifically, catering business operators were required to ensure six or more air changes per hour in the seating area. Otherwise, air purifiers were to be installed as an alternative measure to reduce the risk of airborne transmission of SARS-CoV-2 [165].

7. The Way Forward

Although the stringent control measures of universal masking, social distancing, testing for isolation and quarantine, contact tracing, and border testing with quarantine worked well for 2 years with previous SARS-CoV-2 variants or lineages, such measures failed to control the Omicron variant in Hong Kong. The global spread of the more benign Omicron variant might be the beginning of the end for the COVID-19 pandemic [166,167], given that a high proportion of the global population has some immunity from natural infection, vaccination, or hybrid immunity. The risk of COVID-19 related death associated with Omicron was comparatively lower than that of the Delta variant [168,169] and was also lower than the previous variants [170], even during the pre-vaccination era [171]. Paradoxically, the crude population mortality rate for COVID-19 in Hong Kong (37.7 per million) during the peak level of the fifth wave, caused by the BA.2 sub-lineage, was among the highest reported worldwide because of the low COVID-19 vaccination coverage in the elderly population [172]. The severe community outbreak of Omicron BA.2 could be attributed to the far higher airborne transmissibility as well as the lower population immunity [173]. The overall seropositive rate of IgG against the receptor-binding domain of SARS-CoV-2 increased from 52% in December 2021 to 89% in May 2022, at the end of the fifth wave of COVID-19 in Hong Kong [174]. Hybrid immunity has been established locally and the case fatality rate of COVID-19 during the secondary Omicron wave, predominantly caused by Omicron BA.5, was about 0.1%, which was comparable to the case fatality rate of seasonal influenza. This could be attributable to the marked increase in vaccination rate and early antiviral treatment. However, the case fatality rate among unvaccinated persons still remained higher at 0.6% overall and 9% among elderly aged 80 years or above. After analyzing scientific data and striking a balance among factors such as transmission risks, the public health policy of compulsory quarantine in designated quarantine hotels for all inbound travelers was no longer required from 26 September 2022 (day 1001), facilitating international travel and recovery of the economy while being able to contain the spread of infection. In Hong Kong, the healthcare services have gradually resumed normalcy. Healthcare workers were fully vaccinated and are well trained for COVID-19. The modalities of pharmaceutical intervention, including remdesivir, nirmatrelvir/ritonavir, molnupiravir, baricitinib, tocilizumab, and tixagevimab co-packaged with cilgavimab were made available to patients according to local studies [175–180] and in accordance with the evolving treatment guidelines [181,182]. The government will continue to adjust the anti-pandemic strategy based on scientific data and evidence [183].

Analogous to the influenza virus with the phenomenon of antigenic drift and shift [184], new variants of SARS-CoV-2 are expected to develop with time. Since September 2022, Omicron sub-lineages BA.2.75, BF.7, and BA.4.6 have been reported to be increasing in Asia (India, Nepal, Singapore), Europe (Belgium, France, German), and the United States, respectively [185]. These three sub-lineages could substantially escape neutralizing antibodies induced by vaccination or previous infection, or both [186]. International collaboration is required to closely monitor the evolving SARS-CoV-2 variants and see whether the T lymphocyte response induced by infection, vaccination, or hybrid immunity will continue to turn this virulent pandemic virus into an endemic benign respiratory virus. Infection control and public health measures remain the most important non-pharmaceutical interventions against COVID-19 and any other emerging infectious diseases.

Author Contributions: Conceptualization, S.-C.W. and V.C.-C.C.; methodology, S.-C.W. and V.C.-C.C.; formal analysis, S.-C.W., P.-L.H. and V.C.-C.C.; investigation, S.-C.W., A.K.-W.A., J.Y.-C.L., P.-L.H., K.K.-W.T. and V.C.-C.C.; resources, A.K.-W.A., J.Y.-C.L., I.F.-N.H. and K.K.-W.T.; data curation, S.-C.W. and V.C.-C.C.; writing—original draft preparation, S.-C.W. and V.C.-C.C.; writing—review and editing, S.-C.W., A.K.-W.A., J.Y.-C.L., P.-L.H., K.K.-W.T., K.-Y.Y. and V.C.-C.C.; supervision, K.-Y.Y. and V.C.-C.C.; project administration, S.-C.W. and V.C.-C.C.; funding acquisition, K.-Y.Y. All authors have read and agreed to the published version of the manuscript.

Funding: This study was supported by the Health and Medical Research Fund (HMRF) Commissioned Research on the Control of Infectious Disease (Phase IV), CID-HKU1-16, Health Bureau, Hong Kong Special Administrative Regional Government.

Institutional Review Board Statement: Not applicable.

Informed Consent Statement: Not applicable.

Data Availability Statement: All the data are publicly available.

Acknowledgments: We are grateful to the contribution of our frontline staff and laboratory staff in enforcing the infection control measures and undertaking laboratory work, respectively, in Hong Kong.

Conflicts of Interest: All authors report no conflict of interest relevant to this article.

References

1. Morens, D.M.; Fauci, A.S. Emerging Pandemic Diseases: How We Got to COVID-19. *Cell* **2020**, *182*, 1077–1092. [CrossRef] [PubMed]
2. Raoult, D.; Mouffok, N.; Bitam, I.; Piarroux, R.; Drancourt, M. Plague: History and contemporary analysis. *J. Infect.* **2013**, *66*, 18–26. [CrossRef] [PubMed]
3. Butler, T. Plague into the 21st century. *Clin. Infect. Dis.* **2009**, *49*, 736–742. [CrossRef]
4. Taubenberger, J.K.; Morens, D.M. 1918 Influenza: The mother of all pandemics. *Emerg. Infect. Dis.* **2006**, *12*, 15–22. [CrossRef]
5. Bootsma, M.C.; Ferguson, N.M. The effect of public health measures on the 1918 influenza pandemic in U.S. cities. *Proc. Natl. Acad. Sci. USA* **2007**, *104*, 7588–7593. [CrossRef] [PubMed]

6. Honigsbaum, M. Revisiting the 1957 and 1968 influenza pandemics. *Lancet* **2020**, *395*, 1824–1826. [CrossRef]
7. Jester, B.J.; Uyeki, T.M.; Jernigan, D.B. Fifty Years of Influenza A(H3N2) Following the Pandemic of 1968. *Am. J. Public Health* **2020**, *110*, 669–676. [CrossRef]
8. Cheng, V.C.; To, K.K.; Tse, H.; Hung, I.F.; Yuen, K.-Y. Two years after pandemic influenza A/2009/H1N1: What have we learned? *Clin. Microbiol. Rev.* **2012**, *25*, 223–263. [CrossRef]
9. Dawood, F.S.; Iuliano, A.D.; Reed, C.; Meltzer, M.I.; Shay, D.K.; Cheng, P.Y.; Bandaranayake, D.; Breiman, R.F.; Brooks, W.A.; Buchy, P.; et al. Estimated global mortality associated with the first 12 months of 2009 pandemic influenza A H1N1 virus circulation: A modelling study. *Lancet Infect. Dis.* **2012**, *12*, 687–695. [CrossRef]
10. Iuliano, A.D.; Roguski, K.M.; Chang, H.H.; Muscatello, D.J.; Palekar, R.; Tempia, S.; Cohen, C.; Gran, J.M.; Schanzer, D.; Cowling, B.J.; et al. Estimates of global seasonal influenza-associated respiratory mortality: A modelling study. *Lancet* **2018**, *391*, 1285–1300. [CrossRef]
11. Rajendran, M.; Nachbagauer, R.; Ermler, M.E.; Bunduc, P.; Amanat, F.; Izikson, R.; Cox, M.; Palese, P.; Eichelberger, M.; Krammer, F. Analysis of Anti-Influenza Virus Neuraminidase Antibodies in Children, Adults, and the Elderly by ELISA and Enzyme Inhibition: Evidence for Original Antigenic Sin. *mBio* **2017**, *8*, e02281-16. [CrossRef] [PubMed]
12. Mulabbi, E.N.; Tweyongyere, R.; Byarugaba, D.K. The history of the emergence and transmission of human coronaviruses. *Onderstepoort. J. Vet. Res.* **2021**, *88*, e1–e8. [CrossRef] [PubMed]
13. van der Hoek, L.; Pyrc, K.; Jebbink, M.F.; Vermeulen-Oost, W.; Berkhout, R.J.; Wolthers, K.C.; Wertheim-van Dillen, P.M.; Kaandorp, J.; Spaargaren, J.; Berkhout, B. Identification of a new human coronavirus. *Nat. Med.* **2004**, *10*, 368–373. [CrossRef] [PubMed]
14. Woo, P.C.; Lau, S.K.; Chu, C.-M.; Chan, K.-H.; Tsoi, H.-W.; Huang, Y.; Wong, B.H.; Poon, R.W.; Cai, J.J.; Luk, W.-K.; et al. Characterization and complete genome sequence of a novel coronavirus, coronavirus HKU1, from patients with pneumonia. *J. Virol.* **2005**, *79*, 884–895. [CrossRef]
15. Cheng, V.C.; Lau, S.K.; Woo, P.C.; Yuen, K.-Y. Severe acute respiratory syndrome coronavirus as an agent of emerging and reemerging infection. *Clin. Microbiol. Rev.* **2007**, *20*, 660–694. [CrossRef]
16. Chan, J.F.; Lau, S.K.; To, K.K.; Cheng, V.C.; Woo, P.C.; Yuen, K.-Y. Middle East respiratory syndrome coronavirus: Another zoonotic betacoronavirus causing SARS-like disease. *Clin. Microbiol. Rev.* **2015**, *28*, 465–522. [CrossRef]
17. Bloom, D.E.; Cadarette, D. Infectious Disease Threats in the Twenty-First Century: Strengthening the Global Response. *Front. Immunol.* **2019**, *10*, 549. [CrossRef]
18. Honigsbaum, M. Disease X and other unknowns. *Lancet* **2019**, *393*, 1496–1497. [CrossRef]
19. Van Kerkhove, M.D.; Ryan, M.J.; Ghebreyesus, T.A. Preparing for "Disease X". *Science* **2021**, *374*, 377. [CrossRef]
20. Jiang, S.; Shi, Z.L. The First Disease X is Caused by a Highly Transmissible Acute Respiratory Syndrome Coronavirus. *Virol. Sin.* **2020**, *35*, 263–265. [CrossRef]
21. The Centre for Health Protection Closely Monitors Cluster of Pneumonia Cases on Mainland. Press Release of the Department of Health, Hong Kong Special Administrative Region. Available online: https://www.info.gov.hk/gia/general/201912/31/P2019123100667.htm (accessed on 1 October 2022).
22. Cheng, V.C.-C.; Wong, S.-C.; To, K.K.-W.; Ho, P.-L.; Yuen, K.-Y. Preparedness and proactive infection control measures against the emerging novel coronavirus in China. *J. Hosp. Infect.* **2020**, *104*, 254–255. [CrossRef] [PubMed]
23. Lu, R.; Zhao, X.; Li, J.; Niu, P.; Yang, B.; Wu, H.; Wang, W.; Song, H.; Huang, B.; Zhu, N.; et al. Genomic characterisation and epidemiology of 2019 novel coronavirus: Implications for virus origins and receptor binding. *Lancet* **2020**, *395*, 565–574. [CrossRef]
24. Delaune, D.; Hul, V.; Karlsson, E.A.; Hassanin, A.; Ou, T.P.; Baidaliuk, A.; Gámbaro, F.; Prot, M.; Tu, V.T.; Chea, S.; et al. A novel SARS-CoV-2 related coronavirus in bats from Cambodia. *Nat. Commun.* **2021**, *12*, 6563. [CrossRef]
25. Lam, T.T.; Jia, N.; Zhang, Y.W.; Shum, M.H.; Jiang, J.F.; Zhu, H.C.; Tong, Y.G.; Shi, Y.X.; Ni, X.B.; Liao, Y.S.; et al. Identifying SARS-CoV-2-related coronaviruses in Malayan pangolins. *Nature* **2020**, *583*, 282–285. [CrossRef] [PubMed]
26. Temmam, S.; Vongphayloth, K.; Baquero, E.; Munier, S.; Bonomi, M.; Regnault, B.; Douangboubpha, B.; Karami, Y.; Chrétien, D.; Sanamxay, D.; et al. Bat coronaviruses related to SARS-CoV-2 and infectious for human cells. *Nature* **2022**, *604*, 330–336. [CrossRef] [PubMed]
27. Lau, S.K.; Woo, P.C.; Li, K.S.; Huang, Y.; Tsoi, H.-W.; Wong, B.H.; Wong, S.S.; Leung, S.-Y.; Chan, K.-H.; Yuen, K.-Y. Severe acute respiratory syndrome coronavirus-like virus in Chinese horseshoe bats. *Proc. Natl. Acad. Sci. USA* **2005**, *102*, 14040–14045. [CrossRef]
28. Lau, S.K.; Chan, J.F. Coronaviruses: Emerging and re-emerging pathogens in humans and animals. *Virol. J.* **2015**, *12*, 209. [CrossRef]
29. Coronavirus Disease (COVID-19) Pandemic, World Health Organization. Available online: https://www.who.int/europe/emergencies/situations/covid-19 (accessed on 1 October 2022).
30. World Health Organization. Coronavirus Disease 2019 (COVID-19) Situation Report-51. Available online: https://www.who.int/docs/default-source/coronaviruse/situation-reports/20200311-sitrep-51-covid-19.pdf?sfvrsn=1ba62e57_10 (accessed on 1 October 2022).
31. COVID-19 Weekly Epidemiological Update, Edition 111 Published 28 September 2022, World Health Organization. Available online: https://www.who.int/publications/m/item/weekly-epidemiological-update-on-covid-19---28-september-2022 (accessed on 1 October 2022).

32. Lockdowns Compared: Tracking Governments' Coronavirus Responses. Financial Times. Available online: https://ig.ft.com/coronavirus-lockdowns/ (accessed on 1 October 2022).
33. Onyeaka, H.; Anumudu, C.K.; Al-Sharify, Z.T.; Egele-Godswill, E.; Mbaegbu, P. COVID-19 pandemic: A review of the global lockdown and its far-reaching effects. *Sci. Prog.* **2021**, *104*, 368504211019854. [CrossRef]
34. Countries by Population Density. Available online: https://worldpopulationreview.com/country-rankings/countries-by-density (accessed on 1 October 2022).
35. Tracking SARS-CoV-2 Variants, World Health Organization. Available online: https://www.who.int/activities/tracking-SARS-CoV-2-variants (accessed on 1 October 2022).
36. Coronavirus Disease (COVID-19) Weekly Epidemiological Update and Weekly Operational Update, World Health Organization. Available online: https://www.who.int/emergencies/diseases/novel-coronavirus-2019/situation-reports (accessed on 2 October 2022).
37. COVID-19 Situation Update Worldwide. European Centre for Disease Prevention and Control. Available online: https://www.ecdc.europa.eu/en/geographical-distribution-2019-ncov-cases (accessed on 2 October 2022).
38. COVID-19 Dashboard by the Center for Systems Science and Engineering at Johns Hopkins University. Available online: https://coronavirus.jhu.edu/map.html (accessed on 2 October 2022).
39. COVID-19 Coronavirus Pandemic, Worldmeter. Available online: https://www.worldometers.info/coronavirus/ (accessed on 2 October 2022).
40. Coronavirus Pandemic (COVID-19), Our World in Data. Available online: https://ourworldindata.org/coronavirus (accessed on 2 October 2022).
41. Coronavirus World Map: Tracking the Global Outbreak, The New York Times. Available online: https://www.nytimes.com/interactive/2021/world/covid-cases.html (accessed on 2 October 2022).
42. Covid Map: Coronavirus Cases, Deaths, Vaccinations by Country, BBC News. Available online: https://www.bbc.com/news/world-51235105 (accessed on 2 October 2022).
43. COVID-19 (SARS-CoV-2 Coronavirus), American Society for Microbiology. Available online: https://asm.org/Resource-Pages/COVID-19-Resources (accessed on 2 October 2022).
44. Davies, N.G.; Abbott, S.; Barnard, R.C.; Jarvis, C.I.; Kucharski, A.J.; Munday, J.D.; Pearson, C.A.B.; Russell, T.W.; Tully, D.C.; Washburne, A.D.; et al. Estimated transmissibility and impact of SARS-CoV-2 lineage B.1.1.7 in England. *Science* **2021**, *372*, eabg3055. [CrossRef]
45. Volz, E.; Mishra, S.; Chand, M.; Barrett, J.C.; Johnson, R.; Geidelberg, L.; Hinsley, W.R.; Laydon, D.J.; Dabrera, G.; O'Toole, Á.; et al. Assessing transmissibility of SARS-CoV-2 lineage B.1.1.7 in England. *Nature* **2021**, *593*, 266–269. [CrossRef]
46. Earnest, R.; Uddin, R.; Matluk, N.; Renzette, N.; Turbett, S.E.; Siddle, K.J.; Loreth, C.; Adams, G.; Tomkins-Tinch, C.H.; Petrone, M.E.; et al. Comparative transmissibility of SARS-CoV-2 variants Delta and Alpha in New England, USA. *Cell. Rep. Med.* **2022**, *3*, 100583. [CrossRef] [PubMed]
47. Ito, K.; Piantham, C.; Nishiura, H. Relative instantaneous reproduction number of Omicron SARS-CoV-2 variant with respect to the Delta variant in Denmark. *J. Med. Virol.* **2022**, *94*, 2265–2268. [CrossRef] [PubMed]
48. Jalali, N.; Brustad, H.K.; Frigessi, A.; MacDonald, E.A.; Meijerink, H.; Feruglio, S.L.; Nygård, K.M.; Rø, G.; Madslien, E.H.; de Blasio, B.F. Increased household transmission and immune escape of the SARS-CoV-2 Omicron compared to Delta variants. *Nat. Commun.* **2022**, *13*, 5706. [CrossRef] [PubMed]
49. COVID-19 Thematic Website. Centre for Health Protection. Department of Health, The Government of the Hong Kong Special Administrative Region. Available online: https://www.coronavirus.gov.hk/eng/index.html (accessed on 2 October 2022).
50. Cheng, V.C.; Wong, S.-C.; Au, A.K.; Zhang, C.; Chen, J.H.; So, S.Y.; Li, X.; Wang, Q.; Lu, K.K.; Lung, D.C.; et al. Explosive outbreak of SARS-CoV-2 Omicron variant is associated with vertical transmission in high-rise residential buildings in Hong Kong. *Build. Environ.* **2022**, *221*, 109323. [CrossRef]
51. Gu, H.; Chu, D.K.-W.; Chang, L.D.-J.; Cheuk, S.S.-Y.; Gurung, S.; Krishnan, P.; Ng, D.Y.-M.; Liu, G.Y.-Z.; Wan, C.K.-C.; Xie, R.; et al. Genetic Diversity of SARS-CoV-2 among Travelers Arriving in Hong Kong. *Emerg. Infect. Dis.* **2021**, *27*, 2666–2668. [CrossRef]
52. Leung, K.S.; Ng, T.T.; Wu, A.K.; Yau, M.C.; Lao, H.-Y.; Choi, M.-P.; Tam, K.K.; Lee, L.-K.; Wong, B.K.; Ho, A.Y.; et al. Territorywide Study of Early Coronavirus Disease Outbreak, Hong Kong, China. *Emerg. Infect. Dis.* **2021**, *27*, 196–204. [CrossRef]
53. Chan, W.-M.; Ip, J.D.; Chu, A.W.; Tse, H.; Tam, A.R.; Li, X.; Kwan, M.Y.; Yau, Y.-S.; Leung, W.-S.; Chik, T.S.; et al. Phylogenomic analysis of COVID-19 summer and winter outbreaks in Hong Kong: An observational study. *Lancet Reg. Health West Pac.* **2021**, *10*, 100130. [CrossRef]
54. Gu, H.; Cheng, S.S.-M.; Krishnan, P.; Ng, D.Y.-M.; Chang, L.D.-J.; Liu, G.Y.-Z.; Cheuk, S.S.-Y.; Hui, M.M.-Y.; Fan, M.C.-Y.; Wan, J.H.-L.; et al. Monitoring International Travelers Arriving in Hong Kong for Genomic Surveillance of SARS-CoV-2. *Emerg. Infect. Dis.* **2022**, *28*, 247–250. [CrossRef]
55. Cheng, V.C.-C.; Ip, J.D.; Chu, A.W.-H.; Tam, A.R.; Chan, W.-M.; Abdullah, S.M.-U.; Chan, B.P.-C.; Wong, S.-C.; Kwan, M.Y.-W.; Chua, G.T.; et al. Rapid Spread of Severe Acute Respiratory Syndrome Coronavirus 2 (SARS-CoV-2) Omicron Subvariant BA.2 in a Single-Source Community Outbreak. *Clin. Infect. Dis.* **2022**, *75*, e44–e49. [CrossRef]

56. Chan, J.F.-W.; Siu, G.K.-H.; Yuan, S.; Ip, J.D.; Cai, J.-P.; Chu, A.W.-H.; Chan, W.-M.; Abdullah, S.M.-U.; Luo, C.; Chan, B.P.-C.; et al. Probable Animal-to-Human Transmission of Severe Acute Respiratory Syndrome Coronavirus 2 (SARS-CoV-2) Delta Variant AY.127 Causing a Pet Shop-Related Coronavirus Disease 2019 (COVID-19) Outbreak in Hong Kong. *Clin. Infect. Dis.* **2022**, *75*, e76–e81. [CrossRef]
57. Kok, K.-H.; Wong, S.-C.; Chan, W.-M.; Wen, L.; Chu, A.W.; Ip, J.D.; Lee, L.K.; Wong, I.T.; Lo, H.W.; Cheng, V.C.; et al. Co-circulation of two SARS-CoV-2 variant strains within imported pet hamsters in Hong Kong. *Emerg. Microbes Infect.* **2022**, *11*, 689–698. [CrossRef] [PubMed]
58. Summary of Probable SARS Cases with Onset of Illness from 1 November 2002 to 31 July 2003 (Based on Data as of the 31 December 2003). World Health Organization. Available online: https://www.who.int/csr/sars/country/table2004_04_21/en/ (accessed on 4 October 2022).
59. Report of the Select Committee to Inquire into the Handling of the Severe Acute Respiratory Syndrome Outbreak by the Government and the Hospital Authority July 2004. Available online: https://www.legco.gov.hk/yr03-04/english/sc/sc_sars/reports/sars_rpt.htm (accessed on 4 October 2022).
60. SARS Expert Committee. Available online: https://www.sars-expertcom.gov.hk/english/reports/reports.html (accessed on 4 October 2022).
61. Benitez, M.A. Expert committee reviews SARS outbreak in Hong Kong. *Lancet* **2003**, *362*, 624. [CrossRef]
62. Message from the Controller, Centre for Health Protection, Department of Health, The Government of the Hong Kong Special Administrative Region. Available online: https://www.chp.gov.hk/en/static/23993.html (accessed on 4 October 2022).
63. Hu, B.; Guo, H.; Zhou, P.; Shi, Z.L. Characteristics of SARS-CoV-2 and COVID-19. *Nat. Rev. Microbiol.* **2021**, *19*, 141–154. [CrossRef] [PubMed]
64. Cheng, V.C.-C.; Wong, S.-C.; Chen, J.H.-K.; Yip, C.C.-Y.; Chuang, V.W.-M.; Tsang, O.T.-Y.; Sridhar, S.; Chan, J.F.-W.; Ho, P.-L.; Yuen, K.-Y. Escalating infection control response to the rapidly evolving epidemiology of the coronavirus disease 2019 (COVID-19) due to SARS-CoV-2 in Hong Kong. *Infect. Control Hosp. Epidemiol.* **2020**, *41*, 493–498. [CrossRef]
65. Cheng, V.C.; Wong, S.-C.; Tong, D.W.; Chuang, V.W.; Chen, J.H.; Lee, L.L.; To, K.K.; Hung, I.F.; Ho, P.-L.; Yeung, D.T.; et al. Multipronged infection control strategy to achieve zero nosocomial coronavirus disease 2019 (COVID-19) cases among Hong Kong healthcare workers in the first 300 days of the pandemic. *Infect. Control Hosp. Epidemiol.* **2022**, *43*, 334–343. [CrossRef]
66. To, K.K.; Tsang, O.T.; Yip, C.C.; Chan, K.-H.; Wu, T.-C.; Chan, J.M.; Leung, W.-S.; Chik, T.S.; Choi, C.Y.; Kandamby, D.H.; et al. Consistent Detection of 2019 Novel Coronavirus in Saliva. *Clin. Infect. Dis.* **2020**, *71*, 841–843. [CrossRef] [PubMed]
67. To, K.K.; Tsang, O.T.; Leung, W.-S.; Tam, A.R.; Wu, T.-C.; Lung, D.C.; Yip, C.C.; Cai, J.-P.; Chan, J.M.; Chik, T.S.; et al. Temporal profiles of viral load in posterior oropharyngeal saliva samples and serum antibody responses during infection by SARS-CoV-2: An observational cohort study. *Lancet Infect. Dis.* **2020**, *20*, 565–574. [CrossRef]
68. Chen, J.H.; Yip, C.C.; Poon, R.W.; Chan, K.-H.; Cheng, V.C.; Hung, I.F.; Chan, J.F.; Yuen, K.-Y.; To, K.K. Evaluating the use of posterior oropharyngeal saliva in a point-of-care assay for the detection of SARS-CoV-2. *Emerg. Microbes Infect.* **2020**, *9*, 1356–1359. [CrossRef]
69. Butler-Laporte, G.; Lawandi, A.; Schiller, I.; Yao, M.; Dendukuri, N.; McDonald, E.G.; Lee, T.C. Comparison of Saliva and Nasopharyngeal Swab Nucleic Acid Amplification Testing for Detection of SARS-CoV-2: A Systematic Review and Meta-analysis. *JAMA Intern. Med.* **2021**, *181*, 353–360. [CrossRef]
70. Hung, D.L.; Li, X.; Chiu, K.H.; Yip, C.C.; To, K.K.; Chan, J.F.; Sridhar, S.; Chung, T.W.; Lung, K.-C.; Liu, R.W.; et al. Early-Morning vs Spot Posterior Oropharyngeal Saliva for Diagnosis of SARS-CoV-2 Infection: Implication of Timing of Specimen Collection for Community-Wide Screening. *Open Forum Infect. Dis.* **2020**, *7*, ofaa210. [CrossRef]
71. Lynch, J.B.; Davitkov, P.; Anderson, D.J.; Bhimraj, A.; Cheng, V.C.; Guzman-Cottrill, J.; Dhindsa, J.; Duggal, A.; Jain, M.K.; Lee, G.M.; et al. Infectious Diseases Society of America Guidelines on Infection Prevention for Health Care Personnel Caring for Patients with Suspected or Known COVID-19. *Clin. Infect. Dis.* **2020**, ciaa1063. [CrossRef] [PubMed]
72. Lynch, J.B.; Davitkov, P.; Anderson, D.J.; Bhimraj, A.; Cheng, V.C.; Guzman-Cottrill, J.; Dhindsa, J.; Duggal, A.; Jain, M.K.; Lee, G.M.; et al. Infectious Diseases Society of America Guidelines on Infection Prevention for Healthcare Personnel Caring for Patients with Suspected or Known COVID-19. *Clin. Infect. Dis.* **2021**, ciab953. [CrossRef] [PubMed]
73. Wong, S.-C.; AuYeung, C.H.; Lam, G.K.; Leung, E.Y.; Chan, V.W.; Yuen, K.-Y.; Cheng, V.C. Is it possible to achieve 100 percent hand hygiene compliance during the coronavirus disease 2019 (COVID-19) pandemic? *J. Hosp. Infect.* **2020**, *105*, 779–781. [CrossRef] [PubMed]
74. Wong, S.-C.; Lam, G.K.; AuYeung, C.H.; Chan, V.W.; Wong, N.L.; So, S.Y.; Chen, J.H.; Hung, I.F.; Chan, J.F.; Yuen, K.-Y.; et al. Absence of nosocomial influenza and respiratory syncytial virus infection in the coronavirus disease 2019 (COVID-19) era: Implication of universal masking in hospitals. *Infect. Control Hosp. Epidemiol.* **2021**, *42*, 218–221. [CrossRef]
75. Cheng, V.C.; Wong, S.-C.; Chan, V.W.; So, S.Y.; Chen, J.H.; Yip, C.C.; Chan, K.-H.; Chu, H.; Chung, T.W.; Sridhar, S.; et al. Air and environmental sampling for SARS-CoV-2 around hospitalized patients with coronavirus disease 2019 (COVID-19). *Infect. Control Hosp. Epidemiol.* **2020**, *41*, 1258–1265. [CrossRef]
76. Wu, S.; Wang, Y.; Jin, X.; Tian, J.; Liu, J.; Mao, Y. Environmental contamination by SARS-CoV-2 in a designated hospital for coronavirus disease 2019. *Am. J. Infect. Control.* **2020**, *48*, 910–914. [CrossRef]
77. Ryu, B.H.; Cho, Y.; Cho, O.H.; Hong, S.I.; Kim, S.; Lee, S. Environmental contamination of SARS-CoV-2 during the COVID-19 outbreak in South Korea. *Am. J. Infect. Control.* **2020**, *48*, 875–879. [CrossRef]

78. Dancer, S.J. Reducing the risk of COVID-19 transmission in hospitals: Focus on additional infection control strategies. *Surgery* **2021**, *39*, 752–758. [CrossRef]
79. Ong, S.W.X.; Tan, Y.K.; Chia, P.Y.; Lee, T.H.; Ng, O.T.; Wong, M.S.Y.; Marimuthu, K. Air, surface environmental, and personal protective equipment contamination by severe acute respiratory syndrome coronavirus 2 (SARS-CoV-2) from a symptomatic patient. *JAMA* **2020**, *323*, 1610–1612. [CrossRef]
80. Li, Y.H.; Fan, Y.Z.; Jiang, L.; Wang, H.B. Aerosol and environmental surface monitoring for SARS-CoV-2 RNA in a designated hospital for severe COVID-19 patients. *Epidemiol. Infect.* **2020**, *148*, e154. [CrossRef]
81. Kim, U.J.; Lee, S.Y.; Lee, J.Y.; Lee, A.; Kim, S.E.; Choi, O.J.; Lee, J.S.; Kee, S.J.; Jang, H.C. Air and Environmental Contamination Caused by COVID-19 Patients: A Multi-Center Study. *J. Korean Med. Sci.* **2020**, *35*, e332. [CrossRef] [PubMed]
82. Guo, Z.D.; Wang, Z.Y.; Zhang, S.F.; Li, X.; Li, L.; Li, C.; Cui, Y.; Fu, R.B.; Dong, Y.Z.; Chi, X.Y.; et al. Aerosol and Surface Distribution of Severe Acute Respiratory Syndrome Coronavirus 2 in Hospital Wards, Wuhan, China, 2020. *Emerg. Infect. Dis.* **2020**, *26*, 1583–1591. [CrossRef] [PubMed]
83. Chia, P.Y.; Coleman, K.K.; Tan, Y.K.; Ong, S.W.X.; Gum, M.; Lau, S.K.; Lim, X.F.; Lim, A.S.; Sutjipto, S.; Lee, P.H.; et al. Detection of air and surface contamination by SARS-CoV-2 in hospital rooms of infected patients. *Nat. Commun.* **2020**, *11*, 2800. [CrossRef] [PubMed]
84. Wong, S.-C.; Chan, V.W.; AuYeung, C.H.; Chen, J.H.; Yip, C.C.; So, S.Y.; Li, X.; Lung, D.C.; Tsang, A.M.; To, K.K.; et al. Air dispersal of respiratory viruses other than severe acute respiratory coronavirus virus 2 (SARS-CoV-2) and the implication on hospital infection control. *Infect. Control Hosp. Epidemiol.* **2022**, 1–6. [CrossRef]
85. Wong, S.-C.; Yuen, L.L.; Chan, V.W.; Chen, J.H.; To, K.K.; Yuen, K.-Y.; Cheng, V.C. Airborne transmission of severe acute respiratory syndrome coronavirus 2 (SARS-CoV-2): What is the implication of hospital infection control? *Infect. Control Hosp. Epidemiol.* **2022**, *43*, 1522–1523. [CrossRef]
86. Wong, S.-C.; Chan, V.W.; Yuen, L.L.; Auyeung, C.H.; Leung, J.O.; Li, C.-K.; Kwok, M.O.; So, S.Y.; Chen, J.H.; Tam, A.R.; et al. Air dispersal of SARS-CoV-2 in general wards: Implication on hospital infection control during the fifth wave of COVID-19 due to Omicron variant in Hong Kong. *Infect. Control Hosp. Epidemiol.* 2022; *accepted*; *in press*.
87. Wong, S.-C.; Lam, G.K.; Chen, J.H.; Li, X.; Ip, F.T.; Yuen, L.L.; Chan, V.W.; AuYeung, C.H.; So, S.Y.; Ho, P.-L.; et al. Air dispersal of multidrug-resistant Acinetobacter baumannii: Implications for nosocomial transmission during the COVID-19 pandemic. *J. Hosp. Infect.* **2021**, *116*, 78–86. [CrossRef]
88. Wong, S.-C.; Chen, J.H.; Yuen, L.L.; Chan, V.W.; AuYeung, C.H.; Leung, S.S.; So, S.Y.; Chan, B.W.; Li, X.; Leung, J.O.; et al. Air dispersal of meticillin-resistant Staphylococcus aureus in residential care homes for the elderly: Implications for transmission during the COVID-19 pandemic. *J. Hosp. Infect.* **2022**, *123*, 52–60. [CrossRef]
89. Hung, I.F.; Cheng, V.C.; Li, X.; Tam, A.R.; Hung, D.L.; Chiu, K.H.; Yip, C.C.; Cai, J.-P.; Ho, D.T.; Wong, S.-C.; et al. SARS-CoV-2 shedding and seroconversion among passengers quarantined after disembarking a cruise ship: A case series. *Lancet Infect. Dis.* **2020**, *20*, 1051–1060. [CrossRef]
90. To, K.K.; Cheng, V.C.; Cai, J.-P.; Chan, K.-H.; Chen, L.L.; Wong, L.-H.; Choi, C.Y.; Fong, C.H.; Ng, A.C.; Lu, L.; et al. Seroprevalence of SARS-CoV-2 in Hong Kong and in residents evacuated from Hubei province, China: A multicohort study. *Lancet Microbe.* **2020**, *1*, e111–e118. [CrossRef]
91. Temporary Test Centres Speed up Tests for People Upon Arrival, Press Releases, The Government of the Hong Kong Special Administrative Region. Available online: https://www.info.gov.hk/gia/general/202003/19/P2020031900664.htm (accessed on 4 October 2022).
92. Wong, S.-C.; Leung, M.; Lee, L.L.; Chung, K.-L.; Cheng, V.C. Infection control challenge in setting up a temporary test centre at Hong Kong International Airport for rapid diagnosis of COVID-19 due to SARS-CoV-2. *J. Hosp. Infect.* **2020**, *105*, 571–573. [CrossRef]
93. Wong, S.-C.; Leung, M.; Tong, D.W.; Lee, L.L.; Leung, W.L.; Chan, F.W.; Chen, J.H.; Hung, I.F.; Yuen, K.-Y.; Yeung, D.T.; et al. Infection control challenges in setting up community isolation and treatment facilities for patients with coronavirus disease 2019 (COVID-19): Implementation of directly observed environmental disinfection. *Infect. Control Hosp. Epidemiol.* **2021**, *42*, 1037–1045. [CrossRef] [PubMed]
94. Chen, S.; Zhang, Z.; Yang, J.; Wang, J.; Zhai, X.; Bärnighausen, T.; Wang, C. Fangcang shelter hospitals: A novel concept for responding to public health emergencies. *Lancet* **2020**, *395*, 1305–1314. [CrossRef]
95. Fang, D.; Pan, S.; Li, Z.; Yuan, T.; Jiang, B.; Gan, D.; Sheng, B.; Han, J.; Wang, T.; Liu, Z. Large-scale public venues as medical emergency sites in disasters: Lessons from COVID-19 and the use of Fangcang shelter hospitals in Wuhan, China. *BMJ Glob. Health* **2020**, *5*, e002815. [CrossRef] [PubMed]
96. Coronavirus: Singapore Expo to House First Batch of Mild Cases Tomorrow to Free Up Hospitals, Straits Times Website, Published 9 April 2020. Available online: https://www.straitstimes.com/singapore/health/spore-expo-to-house-first-batch-of-mild-cases-tomorrow-to-free-up-hospitals (accessed on 5 October 2022).
97. UK Opens 4000-Bed Coronavirus Field Hospital, Deutsche Welle Website, Published 3 April 2020. Available online: https://www.dw.com/en/uk-opens-4000-bed-coronavirus-field-hospital/a-53005844 (accessed on 5 October 2022).
98. Coronavirus: Wales' 19 field Hospitals Cost £166m to Build, BBC News Website, Published 21 May 2020. Available online: https://www.bbc.com/news/uk-wales-52752976 (accessed on 5 October 2022).

99. Naganathan, S.; Meehan-Coussee, K.; Pasichow, S.; Rybasack-Smith, H.; Binder, W.; Beaudoin, F.; Musits, A.N.; Sutton, E.; Petrone, G.; Levine, A.C.; et al. From Concerts to COVID: Transforming the RI Convention Center into an Alternate Hospital Site in under a Month. *Rhode Isl. Med. J.* **2020**, *103*, 8–13.
100. Cheng, V.C.; Lung, D.C.; Wong, S.-C.; Au, A.K.; Wang, Q.; Chen, H.; Li, X.; Chu, A.W.; Ip, J.D.; Chan, W.-M.; et al. Outbreak investigation of airborne transmission of Omicron (B.1.1.529)-SARS-CoV-2 variant of concern in a restaurant: Implication for enhancement of indoor air dilution. *J. Hazard. Mater.* **2022**, *430*, 128504. [CrossRef]
101. CE Inspects Handover of Community Isolation Facility in Penny's Bay, Press Releases, The Government of the Hong Kong Special Administrative Region. Available online: https://www.info.gov.hk/gia/general/202204/19/P2022041900544.htm (accessed on 6 October 2022).
102. Kai Tak Community Isolation Facility Commences Operation. Press Releases, The Government of the Hong Kong Special Administrative Region. Available online: https://www.info.gov.hk/gia/general/202208/19/P2022081900552.htm?fontSize=1 (accessed on 6 October 2022).
103. COVID-19 has Infected Some 570,000 health Workers and Killed 2500 in the Americas, Pan American Health Organization, World Health Organization. Available online: https://www.paho.org/en/news/2-9-2020-covid-19-has-infected-some-570000-health-workers-and-killed-2500-americas-paho (accessed on 5 October 2022).
104. Bandyopadhyay, S.; Baticulon, R.E.; Kadhum, M.; Alser, M.; Ojuka, D.K.; Badereddin, Y.; Kamath, A.; Parepalli, S.A.; Brown, G.; Iharchane, S.; et al. Infection and mortality of healthcare workers worldwide from COVID-19: A systematic review. *BMJ Glob. Health.* **2020**, *5*, e003097. [CrossRef]
105. Health and Care Worker Deaths during COVID-19, World Health Organization. Available online: https://www.who.int/news/item/20-10-2021-health-and-care-worker-deaths-during-covid-19 (accessed on 5 October 2022).
106. Cheng, V.C.; Wong, S.-C.; Yuen, K.-Y. Estimating Coronavirus Disease 2019 Infection Risk in Health Care Workers. *JAMA Netw. Open* **2020**, *3*, e209687. [CrossRef]
107. Cheng, V.C.-C.; Wong, S.-C.; Chuang, V.W.-M.; So, S.Y.-C.; Chen, J.H.-K.; Sridhar, S.; To, K.K.-W.; Chan, J.F.-W.; Hung, I.F.-N.; Ho, P.-L.; et al. Absence of nosocomial transmission of coronavirus disease 2019 (COVID-19) due to SARS-CoV-2 in the prepandemic phase in Hong Kong. *Am. J. Infect. Control* **2020**, *48*, 890–896. [CrossRef]
108. Cheng, V.C.; Fung, K.S.; Siu, G.K.; Wong, S.-C.; Cheng, L.S.; Wong, M.-S.; Lee, L.-K.; Chan, W.-M.; Chau, K.-Y.; Leung, J.S.; et al. Nosocomial Outbreak of Coronavirus Disease 2019 by Possible Airborne Transmission Leading to a Superspreading Event. *Clin. Infect. Dis.* **2021**, *73*, e1356–e1364. [CrossRef]
109. Over 12,500 Staff at Hong Kong's Overloaded Public Hospitals Contracted COVID During 5th Wave, Hong Kong Free Press. Available online: https://hongkongfp.com/2022/03/11/over-12500-staff-at-hong-kongs-overloaded-pubic-hospitals-contracted-covid-during-5th-wave-ae-chief-in-tears/ (accessed on 5 October 2022).
110. Wong, S.-C.; Chan, V.W.; Yuen, L.L.; Auyeung, C.H.; Leung, J.O.; Li, C.-K.; Kwok, M.O.; So, S.Y.; Chen, J.H.; Chiu, K.H.; et al. Infection of healthcare workers despite a high vaccination rate during the fifth wave of COVID-19 due to Omicron variant in Hong Kong. 2022; *submitted, under review* .
111. Nguyen, L.H.; Drew, D.A.; Graham, M.S.; Joshi, A.D.; Guo, C.G.; Ma, W.; Mehta, R.S.; Warner, E.T.; Sikavi, D.R.; Lo, C.H.; et al. Risk of COVID-19 among front-line health-care workers and the general community: A prospective cohort study. *Lancet Public Health* **2020**, *5*, e475–e483. [CrossRef]
112. Bai, D.-S.; Geng, P.; Wang, Z.-D.; Wang, X.-L.; Xu, G.-R.; Ye, Q.; Guo, N.; Zhao, Y.; Yang, C.; Song, H.; et al. Practice and experience of regional medical center entrance linkage and closed-loop management under the wartime situation of the COVID-19 in China. *Ann. Transl. Med.* **2022**, *10*, 112. [CrossRef] [PubMed]
113. Hale, T.; Angrist, N.; Goldszmidt, R.; Kira, B.; Petherick, A.; Phillips, T.; Webster, S.; Cameron-Blake, E.; Hallas, L.; Majumdar, S.; et al. A global panel database of pandemic policies (Oxford COVID-19 Government Response Tracker). *Nat. Hum. Behav.* **2021**, *5*, 529–538. [CrossRef]
114. COVID-19 Government Response Tracker. Available online: https://www.bsg.ox.ac.uk/research/research-projects/covid-19-government-response-tracker (accessed on 7 October 2022).
115. Variation in Government Responses to COVID-19. Blavatnik School Working Paper. BSG-WP-2020/032, version 14.1. Updated 12 August 2022. Available online: https://www.bsg.ox.ac.uk/research/publications/variation-government-responses-covid-19 (accessed on 7 October 2022).
116. Gibney, E. Whose coronavirus strategy worked best? Scientists hunt most effective policies. *Nature* **2020**, *581*, 15–16. [CrossRef]
117. Cheng, V.C.; Wong, S.-C.; Chuang, V.W.; So, S.Y.; Chen, J.H.; Sridhar, S.; To, K.K.; Chan, J.F.; Hung, I.F.; Ho, P.-L.; et al. The role of community-wide wearing of face mask for control of coronavirus disease 2019 (COVID-19) epidemic due to SARS-CoV-2. *J. Infect.* **2020**, *81*, 107–114. [CrossRef]
118. Cap. 599I Prevention and Control of Disease (Wearing of Mask) Regulation. Available online: https://www.elegislation.gov.hk/hk/cap599I (accessed on 7 October 2022).
119. Lam, H.Y.; Lam, T.S.; Wong, C.H.; Lam, W.H.; Leung, C.M.E.; Au, K.W.A.; Lam, C.K.Y.; Lau, T.W.W.; Chan, Y.W.D.; Wong, K.H.; et al. The epidemiology of COVID-19 cases and the successful containment strategy in Hong Kong-January to May 2020. *Int. J. Infect. Dis.* **2020**, *98*, 51–58. [CrossRef]
120. Li, K.K.F.; Jarvis, S.A.; Minhas, F. Elementary effects analysis of factors controlling COVID-19 infections in computational simulation reveals the importance of social distancing and mask usage. *Comput. Biol. Med.* **2021**, *134*, 104369. [CrossRef]

121. Government to Impose Mandatory Quarantine on People Entering Hong Kong from Mainland, Press Releases, The Government of the Hong Kong Special Administrative Region. Available online: https://www.info.gov.hk/gia/general/202002/05/P2020020500793.htm (accessed on 7 October 2022).
122. DH to Strengthen Health Quarantine Arrangements on Inbound Travellers from Italy's Three Regions and Iran, Press Releases, The Government of the Hong Kong Special Administrative Region. Available online: https://www.info.gov.hk/gia/general/202002/28/P2020022800820.htm (accessed on 7 October 2022).
123. DH to Further Strengthen Health Quarantine Arrangements on Inbound Travellers from Overseas, Press Releases. The Government of the Hong Kong Special Administrative Region. Available online: https://www.info.gov.hk/gia/general/202003/10/P2020031000663.htm (accessed on 7 October 2022).
124. DH to Further Extend and Adjust Health Quarantine Arrangements on Inbound Travellers from overseas, Press releases, The Government of the Hong Kong Special Administrative Region. Available online: https://www.info.gov.hk/gia/general/202003/13/P2020031300832.htm (accessed on 7 October 2022).
125. Specifications Under the Prevention and Control of Disease (Regulation of Cross-boundary Conveyances and Travellers) Regulation Gazette, Press releases, The Government of the Hong Kong Special Administrative Region. Available online: https://www.info.gov.hk/gia/general/202007/18/P2020071800038.htm?fontSize=3 (accessed on 7 October 2022).
126. COVID-19: All Travellers to Hong Kong Must Quarantine in Hotels, but Some China Arrivals Exempt, Hong Kong Free Press. Available online: https://hongkongfp.com/2020/11/12/covid-19-all-travellers-to-hong-kong-must-quarantine-in-hotels-but-some-china-arrivals-exempt/ (accessed on 7 October 2022).
127. Hong Kong, China, Extends Quarantine Period for Most Inbound Travelers Effective Dec. 25. Available online: https://crisis24.garda.com/alerts/2020/12/china-officials-extend-quarantine-period-for-most-inbound-travelers-effective-dec-25-entry-banned-for-travelers-from-south-africa-update-45 (accessed on 7 October 2022).
128. Hong Kong to Cut Quarantine for Arrivals to 14 Days from Next Month. Available online: https://www.reuters.com/world/china/hong-kong-shorten-21-day-quarantine-requirement-arrivals-2022-01-27/ (accessed on 7 October 2022).
129. Government Adjusts Boarding, Quarantine and Testing Arrangements for Inbound Travellers on Risk-Based Principle, Press Releases, The Government of the Hong Kong Special Administrative Region. Available online: https://www.info.gov.hk/gia/general/202203/27/P2022032700005.htm?fontSize=1 (accessed on 7 October 2022).
130. Government Announces Adjustments to Quarantine Arrangements for Inbound Persons, Press Releases, The Government of the Hong Kong Special Administrative Region. Available online: https://www.info.gov.hk/gia/general/202208/08/P2022080800803.htm (accessed on 7 October 2022).
131. Government Announces Lifting of Compulsory Quarantine Requirement on Arrival at Hong Kong, Press Releases, The Government of the Hong Kong Special Administrative Region. Available online: https://www.info.gov.hk/gia/general/202209/24/P2022092400048.htm?fontSize=1 (accessed on 7 October 2022).
132. Wong, S.-C.; Chen, H.; Lung, D.C.; Ho, P.-L.; Yuen, K.-Y.; Cheng, V.C. To prevent SARS-CoV-2 transmission in designated quarantine hotel for travelers: Is the ventilation system a concern? *Indoor Air* **2021**, *31*, 1295–1297. [CrossRef]
133. Wong, S.-C.; Au, A.K.; Chen, H.; Yuen, L.L.; Li, X.; Lung, D.C.; Chu, A.W.; Ip, J.D.; Chan, W.-M.; Tsoi, H.-W.; et al. Transmission of Omicron (B.1.1.529)-SARS-CoV-2 Variant of Concern in a designated quarantine hotel for travelers: A challenge of elimination strategy of COVID-19. *Lancet Reg. Health West Pac.* **2022**, *18*, 100360. [CrossRef]
134. Cheng, V.C.; Siu, G.K.; Wong, S.-C.; Au, A.K.; Ng, C.S.; Chen, H.; Li, X.; Lee, L.-K.; Leung, J.S.; Lu, K.K.; et al. Complementation of contact tracing by mass testing for successful containment of beta COVID-19 variant (SARS-CoV-2 VOC B.1.351) epidemic in Hong Kong. *Lancet Reg. Health West Pac.* **2021**, *17*, 100281. [CrossRef] [PubMed]
135. Cap. 599J Prevention and Control of Disease (Compulsory Testing for Certain Persons) Regulation. Available online: https://www.elegislation.gov.hk/hk/cap599J (accessed on 7 October 2022).
136. Restriction-testing declaration, Centre for Health Protection, Department of Health, The Government of the Hong Kong Special Administrative Region. Available online: https://www.coronavirus.gov.hk/eng/compulsory-testing.html#Restriction-testing-declaration (accessed on 7 October 2022).
137. Wong, S.-C.; Yuen, L.L.-H.; Chen, J.H.-K.; Yuen, K.-Y.; Cheng, V.C.-C. Infection control challenges in handling recurrent blockage of sewage pipes in isolation facility designated for patients with COVID-19. *J. Hosp. Infect.* **2021**, *114*, 187–189. [CrossRef] [PubMed]
138. Wong, S.-C.; Chan, V.W.; Lam, G.K.-M.; Yuen, L.L.; AuYeung, C.H.; Li, X.; Chen, J.H.; Chau, P.-H.; Yuen, K.-Y.; Cheng, V.C. The impact of personal coaching on influenza vaccination among healthcare workers before and during COVID-19 pandemic. *Vaccine* **2022**, *40*, 4905–4910. [CrossRef] [PubMed]
139. Early Vaccination for All. COVID-19 Vaccination Programme, The Government of the Hong Kong Special Administrative Region. Available online: https://www.covidvaccine.gov.hk/en/dashboard (accessed on 8 October 2022).
140. Ritchie, H.; Mathieu, E.; Rodés-Guirao, L.; Appel, C.; Giattino, C.; Ortiz-Ospina, E.; Hasell, J.; Macdonald, B.; Beltekian, D.; Roser, M. Coronavirus Pandemic (COVID-19). Available online: https://ourworldindata.org/covid-vaccinations?country=OWID_WRL (accessed on 8 October 2022).
141. Archive of Statistics on 5th Wave of COVID-19. Centre for Health Protection, Department of Health, The Government of the Hong Kong Special Administrative Region. Available online: https://www.coronavirus.gov.hk/eng/5th-wave-statistics.html (accessed on 8 October 2022).

142. Luk, T.T.; Zhao, S.; Wu, Y.; Wong, J.Y.; Wang, M.P.; Lam, T.H. Prevalence and determinants of SARS-CoV-2 vaccine hesitancy in Hong Kong: A population-based survey. *Vaccine* **2021**, *39*, 3602–3607. [CrossRef] [PubMed]
143. Yu, B.Y.; Lam, J.C.; Lam, S.C.; Li, Y.; Chen, S.; Lam, M.Y.; Yeung, W.F. COVID-19 vaccine hesitancy and resistance in an urban Chinese population of Hong Kong: A cross-sectional study. *Hum. Vaccin. Immunother.* **2022**, *18*, 2072144. [CrossRef]
144. Yasmin, F.; Najeeb, H.; Moeed, A.; Naeem, U.; Asghar, M.S.; Chughtai, N.U.; Yousaf, Z.; Seboka, B.T.; Ullah, I.; Lin, C.Y.; et al. COVID-19 Vaccine Hesitancy in the United States: A Systematic Review. *Front. Public Health* **2021**, *9*, 770985. [CrossRef]
145. Cénat, J.M.; Noorishad, P.G.; Moshirian Farahi, S.M.M.; Darius, W.P.; Mesbahi El Aouame, A.; Onesi, O.; Broussard, C.; Furyk, S.E.; Yaya, S.; Caulley, L.; et al. Prevalence and factors related to COVID-19 vaccine hesitancy and unwillingness in Canada: A systematic review and meta-analysis. *J. Med. Virol.* **2022**. [CrossRef]
146. Ackah, B.B.B.; Woo, M.; Stallwood, L.; Fazal, Z.A.; Okpani, A.; Ukah, U.V.; Adu, P.A. COVID-19 vaccine hesitancy in Africa: A scoping review. *Glob. Health Res. Policy* **2022**, *7*, 21. [CrossRef]
147. Aw, J.; Seah, S.S.Y.; Seng, B.J.J.; Low, L.L. COVID-19-Related Vaccine Hesitancy among Community Hospitals' Healthcare Workers in Singapore. *Vaccines* **2022**, *10*, 537. [CrossRef]
148. Caiazzo, V.; Witkoski Stimpfel, A. Vaccine hesitancy in American healthcare workers during the COVID-19 vaccine roll out: An integrative review. *Public Health* **2022**, *207*, 94–104. [CrossRef]
149. Venkatesan, K.; Menon, S.; Haroon, N.N. COVID-19 vaccine hesitancy among medical students: A systematic review. *J. Educ. Health Promot.* **2022**, *11*, 218. [PubMed]
150. Pires, C. Global Predictors of COVID-19 Vaccine Hesitancy: A Systematic Review. *Vaccines* **2022**, *10*, 1349. [CrossRef] [PubMed]
151. Chen, X.; Lee, W.; Lin, F. Infodemic, Institutional Trust, and COVID-19 Vaccine Hesitancy: A Cross-National Survey. *Int. J. Environ. Res. Public Health* **2022**, *19*, 8033. [CrossRef] [PubMed]
152. Chau, C.Y.C. COVID-19 vaccination hesitancy and challenges to mass vaccination. *Hong Kong Med. J.* **2021**, *27*, 377–379. [CrossRef] [PubMed]
153. Kwok, K.O.; Li, K.K.; Wei, W.I.; Tang, A.; Wong, S.Y.S.; Lee, S.S. Influenza vaccine uptake, COVID-19 vaccination intention and vaccine hesitancy among nurses: A survey. *Int. J. Nurs. Stud.* **2021**, *114*, 103854. [CrossRef] [PubMed]
154. Kwok, K.O.; Li, K.K.; Tang, A.; Tsoi, M.T.F.; Chan, E.Y.Y.; Tang, J.W.T.; Wong, A.; Wei, W.I.; Wong, S.Y.S. Psychobehavioral Responses and Likelihood of Receiving COVID-19 Vaccines during the Pandemic, Hong Kong. *Emerg. Infect. Dis.* **2021**, *27*, 1802–1810. [CrossRef]
155. Silver, A. 'COVID zero' regions struggle with vaccine complacency. *Nature* **2022**. [CrossRef]
156. LeaveHomeSafe Moblie App. Available online: https://www.leavehomesafe.gov.hk/en/ (accessed on 10 November 2022).
157. Government announces implementation arrangements for Vaccine Pass. Press releases, The Government of the Hong Kong Special Administrative Region. Available online: https://www.info.gov.hk/gia/general/202202/21/P2022022100781.htm (accessed on 8 October 2022).
158. Government adjusts vaccination requirements of Vaccine Pass, Press releases, The Government of the Hong Kong Special Administrative Region. Available online: https://www.info.gov.hk/gia/general/202203/20/P2022032000438.htm (accessed on 8 October 2022).
159. McMenamin, M.E.; Nealon, J.; Lin, Y.; Wong, J.Y.; Cheung, J.K.; Lau, E.H.Y.; Wu, P.; Leung, G.M.; Cowling, B.J. Vaccine effectiveness of one, two, and three doses of BNT162b2 and CoronaVac against COVID-19 in Hong Kong: A population-based observational study. *Lancet Infect. Dis.* **2022**, *22*, 1435–1443. [CrossRef]
160. Kang, W.; Shami, J.J.P.; Yan, V.K.C.; Ye, X.; Blais, J.E.; Li, X.; Lee, V.H.F.; Chui, C.S.L.; Lai, F.T.T.; Wan, E.Y.F.; et al. Safety of two-dose COVID-19 vaccination (BNT162b2 and CoronaVac) in adults with cancer: A territory-wide cohort study. *J. Hematol. Oncol.* **2022**, *15*, 66. [CrossRef]
161. Ye, X.; Ma, T.; Blais, J.E.; Yan, V.K.C.; Kang, W.; Chui, C.S.L.; Lai, F.T.T.; Li, X.; Wan, E.Y.F.; Wong, C.K.H.; et al. Association between BNT162b2 or CoronaVac COVID-19 vaccines and major adverse cardiovascular events among individuals with cardiovascular disease. *Cardiovasc. Res.* **2022**, *118*, 2329–2338. [CrossRef]
162. Cheng, F.W.T.; Fan, M.; Wong, C.K.H.; Chui, C.S.L.; Lai, F.T.T.; Li, X.; Wan, E.Y.F.; Tang, S.C.W.; Chan, E.W.Y.; Wong, I.C.K. The effectiveness and safety of mRNA (BNT162b2) and inactivated (CoronaVac) COVID-19 vaccines among individuals with chronic kidney diseases. *Kidney Int.* **2022**, *102*, 922–925. [CrossRef] [PubMed]
163. Wan, E.Y.F.; Wang, Y.; Chui, C.S.L.; Mok, A.H.Y.; Xu, W.; Yan, V.K.C.; Lai, F.T.T.; Li, X.; Wong, C.K.H.; Chan, E.W.Y.; et al. Safety of an inactivated, whole-virion COVID-19 vaccine (CoronaVac) in people aged 60 years or older in Hong Kong: A modified self-controlled case series. *Lancet Healthy Longev.* **2022**, *3*, e491–e500. [CrossRef]
164. Air Change/Installation of Air Purifier in Catering Premises. Food and Environmental Hygiene Department, The Government of Hong Kong Special Administrative Region. Available online: https://www.fehd.gov.hk/english/licensing/guide_general_reference/report_air-changes_purification.html (accessed on 10 November 2022).
165. Final Report on Compliance with the Requirement on Air Change or Air Purifiers in Seating Areas of Dine-in Catering Premises under Cap. 599F, Food and Environmental Hygiene Department, The Government of Hong Kong Special Administrative Region. Available online: https://www.fehd.gov.hk/english/licensing/guide_general_reference/air-changes_report2021.pdf (accessed on 10 November 2022).
166. Ioannidis, J.P.A. The end of the COVID-19 pandemic. *Eur. J. Clin. Investig.* **2022**, *52*, e13782. [CrossRef] [PubMed]

167. Del Rio, C.; Malani, P.N. COVID-19 in 2022-The Beginning of the End or the End of the Beginning? *JAMA* **2022**, *327*, 2389–2390. [CrossRef]
168. Nyberg, T.; Ferguson, N.M.; Nash, S.G.; Webster, H.H.; Flaxman, S.; Andrews, N.; Hinsley, W.; Bernal, J.L.; Kall, M.; Bhatt, S.; et al. Comparative analysis of the risks of hospitalisation and death associated with SARS-CoV-2 omicron (B.1.1.529) and delta (B.1.617.2) variants in England: A cohort study. *Lancet* **2022**, *399*, 1303–1312. [CrossRef]
169. Ward, I.L.; Bermingham, C.; Ayoubkhani, D.; Gethings, O.J.; Pouwels, K.B.; Yates, T.; Khunti, K.; Hippisley-Cox, J.; Banerjee, A.; Walker, A.S.; et al. Risk of covid-19 related deaths for SARS-CoV-2 omicron (B.1.1.529) compared with delta (B.1.617.2): Retrospective cohort study. *BMJ* **2022**, *378*, e070695. [CrossRef]
170. Liu, Y.; Yu, Y.; Zhao, Y.; He, D. Reduction in the infection fatality rate of Omicron variant compared with previous variants in South Africa. *Int. J. Infect. Dis.* **2022**, *120*, 146–149. [CrossRef]
171. COVID-19 Forecasting Team. Variation in the COVID-19 infection-fatality ratio by age, time, and geography during the pre-vaccine era: A systematic analysis. *Lancet* **2022**, *399*, 1469–1488. [CrossRef]
172. Smith, D.J.; Hakim, A.J.; Leung, G.M.; Xu, W.; Schluter, W.W.; Novak, R.T.; Marston, B.; Hersh, B.S. COVID-19 Mortality and Vaccine Coverage-Hong Kong Special Administrative Region, China, January 6, 2022-March 21, 2022. *MMWR Morb. Mortal. Wkly. Rep.* **2022**, *71*, 545–548. [CrossRef]
173. Chen, L.L.; Abdullah, S.M.U.; Chan, W.-M.; Chan, B.P.; Ip, J.D.; Chu, A.W.; Lu, L.; Zhang, X.; Zhao, Y.; Chuang, V.W.; et al. Contribution of low population immunity to the severe Omicron BA.2 outbreak in Hong Kong. *Nat. Commun.* **2022**, *13*, 3618. [CrossRef]
174. Poon, R.W.; Chan, B.P.; Chan, W.-M.; Fong, C.H.; Zhang, X.; Lu, L.; Chen, L.L.; Lam, J.Y.; Cheng, V.C.; Wong, S.S.-Y.; et al. SARS-CoV-2 IgG seropositivity after the severe Omicron wave of COVID-19 in Hong Kong. *Emerg. Microbes Infect.* **2022**, *11*, 2116–2119. [CrossRef] [PubMed]
175. Hung, I.F.; Lung, K.-C.; Tso, E.Y.; Liu, R.; Chung, T.W.; Chu, M.-Y.; Ng, Y.-Y.; Lo, J.; Chan, J.; Tam, A.R.; et al. Triple combination of interferon beta-1b, lopinavir-ritonavir, and ribavirin in the treatment of patients admitted to hospital with COVID-19: An open-label, randomised, phase 2 trial. *Lancet* **2020**, *395*, 1695–1704. [CrossRef]
176. Tam, A.R.; Zhang, R.R.; Lung, K.C.; Liu, R.; Leung, K.Y.; Liu, D.; Fan, Y.; Lu, L.; Lam, A.H.Y.; Chung, T.W.H.; et al. Early treatment of high-risk hospitalized COVID-19 patients with a combination of interferon beta-1b and remdesivir: A phase 2 open-label randomized controlled trial. *Clin. Infect. Dis.* **2022**, ciac523. [CrossRef] [PubMed]
177. Wong, C.K.H.; Lau, K.T.K.; Au, I.C.H.; Xiong, X.; Lau, E.H.Y.; Cowling, B.J. Clinical Improvement, Outcomes, Antiviral Activity, and Costs Associated with Early Treatment with Remdesivir for Patients with Coronavirus Disease 2019 (COVID-19). *Clin. Infect. Dis.* **2022**, *74*, 1450–1458. [CrossRef]
178. Wong, C.K.H.; Lau, K.T.K.; Au, I.C.H.; Xiong, X.; Chung, M.S.H.; Lau, E.H.Y.; Cowling, B.J. Optimal Timing of Remdesivir Initiation in Hospitalized Patients with Coronavirus Disease 2019 (COVID-19) Administered with Dexamethasone. *Clin. Infect. Dis.* **2022**, *75*, e499–e508. [CrossRef]
179. Wong, C.K.H.; Lau, K.T.K.; Au, I.C.H.; Xiong, X.; Chung, M.S.H.; Leung, B.Y.C.; Lau, E.H.Y.; Cowling, B.J. Initiation of Tocilizumab or Baricitinib Were Associated with Comparable Clinical Outcomes Among Patients Hospitalized With COVID-19 and Treated with Dexamethasone. *Front. Pharmacol.* **2022**, *13*, 866441. [CrossRef]
180. Wong, C.K.H.; Au, I.C.H.; Lau, K.T.K.; Lau, E.H.Y.; Cowling, B.J.; Leung, G.M. Real-world effectiveness of early molnupiravir or nirmatrelvir-ritonavir in hospitalised patients with COVID-19 without supplemental oxygen requirement on admission during Hong Kong's omicron BA.2 wave: A retrospective cohort study. *Lancet Infect. Dis.* **2022**. [CrossRef]
181. Bhimraj, A.; Morgan, R.L.; Shumaker, A.H.; Lavergne, V.; Baden, L.; Cheng, V.C.; Edwards, K.M.; Gandhi, R.; Muller, W.J.; O'Horo, J.C.; et al. Infectious Diseases Society of America Guidelines on the Treatment and Management of Patients with COVID-19. *Clin. Infect. Dis.* **2020**, ciaa478. [CrossRef]
182. Bhimraj, A.; Morgan, R.L.; Shumaker, A.H.; Baden, L.; Cheng, V.C.C.; Edwards, K.M.; Gallagher, J.C.; Gandhi, R.T.; Muller, W.J.; Nakamura, M.M.; et al. Infectious Diseases Society of America Guidelines on the Treatment and Management of Patients with COVID-19. *Clin. Infect. Dis.* **2022**, ciac724. [CrossRef]
183. Response to Media Enquiries about Quarantine Arrangements, Press Releases, The Government of the Hong Kong Special Administrative Region. Available online: https://www.info.gov.hk/gia/general/202209/01/P2022090100815.htm?fontSize=3 (accessed on 15 October 2022).
184. Kim, H.; Webster, R.G.; Webby, R.J. Influenza Virus: Dealing with a Drifting and Shifting Pathogen. *Viral Immunol.* **2018**, *31*, 174–183. [CrossRef]
185. SARS-CoV-2 Variants of Concern Update. Available online: https://www.health.govt.nz/system/files/documents/pages/sars-cov-2-variant-of-concern-update-27sep22.pdf (accessed on 11 October 2022).
186. Tan, C.W.; Lim, B.L.; Young, B.E.; Yeoh, A.Y.; Yung, C.F.; Yap, W.C.; Althaus, T.; Chia, W.N.; Zhu, F.; Lye, D.C.; et al. Comparative neutralisation profile of SARS-CoV-2 omicron subvariants BA.2.75 and BA.5. *Lancet Microbe* **2022**. [CrossRef]

viruses MDPI

Article

Structural Characteristics of Heparin Binding to SARS-CoV-2 Spike Protein RBD of Omicron Sub-Lineages BA.2.12.1, BA.4 and BA.5

Deling Shi [1,2], Changkai Bu [1], Peng He [2], Yuefan Song [2], Jonathan S. Dordick [2], Robert J. Linhardt [2], Lianli Chi [1,*] and Fuming Zhang [2,*]

1 National Glycoengineering Research Center, Shandong University, Qingdao 266237, China
2 Center for Biotechnology and Interdisciplinary Studies, Rensselaer Polytechnic Institute, Troy, NY 12180, USA
* Correspondence: lianlichi@sdu.edu.cn (L.C.); zhangf2@rpi.edu (F.Z.)

Abstract: The now prevalent Omicron variant and its subvariants/sub-lineages have led to a significant increase in COVID-19 cases and raised serious concerns about increased risk of infectivity, immune evasion, and reinfection. Heparan sulfate (HS), located on the surface of host cells, plays an important role as a co-receptor for virus–host cell interaction. The ability of heparin and HS to compete for binding of the SARS-CoV-2 spike (S) protein to cell surface HS illustrates the therapeutic potential of agents targeting protein–glycan interactions. In the current study, phylogenetic tree of variants and mutations in S protein receptor-binding domain (RBD) of Omicron BA.2.12.1, BA.4 and BA.5 were described. The binding affinity of Omicron S protein RBD to heparin was further investigated by surface plasmon resonance (SPR). Solution competition studies on the inhibitory activity of heparin oligosaccharides and desulfated heparins at different sites on S protein RBD–heparin interactions revealed that different sub-lineages tend to bind heparin with different chain lengths and sulfation patterns. Furthermore, blind docking experiments showed the contribution of basic amino acid residues in RBD and sulfo groups and carboxyl groups on heparin to the interaction. Finally, pentosan polysulfate and mucopolysaccharide polysulfate were evaluated for inhibition on the interaction of heparin and S protein RBD of Omicron BA.2.12.1, BA.4/BA.5, and both showed much stronger inhibition than heparin.

Keywords: SARS-CoV-2; Omicron; spike protein RBD; heparin; pentosan polysulfate; mucopolysaccharide polysulfate

Citation: Shi, D.; Bu, C.; He, P.; Song, Y.; Dordick, J.S.; Linhardt, R.J.; Chi, L.; Zhang, F. Structural Characteristics of Heparin Binding to SARS-CoV-2 Spike Protein RBD of Omicron Sub-Lineages BA.2.12.1, BA.4 and BA.5. *Viruses* **2022**, *14*, 2696. https://doi.org/10.3390/v14122696

Academic Editors: Ahmed El-Shamy and Mohamed Ibrahim

Received: 14 October 2022
Accepted: 29 November 2022
Published: 1 December 2022

1. Introduction

Since the beginning of the COVID-19 pandemic, numerous mutations of severe acute respiratory syndrome coronavirus 2 (SARS-CoV-2) have been identified and shared on GISAID (Global Initiative on Sharing Avian Influenza Data). A variant is recognized as a Variant of concern (VOC) by the World Health Organization (WHO) if it demonstrates: (i) increased transmissibility; (ii) detrimental change; (iii) increased in virulence; (iv) change in clinical disease presentation; (v) decreased effectiveness of public health and social measures of available diagnostics, vaccines and therapeutics. Previously circulating VOCs include Alpha, Beta, Gamma and Delta, while Omicron is currently the dominant variant circulating globally with greatly increased transmissibility [1]. Emergence of the Omicron variant has raised serious concerns about the increased risk of infectivity, immune evasion and reinfection.

The Omicron variants include BA.1, BA.2, BA.3, BA.4, BA.5 and descendent lineages, but also BA.1/BA.2 circulating recombinant forms such as XE [1]. The genome of SARS-CoV-2 (~30 kb) encodes 16 non-structural proteins (NSPs) and 4 main structural proteins, including spike (S), envelope (E), core membrane (M), and nucleocapsid (N), and other accessory proteins [2]. Genome sequenced data of the Omicron variant demonstrated that the Omicron variant was the most highly mutated strain compared with the other VOCs, with 50 mutations accumulated throughout the genome and 26–32 mutations in the S protein [3]. Analysis of the mutations data shows that Omicron also carries several mutations found before, which were associated with increased infectivity and the chance of transmission by evading the immune response [4]. The current used COVID-19 vaccines mainly target the S protein [5]. Although several vaccines offering protection for COVID-19, studies showed a marked reduction the neutralizing capacity of vaccine induced immunity against the Omicron variant, especially the sub-lineages BA.4, and BA.5 [4]. Other therapeutics, including various monoclonal antibodies [6], remdesivir [7,8], tocilizumab [9], favipiravir [10], nirmatrelvir plus ritonavir (Paxlovid™) [11], molnupiravir and other approved drugs [12] have been used with different treatments [13]. With the virus mutation occurring so rapidly, alternative, or complementary approaches, need to be considered that require durable therapeutic effects and reduced adverse events and facilitate rapid development and large-scale production.

Glycosaminoglycans (GAGs) are a class of linear polysaccharides, including heparin/heparan sulfate (HS), keratan sulfate (KS), chondroitin sulfate (CS)/dermatan sulfate (DS), and hyaluronan (HA), and commonly expressed in the interior, cell surface, and extracellular environment of many cell types [14]. Pathogens exploit fundamental biological activities of GAGs, such as serving as cell adhesion and internalization receptors, inducing conformational changes, activating signaling pathways, to promote their attachment and invasion of host cells and to protect themselves from immune attack [15,16]. These activities suggest that GAGs are potential targets for the development of specific and effective antipathogen therapies. Studies have confirmed that SARS-CoV-2 interacts with both cellular HS and angiotensin-converting enzyme 2 (ACE2) through its receptor-binding domain (RBD) in the S1 subunit of the S protein. Binding of HS to S protein shifts the structure to favor the RBD open conformation that binds ACE2 [17,18]. Cellular HS acts as a co-factor for SARS-CoV-2 infection, this emphasizes the new therapeutic opportunities for targeting S protein–HS interactions. Studies have also shown that heparin may inhibit the activity of SARS-CoV-2 M^{pro} protein, thereby inhibiting virus replication and transcription, and heparin also reduces the activity of excessive heparanase, thereby inhibiting glycocalyx shedding and redox balance disturbance [19].

In previous work, we and others have shown that sulfated glycans, including heparin/HS, heparin derivatives, fucoidans, fucosylated chondroitin sulfate, and rhamnan sulfate inhibit the interaction between HS and the S protein RBD of wild type (WT), Delta variant and Omicron (B.1.1.529) [17,18,20–24]. Mutations occur in the RBD region of S protein may influence the binding to ACE2 or HS, for example, nine of the 15 RBD mutations in the Omicron (BA.1.1.529) Spike region belong to the binding footprint of the virus' primary entry receptor [25]. Therefore, comparation of mutations in the emerging Omicron sub-lineages and the binding between their RBD and HS requires further analysis. It cannot be ignored that despite the positive effect of heparin on reducing the risk of venous thromboembolism and coagulopathy in COVID-19 patients, the risk of bleeding is increased [19]. In clinical use, heparin also has other side effect, such as heparin-induced thrombocytopenia [26]. Therefore, discovering the structural characteristic of heparin binding to the S-protein is critical for the development of therapeutics targeting S-protein–heparin interaction while reducing adverse effects. In this work, we examined the binding of the S protein RBD in Omicron sub-lineages BA.2.12.1, BA.4 and BA.5 with heparin, heparin oligosaccharides of different lengths, and chemically modified heparins using surface plasmon resonance (SPR) to elucidate the importance of size and sulfo group position for heparin/HS binding. We also performed blind docking experiments to objectively identify the preferred binding residues of heparin/HS and the associated amino acids on RBD region. Finally, highly negative compounds, including pentosan polysulfate (PPS) and mucopolysaccharide polysulfate (MPS) were evaluated for their inhibition of the RBD–heparin interaction.

2. Materials and Methods

2.1. Materials

S protein RBD of Omicron sub-lineages BA.2.12.1 (Cat: 40592-V08H132), BA.4/BA.5 (Cat: 40592-V08H130) were purchased from Sino Biological Inc. (Beijing, China).The proteins were constructed as follows: (1) a DNA sequence encoding the SARS-CoV-2 (BA.2.12.1) Spike RBD (YP_009724390.1, with mutations G339D, S371F, S373P, S375F, T376A, D405N, R408S, K417N, N440K, L452Q, S477N, T478K, E484A, Q493R, Q498R, N501Y, Y505H) (Arg319–Phe541) was expressed with a polyhistidine tag at the C-terminus; and (2) a DNA sequence encoding the SARS-CoV-2 (BA.4/BA.5) Spike RBD (YP_009724390.1, with mutations G339D, S371F, S373P, S375F, T376A, D405N, R408S, K417N, N440K, L452R, S477N, T478K, E484A, F486V, Q498R, N501Y, Y505H) (Arg319–Phe541) was expressed with a polyhistidine tag at the C-terminus. Unfractionated heparin (15 kDa) was purchased from Celsus Laboratories (Cincinnati, OH, USA). Desulfated heparins including *N*-desulfated heparin (14 kDa), 2-*O*-desulfated IdoA heparin (13 kDa), 6-*O*-desulfated heparin (13 kDa) and heparin oligosaccharides from tetrasaccharide (dp4) to octadecasaccharide (dp18) were purchased from Iduron (Manchester, UK). Pentosan polysulfate (PPS; 6.5 kDa) was from Bene Pharma (Munich, Germany). Mucopolysaccharide polysulfate (MPS; 14.5 kDa) was from Luitpold Pharma (Munich, Germany). Sensor SA chips were from Cytiva (Uppsala, Sweden). SPR experiments were performed using a BIAcore 3000 or T200 SPR (Cytiva, Uppsala, Sweden) with Biaevaluation software (version 4.0.1 or 3.2, Cytiva, Uppsala, Sweden).

2.2. Preparation of Heparin Biochips

The preparation of biotinylated heparin was as follows: heparin (2 mg) and amine-PEG3-Biotin (2 mg, Thermo Scientific, Waltham, MA, USA) were dissolved in 200 μL H_2O added with 10 mg $NaCNBH_3$, and reacted at 70 °C for 24 h, followed by additional 10 mg $NaCNBH_3$ and reacted for another 24 h. The desalted biotinylated heparin was immobilized onto streptavidin (SA) chips based on the manufacturer's protocol, as previously described [27].

2.3. Binding Kinetics and Affinity Measurement

S protein RBD of BA.2.12.1 and BA.4/BA.5 were diluted into HBS-EP+ buffer at concentrations of 1000, 500, 250, 125, and 63 nM, respectively. Diluted protein samples were injected at a flow rate of 30 μL/min for 3 min at 25 °C, followed by dissociation with HBS-EP+ buffer for 3 min. The sensor surface was regenerated by 2 M NaCl (30 μL) after each binding measurement.

2.4. Evaluation of the Inhibition Activity of Heparin Oligosaccharides and Chemically Modified Heparins on S Protein RBD–Heparin Interaction Using Solution Competition SPR

Competition studies between surface-immobilized heparin and heparin analogues (heparin oligosaccharides and desulfated heparins) in solution mixed with S protein RBD, were performed as previously described [24]. S protein RBD samples (250 nM) were premixed individually with 1000 nM oligosaccharides (dp4–dp18) or desulfated heparins and injected at a flow rate of 30 μL/min for 3 min at 25 °C. After dissociation, the sensor was regenerated by 2 M NaCl (30 μL). A control experiment (only S protein RBD) was used to test the complete regeneration.

2.5. Model Building and Molecular Docking

Molecular docking and modeling of the S protein RBD with heparin dodecasaccharide were performed using AutoDock Vina. S protein RBD of Omicron (BA.2.12.1) was derived from the PDB library under code 7XNS, and the structure of dodecasaccharide, IdoA2S-GlcNS6S-IdoA2S-GlcNS6S-IdoA2S-GlcNS6S-IdoA2S-GlcNS6S-IdoA2S-GlcNS6S-IdoA2S-GlcNS6S, was derived from the NMR structure (PDB:1HPN). The structure of BA.4/BA.5 S protein RBD was derived from the mutation of BA.2.12.1 (PDB:7XNS) and optimized by the CHARMm force field. All hydrogen atoms were added to S protein RBD and charged using Gasteiger. A box of size (100,77,122) and grid center (42.110,182.058,131.292) was built for ligand docking. The RBD was in the center of the box and covered completely. The dodecaccharide was allowed to move freely in the box. During blind docking, all monosaccharide rings can rotate freely, and ring substituents (such as the sulfonic acid group) are defined as flexible and free to rotate. After molecular docking simulations, the binding poses were assessed based on binding energy in kcal/mol, and the low energy binding pose (more stable conformer) was chosen.

2.6. Evaluation of the Inhibition Activity of PPS and MPS on S Protein RBD–Heparin Interaction Using Solution Competition SPR

Likewise, SARS-CoV-2 S protein RBD samples (250 nM) pre-mixed with 1000 nM PPS or MPS were injected at a flow rate of 30 μL/min. The signal (RU) decreased when the binding sites on the S protein RBD were occupied by PPS or MPS instead of the surface-immobilized heparin.

3. Results and Discussion

3.1. SARS-CoV-2 Variants and Omicron S Protein RBD Mutations

On November 2021, the WHO defined B.1.1.529 as the fifth VOC and named it Omicron. Multiple new subvariants/sub-lineages of Omicron have now emerged causing a significant increase in COVID-19 cases. High-throughput sequencing technologies enabled rapid identification of SARS-CoV-2 variants. The overall relationships of SARS-CoV-2 variants over time and the VOCs are shown in Figure 1A. Orange and red nodes are Omicron and its sub-lineages, where the red nodes are the three sub-lineages studied in this work, BA.2.12.1, BA.4 and BA.5. Sequence comparison showed that mutations in Omicron were mostly restricted to the S and N proteins, while other viral proteins were generally conserved [28]. The sequence alignment of the S protein RBD (Arg319–Phe541) of Omicron sub-lineages is shown in Figure 1B. An * (asterisk) represents shared mutations, and all the three sub-lineages have 13 amino acid mutations in RBD compared to WT, although the mutation positions are not identical. Notably, the mutations in BA4 and BA5 RBD are the same. The positively charged mutations in BA.2.12.1 RBD are N440K, T478K, Q493R, Q498R and Y505H, while in BA.4/BA.5 RBD are N440K, L452R, T478K, Q498R and Y505H. Persistent amino acid mutations will make it more difficult to provide rapid and reliable diagnosis and treatment.

A

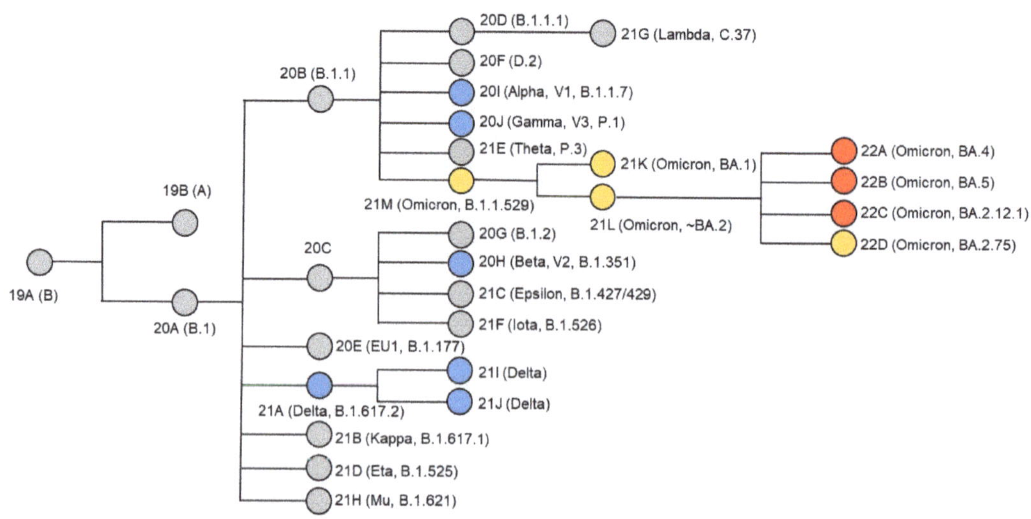

Figure 1. *Cont.*

B

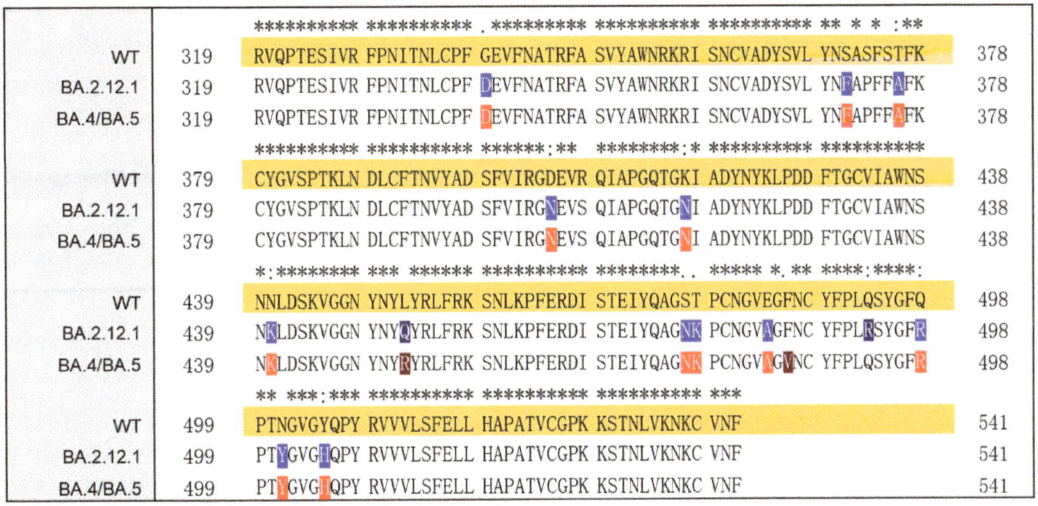

Figure 1. Phylogenetic tree and multiple sequence alignment. (**A**) Phylogenetic relationships of Nextstrain SARS-CoV-2 clades. The phylogenetic tree was adapted from figure provided by Nextstrain and CoVariants [29]. VOCs are represented by colored nodes. (**B**) Mutation profile of S protein RBD of Omicron BA.2.12.1, BA.4/BA.5 compared with WT. Multiple sequence alignment was performed by Clustal Omega (1.2.4). An * (asterisk) indicates positions which have a single, fully conserved residue.

3.2. Binding Affinity and Kinetics Measurement on S Protein RBD–Heparin Interactions

Heparin/HS has variable repeating units, L-iduronic acid (IdoA) or D-glucuronic acid (GlcA) linked to *N*-sulfoglucosamine (GlcNS) or *N*-acetylglucosamine (GlcNAc), with different sulfo group modification [30]. Heparin/HS and other GAGs interact with proteins mainly through their highly negatively charged groups (sulfo groups and carboxyl groups) in polysaccharide chains binding to basic amino acid residues of proteins, for which a limited number of specific binding cases have, thus far, been discovered [31]. HS interacts with SARS-CoV-2 S protein and facilitates host cell entry of SARS-CoV-2 as a co-receptor of ACE2 [17]. Destabilizing and stabilizing mutations may have a large impact on the structure and pathogenesis of the virus. Since the Omicron S protein RBD have a more positive electrostatic potential than both WT and Delta [32], the binding of Omicron BA.2.12.1 and BA.4/BA.5 to heparin/HS is further investigated in the current study.

SPR was used to measure the kinetics and binding affinity of SARS-CoV-2 S protein RBD interaction with heparin, a highly sulfated version of HS. Sensorgrams of S-protein RBD (BA.2.12.1 and BA.4/BA.5) interactions with immobilized heparin are shown in Figure 2. The sensorgrams were used to determine kinetics and binding affinity (i.e., association rate constant, k_a; dissociation rate constant, k_d; and binding equilibrium dissociation constant, K_D, where $K_D = k_d/k_a$) by globally fitting the entire association and dissociation phases using a 1:1 Langmuir-binding model (Table 1). The binding affinities of S protein RBDs (BA.2.12.1 and BA.4/BA.5) were all nanomolar, which were slightly stronger than that of WT measured in our previous work (K_D = 400 nM) and comparable to that of Delta (K_D = 140 nM) and Omicron B.1.1.529 (K_D = 100 nM) [24]. Interestingly, Omicron BA.4/BA.4 RBD had a lower affinity (K_D = 230 nM) for heparin than Omicron BA.2.12.1 RBD (K_D = 140 nM), although they carry the same number of basic amino acids, albeit in different sequences.

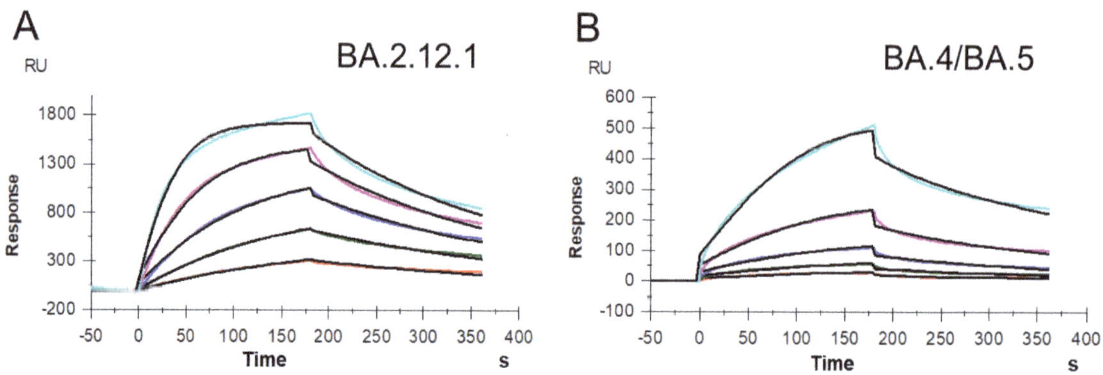

Figure 2. SPR sensorgrams of S protein RBD of BA.2.12.1 and BA.4/BA.5 binding with heparin. (**A**) SPR sensorgrams of S protein RBD of BA.2.12.1 binding with heparin. Concentrations of RBD (from top to bottom) are 1000, 500, 250, 125, and 63 nM, respectively. (**B**) SPR sensorgrams of S protein RBD of BA.4/BA.5 binding with heparin. Concentrations of RBD (from top to bottom) are 1000, 500, 250, 125, and 63 nM, respectively.

Table 1. Summary of kinetic data of S protein RBD of BA.2.12.1 and BA.4/BA.5 binding with heparin.

	k_a ($M^{-1}s^{-1}$)	k_d (s^{-1})	K_D (M)
BA.2.12.1	3.4×10^4 ($\pm$ 270) *	4.7×10^{-3} ($\pm 2.1 \times 10^{-5}$) *	1.4×10^{-7} ($\pm 5.8 \times 10^{-9}$) **
BA.4/BA.5	3.4×10^4 ($\pm$ 630)	7.9×10^{-3} ($\pm 8.7 \times 10^{-5}$)	2.3×10^{-7} ($\pm 2.6 \times 10^{-8}$)

* The data with ($\pm$) in parentheses are the standard deviations (SD) from global fitting of five injections. ** Standard deviation (SD) on triplicated experiments.

3.3. Solution Competition Study on the Inhibition Activity of Heparin Oligosaccharides and Chemically Modified Heparins on S Protein RBD–Heparin Interaction

The molecular and biophysical properties of S protein–GAG binding is currently being investigated. For previous SARS-CoV-2 variants, either the sulfation pattern or chain length of HS/heparin showed an effect on binding affinity. HS with a higher degree of sulfation showed a higher affinity toward SARS-CoV-2 S protein subunits, a full-length molecule and its trimer, and the binding was also positively related to the 6-*O*-sulfation level [33]. Kim et al. showed the level of sulfation had critical impact on the SARS-CoV-2–GAG interaction, and the binding preferred long, highly sulfated structures [18]. For Omicron (BA.1.1.529), heparin showed size-dependent inhibition on the binding to S protein RBD, while higher sulfation level in heparin may not be that important for binding Omicron S protein RBD [24].

Solution competition was applied to test the effect of the chain length and sulfation pattern of heparin on the heparin interactions with RBD of Omicron BA.2.12.1, BA.4/BA.5 S proteins. S protein RBD was pre-mixed with different concentrations of heparin oligosaccharides or chemically desulfated heparins, and then injected onto the heparin chip. The signal (RU) decreased when the binding sites on the S protein RBD were occupied by PPS or MPS instead of the surface-immobilized heparin. Different heparin oligosaccharides at 1000 nM were applied in the competition analysis (Figures 3A,B and 4A,B). For Omicron BA.2.12.1 S protein RBD, heparin oligosaccharides (from dp4 to dp18) showed a weak (2–27% reduction) and size-dependent inhibition on the binding. In the case of Omicron BA.4/BA.5, heparin oligosaccharides in solution competed more effectively (9–35% reduction) against S protein RBD binding to the heparin chip and was independent of chain length. The ability of different chemically desulfated heparins to inhibit the interaction of S protein with surface-immobilized heparin was also measured. All three chemically desulfated heparins reduced the binding of all three S-protein RBDs to surface-immobilized heparin (Figures 3C,D and 4C,D). To our surprise, the binding biases of BA.2.12.1 and BA.4/BA.5 for sulfation patterns were quite different, with BA.4/BA.5 showed stronger charge-dependent and preference for 6-*O*-sulfation groups, possibly due to three amino acids differ in the RBD.

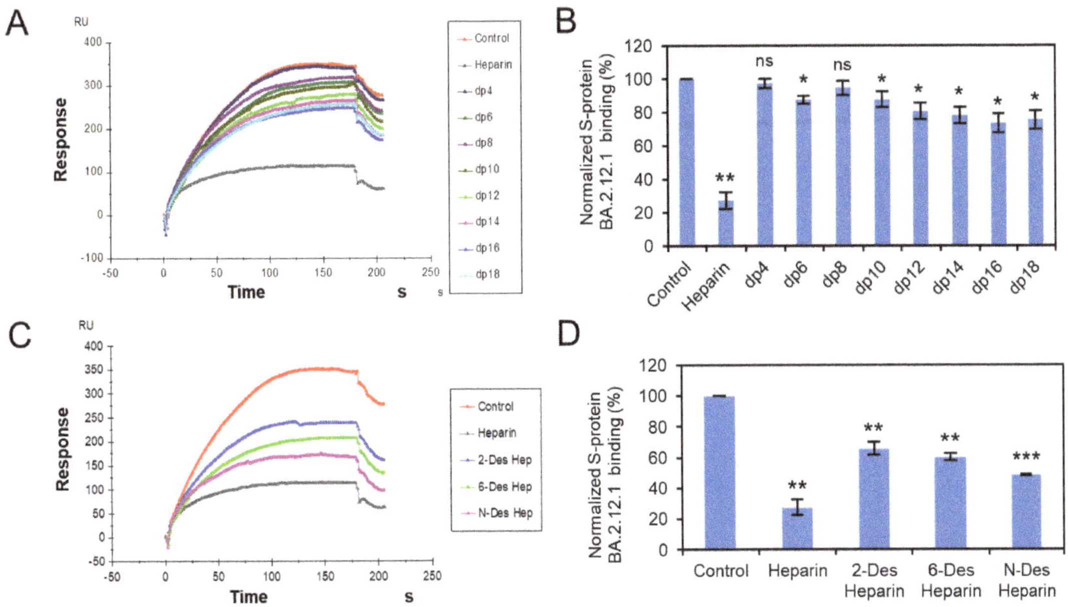

Figure 3. S protein RBD (BA.2.12.1)–heparin interaction inhibited by heparin oligosaccharides/desulfated heparins using solution competition. (**A**) SPR sensorgrams of S protein RBD (BA.2.12.1)–heparin interaction competing with different heparin oligosaccharides. Concentration of S-protein RBD (BA.2.12.1) is 250 nM mixed with 1 μM of different heparin oligosaccharides. (**B**) Bar graphs (based on triplicate experiments with standard deviation) of normalized S-protein RBD (BA.2.12.1) binding preference to surface heparin by competing with different heparin oligosaccharides. (**C**) SPR sensorgrams of S protein RBD (BA.2.12.1)–heparin interaction competing with different desulfated heparins. Concentration of S-protein RBD (BA.2.12.1) is 250 nM mixed with 1 μM of different desulfated heparins. (**D**) Bar graphs (based on triplicate experiments with standard deviation) of normalized S-protein RBD (BA.2.12.1) binding preference to surface heparin by competing with different desulfated heparins. Statistical analysis was performed using unpaired two-tailed *t*-test (ns: $p > 0.05$ compared to the control, *: $p \leq 0.05$ compared to the control, **: $p \leq 0.01$ compared to the control, ***: $p \leq 0.001$ compared to the control).

3.4. Molecular Modeling of the SARS-CoV-2 Spike RBD Interaction with Heparin

AutoDock Vina was used to construct a theoretical binding model of the Omicron S protein RBD and heparin oligosaccharides. Omicron BA.4 variant S protein, derived from the PDB library (PDB:7XNS), was shown in Figure 5A, with RBD domain in red. The RBD in Omicron BA.2.12.1 S protein was derived from the structure of Omicron BA.4/BA.5 RBD (PDB: 7XNS) and the dodecasaccharide, IdoA2S-GlcNS6S-IdoA2S-GlcNS6S-IdoA2S-GlcNS6S-IdoA2S-GlcNS6S-IdoA2S-GlcNS6S-IdoA2S-GlcNS6S, was derived from the NMR structure (PDB: 1HPN). The ranking of the binding poses was based on affinity energy, and the conformation with the lowest energy was selected for subsequent analysis of the S-protein RBD–heparin interaction. We report the best binding pose for BA.2.12.1 based upon lowest binding energy value (affinity energy = −4.2 kcal/mol, RMSD u.b. = 10.879), and the best binding pose for BA.4/BA5 was also selected by the same principle (affinity energy = −5.8 kcal/mol, RMSD u. b. = 31.459). The optimal binding model of the dodecasaccharide binding to BA.2.12.1 or BA.4/BA.5 S protein RBD was displayed using Pymol, and the electrostatic potential map of the binding conformation is shown in Figure 5B. The binding sites of heparin dodecasaccharide to BA.2.12.1 and BA.4/BA.5 were both located at or near the basic amino acid-rich domain, although different amino acid sites were chosen for binding. As shown in Figure 5C, the interaction of dodecasaccharide with BA.2.12.1 is dominated by electrostatic forces, assisted by hydrogen bonds, and both 6-*O*-sulfation and carboxyl groups on heparin chain contribute to the binding. The interaction of the dodecasaccharide and BA.4/BA.5 is governed by a combination of electrostatic forces and hydrogen bonding. The sulfate groups and carboxyl groups on heparin strengthen the interaction. The amino acid residues, such as R355, R577 and R357 in BA.2.12.1 RBD and R346, K440, K444 in BA.4/BA.5 RBD, make up a potential binding site for heparin and heparan sulfate, and other amino acids residues shown in Figure 5C also further strengthen the interaction.

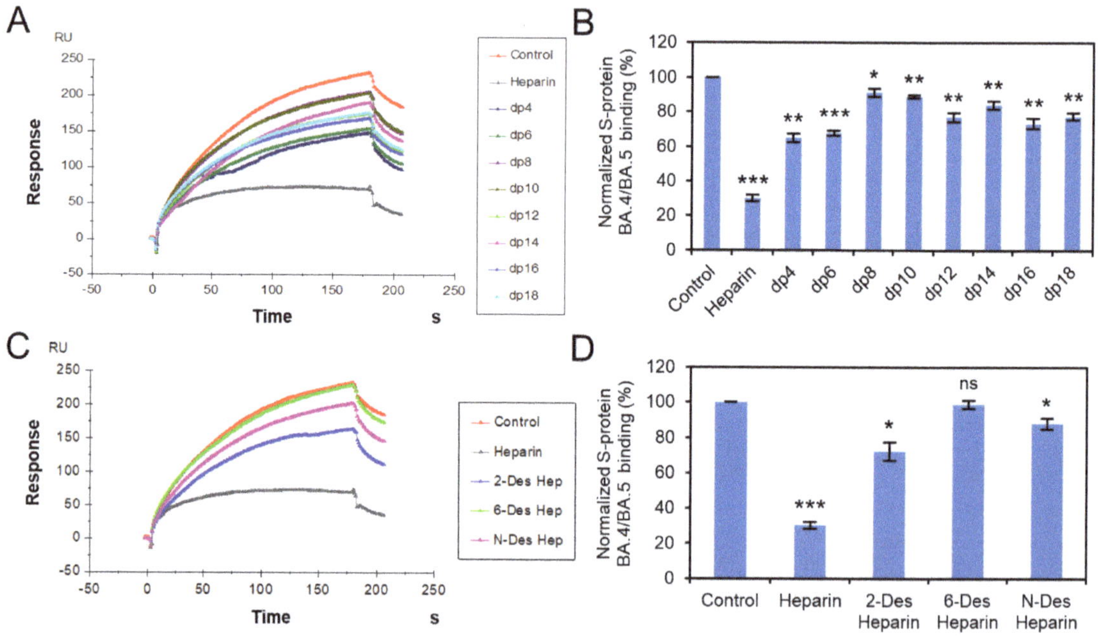

Figure 4. S protein RBD (BA.4/BA.5)–heparin interaction inhibited by heparin oligosaccharides/desulfated heparins using solution competition. (**A**) SPR sensorgrams of S protein RBD (BA.4/BA.5)–heparin interaction competing with different heparin oligosaccharides. Concentration of S-protein RBD (BA.4/BA.5) is 250 nM mixed with 1 μM of different heparin oligosaccharides. (**B**) Bar graphs (based on triplicate experiments with standard deviation) of normalized S-protein RBD (BA.4/BA.5) binding preference to surface heparin by competing with different heparin oligosaccharides. (**C**) SPR sensorgrams of S protein RBD (BA.4/BA.5)–heparin interaction competing with different desulfated heparins. Concentration of S-protein RBD (BA.4/BA.5) is 250 nM mixed with 1 μM of different desulfated heparins. (**D**) Bar graphs (based on triplicate experiments with standard deviation) of normalized S-protein RBD (BA.4/BA.5) binding preference to surface heparin by competing with different desulfated heparins. Statistical analysis was performed using unpaired two-tailed *t*-test (ns: $p > 0.05$ compared to the control, *: $p \leq 0.05$ compared to the control, **: $p \leq 0.01$ compared to the control, ***: $p \leq 0.001$ compared to the control).

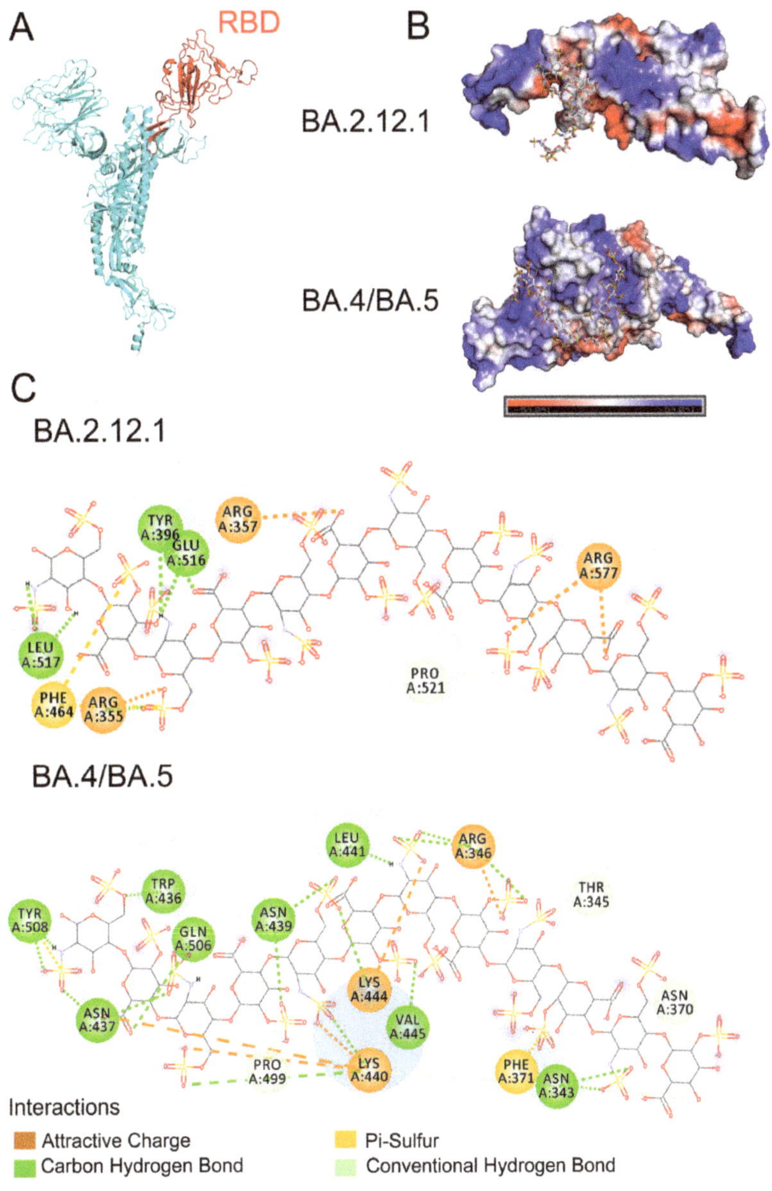

Figure 5. Molecular Docking and modeling simulation. (**A**) Structure of Omicron S protein (PDB: 7XNS) with the RBD domain in red. (**B**) Model electrostatic potential map for docking binding of BA.2.12.1 and BA.4/BA.5 S protein RBD to heparin dodecasaccharide (PDB:1HPN). (**C**) 2D diagram of the interaction of BA.2.12.1 and BA.4/BA.5 S protein RBDs with heparin dodecasaccharide.

3.5. Potential Anti-SARS-CoV-2 Activity of Pentosan Polysulfate and Mucopolysaccharide Polysulfate

Previously work showed that the structural heparin analogues, PPS and MPS, showed strong binding affinities to S protein in isothermal fluorescence titration and surface plasmon resonance (SPR) experiments [27,34,35]. PPS exhibits reduced anticoagulant potential [35] and is less likely to induce bleeding complications for long-term and high-dose use.

Based on these, we investigated the ability of PPS and MPS to inhibit Omicron S protein RBD–heparin binding. The structure of PPS and MPS are shown in Figure 6A, the high level of sulfo groups enable strong interaction with S protein RBD. Both PPS and MPS in solution showed remarkable inhibition activity against surface-immobilized heparin binding with the Omicron S protein RBD, stronger than that of heparin in solution (positive control). PPS and MPS potently inhibited the S protein RBD (BA.2.12.1)–heparin interaction by 99% and 89% (Figure 6B,C), respectively, while inhibiting the S protein RBD (BA.4/BA.5)–heparin interaction by 92% and 80% (Figure 6D,E), respectively. PPS and MPS showed promise for potential therapeutic or preventive agents against COVID-19.

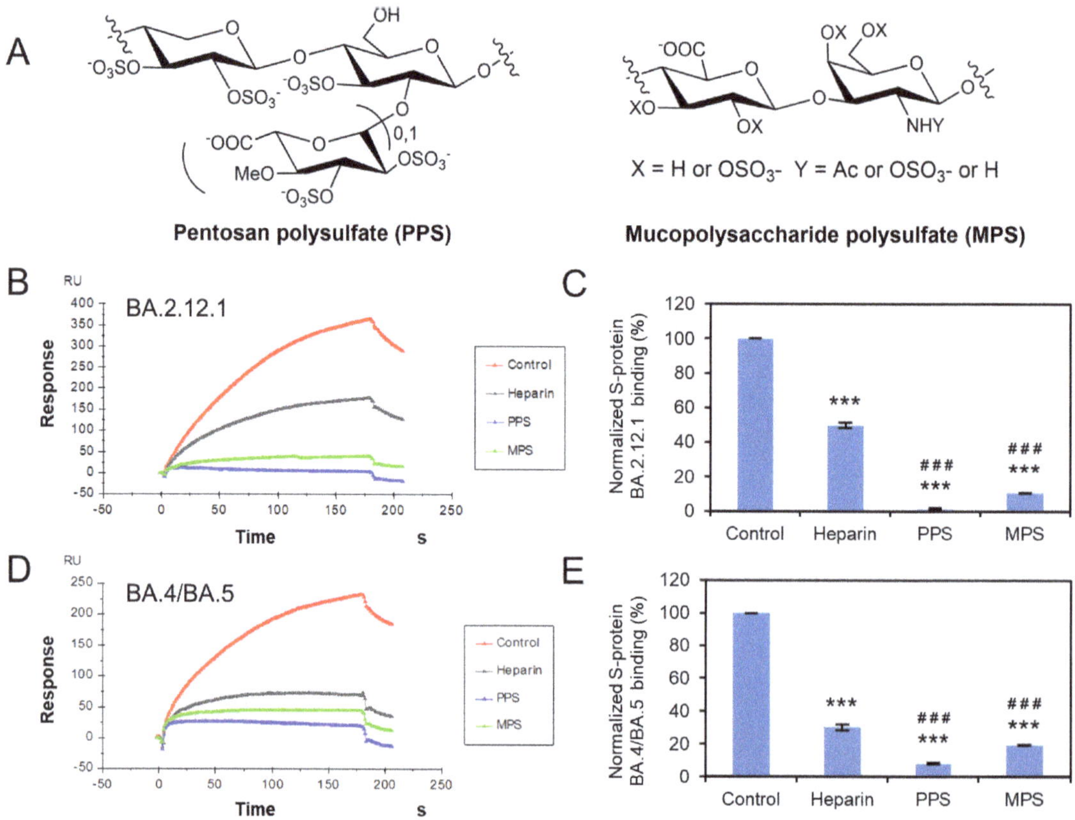

Figure 6. Solution competition between heparin and PPS or MPS. (**A**) Structure of PPS and MPS. (**B**) SPR sensorgrams of S protein RBD (BA.2.12.1)–heparin interaction competing with PPS or MPS. Concentration of S-protein RBD (BA.2.12.1) is 250 nM mixed with 1 μM of PPS or MPS. (**C**) Bar graphs (based on triplicate experiments with standard deviation) of normalized S-protein RBD (BA.2.12.1) binding preference to surface heparin by competing with PPS or MPS. (**D**) SPR sensorgrams of S protein RBD (BA.4/BA.5)–heparin interaction competing with PPS or MPS. Concentration of S-protein RBD (BA.4/BA.5) is 250 nM mixed with 1 μM of PPS or MPS. (**E**) Bar graphs (based on triplicate experiments with standard deviation) of normalized S-protein RBD (BA.4/BA.5) binding preference to surface heparin by competing with PPS or MPS. Statistical analysis was performed using unpaired two-tailed *t*-test (***: $p \leq 0.001$ compared to the control, ###: $p < 0.001$ compared to the heparin).

Solution competition dose response analysis was performed to calculate IC_{50} values to examine the ability of PPS and MPS to inhibit the interaction between surface-immobilized heparin with the S-protein RBD of Omicron variants. Once the active binding site on the S-protein RBD is occupied by glycan in solution, its binding to the surface-immobilized heparin is reduced, resulting in a concentration-dependent decrease in signal. IC_{50} measurement of the inhibition of S-protein RBD of Omicron BA.2.12.1 binding to surface-immobilized heparin using solution competition SPR by PPS and MPS were shown in Figure 7. Measured IC_{50} (concentration of competing analyte resulting in a 50% decrease in RU) = 228.9 nM, 19.3 nM, and 9.7 nM for heparin, PPS, and MPS, respectively. The IC_{50} values of the inhibition of S-protein RBD of Omicron BA.4/BA.5 binding to surface-immobilized heparin were also measured using solution competition SPR by PPS and MPS (Figure 8). The IC_{50} values of heparin, PPS and MPS were 680.1 nM, 84.4 nM, and 124.9 nM, respectively. Although both BA2.12.1 and BA.4/BA.5 were strongly inhibited by PPS and MPS, the activity of BA.4/BA.5 appeared to be more difficult to be inhibited by sulfated glycans.

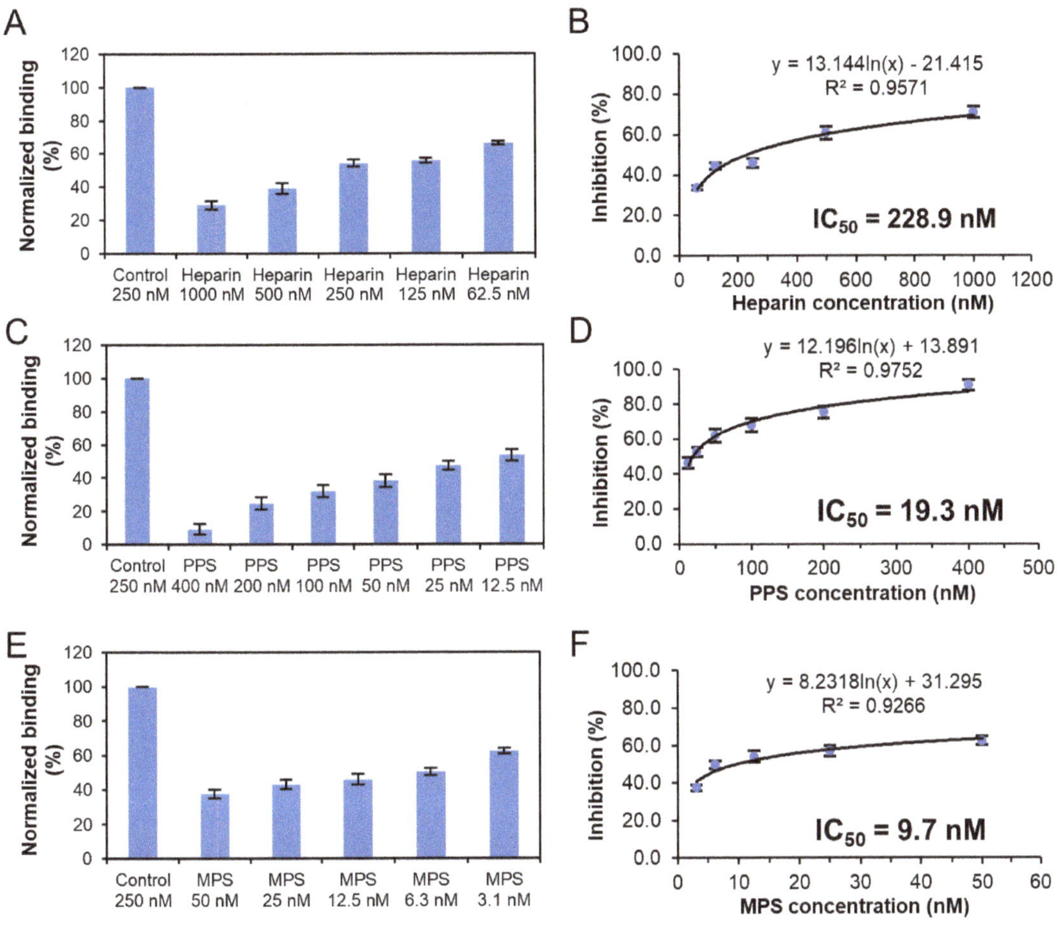

Figure 7. IC_{50} measurement of the inhibition of S-protein RBD (BA.2.12.1) binding to heparin using solution competition SPR by sulfated glycans (heparin, PPS, and MPS). S-protein RBD concentration was 250 nM. Error bars represent standard deviations from triplicate tests. (**A**,**B**) = heparin; (**C**,**D**) = PPS; (**E**,**F**) = MPS.

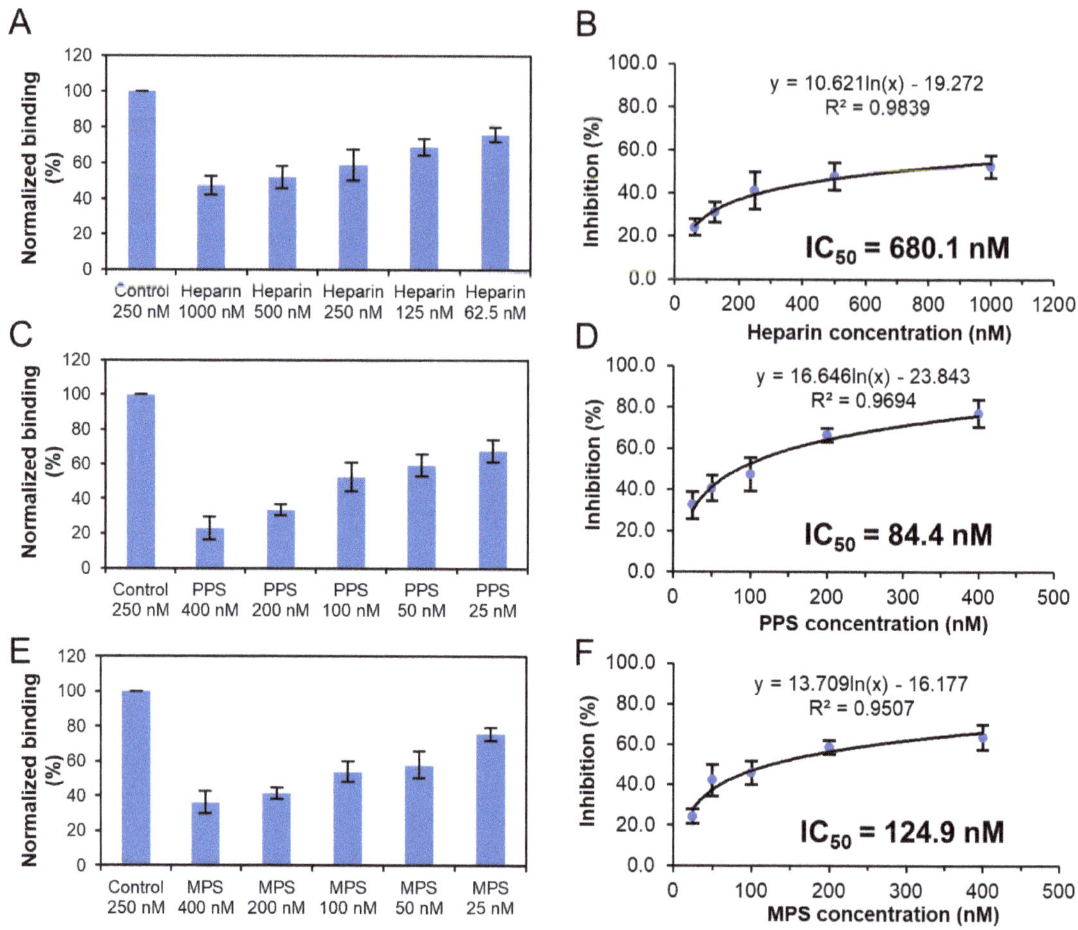

Figure 8. IC_{50} measurement of the inhibition of S-protein RBD (BA.4/BA.5) binding to heparin using solution competition SPR by sulfated glycans (heparin, PPS, and MPS). S-protein RBD concentration was 250 nM. Error bars represent standard deviations from triplicate tests. (**A**,**B**) = heparin; (**C**,**D**) = PPS; (**E**,**F**) = MPS.

4. Conclusions

Mutations occurring in the RBD region of SARS-CoV-2 S protein may influence the binding to HS/heparin. SPR analysis revealed that the binding affinity (K_D) of Omicron BA.2.12.1 and BA.4/BA.5 were all at nanomolar concentrations, which were slightly stronger than that of WT and comparable to that of Delta and Omicron B.1.1.529. Solution competition studies indicated that efficient binding of Omicron BA.2.12.1 requires longer chain length, which is not that necessary for BA.4/BA.5. Competition assays also demonstrated that all the sulfation sites are important for interaction between the S protein RBDs and heparin, although higher sulfation level of HS/heparin are required for binding to BA.4/BA.5. The three sub-lineages showed differences in binding models for heparin dodecasaccharide, suggesting that mutations in RBD have an important effect on viral attachment, possibly explaining differences in the SARS-CoV-2 infection. Binding of human ACE2 to S protein RBD from Omicron and Delta was studied by Han and co-workers, showing that Omicron, Delta, and WT SARS-CoV-2 RBDs have similar binding strengths to hACE2 [32]. Therefore, it is also promising to investigate the effect of different glycans on the binding of ACE2 to different RBDs. Although, persistent amino acid mutations will make it more difficult to provide rapid and reliable diagnosis and treatment, PPS and MPS show promise as therapeutic and/or preventative antiviral drugs against COVID-19.

Author Contributions: Conceptualization, F.Z. and R.J.L.; methodology, D.S., C.B., Y.S., P.H. and F.Z.; formal analysis, D.S. and F.Z.; writing—original draft preparation, D.S.; writing—review and editing, F.Z., J.S.D. and R.J.L.; funding acquisition, L.C., F.Z. and R.J.L. All authors have read and agreed to the published version of the manuscript.

Funding: This work was supported by the NIH (S10OD028523 and 5R21AI156573-02 to F.Z. and R.J.L.) and GlycoMIP, a National Science Foundation Materials Innovation Platform funded through Cooperative Agreement DMR-1933525 (to F.Z., J.S.D. and R.J.L.); New York State Biodefense Commercialization Fund (to F.Z., J.S.D. and R.J.L.).

Institutional Review Board Statement: Not applicable.

Informed Consent Statement: Not applicable.

Data Availability Statement: Not applicable.

Conflicts of Interest: The authors declare no conflict of interest.

References

1. Tracking SARS-CoV-2 Variants. Available online: https://www.who.int/activities/tracking-SARS-CoV-2-variants (accessed on 10 October 2022).
2. Wu, A.; Peng, Y.; Huang, B.; Ding, X.; Wang, X.; Niu, P.; Meng, J.; Zhu, Z.; Zhang, Z.; Wang, J.; et al. Genome Composition and Divergence of the Novel Coronavirus (2019-nCoV) Originating in China. *Cell Host Microbe* **2020**, *27*, 325–328. [CrossRef] [PubMed]
3. Tian, D.; Sun, Y.; Xu, H.; Ye, Q. The emergence and epidemic characteristics of the highly mutated SARS-CoV-2 Omicron variant. *J. Med. Virol.* **2022**, *94*, 2376–2383. [CrossRef] [PubMed]
4. Shrestha, L.B.; Foster, C.; Rawlinson, W.; Tedla, N.; Bull, R.A. Evolution of the SARS-CoV-2 omicron variants BA.1 to BA.5: Implications for immune escape and transmission. *Rev. Med. Virol.* **2022**, *32*, e2381. [CrossRef] [PubMed]
5. Dai, L.; Gao, G.F. Viral targets for vaccines against COVID-19. *Nat. Rev. Immunol.* **2021**, *21*, 73–82. [CrossRef] [PubMed]
6. Corti, D.; Purcell, L.A.; Snell, G.; Veesler, D. Tackling COVID-19 with neutralizing monoclonal antibodies. *Cell* **2021**, *184*, 3086–3108. [CrossRef]
7. Kokic, G.; Hillen, H.S.; Tegunov, D.; Dienemann, C.; Seitz, F.; Schmitzova, J.; Farnung, L.; Siewert, A.; Hobartner, C.; Cramer, P. Mechanism of SARS-CoV-2 polymerase stalling by remdesivir. *Nat. Commun.* **2021**, *12*, 279. [CrossRef]
8. Beigel, J.H.; Tomashek, K.M.; Dodd, L.E.; Mehta, A.K.; Zingman, B.S.; Kalil, A.C.; Hohmann, E.; Chu, H.Y.; Luetkemeyer, A.; Kline, S.; et al. Remdesivir for the Treatment of COVID-19—Final Report. *N. Engl. J. Med.* **2020**, *383*, 1813–1826. [CrossRef]
9. Luo, P.; Liu, Y.; Qiu, L.; Liu, X.; Liu, D.; Li, J. Tocilizumab treatment in COVID-19: A single center experience. *J. Med. Virol.* **2020**, *92*, 814–818. [CrossRef]
10. Cai, Q.; Yang, M.; Liu, D.; Chen, J.; Shu, D.; Xia, J.; Liao, X.; Gu, Y.; Cai, Q.; Yang, Y.; et al. Experimental Treatment with Favipiravir for COVID-19: An Open-Label Control Study. *Engineering* **2020**, *6*, 1192–1198. [CrossRef]
11. Syed, Y.Y. Molnupiravir: First Approval. *Drugs* **2022**, *82*, 455–460. [CrossRef]
12. Ashour, N.A.; Elmaaty, A.A.; Sarhan, A.A.; Elkaeed, E.B.; Moussa, A.M.; Erfan, I.A.; Al-Karmalawy, A.A. A Systematic Review of the Global Intervention for SARS-CoV-2 Combating: From Drugs Repurposing to Molnupiravir Approval. *Drug Des. Dev. Ther.* **2022**, *16*, 685–715. [CrossRef] [PubMed]
13. Bartoletti, M.; Azap, O.; Barac, A.; Bussini, L.; Ergonul, O.; Krause, R.; Pano-Pardo, J.R.; Power, N.R.; Sibani, M.; Szabo, B.G.; et al. ESCMID COVID-19 living guidelines: Drug treatment and clinical management. *Clin. Microbiol. Infect.* **2022**, *28*, 222–238. [CrossRef] [PubMed]
14. Gandhi, N.S.; Mancera, R.L. The structure of glycosaminoglycans and their interactions with proteins. *Chem. Biol. Drug Des.* **2008**, *72*, 455–482. [CrossRef] [PubMed]
15. Aquino, R.S.; Park, P.W. Glycosaminoglycans and infection. *Front. Biosci.* **2016**, *21*, 1260–1277.
16. Kamhi, E.; Joo, E.J.; Dordick, J.S.; Linhardt, R.J. Glycosaminoglycans in infectious disease. *Biol. Rev. Camb. Philos. Soc.* **2013**, *88*, 928–943. [CrossRef]
17. Clausen, T.M.; Sandoval, D.R.; Spliid, C.B.; Pihl, J.; Perrett, H.R.; Painter, C.D.; Narayanan, A.; Majowicz, S.A.; Kwong, E.M.; McVicar, R.N.; et al. SARS-CoV-2 Infection Depends on Cellular Heparan Sulfate and ACE2. *Cell* **2020**, *183*, 1043–1057.e15. [CrossRef]
18. Kim, S.Y.; Jin, W.; Sood, A.; Montgomery, D.W.; Grant, O.C.; Fuster, M.M.; Fu, L.; Dordick, J.S.; Woods, R.J.; Zhang, F.; et al. Characterization of heparin and severe acute respiratory syndrome-related coronavirus 2 (SARS-CoV-2) spike glycoprotein binding interactions. *Antivir. Res.* **2020**, *181*, 104873. [CrossRef]
19. Mangiafico, M.; Caff, A.; Costanzo, L. The Role of Heparin in COVID-19: An Update after Two Years of Pandemics. *J. Clin. Med.* **2022**, *11*, 3099. [CrossRef]

20. Kwon, P.S.; Oh, H.; Kwon, S.J.; Jin, W.; Zhang, F.; Fraser, K.; Hong, J.J.; Linhardt, R.J.; Dordick, J.S. Sulfated polysaccharides effectively inhibit SARS-CoV-2 in vitro. *Cell Discov.* **2020**, *6*, 50. [CrossRef]
21. Tandon, R.; Sharp, J.S.; Zhang, F.; Pomin, V.H.; Ashpole, N.M.; Mitra, D.; McCandless, M.G.; Jin, W.; Liu, H.; Sharma, P.; et al. Effective Inhibition of SARS-CoV-2 Entry by Heparin and Enoxaparin Derivatives. *J. Virol.* **2020**, *95*, e01987-20. [CrossRef]
22. Dwivedi, R.; Samanta, P.; Sharma, P.; Zhang, F.; Mishra, S.K.; Kucheryavy, P.; Kim, S.B.; Aderibigbe, A.O.; Linhardt, R.J.; Tandon, R.; et al. Structural and kinetic analyses of holothurian sulfated glycans suggest potential treatment for SARS-CoV-2 infection. *J. Biol. Chem.* **2021**, *297*, 101207.
23. Song, Y.; He, P.; Rodrigues, A.L.; Datta, P.; Tandon, R.; Bates, J.T.; Bierdeman, M.A.; Chen, C.; Dordick, J.; Zhang, F.; et al. Anti-SARS-CoV-2 Activity of Rhamnan Sulfate from Monostroma nitidum. *Mar. Drugs* **2021**, *19*, 685. [CrossRef] [PubMed]
24. Gelbach, A.L.; Zhang, F.; Kwon, S.J.; Bates, J.T.; Farmer, A.P.; Dordick, J.S.; Wang, C.; Linhardt, R.J. Interactions between heparin and SARS-CoV-2 spike glycoprotein RBD from omicron and other variants. *Front. Mol. Biosci.* **2022**, *9*, 912887. [CrossRef] [PubMed]
25. Dejnirattisai, W.; Huo, J.; Zhou, D.; Zahradnik, J.; Supasa, P.; Liu, C.; Duyvesteyn, H.M.E.; Ginn, H.M.; Mentzer, A.J.; Tuekprakhon, A.; et al. SARS-CoV-2 Omicron-B.1.1.529 leads to widespread escape from neutralizing antibody responses. *Cell* **2022**, *185*, 467–484.e15. [CrossRef] [PubMed]
26. Zhang, F.; Datta, P.; Dordick, J.S.; Linhardt, R.J. Evaluating Heparin Products for Heparin-Induced Thrombocytopenia Using Surface Plasmon Resonance. *J. Pharm. Sci.* **2020**, *109*, 975–980. [CrossRef] [PubMed]
27. Zhang, F.; He, P.; Rodrigues, A.L.; Jeske, W.; Tandon, R.; Bates, J.T.; Bierdeman, M.A.; Fareed, J.; Dordick, J.; Linhardt, R.J. Potential Anti-SARS-CoV-2 Activity of Pentosan Polysulfate and Mucopolysaccharide Polysulfate. *Pharmaceuticals* **2022**, *15*, 258. [CrossRef]
28. Wiegand, T.; Nemudryi, A.; Nemudraia, A.; McVey, A.; Little, A.; Taylor, D.N.; Walk, S.T.; Wiedenheft, B. The Rise and Fall of SARS-CoV-2 Variants and Ongoing Diversification of Omicron. *Viruses* **2022**, *14*, 2009. [CrossRef]
29. Hodcroft, E. CoVariants: Phylogenetic Relationships of Nextstrain SARS-CoV-2 Clades. Available online: https://covariants.org/ (accessed on 10 October 2022).
30. Shriver, Z.; Capila, I.; Venkataraman, G.; Sasisekharan, R. Heparin and Heparan Sulfate: Analyzing Structure and Microheterogeneity. In *Handbook of Experimental Pharmacology*; Springer: Berlin/Heidelberg, Germany, 2012; Volume 207, pp. 159–176.
31. Kjellen, L.; Lindahl, U. Specificity of glycosaminoglycan-protein interactions. *Curr. Opin. Struct. Biol.* **2018**, *50*, 101–108. [CrossRef]
32. Han, P.; Li, L.; Liu, S.; Wang, Q.; Zhang, D.; Xu, Z.; Han, P.; Li, X.; Peng, Q.; Su, C.; et al. Receptor binding and complex structures of human ACE2 to spike RBD from omicron and delta SARS-CoV-2. *Cell* **2022**, *185*, 630–640.e10. [CrossRef]
33. Hao, W.; Ma, B.; Li, Z.; Wang, X.; Gao, X.; Li, Y.; Qin, B.; Shang, S.; Cui, S.; Tan, Z. Binding of the SARS-CoV-2 spike protein to glycans. *Sci. Bull.* **2021**, *66*, 1205–1214. [CrossRef]
34. Ennemoser, M.; Rieger, J.; Muttenthaler, E.; Gerlza, T.; Zatloukal, K.; Kungl, A.J. Enoxaparin and Pentosan Polysulfate Bind to the SARS-CoV-2 Spike Protein and Human ACE2 Receptor, Inhibiting Vero Cell Infection. *Biomedicines* **2021**, *10*, 49. [CrossRef] [PubMed]
35. Bertini, S.; Alekseeva, A.; Elli, S.; Pagani, I.; Zanzoni, S.; Eisele, G.; Krishnan, R.; Maag, K.P.; Reiter, C.; Lenhart, D.; et al. Pentosan Polysulfate Inhibits Attachment and Infection by SARS-CoV-2 In Vitro: Insights into Structural Requirements for Binding. *Thromb. Haemost.* **2022**, *122*, 984–997. [CrossRef] [PubMed]

viruses

MDPI

Article

Severity of COVID-19 among Hospitalized Patients: Omicron Remains a Severe Threat for Immunocompromised Hosts

Louis Nevejan [1,*], Sien Ombelet [1,2], Lies Laenen [1,2], Els Keyaerts [2], Thomas Demuyser [3,4], Lucie Seyler [5], Oriane Soetens [3], Els Van Nedervelde [5], Reinout Naesens [6], Dieter Geysels [6], Walter Verstrepen [6], Lien Cattoir [7], Steven Martens [7], Charlotte Michel [8], Elise Mathieu [8], Marijke Reynders [9], Anton Evenepoel [9], Jorn Hellemans [9], Merijn Vanhee [9], Koen Magerman [10], Justine Maes [10], Veerle Matheeussen [11], Hélène Boogaerts [11], Katrien Lagrou [1,2], Lize Cuypers [1,2,*,†] and Emmanuel André [1,2,*,†]

1 Department of Laboratory Medicine, National Reference Center for Respiratory Pathogens, UZ Leuven—University Hospitals Leuven, Herestraat 49, 3000 Leuven, Belgium
2 KU Leuven Department of Microbiology, Immunology and Transplantation, Laboratory of Clinical Microbiology, 3000 Leuven, Belgium
3 Department of Microbiology and Infection Control, Vrije Universiteit Brussel (VUB), UZ Brussel—University Hospitals Brussels, 1090 Brussels, Belgium
4 Center for Neurosciences, Faculty of Medicine and Pharmacy, Vrije Universiteit Brussel (VUB), 1090 Brussels, Belgium
5 Department of Internal Medicine and Infectious Diseases, Vrije Universiteit Brussel (VUB), UZ Brussel—University Hospitals Brussels, 1090 Brussels, Belgium
6 Department of Medical Microbiology, Department of infection prevention and control, ZNA Middelheim, 2020 Antwerp, Belgium
7 Clinical Laboratory of Microbiology, OLV Hospital, 9300 Aalst, Belgium
8 Department of Microbiology, Laboratoire Hospitalier Universitaire de Bruxelles, Université Libre de Bruxelles, 1000 Brussels, Belgium
9 Department of Laboratory Medicine—Medical Microbiology, AZ Sint Jan Brugge-Oostende, 8000 Brugge, Belgium
10 Clinical Laboratory, Jessa Hospital, 3500 Hasselt, Belgium
11 Department of Microbiology, University Hospital Antwerp, 2650 Antwerp, Belgium
* Correspondence: louis.nevejan@uzleuven.be (L.N.); lize.cuypers@uzleuven.be (L.C.); emmanuel.andre@uzleuven.be (E.A.)
† These authors contributed equally to this work.

Citation: Nevejan, L.; Ombelet, S.; Laenen, L.; Keyaerts, E.; Demuyser, T.; Seyler, L.; Soetens, O.; Van Nedervelde, E.; Naesens, R.; Geysels, D.; et al. Severity of COVID-19 among Hospitalized Patients: Omicron Remains a Severe Threat for Immunocompromised Hosts. *Viruses* **2022**, *14*, 2736. https://doi.org/10.3390/v14122736

Academic Editors: Ahmed El-Shamy and Mohamed Ibrahim

Received: 14 October 2022
Accepted: 6 December 2022
Published: 8 December 2022

Abstract: The Omicron variant of severe acute respiratory syndrome coronavirus 2 (SARS-CoV-2) emerged in the general population in the context of a relatively high immunity gained through the early waves of coronavirus disease 19 (COVID-19), and vaccination campaigns. Despite this context, a significant number of patients were hospitalized, and identifying the risk factors associated with severe disease in the Omicron era is critical for targeting further preventive, and curative interventions. We retrospectively analyzed the individual medical records of 1501 SARS-CoV-2 positive hospitalized patients between 13 December 2021, and 13 February 2022, in Belgium, of which 187 (12.5%) were infected with Delta, and 1036 (69.0%) with Omicron. Unvaccinated adults showed an increased risk of moderate/severe/critical/fatal COVID-19 (crude OR 1.54; 95% CI 1.09–2.16) compared to vaccinated patients, whether infected with Omicron or Delta. In adults infected with Omicron and moderate/severe/critical/fatal COVID-19 (n = 323), immunocompromised patients showed an increased risk of in-hospital mortality related to COVID-19 (adjusted OR 2.42; 95% CI 1.39–4.22), compared to non-immunocompromised patients. The upcoming impact of the pandemic will be defined by evolving viral variants, and the immune system status of the population. The observations support that, in the context of an intrinsically less virulent variant, vaccination and underlying patient immunity remain the main drivers of severe disease.

Keywords: COVID-19; omicron; delta; immunocompromised host; hospitalization

1. Introduction

Compared to the ancestral severe acute respiratory syndrome coronavirus 2 (SARS-CoV-2) Wuhan-Hu-1 strain and to the SARS-CoV-2 lineage B.1.617.2 (Delta variant), the SARS-CoV-2 lineage B.1.1.529 (Omicron variant) is characterized by an antigenic drift of 30, respectively, 28 mutations in the viral spike protein [1]. As a result, Omicron showed more ability than Delta to cause infections in individuals with acquired immunity from either prior infection or vaccination [2]. Omicron outcompeted the Delta variant almost completely in only four weeks in the area of detected emergence, South Africa [3], and caused a nearly complete viral population replacement in a matter of weeks with >3 million daily new confirmed cases worldwide by mid-January 2022 [4]. Initially, the Omicron variant was divided into sub-lineages aliased BA.1 (the main clade), BA.2, and BA.3 [2]. Two distinct sub-lineages BA.4 and BA.5 were designated later on [5].

Increased transmissibility had already been observed with previous variants in the two-year coronavirus disease 2019 (COVID-19) pandemic. The emergence of the SARS-CoV-2 lineage B.1.1.7 (Alpha variant), first detected in the United Kingdom (UK) in September 2020, was associated with a higher transmissibility, but also with an increased severity of disease [6] and risk of death [7], compared to the first COVID-19 pandemic wave. In March 2021, the Delta variant rapidly spread in the UK and showed an even higher risk of hospital admission due to COVID-19 in the unvaccinated population, compared to the patients infected with the Alpha variant [8]. The higher intrinsic severity of the preceding SARS-CoV-2 variants with enhanced transmissibility, compared to the previous ancestor or previous dominant variant, urged studies to investigate the severity of COVID-19 caused by the even more infective Omicron variant [9].

However, early data from South Africa were encouraging. Patients infected with the Omicron variant showed a reduced risk of hospitalization, a shorter length of hospital stay [10], and reduced risk of severe disease compared to patients infected with the Delta variant [3], but also compared to patients infected with the ancestral strain (with D614G mutation), and patients infected with the SARS-CoV-2 lineage B.1.351 (Beta variant) [11,12]. This less severe disease remained after adjusting for both vaccination and prior diagnosed infection indicating a possible intrinsically reduced virulence of Omicron [13].

Extrapolations of the South African studies to other populations should be carried out with caution. Indeed, the age distribution may be different, with a predominantly younger population in South Africa. The vaccination status may also differ, with +/− 35% of the adult population in South Africa having received a basic COVID-19 vaccination (messenger RNA (mRNA) or viral vector vaccine) at the time of Omicron's first detection. In addition, South Africa had high levels of prior infections, with about 70% unvaccinated patients in the province of Gauteng showing SARS-CoV-2 anti-S or anti-N IgG seropositivity at that time [11,12]. Yet, preliminary data retrieved the during the first Omicron emergence from Europe [14,15], the UK [16] and North America [17–20] confirmed these early observations of reduced risk of hospitalization, shorter duration of hospital stay, and fewer severe outcomes, despite an increased incidence rate of Omicron infections.

Despite the overall lower virulence of Omicron, the identification of patient groups remaining vulnerable to severe COVID-19 and prolonged hospitalizations is important to guide health policies when new waves of infections emerge. Data on COVID-19 outcomes, obtained during previous waves in mainly unvaccinated and unexposed patients, showed an association between severity of COVID-19 and increasing age, masculine gender, and a wide range of pre-existing medical conditions, such as diabetes, obesity, heart failure, renal disease, chronic obstructive pulmonary disease (COPD), and immunocompromised status [21]. However, with the emergence of Omicron, evaluating COVID-19 severity is more challenging, given vaccination coverage and a global exposition to previous variants.

It, therefore, seemed interesting to look at the situation in Belgium, characterized by a high vaccination coverage at the time of Omicron's first infection: 88% of adults were fully vaccinated, and 17% of adults had already received a first booster dose [22]. In this multicenter retrospective study, we compared hospitalized patients infected with the Delta variant and with the Omicron variant looking at COVID-19 severity and short-term clinical outcome, taking vaccination status and immune status into account.

2. Materials and Methods

All hospitalized patients with a PCR confirmed SARS-CoV-2 infection from four university and four general hospitals in Belgium were included retrospectively, if admitted between 13 December 2021, and 13 February 2022 (n = 9 weeks). During this period, all newly hospitalized patients were screened for SARS-CoV-2 positivity, regardless of COVID-19 symptoms, in all participating centers in accordance with national testing guidelines at that time. This period was characterized by the full viral replacement of Delta by Omicron BA.1, followed by the emergence of Omicron BA.2, causing a fifth wave of COVID-19 hospital admissions in Belgium, with the highest new number of hospital admissions on 2 February 2022 (n = 2576) [4,23]. Exclusion criteria were admission to the emergency department solely; to a psychiatric or rehabilitation ward; admission to ambulatory care/day surgery; or a positive SARS-CoV-2 screening result, interpreted as a viral remnant of a previous infection (based on viral load <3.0 log copies/mL and/or clinical context).

The following demographic and clinical information were collected: age at admission, sex, date of hospitalization, ward of hospitalization (intensive care unit (ICU) or non-ICU including potential date of transfer to ICU), and mortality during hospitalization (patients still hospitalized on 3 May 2022 were considered non-deceased). COVID-19 symptoms at time of hospital admission (based on National Institutes of Health (NIH) classification [24] (Supplemental Material and Methods A)) were registered, as well as the most severe COVID-19 symptoms during hospitalization. If available, SARS-CoV-2 serology data before or during the first days of admission were collected (SARS-CoV-2 IgG anti-S and/or anti-N). The immune status of patients at time of admission, according to CDC criteria [25], were registered, and immunocompromised status was defined as (1) receiving active cancer treatment for tumors or cancers of the blood; (2) receiving immunosuppressants after organ transplant; (3) recipient of a stem cell transplant within the last 2 years or taking immunosuppressants; (4) moderate or severe primary immunodeficiency; (5) advanced or untreated HIV infection; (6) receiving high-dose corticosteroids (Supplemental Material and Methods B). COVID-19 vaccination status (including date of last vaccine) of the patient on admission was registered, further [25] described in Supplemental Material and Methods C. Regarding SARS-CoV-2 screening results, both the viral load and the variant-of-concern (VOC) were registered, with typing of the VOC being based on whole-genome sequencing results (divided into Delta variant, Omicron BA.1 variant, Omicron BA.1.1 variant, Omicron BA.2 variant, Omicron BA.3 variant, other variant or untypable) or based on a genotyping PCR when WGS could not be performed. Previous SARS-CoV-2 infections were registered if detected in the same hospital as the current admission. An overview of the SARS-CoV-2 serology, PCR and WGS assays, and analyzers is available in Supplemental Material and Methods D.

This study was approved by the Ethics Committee Research UZ/KU Leuven (reference S66037, date of approval 28 February 2022). After signing a Data Transfer Agreement for non-commercial studies, the participating centers sent pseudonymized patients' data to the principal investigator via a secured Excel file. Data were processed using Microsoft® Excel® Version 2204 (Microsoft, Redmond, WA, USA) and R 4.1.3 (RStudio version 2022.02.1) [26].

Statistical analyses comprised bivariate analyses, and multivariate logistic regression analyses. Bivariate analyses were performed by using a Chi-squared test in case of categorical data (reported as absolute number, n, and relative frequency, %). Continuous variables (reported as median and interquartile range (IQR)) were compared using Wilcoxon rank-sum test (sample size $\leq$ 30 and non-paired non-normally distributed data), or two-sample, unpaired t-test (sample size $\leq$ 30 and normally distributed data or sample size > 30). Multivariable logistic regression analysis was performed to assess the effect of independent variables on binary outcomes, such as severity of disease during hospitalization (asymptomatic or mild versus moderate, severe, critical or fatal COVID-19), mortality (discharged or deceased not due to COVID-19 versus deceased due to COVID-19), and need for ICU admission (admission to a non-ICU ward or admission to ICU not due to COVID-19 versus ICU admission due to COVID-19). Independent variables included age at admission, sex, VOC (Delta versus Omicron), immune status at admission (immunocompromised versus non-immunocompromised), vaccination status (unvaccinated versus vaccinated; non-boosted versus boosted; viral vector versus mRNA vaccines) and time since last vaccination (<2/4/6 months before admission versus >2/4/6 months before admission). Because of the multicenter nature of the study, the variable 'hospital' was included as a random effect, using a mixed effects logistic regression model (lme4 package in R). However, due to high numbers of missing data regarding vaccination data, neither multivariable nor mixed-effects models could be run; a bivariate regression model was used instead for these variables. All p-values <0.05 were considered statistically significant.

3. Results

3.1. Patient Demographics

Between 13 December 2021, and 13 February 2022, 1501 newly hospitalized patients with positive SARS-CoV-2 test result were included. In total, 187 patients (12.5%) were infected with Delta, 1036 patients (69.0%) with Omicron, in 278 patients (18.5%) no VOC could be typed (Supplemental Figure S1). Clinical characteristics of all patients are shown in Supplemental Tables S1 and S2. No or mild COVID-19 symptoms were developed in 160/210 (76.2%) children and 753/1291 (58.3%) adults during hospitalization (Supplemental Figure S2). ICU hospitalization accounted for 20/210 (9.5%) children and 178/1291 (13.8%) adults of which, respectively, 8/20 (40.0%) and 49/178 (27.5%) had no or mild COVID-19 symptoms. No children, and 139/1291 (10.8%) adult patients died during hospitalization, however, 40/139 (29%) of them from non-COVID-19 related causes.

3.2. Hospitalized Adults Infected with Omicron Showed a Reduced Intrinsic COVID-19 Severity Compared to Those Infected with Delta

Looking at adult patients with known VOC (Delta and Omicron together, n = 1071/1291, 83.0%), 449/1071 (41.9%) suffered from moderate/severe/critical/fatal COVID-19. Patients infected with Omicron had reduced odds to suffer from these symptoms, compared to patients infected with Delta, after correcting for age, sex, and immune status (odds ratio (OR) 0.14; 95% confidence interval (CI) 0.09–0.22; $p < 0.001$) (Table 1, Figure 1A). Next, the risk of ICU admission due to COVID-19 (total on ICU, n = 112/1071, 10.5%) was reduced for patients with Omicron infection, compared to patients infected with Delta, corrected for using the same variables as above (OR 0.26, 95% CI 0.17–0.40; $p < 0.001$). Finally, regarding the risk of in-hospital mortality related to COVID-19 (total mortality, n = 89/1071, 8.3%), patients infected with Omicron showed reduced odds, compared to patients infected with Delta, corrected for the same variables as above (OR 0.24; 95% CI 0.14–0.40; $p < 0.001$).

Using bivariate logistic regression, unvaccinated patients were more likely than vaccinated patients ([1/2] basic or boosted vaccination scheme) to develop moderate/severe/critical/fatal COVID-19 (crude OR 1.54; 95% CI 1.09–2.16; $p = 0.01$). Similarly, recently vaccinated (<2 months before admission) were less likely than patients vaccinated >2 months before admission to develop moderate/severe/critical/fatal COVID-19 (crude OR 0.55; 95% CI 0.38–0.78; $p = 0.001$). Moreover, unvaccinated patients showed increased odds, compared to vaccinated patients, for admission to ICU due to COVID-19 (crude OR 1.80; 95% CI 1.10–2.87; $p = 0.02$). No significant differences were noticed in vaccination status between patients deceased due to COVID-19 and patients discharged or deceased not due to COVID-19 (crude OR 1.15; 95% CI 0.56–2.19; $p = 0.68$, unadjusted for age, sex, and immune status) (Table 1).

A similar comparative analysis in children with known VOC (n = 152/210, 72.4%) was not possible due to a low number of Delta-infected patients (Supplemental Table S2).

Table 1. Mixed-model logistic regression analysis comparing severity of COVID-19 outcomes between hospitalized adult patients infected with Delta (n = 166) and Omicron (n = 905) (excluding patients with untyped VOC, n = 210). Covariates: 1. versus Delta; 2. odds increase per year; 3. versus female; 4. versus non-immunocompromised; 5. versus any vaccination; 6. versus unvaccinated or vaccinated >2 months before admission; 7. versus unvaccinated or (1/2) basic vaccination; 8. versus vaccinated with mRNA solely. Abbreviations: OR; odds ratio; CI; confidence interval.

	Crude OR	95% CI	*p* value	Adjusted OR	95% CI	*p* Value
Outcome 1—disease severity: odds of moderate/severe/critical/fatal COVID-19 (n = 449/1071, 41.9%)						
Omicron [1]	0.17	0.11–0.26	<0.001	0.14	0.09–0.22	<0.001
Age at admission [2]	1.01	1.01–1.02	<0.001	1.02	1.01–1.03	<0.001
Male [3]	2.07	1.60–2.67	<0.001	1.92	1.47–2.53	<0.001
Immunocompromised [4]	2.12	1.57–2.88	<0.001	2.06	1.48–2.86	<0.001
Unvaccinated [5]	1.54	1.09–2.16	0.01	-	-	-
Vaccinated <2 months before admission [6]	0.55	0.38–0.78	0.001	-	-	-
Boosted [7]	0.82	0.64–1.07	0.14	-	-	-
Viral vector vaccination [8]	1.35	0.99–1.85	0.06	-	-	-
Outcome 2—ICU admission: admission or transferred to ICU due to COVID-19 (n = 112/1071, 10.5%)						
Omicron [1]	0.22	0.14–0.34	<0.001	0.26	0.17–0.40	<0.001
Age at admission [2]	0.99	0.98–1.00	0.01	0.99	0.98–1.00	0.02
Male [3]	1.97	1.32–2.98	0.001	1.99	1.31–3.05	0.001
Immunocompromised [4]	2.59	1.68–3.89	<0.01	2.34	1.50–3.56	<0.001
Unvaccinated [5]	1.80	1.10–2.87	0.02	-	-	-
Vaccinated <2 months before admission [6]	0.75	0.38–1.36	0.36	-	-	-
Outcome 3—mortality: odds of in-hospital mortality related to COVID-19 (n = 89/1071, 8.3%)						
Omicron [1]	0.30	0.19–0.48	<0.001	0.24	0.14–0.40	<0.001
Age at admission [2]	1.03	1.02–1.04	<0.001	1.05	1.03–1.07	<0.001
Male [3]	2.02	1.30–3.20	0.002	1.85	1.16–2.99	0.01
Immunocompromised [4]	2.80	1.75–4.47	<0.001	3.02	1.82–5.05	<0.001
Unvaccinated [5]	1.15	0.56–2.19	0.68	-	-	-
Vaccinated <2 months before admission [6]	0.51	0.15–1.34	0.22	-	-	-

3.3. Immunocompromised Adults HOSPITALIZED with Omicron Infections Had an Increased Risk of Severe COVID-19 Outcomes

Looking at the clinical course of Delta-infected adults separately (n = 166), immunocompromised status was not significantly associated with the development of moderate/severe/critical/fatal COVID-19, after correcting for age and sex; nor the need for ICU admission; nor in-hospital mortality related to COVID-19 (Supplemental Table S3). However, in Omicron-infected adults (n = 905), both an older age, male sex, and immunocompromised status were associated with significantly increased odds for moderate/severe/critical/fatal COVID-19, and in-hospital mortality related to COVID-19 (Supplemental Table S4). Within the Omicron subgroup, recent vaccination (<2 months before admission) was associated with lower odds for moderate/severe/critical/fatal COVID-19 symptoms (crude OR 0.57; 95% CI 0.38–0.82; p = 0.003; unadjusted for age, sex, immune status). Neither ICU admission nor in-hospital mortality related to COVID-19 were significantly associated with vaccination in bivariate analysis in the Omicron subgroup (Supplemental Table S4).

Next, COVID-19 outcomes were investigated in Omicron-infected adults with moderate/severe/critical/fatal COVID-19 only (n = 323) i.e., excluding patients with asymptomatic/mild symptoms. As shown in Figure 1B and Supplemental Table S5, immunocompromised patients had a significantly increased risk of all defined outcomes, i.e., risk of critical/fatal COVID-19 (OR 2.42; 95% CI 1.39–4.22; p = 0.002); risk of admission to ICU due to COVID-19 (OR 1.96; 95% CI 1.06–3.57; p = 0.03); and risk of in-hospital mortality related to COVID-19 (OR 3.13; 95% CI 1.52–6.62; p = 0.002), all corrected for age and sex. Neither the vaccination status—unadjusted for age, sex, immune status (vaccinated versus unvaccinated, boosted versus non-boosted, mRNA vaccination versus viral vector vaccination)—nor the time since last vaccination (<2, <4 or <6 months before admission) showed a significant effect on the defined outcomes using bivariate analysis (Supplemental Table S5).

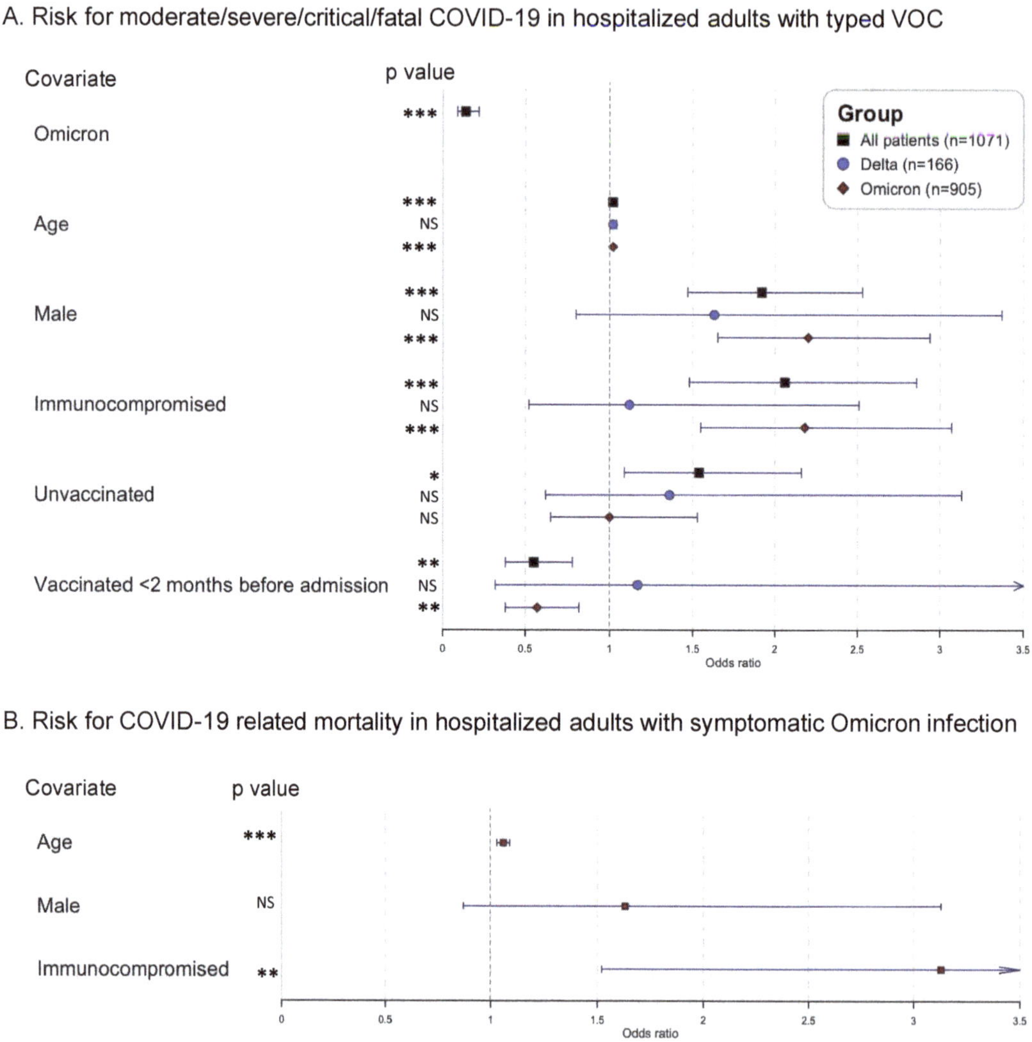

Figure 1. Forest plots of mixed-model logistic regression analysis (**A**). comparing severity of COVID-19 symptoms between hospitalized adult patients infected with Delta ($n = 166$) and Omicron ($n = 905$) (excluding patients with untyped VOC, $n = 210$) and (**B**). identifying risk factors for in-hospital mortality related to COVID-19 in adult patients infected with Omicron and moderate/severe/critical/fatal COVID-19 ($n = 323$) (i.e., excluding patients with asymptomatic/mild symptoms). Covariates: Omicron versus Delta; age odds increase per year; male versus female; immunocompromised versus non-immunocompromised; unvaccinated versus any vaccination; vaccinated <2 months before admission versus unvaccinated or vaccinated >2 months before admission. Abbreviations: NS, non-significant; * $p < 0.05$; ** $p < 0.01$; *** $p < 0.001$.

3.4. Neither Vaccination Status Nor Immune Status Determine Total Hospital Length-of-Stay in Omicron-Infected Adults

Within the subgroup of Omicron-infected adults with moderate/severe/critical/fatal COVID-19 solely i.e., excluding patients with asymptomatic/mild symptoms, no significant association was found between the patients' vaccination status or immune status, and the total hospital length-of-stay. Moreover, when comparing total hospital length-of-stay between patients with different vaccination status or immune status, regardless of the category of symptoms, no difference was noticed (Supplemental Table S6).

4. Discussion

In this multicenter observational cohort study, we investigated the clinical differences between patients infected with the SARS-CoV-2 Delta variant and patients infected with the SARS-CoV-2 Omicron variant in hospitalized patients. We showed that patients infected with Omicron had significantly reduced odds (1) to develop moderate/severe/critical/fatal COVID-19, (2) for COVID-19 associated ICU admission, and (3) for in-hospital mortality related to COVID-19, compared to patients infected with Delta (corrected for age, sex, and immune status at admission). These data confirm previous reports of an inherently lower severity of Omicron, compared to Delta, resulting in a reduced risk of hospitalization [3,14,16,27,28], less severe illness [11,17,18,20,29], and a lower case-fatality ratio [11,19].

However, as the Delta variant is no longer circulating [23], the important considerations for potential COVID-19 waves in the future—with Omicron or another VOC—are the identification of groups requiring additional protection against infection, and severe outcomes. First, our data confirmed the crucial importance of vaccination. With substantially higher vaccination rates in the West, compared to South Africa, at the time of emergence of Omicron (+/− 60% in European and North American versus +/− 30% in South African adults [4]), it has been clearly shown that unvaccinated Omicron-infected patients have a higher risk of hospitalization compared to patients who received two or three vaccine doses [27]. Although our study did not investigate the actual risk of hospitalization, we could show that unvaccinated adult patients have a higher risk of moderate/severe/critical/fatal COVID-19 (OR 1.54; 95% CI 1.09–2.16) and of COVID-19 associated mortality (OR 1.80; 95% CI 1.10–2.87) compared to vaccinated patients (Table 1). Additionally, looking at the subgroup of Omicron-infected adult patients (Supplemental Table S3), we found a reduced risk of moderate/severe/critical/fatal COVID-19 of recently vaccinated patients (<2 months before admission) (crude OR 0.57; 95% CI 0.38–0.82). However, in bivariate analysis, no difference in need for ICU admission nor risk of in-hospital mortality was found between vaccinated and unvaccinated patients in the Omicron subgroup. This could be explained by the lack of adjustment for age—due to low sample size—but also by the lack of reinfection data, since unvaccinated patients with earlier documented SARS-CoV-2 infection gain moderate protection against COVID-19 associated hospitalization [28].

Secondly, immunocompromised adult patients showed an increased risk of moderate/severe/critical/fatal COVID-19, ICU admission, and in-hospital mortality related to COVID-19, compared to non-immunocompromised patients (Table 1). This risk factor was only significant in Omicron-infected adults, not in Delta-infected adults (however, the statistical power in Delta-infected adults was limited due to a smaller sample size). In addition, excluding asymptomatic/mild infections caused by Omicron, immunocompromised adult patients showed an increased risk of critical/fatal COVID-19 (Supplemental Tables S3–S5). A possible underreporting of immunocompromised patients infected with Delta cannot be ruled out, since data were collected during a higher circulation of Omicron in the general community, causing an increased exposure of the virus in this specific patient group. Additionally, considering the higher intrinsic virulence of Delta, more non-immunocompromised patients were at risk of severe outcomes during these Delta waves. Studies on the hospitalization course of COVID-19 in immunocompromised patients are scarce. Mahale et al. [30] showed a high requirement of hospital admission (20%) with substantial morbidity in 114 immunocompromised patients, despite a high vaccination coverage in these patients. Our data confirm that despite a decreased intrinsic virulence of Omicron, humoral and cellular immune responses after vaccination in these patients offer insufficient protection against the most severe COVID-19 outcomes, requiring additional protection measurements (e.g., monoclonal antibodies, directly acting antiviral drugs, modified COVID-19 vaccines) [30–33].

Considering the apparent decrease in virulence of the most recent SARS-CoV-2 VOC, it remains of utmost importance to continue monitoring clinical outcomes in different patient populations, especially when new VOCs emerge. The combination of high incidence rates and lower virulence of Omicron highlights the importance of distinctive reporting of incidental detection of asymptomatic SARS-CoV-2 infections, and symptomatic COVID-19 infections. Our data showed that 160/210 (76.2%) hospitalized children and 753/1291 (58.3%) hospitalized adults experienced no or only mild COVID-19 symptoms; thus concluding that these patients were admitted for other pathologies that were not COVID-19 (nearly identical numbers as reported in South Africa [10]). When looking at patients admitted to ICU, 8/20 (40%) children and 49/178 (28%) adults were not admitted to ICU for COVID-19 symptoms. In the current situation of the pandemic, simultaneous registration of hospital occupancy, and of concise clinical information, is warranted to organize hospital wards adequately (need for exclusive COVID-19 wards versus admitting SARS-CoV-2 positive patients to other pathology-specific wards) [29,34].

By reviewing clinical data using individual patient records, strengths of our study include an accurate classification of a large cohort of patients (n = 1501) according to COVID-19 symptoms. In addition, this is one of the largest studies showing the increased vulnerability of immunocompromised hospitalized patients when infected with Omicron. The limitations of this multicenter study must be addressed too.

First, some crucial clinical information was lacking during data collection. Documented previous infections were highly underreported caused by an inaccessibility to the national SARS-CoV-2 sampling database. However, the emergence of Omicron was associated with a significant increase in reinfection cases [29,35]. Therefore, the effects of acquired immunity due to previous infection could not be included in the multivariate analyses. Other important missing data were due to inaccessibility of the national vaccination database of deceased patients, leading to unknown vaccination status for deceased patients when the vaccination data were not explicitly mentioned in the medical files (i.e., in 30/99 (30.3%) patients deceased from COVID-19, vaccination status was unknown). Additionally, data on general comorbidities (e.g., diabetes, renal insufficiency) were not collected. However, immunocompromised patients very often have additional comorbidities (e.g., steroids-induced diabetes, chronic kidney disease in nonrenal organ transplant recipients), so the exact attributable risk of being immunocompromised could potentially be outweighed in the current analysis. Next, data on the specific subgroup of immunocompromised status (according to CDC criteria [25]) were not collected, hampering additional sub-analyses in these patient groups. It should be noted that CDC criteria defining immunocompromised patients are very broad, as it has been recently reported that seroconversion after a booster vaccine is not equal between patients with solid organ malignancies receiving recent systemic anticancer therapy, and patients with hematological malignancies receiving B-cell depleting therapies [36].

Second, assigning COVID-19 patients into the classification of symptoms we used was not always straightforward. For example, a geriatric patient admitted with general deterioration with positive SARS-CoV-2 screening but no clinical nor radiological evidence of a lower respiratory infection was classified in the 'mild symptoms' category, even if residual viral infection might have contributed to her hospitalization. In contrast, in SARS-CoV-2 positive children with severe respiratory symptoms and concomitant respiratory syncytial virus (RSV) infection, the causative role for SARS-CoV-2 is questionable. In general, the lack of clinical information about potential respiratory co-infections, in both adults and children, could have resulted in a biased classification of symptoms. Additionally, the number of children infected with Omicron and with severe COVID-19 outcomes was too low for further analyses.

Third, data collection started at a time of decreasing viral circulation of the Delta variant, at the 'tail' of infections and hospitalization of patients infected with the Delta variant, compared to the Omicron variant. However, this was also a deliberate choice to minimize differences in vaccination status between the two patient groups. Indeed, a lower vaccination coverage of hospitalized patients in the previous Delta waves could potentially introduce bias, when comparing their outcomes with those of a higher vaccinated population during the Omicron wave. Lastly, the use of monoclonal antibodies was not registered in the database, and this may be of importance when comparing clinical outcomes in immunocompromised patients, with or without previous treatment with monoclonal antibodies.

5. Conclusions

The antibody-evasive nature of the Omicron SARS-CoV-2 variant has led to a worldwide increased incidence of COVID-19, and rapid displacement of the Delta variant. In this multicenter study, the lower intrinsic virulence of Omicron was demonstrated by reduced odds of moderate/severe/critical/fatal COVID-19, need for ICU admission, and COVID-19 associated in-hospital mortality in Omicron-infected patients, compared to patients infected with Delta. Additionally, of importance was the high number of hospitalized children and adults with no or solely mild COVID-19 symptoms (respectively, 76.2% and 58.3%). This highlights the need for a symptomatic registration of hospitalized COVID-19 patients to plan healthcare capacity needs adequately, and more efficiently. In symptomatic Omicron-infected patients, immunocompromised patients had a significantly increased risk of suffering from the most severe COVID-19 outcomes. Despite the lower intrinsic virulence of Omicron, additional precautions are still warranted to protect the vulnerable immunocompromised patients especially when viral transmission in the community is high.

Supplementary Materials: The following supporting information can be downloaded at: https://www.mdpi.com/article/10.3390/v14122736/s1. References [37–39] are cited in Supplementary Materials.

Author Contributions: Conceptualization: L.N., L.C. (Lize Cuypers) and E.A.; methodology: L.N., S.O., L.C. (Lize Cuypers) and E.A.; resources: L.L., E.K., T.D., L.S., O.S., E.V.N., R.N., D.G., W.V., L.C. (Lien Cattoir), S.M., C.M., E.M., M.R., A.E., J.H., M.V., K.M., J.M., V.M. and H.B.; formal analysis: L.N. and S.O.; writing—original draft: L.N.; writing—reviewing and editing: S.O., L.L., E.K., T.D., L.S., O.S., R.N., D.G., L.C. (Lien Cattoir), S.M., C.M., M.R., J.M., V.M., K.L., L.C. (Lize Cuypers) and E.A.; supervision: L.C. (Lize Cuypers) and E.A. All authors have read and agreed to the published version of the manuscript.

Funding: This research received no external funding.

Institutional Review Board Statement: This study was conducted in accordance with the Declaration of Helsinki and approved by the Ethics Committee Research UZ/KU Leuven (reference S66037, date of approval 28 February 2022).

Informed Consent Statement: Patient consent was waived based on art. 6 and 9 of the GDPR.

Data Availability Statement: The data presented in this study are available on request from the corresponding author.

Acknowledgments: UZ Leuven, as national reference center for respiratory pathogens, is supported by Sciensano, the Belgian national institute of health, which is gratefully acknowledged. All members of the Belgian SARS-CoV-2 genomic consortium are gratefully acknowledged.

Conflicts of Interest: The authors declare no conflict of interest.

References

1. Kumar, S.; Thambiraja, T.S.; Karuppanan, K.; Subramaniam, G. Omicron and Delta variant of SARS-CoV-2: A comparative computational study of spike protein. *J. Med. Virol.* **2021**, *94*, 1641–1649. [CrossRef]
2. Viana, R.; Moyo, S.; Amoako, D.G.; Tegally, H.; Scheepers, C.; Althaus, C.L.; Anyaneji, U.J.; Bester, P.A.; Boni, M.F.; Chand, M.; et al. Rapid epidemic expansion of the SARS-CoV-2 Omicron variant in southern Africa. *Nature* **2022**, *603*, 679–686. [CrossRef] [PubMed]
3. Wolter, N.; Jassat, W.; Walaza, S.; Welch, R.; Moultrie, H.; Groome, M.; Amoako, D.G.; Everatt, J.; Bhiman, J.N.; Scheepers, C.; et al. Early assessment of the clinical severity of the SARS-CoV-2 omicron variant in South Africa: A data linkage study. *Lancet* **2022**, *399*, 437–446. [CrossRef] [PubMed]
4. Ritchie, H.; Mathieu, E.; Rodés-Guirao, L.; Appel, C.; Gavrilov, D.; Giattino, C.; Hasell, J.; Macdonald, B.; Dattani, S.; Beltekian, D.; et al. 'Coronavirus Pandemic (COVID-19)'—Published Online at OurWorldInData.org. Available online: https://ourworldindata.org/coronavirus (accessed on 1 June 2022).
5. Tegally, H.; Moir, M.; Everatt, J.; Giovanetti, M.; Scheepers, C.; Wilkinson, E.; Subramoney, K.; Makatini, Z.; Moyo, S.; Amoako, D.G.; et al. Emergence of SARS-CoV-2 Omicron lineages BA.4 and BA.5 in South Africa. *Nat. Med.* **2022**, *28*, 1785–1790. [CrossRef] [PubMed]
6. Snell, L.B.; Wang, W.; Alcolea-Medina, A.; Charalampous, T.; Batra, R.; de Jongh, L.; Higgins, F.; Nebbia, G.; Wang, Y.; Edgeworth, J.; et al. Descriptive comparison of admission characteristics between pandemic waves and multivariable analysis of the association of the Alpha variant (B.1.1.7 lineage) of SARS-CoV-2 with disease severity in inner London. *BMJ Open* **2022**, *12*, e055474. [CrossRef]
7. Davies, N.G.; Jarvis, C.I.; Edmunds, W.J.; Jewell, N.P.; Diaz-Ordaz, K.; Keogh, R.H. Increased mortality in community-tested cases of SARS-CoV-2 lineage B.1.1.7. *Nature* **2021**, *593*, 270–274. [CrossRef]
8. Twohig, K.A.; Nyberg, T.; Zaidi, A.; Thelwall, S.; Sinnathamby, M.A.; Aliabadi, S.; Seaman, S.R.; Harris, R.J.; Hope, R.; Lopez-Bernal, J.; et al. Hospital admission and emergency care attendance risk for SARS-CoV-2 delta (B.1.617.2) compared with alpha (B.1.1.7) variants of concern: A cohort study. *Lancet Infect. Dis.* **2022**, *22*, 35–42. [CrossRef]
9. Bhattacharyya, R.P.; Hanage, W.P. Challenges in Inferring Intrinsic Severity of the SARS-CoV-2 Omicron Variant. *N. Engl. J. Med.* **2022**, *386*, e14. [CrossRef]
10. Abdullah, F.; Myers, J.; Basu, D.; Tintinger, G.; Ueckermann, V.; Mathebula, M.; Ramlall, R.; Spoor, S.; de Villiers, T.; Van der Walt, Z.; et al. Decreased severity of disease during the first global omicron variant covid-19 outbreak in a large hospital in tshwane, south africa. *Int. J. Infect. Dis.* **2021**, *116*, 38–42. [CrossRef]
11. Jassat, W.; Karim, S.S.A.; Mudara, C.; Welch, R.; Ozougwu, L.; Groome, M.J.; Govender, N.; von Gottberg, A.; Wolter, N.; Wolmarans, M.; et al. Clinical severity of COVID-19 in patients admitted to hospital during the omicron wave in South Africa: A retrospective observational study. *Lancet Glob. Health* **2022**, *10*, e961–e969. [CrossRef]
12. Madhi, S.A.; Kwatra, G.; Myers, J.E.; Jassat, W.; Dhar, N.; Mukendi, C.K.; Nana, A.J.; Blumberg, L.; Welch, R.; Ngorima-Mabhena, N.; et al. Population Immunity and Covid-19 Severity with Omicron Variant in South Africa. *N. Engl. J. Med.* **2022**, *386*, 1314–1326. [CrossRef] [PubMed]
13. Davies, M.-A.; Kassanjee, R.; Rousseau, P.; Morden, E.; Johnson, L.; Solomon, W.; Hsiao, N.-Y.; Hussey, H.; Meintjes, G.; Paleker, M.; et al. Outcomes of laboratory-confirmed SARS-CoV-2 infection in the Omicron-driven fourth wave compared with previous waves in the Western Cape Province, South Africa. *Trop. Med. Int. Health* **2022**, *27*, 564–573. [CrossRef] [PubMed]
14. Veneti, L.; Bøås, H.; Kristoffersen, A.B.; Stålcrantz, J.; Bragstad, K.; Hungnes, O.; Storm, M.L.; Aasand, N.; Rø, G.; Starrfelt, J.; et al. Reduced risk of hospitalisation among reported COVID-19 cases infected with the SARS-CoV-2 Omicron BA.1 variant compared with the Delta variant, Norway, December 2021 to January 2022. *Eurosurveillance* **2022**, *27*, 2200077. [CrossRef] [PubMed]
15. Dinh, A.; Dahmane, L.; Dahoumane, M.; Masingue, X.; Jourdain, P.; Lescure, F.-X. Impact of Omicron surge in community setting in greater Paris area. *Clin. Microbiol. Infect.* **2022**, *28*, 897–899. [CrossRef]
16. Ferguson, N.; Ghani, A.; Hinsley, W.; Volz, E. *Report 50: Hospitalisation Risk for Omicron Cases in England*; Imperial College London: London, UK, 2021. [CrossRef]
17. Wang, D.; Hu, B.; Hu, C.; Zhu, F.; Liu, X.; Zhang, J.; Wang, B.; Xiang, H.; Cheng, Z.; Xiong, Y.; et al. Clinical Characteristics of 138 Hospitalized Patients With 2019 Novel Coronavirus—Infected Pneumonia in Wuhan, China. *JAMA-J. Am. Med. Assoc.* **2020**, *323*, 1061–1069. [CrossRef]
18. Christensen, P.A.; Olsen, R.J.; Long, S.W.; Snehal, R.; Davis, J.J.; Saavedra, M.O.; Reppond, K.; Shyer, M.N.; Cambric, J.; Gadd, R.; et al. Signals of Significantly Increased Vaccine Breakthrough, Decreased Hospitalization Rates, and Less Severe Disease in Patients with Coronavirus Disease 2019 Caused by the Omicron Variant of Severe Acute Respiratory Syndrome Coronavirus 2 in Houston, Texas. *Am. J. Pathol.* **2022**, *192*, 642–652. [CrossRef]
19. Ulloa, A.C.; Buchan, S.A.; Daneman, N.; Brown, K.A. Estimates of SARS-CoV-2 Omicron Variant Severity in Ontario, Canada. *JAMA-J. Am. Med. Assoc.* **2022**, *327*, 1286–1288. [CrossRef]
20. Lewnard, J.A.; Hong, V.X.; Patel, M.M.; Kahn, R.; Lipsitch, M.; Tartof, S.Y. Clinical Outcomes among Patients Infected with Omicron (B.1.1.529) SARS-CoV-2 Variant in Southern California. *MedRxiv* **2022**. [CrossRef]
21. European Centre for Disease Prevention and Control. COVID-19—Risk Factors and Risk Groups. 2022. Available online: https://www.ecdc.europa.eu/en/covid-19/latest-evidence/risk-factors-risk-groups (accessed on 6 August 2022).
22. Vaesen, J. Covid Vaccinaties België. 2021. Available online: https://covid-vaccinatie.be/nl (accessed on 6 August 2022).

23. Cuypers, L.; Baele, G.; Dellicour, S.; Maes, P.; André, E. *Genomic Surveillance of SARS-CoV-2 in Belgium*; UZ Leuven: Leuven, Belgium, 2022.
24. NIH. Clinical Spectrum of SARS-CoV-2 Infection. 2020. Available online: https://www.covid19treatmentguidelines.nih.gov/overview/clinical-spectrum/ (accessed on 28 January 2021).
25. CDC. COVID-19 Vaccines for Moderately or Severely Immunocompromised People. 2022. Available online: https://www.cdc.gov/coronavirus/2019-ncov/vaccines/recommendations/immuno.html (accessed on 9 February 2022).
26. Rstudio Team. RStudio: Integrated Development for R. 2020. Available online: https://www.rstudio.com/ (accessed on 9 February 2022).
27. Bager, P.; Wohlfahrt, J.; Bhatt, S.; Stegger, M.; Legarth, R.; Møller, C.H.; Skov, R.L.; Valentiner-Branth, P.; Voldstedlund, M.; Fischer, T.K.; et al. Risk of hospitalisation associated with infection with SARS-CoV-2 omicron variant versus delta variant in Denmark: An observational cohort study. *Lancet Infect. Dis.* **2022**, *22*, 967–976. [CrossRef]
28. Nyberg, T.; Ferguson, N.M.; Nash, S.G.; Webster, H.H.; Flaxman, S.; Andrews, N.; Hinsley, W.; Bernal, J.L.; Kall, M.; Bhatt, S.; et al. Comparative analysis of the risks of hospitalisation and death associated with SARS-CoV-2 omicron (B.1.1.529) and delta (B.1.617.2) variants in England: A cohort study. *Lancet* **2022**, *399*, 1303–1312. [CrossRef]
29. Van Goethem, N.; Chung, P.Y.J.; Meurisse, M.; Vandromme, M.; De Mot, L.; Brondeel, R.; Stouten, V.; Klamer, S.; Cuypers, L.; Braeye, T.; et al. Clinical Severity of SARS-CoV-2 Omicron Variant Compared with Delta among Hospitalized COVID-19 Patients in Belgium during Autumn and Winter Season 2021–2022. *Viruses* **2022**, *14*, 1297. [CrossRef] [PubMed]
30. Malahe, S.R.K.; Hoek, R.A.S.; Dalm, V.A.S.H.; Broers, A.E.C.; Hoed, C.M.D.; Manintveld, O.C.; Baan, C.C.; van Deuzen, C.M.; Papageorgiou, G.; Bax, H.I.; et al. Clinical Characteristics and Outcomes of Immunocompromised Patients With Coronavirus Disease 2019 Caused by the Omicron Variant: A Prospective, Observational Study. *Clin. Infect. Dis.* **2022**. [CrossRef] [PubMed]
31. Chenchula, S.; Karunakaran, P.; Sharma, S.; Chavan, M. Current evidence on efficacy of COVID-19 booster dose vaccination against the Omicron variant: A systematic review. *J. Med. Virol.* **2022**, *94*, 2969–2976. [CrossRef] [PubMed]
32. Guo, Y.; Han, J.; Zhang, Y.; He, J.; Yu, W.; Zhang, X.; Wu, J.; Zhang, S.; Kong, Y.; Guo, Y.; et al. SARS-CoV-2 Omicron Variant: Epidemiological Features, Biological Characteristics, and Clinical Significance. *Front. Immunol.* **2022**, *13*, 877101. [CrossRef]
33. Focosi, D.; McConnell, S.; Casadevall, A.; Cappello, E.; Valdiserra, G.; Tuccori, M. Monoclonal antibody therapies against SARS-CoV-2. *Lancet Infect. Dis.* **2022**, *22*, e311–e326. [CrossRef]
34. Van Goethem, N.; Vilain, A.; Wyndham-Thomas, C.; Deblonde, J.; Bossuyt, N.; Lernout, T.; Gonzalez, J.R.; Quoilin, S.; Melis, V.; Van Beckhoven, D. Rapid establishment of a national surveillance of COVID-19 hospitalizations in Belgium. *Arch. Public Health* **2020**, *78*, 121. [CrossRef]
35. Pulliam, J.R.C.; van Schalkwyk, C.; Govender, N.; von Gottberg, A.; Cohen, C.; Groome, M.J.; Dushoff, J.; Mlisana, K.; Moultrie, H. Increased risk of SARS-CoV-2 reinfection associated with emergence of Omicron in South Africa. *Science* **2022**, *376*, eabn4947. [CrossRef]
36. Al Hajji, Y.; Taylor, H.; Starkey, T.; Lee, L.Y.W.; Tilby, M. Antibody response to a third booster dose of SARS-CoV-2 vaccination in adults with haematological and solid cancer: A systematic review. *Br. J. Cancer* **2022**, *127*, 1827–1836. [CrossRef]
37. CDC Health Alert Network. MIS-C associated with COVID-19. Available online: https://emergency.cdc.gov/han/2020/han00432.asp (accessed on 10 February 2022).
38. Sullivan, K.E.; Jyonouchi, S.C. Inborn errors of immunity (primary immunodeficiencies): Classification. Available online: https://www.uptodate.com/contents/inborn-errors-of-immunity primary-immunodeficiencies- (accessed on 15 December 2021).
39. Shroff, A.; Mertz, D. Infectious Diseases Risk Whole in Chronic, High-Dose Corticosteroids. *Can. J. Gen. Internal Med.* **2017**, *12*, 10–13. [CrossRef]

viruses

MDPI

Review

A Glimpse on the Evolution of RNA Viruses: Implications and Lessons from SARS-CoV-2

Petra Šimičić and Snježana Židovec-Lepej *

Department of Immunological and Molecular Diagnostics, University Hospital for Infectious Diseases "Dr. Fran Mihaljević", HR-10000 Zagreb, Croatia
* Correspondence: szidovec@gmail.com; Tel.: +385-1-2826-625

Abstract: RNA viruses are characterised by extremely high genetic variability due to fast replication, large population size, low fidelity, and (usually) a lack of proofreading mechanisms of RNA polymerases leading to high mutation rates. Furthermore, viral recombination and reassortment may act as a significant evolutionary force among viruses contributing to greater genetic diversity than obtainable by mutation alone. The above-mentioned properties allow for the rapid evolution of RNA viruses, which may result in difficulties in viral eradication, changes in virulence and pathogenicity, and lead to events such as cross-species transmissions, which are matters of great interest in the light of current severe acute respiratory syndrome coronavirus 2 (SARS-CoV-2) pandemics. In this review, we aim to explore the molecular mechanisms of the variability of viral RNA genomes, emphasising the evolutionary trajectory of SARS-CoV-2 and its variants. Furthermore, the causes and consequences of coronavirus variation are explored, along with theories on the origin of human coronaviruses and features of emergent RNA viruses in general. Finally, we summarise the current knowledge on the circulating variants of concern and highlight the many unknowns regarding SARS-CoV-2 pathogenesis.

Keywords: RNA virus; evolution; quasispecies; mutation; recombination; coronavirus; SARS-CoV-2

Citation: Šimičić, P.; Židovec-Lepej, S. A Glimpse on the Evolution of RNA Viruses: Implications and Lessons from SARS-CoV-2. *Viruses* **2023**, *15*, 1. https://doi.org/10.3390/v15010001

Academic Editors: Ahmed El-Shamy and Mohamed Ibrahim

Received: 1 December 2022
Revised: 15 December 2022
Accepted: 16 December 2022
Published: 20 December 2022

1. Introduction

RNA viruses are obligate intracellular parasites characterised by extremely high genetic variability and phenotypic diversity, facilitating infection of an extensive range of hosts [1–3]. Significantly different genome structures and replication strategies allow for great adaptability and exploitation of various host cellular mechanisms [4–7]. One of the prerequisites of successful adaptation to various hosts and environments is the ability to efficiently introduce genetic change in a short amount of time [8,9]. Alterations in the virus's genetic material may result in changes in the virus phenotype and population dynamics, leading to events such as a change in virulence or host tropism, potentially resulting in new emergent pathogens [10–12].

The outbreak of severe acute respiratory syndrome coronavirus 2 (SARS-CoV-2) in 2019 and the ongoing pandemics once again emphasised the importance of elucidating molecular mechanisms of RNA genome evolution and potential viral emergence [13–15]. Coronaviruses, which constitute the subfamily *Orthocoronaviridae* among the *Coronaviridae* family, were shown to be distributed among various animals and capable of causing diseases with a broad range of symptoms and degrees of pathology [16–18]. However, many of the evolutionary, structural, and functional features are shared and sometimes intertwined among the coronaviruses, enabling the analysis of the genetic structure change in entire populations [12,19–21].

The first part of this review article aims to summarise the main RNA virus characteristics necessary for understanding the mechanisms and causal agents that govern RNA virus variation. The second part applies the above-mentioned concepts to the evolution of human coronaviruses with an emphasis on the emergence of SARS-CoV-2. Finally, the unresolved questions and long-term consequences of the rapid RNA virus evolution are discussed as we witness the aftermath of one of the largest globally known pandemics.

2. RNA Virus Classification

2.1. Baltimore Classification

In 1971 David Baltimore published an article titled *Expression of Animal Virus Genomes*, suggesting a division of viruses into six classes according to the method of transmission of their genetic information from one generation to another (mRNA synthesis) and the style of expression of their genetic information (structure of nucleic acid) [4]. The original Baltimore classification placed viruses into one of six classes—class I (double stranded DNA genome, (+/−) dsDNA), class II (single-stranded DNA genome, (+) ssDNA), class III (double-stranded RNA genome, (+/−) dsRNA), class IV (positive sense single-stranded RNA genome, (+) ssRNA), class V (negative-sense single-stranded RNA genome, (−) ssRNA), and class VI (reverse-transcribing single-stranded RNA genome, (+) ssRNA-RT). The polarity of the genome is designated as "+" and "−" strands, or strands of both polarities ("+/−") [4]. This classification was later updated with class VII to accommodate the viruses with a double-stranded DNA genome with an RNA intermediate in its replication cycle ((+/−) dsDNA-RT) [22] (Figure 1).

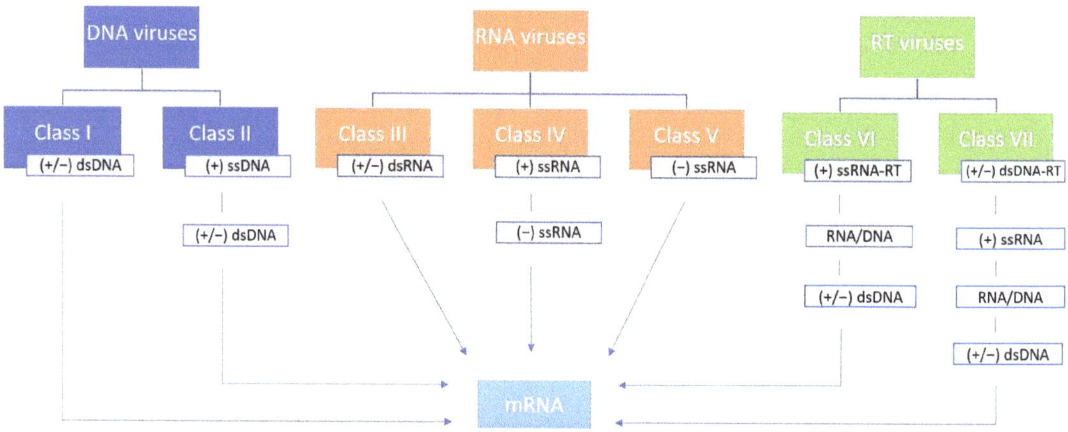

Figure 1. Baltimore classification of viruses. Seven classes of viruses are depicted based on the genome structure and mRNA synthesis strategy. The polarity of the genome is designated as "+" and "−" strands, or strands of both polarities ("+/−"), reverse transcribing characteristic is abbreviated as "RT", while "ds" and "ss" indicate double-stranded and single-stranded genomes, respectively. Data are based on the original and updated Baltimore classification [4,22].

This simple yet functional and intuitive classification system is still used today since it nicely complements virus taxonomy [18]. Unlike cellular life forms, which are strictly dependent on the replication of the dsDNA genome, the abundance of viral RNA and DNA replication–expression strategies enables their stark genetic and phenotypic diversity [4,23]. True RNA viruses with no DNA intermediate in their replication cycle comprise 3 out of 7 Baltimore classes (III, IV, and V) and could be considered one of the earliest descendants from the primordial genetic pool [24,25]. All three of the above-mentioned Baltimore classes include viruses with non-segmented and segmented genomes. A segmented genome can be packed into single virions resulting in true segmented viruses, or into multiple virions, resulting in multipartite viruses [26,27]. It is important to note that recent advances in evolutionary dynamics analyses confirmed that Baltimore classes do not accurately reflect evolutionary relationships among viruses, discerning the implicit presumption that each class is of monophyletic origin [28].

2.2. *The International Committee on Taxonomy of Viruses Classification*

In 2019 and 2020, the International Committee on Taxonomy of Viruses (ICTV), a global organisation responsible for developing virus taxonomy and nomenclature, introduced a new fifteen-rank classification hierarchy of virus taxonomy to replace the previously used five-rank hierarchy of species, genus, subfamily, family, and order. The 15-rank classification includes eight primary ranks (realm, kingdom, phylum, class, order, family, genus, and species) and seven secondary ranks, which resemble those used in Linnean systems [18]. Realm *Riboviria* was established as a likely monophyletic clade of RNA viruses that use RNA-directed RNA polymerase for genome replication, which includes viruses from 3 out of 7 Baltimore classes (III, IV, and V) [18,29]. The realm *Riboviria* was later extended to include nearly all RNA viruses and reverse-transcribing viruses, with a total of six realms recognised up to this day [30,31]. Even though some of the Baltimore classes do seem to be strictly monophyletic (such as class V, (−) ssRNA viruses) and others are clearly polyphyletic (such as class I, (+/−) dsDNA viruses), some form paraphyletic taxons with respect to other classes and therefore cannot be considered traditionally either (such as class III, dsRNA, with respect to class V, (−) ssRNA) [28]. With time, new discoveries obtained by metaviromics will inevitably change the hierarchical taxonomy of viruses. It is important to note that even though it was shown that Baltimore classes could not be considered taxonomic categories, both classification systems are helpful and complement each other well, especially for deciphering the means of viral evolution.

3. RNA Virus Characteristics

3.1. *Host Range*

RNA viruses are primarily infectious agents of Eukarya [32]. No known RNA viruses that infect Archaea were known until the last decade. At the same time, recent findings suggest that RNA viruses detected in archaeal hosts could be direct ancestors of eukaryotic RNA viruses [1]. RNA viruses of bacteria most commonly have a dsRNA genome; however, the taxonomy of bacteria-infecting ssRNA viruses was recently expanded [2]. Nevertheless, the majority of the viruses infecting prokaryotes have DNA genomes, particularly Baltimore class I ((+/−) dsDNA), with RNA viruses accounting for a minority of virome diversity [33]. In contrast to Archaea and Bacteria, eukaryotes are a host for the vast majority of RNA viruses, particularly of Baltimore class IV ((+) ssRNA) [3].

3.2. *Structural Genome Features*

Even though RNA viruses are known for their large structural and functional diversity, their genome sizes exhibit relatively small variations with differences within only one order of magnitude. While the smallest RNA viruses measure around 2 kb in size, the maximum observed genome length of 30–40 kb is reached among the *Coronaviridae* family [34,35]. This is primarily due to physical and structural constraints on RNA genome stability and RNA polymerase fidelity limitations with larger genomes consisting exclusively of class I dsDNA viruses [22,23,34,36,37]. One of the consequences of the small genome size is genome compression, which commonly manifests as gene overlap [38,39]. A universally shared structural characteristic of RNA viruses is the presence of untranslated regions (UTR) surrounding one or more open reading frames (ORF) at both 5′ and 3′ ends, usually containing sequences or conserved structures needed for replication regulation [40].

3.3. *RNA Regulatory Processes*

RNA-dependent RNA polymerase is an essential enzyme for viral RNA replication because it catalyses RNA replication from an RNA template. It is a process generally not typical in eukaryotic cells, except in events such as RNA-mediated silencing pathways and telomere formation [41]. Therefore, RNA viruses from Baltimore classes III and V encode their own RNA-dependent RNA polymerase and incorporate it into virions. The exception is viruses from class IV, which carry a positive single-stranded RNA genome that cellular machinery can directly transcribe upon viral entry [22]. Regulation of RNA replication, a fundamental step in the virus replication cycle, primarily depends on the structure of the viral RNA genome. Viruses with dsRNA genomes must synthesise both mRNA (which can occur as a direct transcription of the viral genome) and viral RNA (which is generated via (−) RNA strand intermediate). Similarly, (+) ssRNA genomes may already act as mRNA but still require (−) ssRNA intermediate for generating viral RNA. On the other hand, viruses with (−) ssRNA genomes, which are complementary in base sequence to the mRNA, first require its transcription to produce the (+) ssRNA, which may act as both mRNA and a template for genome replication. A delicate regulation balance between replication and transcription processes is vital for the successful synthesis of new viral progeny [4,22,42].

Canonical posttranscriptional processing necessary for the maturation of the primary transcript in cellular organisms includes obtaining a 5′-cap, splicing, and generation of a 3′ poly-A tail [43]. By mimicking the mechanisms mentioned above, viruses could increase the probability of perseverance of their genetic information during replication. Some RNA viruses use cap-snatching or 'stealing' of the cap from cellular mRNA; some use viral replication machinery to synthesise cap equivalent, while others disregard the use of 5′-cap completely, as reviewed in Decroly et al. (2012) [44]. Alternative polyadenylation signals, such as short poly-U or poly-A stretches, or the presence of highly conserved secondary structures, were observed [45–47]. The occurrence of splicing is generally not common in RNA viruses whose genome rarely contains introns [48]. The above-mentioned modifications may be applied to both whole-genome transcripts and nested sets of subgenomic mRNAs, which result from discontinuous transcription characteristic for members of the *Coronaviridae* family [49,50].

While the degree of independence of transcription and posttranscriptional processing in RNA viruses is a result of the intrinsic viral characteristics, translation is ultimately dependent on the host cell translation machinery. No matter if their genome may be used directly as mRNA or if the transcription of the minus strand must occur first, the translation of the resulting viral mRNA is completely dependent on cellular ribosomes. Since the cellular translation machinery is rather complex, viruses exhibit numerous tactics of manipulation, from an attack on regulatory signalling pathways to targeting specific steps of the translation process itself, comprehensively reviewed by Jaafar et al. (2018) and Jan et al. (2016) [51,52]. Furthermore, many viruses developed specific mechanisms in order to bypass the canonical phases of translation, such as the utilisation of the internal ribosome entry sites (IRES), the use of ribosomal frameshifting signals, and leaky scanning, which simultaneously contribute to the increase in the coding capacity of the rather small RNA virus genomes [53,54]. All of the properties and strategies mentioned above result in the successful hijacking of host cell machinery for the optimisation of the production of a large amount of viral progeny in a very short time, which is another hallmark of virus replication [55].

3.4. Quasispecies Concept

The concept of quasispecies was first developed by Manfred Eigen and Peter Schuster in the search of an adequate model for the origin and evolution of early life forms, defining quasispecies as a distribution of closely related replicative units centred around one or several (degenerate) master copies [56–58]. Before the introduction of the quasispecies concept, the main target of the (genetic) selection was considered to be the 'wild type' population. The quasispecies concept emphasises that selection does not apply to a single individual 'wild type' genome but rather within the 'wild type' distribution of closely related mutant genomes [59]. The error threshold seems to be a consequence of quasispecies existence and simultaneously the constraint of quasispecies evolution [59,60]. Viruses operate closely to their error threshold, which enables the greatest genetic variation; however, a slight change, such as a decrease in replication fidelity, can push the whole viral population beyond the error threshold into extinction [59–61]. The concept of quasispecies was further developed in recent years, emphasising real quasispecies in contrast to the theoretical model. The first experimental evidence of quasispecies was observed by analysing nucleotide sequences of bacteriophage Qβ. It was concluded that no unique structure of the 'wild type' bacteriophage Qβ genome existed, but rather, a large number of individual mutant sequences in dynamic equilibrium [62]. Furthermore, since the scope of viral pathogenesis could hardly be predicted due to the everchanging and fleeting quasispecies dynamics, possible medical implications of quasispecies existence were proposed [63]. The quasispecies nature of the virus population was soon demonstrated for various RNA virus populations, such as vesicular stomatitis virus and hepatitis C virus [63–65]. The scope of quasispecies was later broadened to refer to "dynamic distributions of non-identical but closely related mutant and recombinant viral genomes subjected to a continuous process of genetic variation, competition and selection, and which act as a unit of selection" [66].

4. Mechanisms of RNA Virus Variation

The most important mechanisms responsible for genetic change and, therefore, the rapid evolution and quasispecies nature of RNA viruses are mutations, recombination, and gene segment reassortment [67–69]. In addition to the enhancement of viral diversity by introducing genetic change on a greater level than by mutation alone, viral recombination and reassortment can also be one of the mechanisms of repair of genomic molecules [70]. However, for both viral recombination and reassortment to occur successfully, several predispositions have to be met: at least two different viral genomes must be present in the same intracellular area, the resulting recombinant or reassortant must be replication-competent and form infectious viral particles, and finally, the recombinant and reassortant virions must have features that favour their selection among the viral population [71,72]. Other less commonly observed genetic change mechanisms among RNA viruses are gene duplication and gene transfers [73].

4.1. Mutation

RNA-dependent RNA polymerase was identified as one of the viral hallmark genes responsible for critical functions in virion structure and genome replication, but is missing from the cellular genome [3,24,74,75]. The origin of RNA-dependent polymerase is intrinsically associated with the origin and evolution of life in general because RNA molecules were considered self-sufficient in the primordial RNA world, with elements such as self-splicing introns speeding up the evolution of hypothetical ribozymes [76]. It is important to note that one of the most conserved protein domains in RNA biogenesis in all kingdoms of life—the RNA recognition motif is related to the structural fold of the catalytic domain of RNA-dependent RNA polymerase [77]. Eukaryotic RNA-dependent RNA polymerases, however, do not share an evolutionary relationship with viral enzymes [78]. Viral RNA polymerases exhibit a relatively low fidelity of 10^{-3} to 10^{-5} per nucleotide polymerised, leading to the introduction of many mutations at every replication cycle due to the absence of 3′ to 5′ exonucleolytic proofreading activity [66]. However, new revelations suggest that early RNA polymerases had a feature of both proofreading and repair due to the need for extensive information contained within the relatively small genome size of the last universal common ancestor (LUCA) [37]. Low replication fidelity is thought to be one of the factors contributing to the RNA virus's vast diversity. It is worth mentioning that both an increase and decrease in replication fidelity greater than 4-fold have a negative impact on viral phenotype [79].

4.2. Recombination

Recombination is a process of exchange of genetic material between different non-segmented genomes, which is one of the main drivers of RNA virus heterogeneity [66]. The role of recombination in RNA viruses was first observed among the *Picornaviridae* family (such as the poliovirus) as an 'exception', but it was soon found to be relatively common, especially among Baltimore class IV (+) ssRNA viruses [80–83]. Recombination may act as a significant evolutionary force among viruses by creating new potential genome combinations [66]. However, the threshold for the selection of recombinants seems relatively high, with the majority of recombinant viruses being pruned away due to unfavourable properties [84]. Two mechanisms of recombination observed among RNA viruses are replicative recombination, which occurs during RNA synthesis via template switching of the viral polymerase according to the widely accepted copy-choice mechanism, and nonreplicative recombination by breaking and rejoining of the RNA molecules [68,82,85].

On the other hand, homologous recombination occurs between two similar RNA molecules with substantial sequence homology at the same or comparable sites on the paternal strains, while nonhomologous recombination is used to refer to genetic crossover between RNA molecules with no sequence homology [68]. It should be emphasised that even though replicative and nonreplicative recombination mechanisms can theoretically result in both homologous and nonhomologous recombinants, which are ultimately indistinguishable as end products, the vast majority of homologous RNA recombination occurs due to replicative mechanisms because sequence similarity acts as a guideline for the copy-choice mechanism [71,72,85,86]. The fidelity of RNA-dependent RNA polymerase was shown to be a determining factor inversely correlated with replicative recombination frequency [87,88]. Secondary structures of the RNA template, such as hairpins and loops, were also linked to the frequency of replicative recombination; however, data regarding various viruses are conflicting, and no universal control mechanism was established [89–91]. The frequency of recombination varies greatly among RNA viruses, with no apparent relation to genome type or replication strategy [83,92].

4.3. Reassortment

Another form of genetic exchange sometimes referred to as pseudo-recombination, a feature of all segmented RNA viruses, is gene segment reassortment. Reassortment is a consequence of the co-infection of a single host with two or more segmented viruses, which may result in progeny that contains novel genome combinations derived from both parental genomes [69,93]. Early reassortment models favour random packaging of segments into virions, with the probability of the generation of viable reassortants being determined entirely by chance [94]. Further research suggested that, while this stochastic model probably accurately describes reassortment among multipartite viruses who pack their segmented genome into multiple virus particles, genome segment exchange among viruses who pack their segmented genome into single virions is most likely guided by specific packaging signals [26,95,96]. Unlike recombination, which can in theory occur on any given genome segment and lead to the production of deleterious or nonfunctional proteins, the exchange of entire segments during reassortment ensures the maintenance of functional gene products [26,68]. The frequency of reassortment varies among segmented viruses—while it is relatively common in some viruses, such as influenza A, in other viruses, such as hantaviruses, it seems to occur less frequently [93,97]. It is worth mentioning that non-multipartite segmented viruses usually have meagre recombination rates, making reassortment particularly important for exploring available sequence space [98,99].

5. Causes and Consequences of RNA Virus Variation

High mutation rates observed in RNA viruses can be considered pillars of fast RNA virus evolution [66]. However, it should be emphasised that high mutation rates may simultaneously hinder virus emergence and adaptability by not allowing genotypes with potentially advantageous mutations to linger long enough to become fixed in a viral population due to strong selection against deleterious changes [100,101]. Therefore, the notion that a high mutation rate unequivocally enables faster viral adaptation was questioned, and other causes for high mutation rates were explored, such as selection for the robustness of the viral population and selection for fast replication [102]. On the other hand, several theories were proposed to explain the evolutionary reasons for recombination occurrence in RNA viruses. Even though recombination may serve a role in the repair of defective genomes, it does not seem likely that recombination evolved purely for purging detrimental mutations [103]. One theory suggests that recombination is a consequence of genome organisation and the means of RNA replication. Rarely observed recombination events in (−) ssRNA viruses could be attributed to the existence of the ribonucleoprotein complex as a unit of replication, lowering the probability of template switching required for recombination [104,105]. On the other hand, some transcription strategies observed among (+) ssRNA viruses, such as subgenomic RNA transcription, may be responsible for high recombination rates due to the more frequent occurrence of template switching events [79]. However, significant differences in recombination frequency were observed in many viruses with similar genome organisation, such as members of the *Flaviviridae* family [86,106,107]. Nevertheless, the ability to recombine provides many RNA viruses with rapid phenotype change, which can potentially enable further alterations in host tropism, evasion of host immune response, development of resistance to antiviral drugs, or changes in virulence and pathogenicity potentially leading to cross-species transmission [7,12,108].

6. Evolution of Human Coronaviruses

6.1. Classification of Coronaviruses

Coronaviruses are a group of viruses with the (+) ssRNA genome of Baltimore class IV under the realm of *Riboviria,* which belong to the order *Nidovirales,* family *Coronaviridae,* and subfamily *Orthocoronavirine* [18]. Phylogenetic analyses showed the division of the *Orthocoronavirinae* subfamily into four major genera (*Alphacoronavirus,* Alpha-CoV; *Betacoronavirus,* Beta-CoV, *Gammacoronavirus,* Gamma-CoV, and *Deltacoronavirus,* Delta-CoV) and 26 subgenera, with 52 currently recognised species [109]. Coronaviruses infect a wide variety of vertebrates, with bats and birds being notable reservoirs for Alpha-CoV to Beta-CoV and Gamma-CoV to Delta-CoV, respectively [16]. Even though coronaviruses in animals may cause a wide variety of symptoms, most infections typically result in respiratory or gastrointestinal illnesses [17]. However, due to the sometimes severe and lethal consequences, coronaviruses were more of interest in veterinary medicine, whereas recent epidemics in the 21st century shifted the focus to human coronaviruses (HCoV) [110]. There are currently seven known species or strains of coronaviruses which infect humans, all from Alpha-CoV or Beta-CoV genera, including *Human coronavirus 229E* (HCoV-229E), human coronavirus OC43 (HCoV-OC43), *Human coronavirus NL63* (HCoV-NL63), *Human coronavirus HKU1* (HCoV-HKU1), *Middle East respiratory syndrome-related coronavirus* (MERS-CoV), severe acute respiratory syndrome coronavirus (SARS-CoV) and, as of most recently, severe acute respiratory syndrome coronavirus 2 (SARS-CoV-2) [32,109]. Major epidemics of novel human coronaviruses in the last two decades resulted in multiple changes in *Coronaviridae* taxonomy to accommodate emergent viruses (Figure 2).

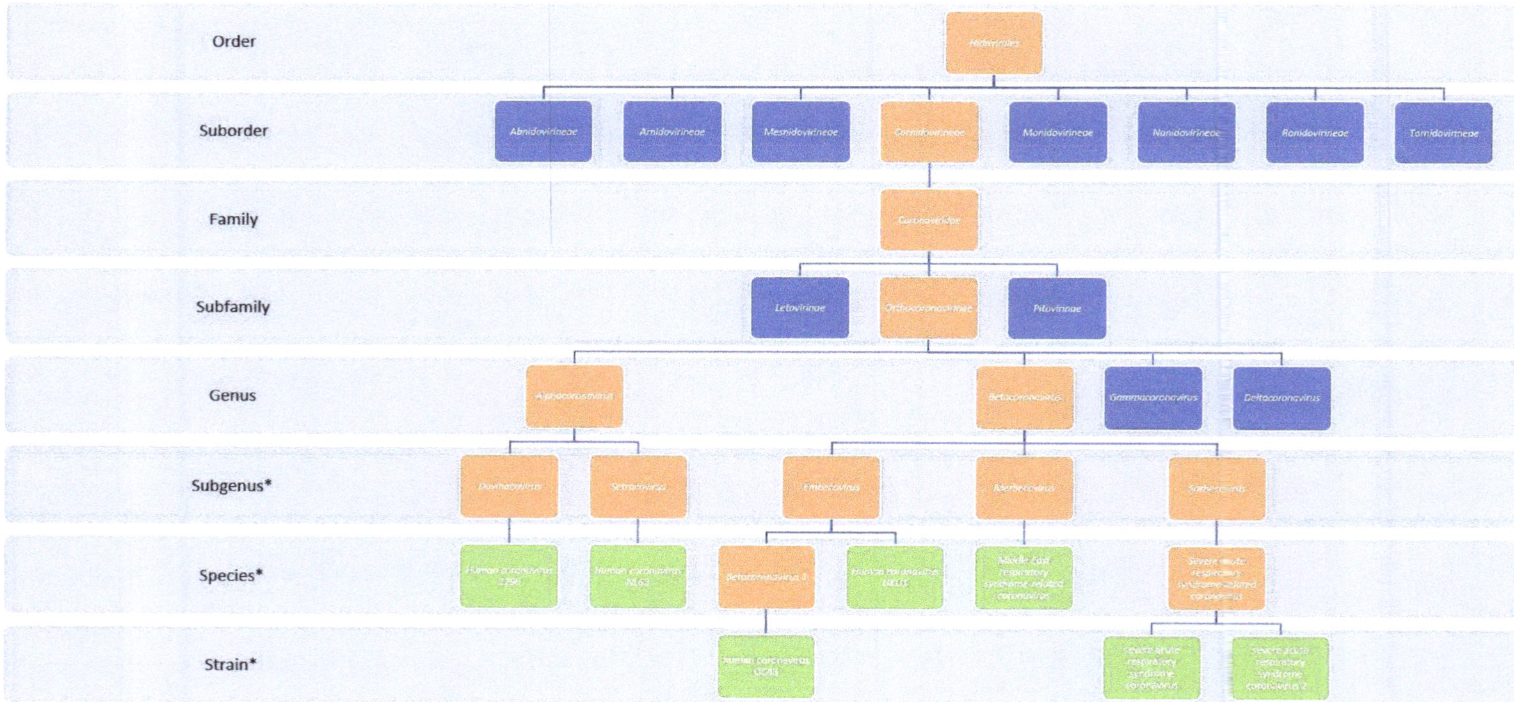

Figure 2. Classification of human coronaviruses. Seven currently recognised human coronavirus species are shown in green, their higher taxonomical categories are shown in orange, and other taxonomical categories are shown in blue. *: not all members of the particular taxonomical category are depicted. Data are based on the International Committee on Taxonomy of Viruses: Current Taxonomy Release 2021 and Master Species List 2021 v3 [32,109].

The first identified human coronaviruses were HCoV-229E from the genus Alpha-CoV and HCoV-OC43 from the genus Beta-CoV isolated during the 1960s, which were later found to be distributed globally and manifest primarily with respiratory symptoms, such as a common cold [111,112]. An outbreak of severe acute respiratory syndrome (SARS) was reported in 2002–2003, originating in Guangdong province in China and later spreading to other Asian countries, Europe, and North America, with the causative agent identified to be SARS-CoV having more than 8000 confirmed cases and mortality of around 10% [113–115]. Subsequent research on coronaviruses led to the discovery of HCoV-NL63 from the genus Alpha-CoV in 2004 and HCoV-HKU1 Beta-CoV in 2005 in patients with respiratory illnesses [116,117]. Both viruses are distributed globally and, together with HCoV-229E and HCoV-OC43, are responsible for 10–29% of common colds [118]. Another ongoing coronavirus outbreak started in 2012 in the Arabian Peninsula and was found to be caused by MERS-CoV, with more than 2600 cases reported as of November 2022 and high mortality exceeding 30% [119–121]. However, none of the previous outbreaks were even remotely comparable in scale, as was the third epidemic introduction of coronavirus in the human population, which started in late 2019 in Wuhan, China, and was later attributed to the novel SARS-CoV-2 [14,16]. SARS-CoV-2 is now the seventh coronavirus known to infect humans and was designated as a sister strain of SARS-CoVs of the species *Severe acute respiratory syndrome-related coronavirus* (Figure 2) [122]. As of November 2022, more than 630,000,000 confirmed cases of SARS-CoV-2, including more than 6,600,000 deaths, were observed with enormous socioeconomic consequences [123].

6.2. Structure and Genome of Coronaviruses

Coronaviruses have one of the largest genomes among RNA viruses, ranging from 26 to 32 kb, which is almost double the size of the other RNA viruses with large genomes [20,124]. The highly conserved genomic organisation is characteristic of all coronaviruses whose (+) ssRNA genomes end with 5′-cap and 3′ poly-A tail and consist of multiple ORFs often preceded by transcriptional regulatory sequences (TRS). ORFs are surrounded by terminal untranslated regions (UTR) rich in secondary structures with critical regulatory functions necessary for viral replication and transcription [19]. The first two-thirds of the coronavirus genome consist of two major ORFs, ORF1a, and ORF1b, and encode for large replicase polyproteins 1a (PP1a) and C-terminally extended 1b (PP1ab), which are processed into 16 nonstructural proteins (nsp) crucial for coordinating various intercellular aspects of coronavirus replication, including cleavage of polyproteins, vesicle membrane formation, excision of misincorporated nucleotides, mRNA processing, and modulation of host responses [125,126]. The final third of the coronavirus genome encodes for (depending on the coronavirus species) at least four structural proteins (spike, envelope, membrane, and nucleocapsid), which are a part of the infectious viral particle, interspersed with a various number of ORFs encoding for accessory proteins [19,125]. Translation of nonstructural proteins is enabled by ribosome frameshifting, while the expression of structural and accessory proteins is initiated after the generation of 3′ nested subgenomic mRNAs, which all contain a typical leader sequence on the 5′ end by discontinuous transcription guided by TRS [49,50].

6.3. Mutation in Coronaviruses

Replication fidelity is one of the main constraints on RNA virus genome size since large genomes generally introduce more mutations in each replication cycle, which threatens to push the virus over the error threshold [59,60]. One of the unique features of coronaviruses is the presence of 3′–5′ exonuclease (ExoN) in nsp14, which was shown to be indispensable for their RNA synthesis [21,126–128]. It was suggested that ExoN function was acquired by an ancestral virus that already had a means of successfully replicating the above-average size genome and was critical for its maintenance and eventual enlargement [115,124]. Engineering of CoV nsp14-ExoN viable mutants by substitutions at the ExoN motifs I, II, and III of the nsp14 demonstrated more than a 20× increase in mutation frequency during replication in vitro and a drastic reduction in the accumulation of viral RNA along with defects in sgRNA synthesis [126,127,129]. Exon I motifs (DE), II (D), and III (D) are conserved in the DEDD family of 3′–5′ exonuclease, named after four invariant acidic residues present in cellular organisms that catalyse DNA proofreading [130]. The contribution of nsp14-ExoN to the increase in coronavirus replication fidelity and faithful genome replication is undeniable; however, precise mechanisms still need to be completely elucidated. Several models were proposed, such as direct 3′–5′ proofreading of the nsp14-exoN analogous to cellular DNA mechanisms, direct or indirect stimulation of the intrinsic 3′–5′ activity of RdRp, or potential regulation of RNA recombination [79]. Several studies showed a significant variation in the missense mutation rate along the SARS-CoV-2 genome, with the minimum mutation rate observed in essential regions of SARS-CoV-2 proteins, such as RNA replication machinery and with higher rates of mutation corresponding to structurally more relaxed regions such as spike protein, which was found to be a target of both purifying and positive selection [131,132]. Mutations in the receptor binding domain of spike proteins may lead to the emergence of new variants with changes in fitness, transmission efficacy, or influence on host response and virus neutralisation [133]. The mutation rate for SARS-CoV-2 was found to be between 1 and 5×10^{-6} per nucleotide polymerised or of the order of 0.1 per genome per infection cycle [134].

6.4. Recombination in Coronaviruses

One of the first occurrences of RNA recombination in viruses with non-segmented RNA genomes was observed on murine coronaviruses. A high frequency of recombination (up to 25%) among RNA genomes of different strains of coronaviruses was shown during mixed infection of susceptible cells, with some of the recombinants appearing to go through multiple crossover events [68,135,136]. Even though mechanisms of coronavirus recombination were not fully understood at the time, it was evident that RNA recombination must have an essential role in virus evolution. It was shown that some recombinants could have evolutionary advantages allowing them to survive or even become dominant in a mixed virus population [68,136]. Early studies also showed that some coronavirus genome regions could be more prone to recombination than others, the so-called recombination 'hot-spots' [89]. The large genome size of coronaviruses prone to accumulation of deleterious mutations was considered one of the main reasons for the observed RNA recombination frequency as a means of genome diversification and repair [68,137].

On the other hand, the discontinuous transcription and generation of subgenomic mRNAs by RNA-dependent RNA polymerase used by coronaviruses as a strategy of gene expression resembles the mechanism of copy-choice RNA recombination via template switching and creates a predisposition for such events to occur [49,72]. Therefore, the rate of recombination observed among coronaviruses was found to be the highest among non-segmented (+) ssRNA viruses, with the frequency of mutants being 25% or more during mixed infection with different strains [136,138]. Such high recombination frequency could be partially attributed to the mutational constraint imposed by the proofreading capacity of nsp14-ExoN and the consequentially lower mutation rate, making recombination one of the main mechanisms of exploring available sequence space [139]. Spike glycoprotein gene, one of the main determinants of coronavirus host range and principal factors in viral entry, was identified as a 'hot-spot' for recombination among Alpha-CoV and Beta-CoV, allowing for the robust interchange of protein-coding sequences without the loss of infectivity [12,139]. Such changes in spike protein may alter spike–receptor interaction and influence host range, leading to cross-species transmission. The highly recombinogenic nature of coronaviruses from different hosts seems to be one of the critical factors in the origin of SARS-CoV-2, the mechanisms of which are further described in the following chapter [140].

6.5. Origin of Human Coronaviruses

With current evidence supporting the hypothesis that bat coronaviruses are gene sources for Alpha-CoV and Beta-CoV, while bird coronaviruses are gene sources for Gamma-

CoV and Delta-CoV, recent studies showed that the most recent common ancestor of all coronaviruses dates to around 8100 BC, with no concluding proof of whether it first occurred in bats and later jumped to birds or vice versa [16,141]. All seven coronaviruses currently present in humans emerged from animal reservoirs; however, it is clear that some of the common human coronaviruses, such as HCoV-229E, HCoV-OC43, HCoV-NL63, and HCoV-HKU1 adapted to their new host quite successfully [118,142]. Screening of bats showed the presence of closely related sequences to HCoV-NL63 and HCoV-229E, which suggests they have a bat origin [143,144]. Furthermore, zoonotic recombination of African bat NL63-like viruses and 229E-like viruses in spike protein genes most likely resulted in the ancestor of the HCoV-NL63, while it seems the ancestor of HCoV-229E infected camelids as intermediate hosts [144,145]. Phylogenetic analyses showed the role of rodents in the emergence of HCoV-OC43 and HCov-HKU1 [146]. On the other hand, SARS-CoV, MERS-CoV, and SARS-CoV-2 do not seem to be that well adapted to humans, but rather their animal hosts with the occasional spillover to the human population [147]. It was also shown that MERS-CoV originated in bats, with dromedary camels being the main intermediate reservoir hosts [148,149]. Similarly, SARS-CoV and SARS-CoV-2 were shown to be of probable bat origin, with other wild animals acting as intermediate hosts, such as masked palm civets and pangolins, respectively [15,150,151]. However, these animals most likely represent only transient and accidental hosts, while the circulation of MERS in dromedary camels seems to be long-term, with cross-species transmission occurring >30 years ago [152,153].

The most likely natural reservoirs of sarbecoviruses are various horseshoe bats from the *Rhinolophidae* family, which are hosts to bat SARS-CoV [15,141,144]. However, none of the bat SARS-CoVs seem to be a direct ancestor of the SARS-CoV and SARS-CoV-2 due to relatively distant phylogenetic relationships [154,155]. It should be emphasised that both SARS-CoV and SARS-CoV-2 use angiotensin-converting enzyme II (ACE2) as a cell entry receptor [15,156]. Even though this is not a universal characteristic of the *Sarbecovirus* subgenus, several of the bat SARS-like coronaviruses, such as WIV1 and WIV16, can bind human and bat ACE2 alike [157,158]. While there is no consensus on whether the ACE2 binding ability of bat-CoV was a loss- or gain-of-function mutation, the use of this receptor seems to be a highly evolvable characteristic with single mutations in receptor binding domains of spike gene leading to severe differences in binding efficiency [140,159]. One of the closest animal viral genomes to SARS-CoV-2 was initially shown to be Bat-CoV-RaTG13, with the second closest relative being Pangolin-CoV [147,151]. The sequence identity between Bat-CoV-RaTG13 and SARS-CoV-2 is 96.2% in the overall genome and 93.1% in the S gene region [15]. On the other hand, Pangolin-CoV genome comparison showed high sequence identity with both SARS-CoV-2 (91.0%) and Bat-CoV-RaTG13 (90.6%), along even higher amino acid sequence identity with SARS-CoV-2 S protein genes (97.5%) than Bat-CoV-RaTG13 S protein genes (95.4%) [151]. Therefore, one of the putative evolutionary patterns of SARS-CoV-2 origin could be the integration of the RNA fragment of the Pangolin-CoV-2019-related strain into the Bat-CoV-RaTG13 spike protein gene [147,154]. Furthermore, even though gene transfer is rarely observed in RNA viruses, gene-by-gene horizontal transfer and recombinational analysis of SARS-CoV-2-related viruses suggest that SARS-CoV-2 could also be a close relative of the bat-CoV ZC45 and ZXC21 strains [160]. A unique feature of SARS-CoV-2, which distinguishes it from other sarbecoviruses, is the presence of a furin cleavage site between S1 and S2 subunits of spike protein beneficial for its priming and increasing the affinity for the ACE2 receptor, most likely acquired during the frequent recombination events between SARS-CoV-2 and SARS-related CoV in bat co-infection, followed by a spillover to the human population [161–163].

6.6. SARS-CoV-2 Variant Evolution

The emergence of SARS-CoV-2 in December 2019 was followed by its swift spread in the human population and continuously increasing transmissibility leading to the evolution of newly evolved variants. Mutations causing single amino acid change, such as

D614G in spike protein gene and P323L in RNA-dependent RNA polymerase protein nsp12, were shown early on to be associated with higher viral loads and higher infectivity of the haplotypes that harbour them, leading to enhanced viral fitness but no change in pathogenesis [164,165]. Therefore, the evolution of SARS-CoV-2 during the first 11 months (the so-called first wave) of the pandemic was nearly neutral, requiring minimal adaptation of SARS-CoV-2 to humans [166]. However, by late 2020, a rapidly growing genomic cluster with a larger than the usual number of genetic changes was observed in the UK and labelled as the first variant of concern (VOC), Alpha (B.1.1.7), starting the second wave of the pandemic [167]. Variants of concern are defined as SARS-CoV-2 variants shown to be associated with one of the following: increase in transmissibility, increase in virulence, change in clinical disease presentation, detrimental change in epidemiology or reduced effectiveness in available therapeutics, vaccines, or diagnostics [168]. VOC Alpha was associated with multiple mutations in the spike protein gene, the most significant being N501Y within the receptor binding domain associated with the increased binding ability to ACE2 receptor, deletion H69, responsible for immune response evasion, and P681H in the furin cleavage site, which enhances spike cleavage [167,169]. Almost simultaneously, two more lineages with characteristic N501Y mutation were identified as VOC, Beta (B.1.351), and Gamma (P.1), first reported in South Africa and Brazil, respectively [170,171]. The third pandemic wave, initiated by the Delta variant (B.1.617.2), was first identified in India in late 2020 and was declared VOC in May 2021 [172]. The Delta variant seemed to lack the N501Y mutation observed in previous variants; however, multiple other mutations in spike protein resulted in a variant characterised by higher transmissibility and mortality [173,174]. The currently circulating variant of concern is Omicron (B.1.1.529), which was designated as such in November 2021 due to more than 30 mutations in the spike protein that were rarely seen in previous SARS-CoV-2 genomes. These mutations were shown to act as highly adaptive mutations and cooperatively interact, leading to an additional increase in viral infectivity and driving the fourth wave of the pandemic [175,176]. Omicron subvariants, such as BA.1, BA.2, BA.3, BA.4, and BA.5, were since identified [177,178]. It should be emphasised that all VOC originated from the initial SARS-CoV-2 genotype [179]. Visualisation of SARS-CoV-2 evolution and spread since the start of the pandemic, performed using Nextstrain, a publicly accessible bioinformatic tool for real-life tracking of pathogen evolution, shows global dominance of Omicron strains that almost completely replaced Delta and all previous variants (Figure 3) [180,181].

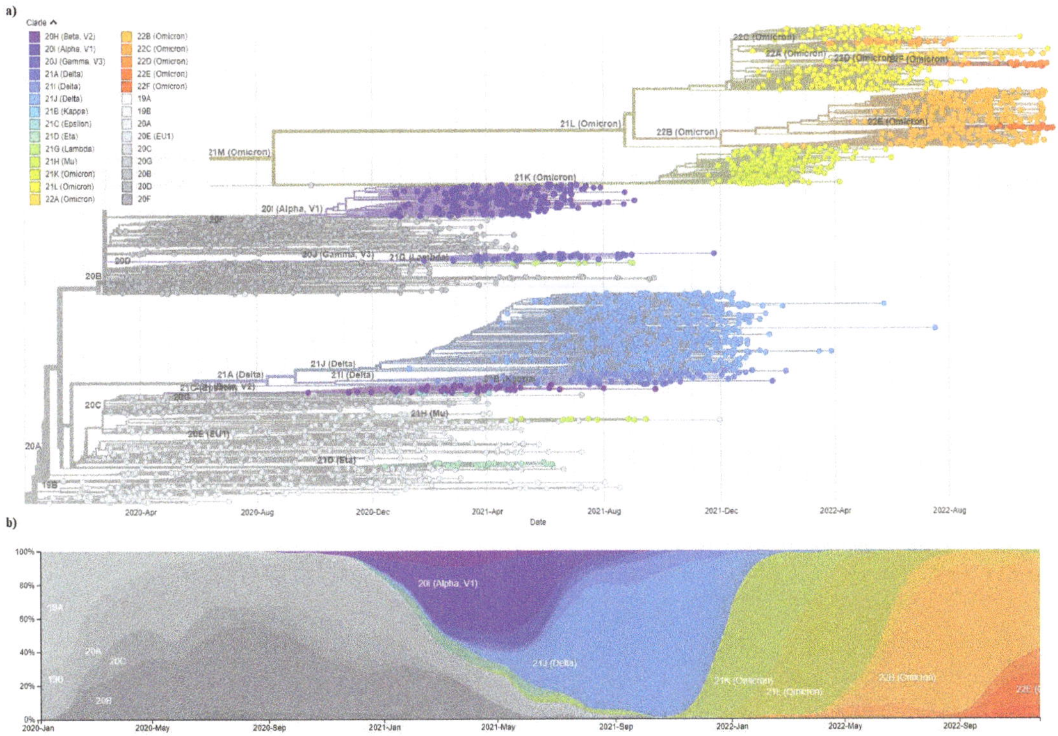

Figure 3. Global genomic epidemiology of severe acute respiratory syndrome coronavirus 2 (SARS-CoV-2) since the start of the coronavirus disease 2019 (COVID-19) pandemic. Visualisation of SARS-CoV-2 evolution and spread since the start of the pandemic was performed using Nextstrain (nextstrain.org), a publicly accessible bioinformatic tool for real-life tracking of pathogen evolution. A depiction of a subsample of 3059 genomes obtained between December 2019 and November 2022 from the GISAID database, with currently recognised SARS-CoV-2 variants indicated by different coloured branches on (**a**) a time-resolved phylogenetic tree. (**b**) Frequency visualisation by clade. Isolate Wuhan-Hu-1 (accession number: MN908947.3) was used as a reference. Vector images and live display can be found at: https://nextstrain.org/ncov/gisaid/global/all-time?d=tree,frequencies&p=full&showBranchLabels=all (accessed on 29 November 2022). In addition, a complete list of 3059 sequence authors was downloaded and shown in Table S1 as a tabs-separated value (TSV) file. Vector images are licensed with Attribution 4.0. International (CC BY 4.0) license [180,181].

7. Conclusions and Future Perspectives

Even though evolutionary processes are generally rather slow and long-lasting when observed on a level of cellular organisms, characteristics of the RNA viruses described in this review article enable real-time monitoring of some of the key evolutionary mechanisms. The increase in genetic variability could be considered one of the prerequisites for exploring new niches, while the plasticity of RNA viruses is best described by the multitude of ways utilised for processes such as genetic recombination and mutation. The emergence of three highly pathogenic RNA viruses from the *Coronaviridae* family showed that there are still many unknowns regarding viral pathogenesis, especially events leading up to potential cross-species transmission. Furthermore, the evolution of SARS-CoV-2 and its variants demonstrated the importance of the quasispecies concept and the need to revise rather old-fashioned constructs, such as the existence of the 'wild type' genome, when it comes to RNA viruses. There are many unknowns regarding the further trajectory of the ongoing SARS-CoV-2 pandemic. Despite the relative success of the SARS-CoV-2 vaccines (which is out of the scope of this review), it must be emphasised that the expectations for vaccine efficiency were a lot higher during 2020 based on then-observed limited diversity between viral genomes [182]. On the other hand, a considerable boost in transmissivity and host adaptation led to a drop in infection severity distinctive for Omicron variants [183]. The highly recombinogenic nature of coronaviruses emphasises the importance of surveillance of both intermediate and reservoir host animals. Furthermore, elucidating the exact mechanisms of cross-species transfer along with deciphering the complete mutational profile of the significant VOC could be highly beneficial in predicting the trajectory of the SARS-CoV-2 pandemic and preparing for the inevitable occurrence of the next one.

Supplementary Materials: The following supporting information can be downloaded at: https://www.mdpi.com/article/10.3390/v15010001/s1, Table S1: Complete list of 3059 sequence authors used by Nextstrain software.

Author Contributions: Conceptualisation, P.Š. and S.Ž.-L.; software, P.Š.; formal analysis, P.Š. and S.Ž.-L.; writing—original draft preparation, P.Š.; writing—review and editing, S.Ž.-L.; visualisation, P.Š.; supervision, S.Ž.-L.; project administration, P.Š. and S.Ž.-L. All authors have read and agreed to the published version of the manuscript.

Funding: This research was supported by the grant, "Strengthening the capacity of CerVirVac for research in virus immunology and vaccinology", KK.01.1.1.01.0006, awarded to the Scientific Centre of Excellence for Virus Immunology and Vaccines and co-financed by the European Regional Development Fund.

Institutional Review Board Statement: Not applicable.

Informed Consent Statement: Not applicable.

Data Availability Statement: The data presented in this study are available in Table S1.

Conflicts of Interest: The authors declare no conflict of interest.

References

1. Bolduc, B.; Shaughnessy, D.P.; Wolf, Y.I.; Koonin, E.V.; Roberto, F.F.; Young, M. Identification of Novel Positive-Strand RNA Viruses by Metagenomic Analysis of Archaea-Dominated Yellowstone Hot Springs. *J. Virol.* **2012**, *86*, 5562–5573. [CrossRef] [PubMed]
2. Callanan, J.; Stockdale, S.R.; Adriaenssens, E.M.; Kuhn, J.H.; Rumnieks, J.; Pallen, M.J.; Shkoporov, A.N.; Draper, L.A.; Ross, R.P.; Hill, C. Leviviricetes: Expanding and Restructuring the Taxonomy of Bacteria-Infecting Single-Stranded RNA Viruses. *Microb. Genom.* **2021**, *7*, 686. [CrossRef] [PubMed]
3. Koonin, E.V.; Dolja, V.V.; Krupovic, M. Origins and Evolution of Viruses of Eukaryotes: The Ultimate Modularity. *Virology* **2015**, *479–480*, 2–25. [CrossRef] [PubMed]
4. Baltimore, D. Expression of Animal Virus Genomes. *Bacteriol. Rev.* **1971**, *35*, 235–241. [CrossRef] [PubMed]
5. Walsh, D.; Mohr, I. Viral Subversion of the Host Protein Synthesis Machinery. *Nat. Rev. Microbiol.* **2011**, *9*, 860–875. [CrossRef]
6. Payne, S. Virus Interactions with the Cell. In *Viruses: From Understanding to Investigation*; Elsevier: Amsterdam, The Netherlands, 2017; pp. 23–35. [CrossRef]
7. Becher, P.; Tautz, N. RNA Recombination in Pestiviruses: Cellular RNA Sequences in Viral Genomes Highlight the Role of Host Factors for Viral Persistence and Lethal Disease. *RNA Biol.* **2011**, *8*, 216–224. [CrossRef]
8. Sanjuán, R.; Domingo-Calap, P. Mechanisms of Viral Mutation. *Cell. Mol. Life Sci.* **2016**, *73*, 4433–4448. [CrossRef]
9. Mattenberger, F.; Vila-Nistal, M.; Geller, R. Increased RNA Virus Population Diversity Improves Adaptability. *Sci. Rep.* **2021**, *11*, 6824. [CrossRef]
10. Fleischmann, W.R., Jr. Viral Genetics. In *Medical Microbiology*, 4th ed.; Baron, S., Ed.; University of Texas Medical Branch at Galveston: Galveston, TX, USA, 1996; Chapter 43.
11. Holland, D.J. Emerging Viruses. *Curr. Opin. Pediatr.* **1998**, *10*, 34–40. [CrossRef]
12. Graham, R.L.; Baric, R.S. Recombination, Reservoirs, and the Modular Spike: Mechanisms of Coronavirus Cross-Species Transmission. *J. Virol.* **2009**, *84*, 3134–3146. [CrossRef]
13. Wu, F.; Zhao, S.; Yu, B.; Chen, Y.-M.; Wang, W.; Song, Z.-G.; Hu, Y.; Tao, Z.-W.; Tian, J.-H.; Pei, Y.-Y.; et al. A New Coronavirus Associated with Human Respiratory Disease in China. *Nature* **2020**, *579*, 265–269. [CrossRef]
14. Zhu, N.; Zhang, D.; Wang, W.; Li, X.; Yang, B.; Song, J.; Zhao, X.; Huang, B.; Shi, W.; Lu, R.; et al. A Novel Coronavirus from Patients with Pneumonia in China, 2019. *N. Engl. J. Med.* **2020**, *382*, 727–733. [CrossRef]
15. Zhou, P.; Yang, X.-L.; Wang, X.-G.; Hu, B.; Zhang, L.; Zhang, W.; Si, H.-R.; Zhu, Y.; Li, B.; Huang, C.-L.; et al. A Pneumonia Outbreak Associated with a New Coronavirus of Probable Bat Origin. *Nature* **2020**, *579*, 270–273. [CrossRef]

16. Woo, P.C.Y.; Lau, S.K.P.; Lam, C.S.F.; Lau, C.C.Y.; Tsang, A.K.L.; Lau, J.H.N.; Bai, R.; Teng, J.L.L.; Tsang, C.C.C.; Wang, M.; et al. Discovery of Seven Novel Mammalian and Avian Coronaviruses in the Genus Deltacoronavirus Supports Bat Coronaviruses as the Gene Source of Alphacoronavirus and Betacoronavirus and Avian Coronaviruses as the Gene Source of Gammacoronavirus and Deltacoronavirus. *J. Virol.* **2012**, *86*, 3995–4008.
17. Haake, C.; Cook, S.; Pusterla, N.; Murphy, B. Coronavirus Infections in Companion Animals: Virology, Epidemiology, Clinical and Pathologic Features. *Viruses* **2020**, *12*, 1023. [CrossRef]
18. The New Scope of Virus Taxonomy: Partitioning the Virosphere into 15 Hierarchical Ranks. *Nat. Microbiol.* **2020**, *5*, 668–674. [CrossRef]
19. Romano, M.; Ruggiero, A.; Squeglia, F.; Maga, G.; Berisio, R. A Structural View of SARS-CoV-2 RNA Replication Machinery: RNA Synthesis, Proofreading and Final Capping. *Cells* **2020**, *9*, 1267. [CrossRef]
20. Gorbalenya, A.E.; Enjuanes, L.; Ziebuhr, J.; Snijder, E.J. Nidovirales: Evolving the Largest RNA Virus Genome. *Virus Res.* **2006**, *117*, 17–37. [CrossRef]
21. Eckerle, L.D.; Lu, X.; Sperry, S.M.; Choi, L.; Denison, M.R. High Fidelity of Murine Hepatitis Virus Replication Is Decreased in Nsp14 Exoribonuclease Mutants. *J. Virol.* **2007**, *81*, 12135–12144. [CrossRef]
22. Koonin, E.V.; Krupovic, M.; Agol, V.I. The Baltimore Classification of Viruses 50 Years Later: How Does It Stand in the Light of Virus Evolution? *Microbiol. Mol. Biol. Rev.* **2021**, *85*, e0005321. [CrossRef]
23. Koonin, E.V.; Dolja, V.V. A Virocentric Perspective on the Evolution of Life. *Curr. Opin. Virol.* **2013**, *3*, 546–557. [CrossRef] [PubMed]
24. Koonin, E.V.; Senkevich, T.G.; Dolja, V.V. The Ancient Virus World and Evolution of Cells. *Biol. Direct* **2006**, *1*, 29.
25. Wolf, Y.I.; Kazlauskas, D.; Iranzo, J.; Lucía-Sanz, A.; Kuhn, J.H.; Krupovic, M.; Dolja, V.V.; Koonin, E.V. Origins and Evolution of the Global RNA Virome. *mBio* **2018**, *9*, e02329. [CrossRef] [PubMed]
26. McDonald, S.M.; Nelson, M.I.; Turner, P.E.; Patton, J.T. Reassortment in Segmented RNA Viruses: Mechanisms and Outcomes. *Nat. Rev. Microbiol.* **2016**, *14*, 448–460. [CrossRef] [PubMed]
27. Newburn, L.R.; White, K.A. Trans-Acting RNA–RNA Interactions in Segmented RNA Viruses. *Viruses* **2019**, *11*, 751. [CrossRef]
28. Koonin, E.V.; Dolja, V.V.; Krupovic, M.; Varsani, A.; Wolf, Y.I.; Yutin, N.; Zerbini, F.M.; Kuhn, J.H. Global Organization and Proposed Megataxonomy of the Virus World. *Microbiol. Mol. Biol. Rev.* **2020**, *84*, e00061. [CrossRef]
29. Walker, P.J.; Siddell, S.G.; Lefkowitz, E.J.; Mushegian, A.R.; Dempsey, D.M.; Dutilh, B.E.; Harrach, B.; Harrison, R.L.; Hendrickson, R.C.; Junglen, S.; et al. Changes to Virus Taxonomy and the International Code of Virus Classification and Nomenclature Ratified by the International Committee on Taxonomy of Viruses (2019). *Arch. Virol.* **2019**, *164*, 2417–2429. [CrossRef]
30. Walker, P.J.; Siddell, S.G.; Lefkowitz, E.J.; Mushegian, A.R.; Adriaenssens, E.M.; Dempsey, D.M.; Dutilh, B.E.; Harrach, B.; Harrison, R.L.; Hendrickson, R.C.; et al. Changes to Virus Taxonomy and the Statutes Ratified by the International Committee on Taxonomy of Viruses (2020). *Arch. Virol.* **2020**, *165*, 2737–2748. [CrossRef]
31. Walker, P.J.; Siddell, S.G.; Lefkowitz, E.J.; Mushegian, A.R.; Adriaenssens, E.M.; Alfenas-Zerbini, P.; Dempsey, D.M.; Dutilh, B.E.; García, M.L.; Curtis Hendrickson, R.; et al. Recent Changes to Virus Taxonomy Ratified by the International Committee on Taxonomy of Viruses (2022). *Arch. Virol.* **2022**, *167*, 2429–2440. [CrossRef]
32. Current ICTV Taxonomy Release | ICTV. Available online: https://ictv.global/taxonomy (accessed on 6 October 2022).
33. Krupovic, M.; Prangishvili, D.; Hendrix, R.W.; Bamford, D.H. Genomics of Bacterial and Archaeal Viruses: Dynamics within the Prokaryotic Virosphere. *Microbiol. Mol. Biol. Rev.* **2011**, *75*, 610–635. [CrossRef]
34. Saberi, A.; Gulyaeva, A.A.; Brubacher, J.L.; Newmark, P.A.; Gorbalenya, A.E. A Planarian Nidovirus Expands the Limits of RNA Genome Size. *PLoS Pathog.* **2018**, *14*, e1007314. [CrossRef]
35. Rodríguez-Cousiño, N.; Maqueda, M.; Ambrona, J.; Zamora, E.; Esteban, R.; Ramírez, M. A New Wine *Saccharomyces Cerevisiae* Killer Toxin (Klus), Encoded by a Double-Stranded RNA Virus, with Broad Antifungal Activity Is Evolutionarily Related to a Chromosomal Host Gene. *Appl. Environ. Microbiol.* **2011**, *77*, 1822–1832. [CrossRef]
36. Holmes, E.C. Error Thresholds and the Constraints to RNA Virus Evolution. *Trends Microbiol.* **2003**, *11*, 543–546. [CrossRef]
37. Poole, A.M.; Logan, D.T. Modern MRNA Proofreading and Repair: Clues That the Last Universal Common Ancestor Possessed an RNA Genome? *Mol. Biol. Evol.* **2005**, *22*, 1444–1455. [CrossRef]
38. Chirico, N.; Vianelli, A.; Belshaw, R. Why Genes Overlap in Viruses. *Proc. R. Soc. B Biol. Sci.* **2010**, *277*, 3809–3817. [CrossRef]
39. Belshaw, R.; Pybus, O.G.; Rambaut, A. The Evolution of Genome Compression and Genomic Novelty in RNA Viruses. *Genome Res.* **2007**, *17*, 1496–1504. [CrossRef]
40. Dreher, T.W. Functions of the 3′-untranslated regions of positive strand rna viral genomes. *Annu. Rev. Phytopathol.* **1999**, *37*, 151–174. [CrossRef]
41. Zong, J.; Yao, X.; Yin, J.; Zhang, D.; Ma, H. Evolution of the RNA-Dependent RNA Polymerase (RdRP) Genes: Duplications and Possible Losses before and after the Divergence of Major Eukaryotic Groups. *Gene* **2009**, *447*, 29–39. [CrossRef]
42. Rampersad, S.; Tennant, P. Replication and Expression Strategies of Viruses. In *Viruses: From Understanding to Investigation*; Elsevier: Amsterdam, The Netherlands, 2018; pp. 55–82.
43. Vanden Heuvel, J.P. Posttranscriptional Processing of Messenger RNA. In *Regulation of Gene Expression*; Perdew, G.H., Vanden Heuvel, J.P., Peters, J.M., Eds.; Humana Press: Totowa, NJ, USA, 2007; pp. 105–106.
44. Decroly, E.; Ferron, F.; Lescar, J.; Canard, B. Conventional and Unconventional Mechanisms for Capping Viral MRNA. *Nat. Rev. Microbiol.* **2012**, *10*, 51–65. [CrossRef]

45. Li, X.; Palese, P. Characterization of the Polyadenylation Signal of Influenza Virus RNA. *J. Virol.* **1994**, *68*, 1245–1249. [CrossRef]
46. Kempf, B.J.; Barton, D.J. Picornavirus RNA Polyadenylation by 3Dpol, the Viral RNA-Dependent RNA Polymerase. *Virus Res.* **2015**, *206*, 3–11. [CrossRef] [PubMed]
47. Piñeiro, D.; Martinez-Salas, E. RNA Structural Elements of Hepatitis c Virus Controlling Viral RNA Translation and the Implications for Viral Pathogenesis. *Viruses* **2012**, *4*, 2233–2250. [CrossRef] [PubMed]
48. Chua, M.A.; Schmid, S.; Perez, J.T.; Langlois, R.A.; Oever, B.R. Influenza a Virus Utilizes Suboptimal Splicing to Coordinate the Timing of Infection. *Cell Rep.* **2013**, *3*, 23–29. [CrossRef] [PubMed]
49. Sawicki, S.G.; Sawicki, D.L.; Siddell, S.G. A Contemporary View of Coronavirus Transcription. *J. Virol.* **2007**, *81*, 20–29. [CrossRef] [PubMed]
50. Sola, I.; Almazán, F.; Zúñiga, S.; Enjuanes, L. Continuous and Discontinuous RNA Synthesis in Coronaviruses. *Annu. Rev. Virol.* **2015**, *2*, 265–288. [CrossRef] [PubMed]
51. Jaafar, Z.A.; Kieft, J.S. Viral RNA Structure-Based Strategies to Manipulate Translation. *Nat. Rev. Microbiol.* **2018**, *17*, 110–123. [CrossRef]
52. Jan, E.; Mohr, I.; Walsh, D. A Cap-To-Tail Guide to MRNA Translation Strategies in Virus-Infected Cells. *Annu. Rev. Virol.* **2016**, *3*, 283–307. [CrossRef]
53. Lukavsky, P.J. Structure and Function of HCV IRES Domains. *Virus Res.* **2009**, *139*, 166–171. [CrossRef]
54. Ishimaru, D.; Plant, E.P.; Sims, A.C.; Yount, B.L.; Roth, B.M.; Eldho, N.V.; Pérez-Alvarado, G.C.; Armbruster, D.W.; Baric, R.S.; Dinman, J.D.; et al. RNA Dimerization Plays a Role in Ribosomal Frameshifting of the SARS Coronavirus. *Nucleic Acids Res.* **2012**, *41*, 2594–2608. [CrossRef]
55. Sender, R.; Bar-On, Y.M.; Gleizer, S.; Bernshtein, B.; Flamholz, A.; Phillips, R.; Milo, R. The Total Number and Mass of SARS-CoV-2 Virions. *Proc. Natl. Acad. Sci. USA* **2021**, *118*, e2024815118. [CrossRef]
56. Eigen, M.; Schuster, P. A Principle of Natural Self-Organization. *Naturwissenschaften* **1977**, *64*, 541–565. [CrossRef]
57. Eigen, M.; Schuster, P. The Hypercycle. *Naturwissenschaften* **1978**, *65*, 7–41. [CrossRef]
58. Eigen, M.; Schuster, P. *The Hypercycle*; Springer: Berlin/Heidelberg, Germany, 1979.
59. Eigen, M. On the Nature of Virus Quasispecies. *Trends Microbiol.* **1996**, *4*, 216–218. [CrossRef]
60. Biebricher, C.K.; Eigen, M. The Error Threshold. *Virus Res.* **2005**, *107*, 117–127. [CrossRef]
61. Eigen, M. The Origin of Genetic Information: Viruses as Models. *Gene* **1993**, *135*, 37–47. [CrossRef]
62. Domingo, E.; Sabo, D.; Taniguchi, T.; Weissmann, C. Nucleotide Sequence Heterogeneity of an RNA Phage Population. *Cell* **1978**, *13*, 735–744. [CrossRef]
63. Holland, J.; Spindler, K.; Horodyski, F.; Grabau, E.; Nichol, S.; VandePol, S. Rapid Evolution of RNA Genomes. *Science* **1982**, *215*, 1577–1585. [CrossRef]
64. Duarte, E.A.; Novella, I.S.; Ledesma, S.; Clarke, D.K.; Moya, A.; Elena, S.F.; Domingo, E.; Holland, J.J. Subclonal Components of Consensus Fitness in an RNA Virus Clone. *J. Virol.* **1994**, *68*, 4295–4301. [CrossRef]
65. Martell, M.; Esteban, J.I.; Quer, J.; Genescà, J.; Weiner, A.; Esteban, R.; Guardia, J.; Gómez, J. Hepatitis c Virus (HCV) Circulates as a Population of Different but Closely Related Genomes: Quasispecies Nature of HCV Genome Distribution. *J. Virol.* **1992**, *66*, 3225–3229. [CrossRef]
66. Domingo, E. Quasispecies. *Encycl. Virol.* **1999**, 1431–1436.
67. Domingo, E.; Holland, J.J. RNA VIRUS MUTATIONS and FITNESS for SURVIVAL. *Annu. Rev. Microbiol.* **1997**, *51*, 151–178. [CrossRef] [PubMed]
68. Lai, M.M. RNA Recombination in Animal and Plant Viruses. *Microbiol. Rev.* **1992**, *56*, 61–79. [CrossRef] [PubMed]
69. Vijaykrishna, D.; Mukerji, R.; Smith, G.J.D. RNA Virus Reassortment: An Evolutionary Mechanism for Host Jumps and Immune Evasion. *PLOS Pathog.* **2015**, *11*, e1004902. [CrossRef] [PubMed]
70. Jackowiak, P.; Kuls, K.; Budzko, L.; Mania, A.; Figlerowicz, M.; Figlerowicz, M. Phylogeny and Molecular Evolution of the Hepatitis c Virus. *Infect. Genet. Evol. J. Mol. Epidemiol. Evol. Genet. Infect. Dis.* **2014**, *21*, 67–82. [CrossRef] [PubMed]
71. Bentley, K.; Evans, D.J. Mechanisms and Consequences of Positive-Strand RNA Virus Recombination. *J. Gen. Virol.* **2018**, *99*, 1345–1356. [CrossRef]
72. Simon-Loriere, E.; Holmes, E.C. Why Do RNA Viruses Recombine? *Nat. Rev. Microbiol.* **2011**, *9*, 617–626. [CrossRef]
73. Simon-Loriere, E.; Holmes, E.C. Gene Duplication Is Infrequent in the Recent Evolutionary History of RNA Viruses. *Mol. Biol. Evol.* **2013**, *30*, 1263–1269. [CrossRef]
74. Koonin, E.V.; Dolja, V.V.; Morris, T.J. Evolution and Taxonomy of Positive-Strand RNA Viruses: Implications of Comparative Analysis of Amino Acid Sequences. *Crit. Rev. Biochem. Mol. Biol.* **1993**, *28*, 375–430. [CrossRef]
75. Iranzo, J.; Krupovic, M.; Koonin, E.V. The Double-Stranded DNA Virosphere as a Modular Hierarchical Network of Gene Sharing. *mBio* **2016**, *7*, e00978. [CrossRef]
76. Gilbert, W. Origin of Life: The RNA World. *Nature* **1986**, *319*, 618. [CrossRef]
77. Krupovic, M.; Dolja, V.V.; Koonin, E.V. Origin of Viruses: Primordial Replicators Recruiting Capsids from Hosts. *Nat. Rev. Microbiol.* **2019**, *17*, 449–458. [CrossRef]
78. Iyer, L.M.; Koonin, E.V.; Aravind, L. Evolutionary Connection between the Catalytic Subunits of DNA-Dependent RNA Polymerases and Eukaryotic RNA-Dependent RNA Polymerases and the Origin of RNA Polymerases. *BMC Struct. Biol.* **2003**, *3*, 1. [CrossRef]

79. Denison, M.R.; Graham, R.L.; Donaldson, E.F.; Eckerle, L.D.; Baric, R.S. Coronaviruses. *RNA Biol.* **2011**, *8*, 270–279. [CrossRef]
80. Hirst, G.K. Genetic Recombination with Newcastle Disease Virus, Polioviruses, and Influenza. *Cold Spring Harb. Symp. Quant. Biol.* **1962**, *27*, 303–309. [CrossRef]
81. Ledinko, N. Genetic Recombination with Poliovirus Type 1. *Virology* **1963**, *20*, 107–119. [CrossRef]
82. Cooper, P.D.; Steiner-Pryor, A.; Scotti, P.D.; Delong, D. On the Nature of Poliovirus Genetic Recombinants. *J. Gen. Virol.* **1974**, *23*, 41–49. [CrossRef]
83. Savolainen-Kopra, C.; Blomqvist, S. Mechanisms of Genetic Variation in Polioviruses. *Rev. Med. Virol.* **2010**, *20*, 358–371. [CrossRef]
84. Banner, L.R.; Mc Lai, M. Random Nature of Coronavirus RNA Recombination in the Absence of Selection Pressure. *Virology* **1991**, *185*, 441–445. [CrossRef]
85. Gmyl, A.P.; Belousov, E.V.; Maslova, S.V.; Khitrina, E.V.; Chetverin, A.B.; Agol, V.I. Nonreplicative RNA Recombination in Poliovirus. *J. Virol.* **1999**, *73*, 8958–8965. [CrossRef]
86. Scheel, T.K.H.; Galli, A.; Li, Y.-P.; Mikkelsen, L.S.; Gottwein, J.M.; Bukh, J. Productive Homologous and Non-Homologous Recombination of Hepatitis c Virus in Cell Culture. *PLoS Pathog.* **2013**, *9*, e1003228. [CrossRef]
87. Poirier, E.Z.; Mounce, B.C.; Rozen-Gagnon, K.; Hooikaas, P.J.; Stapleford, K.A.; Moratorio, G.; Vignuzzi, M. Low-Fidelity Polymerases of Alphaviruses Recombine at Higher Rates to Overproduce Defective Interfering Particles. *J. Virol.* **2016**, *90*, 2446–2454. [CrossRef] [PubMed]
88. Woodman, A.; Arnold, J.J.; Cameron, C.E.; Evans, D.J. Biochemical and Genetic Analysis of the Role of the Viral Polymerase in Enterovirus Recombination. *Nucleic Acids Res.* **2016**, *44*, 6883–6895. [CrossRef] [PubMed]
89. Rowe, C.L.; Fleming, J.O.; Nathan, M.J.; Sgro, J.Y.; Palmenberg, A.C.; Baker, S.C. Generation of Coronavirus Spike Deletion Variants by High-Frequency Recombination at Regions of Predicted RNA Secondary Structure. *J. Virol.* **1997**, *71*, 6183–6190. [CrossRef] [PubMed]
90. Runckel, C.; Westesson, O.; Andino, R.; DeRisi, J.L. Identification and Manipulation of the Molecular Determinants Influencing Poliovirus Recombination. *PLoS Pathog.* **2013**, *9*, e1003164. [CrossRef] [PubMed]
91. Lowry, K.; Woodman, A.; Cook, J.; Evans, D.J. Recombination in Enteroviruses Is a Biphasic Replicative Process Involving the Generation of Greater-than Genome Length "Imprecise" Intermediates. *PLoS Pathog.* **2014**, *10*, e1004191. [CrossRef] [PubMed]
92. Taucher, C.; Berger, A.; Mandl, C.W. A Trans -Complementing Recombination Trap Demonstrates a Low Propensity of Flaviviruses for Intermolecular Recombination. *J. Virol.* **2010**, *84*, 599–611. [CrossRef] [PubMed]
93. Marshall, N.; Priyamvada, L.; Ende, Z.; Steel, J.; Lowen, A.C. Influenza Virus Reassortment Occurs with High Frequency in the Absence of Segment Mismatch. *PLoS Pathog.* **2013**, *9*, e1003421. [CrossRef]
94. Bancroft, C.T.; Parslow, T.G. Evidence for Segment-Nonspecific Packaging of the Influenza a Virus Genome. *J. Virol.* **2002**, *76*, 7133–7139. [CrossRef]
95. Essere, B.; Yver, M.; Gavazzi, C.; Terrier, O.; Isel, C.; Fournier, E.; Giroux, F.; Textoris, J.; Julien, T.; Socratous, C.; et al. Critical Role of Segment-Specific Packaging Signals in Genetic Reassortment of Influenza a Viruses. *Proc. Natl. Acad. Sci. USA* **2013**, *110*, E3840–E3848. [CrossRef]
96. Baker, S.F.; Nogales, A.; Finch, C.; Tuffy, K.M.; Domm, W.; Perez, D.R.; Topham, D.J.; Martínez-Sobrido, L. Influenza a and B Virus Intertypic Reassortment through Compatible Viral Packaging Signals. *J. Virol.* **2014**, *88*, 10778–10791. [CrossRef]
97. Klempa, B. Reassortment Events in the Evolution of Hantaviruses. *Virus Genes* **2018**, *54*, 638–646. [CrossRef]
98. Boni, M.F.; Zhou, Y.; Taubenberger, J.K.; Holmes, E.C. Homologous Recombination Is Very Rare or Absent in Human Influenza a Virus. *J. Virol.* **2008**, *82*, 4807–4811. [CrossRef]
99. Varsani, A.; Lefeuvre, P.; Roumagnac, P.; Martin, D. Notes on Recombination and Reassortment in Multipartite/Segmented Viruses. *Curr. Opin. Virol.* **2018**, *33*, 156–166. [CrossRef]
100. Johnson, T.; Barton, N.H. The Effect of Deleterious Alleles on Adaptation in Asexual Populations. *Genetics* **2002**, *162*, 395–411. [CrossRef]
101. Peck, J.R. A Ruby in the Rubbish: Beneficial Mutations, Deleterious Mutations and the Evolution of Sex. *Genetics* **1994**, *137*, 597–606. [CrossRef]
102. Elena, S.F.; Sanjuan, R. Adaptive Value of High Mutation Rates of RNA Viruses: Separating Causes from Consequences. *J. Virol.* **2005**, *79*, 11555–11558. [CrossRef]
103. Bonhoeffer, S.; Chappey, C.; Parkin, N.T.; Whitcomb, J.M.; Petropoulos, C.J. Evidence for Positive Epistasis in HIV-1. *Science* **2004**, *306*, 1547–1550. [CrossRef]
104. Chare, E.R.; Gould, E.A.; Holmes, E.C. Phylogenetic Analysis Reveals a Low Rate of Homologous Recombination in Negative-Sense RNA Viruses. *J. Gen. Virol.* **2003**, *84*, 2691–2703. [CrossRef]
105. Han, G.-Z.; Worobey, M. Homologous Recombination in Negative Sense RNA Viruses. *Viruses* **2011**, *3*, 1358–1373. [CrossRef]
106. Twiddy, S.S. The Extent of Homologous Recombination in Members of the Genus Flavivirus. *J. Gen. Virol.* **2003**, *84*, 429–440. [CrossRef]
107. Pickett, B.E.; Lefkowitz, E.J. Recombination in West Nile Virus: Minimal Contribution to Genomic Diversity. *Virol. J.* **2009**, *6*, 165. [CrossRef] [PubMed]
108. González-Candelas, F.; López-Labrador, F.X.; Bracho, M.A. Recombination in Hepatitis c Virus. *Viruses* **2011**, *3*, 2006–2024. [CrossRef] [PubMed]

109. Master Species Lists | ICTV. Available online: https://ictv.global/msl (accessed on 5 November 2022).
110. Bonilauri, P.; Rugna, G. Animal Coronaviruses and SARS-COV-2 in Animals, What Do We Actually Know? *Life* **2021**, *11*, 123. [CrossRef] [PubMed]
111. Hamre, D.; Procknow, J.J. A New Virus Isolated from the Human Respiratory Tract. *Exp. Biol. Med.* **1966**, *121*, 190–193. [CrossRef] [PubMed]
112. McIntosh, K.; Dees, J.H.; Becker, W.B.; Kapikian, A.Z.; Chanock, R.M. Recovery in Tracheal Organ Cultures of Novel Viruses from Patients with Respiratory Disease. *Proc. Natl. Acad. Sci. USA* **1967**, *57*, 933–940. [CrossRef]
113. Peiris, J.; Lai, S.; Poon, L.; Guan, Y.; Yam, L.; Lim, W.; Nicholls, J.; Yee, W.; Yan, W.; Cheung, M.; et al. Coronavirus as a Possible Cause of Severe Acute Respiratory Syndrome. *Lancet* **2003**, *361*, 1319–1325. [CrossRef]
114. Peiris, J.S.M.; Yuen, K.Y.; Osterhaus, A.D.M.E.; Stöhr, K. The Severe Acute Respiratory Syndrome. *N. Engl. J. Med.* **2003**, *349*, 2431–2441. [CrossRef]
115. Snijder, E.J.; Bredenbeek, P.J.; Dobbe, J.C.; Thiel, V.; Ziebuhr, J.; Poon, L.L.M.; Guan, Y.; Rozanov, M.; Spaan, W.J.M.; Gorbalenya, A.E. Unique and Conserved Features of Genome and Proteome of SARS-Coronavirus, an Early Split-off from the Coronavirus Group 2 Lineage. *J. Mol. Biol.* **2003**, *331*, 991–1004. [CrossRef]
116. van der Hoek, L.; Pyrc, K.; Berkhout, B. Human Coronavirus NL63, a New Respiratory Virus. *FEMS Microbiol. Rev.* **2006**, *30*, 760–773. [CrossRef]
117. Woo, P.C.Y.; Lau, S.K.P.; Chu, C.; Chan, K.; Tsoi, H.; Huang, Y.; Wong, B.H.L.; Poon, R.W.S.; Cai, J.J.; Luk, W.; et al. Characterization and Complete Genome Sequence of a Novel Coronavirus, Coronavirus HKU1, from Patients with Pneumonia. *J. Virol.* **2004**, *79*, 884–895. [CrossRef]
118. Monto, A.S. Medical Reviews. Coronaviruses. *Yale J. Biol. Med.* **1974**, *47*, 234–251.
119. Zaki, A.M.; van Boheemen, S.; Bestebroer, T.M.; Osterhaus, A.D.M.E.; Fouchier, R.A.M. Isolation of a Novel Coronavirus from a Man with Pneumonia in Saudi Arabia. *N. Engl. J. Med.* **2012**, *367*, 1814–1820. [CrossRef]
120. MERS-CoV Worldwide Overview. Available online: https://www.ecdc.europa.eu/en/middle-east-respiratory-syndrome-coronavirus-mers-cov-situation-update (accessed on 10 November 2022).
121. World Health Organization (WHO). Middle East Respiratory Syndrome Coronavirus (MERS-CoV). Available online: https://www.who.int/news-room/fact-sheets/detail/middle-east-respiratory-syndrome-coronavirus-(mers-cov) (accessed on 10 November 2022).
122. Gorbalenya, A.E.; Baker, S.C.; Baric, R.S.; de Groot, R.J.; Drosten, C.; Gulyaeva, A.A.; Haagmans, B.L.; Lauber, C.; Leontovich, A.M.; Neuman, B.W.; et al. The Species Severe Acute Respiratory Syndrome-Related Coronavirus: Classifying 2019-NCoV and Naming It SARS-CoV-2. *Nat. Microbiol.* **2020**, *5*, 1–9.
123. World Health Organization. WHO COVID-19 Dashboard. Available online: https://covid19.who.int/ (accessed on 2 November 2022).
124. Nayak, B.; Kumar, S.; Collins, P.L.; Samal, S.K. Molecular Characterization and Complete Genome Sequence of Avian Paramyxovirus Type 4 Prototype Strain Duck/Hong Kong/D3/75. *Virol. J.* **2008**, *5*, 124. [CrossRef]
125. Chen, Y.; Liu, Q.; Guo, D. Emerging Coronaviruses: Genome Structure, Replication, and Pathogenesis. *J. Med. Virol.* **2020**, *92*, 418–423. [CrossRef]
126. Tahir, M. Coronavirus Genomic Nsp14-ExoN, Structure, Role, Mechanism, and Potential Application as a Drug Target. *J. Med. Virol.* **2021**, *93*, 4258–4264. [CrossRef]
127. Minskaia, E.; Hertzig, T.; Gorbalenya, A.E.; Campanacci, V.; Cambillau, C.; Canard, B.; Ziebuhr, J. Discovery of an RNA Virus 3′->5′ Exoribonuclease That Is Critically Involved in Coronavirus RNA Synthesis. *Proc. Natl. Acad. Sci. USA* **2006**, *103*, 5108–5113. [CrossRef]
128. Robson, F.; Khan, K.S.; Le, T.K.; Paris, C.; Demirbag, S.; Barfuss, P.; Rocchi, P.; Ng, W.-L. Coronavirus RNA Proofreading: Molecular Basis and Therapeutic Targeting. *Mol. Cell* **2020**, *79*, 710–727. [CrossRef]
129. Eckerle, L.D.; Becker, M.M.; Halpin, R.A.; Li, K.; Venter, E.; Lu, X.; Scherbakova, S.; Graham, R.L.; Baric, R.S.; Stockwell, T.B.; et al. Infidelity of SARS-CoV Nsp14-Exonuclease Mutant Virus Replication Is Revealed by Complete Genome Sequencing. *PLoS Pathog.* **2010**, *6*, e1000896. [CrossRef]
130. Zuo, Y. Exoribonuclease Superfamilies: Structural Analysis and Phylogenetic Distribution. *Nucleic Acids Res.* **2001**, *29*, 1017–1026. [CrossRef]
131. Berrio, A.; Gartner, V.; Wray, G.A. Positive Selection within the Genomes of SARS-CoV-2 and Other Coronaviruses Independent of Impact on Protein Function. *PeerJ* **2020**, *8*, e10234. [CrossRef] [PubMed]
132. Jaroszewski, L.; Iyer, M.; Alisoltani, A.; Sedova, M.; Godzik, A. The Interplay of SARS-CoV-2 Evolution and Constraints Imposed by the Structure and Functionality of Its Proteins. *PLoS Comput. Biol.* **2021**, *17*, e1009147. [CrossRef] [PubMed]
133. Magazine, N.; Zhang, T.; Wu, Y.; McGee, M.C.; Veggiani, G.; Huang, W. Mutations and Evolution of the SARS-CoV-2 Spike Protein. *Viruses* **2022**, *14*, 640. [CrossRef] [PubMed]
134. Amicone, M.; Borges, V.; Alves, M.J.; Isidro, J.; Zé-Zé, L.; Duarte, S.; Vieira, L.; Guiomar, R.; Gomes, J.P.; Gordo, I. Mutation Rate of SARS-CoV-2 and Emergence of Mutators during Experimental Evolution. *Evol. Med. Public Health* **2022**, *10*, 142–155. [CrossRef] [PubMed]
135. Lai, M.M.; Baric, R.S.; Makino, S.; Keck, J.G.; Egbert, J.; Leibowitz, J.L.; Stohlman, S.A. Recombination between Nonsegmented RNA Genomes of Murine Coronaviruses. *J. Virol.* **1985**, *56*, 449–456. [CrossRef]

136. Makino, S.; Keck, J.G.; Stohlman, S.A.; Lai, M.M. High-Frequency RNA Recombination of Murine Coronaviruses. *J. Virol.* **1986**, *57*, 729–737. [CrossRef]
137. Koetzner, C.A.; Parker, M.M.; Ricard, C.S.; Sturman, L.S.; Masters, P.S. Repair and Mutagenesis of the Genome of a Deletion Mutant of the Coronavirus Mouse Hepatitis Virus by Targeted RNA Recombination. *J. Virol.* **1992**, *66*, 1841–1848. [CrossRef]
138. Baric, R.S.; Fu, K.; Schaad, M.C.; Stohlman, S.A. Establishing a Genetic Recombination Map for Murine Coronavirus Strain A59 Complementation Groups. *Virology* **1990**, *177*, 646–656. [CrossRef]
139. Goldstein, S.A.; Brown, J.; Pedersen, B.S.; Quinlan, A.R.; Elde, N.C. Extensive Recombination-Driven Coronavirus Diversification Expands the Pool of Potential Pandemic Pathogens. *Genome Biol. Evol.* **2021**, *14*, evac161. [CrossRef]
140. Boni, M.F.; Lemey, P.; Jiang, X.; Lam, T.T.-Y.; Perry, B.W.; Castoe, T.A.; Rambaut, A.; Robertson, D.L. Evolutionary Origins of the SARS-CoV-2 Sarbecovirus Lineage Responsible for the COVID-19 Pandemic. *Nat. Microbiol.* **2020**, *5*, 1408–1417. [CrossRef]
141. Woo, P.C.Y.; Lau, S.K.P.; Huang, Y.; Yuen, K.-Y. Coronavirus Diversity, Phylogeny and Interspecies Jumping. *Exp. Biol. Med.* **2009**, *234*, 1117–1127. [CrossRef]
142. Liu, D.X.; Liang, J.Q.; Fung, T.S. Human Coronavirus-229E, -OC43, -NL63, and -HKU1 (Coronaviridae). In *Encyclopedia of Virology*; Elsevier: Amsterdam, The Netherlands, 2021; pp. 428–440.
143. Pfefferle, S.; Oppong, S.; Drexler, J.F.; Gloza-Rausch, F.; Ipsen, A.; Seebens, A.; Müller, M.A.; Annan, A.; Vallo, P.; Adu-Sarkodie, Y.; et al. Distant Relatives of Severe Acute Respiratory Syndrome Coronavirus and Close Relatives of Human Coronavirus 229E in Bats, Ghana. *Emerg. Infect. Dis.* **2009**, *15*, 1377–1384. [CrossRef]
144. Tao, Y.; Shi, M.; Chommanard, C.; Queen, K.; Zhang, J.; Markotter, W.; Kuzmin, I.V.; Holmes, E.C.; Tong, S. Surveillance of Bat Coronaviruses in Kenya Identifies Relatives of Human Coronaviruses NL63 and 229E and Their Recombination History. *J. Virol.* **2017**, *91*, e01953. [CrossRef]
145. Corman, V.M.; Baldwin, H.J.; Tateno, A.F.; Zerbinati, R.M.; Annan, A.; Owusu, M.; Nkrumah, E.E.; Maganga, G.D.; Oppong, S.; Adu-Sarkodie, Y.; et al. Evidence for an Ancestral Association of Human Coronavirus 229E with Bats. *J. Virol.* **2015**, *89*, 11858–11870. [CrossRef]
146. Stout, A.E.; Millet, J.K.; Stanhope, M.J.; Whittaker, G.R. Furin Cleavage Sites in the Spike Proteins of Bat and Rodent Coronaviruses: Implications for Virus Evolution and Zoonotic Transfer from Rodent Species. *One Health* **2021**, *13*, 100282. [CrossRef]
147. Zhu, Z.; Meng, K.; Meng, G. Genomic Recombination Events May Reveal the Evolution of Coronavirus and the Origin of SARS-CoV-2. *Sci. Rep.* **2020**, *10*, 21617. [CrossRef]
148. Khalafalla, A.I.; Lu, X.; Al-Mubarak, A.I.A.; Dalab, A.H.S.; Al-Busadah, K.A.S.; Erdman, D.D. MERS-CoV in Upper Respiratory Tract and Lungs of Dromedary Camels, Saudi Arabia, 2013–2014. *Emerg. Infect. Dis.* **2015**, *21*, 1153–1158. [CrossRef]
149. Memish, Z.A.; Mishra, N.; Olival, K.J.; Fagbo, S.F.; Kapoor, V.; Epstein, J.H.; AlHakeem, R.; Durosinloun, A.; Al Asmari, M.; Islam, A.; et al. Middle East Respiratory Syndrome Coronavirus in Bats, Saudi Arabia. *Emerg. Infect. Dis.* **2013**, *19*, 1819–1823. [CrossRef]
150. Song, H.-D.; Tu, C.-C.; Zhang, G.-W.; Wang, S.-Y.; Zheng, K.; Lei, L.-C.; Chen, Q.-X.; Gao, Y.-W.; Zhou, H.-Q.; Xiang, H.; et al. Cross-Host Evolution of Severe Acute Respiratory Syndrome Coronavirus in Palm Civet and Human. *Proc. Natl. Acad. Sci. USA* **2005**, *102*, 2430–2435. [CrossRef]
151. Zhang, T.; Wu, Q.; Zhang, Z. Probable Pangolin Origin of SARS-CoV-2 Associated with the COVID-19 Outbreak. *Curr. Biol.* **2020**, *30*, 1346–1351. [CrossRef]
152. Poon, L.L.M.; Chu, D.K.W.; Chan, K.H.; Wong, O.K.; Ellis, T.M.; Leung, Y.H.C.; Lau, S.K.P.; Woo, P.C.Y.; Suen, K.Y.; Yuen, K.Y.; et al. Identification of a Novel Coronavirus in Bats. *J. Virol.* **2005**, *79*, 2001–2009. [CrossRef] [PubMed]
153. Müller, M.A.; Corman, V.M.; Jores, J.; Meyer, B.; Younan, M.; Liljander, A.; Bosch, B.-J.; Lattwein, E.; Hilali, M.; Musa, B.E.; et al. MERS Coronavirus Neutralizing Antibodies in Camels, Eastern Africa, 1983–1997. *Emerg. Infect. Dis.* **2014**, *20*, 2093–2095. [CrossRef] [PubMed]
154. Hu, B.; Zeng, L.-P.; Yang, X.-L.; Ge, X.-Y.; Zhang, W.; Li, B.; Xie, J.-Z.; Shen, X.-R.; Zhang, Y.-Z.; Wang, N.; et al. Discovery of a Rich Gene Pool of Bat SARS-Related Coronaviruses Provides New Insights into the Origin of SARS Coronavirus. *PLOS Pathog.* **2017**, *13*, e1006698. [CrossRef] [PubMed]
155. Hon, C.-C.; Lam, T.-Y.; Shi, Z.-L.; Drummond, A.J.; Yip, C.-W.; Zeng, F.; Lam, P.-Y.; Leung, F.C.-C. Evidence of the Recombinant Origin of a Bat Severe Acute Respiratory Syndrome (SARS)-like Coronavirus and Its Implications on the Direct Ancestor of SARS Coronavirus. *J. Virol.* **2007**, *82*, 1819–1826. [CrossRef] [PubMed]
156. Walls, A.C.; Park, Y.-J.; Tortorici, M.A.; Wall, A.; McGuire, A.T.; Veesler, D. Structure, Function, and Antigenicity of the Sars-Cov-2 Spike Glycoprotein. *Cell* **2020**, *181*, 281–292. [CrossRef]
157. Ge, X.-Y.; Li, J.-L.; Yang, X.-L.; Chmura, A.A.; Zhu, G.; Epstein, J.H.; Mazet, J.K.; Hu, B.; Zhang, W.; Peng, C.; et al. Isolation and Characterization of a Bat SARS-like Coronavirus That Uses the ACE2 Receptor. *Nature* **2013**, *503*, 535–538. [CrossRef]
158. Yang, X.-L.; Hu, B.; Wang, B.; Wang, M.-N.; Zhang, Q.; Zhang, W.; Wu, L.-J.; Ge, X.-Y.; Zhang, Y.-Z.; Daszak, P.; et al. Isolation and Characterization of a Novel Bat Coronavirus Closely Related to the Direct Progenitor of Severe Acute Respiratory Syndrome Coronavirus. *J. Virol.* **2015**, *90*, 3253–3256. [CrossRef]
159. Starr, T.N.; Zepeda, S.K.; Walls, A.C.; Greaney, A.J.; Alkhovsky, S.; Veesler, D.; Bloom, J.D. ACE2 Binding Is an Ancestral and Evolvable Trait of Sarbecoviruses. *Nature* **2022**, *603*, 913–918. [CrossRef]
160. Makarenkov, V.; Mazoure, B.; Rabusseau, G.; Legendre, P. Horizontal Gene Transfer and Recombination Analysis of SARS-CoV-2 Genes Helps Discover Its Close Relatives and Shed Light on Its Origin. *BMC Ecol. Evol.* **2021**, *21*, 5. [CrossRef]

161. Coutard, B.; Valle, C.; de Lamballerie, X.; Canard, B.; Seidah, N.G.; Decroly, E. The Spike Glycoprotein of the New Coronavirus 2019-NCoV Contains a Furin-like Cleavage Site Absent in CoV of the Same Clade. *Antivir. Res.* **2020**, *176*, 104742. [CrossRef]
162. Kadam, S.B.; Sukhramani, G.S.; Bishnoi, P.; Pable, A.A.; Barvkar, V.T. SARS-CoV-2, the Pandemic Coronavirus: Molecular and Structural Insights. *J. Basic Microbiol.* **2021**, *61*, 180–202. [CrossRef]
163. The Sarbecovirus Origin of SARS-CoV-2′s Furin Cleavage Site. Available online: https://virological.org/t/the-sarbecovirus-origin-of-sars-cov-2-s-furin-cleavage-site/536 (accessed on 15 November 2022).
164. Korber, B.; Fischer, W.M.; Gnanakaran, S.; Yoon, H.; Theiler, J.; Abfalterer, W.; Hengartner, N.; Giorgi, E.E.; Bhattacharya, T.; Foley, B.; et al. Tracking Changes in SARS-CoV-2 Spike: Evidence That D614G Increases Infectivity of the COVID-19 Virus. *Cell* **2020**, *182*, 812–827. [CrossRef]
165. Garvin, M.R.; Prates, E.T.; Pavicic, M.; Jones, P.; Amos, B.K.; Geiger, A.; Shah, M.B.; Streich, J.; Felipe Machado Gazolla, J.G.; Kainer, D.; et al. Potentially Adaptive SARS-CoV-2 Mutations Discovered with Novel Spatiotemporal and Explainable AI Models. *Genome Biol.* **2020**, *21*, 304. [CrossRef]
166. MacLean, O.A.; Lytras, S.; Weaver, S.; Singer, J.B.; Boni, M.F.; Lemey, P.; Kosakovsky Pond, S.L.; Robertson, D.L. Natural Selection in the Evolution of SARS-CoV-2 in Bats Created a Generalist Virus and Highly Capable Human Pathogen. *PLOS Biol.* **2021**, *19*, e3001115. [CrossRef]
167. Hill, V.; Du Plessis, L.; Peacock, T.P.; Aggarwal, D.; Colquhoun, R.; Carabelli, A.M.; Ellaby, N.; Gallagher, E.; Groves, N.; Jackson, B.; et al. The Origins and Molecular Evolution of SARS-CoV-2 Lineage B.1.1.7 in the UK. *Virus Evol.* **2022**, *8*, e73896. [CrossRef]
168. WHO. Tracking SARS-CoV-2 variants. Available online: https://www.who.int/activities/tracking-SARS-CoV-2-variants (accessed on 20 November 2022).
169. Preliminary Genomic Characterisation of an Emergent SARS-CoV-2 Lineage in the UK Defined by a Novel set of Spike Mutations. Available online: https://virological.org/t/preliminary-genomic-characterisation-of-an-emergent-sars-cov-2-lineage-in-the-uk-defined-by-a-novel-set-of-spike-mutations/563 (accessed on 20 November 2022).
170. Tegally, H.; Wilkinson, E.; Giovanetti, M.; Iranzadeh, A.; Fonseca, V.; Giandhari, J.; Doolabh, D.; Pillay, S.; San, E.J.; Msomi, N.; et al. Detection of a SARS-CoV-2 Variant of Concern in South Africa. *Nature* **2021**, *592*, 438–443. [CrossRef]
171. Faria, N.R.; Mellan, T.A.; Whittaker, C.; Claro, I.M.; da, S. Candido, D.; Mishra, S.; Crispim, M.A.E.; Sales, F.C.S.; Hawryluk, I.; McCrone, J.T.; et al. Genomics and Epidemiology of the P.1 SARS-CoV-2 Lineage in Manaus, Brazil. *Science* **2021**, *372*, 815–821. [CrossRef]
172. Singh, J.; Rahman, S.A.; Ehtesham, N.Z.; Hira, S.; Hasnain, S.E. SARS-CoV-2 Variants of Concern Are Emerging in India. *Nat. Med.* **2021**, *27*, 1131–1133. [CrossRef]
173. Tian, D.; Sun, Y.; Zhou, J.; Ye, Q. The Global Epidemic of the SARS-CoV-2 Delta Variant, Key Spike Mutations and Immune Escape. *Front. Immunol.* **2021**, *12*, 751778. [CrossRef]
174. Novelli, G.; Colona, V.; Pandolfi, P. A Focus on the Spread of the Delta Variant of SARS-CoV-2 in India. *Indian J. Med. Res.* **2021**, *153*, 537–541. [CrossRef]
175. Cameroni, E.; Bowen, J.E.; Rosen, L.E.; Saliba, C.; Zepeda, S.K.; Culap, K.; Pinto, D.; VanBlargan, L.A.; De Marco, A.; di Iulio, J.; et al. Broadly Neutralizing Antibodies Overcome SARS-CoV-2 Omicron Antigenic Shift. *Nature* **2021**, *602*, 664–670. [CrossRef] [PubMed]
176. Martin, D.P.; Lytras, S.; Lucaci, A.G.; Maier, W.; Grüning, B.; Shank, S.D.; Weaver, S.; MacLean, O.A.; Orton, R.J.; Lemey, P.; et al. Selection Analysis Identifies Clusters of Unusual Mutational Changes in Omicron Lineage BA.1 That Likely Impact Spike Function. *Mol. Biol. Evol.* **2022**, *39*, msac061.
177. Chen, J.; Wei, G.W. Omicron BA.2 (B.1.1.529.2): High potential to becoming the next dominating variant. *arXiv* **2022**, arXiv:2202.05031v1.
178. Wang, Q.; Guo, Y.; Iketani, S.; Nair, M.S.; Li, Z.; Mohri, H.; Wang, M.; Yu, J.; Bowen, A.D.; Chang, J.Y.; et al. Antibody Evasion by SARS-CoV-2 Omicron Subvariants BA.2.12.1, BA.4 and BA.5. *Nature* **2022**, *608*, 603–608. [CrossRef]
179. Telenti, A.; Hodcroft, E.B.; Robertson, D.L. The Evolution and Biology of SARS-CoV-2 Variants. *Cold Spring Harb. Perspect. Med.* **2022**, *12*, a041390. [CrossRef]
180. Hadfield, J.; Megill, C.; Bell, S.M.; Huddleston, J.; Potter, B.; Callender, C.; Sagulenko, P.; Bedford, T.; Neher, R.A. Nextstrain: Real-Time Tracking of Pathogen Evolution. *Bioinformatics* **2018**, *34*, 4121–4123. [CrossRef]
181. Sagulenko, P.; Puller, V.; Neher, R.A. TreeTime: Maximum-Likelihood Phylodynamic Analysis. *Virus Evol.* **2018**, *4*, vex042. [CrossRef]

182. Dearlove, B.; Lewitus, E.; Bai, H.; Li, Y.; Reeves, D.B.; Joyce, M.G.; Scott, P.T.; Amare, M.F.; Vasan, S.; Michael, N.L.; et al. A SARS-CoV-2 Vaccine Candidate Would Likely Match All Currently Circulating Variants. *Proc. Natl. Acad. Sci. USA* **2020**, *117*, 23652–23662. [CrossRef]
183. Nealon, J.; Cowling, B.J. Omicron Severity: Milder but Not Mild. *Lancet* **2022**, *399*, 412–413. [CrossRef]

viruses

MDPI

Brief Report

Alpha and Omicron SARS-CoV-2 Adaptation in an Upper Respiratory Tract Model

Gregory Mathez, Trestan Pillonel, Claire Bertelli and Valeria Cagno *

Institute of Microbiology, Lausanne University Hospital, University of Lausanne, 1011 Lausanne, Switzerland
* Correspondence: valeria.cagno@chuv.ch

Abstract: Severe acute respiratory syndrome coronavirus 2 (SARS-CoV-2) is currently causing an unprecedented pandemic. Although vaccines and antivirals are limiting the spread, SARS-CoV-2 is still under selective pressure in human and animal populations, as demonstrated by the emergence of variants of concern. To better understand the driving forces leading to new subtypes of SARS-CoV-2, we infected an ex vivo cell model of the human upper respiratory tract with Alpha and Omicron BA.1 variants for one month. Although viral RNA was detected during the entire course of the infection, infectious virus production decreased over time. Sequencing analysis did not show any adaptation in the spike protein, suggesting a key role for the adaptive immune response or adaptation to other anatomical sites for the evolution of SARS-CoV-2.

Keywords: SARS-CoV-2; virus adaptation; persistent infection

Citation: Mathez, G.; Pillonel, T.; Bertelli, C.; Cagno, V. Alpha and Omicron SARS-CoV-2 Adaptation in an Upper Respiratory Tract Model. *Viruses* **2023**, *15*, 13. https://doi.org/10.3390/v15010013

Academic Editors: Ahmed El-Shamy and Mohamed Ibrahim

Received: 14 October 2022
Revised: 16 December 2022
Accepted: 19 December 2022
Published: 20 December 2022

1. Introduction

Almost three years after the identification of a novel virus which causes pneumonia, severe acute respiratory syndrome coronavirus 2 (SARS-CoV-2), more than 600 million people have been infected [1–3]. Probably originating from bat coronaviruses, SARS-CoV-2 adapted to infect humans [1,4–7]. The first human infections likely occurred in a Wuhan market in China in December 2019 [1,8–10]. From the beginning of 2020, the virus spread worldwide, leading to huge efforts to better understand this new virus and to keep it under control.

One of the major issues in the spread of SARS-CoV-2 is the multiple waves of infections, linked to the continuous emergence of variants (such as Alpha, Beta, Delta, and Omicron) and subvariants. For instance, the Alpha variant appeared in September 2020 and possibly emerged from a chronic infection, although these patients were generally unlikely to spread the infection [11,12]. The emergence of the Omicron variant, in November 2021, is still under investigation. A passage to animals and back to humans, or the infection of a hidden population, are possibilities under scrutiny [13–15].

SARS-CoV-2 variants differ primarily due to mutations in the spike protein, resulting in either increased binding for angiotensin-converting enzyme 2 (ACE-2), the entry receptor, or immune evasion. For instance, the mutation D614G, which became dominant in February 2020 [4,16], confers enhanced binding to ACE-2 and increased transmissibility [17]. Multiple mutations (e.g., K417N, E484A, and N501Y) were shown to lead to decreased antibody neutralization, as demonstrated by the augmented risk of re-infection with the Omicron variant [4,14,17–22]. However, mutations can also occur in other structural and non-structural proteins, resulting in innate immune evasion and increased viral replication: R203K and G204R in the nucleocapsid enhanced viral fitness [23] and Δ500−532 in the non-structural protein 1 (NSP1) are associated with decrease in interferon type I [24].

Importantly, SARS-CoV-2 was also reported to adapt rapidly to cell lines and different hosts, with the appearance of some mutations currently present in circulating variants. For instance, in Vero E6 cell lines, the furin cleavage site is lost, whereas it has been shown to be re-acquired in cells expressing TMPRSS2 or in vivo [25,26]. Studies in primates and mice with isolates from 2020 observed the apparition of the H665Y mutation now present in Omicron subvariants [26,27]. Moreover, mouse-adapted SARS-CoV-2 harbored N501Y which was then associated with almost all variants of concern [27]. However, animal models present important differences with humans, such as difference in receptors, leading to a need for viral adaptation in mice, or the lack of severe symptoms in non-human primates.

Therefore, we focused on suitable human surrogate models to study SARS-CoV-2 evolution. Nevertheless, cell lines were not suitable because SARS-CoV-2 enters through different routes in cell lines in vitro and in vivo [17].

To overcome this limitation, we used MucilAir tissues, which are human-derived pseudostratified epithelia, derived from human donor biopsies, mimicking the human upper respiratory tract. These tissues are composed of secretory, ciliated, and basal cells [28,29], although we do not have detailed information about the presence of rare cell types present in the natural landscape of respiratory tract such as the pulmonary neuroendocrine, suprabasal, ionocyte, tuft, or even deuterosomal cells [30]. Unfortunately, these types of cells have similar gene expression to other cells, and their presence has not been evaluated in previous characterizations [30,31]. However, the main difference between the human respiratory tract and MucilAir is the lack of immune cells. In previous experiments with respiratory viruses, it was demonstrated that the tissues were unable to clear infections [32]. For instance, the addition of macrophages to the system has been shown to limit the propagation of the virus [33]. Although the adaptive immune system is not present, this respiratory tract model can still produce cytokines and chemokines [32–34]. In addition to immune cells and rare cell types previously cited, neurons, smooth muscles, and fibroblasts are missing compared with the natural human respiratory tract and could somehow play a role during viral infection [30]. However, to the best of our knowledge, this model is still one of the best in vitro options to study viral growth in the upper respiratory tract.

In our study, we investigated whether the human respiratory tract per se can drive adaptations of SARS-CoV-2. We infected human-derived respiratory tissues with two variants of concern (Alpha and Omicron BA.1) for one month. During the infection, we monitored the level of RNA and the infectiousness of apically released viruses. At the end of this long infection, SARS-CoV-2 viruses were sequenced to reveal any adaptation.

2. Materials and Methods

2.1. Cells and Viruses

Vero E6 given by Prof. Gary Kobinger originally derived from ATCC CRL-1586 were maintained in Dulbecco's modified Eagle medium (DMEM), high glucose, and Glutamax (Thermo Fisher Scientific, Waltham, MA, USA) supplemented by 10% fetal bovine serum (FBS) (Pan Biotech, Aidenbach, Germany) and 1% penicillin/streptavidin (Sigma, St. Louis, MI, USA). MucilAir (human upper respiratory tissues) were purchased from Epithelix (Plan-les-Ouates, Switzerland). They were maintained according to the manufacturer's protocol. The specimens of the two variants of concern were collected at the University Hospital of Lausanne (CHUV). Their consensus genome sequences and raw reads were submitted to GISAID (B.1.1.7: EPI_ISL_2359887, BA.1: EPI_ISL_7681695) and the ENA (B.1.1.7: ERR6094646, BA.1: ERR8526519).

2.2. Infection of MucilAir

PBS with calcium and magnesium (Thermo Fisher Scientific, Waltham, MA, USA) was added to the apical side of MucilAir before infection with SARS-CoV-2. Of the nasopharyngeal sample, 50 µL testing positive for SARS-CoV-2 (B.1.1.7 or BA.1) was used to infect the apical side of the respiratory tissue. After infection, MucilAir was washed with MucilAir medium. Respiratory tissues were maintained at 33 °C. In the first week of infection, SARS-CoV-2 viruses apically released were collected every day. The following week, until the 25 days post-infection, they were collected twice per week. For the collection, 200 µL MucilAir medium was added on the apical side for 20 min at 33 °C and then stored at −80 °C for further analysis. The basal medium was changed twice per week. Viral RNAs were quantified by RT-qPCR and plaque assay of the apical wash were performed on Vero E6 (see below). On the last day of infection, MucilAir was fixed with 4% formaldehyde. The last apical washes collected were sequenced by next-generation sequencing.

2.3. Infection of Vero E6

Infection of Vero E6 with several passages was performed as described previously [35]. Briefly, Vero E6 cells (3.5×10^5 cells per well) were seeded on a 6-well plate. Cells were infected with B.1.1.7 SARS-CoV-2 (MOI 0.01) for 1 h at 37 °C. The inoculum was removed, and DMEM with 2.5% FBS was added. Three days post-infection, viruses were collected and titred by plaque assay. For the following passage, viruses collected in the previous passage were used for the infection (MOI 0.01). After 11 passages, viruses were sequenced by next-generation sequencing.

2.4. Plaque Assay

Plaque assays on Vero E6 cells were performed as described by Mathez and Cagno, 2021 [36]. Briefly, Vero E6 (10^4 cells per well) was seeded in a 24-well plate. The apical washes of MucilAir were serially diluted and used to infect Vero E6 cells. After one hour of incubation at 37 °C, an overlay of 0.6% avicel gp3515 (SelectChemie, DuPont, Zurich, Switzerland) in 2.5% FBS DMEM was added. Three days post-infection, cells were fixed with 4% formaldehyde and crystal violet. Plaques were counted manually.

2.5. Immunofluorescence

MucilAir tissues were stained as described previously [32,34], with minor modifications. Tissues were blocked with 5% FBS and Tween 0.5% for 30 min at room temperature. J2 (1:500) (Nordic-MUbio, BC Susteren, The Netherlands) and β-tubulin IV (1:250) (Abcam, Cambridge, United Kingdom) were added for 1 h at 37 °C. After three washes with PBS Tween 0.05%, tissues were incubated with 1:2000 Alexa Fluor 488 anti-mouse and Alexa Fluor 594 anti-rabbit (Thermo Fisher Scientific, Waltham, MA, USA) for 1 h at 37 °C. Then, 4,6′-diamidino-2-phenylindole (DAPI) (0.5 μg/mL) was added for 5 min. Tissues were mounted on a slide with Fluoromount g (Thermo Fisher Scientific, Waltham, MA, USA). A Zeiss LSM-900 confocal microscope was used to visualize the tissue with a Plan-Apochromat 63×/1.40 objective.

2.6. RT-qPCR and Next-Generation Sequencing

Supernatants were lysed with TRK lysis and RNA was extracted with the E.Z.N.A Total RNA kit (Omega Biotech, Norcross, GA, USA). RT-qPCRs of SARS-CoV-2 were performed with TaqPath 1-Step RT-qPCR (Thermo Fisher Scientific, Waltham, MA, USA) with primers and probe targeting the E gene (Fwd: 5′-ACAGGTACGTTAATAGTTAATAGCGT-3′, Rev: 5′-ATATTGCAGCAGTACGCACACA-3′, probe: 5′-ACACTAGCCATCCTTACTGCGCTTCG-3′) [37]. Next-generation sequencing was performed as described previously [35,38]. The consensus sequences represent mutations observed in more than 70% of the reads. Due to a technical issue, BA.1 sequencing was not performed between 22,916 and 23,012 (according to the reference sequence of Wuhan).

2.7. Statistics and Analysis

Experiments were performed with two or three MucilAir samples for each variant and in duplicate for in vitro experiments. Results are shown as the mean and SEM. The area under the curve analysis followed by an unpaired *t*-test was performed to compare curves. GraphPad Prism version 9.1 (San Diego, CA, USA) and Geneious Prime (San Diego, CA, USA) version 2022.01 software were used for the analyses.

3. Results and Discussion

Viruses, including SARS-CoV-2, continuously adapt themselves to their environment, including new hosting organisms, cells, and their surroundings. To observe the respiratory tissue adaptations of SARS-CoV-2, we infected human-derived upper respiratory tract tissue with two nasopharyngeal samples that tested positive for SARS-CoV-2 for one month. Samples were never grown on cells or in tissues and were fully sequenced before experiments.

The first phase of the infection showed an increase in viral RNA released that peaked on the third day of infection (Figure 1A). The Alpha variant (B.1.1.7) had a higher fold change (3.58 log increase) during the same period compared with Omicron (BA.1) (1.12 log increase). After this initial phase of infection, viral RNA slightly decreased and stabilized. After 25 days of infection, 9.73×10^8 RNA copies were detected for B.1.1.7 compared with 3.96×10^7 for BA.1.

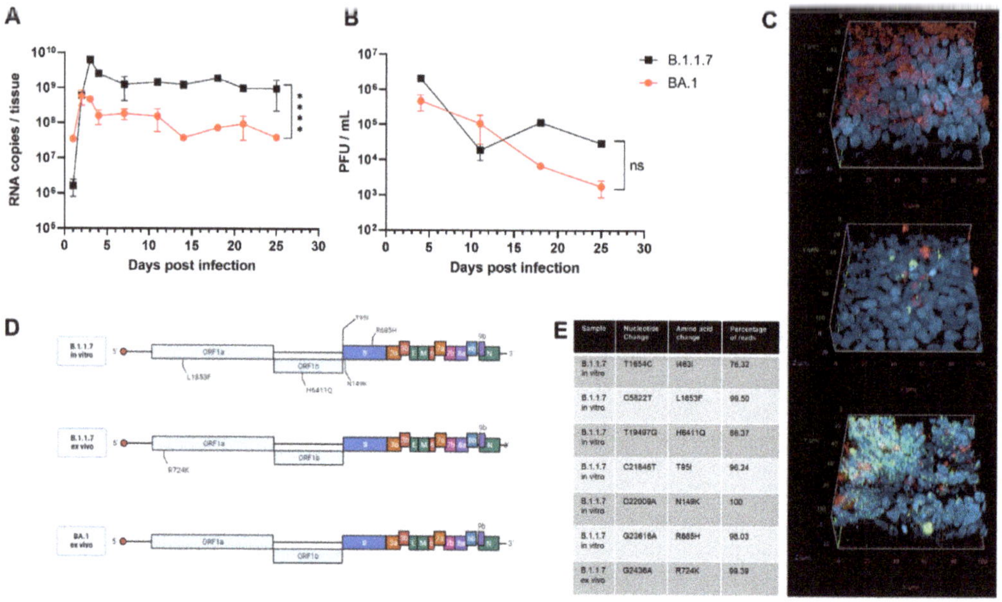

Figure 1. Evolution of SARS-CoV-2 variants ex vivo and in vitro after 25 days of infection. MucilAir tissues were infected with Alpha (B.1.1.7) or Omicron (BA.1) SARS-CoV-2. For 25 days, tissues were maintained at 33 °C and apical washes were performed to collect released SARS-CoV-2 viruses, quantified by RT-qPCR (**A**) and plaque assay (**B**). (**C**) Tissues were fixed and stained with J2 (dsRNA: green), β-tubulin IV (ciliated cells: red), and DAPI (nuclei: blue) at the end of infection. (**D**) Viruses were deep-sequenced at the end of infection ex vivo or after 11 passages in Vero E6. Mutations that were not present in the original sample are shown (created with biorender.com). (**E**) Nucleotide changes and percentage of reads present in the different samples. The results are the mean and SEM of two (B.1.1.7) or three (BA.1) MucilAir. p-value < 0.0001 (****).

Omicron presented lower viral RNA copies and produced fewer infectious viruses than Alpha (Figure 1B), but the difference was not statistically significant. Previous results showed a better fitness of Omicron in the bronchi if compared with Delta and an opposite trend in the lung [39–41]. The results in nasal cell lines, however, were more ambiguous. In one study, Omicron was shown to have a faster replication but lower viral peak [42], whereas in another, higher replication with comparable viral peak has been observed [41]. In the comparison of infectivity, it is important to take into consideration the possible under estimation of Omicron due to its reduced plaque-forming ability [41,43].

The discrepancy between the high RNA level and the low level of infectious virus is in line with reports of human infections, which highlighted the presence of viral RNA in nasopharyngeal samples, although there was a lack of viral isolation one week after the first positive PCR test [44–46]. In our case, infectious viruses could be detected until the end of infection, but a drop (2.04 log and 0.64 log decrease for B.1.1.7 and BA.1, respectively) was observed in the same period.

After 25 days of infection, through immunostaining, we evidenced that both variants caused a reduction in the number of ciliated cells and damaged the tissues (Figure 1C). This finding is in line with previous reports showing that SARS-CoV-2 principally infects this type of cells [47,48]. We could identify, for both variants, cells with active viral replication (through double-stranded RNA staining), although in larger numbers for Omicron than for Alpha (Figure 1C). If combined with the infectivity and genome quantification, these results could suggest a higher replication per cell for Alpha compared with Omicron; however, no single-cell measurements of the two variants are available to further confirm this hypothesis.

We then performed next-generation sequencing on the virus collected 25 days post-infection in the apical wash to identify any potential adaptations to our ex vivo model and compared the consensus sequences with the initial sequence of the original clinical sample. Surprisingly, BA.1 did not harbor additional fixed mutations, whereas after B.1.1.7 infection, a single mutation, R724K, was observed in all the replicates (Figure 1D,E). No minority variants were observed for both BA.1 and BA.1.1.7.

The R724K mutation located in the NSP2 was previously reported to be characteristic of a Beta variant detected in Brazil [49]. Although the function of NSP2 is still unclear, it was associated with inhibition of the antiviral state of cells [50,51]. The mutation of arginine to lysine does not change the charge of the amino acid, with a probable minor impact on NSP2; however, further research should clarify the role of this mutation.

We cannot exclude that mutations arose at shorter time points but were not selected, or could arise with longer infections; however, the decreased amount of viruses released over time and the altered architecture of the tissues are not suitable to evaluate this possibility.

In contrast, when we infected Vero E6 cells with B.1.1.7, five mutations were present after several passages (Figure 1D,E): L1853F was predicted to affect the stability of NSP3 [52]; T95I was found in several variants including BA.1 and B.1.526 [53,54]; N149 is a possible site of N-glycosylation that might be associated with increased sensitivity to antibodies [55]; R685H results in a loss of furin cleavage site and is a mark of in vitro adaptation [25,56]. These results further confirm the possible adaptation to the host cell as a driver of evolution in the absence of adaptive immunity and emphasize that MucilAir models more faithfully mimic natural infection conditions than Vero E6.

In conclusion, in our study, we infected a representative model of the upper respiratory tract with two SARS-CoV-2 variants of concern (B.1.1.7 and BA.1) for one month. We highlighted that the viral RNA can be still detected but with a decrease in infectivity of viruses released over time, possibly linked to the loss of ciliated cells observed. Both variants did not show adaptations to the upper respiratory tract model. Therefore, in our experimental setting, the emergence of new variants was not due to the adaptation to the upper respiratory tract. Further work could be performed to assess the adaptation of clinical samples in different tissue models (e.g., brain, intestinal, etc.) or in the presence of suboptimal concentrations of immune sera to assess the role of other tissue tropism and adaptative immunity in SARS-CoV-2 evolution.

Author Contributions: Conceptualization, V.C. and G.M.; methodology, V.C., G.M., C.B. and T.P.; investigation, V.C. and G.M.; data curation, G.M. and T.P.; writing—original draft preparation, G.M. and V.C.; writing—review and editing, G.M., V.C., T.P. and C.B.; visualization, G.M.; supervision, V.C. and C.B.; project administration, V.C.; funding acquisition, V.C. All authors have read and agreed to the published version of the manuscript.

Funding: This study was supported by the Swiss National Science Foundation [PZ00P3_193289 to V.C.].

Institutional Review Board Statement: Not applicable.

Informed Consent Statement: Not applicable.

Data Availability Statement: Raw data are available at 10.6084/m9.figshare.20596845 and sequenced data with accession number: ERS12330501, ERS12330502, ERS12330504, ERS12330505, ERS12330506, ERS12330507 and ERS12330508.

Conflicts of Interest: The authors declare no conflict of interest.

References

1. Zhou, P.; Yang, X.L.; Wang, X.G.; Hu, B.; Zhang, L.; Zhang, W.; Si, H.R.; Zhu, Y.; Li, B.; Huang, C.L.; et al. A Pneumonia Outbreak Associated with a New Coronavirus of Probable Bat Origin. *Nature* **2020**, *579*, 270–273. [CrossRef] [PubMed]
2. Who Coronavirus (COVID-19) Dashboard. Available online: https://covid19.who.int/ (accessed on 23 August 2022).
3. COVID-19 Coronavirus Pandemic. Available online: https://www.worldometers.info/coronavirus/ (accessed on 25 August 2022).
4. Tosta, E. The Adaptation of SARS-CoV-2 to Humans. *Mem. Inst. Oswaldo Cruz.* **2022**, *116*, e210127. [CrossRef] [PubMed]
5. Li, L.L.; Wang, J.L.; Ma, X.H.; Sun, X.M.; Li, J.S.; Yang, X.F.; Shi, W.F.; Duan, Z.J. A Novel SARS-CoV-2 Related Coronavirus with Complex Recombination Isolated from Bats in Yunnan Province, China. *Emerg. Microbes Infect.* **2021**, *10*, 1683–1690. [CrossRef] [PubMed]
6. Li, X.; Giorgi, E.E.; Marichannegowda, M.H.; Foley, B.; Xiao, C.; Kong, X.P.; Chen, Y.; Gnanakaran, S.; Korber, B.; Gao, F. Emergence of SARS-CoV-2 through Recombination and Strong Purifying Selection. *Sci. Adv.* **2020**, *6*. [CrossRef]
7. Lytras, S.; Hughes, J.; Martin, D.; Swanepoel, P.; de Klerk, A.; Lourens, R.; Pond, S.L.K.; Xia, W.; Jiang, X.; Robertson, D.L. Exploring the Natural Origins of SARS-CoV-2 in the Light of Recombination. *Genome Biol. Evol.* **2022**, *14*. [CrossRef]
8. Pekar, J.; Worobey, M.; Moshiri, N.; Scheffler, K.; Wertheim, J.O. Timing the SARS-CoV-2 Index Case in Hubei Province. *Science* **2021**, *372*, 412–417. [CrossRef]
9. Fongaro, G.; PStoco, H.; Souza, D.S.M.; Grisard, E.C.; Magri, M.E.; Rogovski, P.; Schorner, M.A.; Barazzetti, F.H.; Christoff, A.P.; de Oliveira, L.F.V.; et al. The Presence of SARS-CoV-2 Rna in Human Sewage in Santa Catarina, Brazil, November 2019. *Sci. Total Environ.* **2021**, *778*, 146198. [CrossRef]
10. Pekar, J.E.; Magee, A.; Parker, E.; Moshiri, N.; Izhikevich, K.; Havens, J.L.; Gangavarapu, K.; Serrano, L.M.M.; Crits-Christoph, A.; Matteson, N.L.; et al. The Molecular Epidemiology of Multiple Zoonotic Origins of SARS-CoV-2. *Science* **2022**, *377*, eabp8337. [CrossRef]
11. Harari, S.; Tahor, M.; Rutsinsky, N.; Meijer, S.; Miller, D.; Henig, O.; Halutz, O.; Levytskyi, K.; Ben-Ami, R.; Adler, A.; et al. Drivers of Adaptive Evolution during Chronic SARS-CoV-2 Infections. *Nat. Med.* **2022**. [CrossRef]
12. Preliminary Genomic Characterisation of an Emergent SARS-CoV-2 Lineage in the UK Defined by a Novel Set of Spike Mutations. Available online: https://virological.org/t/preliminary-genomic-characterisation-of-an-emergent-SARS-CoV-2-lineage-in-the-uk-defined-by-a-novel-set-of-spike-mutations/563 (accessed on 24 August 2022).
13. Du, P.; Gao, G.F.; Wang, Q. The Mysterious Origins of the Omicron Variant of SARS-CoV-2. *Innovation* **2022**, *3*, 100206. [CrossRef]
14. Variant: 21k (Omicron). Available online: https://covariants.org/variants/21K.Omicron (accessed on 24 August 2022).
15. Classification of Omicron (B.1.1.529): SARS-CoV-2 Variant of Concern. Available online: https://www.who.int/news/item/26-11-2021-classification-of-omicron-(b.1.1.529)-sars-cov-2-variant-of-concern (accessed on 24 August 2022).
16. Yurkovetskiy, L.; Wang, X.; Pascal, K.E.; Tomkins-Tinch, C.; Nyalile, T.P.; Wang, Y.; Baum, A.; Diehl, W.E.; Dauphin, A.; Carbone, C.; et al. Structural and Functional Analysis of the D614g SARS-CoV-2 Spike Protein Variant. *Cell* **2020**, *183*, 739–751.e8. [CrossRef] [PubMed]
17. Jackson, C.B.; Farzan, M.; Chen, B.; Choe, H. Mechanisms of SARS-CoV-2 Entry into Cells. *Nat. Rev. Mol. Cell Biol.* **2022**, *23*, 3–20. [CrossRef] [PubMed]
18. Cao, Y.; Wang, J.; Jian, F.; Xiao, T.; Song, W.; Yisimayi, A.; Huang, W.; Li, Q.; Wang, P.; An, R.; et al. Omicron Escapes the Majority of Existing SARS-CoV-2 Neutralizing Antibodies. *Nature* **2022**, *602*, 657–663. [CrossRef] [PubMed]
19. Planas, D.; Saunders, N.; Maes, P.; Guivel-Benhassine, F.; Planchais, C.; Buchrieser, J.; Bolland, W.H.; Porrot, F.; Staropoli, I.; Lemoine, F.; et al. Considerable Escape of SARS-CoV-2 Omicron to Antibody Neutralization. *Nature* **2022**, *602*, 671–675. [CrossRef] [PubMed]
20. Hu, J.; Peng, P.; Cao, X.; Wu, K.; Chen, J.; Wang, K.; Tang, N.; Huang, A.L. Increased Immune Escape of the New SARS-CoV-2 Variant of Concern Omicron. *Cell Mol. Immunol.* **2022**, *19*, 293–295. [CrossRef]
21. Pulliam, J.R.C.; van Schalkwyk, C.; Govender, N.; von Gottberg, A.; Cohen, C.; Groome, M.J.; Dushoff, J.; Mlisana, K.; Moultrie, H. Increased Risk of SARS-CoV-2 Reinfection Associated with Emergence of Omicron in South Africa. *Science* **2022**, *376*, eabn4947. [CrossRef]
22. Subissi, L.; von Gottberg, A.; Thukral, L.; Worp, N.; Munnink, B.B.O.; Rathore, S.; Abu-Raddad, L.J.; Aguilera, X.; Alm, E.; Archer, B.N.; et al. An Early Warning System for Emerging SARS-CoV-2 Variants. *Nat. Med.* **2022**, *28*, 1110–1115. [CrossRef]
23. Johnson, B.A.; Zhou, Y.; Lokugamage, K.G.; Vu, M.N.; Bopp, N.; Crocquet-Valdes, P.A.; Kalveram, B.; Schindewolf, C.; Liu, Y.; Scharton, D.; et al. Nucleocapsid Mutations in SARS-CoV-2 Augment Replication and Pathogenesis. *PLoS Pathog.* **2022**, *18*, e1010627. [CrossRef]
24. Lin, J.W.; Tang, C.; Wei, H.C.; Du, B.; Chen, C.; Wang, M.; Zhou, Y.; Yu, M.X.; Cheng, L.; Kuivanen, S.; et al. Genomic Monitoring of SARS-CoV-2 Uncovers an Nsp1 Deletion Variant That Modulates Type I Interferon Response. *Cell Host Microbe* **2021**, *29*, 489–502.e8. [CrossRef]
25. Chaudhry, M.Z.; Eschke, K.; Hoffmann, M.; Grashoff, M.; Abassi, L.; Kim, Y.; Brunotte, L.; Ludwig, S.; Kroger, A.; Klawonn, F.; et al. Rapid SARS-CoV-2 Adaptation to Available Cellular Proteases. *J. Virol.* **2022**, *96*, e0218621. [CrossRef]
26. Rowe, L.A.; Beddingfield, B.J.; Goff, K.; Killeen, S.Z.; Chirichella, N.R.; Melton, A.; Roy, C.J.; Maness, N.J. Intra-Host SARS-CoV-2 Evolution in the Gut of Mucosally-Infected Chlorocebus Aethiops (African Green Monkeys). *Viruses* **2022**, *14*, 77. [CrossRef] [PubMed]

27. Rathnasinghe, R.; Jangra, S.; Ye, C.; Cupic, A.; Singh, G.; Martinez-Romero, C.; Mulder, L.C.F.; Kehrer, T.; Yildiz, S.; Choi, A.; et al. Characterization of SARS-CoV-2 Spike Mutations Important for Infection of Mice and Escape from Human Immune Sera. *Nat. Commun.* **2022**, *13*, 3921. [CrossRef] [PubMed]
28. Epithelix. Mucilair™: In Vitro 3d Human upper Airway Epithelium. Available online: https://www.epithelix.com/products/mucilair (accessed on 6 October 2022).
29. Mercier, C.; Jacqueroux, E.; He, Z.; Hodin, S.; Constant, S.; Perek, N.; Boudard, D.; Delavenne, X. Pharmacological Characterization of the 3d Mucilair Nasal Model. *Eur. J. Pharm. Biopharm.* **2019**, *139*, 186–196. [CrossRef]
30. Hewitt, R.J.; Lloyd, C.M. Regulation of Immune Responses by the Airway Epithelial Cell Landscape. *Nat. Rev. Immunol.* **2021**, *21*, 347–362. [CrossRef]
31. Medaglia, C.; Kolpakov, I.; Zwygart, A.C.A.; Zhu, Y.; Constant, S.; Huang, S.; Cagno, V.; Dermitzakis, E.T.; Stellacci, F.; Xenarios, I.; et al. An anti-influenza combined therapy assessed by single cell RNA-sequencing. *Commun. Biol.* **2022**, *5*, 1075. [CrossRef] [PubMed]
32. Essaidi-Laziosi, M.; Brito, F.; Benaoudia, S.; Royston, L.; Cagno, V.; Fernandes-Rocha, M.; Piuz, I.; Zdobnov, E.; Huang, S.; Constant, S.; et al. Propagation of Respiratory Viruses in Human Airway Epithelia Reveals Persistent Virus-Specific Signatures. *J. Allergy Clin. Immunol.* **2018**, *141*, 2074–2084. [CrossRef]
33. Barreto-Duran, E.; Artur, S.; Adrianna, G.-B.; Marcin, S.; Maciej, S.; Marek, S.; Zenon, R.; Aleksandra, M.; Marzena, L.; Krzysztof, P. The Interplay between the Airway Epithelium and Tissue Macrophages During the SARS-CoV-2 Infection. *Front. Immunol.* **2022**, *13*, 6017. [CrossRef]
34. Essaidi-Laziosi, M.; Alvarez, C.; Puhach, O.; Sattonnet-Roche, P.; Torriani, G.; Tapparel, C.; Kaiser, L.; Eckerle, I. Sequential Infections with Rhinovirus and Influenza Modulate the Replicative Capacity of SARS-CoV-2 in the Upper Respiratory Tract. *Emerg. Microbes Infect.* **2022**, *11*, 412–423. [CrossRef]
35. Varricchio, C.; Gregory, M.; Trestan, P.; Claire, B.; Laurent, K.; Caroline, T.; Andrea, B.; Valeria, C. Geneticin Shows Selective Antiviral Activity against SARS-CoV-2 by Interfering with Programmed −1 Ribosomal Frameshifting. *Antivir. Res.* **2022**, *208*, 105452. [CrossRef]
36. Mathez, G.; Cagno, V. Clinical Severe Acute Respiratory Syndrome Coronavirus 2 Isolation and Antiviral Testing. *Antivir. Chem. Chemother.* **2021**, *29*, 20402066211061063. [CrossRef] [PubMed]
37. Corman, V.M.; Landt, O.; Kaiser, M.; Molenkamp, R.; Meijer, A.; Chu, D.K.; Bleicker, T.; Brunink, S.; Schneider, J.; Schmidt, M.L.; et al. Detection of 2019 Novel Coronavirus (2019-Ncov) by Real-Time Rt-Pcr. *Eurosurveillance* **2020**, *25*, 2000045. [CrossRef] [PubMed]
38. Jacot, D.; Pillonel, T.; Greub, G.; Bertelli, C. Assessment of SARS-CoV-2 Genome Sequencing: Quality Criteria and Low-Frequency Variants. *J. Clin. Microbiol.* **2021**, *59*, e0094421. [CrossRef] [PubMed]
39. Hui, K.P.Y.; Ho, J.C.W.; Cheung, M.C.; Ng, K.C.; Ching, R.H.H.; Lai, K.L.; Kam, T.T.; Gu, H.; Sit, K.Y.; Hsin, M.K.Y.; et al. SARS-CoV-2 Omicron Variant Replication in Human Bronchus and Lung Ex Vivo. *Nature* **2022**, *603*, 715–720. [CrossRef] [PubMed]
40. Mache, C.; Schulze, J.; Holland, G.; Bourquain, D.; Gensch, J.M.; Oh, D.Y.; Nitsche, A.; Durrwald, R.; Laue, M.; Wolff, T. SARS-CoV-2 Omicron Variant Is Attenuated for Replication in a Polarized Human Lung Epithelial Cell Model. *Commun. Biol.* **2022**, *5*, 1138. [CrossRef] [PubMed]
41. Willett, B.J.; Grove, J.; MacLean, O.A.; Wilkie, C.; de Lorenzo, G.; Furnon, W.; Cantoni, D.; Scott, S.; Logan, N.; Ashraf, S.; et al. SARS-CoV-2 Omicron Is an Immune Escape Variant with an Altered Cell Entry Pathway. *Nat. Microbiol.* **2022**, *7*, 1161–1179. [CrossRef]
42. Peacock, T.P.; Jonathan, C.B.; Jie, Z.; Nazia, T.; Ksenia, S.; Joseph, N.; Ruthiran, K.; Yan, A.W.C.; Wilhelm, F.; Giuditta, D.L.; et al. The Altered Entry Pathway and Antigenic Distance of the SARS-CoV-2 Omicron Variant Map to Separate Domains of Spike Protein. *Nat. Microbiol.* **2022**, *7*, 1161–1179.
43. Pia, L.; Rowland-Jones, S. Omicron Entry Route. *Nat. Rev. Immunol.* **2022**, *22*, 144. [CrossRef]
44. Owusu, D.; Pomeroy, M.A.; Lewis, N.M.; Wadhwa, A.; Yousaf, A.R.; Whitaker, B.; Dietrich, E.; Hall, A.J.; Chu, V.; Thornburg, N.; et al. Persistent SARS-CoV-2 Rna Shedding without Evidence of Infectiousness: A Cohort Study of Individuals with COVID-19. *J. Infect. Dis.* **2021**, *224*, 1362–1371. [CrossRef]
45. Takahashi, K.; Ishikane, M.; Ujiie, M.; Iwamoto, N.; Okumura, N.; Sato, T.; Nagashima, M.; Moriya, A.; Suzuki, M.; Hojo, M.; et al. Duration of Infectious Virus Shedding by SARS-CoV-2 Omicron Variant-Infected Vaccinees. *Emerg. Infect. Dis.* **2022**, *28*, 998–1001. [CrossRef]
46. Boucau, J.; Marino, C.; Regan, J.; Uddin, R.; Choudhary, M.C.; Flynn, J.P.; Chen, G.; Stuckwisch, A.M.; Mathews, J.; Liew, M.Y.; et al. Duration of Shedding of Culturable Virus in SARS-CoV-2 Omicron (Ba.1) Infection. *N. Engl. J. Med.* **2022**, *387*, 275–277. [CrossRef]
47. Ahn, J.H.; Kim, J.; Hong, S.P.; Choi, S.Y.; Yang, M.J.; Ju, Y.S.; Kim, Y.T.; Kim, H.M.; Rahman, M.D.T.; Chung, M.K.; et al. Nasal Ciliated Cells Are Primary Targets for SARS-CoV-2 Replication in the Early Stage of COVID-19. *J. Clin. Investig.* **2021**, 131. [CrossRef] [PubMed]
48. Robinot, R.; Hubert, M.; de Melo, G.D.; Lazarini, F.; Bruel, T.; Smith, N.; Levallois, S.; Larrous, F.; Fernandes, J.; Gellenoncourt, S.; et al. SARS-CoV-2 Infection Induces the Dedifferentiation of Multiciliated Cells and Impairs Mucociliary Clearance. *Nat. Commun.* **2021**, *12*, 4354. [CrossRef] [PubMed]
49. Slavov, S.N.; Patane, J.S.L.; Bezerra, R.D.S.; Giovanetti, M.; Fonseca, V.; Martins, A.J.; Viala, V.L.; Rodrigues, E.S.; Santos, E.V.; Barros, C.R.S.; et al. Genomic Monitoring Unveil the Early Detection of the SARS-CoV-2 B.1.351 (Beta) Variant (20h/501y.V2) in Brazil. *J. Med. Virol.* **2021**, *93*, 6782–6787. [CrossRef]
50. Xu, Z.; Choi, J.H.; Dai, D.L.; Luo, J.; Ladak, R.J.; Li, Q.; Wang, Y.; Zhang, C.; Wiebe, S.; Liu, A.C.H.; et al. SARS-CoV-2 Impairs Interferon Production Via Nsp2-Induced Repression of Mrna Translation. *Proc. Natl. Acad. Sci. USA* **2022**, *119*, e2204539119. [CrossRef]
51. Zou, L.; Moch, C.; Graille, M.; Chapat, C. The SARS-CoV-2 Protein Nsp2 Impairs the Silencing Capacity of the Human 4ehp-Gigyf2 Complex. *iScience* **2022**, *25*, 104646. [CrossRef]

52. Das, J.K.; Sengupta, A.; Choudhury, P.P.; Roy, S. Characterizing Genomic Variants and Mutations in SARS-CoV-2 Proteins from Indian Isolates. *Gene Rep.* **2021**, *25*, 101044. [CrossRef] [PubMed]
53. Viana, R.; Moyo, S.; Amoako, D.G.; Tegally, H.; Scheepers, C.; Althaus, C.L.; Anyaneji, U.J.; Bester, P.A.; Boni, M.F.; Chand, M.; et al. Rapid Epidemic Expansion of the SARS-CoV-2 Omicron Variant in Southern Africa. *Nature* **2022**, *603*, 679–686. [CrossRef]
54. West, A.P., Jr.; Wertheim, J.O.; Wang, J.C.; Vasylyeva, T.I.; Havens, J.L.; Chowdhury, M.A.; Gonzalez, E.; Fang, C.E.; di Lonardo, S.S.; Hughes, S.; et al. Detection and Characterization of the SARS-CoV-2 Lineage B.1.526 in New York. *Nat. Commun.* **2021**, *12*, 4886. [CrossRef]
55. Gong, Y.; Qin, S.; Dai, L.; Tian, Z. The Glycosylation in SARS-CoV-2 and Its Receptor Ace2. *Signal. Transduct Target. Ther.* **2021**, *6*, 396. [PubMed]
56. Mykytyn, A.Z.; Breugem, T.I.; Riesebosch, S.; Schipper, D.; van den Doel, P.B.; Rottier, R.J.; Lamers, M.M.; Haagmans, B.L. SARS-CoV-2 Entry into Human Airway Organoids Is Serine Protease-Mediated and Facilitated by the Multibasic Cleavage Site. *eLife* **2021**, *10*, e64508. [CrossRef]

viruses

MDPI

Review

The Elusive Coreceptors for the SARS-CoV-2 Spike Protein

Reed L. Berkowitz and David A. Ostrov *

Department of Pathology, Immunology and Laboratory Medicine, University of Florida College of Medicine, Gainesville, FL 32610, USA

* Correspondence: ostroda@pathology.ufl.edu; Tel.: +(352)-273-8166

Abstract: Evidence suggests that the N-terminal domain (NTD) of the SARS-CoV-2 spike protein interacts with host coreceptors that participate in viral entry. Resolving the identity of coreceptors has important clinical implications as it may provide the basis for the development of antiviral drugs and vaccine candidates. The majority of characteristic mutations in variants of concern (VOCs) have occurred in the NTD and receptor binding domain (RBD). Unlike the RBD, mutations in the NTD have clustered in the most flexible parts of the spike protein. Many possible coreceptors have been proposed, including various sugars such as gangliosides, sialosides, and heparan sulfate. Protein coreceptors, including neuropilin-1 and leucine-rich repeat containing 15 (LRRC15), are also proposed coreceptors that engage the NTD.

Keywords: SARS-CoV-2; COVID-19; N-terminal domain; receptor binding domain; LRRC15; leucine rich repeat containing 15; neuropilin-1; NRP-1; mutations

Citation: Berkowitz, R.L.; Ostrov, D.A. The Elusive Coreceptors for the SARS-CoV-2 Spike Protein. *Viruses* **2023**, *15*, 67. https://doi.org/10.3390/v15010067

Academic Editors: Ahmed El-Shamy and Mohamed Ibrahim

Received: 17 November 2022
Revised: 13 December 2022
Accepted: 22 December 2022
Published: 25 December 2022

1. Introduction to Coronavirus Spike Proteins (Function, Structure, Receptors)

1.1. Function and Structure

The coronavirus spike (S) protein, comprised of S1 and S2 subunits, mediates the entry of coronaviruses into host cells. Generally, in the viral attachment of coronaviruses, the S1 subunit binds to a receptor on the host cell's surface. This binding prompts a conformational change in the spike protein that enables the S2 subunit to fuse together the viral and host membranes [1].

Due to its fundamental role in initiating infection, the S protein is of high relevance to natural immune response as well as clinical therapies. It follows that SARS-CoV-2 infection elicits neutralizing antibodies that bind to the S protein, and that vaccines targeting the S protein provide protection against infection.

Each spike protein monomer is comprised of approximately 1273 residues (depending on deletions in certain variants) which trimerize upon assembly into a clove-shaped trimer with three S1 heads and a trimeric S2 stalk [2]. There are characteristic domains present in all coronavirus spike domains. In the SARS-CoV-2 S1, these include the N-terminal domain (NTD, residues 14–305) and receptor-binding domain (RBD, residues 319–541). In S2, these include the fusion peptide (FP, residues 788–806), heptad repeat sequences 1 (HR1, residues 912–984) and 2 (HR2, residues 1163–1213), transmembrane domain (TM, residues 1213–1237), and intracellular domain (IC, residues 1237–1273) [2].

1.2. Receptors

Different coronaviruses exhibit a diverse range of tropism to different cells and tissues, resulting in varying clinical manifestations ranging from mild to severe disease (e.g., HCoV-229E results in the common cold while MERS-CoV results in approximately 35% mortality) [3]. S1 C-terminal domains (CTDs) are known to bind to protein receptors, such as ACE2 in SARS-CoV and SARS-CoV-2 or dipeptidyl peptidase 4 (DPP4) in MERS-CoV. The S1-NTD has been shown to bind host sugar moieties for the majority of coronaviruses (e.g., TGEV and PEDV) [1]. A subset of coronaviruses exhibits NTDs that bind protein elements (e.g., the NTD of mouse hepatitis virus, MHV, binds carcinoembryonic antigen-related cell adhesion molecule 1, CEACAM1) [1]. Recent studies have suggested a dual-receptor mechanism for MERS-CoV, positing that it binds not only to DPP4 at the CTD but also to host sugars (sialosides) at the NTD [4].

The host cell receptor for the NTD of SARS-CoV-2 is not yet known [5]. Resolving the identity of such a ligand could have a large impact on drug development and COVID treatment strategies. This unidentified ligand could either act as a coreceptor that helps facilitate entry mediated by the primary receptor (e.g., CCR5 and CD4 for HIV) or a distinct alternative receptor (e.g., CD147 for SARS-CoV-2) [6,7]. Since the putative ligands of the SARS-CoV-2 S1-NTD are not clear, it is also not clear how mutations in the S1-NTD might affect virus fitness.

1.3. Mutations in Variants of Concern (VOCs)

In variants of concern (VOC) of SARS-CoV-2, many distinguishing mutations of the spike protein occur at the S1-NTD and S1-RBD (Figure 1). The majority of S1-NTD mutations that distinguish VOCs occur on the solvent-accessible surface at a site available for binding ligands (Figure 2). Additionally, SARS-CoV-2 elicits neutralizing antibodies binding to multiple locations of the NTD [5,8]. These facts suggest that the NTD has some important but unknown functions affecting virus fitness, such as coreceptor binding, avoidance of restriction factors, or antibody-dependent enhancement [8]. Specifically, host cell restriction factors might block virus entry by binding the S1-NTD, so mutations here could help the virus evade this defense.

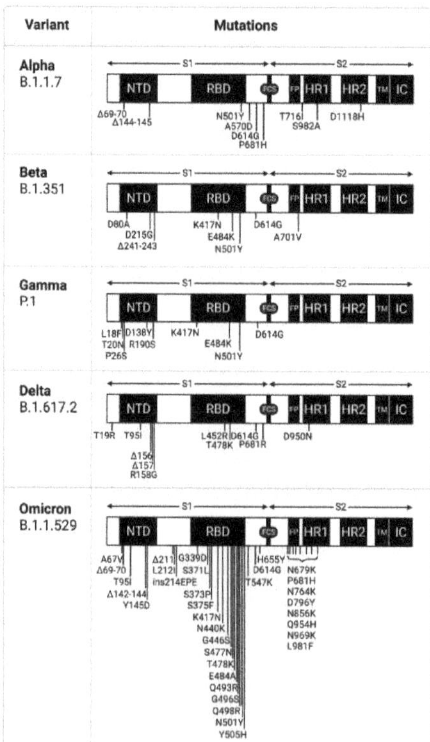

Figure 1. Mutations in the SARS-CoV-2 spike glycoprotein. NTD, N-terminal domain; RBD, receptor binding domain; FCS, furin cleavage site; FP, fusion peptide; HR1, heptad repeat 1; HR2, heptad repeat 2; TM, transmembrane anchor; IC, intracellular tail. The Delta symbol indicates deletion. Mutations as defined by the Centers for Disease Control and Prevention.

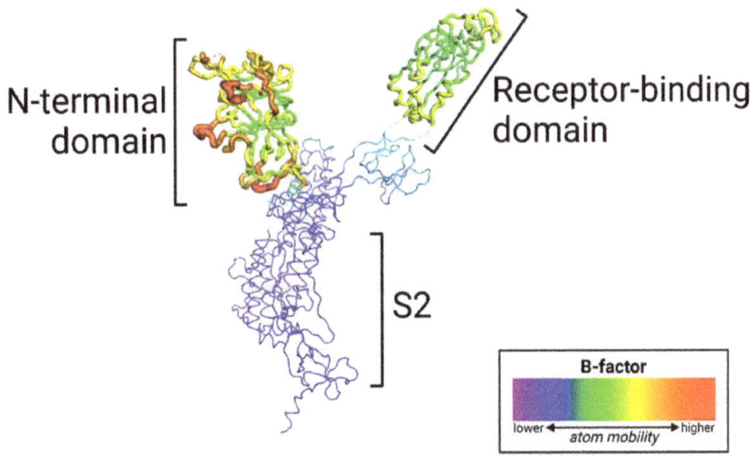

Figure 2. The N-terminal domain (NTD) of the SARS-CoV-2 spike glycoprotein exhibits a high level of flexibility. The cryo-EM structure of the prefusion spike protein is colored by B-factor; the thickness of the wire corresponds to structural variability.

Importantly, these mutations in the NTD occur in the most flexible part of the spike protein (Figures 2–4). The NTD has a far higher B-factor than other parts of the spike, indicating its high structural variability. The functional consequence of mutations that impact flexibility is likely to improve fitness since these mutations distinguish VOCs. For example, increased flexibility of an epitope is likely to reduce antibody binding.

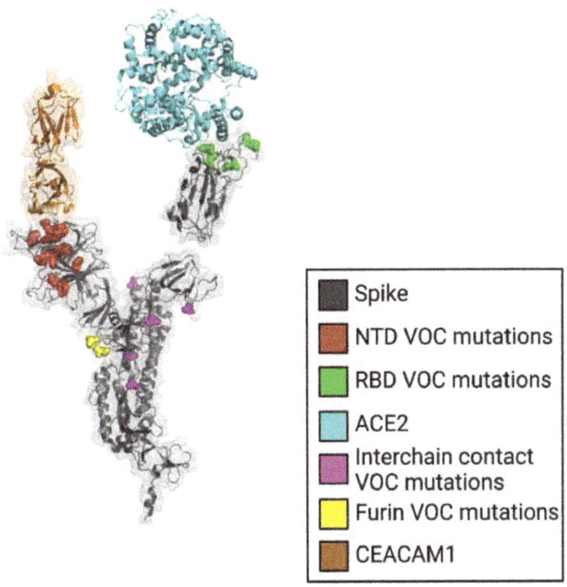

Figure 3. Mutations in the N-terminal domain (NTD) cluster in a manner that likely modulates binding interactions. The spike glycoprotein is shown in gray with mutations present in variants of concern (VOC) shown in red, NTD VOC mutations in red, RBD VOC mutations in green, interchain contact VOC mutations in magenta, furin cleavage site mutations shown in yellow, ACE2 shown in cyan, CEACAM1 shown in orange.

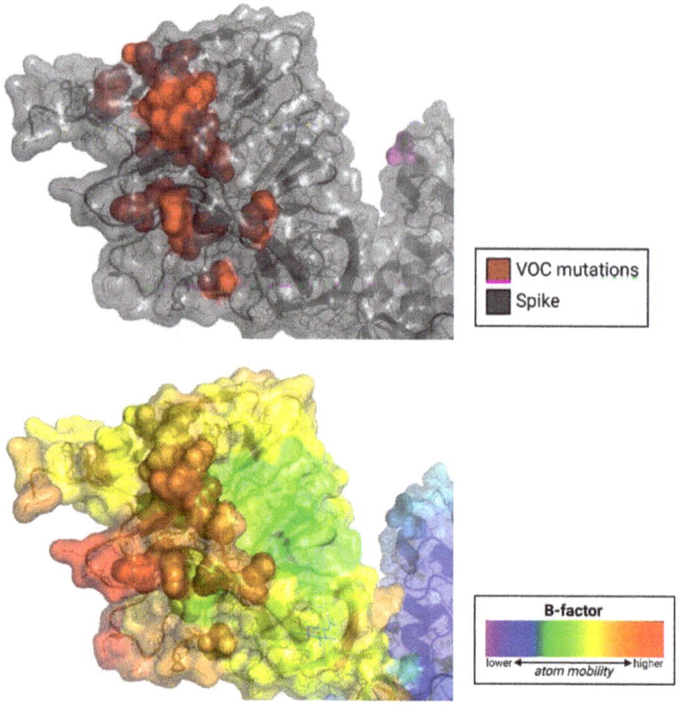

Figure 4. SARS-CoV-2 variants of concern (VOC) acquire mutations in the N-terminal domain (NTD) at the sites of the highest flexibility. The cryo-EM structure of the prefusion spike glycoprotein is shown. Top panel, the NTD is shown in gray; mutations in variants of concern are shown in red; bottom panel, the NTD is colored by B-factor, indicating structural variability.

A separate site of importance is the S1-RBD, which has also mutated extensively in highly contagious variants of concern. Though it is known that the RBD binds to ACE2 to mediate host cell entry, several of these mutations have decreased the RBD's affinity for ACE2 in spite of the increased transmission of more recent variants [9]. Although the functional consequence of mutations in the RBD likely contributes to immune evasion, it is possible that the RBD takes part in other interactions that affect infectivity.

The furin cleavage site (FCS) is another key to infection, as its cleavage promotes or enables virus-host cell fusion. The identity of host cell proteases that facilitate this cleavage is also of interest and is discussed further in this review.

2. Sugars as SARS-CoV-2 Spike Protein Ligands

2.1. *Sialosides*

As mentioned before, a dual-receptor mechanism binding to DPP4 and host sialoside sugars was proposed to be used by MERS-CoV for the infection of host cells. Awasthi et al. noted that the SARS-CoV-2 S1-NTD has three divergent loop regions that structurally resemble the MERS-CoV sialoside-binding pocket, as confirmed by a cryo-EM study [4,10]. Specifically, the β14–β15 loop of the NTD is implicated in sialoside binding [4].

Computational binding studies found that diverse sialosides interacted with and localized to the proposed sialoside-binding pocket in the NTD of both MERS-CoV and SARS-CoV-2. SARS-CoV was not found to interact similarly with sialosides, demonstrating binding in various locations of the NTD, which was the expected result as SARS-CoV is not known to bind to sialosides [4]. Further, Milanetti et al. also predicted a sialoside-binding pocket in the SARS-CoV-2 NTD by surface iso-electron density mapping [11].

Awasthi et al. also propose that the predicted ability of SARS-CoV-2 to engage diverse sialosides could explain its high infectivity with broad tissue tropism [4]. The distribution of sialosides in the respiratory tract could explain why SARS-CoV-2 exhibits high infectivity for these cells despite their limited ACE2 expression.

Another proposed function of sialoside binding is the facilitation of viral surfing over the host cell surface [11–14]. Viral surfing describes the movement of SARS-CoV-2 virions across a cell surface's sialic acid layer to find and attach to ACE2. Similarly, Milanetti et al. proposed a dual or even triple binding of SARS-CoV-2 to ACE2 and gangliosides present in lipid rafts [11]. A glycan-binding domain could bind to certain glycosphingolipids present in lipid rafts. Lipid raft coalescence is proposed to lead to the recruitment of ACE2, and ganglioside expression is higher in epithelial intestinal and brain cells, supporting the potential role of gangliosides in SARS-CoV-2 infectivity.

Sialoglycan microarrays were used to test the spike protein's sialic acid binding capacity. Hao et al. did not determine significant fluorescent signals when recombinant SARS-CoV-2 S protein was incubated with sialic acid-containing oligosaccharides on an array chip [15]. Though, these immobilized sialic acids may not model the native presentation of sialic acids on the cell surface in vivo, where sialic acids cluster in the flexible plasma membrane. Baker et al. used polymer-stabilized gold nanoparticle bearing sialic acids to confirm the binding of single sialic acid to the S protein, but not sialyllactoses [16].

Yang et al. investigated the role of glycans containing sialic acids on the ACE2 receptor for SARS-CoV-2 infection [17]. They found that glycans did not greatly contribute to the binding of the S protein to ACE2 and that ACE2′s sialic acids actually shielded cells from pseudovirus binding. Chu et al. then treated epithelial cells with neuraminidase to remove cell surface sialic acids [18]. They found that this treatment reduced MERS-CoV entry by 86%, which was expected since MERS-CoV uses DPP4 and sialic acids as coreceptors to facilitate binding. However, this sialic acid-removing treatment increased SARS-CoV and SARS-CoV-2 infection by 492% and 80.3%, respectively [19]. These results indicate that the presence of cell-surface sialic acids prevents ACE2-S protein binding, thus inhibiting virus entry.

2.2. *Gangliosides*

Though Yang et al. and Chu et al. reveal a decreasing likelihood that sialic acids potentiate SARS-CoV-2-cell binding, a preprint study by Nguyen et al. suggested that the RBD had an affinity for monosialylated gangliosides [17,18,20]. Using an artificial membrane embedded within gangliosides, they found that three monosialylated gangliosides (GM1, GM2, and GM3) were recognized by the RBD [20]. The affinity was similar to that of the glycan heparan sulfate, another proposed ligand of the RBD. Next, by depleting cell surface sialic acids using sialyltransferase inhibition, genetic knockout of sialic acid biosynthesis, and neuraminidase treatment of ACE2-expressing cells, they demonstrated decreased binding and infection.

Functions of gangliosides supporting their theoretical implication in SARS-CoV-2 infection include their negatively charged flat surface that attracts the electropositive tip of virus envelope proteins, ability to facilitate the recruitment of virus protein receptors from lipid rafts, association with cholesterol to form lipid rafts that could enhance fusion and activation of viral proteins through membrane chaperone properties [19].

Fantini et al. proposed a ganglioside binding domain at the top of the NTD, positing that it allows the S protein to interact with lipid rafts independently of the RBD [21]. The study found that neutralizing antibodies directed against the NTD's tip did, in fact, prevent access of the S protein to lipid rafts independently of RBD-ACE2 interactions [21]. Though, Fantini et al.'s data led to the proposal that hydroxychloroquine and other antimalarials could block the interaction between the S protein and cell surface gangliosides due to their affinity for gangliosides. The long-term lack of clinical validation for this strategy is not supportive of this theory [22].

2.3. Heparan Sulfate

Heparan sulfate is a highly conserved, negatively charged linear polysaccharide and cellular receptor found in almost all mammalian cells. In its proteoglycan form, HS binds to a variety of extracellular proteins, with wide-ranging functions related to development, inflammation, coagulation, angiogenesis, and viral entry. Heparan sulfate proteoglycans (HSPGs) are known to serve as coreceptors for many viruses [23].

Clausen et al. determined via molecular docking that the RBD likely contains a positively charged site adjacent to the ACE2-binding site that binds negatively charged heparan sulfate [24]. Subsequent competition studies, enzymatic removal of HS, and genetic studies showed that recombinant and pseudoviral S proteins, as well as authentic SARS-CoV-2 virions, bind to cell surface HS cooperatively with ACE2 [24].

A ternary complex of ACE2, heparin and the S protein using heparin as a scaffold was observed by Clausen et al. in vitro. Kearns et al. demonstrated a polyanionic HS-binding site that starts at the RBD and runs between the RBD and NTD down to the furin cleavage site [25]. Electron micrographs suggested that binding to heparin enhances the open conformation of the RBD, which supports ACE2 binding. Further, in vitro treatment of cells with heparin lyases that degrade HS significantly reduced infection. Clausen et al. concluded that HS not only supports infection but is an essential factor in infection that enables the RBD to bind ACE2. This proposed role has since been supported by a few studies (including Liu et al. [26]).

HSPGs have previously been proposed as coreceptors for HCoV-NL63 [27] and SARS-CoV [28]. In both cases, HSPGs were also proposed as necessary adhesion molecules for HCoV-NL63 and SARS-CoV infection. In fact, the basis of treatment of SARS with lactoferrin, an innate immunity protein, is lactoferrin's colocalization with HSPGs, blocking S protein-HSPG binding, and thus infection [28,29]. Also, HS's role in coagulation could be responsible for thrombotic complications seen in critically ill COVID-19 patients since the S protein could outcompete antithrombin and heparin cofactor II for HS binding [30].

2.4. Summary of Sugars as SARS-CoV-2 Spike Protein Receptors

Experimental data has yet to reveal anything conclusive regarding the role of sialosides or gangliosides in SARS-CoV-2-cell binding. Studies offer conflicting data regarding the enabling or prevention of virus entry due to these cell surface sugars. Due to the structural basis for the spike protein's proposed affinity for sialosides, including divergent loop regions in the NTD resembling sialoside-binding pockets and locations with an affinity for gangliosides in the NTD and RBD, more investigation into spike-sialoside interactions is warranted. If sialosides or gangliosides do, in fact, play a role in coronavirus host-cell binding, it is likely to use a novel mechanism yet unknown.

More likely is the possibility of heparan sulfate and its proteoglycan form as an important coreceptor for SARS-CoV-2. Strong but limited experimental data supports its role as a necessary or supporting factor for cell-virus adhesion, as its binding to the S protein prompts the open conformation of the RBD to accommodate subsequent ACE2 binding and infection [17].

3. Other Molecules

3.1. LRRC15 as a SARS-CoV-2 Spike Protein Ligand

Leucine-rich repeat (LRR) proteins are a family of functionally unrelated α/β horseshoe-shaped proteins that contain tandem repeats of 20–30 amino acids with an unusually high composition of leucine [31]. Leucine-rich repeat-containing protein 15 (LRRC15) is a member of the LRR family with many known functions, including innate immunity and nervous system development [32]. LRRC15 is known to be involved in the negative regulation of protein localization to the plasma membrane [33].

Multiple late-2021 preprints independently identified LRRC15 as a ligand of the SARS-CoV-2 spike protein (Figure 5). Shilts et al. employed two strategies to determine possible ligands of the spike protein [34]. First, HEK293 cells were individually transfected with one of 2363 genes encoding cell surface membrane proteins, measuring protein-spike binding with flow cytometry. Second, a genome-wide CRISPR activation library was used in RPE1 cells to identify which genes, when upregulated, induced the binding of the spike protein. Both systematic cell-based screens identified LRRC15 as SARS-CoV-2 spike protein ligands, though structural details of this interaction were unclear [34].

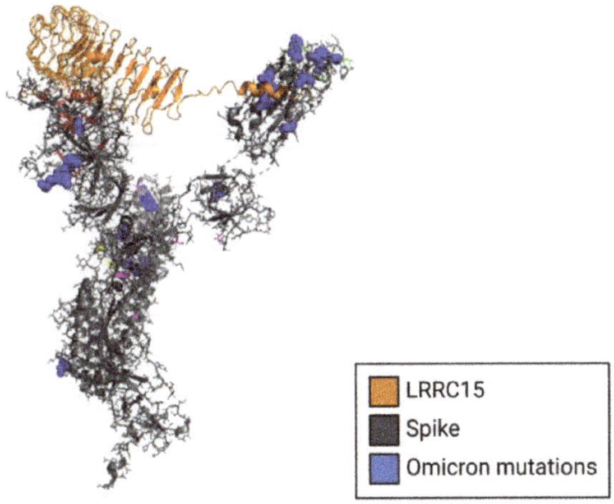

Figure 5. Structural model of LRRC15 complexed to the NTD of the SARS-CoV-2 spike glycoprotein. The cryo-EM structure of the spike protein is shown in gray complexed to LRRC15, shown in gold. Spike mutations from the omicron variant are shown in blue.

Loo et al. also identified LRRC15 as a spike protein ligand using a CRISPR activation strategy similar to Shilts et al. [34,35]. Further, LRRC15 sequestered virions and functioned as a negative receptor that suppressed live SARS-CoV-2 infection in trans [35]. Since LRRC15 and ACE2 expression are mutually exclusive (not expressed in the same cell types) in lung cells, it follows that LRRC15′s virion sequestration may prevent virions from reaching ACE2-expressing cells.

Additionally, LRRC15 expression could regulate the reaction of fibroblasts to infection. It is found in collagen-producing myofibroblasts, where it regulates collagen production. Since it is known to be upregulated by proinflammatory cytokines, including TNFα, IL-1β, and IFNγ [36], LRRC15 could suppress lung fibrosis during virus-induced inflammation, independent of virion sequestration and immobilization. LRRC15 could potentially help fibroblasts transport virions to antigen-presenting cells.

A subsequent decrease in LRRC15 levels following inflammation could promote collagen production to support lung repair. It is posited by Loo et al. that dysregulation of this system caused by chronic lung infection could cause inappropriate collagen production, contributing to lung fibrosis seen in "long-haul" COVID patients [35].

Interestingly, when compared to the earlier D614G variant of SARS-CoV-2, Loo et al. found that the mutations in the Delta variant's spike protein reduced LRRC15′s antiviral activity. This indicates that the numerous mutations of the spike protein in highly contagious variants could be attributed to an adaptation against the anti-SARS-CoV-2 activity of LRRC15.

A third preprint, by Song et al., affirmed the *trans*-inhibition of viral entry of SARS-CoV-2 by LRRC15 [37]. The interaction of LRRC15 with the RBD was not found to compete with the interaction of ACE2 with the RBD. LRRC15 did, in fact, inhibit spike-mediated viral entry in LRRC15-expressing ACE2-negative cells as well as neighboring ACE2-positive cells in trans. LRRC15 had a specific inhibitory effect on multiple variants of SARS-CoV-2 and SARS-CoV-1 but not MERS-CoV, which Song et al. noted as a suggestion of an evolutionary arms race between humans and coronaviruses. Song et al. also confirmed by analysis of human lung single cell RNA sequencing that LRRC15 expression is primarily detected in fibroblasts, including those pathologically enriched in COVID-19 patients.

3.2. *Summary of LRRC15 Relating to SARS-CoV-2*

These preprints display convincing experimental evidence that LRRC15 binds to the spike protein, offering mechanistic theories with plausibility established by experimental observations. The role of LRRC15 is currently best explained as SARS-CoV-2-virion-sequestering, protecting nearby ACE2-positive cells from infection. Further speculation reveals a possible mechanism by which LRRC15′s role in collagen production regulation on fibroblasts could contribute to "long-haul" COVID. These roles have earned the protein its characterization as a "master regulator" of SARS-CoV-2 infection and collagen production associated with "long-haul" COVID-related lung fibrosis.

More experimental investigation into the spike-LRRC15 interaction is warranted to further establish the theories about its potential role in preventing infection. Resolving the structure of the spike-LRRC15 complex could offer insight into the exact binding location of LRRC15 on the spike protein.

3.3. *Neuropilin-1 as a SARS-CoV-2 Spike Protein Ligand*

Neuropilin-1 is a glycoprotein receptor found in neurons that have functions related to angiogenesis, neuronal development, and immune response regulation [38].

Some studies have identified NRP-1 as a coreceptor for SARS-CoV-2 entry [38–40]. The presence of NRP-1 on the host cell membrane has been shown to increase infection and spread of SARS-CoV-2, as shown by Cantuti-Castelvetri et al. [39]. Particularly, NRP-1 is implicated in neurological manifestations of COVID-19, as it is thought to enable SARS-CoV-2 to enter the central nervous system through the respiratory and olfactory epithelia—where NRP-1 is highly expressed—including those of the nasal cavity.

Daly et al. explains that NRP-1 binds furin-cleaved substrates such as the one in the spike protein (Figure 6) [40]. Daly et al. successfully demonstrated that the furin-cleaved spike protein binds directly to cell surface NRP-1. Further, blocking the S1-NRP-1 interaction with small-molecule inhibitors and monoclonal antibodies reduced viral infection in vitro. This finding has significant implications for future antiviral therapeutics.

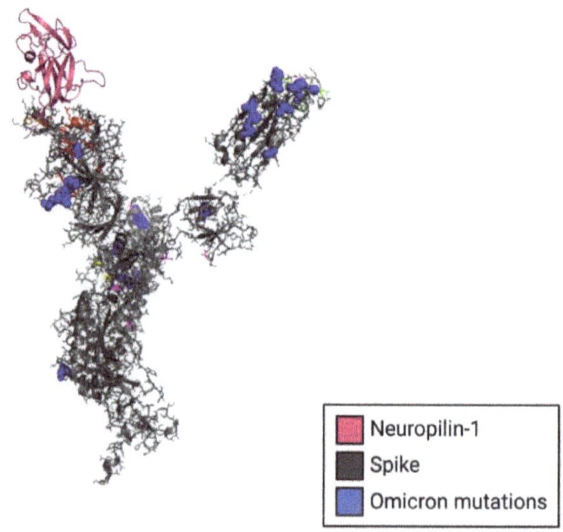

Figure 6. Structural model of Neuropilin-1 complexed to the NTD of the SARS-CoV-2 spike glycoprotein monomer. The cryo-EM structure of the spike protein monomer is shown in gray complex to neuropilin-1, shown in magenta. Spike mutations from the omicron variant are shown in blue.

4. Non-Receptor Spike-Host Cell Protein Interactions

4.1. SARS-CoV-2 Spike Protein Cleavage and Restriction Factors

The SARS-CoV-2 spike protein relies on cleavage by proteases on the cell surface to transform into its infectious conformation that enables cell entry through plasma membrane fusion [41]. Spike protein cleavage/fusion at (or in close proximity to) the cell surface is crucial to successful infection. Spike protein fusion with host cell membranes near the cell surface permits avoidance of restriction factors located in early and late endosomes within the cell. The furin cleavage site where the spike is cleaved is thought to be an advantage of SARS-CoV-2, as viruses that lack such a cleavage site generally must enter cells through the restriction-factor-containing endosome [42].

Restriction factors are proteins that host cells use as the first line of defense against viruses [42]. Some interfere with viral replication and propagation, while others are sensors that trigger innate immune responses. They are generally induced by interferons.

Experimental evidence showed that SARS-CoV-2 is predominantly sensitive to IFITM2, a restriction factor that inhibits virus-host cell fusion [43]. The same study observed decreased infectivity when removing the furin cleavage site from pre-Omicron variants, supporting the theory that this cleavage is an advantage of the virus. Additionally, SARS-CoV-2 was found to be highly sensitive to IFN-β and IFN-γ. IFN-α and IFN-λ were less effective but still demonstrated restrictive properties in various cell types. It is not yet elucidated how changes in Omicron's furin cleavage site (and consequently its protease-based infection cascade) may affect its interactions with restriction factors.

4.2. Membrane-Associated Serine Proteases

The spike protein must undergo a conformational change at the RBD to interact with ACE2. TMPRSS2 has been known as the first host cell-surface protease used by the spike protein to cleave its furin cleavage site prompting the conformational changes necessary to enter the cell [41]. Following cleavage by TMPRSS2, the spike is further cleaved and processed by furin and cathepsin, then fusing with the cell and provoking viral replication.

Currently, Omicron variants inefficiently use TMPRSS2, causing the spike to rely on cell entry through endocytosis [44]. This caused a change in the cellular tropism of SARS-CoV-2, moving away from TMPRSS2-expressing cells. However, BA.5 has recently shown efficient use of TMPRSS2, indicating a possible shift back to pre-Omicron tropism and infectious mechanism [45].

5. Summary

The identity of a putative SARS-CoV-2 coreceptor has yet to be determined. Though, few proposed candidates increasingly demonstrate activity as potential ligands of the spike protein (Table 1). Further experimentation is warranted to elucidate the roles of these candidates in the SARS-CoV-2 infection mechanism. Significant improvements in the prevention and treatment of COVID-19 may be achieved by understanding how to modulate the interactions between the S1-NTD/RBD and coreceptors, antibodies, and restriction factors.

Table 1. Summary of potential coreceptor ligands of the SARS-CoV-2 spike protein.

Ligand	Description	Proposed Virological Function	References
Sialic acid	Sugar chain with nine-carbon backbone found on the surfaces of all vertebrate cells. Implicated in various functions and diseases.	S1-NTD may bind sialic acids in a sialosides-binding pocket that resembles that of the MERS-CoV S1-NTD. Sialic acid may act as a coreceptor to help S1 reach ACE2 via viral surfing, or it may inhibit ACE2-S1 binding.	[4,11,17,46]

Table 1. *Cont.*

Ligand	Description	Proposed Virological Function	References
Monosialylated gangliosides (GM1, GM2, GM3)	Sialic acid-containing glycosphingolipids widely expressed in the nervous system. As part of the plasma membrane's outer leaflet, these sugar chains are key to cell-cell recognition, adhesion, and signal transduction.	Negatively charged flat surface may attract the electropositive tip of virus envelope proteins. Association with cholesterol to form lipid rafts could enhance fusion. Membrane chaperone properties could enhance activation of viral proteins.	[19,21,47]
Heparan sulfate proteoglycans (HSPGs)	Glycoprotein with attached heparan sulfate polysaccharide chains. Found as a cellular receptor in almost all mammalian cells with functions related to development, inflammation, coagulation, angiogenesis, and viral entry.	Binding between HSPGs/HS and a polyanionic path along the S1-RBD, NTD, and FCS may enhance the open conformation of the RBD, supporting RBD-ACE2 binding.	[23–26]
Leucine-rich repeat-containing protein 15 (LRRC15)	Protein family with unrelated functions and characteristic α/β-horseshoe shape with leucine-rich tandem repeats. Functions of LRRC15 include innate immunity, down regulation of protein localization to the plasma membrane, and nervous system development.	LRRC15 may sequester and immobilize SARS-CoV-2 virions. Upregulated by proinflammatory cytokines, LRRC15 could suppress lung fibrosis during virus-induced inflammation. Demonstrated affinity for S1-NTD.	[31–35,37]
Neuropilin-1 (NRP-1)	Glycoprotein receptor found on cell membranes of neurons with functions related to angiogenesis, neuronal development, and immune response regulation.	Potentially enables SARS-CoV-2 virions to enter the nervous system through respiratory and olfactory epithelia. Known to bind furin-cleaved substrates.	[38–40]

6. Methods

6.1. SARS-CoV-2 Mutations in Variants of Concern

Mutations in SARS-CoV-2 variants of concern are shown with the Wuhan-Hu-1 sequence as a reference (GENBANK accession number MN908947). The Centers for Disease Control and Prevention classification and definitions of mutations that distinguish variants of concern were used: https://www.cdc.gov/coronavirus/2019-ncov/variants/variant-classifications.html (accessed on 26 April 2022) [48].

6.2. Mapping Mutations and Structural Variability on Structure of the Spike Protein Monomer

The prefusion SARS-CoV-2 spike glycoprotein with a single receptor-binding domain up (PDB 6VSB) [49] was used for mapping mutations and structural variability (B-factor) in variants of concern.

6.3. Modeling Interactions between LRRC15, Neuropilin-1 and the SARS-CoV-2 Spike Protein NTD

The crystal structure of Neuropilin-1 (PDB 7JJC) [40] was docked to the SARS-CoV-2 spike protein NTD using the program HDOCK [50]. A structural model of human LRRC15 was generated by SWISS-MODEL [51] and docked to the SARS-CoV-2 spike protein NTD using the program HDOCK [50].

Author Contributions: Writing—Original draft preparation, R.L.B.; Writing—Review and editing, D.A.O.; supervision, D.A.O. All authors have read and agreed to the published version of the manuscript.

Funding: This research was funded by the National Institute of Allergy and Infectious Diseases, grant number R01AI170187. The APC was funded by David A. Ostrov.

Institutional Review Board Statement: Not applicable.

Informed Consent Statement: Not applicable.

Data Availability Statement: Not applicable.

Acknowledgments: Figures created using BioRender.com.

Conflicts of Interest: The authors declare no conflict of interest.

References

1. Li, F. Structure, Function, and Evolution of Coronavirus Spike Proteins. *Annu. Rev. Virol.* **2016**, *3*, 237–261. [CrossRef] [PubMed]
2. Huang, Y.; Yang, C.; Xu, X.; Xu, W.; Liu, S. Structural and functional properties of SARS-CoV-2 spike protein: Potential antivirus drug development for COVID-19. *Acta Pharmacol. Sin.* **2020**, *41*, 1141–1149. [CrossRef] [PubMed]
3. Eslami, N.; Aghbash, P.S.; Shamekh, A.; Entezari-Maleki, T.; Nahand, J.S.; Sales, A.J.; Baghi, H.B. SARS-CoV-2: Receptor and Co-receptor Tropism Probability. *Curr. Microbiol.* **2022**, *79*, 133. [CrossRef] [PubMed]
4. Awasthi, M.; Gulati, S.; Sarkar, D.P.; Tiwari, S.; Kateriya, S.; Ranjan, P.; Verma, S.K. The Sialoside-Binding Pocket of SARS-CoV-2 Spike Glycoprotein Structurally Resembles MERS-CoV. *Viruses* **2020**, *12*, 909. [CrossRef]
5. Ostrov, D.A.; Knox, G.W. Emerging mutation patterns in SARS-CoV-2 variants. *Biochem. Biophys. Res. Commun.* **2022**, *586*, 87–92. [CrossRef]
6. Wang, K.; Chen, W.; Zhang, Z.; Deng, Y.; Lian, J.-Q.; Du, P.; Wei, D.; Zhang, Y.; Sun, X.-X.; Gong, L.; et al. CD147-spike protein is a novel route for SARS-CoV-2 infection to host cells. *Signal Transduct. Target. Ther.* **2020**, *5*, 283. [CrossRef]
7. Jasinska, A.J.; Pandrea, I.; Apetrei, C. CCR5 as a Coreceptor for Human Immunodeficiency Virus and Simian Immunodeficiency Viruses: A Prototypic Love-Hate Affair. *Front. Immunol.* **2022**, *13*, 835994. [CrossRef]
8. Liu, Y.; Soh, W.T.; Kishikawa, J.; Hirose, M.; Nakayama, E.E.; Li, S.; Sasai, M.; Suzuki, T.; Tada, A.; Arakawa, A.; et al. An infectivity-enhancing site on the SARS-CoV-2 spike protein targeted by antibodies. *Cell* **2021**, *184*, 3452–3466.e18. [CrossRef]
9. Wu, L.; Zhou, L.; Mo, M.; Liu, T.; Wu, C.; Gong, C.; Lu, K.; Gong, L.; Zhu, W.; Xu, Z. SARS-CoV-2 Omicron RBD shows weaker binding affinity than the currently dominant Delta variant to human ACE2. *Signal Transduct. Target. Ther.* **2022**, *7*, 8. [CrossRef]
10. Yao, H.; Song, Y.; Chen, Y.; Wu, N.; Xu, J.; Sun, C.; Zhang, J.; Weng, T.; Zhang, Z.; Wu, Z.; et al. Molecular Architecture of the SARS-CoV-2 Virus. *Cell* **2020**, *183*, 730–738.e13. [CrossRef]
11. Milanetti, E.; Miotto, M.; Di Rienzo, L.; Nagaraj, M.; Monti, M.; Golbek, T.W.; Gosti, G.; Roeters, S.J.; Weidner, T.; Otzen, D.E.; et al. In-Silico Evidence for a Two Receptor Based Strategy of SARS-CoV-2. *Front. Mol. Biosci.* **2021**, *8*, 690655. [CrossRef] [PubMed]
12. Burckhardt, C.J.; Greber, U.F. Virus Movements on the Plasma Membrane Support Infection and Transmission between Cells. *PLoS Pathog.* **2009**, *5*, e1000621. [CrossRef] [PubMed]
13. Caldas, L.A.; Carneiro, F.A.; Higa, L.M.; Monteiro, F.L.; da Silva, G.P.; da Costa, L.J.; Durigon, E.L.; Tanuri, A.; de Souza, W. Ultrastructural analysis of SARS-CoV-2 interactions with the host cell via high resolution scanning electron microscopy. *Sci. Rep.* **2020**, *10*, 16099. [CrossRef] [PubMed]
14. Seyran, M.; Takayama, K.; Uversky, V.N.; Lundstrom, K.; Palù, G.; Sherchan, S.P.; Attrish, D.; Rezaei, N.; Aljabali, A.A.A.; Ghosh, S.; et al. The structural basis of accelerated host cell entry by SARS-CoV-2. *FEBS J.* **2021**, *288*, 5010–5020. [CrossRef] [PubMed]
15. Hao, W.; Ma, B.; Li, Z.; Wang, X.; Gao, X.; Li, Y.; Qin, B.; Shang, S.; Cui, S.; Tan, Z. Binding of the SARS-CoV-2 Spike Protein to Glycans. *Sci. Bull.* **2021**, *66*, 1205–1214. [CrossRef]
16. Baker, A.N.; Richards, S.-J.; Guy, C.S.; Congdon, T.R.; Hasan, M.; Zwetsloot, A.J.; Gallo, A.; Lewandowski, J.R.; Stansfeld, P.J.; Straube, A.; et al. The SARS-COV-2 Spike Protein Binds Sialic Acids and Enables Rapid Detection in a Lateral Flow Point of Care Diagnostic Device. *ACS Cent. Sci.* **2020**, *6*, 2046–2052. [CrossRef]
17. Yang, Q.; Hughes, T.A.; Kelkar, A.; Yu, X.; Cheng, K.; Park, S.; Huang, W.-C.; Lovell, J.F.; Neelamegham, S. Inhibition of SARS-CoV-2 viral entry upon blocking N- and O-glycan elaboration. *eLife* **2020**, *9*, e61552. [CrossRef]
18. Chu, H.; Hu, B.; Huang, X.; Chai, Y.; Zhou, D.; Wang, Y.; Shuai, H.; Yang, D.; Hou, Y.; Zhang, X.; et al. Host and viral determinants for efficient SARS-CoV-2 infection of the human lung. *Nat. Commun.* **2021**, *12*, 134. [CrossRef]
19. Sun, X.-L. The role of cell surface sialic acids for SARS-CoV-2 infection. *Glycobiology* **2021**, *31*, 1245–1253. [CrossRef]
20. Nguyen, L.; McCord, K.A.; Bui, D.T.; Bouwman, K.M.; Kitova, E.N.; Elaish, M.; Kumawat, D.; Daskhan, G.C.; Tomris, I.; Han, L.; et al. Sialic Acid-Containing Glycolipids Mediate Binding and Viral Entry of SARS-CoV-2. *Nat Chem Biol* **2022**, *18*, 81–90. [CrossRef]
21. Fantini, J.; Di Scala, C.; Chahinian, H.; Yahi, N. Structural and molecular modelling studies reveal a new mechanism of action of chloroquine and hydroxychloroquine against SARS-CoV-2 infection. *Int. J. Antimicrob. Agents* **2020**, *55*, 105960. [CrossRef] [PubMed]
22. Schwartz, I.S.; Boulware, D.R.; Lee, T.C. Hydroxychloroquine for COVID19: The curtains close on a comedy of errors. *Lancet Reg. Health Am.* **2022**, *11*, 100268. [CrossRef] [PubMed]

23. Cagno; Tseligka; Jones; Tapparel Heparan Sulfate Proteoglycans and Viral Attachment: True Receptors or Adaptation Bias? *Viruses* **2019**, *11*, 596. [CrossRef] [PubMed]
24. Clausen, T.M.; Sandoval, D.R.; Spliid, C.B.; Pihl, J.; Perrett, H.R.; Painter, C.D.; Narayanan, A.; Majowicz, S.A.; Kwong, E.M.; McVicar, R.N.; et al. SARS-CoV-2 Infection Depends on Cellular Heparan Sulfate and ACE2. *Cell* **2020**, *183*, 1043–1057.e15. [CrossRef] [PubMed]
25. Kearns, F.L.; Sandoval, D.R.; Casalino, L.; Clausen, T.M.; Rosenfeld, M.A.; Spliid, C.B.; Amaro, R.E.; Esko, J.D. Spike-heparan sulfate interactions in SARS-CoV-2 infection. *Curr. Opin. Struct. Biol.* **2022**, *76*, 102439. [CrossRef]
26. Liu, L.; Chopra, P.; Li, X.; Bouwman, K.M.; Tompkins, S.M.; Wolfert, M.A.; de Vries, R.P.; Boons, G.-J. Heparan Sulfate Proteoglycans as Attachment Factor for SARS-CoV-2. *ACS Cent. Sci.* **2021**, *7*, 1009–1018. [CrossRef]
27. Milewska, A.; Zarebski, M.; Nowak, P.; Stozek, K.; Potempa, J.; Pyrc, K. Human Coronavirus NL63 Utilizes Heparan Sulfate Proteoglycans for Attachment to Target Cells. *J. Virol.* **2014**, *88*, 13221–13230. [CrossRef]
28. Lang, J.; Yang, N.; Deng, J.; Liu, K.; Yang, P.; Zhang, G.; Jiang, C. Inhibition of SARS Pseudovirus Cell Entry by Lactoferrin Binding to Heparan Sulfate Proteoglycans. *PLoS ONE* **2011**, *6*, e23710. [CrossRef]
29. Hu, Y.; Meng, X.; Zhang, F.; Xiang, Y.; Wang, J. The in vitro antiviral activity of lactoferrin against common human coronaviruses and SARS-CoV-2 is mediated by targeting the heparan sulfate co-receptor. *Emerg. Microbes Infect.* **2021**, *10*, 317–330. [CrossRef]
30. Zheng, Y.; Zhao, J.; Li, J.; Guo, Z.; Sheng, J.; Ye, X.; Jin, G.; Wang, C.; Chai, W.; Yan, J.; et al. SARS-CoV-2 spike protein causes blood coagulation and thrombosis by competitive binding to heparan sulfate. *Int. J. Biol. Macromol.* **2021**, *193*, 1124–1129. [CrossRef]
31. Ng, A.; Xavier, R.J. Leucine-rich repeat (LRR) proteins: Integrators of pattern recognition and signaling in immunity. *Autophagy* **2011**, *7*, 1082–1084. [CrossRef] [PubMed]
32. Ray, U.; Pathoulas, C.L.; Thirusangu, P.; Purcell, J.W.; Kannan, N.; Shridhar, V. Exploiting LRRC15 as a Novel Therapeutic Target in Cancer. *Cancer Res.* **2022**, *82*, 1675–1681. [CrossRef] [PubMed]
33. O'Prey, J.; Wilkinson, S.; Ryan, K.M. Tumor Antigen LRRC15 Impedes Adenoviral Infection: Implications for Virus-Based Cancer Therapy. *J. Virol.* **2008**, *82*, 5933–5939. [CrossRef] [PubMed]
34. Shilts, J.; Crozier, T.W.M.; Teixeira-Silva, A.; Gabaev, I.; Greenwood, E.J.D.; Watson, S.J.; Ortmann, B.M.; Gawden-Bone, C.M.; Pauzaite, T.; Hoffmann, M.; et al. LRRC15 mediates an accessory interaction with the SARS-CoV-2 spike protein. *bioRxiv* **2021**.
35. Loo, L.; Waller, M.A.; Cole, A.J.; Stella, A.O.; Moreno, C.L.; Denes, C.E.; Hamoudi, Z.; Chung, F.; Aggarwal, A.; Low, J.K.K.; et al. LRRC15 suppresses SARS-CoV-2 infection and controls collagen production. *bioRxiv* **2021**.
36. Satoh, K.; Hata, M.; Yokota, H. A Novel Member of the Leucine-Rich Repeat Superfamily Induced in Rat Astrocytes by β-Amyloid. *Biochem. Biophys. Res. Commun.* **2002**, *290*, 756–762. [CrossRef]
37. Song, J.; Chow, R.D.; Pena-Hernandez, M.; Zhang, L.; Loeb, S.A.; So, E.-Y.; Liang, O.D.; Wilen, C.B.; Lee, S. LRRC15 is an inhibitory receptor blocking SARS-CoV-2 spike-mediated entry in trans. *bioRxiv* **2021**.
38. Gudowska-Sawczuk, M.; Mroczko, B. The Role of Neuropilin-1 (NRP-1) in SARS-CoV-2 Infection: Review. *J. Clin. Med.* **2021**, *10*, 2772. [CrossRef]
39. Cantuti-Castelvetri, L.; Ojha, R.; Pedro, L.D.; Djannatian, M.; Franz, J.; Kuivanen, S.; van der Meer, F.; Kallio, K.; Kaya, T.; Anastasina, M.; et al. Neuropilin-1 facilitates SARS-CoV-2 cell entry and infectivity. *Science* **2020**, *370*, 856–860. [CrossRef]
40. Daly, J.L.; Simonetti, B.; Klein, K.; Chen, K.-E.; Williamson, M.K.; Antón-Plágaro, C.; Shoemark, D.K.; Simón-Gracia, L.; Bauer, M.; Hollandi, R.; et al. Neuropilin-1 is a host factor for SARS-CoV-2 infection. *Science* **2020**, *370*, 861–865. [CrossRef]
41. Peacock, T.P.; Goldhill, D.H.; Zhou, J.; Baillon, L.; Frise, R.; Swann, O.C.; Kugathasan, R.; Penn, R.; Brown, J.C.; Sanchez-David, R.Y.; et al. The furin cleavage site in the SARS-CoV-2 spike protein is required for transmission in ferrets. *Nat. Microbiol.* **2021**, *6*, 899–909. [CrossRef] [PubMed]
42. Colomer-Lluch, M.; Ruiz, A.; Moris, A.; Prado, J.G. Restriction Factors: From Intrinsic Viral Restriction to Shaping Cellular Immunity Against HIV-1. *Front. Immunol.* **2018**, *9*, 2876. [CrossRef] [PubMed]
43. Winstone, H.; Lista, M.J.; Reid, A.C.; Bouton, C.; Pickering, S.; Galao, R.P.; Kerridge, C.; Doores, K.J.; Swanson, C.M.; Neil, S.J.D. The Polybasic Cleavage Site in SARS-CoV-2 Spike Modulates Viral Sensitivity to Type I Interferon and IFITM2. *J. Virol.* **2021**, *95*, e02422-20. [CrossRef]
44. Meng, B.; Abdullahi, A.; Ferreira, I.A.T.M.; Goonawardane, N.; Saito, A.; Kimura, I.; Yamasoba, D.; Gerber, P.P.; Fatihi, S.; Rathore, S.; et al. Altered TMPRSS2 usage by SARS-CoV-2 Omicron impacts infectivity and fusogenicity. *Nature* **2022**, *603*, 706–714. [CrossRef] [PubMed]
45. Aggarwal, A.; Akerman, A.; Milogiannakis, V.; Silva, M.R.; Walker, G.; Stella, A.O.; Kindinger, A.; Angelovich, T.; Waring, E.; Amatayakul-Chantler, S.; et al. SARS-CoV-2 Omicron BA.5: Evolving tropism and evasion of potent humoral responses and resistance to clinical immunotherapeutics relative to viral variants of concern. *eBioMedicine* **2022**, *84*, 104270. [CrossRef] [PubMed]
46. Varki, A. Sialic acids in human health and disease. *Trends Mol. Med.* **2008**, *14*, 351–360. [CrossRef]
47. Yu, R.K.; Tsai, Y.-T.; Ariga, T.; Yanagisawa, M. Structures, Biosynthesis, and Functions of Gangliosides-an Overview. *J. Oleo Sci.* **2011**, *60*, 537–544. [CrossRef]
48. Centers for Disease Control and Prevention. *SARS-CoV-2 Variant Classifications and Definitions 2022*; Centers for Disease Control and Prevention: Atlanta, GA, USA, 2022.
49. Wrapp, D.; Wang, N.; Corbett, K.S.; Goldsmith, J.A.; Hsieh, C.-L.; Abiona, O.; Graham, B.S.; McLellan, J.S. Cryo-EM structure of the 2019-nCoV spike in the prefusion conformation. *Science* **2020**, *367*, 1260–1263. [CrossRef]

50. Yan, Y.; Zhang, D.; Zhou, P.; Li, B.; Huang, S.-Y. HDOCK: A web server for protein–protein and protein–DNA/RNA docking based on a hybrid strategy. *Nucleic Acids Res.* **2017**, *45*, W365–W373. [CrossRef]
51. Waterhouse, A.; Bertoni, M.; Bienert, S.; Studer, G.; Tauriello, G.; Gumienny, R.; Heer, F.T.; de Beer, T.A.P.; Rempfer, C.; Bordoli, L.; et al. SWISS-MODEL: Homology modelling of protein structures and complexes. *Nucleic Acids Res.* **2018**, *46*, W296–W303. [CrossRef]

viruses

MDPI

Review

On the Evolutionary Trajectory of SARS-CoV-2: Host Immunity as a Driver of Adaptation in RNA Viruses

Jacob Warger [1] **and Silvana Gaudieri** [2,3,4,*]

1 School of Medicine and Pharmacology, University of Western Australia, Crawley, WA 6009, Australia
2 School of Human Sciences, University of Western Australia, Crawley, WA 6009, Australia
3 Institute for Immunology and Infectious Diseases, Murdoch University, Mandurah, WA 6150, Australia
4 Division of Infectious Diseases, Department of Medicine, Vanderbilt University Medical Center, Nashville, TN 37232, USA
* Correspondence: silvana.gaudieri@uwa.edu.au

Abstract: Host immunity can exert a complex array of selective pressures on a pathogen, which can drive highly mutable RNA viruses towards viral escape. The plasticity of a virus depends on its rate of mutation, as well as the balance of fitness cost and benefit of mutations, including viral adaptations to the host's immune response. Since its emergence, SARS-CoV-2 has diversified into genetically distinct variants, which are characterised often by clusters of mutations that bolster its capacity to escape human innate and adaptive immunity. Such viral escape is well documented in the context of other pandemic RNA viruses such as the human immunodeficiency virus (HIV) and influenza virus. This review describes the selection pressures the host's antiviral immunity exerts on SARS-CoV-2 and other RNA viruses, resulting in divergence of viral strains into more adapted forms. As RNA viruses obscure themselves from host immunity, they uncover weak points in their own armoury that can inform more comprehensive, long-lasting, and potentially cross-protective vaccine coverage.

Keywords: RNA viruses; SARS-CoV-2 variants; host immunity; adaptation; viral escape

Citation: Warger, J.; Gaudieri, S. On the Evolutionary Trajectory of SARS-CoV-2: Host Immunity as a Driver of Adaptation in RNA Viruses. *Viruses* **2023**, *15*, 70. https://doi.org/10.3390/v15010070

Academic Editors: Ahmed El-Shamy and Mohamed Ibrahim

Received: 4 December 2022
Revised: 21 December 2022
Accepted: 24 December 2022
Published: 26 December 2022

1. Introduction

The evolutionary trajectory of an organism is oriented toward the maximum likelihood for it to survive encounters with external stressors. The evolution of RNA viruses is influenced by their interaction with host immunity in much the same way. The fitness of a virus depends in part on its ability to circumvent the full artillery of host innate and adaptive immunity. A virus which reaches pandemic levels, such as SARS-CoV-2, does so by evolving into distinct viral strains, or *variants*. These variants can be characterised by specific mutations in the viral genome that may bolster evolutionary fitness—termed *viral adaptations* [1]. The frequency at which new variants emerge depends on the capacity of the virus to acquire adaptations [2,3]. RNA viruses are highly mutable and can quickly adapt to evade immune defence [2,4–6], which has significant bearing on clinical outcome and vaccine efficacy [7]. Mutation is sometimes associated with fitness cost [8], such that the frequency of any given mutation is often skewed in favour of mutations that impart a net fitness advantage. Understanding the pressures by which host immunity influences viral adaptation provides critical insight into the pathways utilised by RNA viruses to maximise replicative efficiency and immune evasion.

2. Viral Plasticity Is an Important Determinant of Fitness

Viral evolution can be thought of as a balancing act between genetic diversity and viral extinction [5,8]. The error threshold of a virus is defined as its maximum rate of mutation, beyond which mutations can result in viral failure [3,8]. This threshold is generally defined by genome length and selective constraints and represents the upper limit for even the most highly mutable viruses.

RNA viruses tend to mutate at rates very close to this threshold [4,5,8] and replicate with relatively low fidelity compared to DNA viruses. As such, RNA viruses can diverge considerably even during a short infection [2–5,9] and generally exist in a host as a collection of genetically similar variants, termed *quasispecies* [10,11]. The ability of RNA viruses to diversify so rapidly represents a fitness advantage when infecting a biologically diverse host population [12]. For example, the spherical virion of the influenza virus is decorated by hemagglutinin glycoproteins (HA), which represent the major surface antigen. During the flu season, mutations accumulate in HA, resulting in antigenic drift [13]. As a result, the influenza vaccine is updated regularly to ensure continued protection [14]. The ability of HIV to rapidly mutate is also of significant clinical concern. The HIV genome is highly malleable and can accumulate mutation at alarming rates, such that it can rapidly develop drug resistance [9].

Mutations often arise during replication, and the high mutation rate of RNA viruses is generally attributable to the RNA-dependent RNA polymerase (RdRp), which is incapable of proofreading [6,15]. Coronaviruses represent one of the few RNA viruses whose viral replicase machinery has the capacity to proofread [15–17]. The proofreading ability of coronaviruses limits the number of genetic changes that are introduced during each round of replication, and protects its genome from excessive deleterious mutation. The coronavirus genome is composed of two large transcription units. The larger of the two units, ORF1ab, occupies more than two thirds of the genome and encodes a large polyprotein (PP1ab), which embeds 16 non-structural proteins (NSP). These proteins make up the viral replicase machinery [15,18,19]. NSP 14 represents the 3′–5′ exonuclease domain of the replicase machinery and is capable of cleaving mismatched nucleotides from the end of the polynucleotide, significantly improving replication fidelity [15,16,18–21]. As a consequence, coronaviruses have considerably larger genomes than other RNA viruses [15]. The remaining third of the coronavirus genome is mostly composed of structural proteins, including its major surface antigen—the spike glycoprotein (*S*). The low rate at which coronaviruses acquire mutations underscores those mutations that do rise to significant frequencies because they may be adaptations that afford sufficient evolutionary advantage. For example, the non-coding 5′ and 3′ terminal regions of betacoronaviruses contain various housekeeping and regulatory genes, and mutations do not accumulate in this region as they are likely to be deleterious [22,23]. Conversely, the highly immunogenic *S* protein can quickly accumulate mutations that can diminish recognition by neutralizing antibodies [24–27].

3. Selection Is an Indicator of Viral Evolution

Selection is an important indicator of genetic evolution and influences the frequency of particular alleles dependent on the replicative fitness of specific genotypes [28]. Indeed, the genomic landscape of SARS-CoV-2 is still actively being shaped by selection imposed by immune pressure, which may reflect the immunogenicity of various components of the virus [29]. While some parts of the genome, such as the replicase machinery, of SARS-CoV-2 are under purifying selection and are relatively conserved, positive selection is actively driving the evolution of certain regions of the viral genome [28,30]. The SARS-CoV-2 genome shows signs of continued adaptation to human immunity after transmission from the non-human reservoir, much like other zoonotic RNA viruses Middle Eastern Respiratory Syndome Coronavirus (MERS-CoV) and influenza [29,31]. Furthermore, the relative fitness of different SARS-CoV-2 variants is heterogenous across different geographic regions, likely reflecting a changing immunological landscape influenced by genetics, as well as infection and vaccination rates [30,32].

One way by which selection can be measured is by evaluating the relationship between the rate of non-synonymous (dN; changes amino acid) to synonymous (dS; does not change the amino acid) changes in a viral protein [1,29,33]. Shortly after the emergence of the novel Alpha (B.1.1.7) variant, we conducted a whole-genome evolutionary analysis of the SARS-CoV-2 viral genome to identify mutations occurring at $\geq$3% of 72,000 captured Alpha genomes from an outbreak period between 1 November 2020 and 1 February 2021 in the UK. All genomes were accessed via the GISAID public SARS-CoV-2 sequence repository, EpiCov (www.epicov.org). We used Single Likelihood Ancestry Counting (SLAC; [34]) to evaluate the dN and dS relationship at single sites across the coding regions of the genome. Our results indicate that a number of these mutations are under the influence of selection. In particular, we detected evidence of positively selected mutations in the coding regions of NSP 4 and NSP 13. A similar study conducted by Hou et al. identified both proteins as regions of strong positive selection [28]. Importantly, NSP13 embeds the viral helicase and is involved in interferon (IFN) suppression [35] pointing to immune pressure as a potential selection force. Further NSP4, in conjunction with ORF9B, is involved in extensive mitochondrial restructuring associated with severe illness, though its role in immune evasion is unclear [36].

4. Adaptation to Innate Interferon Signalling Can Result in Immune Suppression

The innate immune response is the initial call to arms, and type-I/III IFNs are the first line of defence against viral infection [37,38] (Figure 1A). Following viral entry, certain viral pathogen-associated molecular patterns (PAMPs) are recognised by host pattern recognition receptors (PRRs)—termed immune sensing. Intracellular signalling triggered by activation of PRRs leads to the transcription of type I IFNs, which go on to initiate the expression of a network of IFN-stimulated genes (ISG) to induce an antiviral state [37,39,40]. Given that this is a critical first stage of infection control, resistance to type-I IFNs provides a significant fitness advantage for RNA viruses.

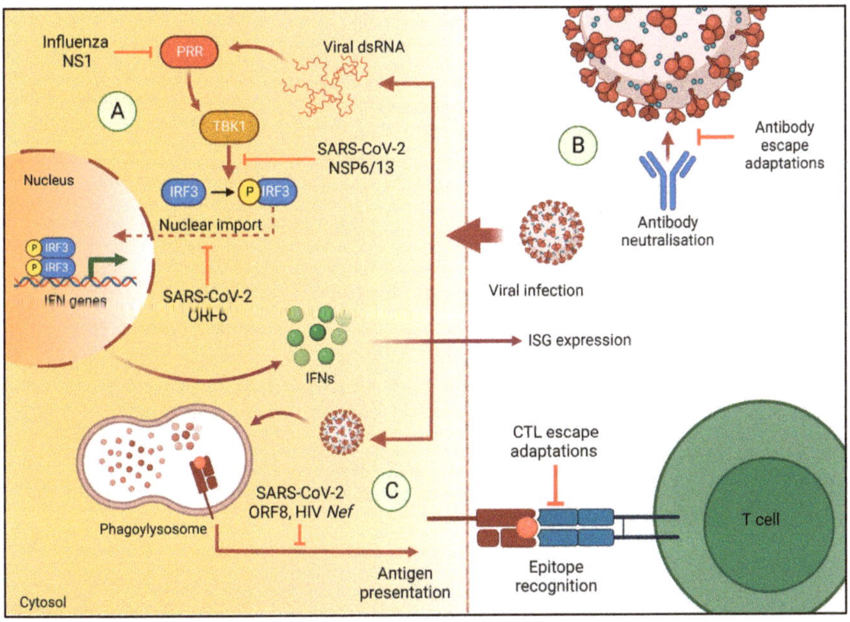

Figure 1. Summary of three sources of immune pressure during viral infection and viral proteins that act is immune modulators: (**A**) viral dsRNA is detected by immune sensing PRRs, triggering expression of IFN genes by phosphorylation and nuclear import of IRF-3. NS1 is a potent inhibitor of IRF-3 activation by binding viral dsRNA and preventing recognition by RIG-I. SARS-CoV-2 NSP6/13 are capable of binding TBK1 to suppress phosphorylation and activation of IRF-3, whereas ORF6 blocks its nuclear import factors; (**B**) neutralization of viral surface antigens by host antibodies, which is attenuated by adaptations in the surface antigen; (**C**) presentation of T cell epitopes by HLA class I molecules on the surface of the infected cell, which can be undermined by adaptations in HLA-restricted epitopes in a viral protein. (Created with BioRender.com).

In donor-recipient HIV transmission pairs, recipient virions are uniformly more resistant to type-I IFN, indicating that the net selective pressure, which determines the transmitted founder virus, is at least partially represented by resistance to type-I IFNs [41]. This resistance is particularly important in the counteraction of IFN-β, which is a potent inhibitor of HIV-1 [42–44]. Furthermore, HIV-1(M), a pandemic strain of HIV, contains a number of amino acid substitutions in its capsid protein and is uniquely capable of evading immune sensing by host innate immunity. Reversing the HIV-1(M) capsid mutations results in induction of type-I IFN genes and suppression of viral replication to levels comparable to non-pandemic HIV-1, suggesting that HIV-1(M) reaches pandemic levels as a consequence of adaptations in its capsid that aid type-I IFN evasion [45].

Suppression of IFN production is a major strategy of immune evasion utilised by SARS-CoV-2. Subjects with COVID-19 exhibit vastly delayed/reduced production of type-I/III IFNs compared with individuals infected with influenza virus [46]. Furthermore, in cells infected with SARS-CoV-2, expression of IFN-β and ISG56, a family of ISGs, peaks much later compared to cells infected with Sendai virus [47]. This pattern of expression may be a result of SARS-CoV-2 viral proteins antagonising IFN production. A number of studies demonstrate that SARS-CoV-2 NSP 6 and 13, as well as ORF6, significantly inhibit activation of the IFN-β gene [47,48]. Specifically, ORF6 blocks nuclear translocation of IFN regulatory factor-3 (IRF3), a critical regulator of the IFN-β gene, by interaction with its upstream nuclear import factors [47]. Similarly, NSP 6 and 13 bind to TANK binding kinase 1 (TBK1) to inhibit phosphorylation of IRF-3 and its subsequent activation [48] (Figure 1A, Table A1).

More recent variants of SARS-CoV-2 show evidence of evolution towards IFN evasion. The Alpha variant is associated with even lower activation of IRF-3 compared to the ancestral strain, as well as reduced activation of nuclear factor kappa B (NF-κB), which is also involved in IFN-β transcription [49,50]. Upregulation of ORF6 in the Alpha variant likely also contributes to its enhanced transmission and infectivity [51,52]. Interestingly, the recently emerged Omicron subvariants BA.2 and BA.4 harbour a D61L mutation in ORF6, which attenuates its capacity to interact with IRF-3 nuclear import factors, reducing IFN evasion [52]. This mutation is no longer present in the currently circulating BA.5 subvariant. Furthermore, a nine-nucleotide deletion in NSP6 has emerged independently in several variants including Alpha, Beta (B.1.351), Gamma, and Iota (B.1.526), as well as in B lineages B.1.620, B.1.1.318, B.1.525 which may enhance IFN antagonism [53–55].

The SARS-CoV-2 subgenome may also play a role in IFN suppression. We previously reported that the Alpha variant produces much higher levels of SARS-CoV-2 ORF9b subgenomic RNA (sgRNA), potentially as a result of a rare triple nucleotide mutation in the nucleocapsid protein (*N*), D3L, resulting in a novel transcription regulatory sequence (TRS) upstream of the ORF9b gene [56]. ORF9b is an accessory protein, which is non-essential for virus replication, but is involved in potent type-I IFN inhibition by interaction with mitochondrial TOM70 [57]. Additionally, two adjacent R203K/G204R substitutions in the *N* protein, another IFN antagonist, result in the introduction of a TRS-like sequence motif, generating a novel sgRNA transcript [58,59]. This novel sgRNA is also present in the Gamma (P.1) and Omicron (BA.1) variants, though its precise function is unclear [51] (Table A1).

IRF-3 suppression is a common evasive manoeuvre utilised by other RNA viruses [60]. Influenza A's non-structural protein NS1 is an important immunomodulatory protein that is capable of inhibiting IRF-3 phosphorylation and subsequent activation [61] (Table A1). The receptor binding domain (RBD) of NS1 binds to short 5′ triphosphate double stranded RNA (dsRNA) produced during infection, preventing recognition by PRRs such as RIG-I and suppressing production of IFN-β [61–65] (Figure 1A, Table A1). Adaptations in NS1 over time contribute to the emergence of highly virulent strains [61,66]. Isolates of the Avian influenza strain H5N1 from the 1997 pandemic harbour a P42S substitution in the RBD of NS1, which is shared with Swine influenza strain H1N1, and may enhance suppression of IRF-3 and NF-κB [66–68].

5. Repeated Exposure of Antigenic Sites to Host Adaptive Immune Memory Gives Rise to Immune Evasion

Memory B cells are responsible for the production of specific antibodies during recall of immunological memory via natural infection or vaccination. Mature memory B cells are generated in the lymphoid tissue germinal centre, where the B cell receptor undergoes immunoglobulin gene somatic hypermutation and selection in a process termed affinity maturation [69]. During affinity maturation, antibodies are selected mostly based on the strength of their affinity to the antigen, rather than their neutralizing capacity or ability to recognise closely related antigens from different viral strains [70]. This renders antibody-mediated immunity vulnerable to antigenic drift and new variants, which are sufficiently distinct from the original infectious insult, can rapidly emerge [3,70,71]. Viruses associated with acute infection are generally slow to escape adaptive immunity, because viral antigens are only transiently exposed to host immune machinery which requires prolonged exposure to influence the composition of the quasispecies [72,73]. As such, viral adaptations to the immune response tend to emerge only as the host population generates immunity [74] or during persistent infection [75].

Antibodies can prevent viral entry by steric hindrance of the surface glycoproteins, which are involved in cell entry (Figure 1B). Influenza's HA is a homotrimer responsible for viral fusion via association with the sialic acid residues on host-cell membranes. The globular head of the protein (HA1) features a RBD surrounded by antigenic sites, which are targeted by neutralizing antibodies and are highly diverse between viral strains [76,77] (Table A1). This facilitates the need for regular vaccine renewal to confer sufficient protection against drifted strains. Antigenic drift also underpins the need for SARS-CoV-2 repeat vaccinations to bolster adaptive immunity against variants whose *S* genes have diverged. The *S* protein accumulates adaptations as a result of a selective drive towards infectivity and transmissibility [78–80], but also towards immune evasion [26]. In the early days of the SARS-CoV-2 pandemic, a prominent substitution in S, D614G, quickly reached fixation in global sequence repositories, supplanting the wild-type *S* (Table A1). This variant enhances infectivity but does not significantly alter disease severity nor attenuate neutralization by host antibodies, suggesting that it is not antigenically distinct [27,79,80]—likely because amino acid 614 lies outside the highly antigenic RBD [24,81].

Adaptations within the RBD can aid immune evasion but must do so without compromising the capacity of the virus for cell entry. As such, they are a result of competing selective pressure [24,81]. One such mutation, N439K, has arisen independently in a number of distinct variants and confers enhanced receptor binding to human ACE2 while affording protection against neutralizing monoclonal and convalescent polyclonal antibodies [25] (Table A1). This variation was one of the most prevalent *S* mutations in the global sequence repositories during 2020–2021 and is indicative of potential viral escape pathways, which are associated with little to no cost to replicative fitness. Indeed, the RBD of the Delta variant (B.1.617) exhibited considerable diversity compared to previous strains such as Alpha and Kappa (B.1.617.1), resulting in higher affinity for ACE2 and evasion of antibody neutralization; characteristics that were bolstered in the later Omicron variant [82–84] whose 50 mutations include 30 *S* mutations, half of which occur in the RBD [85]. Efficacy of convalescent and vaccine-induced antibodies against the Omicron variant is markedly reduced, owing to four novel mutations in the RBD [86]. Omicron sublineages BA.1.1 and BA.2, which are even more resistant to antibody neutralization, each contain novel *S* mutations [87].

Many *S* adaptations have arisen independently in distinct SARS-CoV-2 variants. For example, in late 2020, the *S* N501Y mutation emerged independently in a number of geographically distinct regions in several unrelated strains [53,82,88–90]. This mutation is shown to increase affinity for human ACE2 and is a major determinant of infection and transmission [78] (Table A1). Concerningly, the N501Y mutation emerged contemporaneously with a constellation of other mutations, including a number of additional mutations in the *S* RBD domain, which enhance receptor binding and reduce efficacy of neutralizing antibodies [91–94].

A number of adaptations in the *S* gene occur in its N-terminal domain (NTD) [95,96]. While the NTD was originally thought to be non-immunogenic as a result of its extensive glycan shielding [97], recent studies have identified a subset of ultrapotent monoclonal antibodies, all of which target residues within the same stretch of amino acids in the NTD—termed the NTD antigenic supersite [92,95,98]. Variants Alpha, Beta, and Delta all exhibit mutations in this site [84,92,95].

6. The Cytotoxic T Cell Response Leads to HLA-Restricted Viral Adaptation

The cytotoxic T lymphocyte (CTL) immune response is associated with specific host human leucocyte antigen (HLA) alleles and represents a major immune pressure. HLA molecules are cell-surface proteins that present peptide fragments (epitopes) to CTLs to initiate death of the infected cell (Figure 1C). The genes that encode human HLA molecules are highly polymorphic, such that different hosts may not present the same viral targets to CTL. Thus, the HLA type of the host restricts the repertoire of viral peptides that are presented to the CTL.

HLA-restricted viral escape is a well-documented pathway for immune evasion of RNA viruses such as HIV and expression of certain HLA class I alleles is associated with lower viral load and other clinical outcomes [99,100]. During the acute phase, rapid elimination of vulnerable virus by the CTL-mediated antiviral response constitutes a selective pressure, enriching for mutations in CTL epitopes which abrogate CTL recognition and propagate over the course of chronic infection [101,102]. The Gag protein of HIV is known to generate immunodominant peptides towards which the CTL-mediated immune response is skewed [103] (Table A1). HLA-B*27 is a common allele in Caucasian populations (~8%) and is overrepresented in long-term HIV non-progressors [104]. CTL escape mutations within the HLAB*27-restricted KK10 epitope in Gag, which abrogate CTL recognition, are associated with progression to AIDS [104–106]. Similarly, the Hepatitis C virus (HCV) is often associated with persistent infection. During chronic HCV infection, mutations accumulate in viral T cell epitopes, which drastically attenuate T cell recognition in mutant peptides compared to the initial transmitted sequence [107]. Such mutations occur in known CTL epitopes and are associated with specific host HLA alleles [108]. For both HIV and HCV, HLA allele-associated viral adaptations have been identified across the viral genomes and account for a significant proportion of the overall diversity of these RNA viruses [109,110], and have been shown to accumulate in a host population over time [111] to become in some instances, the consensus sequence [110]. Importantly, the level of T cell-mediated viral adaptation in the autologous virus to the host's HLA repertoire is associated with clinical outcomes [112].

In cases where viral fitness is undermined by CTL escape, mutations often co-occur with separate compensatory mutations. The Gag KK10 mutations impair host DNA integration and compromise the integrity of the viral capsid, significantly reducing replicative efficiency; however, this deleterious effect is rescued by a compensatory mutation in the capsid, S173A [106,113,114]. Similarly, CTL escape variant R384G in the influenza nucleoprotein obscures two CTL epitopes, HLA-B*27-restricted NP383-391 and HLA-B*08-restricted NP380-388, completely abolishing T cell recognition [115,116] (Table A1). Although this mutation quickly reached fixation in influenza strains, it is associated with a detriment to viral fitness and frequently co-occurs with a number of compensatory mutations [117,118].

The extent to which CTL-mediated immunity exerts an evolutionary force on SARS-CoV-2 remains contentious. The SARS-CoV-2 ORF8 protein, which shares the least homology with other coronaviruses, is capable of down-regulating HLA class I molecules, impairing CTL-mediated immunity [119], a characteristic shared by HIV's Nef protein [120] (Figure 1C, Table A1). Nevertheless, a number of CTL epitopes derived from SARS-CoV-2 viral proteins have been identified [121]. Motozono et al. describe a 9-mer peptide in the RBD of the SARS-CoV-2 *S* protein, NF9, which is shown to be an immunodominant epitope presented by common HLA-A alleles [122]. The same study established that SARS-CoV-2 variants such as variant Epsilon (B.1.427/429) and B lineage B.1.1.298 can evade HLA-restricted CTL detection and show significantly enhanced infectivity and virulence. Furthermore, SARS-CoV-2 variants Alpha, Beta, Gamma and Delta all harbour mutations in four known HLA-A-restricted CTL epitopes in *S* and ORF1a, which significantly attenuate CTL activation [123]. However, it is difficult to determine whether these mutations are the result of CTL adaptation or other immune pressures acting on the *S* protein [124]. Many SARS-CoV-2 CTL epitopes occur in internal proteins, which are more conserved and are can consistently elicit a CTL response across several distinct variants [125,126]. CTL escape may not be as extensive for SARS-CoV-2 as observed for the more mutable RNA viruses such as HIV and HCV that are also associated with chronic infection [127].

7. Chronic Infection May Accelerate Viral Adaptation

The global SARS-CoV-2 pandemic can be characterised by the periodic emergence of highly divergent viral variants. Such punctuated evolutionary shifts, termed saltational evolution [128], are not characteristic of stepwise diversification expected from multiple host-host transmissions. Most SARS-CoV-2 infections are acute, and de novo intrahost viral diversity is relatively low [129]. The infection event constitutes a bottleneck, resulting in the transmission of a small number of founding viruses from the donor viral pool. This founding virus is usually a representative of the dominant viral strain from the infectious donor, which frequently corresponds to the dominant circulating strain at the time [75].

Mounting immune pressure over the course of persistent infection can induce viral adaptation—thus, intrahost viral diversity may be considered a function of infection time. Accordingly, the likelihood of a transmitted founder virus harbouring de novo variation increases as infection persists [75]. In cases of incomplete immunity experienced by subjects with primary or acquired immunodeficiencies, or undergoing immunosuppressive therapy, viral clearance is prevented as compared to immunocompetent subjects [130]. As a consequence, replicating virus can persist in the host for longer periods of time. The chronic infection hypothesis of SARS-CoV-2 viral divergence posits that highly divergent viral variants evolve in such subjects before transmitting into the general population [131]. Such mutational jumps are relatively understudied in the context of human viral disease. Nevertheless, several studies report cases of punctuated, extensive adaptation in viral isolates from cases of persistent infection [128,131–137].

A case study published by Avanzato et al. documents persistent infection of a septuagenarian female experiencing immune dysfunction as a result of acquired hypogammaglobulinemia [136]. The subject was admitted in early 2020 with SARS-CoV-2 infection which persisted for 105 days and was treated with two doses of convalescent plasma therapy. Genomic sequencing of viral isolates from day 70 of infection uncovered a 12-nucleotide deletion resulting in deletion of amino acids 141–144 in *S* in 100% of reads. This deletion falls within the NTD antigenic supersite [96,98,136] and overlaps a 3-nucleotide deletion (Δ143-145) observed in Omicron [137] (Table A1). A similar case study describes a septuagenarian male who was experiencing severe combined immunodeficiency and was admitted for persistent SARS-CoV-2 infection [134]. Treatment with convalescent plasma therapy yielded dramatic punctuated shifts in viral diversity towards escape mutants, including deletions in the NTD (Δ69/70), as well as a D796H substitution in the S2 subunit [134] (Table A1). These variations were less sensitive to antibody neutralization. Of note, Δ69/70 has been documented in several distinct and independent lineages and may be an example of evolutionary convergence [53,138,139].

A third study published by Cele et al. describes a subject experiencing persistent SARS-CoV-2 infection as a result of immune suppression associated with HIV infection. This study documents the de novo emergence of adaptations in the RBD, including adaptations which reduce efficacy of Delta neutralizing antibodies despite infection time predating the emergence of the Delta variant [137].

The overall mutational patterns observed in chronic SARS-CoV-2 infection often correspond to those observed in emerging variants of concern, however in many cases, such intrahost viral evolution is host-specific and divergent virions often do not carry key mutations observed in circulating variants. Harari et al. report that mutations associated with higher transmissibility, that are documented in circulating variants, are not detected in high numbers in chronically infected hosts [140]. The authors suggest that there may be insufficient pressure from the chronically infected host to drive evolution of more transmissible variants, rather intrahost viral adaptation is skewed in favour of replicative fitness that is bolstered by immune evasion [140].

8. Cross Protective Immunity Highlights Conserved Targets for Immunisation

While broad immunity is often thwarted by antigenic drift, repeated exposure to the same or similar viruses can provide valuable insight into those peptides that represent a more suitable target; peptides that are antigenic and functionally indispensable. Antibodies against these conserved peptides often impart a broad range of immunity against drifted strains. Such viral cross reactivity can inform the development of putative universal vaccines. The HA protein of the influenza virus is its most highly variable protein and represents a moving target for influenza immunisation, as such, prospective universal influenza vaccines should also target more conserved regions [141]. The membrane proximal stalk domain of HA, which is involved in several structural rearrangements to facilitate internalisation of the influenza virus into the host cell, is functionally indispensable and mutations in this protein are usually not tolerated [142,143]. This region is therefore relatively resistant to antigenic drift, representing an attractive target for universal influenza vaccine design [144]. In support of this approach, anti-stalk antibodies can provide protection against several distinct influenza strains and accumulate after repeated infection [13,14,143].

The emerging SARS-CoV-2 Omicron variant and its sublineages harbour adaptations in previously conserved RBD residues, which aid in evasion even of cross-protective antibodies with broad neutralization capacity against other sarbecoviruses [145–147]. As such, determining a suitable target for vaccination that can protect against emerging variants is of significant interest. Pre-existing immunity for SARS-CoV-2 has been described in human hosts naïve to SARS-CoV-2 [148]. Convalescent SARS-CoV subjects retain robust cross-reactive memory CD4+ T cells against the SARS-CoV-2 *N* protein up to 11 years after infection [149,150]. *N* is the most abundant protein in the SARS-CoV-2 proteome and maintains a high degree of homology between a number of beta coronaviruses [149]. Furthermore, *N* contains an immunodominant T cell epitope recognised by SARS-CoV- and MERS-CoV-infected HLA-DR2 and -DR3 transgenic mice, and provides protection against bat CoVs, which may be an attractive target for pan-coronavirus heterologous immunisation. Interestingly, when comparing CD4+ T cell reactivity in SARS-CoV-2 naïve individuals with historic exposure to the common cold coronaviruses, cross-reactivity against a number of SARS-CoV-2 specific peptides was observed, including NSP14, NSP4 and NSP6, but only marginally against *N* [148].

9. Conclusions

RNA viruses' considerable capacity for adaptation represents the ever-present threat of pandemic viral outbreak of which SARS-CoV-2 is an emphatic reminder. The emergence of a pandemic RNA virus is followed by an evolutionary arms race wherein host immunity must mount a successful defence faster than the virus can adapt. It is critical to understand the vulnerabilities of the RNA virus, which are inevitably influenced by host immunity, in order to mitigate its impact on the global community and inform comprehensive vaccine coverage and pre-empt its evolutionary trajectory.

Adaptation to the innate IFN response represents a critical turning point for an infecting virus. A number of RNA viruses have evolved the capacity to down regulate IFN regulatory elements. Emerging SARS-CoV-2 variants have a similar effect on expression of IFN-β through interaction with IRF-3 as do seasonal influenza viruses, effectively sequestering it and preventing transcription of its downstream genes. This capacity for IFN evasion is renewed with the emergence of adaptations in new variants and highlights the importance of IFN-β regulation in control of infection.

Adaptation to humoral immune memory represents a delicate balancing act wherein the viral surface proteins are frequently the subject of immense selective pressure. Influenza's globular HA domain is its most antigenic asset and can acquire mutation at little-to-no cost to replicative fitness. As such, it can quickly acquire antibody escape characteristics. Similarly, mutations in a SARS-CoV-2 antigenic supersite are the result of considerable immune pressure in a region which is highly exposed and dispensable to function, resulting in escape mutations in this region, which rise in frequency as new variants emerge. Such adaptations can also emerge de novo during persistent infection. This phenomenon is popularly cited as the evolutionary mechanism by which highly divergent SARS-CoV-2 variants first emerged in the host population.

CTL immunity represents a host-specific selective force on many RNA viruses. Such HLA-restricted viral escape is well documented in the context of HIV and HCV but remains controversial in the case of SARS-CoV-2. While a number of variants harbour mutations in known HLA-A restricted CTL epitopes, other epitopes lie within relatively conserved proteins and CTL escape may be limited. However, the importance of CTL immune responses against SARS-CoV-2 is highlighted by the downregulation of HLA class I by the ORF8 protein.

As the viral genomic landscape changes, it becomes clear precisely which genomic regions cannot tolerate extensive mutation. Such viral proteins as influenza's membrane proximal Hemagglutinin stalk or the *N* protein of SARS-CoV and SARS-CoV-2 are suitable candidates for universal immunisation because they are antigenic but functionally indispensable and demonstrate a resilience to evolutionary change that can be relied upon with some degree of certainty.

As the global community begins to emerge from the SARS-CoV-2 pandemic, continued monitoring of emerging variants is critical to better understand the weak-points of the RNA virus and pre-empt future viral outbreaks.

Author Contributions: Writing—original draft preparation, J.W.; writing—review and editing, S.G.; supervision, S.G. All authors have read and agreed to the published version of the manuscript.

Funding: This research received no external funding.

Conflicts of Interest: The authors declare no conflict of interest.

Appendix A

Table A1. Summary of viral proteins involved in defence against host immunity for various RNA viruses, some of which show evidence of adaptation towards viral escape (examples provided).

SARS-CoV-2			
Coding Region	**Role in Viral Defence**	**Adaptations**	**Consequence**
NSP6/13	IFN antagonism (TBK1 inhibition) [48]	9 nucleotide deletion [53]	
ORF6	IFN antagonism (blocking of IRF-3 nuclear import factors) [47]	D61L [52]	IFN susceptibility
Nucleocapsid	IFN antagonism	R203K/G204R [58]	Novel subgenomic transcript
		D3L [56]	Upregulation of ORF9b

Table A1. *Cont.*

SARS-CoV-2			
Coding Region	**Role in Viral Defence**	**Adaptations**	**Consequence**
Spike	Surface antigen	D614G [27]	Increased infectivity
		N439K [25]	Increased infectivity, antibody evasion
		N501Y [53]	Increased infectivity, antibody evasion
		NTD Δ143-145 [136]	Antibody evasion
		NTD Δ141-144 [137]	Antibody evasion
		NTD Δ69/70 [134]	Antibody evasion
		S2 D796H [134]	Antibody evasion
ORF8	HLA-I suppression		CTL escape
Influenza			
NS1	IFN antagonism (sequestration of viral dsRNA) [61]	P42S [67]	IFN evasion
Nucleoprotein	Encapsulates viral genome	R384G [115]	HLA-*27/08 restricted CTL escape
HA	Surface antigen		Antibody evasion
HIV			
Gag	Immunodominant CTL target	HLAB*27 restricted KK10 epitope mutations [106]	CTL escape
Capsid	Encapsulates viral genome	S173A [106]	Compensatory to Gag KK10 escape mutations
Nef	HLA-I suppression		CTL escape

References

1. Singh, D.; Yi, S.V. On the origin and evolution of SARS-CoV-2. *Exp. Mol. Med.* **2021**, *53*, 537–547. [CrossRef] [PubMed]
2. Sanjuán, R.; Domingo-Calap, P. Mechanisms of viral mutation. *Cell. Mol. Life Sci.* **2016**, *73*, 4433–4448. [CrossRef] [PubMed]
3. Duffy, S.; Shackelton, L.A.; Holmes, E.C. Rates of evolutionary change in viruses: Patterns and determinants. *Nat. Rev. Genet.* **2008**, *9*, 267–276. [CrossRef] [PubMed]
4. Sanjuán, R.; Nebot, M.R.; Chirico, N.; Mansky, L.M.; Belshaw, R. Viral mutation rates. *J. Virol.* **2010**, *84*, 9733–9748. [CrossRef] [PubMed]
5. Smith, E.C. The not-so-infinite malleability of RNA viruses: Viral and cellular determinants of RNA virus mutation rates. *PLOS Pathog.* **2017**, *13*, e1006254. [CrossRef] [PubMed]
6. Steinhauer, D.A.; Domingo, E.; Holland, J.J. Lack of evidence for proofreading mechanisms associated with an RNA virus polymerase. *Gene* **1992**, *122*, 281–288. [CrossRef] [PubMed]
7. Nelemans, T.; Kikkert, M. Viral Innate Immune Evasion and the Pathogenesis of Emerging RNA Virus Infections. *Viruses* **2019**, *11*, 961. [CrossRef] [PubMed]
8. Fornés, J.; Tomás Lázaro, J.; Alarcón, T.; Elena, S.F.; Sardanyés, J. Viral replication modes in single-peak fitness landscapes: A dynamical systems analysis. *J. Theor. Biol.* **2019**, *460*, 170–183. [CrossRef]
9. Perelson, A.S. Modelling viral and immune system dynamics. *Nat. Rev. Immunol.* **2002**, *2*, 28–36. [CrossRef]
10. Lauring, A.S.; Andino, R. Quasispecies Theory and the Behavior of RNA Viruses. *PLoS Pathog.* **2010**, *6*, e1001005. [CrossRef]
11. Agius, J.E.; Johnson-Mackinnon, J.C.; Fong, W.; Gall, M.; Lam, C.; Basile, K.; Kok, J.; Arnott, A.; Sintchenko, V.; Rockett, R.J. SARS-CoV-2 Within-Host and in vitro Genomic Variability and Sub-Genomic RNA Levels Indicate Differences in Viral Expression Between Clinical Cohorts and in vitro Culture. *Front. Microbiol.* **2022**, *13*, 824217. [CrossRef] [PubMed]
12. Barrett, R.D.; Schluter, D. Adaptation from standing genetic variation. *Trends Ecol. Evol.* **2008**, *23*, 38–44. [CrossRef] [PubMed]
13. Carrat, F.; Flahault, A. Influenza vaccine: The challenge of antigenic drift. *Vaccine* **2007**, *25*, 6852–6862. [CrossRef] [PubMed]

14. DiazGranados, C.A.; Denis, M.; Plotkin, S. Seasonal influenza vaccine efficacy and its determinants in children and non-elderly adults: A systematic review with meta-analyses of controlled trials. *Vaccine* **2012**, *31*, 49–57. [CrossRef] [PubMed]
15. Gorbalenya, A.E.; Enjuanes, L.; Ziebuhr, J.; Snijder, E.J. Nidovirales: Evolving the largest RNA virus genome. *Virus Res.* **2006**, *117*, 17–37. [CrossRef]
16. Robson, F.; Khan, K.S.; Le, T.K.; Paris, C.; Demirbag, S.; Barfuss, P.; Rocchi, P.; Ng, W.L. Coronavirus RNA Proofreading: Molecular Basis and Therapeutic Targeting. *Mol. Cell* **2020**, *79*, 710–727. [CrossRef]
17. Minskaia, E.; Hertzig, T.; Gorbalenya, A.E.; Campanacci, V.; Cambillau, C.; Canard, B.; Ziebuhr, J. Discovery of an RNA virus 3′->5′ exoribonuclease that is critically involved in coronavirus RNA synthesis. *Proc. Natl. Acad. Sci. USA* **2006**, *103*, 5108–5113. [CrossRef]
18. Romano, M.; Ruggiero, A.; Squeglia, F.; Maga, G.; Berisio, R. A Structural View of SARS-CoV-2 RNA Replication Machinery: RNA Synthesis, Proofreading and Final Capping. *Cells* **2020**, *9*, 1267. [CrossRef]
19. Knoops, K.; Kikkert, M.; Worm, S.H.; Zevenhoven-Dobbe, J.C.; van der Meer, Y.; Koster, A.J.; Mommaas, A.M.; Snijder, E.J. SARS-coronavirus replication is supported by a reticulovesicular network of modified endoplasmic reticulum. *PLoS Biol.* **2008**, *6*, e226. [CrossRef]
20. Yan, L.; Yang, Y.; Li, M.; Zhang, Y.; Zheng, L.; Ge, J.; Huang, Y.C.; Liu, Z.; Wang, T.; Gao, S.; et al. Coupling of N7-methyltransferase and 3′-5′ exoribonuclease with SARS-CoV-2 polymerase reveals mechanisms for capping and proofreading. *Cell* **2021**, *184*, 3474–3485.e11. [CrossRef]
21. Liu, C.; Shi, W.; Becker, S.T.; Schatz, D.G.; Liu, B.; Yang, Y. Structural basis of mismatch recognition by a SARS-CoV-2 proofreading enzyme. *Science* **2021**, *373*, 1142–1146. [CrossRef] [PubMed]
22. Yang, D.; Leibowitz, J.L. The structure and functions of coronavirus genomic 3′ and 5′ ends. *Virus Res.* **2015**, *206*, 120–133. [CrossRef] [PubMed]
23. Chan, A.P.; Choi, Y.; Schork, N.J. Conserved Genomic Terminals of SARS-CoV-2 as Coevolving Functional Elements and Potential Therapeutic Targets. *msphere* **2020**, *5*, e00754-20. [CrossRef] [PubMed]
24. Piccoli, L.; Park, Y.-J.; Tortorici, M.A.; Czudnochowski, N.; Walls, A.C.; Beltramello, M.; Silacci-Fregni, C.; Pinto, D.; Rosen, L.E.; Bowen, J.E.; et al. Mapping Neutralizing and Immunodominant Sites on the SARS-CoV-2 Spike Receptor-Binding Domain by Structure-Guided High-Resolution Serology. *Cell* **2020**, *183*, 1024–1042.e21. [CrossRef]
25. Thomson, E.C.; Rosen, L.E.; Shepherd, J.G.; Spreafico, R.; da Silva Filipe, A.; Wojcechowskyj, J.A.; Davis, C.; Piccoli, L.; Pascall, D.J.; Dillen, J.; et al. Circulating SARS-CoV-2 spike N439K variants maintain fitness while evading antibody-mediated immunity. *Cell* **2021**, *184*, 1171–1187.e20. [CrossRef]
26. Mei, Y.; Jiaqi, L.; Zhaoxia, H.; Heng, L.; Yueming, W.; Xiaoli, W.; Sisi, K.; Xing, H.; Changwen, W.; Tong, L.; et al. Structural Basis of a Human Neutralizing Antibody Specific to the SARS-CoV-2 Spike Protein Receptor-Binding Domain. *Microbiol. Spectr.* **2021**, *9*, e01352-21.
27. Yurkovetskiy, L.; Wang, X.; Pascal, K.E.; Tomkins-Tinch, C.; Nyalile, T.P.; Wang, Y.; Baum, A.; Diehl, W.E.; Dauphin, A.; Carbone, C.; et al. Structural and Functional Analysis of the D614G SARS-CoV-2 Spike Protein Variant. *Cell* **2020**, *183*, 739–751.e8. [CrossRef]
28. Hou, Y.; Zhao, S.; Liu, Q.; Zhang, X.; Sha, T.; Su, Y.; Zhao, W.; Bao, Y.; Xue, Y.; Chen, H. Ongoing positive selection drives the evolution of SARS-CoV-2 genomes. *Genom. Proteom. Bioinform.* **2022**. [CrossRef]
29. Yi, K.; Kim, S.Y.; Bleazard, T.; Kim, T.; Youk, J.; Ju, Y.S. Mutational spectrum of SARS-CoV-2 during the global pandemic. *Exp. Mol. Med.* **2021**, *53*, 1229–1237. [CrossRef]
30. Dorp, C.H.V.; Goldberg, E.E.; Hengartner, N.; Ke, R.; Romero-Severson, E.O. Estimating the strength of selection for new SARS-CoV-2 variants. *Nat. Commun.* **2021**, *12*, 7239. [CrossRef]
31. Wada, Y.; Wada, K.; Iwasaki, Y.; Kanaya, S.; Ikemura, T. Directional and reoccurring sequence change in zoonotic RNA virus genomes visualized by time-series word count. *Sci. Rep.* **2016**, *6*, 36197. [CrossRef] [PubMed]
32. van Dorp, C.; Goldberg, E.; Ke, R.; Hengartner, N.; Romero-Severson, E. Global estimates of the fitness advantage of SARS-CoV-2 variant Omicron. *Virus Evol.* **2022**, *8*, veac089. [CrossRef] [PubMed]
33. Jeffares, D.C.; Tomiczek, B.; Sojo, V.; dos Reis, M. A beginners guide to estimating the non-synonymous to synonymous rate ratio of all protein-coding genes in a genome. *Methods Mol. Biol.* **2015**, *1201*, 65–90. [PubMed]
34. Kosakovsky Pond, S.L.; Frost, S.D.W. Not So Different After All: A Comparison of Methods for Detecting Amino Acid Sites Under Selection. *Mol. Biol. Evol.* **2005**, *22*, 1208–1222. [CrossRef]
35. Fung, S.Y.; Siu, K.L.; Lin, H.; Chan, C.P.; Yeung, M.L.; Jin, D.Y. SARS-CoV-2 NSP13 helicase suppresses interferon signaling by perturbing JAK1 phosphorylation of STAT1. *Cell Biosci.* **2022**, *12*, 36. [CrossRef]
36. Faizan, M.I.; Chaudhuri, R.; Sagar, S.; Albogami, S.; Chaudhary, N.; Azmi, I.; Akhtar, A.; Ali, S.M.; Kumar, R.; Iqbal, J.; et al. NSP4 and ORF9b of SARS-CoV-2 Induce Pro-Inflammatory Mitochondrial DNA Release in Inner Membrane-Derived Vesicles. *Cells* **2022**, *11*, 2969. [CrossRef]
37. Schneider, W.M.; Chevillotte, M.D.; Rice, C.M. Interferon-Stimulated Genes: A Complex Web of Host Defenses. *Annu. Rev. Immunol.* **2014**, *32*, 513–545. [CrossRef]
38. Galani, I.E.; Triantafyllia, V.; Eleminiadou, E.E.; Koltsida, O.; Stavropoulos, A.; Manioudaki, M.; Thanos, D.; Doyle, S.E.; Kotenko, S.V.; Thanopoulou, K.; et al. Interferon-λ Mediates Non-redundant Front-Line Antiviral Protection against Influenza Virus Infection without Compromising Host Fitness. *Immunity* **2017**, *46*, 875–890.e6. [CrossRef]

39. Schoenborn, J.R.; Wilson, C.B. Regulation of Interferon-γ during Innate and Adaptive Immune Responses. In *Advances in Immunology*; Academic Press: Cambridge, MA, USA, 2007; Volume 96, pp. 41–101.
40. González-Navajas, J.M.; Lee, J.; David, M.; Raz, E. Immunomodulatory functions of type I interferons. *Nat. Rev. Immunol.* **2012**, *12*, 125–135. [CrossRef]
41. Iyer, S.S.; Bibollet-Ruche, F.; Sherrill-Mix, S.; Learn, G.H.; Plenderleith, L.; Smith, A.G.; Barbian, H.J.; Russell, R.M.; Gondim, M.V.; Bahari, C.Y.; et al. Resistance to type 1 interferons is a major determinant of HIV-1 transmission fitness. *Proc. Natl. Acad. Sci. USA* **2017**, *114*, E590–E599. [CrossRef]
42. Veazey, R.S.; Pilch-Cooper, H.A.; Hope, T.J.; Alter, G.; Carias, A.M.; Sips, M.; Wang, X.; Rodriguez, B.; Sieg, S.F.; Reich, A.; et al. Prevention of SHIV transmission by topical IFN-β treatment. *Mucosal Immunol.* **2016**, *9*, 1528–1536. [CrossRef] [PubMed]
43. Fenton-May, A.E.; Dibben, O.; Emmerich, T.; Ding, H.; Pfafferott, K.; Aasa-Chapman, M.M.; Pellegrino, P.; Williams, I.; Cohen, M.S.; Gao, F.; et al. Relative resistance of HIV-1 founder viruses to control by interferon-alpha. *Retrovirology* **2013**, *10*, 146. [CrossRef] [PubMed]
44. Tahmina, S.; João, I.M.; Akatsuki, S.; Hirotaka, O.; Kyotaro, N.; Romy, C.; Emi, E.N.; Yasumasa, I.; Masahiro, Y.; Thomas, J.H.; et al. Multiple Pathways to Avoid Beta Interferon Sensitivity of HIV-1 by Mutations in Capsid. *J. Virol.* **2019**, *93*, e00986-19.
45. Zuliani-Alvarez, L.; Govasli, M.L.; Rasaiyaah, J.; Monit, C.; Perry, S.O.; Sumner, R.P.; McAlpine-Scott, S.; Dickson, C.; Rifat Faysal, K.M.; Hilditch, L.; et al. Evasion of cGAS and TRIM5 defines pandemic HIV. *Nat. Microbiol.* **2022**, *7*, 1762–1776. [CrossRef] [PubMed]
46. Liu, G.; Gack, M.U. Insights into pandemic respiratory viruses: Manipulation of the antiviral interferon response by SARS-CoV-2 and influenza A virus. *Curr. Opin. Immunol.* **2022**, *78*, 102252. [CrossRef] [PubMed]
47. Lei, X.; Dong, X.; Ma, R.; Wang, W.; Xiao, X.; Tian, Z.; Wang, C.; Wang, Y.; Li, L.; Ren, L.; et al. Activation and evasion of type I interferon responses by SARS-CoV-2. *Nat. Commun.* **2020**, *11*, 3810. [CrossRef] [PubMed]
48. Xia, H.; Cao, Z.; Xie, X.; Zhang, X.; Chen, J.Y.-C.; Wang, H.; Menachery, V.D.; Rajsbaum, R.; Shi, P.-Y. Evasion of Type I Interferon by SARS-CoV-2. *Cell Rep.* **2020**, *33*, 108234. [CrossRef] [PubMed]
49. Laine, L.; Skön, M.; Väisänen, E.; Julkunen, I.; Österlund, P. SARS-CoV-2 variants Alpha, Beta, Delta and Omicron show a slower host cell interferon response compared to an early pandemic variant. *Front. Immunol.* **2022**, *13*, 1016108. [CrossRef]
50. Randall, R.E.; Goodbourn, S. Interferons and viruses: An interplay between induction, signalling, antiviral responses and virus countermeasures. *J. Gen. Virol.* **2008**, *89 Pt 1*, 1–47.
51. Thorne, L.G.; Bouhaddou, M.; Reuschl, A.K.; Zuliani-Alvarez, L.; Polacco, B.; Pelin, A.; Batra, J.; Whelan, M.V.X.; Hosmillo, M.; Fossati, A.; et al. Evolution of enhanced innate immune evasion by SARS-CoV-2. *Nature* **2022**, *602*, 487–495. [CrossRef]
52. Kehrer, T.; Cupic, A.; Ye, C.; Yildiz, S.; Bouhhadou, M.; Crossland, N.A.; Barrall, E.; Cohen, P.; Tseng, A.; Çağatay, T.; et al. Impact of SARS-CoV-2 ORF6 and its variant polymorphisms on host responses and viral pathogenesis. *bioRxiv* **2022**. [CrossRef]
53. Martin, D.P.; Weaver, S.; Tegally, H.; San, J.E.; Shank, S.D.; Wilkinson, E.; Lucaci, A.G.; Giandhari, J.; Naidoo, S.; Pillay, Y.; et al. The emergence and ongoing convergent evolution of the SARS-CoV-2 N501Y lineages. *Cell* **2021**, *184*, 5189–5200.e7. [CrossRef] [PubMed]
54. Dudas, G.; Hong, S.L.; Potter, B.I.; Calvignac-Spencer, S.; Niatou-Singa, F.S.; Tombolomako, T.B.; Fuh-Neba, T.; Vickos, U.; Ulrich, M.; Leendertz, F.H.; et al. Emergence and spread of SARS-CoV-2 lineage B.1.620 with variant of concern-like mutations and deletions. *Nat. Commun.* **2021**, *12*, 5769. [CrossRef] [PubMed]
55. Annavajhala, M.K.; Mohri, H.; Wang, P.; Nair, M.; Zucker, J.E.; Sheng, Z.; Gomez-Simmonds, A.; Kelley, A.L.; Tagliavia, M.; Huang, Y.; et al. Emergence and Expansion of the SARS-CoV-2 Variant B.1.526 Identified in New York. *medRxiv* **2021**. [CrossRef]
56. Parker, M.D.; Stewart, H.; Shehata, O.M.; Lindsey, B.B.; Shah, D.R.; Hsu, S.; Keeley, A.J.; Partridge, D.G.; Leary, S.; Cope, A.; et al. Altered subgenomic RNA abundance provides unique insight into SARS-CoV-2 B.1.1.7/Alpha variant infections. *Commun. Biol.* **2022**, *5*, 666. [CrossRef] [PubMed]
57. Jiang, H.-w.; Zhang, H.-n.; Meng, Q.-f.; Xie, J.; Li, Y.; Chen, H.; Zheng, Y.-x.; Wang, X.-n.; Qi, H.; Zhang, J.; et al. SARS-CoV-2 Orf9b suppresses type I interferon responses by targeting TOM70. *Cell. Mol. Immunol.* **2020**, *17*, 998–1000. [CrossRef]
58. Leary, S.; Gaudieri, S.; Parker, M.D.; Chopra, A.; James, I.; Pakala, S.; Alves, E.; John, M.; Lindsey, B.B.; Keeley, A.J.; et al. Generation of a Novel SARS-CoV-2 Sub-genomic RNA Due to the R203K/G204R Variant in Nucleocapsid: Homologous Recombination has Potential to Change SARS-CoV-2 at Both Protein and RNA Level. *Pathog. Immunol.* **2021**, *6*, 27–49. [CrossRef]
59. Yong, H.; Wei, L.; Ting, G.; Yan, C.; Yanwen, J.; Ping, L.; Qingjun, M.; Xuan, L.; Cheng, C. The Severe Acute Respiratory Syndrome Coronavirus Nucleocapsid Inhibits Type I Interferon Production by Interfering with TRIM25-Mediated RIG-I Ubiquitination. *J. Virol.* **2017**, *91*, e02143-16.
60. Geiss, G.K.; Salvatore, M.; Tumpey, T.M.; Carter, V.S.; Wang, X.; Basler, C.F.; Taubenberger, J.K.; Bumgarner, R.E.; Palese, P.; Katze, M.G.; et al. Cellular transcriptional profiling in influenza A virus-infected lung epithelial cells: The role of the nonstructural NS1 protein in the evasion of the host innate defense and its potential contribution to pandemic influenza. *Proc. Natl. Acad. Sci. USA* **2002**, *99*, 10736–10741. [CrossRef]
61. Ji, Z.X.; Wang, X.Q.; Liu, X.F. NS1: A Key Protein in the "Game" Between Influenza A Virus and Host in Innate Immunity. *Front. Cell. Infect. Microbiol.* **2021**, *11*, 670177. [CrossRef]
62. Mibayashi, M.; Martínez-Sobrido, L.; Loo, Y.M.; Cárdenas, W.B.; Gale, M., Jr.; García-Sastre, A. Inhibition of retinoic acid-inducible gene I-mediated induction of beta interferon by the NS1 protein of influenza A virus. *J. Virol.* **2007**, *81*, 514–524. [CrossRef] [PubMed]

63. Talon, J.; Horvath, C.M.; Polley, R.; Basler, C.F.; Muster, T.; Palese, P.; García-Sastre, A. Activation of interferon regulatory factor 3 is inhibited by the influenza A virus NS1 protein. *J. Virol.* **2000**, *74*, 7989–7996. [CrossRef] [PubMed]
64. Hale, B.G.; Randall, R.E.; Ortín, J.; Jackson, D. The multifunctional NS1 protein of influenza A viruses. *J. Gen. Virol.* **2008**, *89*, 2359–2376. [CrossRef] [PubMed]
65. García-Sastre, A.; Egorov, A.; Matassov, D.; Brandt, S.; Levy, D.E.; Durbin, J.E.; Palese, P.; Muster, T. Influenza A virus lacking the NS1 gene replicates in interferon-deficient systems. *Virology* **1998**, *252*, 324–330. [CrossRef]
66. Neumann, G. H5N1 influenza virulence, pathogenicity and transmissibility: What do we know? *Future Virol.* **2015**, *10*, 971–980. [CrossRef]
67. Peirong, J.; Guobin, T.; Yanbing, L.; Guohua, D.; Yongping, J.; Chang, L.; Weilong, L.; Zhigao, B.; Yoshihiro, K.; Hualan, C. A Single-Amino-Acid Substitution in the NS1 Protein Changes the Pathogenicity of H5N1 Avian Influenza Viruses in Mice. *J. Virol.* **2008**, *82*, 1146–1154.
68. Cheng, J.; Zhang, C.; Tao, J.; Li, B.; Shi, Y.; Liu, H. Effects of the S42 residue of the H1N1 swine influenza virus NS1 protein on interferon responses and virus replication. *Virol. J.* **2018**, *15*, 57. [CrossRef]
69. Seifert, M.; Küppers, R. Human memory B cells. *Leukemia* **2016**, *30*, 2283–2292. [CrossRef]
70. Victora, G.D.; Wilson, P.C. Germinal Center Selection and the Antibody Response to Influenza. *Cell* **2015**, *163*, 545–548. [CrossRef]
71. Antara, D. Molecular evolution of hemagglutinin gene of Influenza A virus. *FBS* **2018**, *10*, 101–118.
72. Kitphati, R.; Pooruk, P.; Lerdsamran, H.; Poosuwan, S.; Louisirirotchanakul, S.; Auewarakul, P.; Chokphaibulkit, K.; Noisumdaeng, P.; Sawanpanyalert, P.; Puthavathana, P. Kinetics and longevity of antibody response to influenza A H5N1 virus infection in humans. *Clin. Vaccine Immunol.* **2009**, *16*, 978–981. [CrossRef] [PubMed]
73. Ochsenbein, A.F.; Pinschewer, D.D.; Sierro, S.; Horvath, E.; Hengartner, H.; Zinkernagel, R.M. Protective long-term antibody memory by antigen-driven and T help-dependent differentiation of long-lived memory B cells to short-lived plasma cells independent of secondary lymphoid organs. *Proc. Natl. Acad. Sci. USA* **2000**, *97*, 13263–13268. [CrossRef] [PubMed]
74. Yewdell, J.W. Antigenic drift: Understanding COVID-19. *Immunity* **2021**, *54*, 2681–2687. [CrossRef] [PubMed]
75. Lythgoe, K.A.; Hall, M.; Ferretti, L.; de Cesare, M.; MacIntyre-Cockett, G.; Trebes, A.; Andersson, M.; Otecko, N.; Wise, E.L.; Moore, N.; et al. SARS-CoV-2 within-host diversity and transmission. *Science* **2021**, *372*, eabg0821. [CrossRef] [PubMed]
76. Caton, A.J.; Brownlee, G.G.; Yewdell, J.W.; Gerhard, W. The antigenic structure of the influenza virus A/PR/8/34 hemagglutinin (H1 subtype). *Cell* **1982**, *31 Pt 1*, 417–427. [CrossRef] [PubMed]
77. Chambers, B.S.; Li, Y.; Hodinka, R.L.; Hensley, S.E. Recent H3N2 influenza virus clinical isolates rapidly acquire hemagglutinin or neuraminidase mutations when propagated for antigenic analyses. *J. Virol.* **2014**, *88*, 10986–10989. [CrossRef] [PubMed]
78. Liu, Y.; Liu, J.; Plante, K.S.; Plante, J.A.; Xie, X.; Zhang, X.; Ku, Z.; An, Z.; Scharton, D.; Schindewolf, C.; et al. The N501Y spike substitution enhances SARS-CoV-2 infection and transmission. *Nature* **2022**, *602*, 294–299. [CrossRef]
79. Plante, J.A.; Liu, Y.; Liu, J.; Xia, H.; Johnson, B.A.; Lokugamage, K.G.; Zhang, X.; Muruato, A.E.; Zou, J.; Fontes-Garfias, C.R.; et al. Spike mutation D614G alters SARS-CoV-2 fitness. *Nature* **2021**, *592*, 116–121. [CrossRef]
80. Korber, B.; Fischer, W.M.; Gnanakaran, S.; Yoon, H.; Theiler, J.; Abfalterer, W.; Hengartner, N.; Giorgi, E.E.; Bhattacharya, T.; Foley, B.; et al. Tracking Changes in SARS-CoV-2 Spike: Evidence that D614G Increases Infectivity of the COVID-19 Virus. *Cell* **2020**, *182*, 812–827.e19. [CrossRef]
81. Robbiani, D.F.; Gaebler, C.; Muecksch, F.; Lorenzi, J.C.C.; Wang, Z.; Cho, A.; Agudelo, M.; Barnes, C.O.; Gazumyan, A.; Finkin, S.; et al. Convergent antibody responses to SARS-CoV-2 in convalescent individuals. *Nature* **2020**, *584*, 437–442. [CrossRef]
82. Kumar, S.; Thambiraja, T.S.; Karuppanan, K.; Subramaniam, G. Omicron and Delta variant of SARS-CoV-2: A comparative computational study of spike protein. *J. Med. Virol.* **2022**, *94*, 1641–1649. [CrossRef] [PubMed]
83. Mlcochova, P.; Kemp, S.A.; Dhar, M.S.; Papa, G.; Meng, B.; Ferreira, I.A.T.M.; Datir, R.; Collier, D.A.; Albecka, A.; Singh, S.; et al. SARS-CoV-2 B.1.617.2 Delta variant replication and immune evasion. *Nature* **2021**, *599*, 114–119. [CrossRef] [PubMed]
84. Planas, D.; Veyer, D.; Baidaliuk, A.; Staropoli, I.; Guivel-Benhassine, F.; Rajah, M.M.; Planchais, C.; Porrot, F.; Robillard, N.; Puech, J.; et al. Reduced sensitivity of SARS-CoV-2 variant Delta to antibody neutralization. *Nature* **2021**, *596*, 276–280. [CrossRef] [PubMed]
85. Thakur, V.; Ratho, R.K. OMICRON (B.1.1.529): A new SARS-CoV-2 variant of concern mounting worldwide fear. *J. Med. Virol.* **2022**, *94*, 1821–1824. [CrossRef]
86. Liu, L.; Iketani, S.; Guo, Y.; Chan, J.F.W.; Wang, M.; Liu, L.; Luo, Y.; Chu, H.; Huang, Y.; Nair, M.S.; et al. Striking antibody evasion manifested by the Omicron variant of SARS-CoV-2. *Nature* **2022**, *602*, 676–681. [CrossRef] [PubMed]
87. Iketani, S.; Liu, L.; Guo, Y.; Liu, L.; Chan, J.F.W.; Huang, Y.; Wang, M.; Luo, Y.; Yu, J.; Chu, H.; et al. Antibody evasion properties of SARS-CoV-2 Omicron sublineages. *Nature* **2022**, *604*, 553–556. [CrossRef] [PubMed]
88. Galloway, S.E.; Paul, P.; MacCannell, D.R.; Johansson, M.A.; Brooks, J.T.; MacNeil, A.; Slayton, R.B.; Tong, S.; Silk, B.J.; Armstrong, G.L.; et al. Emergence of SARS-CoV-2 B.1.1.7 Lineage-United States, December 29, 2020–January 12, 2021. *MMWR Morb. Mortal. Wkly. Rep.* **2021**, *70*, 95–99. [CrossRef]
89. Claro, I.M.; da Silva Sales, F.C.; Ramundo, M.S.; Candido, D.S.; Silva, C.A.M.; de Jesus, J.G.; Manuli, E.R.; de Oliveira, C.M.; Scarpelli, L.; Campana, G.; et al. Local Transmission of SARS-CoV-2 Lineage B.1.1.7, Brazil, December 2020. *Emerg. Infect. Dis.* **2021**, *27*, 970–972. [CrossRef]
90. Kupferschmidt, K. Fast-spreading U.K. virus variant raises alarms. *Science* **2021**, *371*, 9–10. [CrossRef]

91. Zahradník, J.; Marciano, S.; Shemesh, M.; Zoler, E.; Chiaravalli, J.; Meyer, B.; Rudich, Y.; Dym, O.; Elad, N.; Schreiber, G. SARS-CoV-2 RBD in vitro evolution follows contagious mutation spread, yet generates an able infection inhibitor. *bioRxiv* **2021**. [CrossRef]
92. Wang, P.; Nair, M.S.; Liu, L.; Iketani, S.; Luo, Y.; Guo, Y.; Wang, M.; Yu, J.; Zhang, B.; Kwong, P.D.; et al. Antibody resistance of SARS-CoV-2 variants B.1.351 and B.1.1.7. *Nature* **2021**, *593*, 130–135. [CrossRef] [PubMed]
93. Collier, D.A.; De Marco, A.; Ferreira, I.A.T.M.; Meng, B.; Datir, R.P.; Walls, A.C.; Kemp, S.A.; Bassi, J.; Pinto, D.; Silacci-Fregni, C.; et al. Sensitivity of SARS-CoV-2 B.1.1.7 to mRNA vaccine-elicited antibodies. *Nature* **2021**, *593*, 136–141. [PubMed]
94. Greaney, A.J.; Loes, A.N.; Crawford, K.H.D.; Starr, T.N.; Malone, K.D.; Chu, H.Y.; Bloom, J.D. Comprehensive mapping of mutations in the SARS-CoV-2 receptor-binding domain that affect recognition by polyclonal human plasma antibodies. *Cell Host Microbe* **2021**, *29*, 463–476.e6. [CrossRef] [PubMed]
95. Cerutti, G.; Guo, Y.; Zhou, T.; Gorman, J.; Lee, M.; Rapp, M.; Reddem, E.R.; Yu, J.; Bahna, F.; Bimela, J.; et al. Potent SARS-CoV-2 neutralizing antibodies directed against spike N-terminal domain target a single supersite. *Cell Host Microbe* **2021**, *29*, 819–833.e7. [CrossRef] [PubMed]
96. Lok, S.M. An NTD supersite of attack. *Cell Host Microbe* **2021**, *29*, 744–746. [CrossRef]
97. Walls, A.C.; Park, Y.-J.; Tortorici, M.A.; Wall, A.; McGuire, A.T.; Veesler, D. Structure, Function, and Antigenicity of the SARS-CoV-2 Spike Glycoprotein. *Cell* **2020**, *181*, 281–292.e6. [CrossRef] [PubMed]
98. McCallum, M.; De Marco, A.; Lempp, F.A.; Tortorici, M.A.; Pinto, D.; Walls, A.C.; Beltramello, M.; Chen, A.; Liu, Z.; Zatta, F.; et al. N-terminal domain antigenic mapping reveals a site of vulnerability for SARS-CoV-2. *Cell* **2021**, *184*, 2332–2347.e16. [CrossRef]
99. Fellay, J.; Shianna, K.V.; Ge, D.; Colombo, S.; Ledergerber, B.; Weale, M.; Zhang, K.; Gumbs, C.; Castagna, A.; Cossarizza, A.; et al. A Whole-Genome Association Study of Major Determinants for Host Control of HIV-1. *Science* **2007**, *317*, 944–947. [CrossRef]
100. Kløverpris, H.N.; Leslie, A.; Goulder, P. Role of HLA Adaptation in HIV Evolution. *Front. Immunol.* **2016**, *6*, 665. [CrossRef]
101. Borrow, P.; Lewicki, H.; Wei, X.; Horwitz, M.S.; Peffer, N.; Meyers, H.; Nelson, J.A.; Gairin, J.E.; Hahn, B.H.; Oldstone, M.B.A.; et al. Antiviral pressure exerted by HIV-l-specific cytotoxic T lymphocytes (CTLs) during primary infection demonstrated by rapid selection of CTL escape virus. *Nat. Med.* **1997**, *3*, 205–211. [CrossRef]
102. Fiebig, E.W.; Wright, D.J.; Rawal, B.D.; Garrett, P.E.; Schumacher, R.T.; Peddada, L.; Heldebrant, C.; Smith, R.; Conrad, A.; Kleinman, S.H.; et al. Dynamics of HIV viremia and antibody seroconversion in plasma donors: Implications for diagnosis and staging of primary HIV infection. *AIDS* **2003**, *17*, 1871–1879. [CrossRef] [PubMed]
103. Turk, G.; Ghiglione, Y.; Falivene, J.; Socias, M.E.; Laufer, N.; Coloccini, R.S.; Rodriguez, A.M.; Ruiz, M.J.; Pando, M.; Giavedoni, L.D.; et al. Early Gag immunodominance of the HIV-specific T-cell response during acute/early infection is associated with higher CD8+ T-cell antiviral activity and correlates with preservation of the CD4+ T-cell compartment. *J. Virol.* **2013**, *87*, 7445–7462. [CrossRef] [PubMed]
104. Setiawan, L.C.; Gijsbers, E.F.; van Nuenen, A.C.; Kootstra, N.A. Viral evolution in HLA-B27-restricted CTL epitopes in human immunodeficiency virus type 1-infected individuals. *J. Gen. Virol.* **2015**, *96*, 2372–2380. [CrossRef] [PubMed]
105. Goulder, P.J.; Phillips, R.E.; Colbert, R.A.; McAdam, S.; Ogg, G.; Nowak, M.A.; Giangrande, P.; Luzzi, G.; Morgan, B.; Edwards, A.; et al. Late escape from an immunodominant cytotoxic T-lymphocyte response associated with progression to AIDS. *Nat. Med.* **1997**, *3*, 212–217. [CrossRef] [PubMed]
106. Schneidewind, A.; Brockman, M.A.; Yang, R.; Adam, R.I.; Li, B.; Le Gall, S.; Rinaldo, C.R.; Craggs, S.L.; Allgaier, R.L.; Power, K.A.; et al. Escape from the dominant HLA-B27-restricted cytotoxic T-lymphocyte response in Gag is associated with a dramatic reduction in human immunodeficiency virus type 1 replication. *J. Virol.* **2007**, *81*, 12382–12393. [CrossRef]
107. Cox, A.L.; Mosbruger, T.; Mao, Q.; Liu, Z.; Wang, X.-H.; Yang, H.-C.; Sidney, J.; Sette, A.; Pardoll, D.; Thomas, D.L.; et al. Cellular immune selection with hepatitis C virus persistence in humans. *J. Exp. Med.* **2005**, *201*, 1741–1752. [CrossRef]
108. Bowen, D.G.; Walker, C.M. Mutational escape from CD8+ T cell immunity: HCV evolution, from chimpanzees to man. *J. Exp. Med.* **2005**, *201*, 1709–1714. [CrossRef]
109. Rauch, A.; James, I.; Pfafferott, K.; Nolan, D.; Klenerman, P.; Cheng, W.; Mollison, L.; McCaughan, G.; Shackel, N.; Jeffrey, G.P.; et al. Divergent adaptation of hepatitis C virus genotypes 1 and 3 to human leukocyte antigen-restricted immune pressure. *Hepatology* **2009**, *50*, 1017–1029. [CrossRef]
110. Moore, C.B.; John, M.; James, I.R.; Christiansen, F.T.; Witt, C.S.; Mallal, S.A. Evidence of HIV-1 adaptation to HLA-restricted immune responses at a population level. *Science* **2002**, *296*, 1439–1443. [CrossRef]
111. Alteri, C.; Fabeni, L.; Scutari, R.; Berno, G.; Di Carlo, D.; Gori, C.; Bertoli, A.; Vergori, A.; Mastrorosa, I.; Bellagamba, R.; et al. Genetic divergence of HIV-1 B subtype in Italy over the years 2003–2016 and impact on CTL escape prevalence. *Sci. Rep.* **2018**, *8*, 15739. [CrossRef]
112. Carlson, J.M.; Du, V.Y.; Pfeifer, N.; Bansal, A.; Tan, V.Y.; Power, K.; Brumme, C.J.; Kreimer, A.; DeZiel, C.E.; Fusi, N.; et al. Impact of pre-adapted HIV transmission. *Nat. Med.* **2016**, *22*, 606–613. [CrossRef] [PubMed]
113. Balasubramaniam, M.; Davids, B.O.; Bryer, A.; Xu, C.; Thapa, S.; Shi, J.; Aiken, C.; Pandhare, J.; Perilla, J.R.; Dash, C. HIV-1 mutants that escape the cytotoxic T-lymphocytes are defective in viral DNA integration. *PNAS Nexus* **2022**, *1*, pgac064. [CrossRef] [PubMed]
114. Schommers, P.; Martrus, G.; Matschl, U.; Sirignano, M.; Lütgehetmann, M.; Richert, L.; Hope, T.J.; Fätkenheuer, G.; Altfeld, M. Changes in HIV-1 Capsid Stability Induced by Common Cytotoxic-T-Lymphocyte-Driven Viral Sequence Mutations. *J. Virol.* **2016**, *90*, 7579–7586. [CrossRef] [PubMed]

115. Rimmelzwaan, G.F.; Boon, A.C.; Voeten, J.T.; Berkhoff, E.G.; Fouchier, R.A.; Osterhaus, A.D. Sequence variation in the influenza A virus nucleoprotein associated with escape from cytotoxic T lymphocytes. *Virus Res.* **2004**, *103*, 97–100. [CrossRef]
116. Voeten, J.T.; Bestebroer, T.M.; Nieuwkoop, N.J.; Fouchier, R.A.; Osterhaus, A.D.; Rimmelzwaan, G.F. Antigenic drift in the influenza A virus (H3N2) nucleoprotein and escape from recognition by cytotoxic T lymphocytes. *J. Virol.* **2000**, *74*, 6800–6807. [CrossRef]
117. Rimmelzwaan, G.F.; Berkhoff, E.G.M.; Nieuwkoop, N.J.; Smith, D.J.; Fouchier, R.A.M.; Osterhaus, A.D.M.E. Full restoration of viral fitness by multiple compensatory co-mutations in the nucleoprotein of influenza A virus cytotoxic T-lymphocyte escape mutants. *J. Gen. Virol.* **2005**, *86*, 1801–1805. [CrossRef]
118. Rimmelzwaan, G.F.; Berkhoff, E.G.; Nieuwkoop, N.J.; Fouchier, R.A.; Osterhaus, A.D. Functional compensation of a detrimental amino acid substitution in a cytotoxic-T-lymphocyte epitope of influenza a viruses by comutations. *J. Virol.* **2004**, *78*, 8946–8949. [CrossRef]
119. Zhang, Y.; Chen, Y.; Li, Y.; Huang, F.; Luo, B.; Yuan, Y.; Xia, B.; Ma, X.; Yang, T.; Yu, F.; et al. The ORF8 protein of SARS-CoV-2 mediates immune evasion through down-regulating MHC-I. *Proc. Natl. Acad. Sci. USA* **2021**, *118*, e2024202118. [CrossRef] [PubMed]
120. Staudt, R.P.; Alvarado, J.J.; Emert-Sedlak, L.A.; Shi, H.; Shu, S.T.; Wales, T.E.; Engen, J.R.; Smithgall, T.E. Structure, function, and inhibitor targeting of HIV-1 Nef-effector kinase complexes. *J. Biol. Chem.* **2020**, *295*, 15158–15171. [CrossRef]
121. Grifoni, A.; Sidney, J.; Vita, R.; Peters, B.; Crotty, S.; Weiskopf, D.; Sette, A. SARS-CoV-2 human T cell epitopes: Adaptive immune response against COVID-19. *Cell Host Microbe* **2021**, *29*, 1076–1092. [CrossRef]
122. Motozono, C.; Toyoda, M.; Zahradnik, J.; Saito, A.; Nasser, H.; Tan, T.S.; Ngare, I.; Kimura, I.; Uriu, K.; Kosugi, Y.; et al. SARS-CoV-2 spike L452R variant evades cellular immunity and increases infectivity. *Cell Host Microbe* **2021**, *29*, 1124–1136.e11. [CrossRef] [PubMed]
123. Zhang, H.; Deng, S.; Ren, L.; Zheng, P.; Hu, X.; Jin, T.; Tan, X. Profiling CD8(+) T cell epitopes of COVID-19 convalescents reveals reduced cellular immune responses to SARS-CoV-2 variants. *Cell Rep.* **2021**, *36*, 109708. [CrossRef] [PubMed]
124. Laha, S.; Chakraborty, J.; Das, S.; Manna, S.K.; Biswas, S.; Chatterjee, R. Characterizations of SARS-CoV-2 mutational profile, spike protein stability and viral transmission. *Infect. Genet. Evol.* **2020**, *85*, 104445. [CrossRef] [PubMed]
125. Akira, T.; Masanori, M. Identification of HLA-A*24:02-Restricted CTL Candidate Epitopes Derived from the Nonstructural Polyprotein 1a of SARS-CoV-2 and Analysis of Their Conservation Using the Mutation Database of SARS-CoV-2 Variants. *Microbiol. Spectr.* **2021**, *9*, e01659-21.
126. Tarke, A.; Sidney, J.; Methot, N.; Yu, E.D.; Zhang, Y.; Dan, J.M.; Goodwin, B.; Rubiro, P.; Sutherland, A.; Wang, E.; et al. Impact of SARS-CoV-2 variants on the total CD4(+) and CD8(+) T cell reactivity in infected or vaccinated individuals. *Cell Rep. Med.* **2021**, *2*, 100355. [CrossRef]
127. Geers, D.; Shamier, M.C.; Bogers, S.; den Hartog, G.; Gommers, L.; Nieuwkoop, N.N.; Schmitz, K.S.; Rijsbergen, L.C.; van Osch, J.A.T.; Dijkhuizen, E.; et al. SARS-CoV-2 variants of concern partially escape humoral but not T-cell responses in COVID-19 convalescent donors and vaccinees. *Sci. Immunol.* **2021**, *6*, eabj1750. [CrossRef]
128. Corey, L.; Beyrer, C.; Cohen, M.S.; Michael, N.L.; Bedford, T.; Rolland, M. SARS-CoV-2 Variants in Patients with Immunosuppression. *N. Engl. J. Med.* **2021**, *385*, 562–566. [CrossRef]
129. Braun, K.M.; Moreno, G.K.; Wagner, C.; Accola, M.A.; Rehrauer, W.M.; Baker, D.A.; Koelle, K.; O'Connor, D.H.; Bedford, T.; Friedrich, T.C.; et al. Acute SARS-CoV-2 infections harbor limited within-host diversity and transmit via tight transmission bottlenecks. *PLoS Pathog.* **2021**, *17*, e1009849. [CrossRef]
130. Fung, M.; Babik, J.M. COVID-19 in Immunocompromised Hosts: What We Know So Far. *Clin. Infect. Dis.* **2021**, *72*, 340–350. [CrossRef]
131. Chaguza, C.; Hahn, A.M.; Petrone, M.E.; Zhou, S.; Ferguson, D.; Breban, M.I.; Pham, K.; Peña-Hernández, M.A.; Castaldi, C.; Hill, V.; et al. Accelerated SARS-CoV-2 intrahost evolution leading to distinct genotypes during chronic infection. *medRxiv* **2022**. [CrossRef]
132. Scherer, E.M.; Babiker, A.; Adelman, M.W.; Allman, B.; Key, A.; Kleinhenz, J.M.; Langsjoen, R.M.; Nguyen, P.V.; Onyechi, I.; Sherman, J.D.; et al. SARS-CoV-2 Evolution and Immune Escape in Immunocompromised Patients. *N. Engl. J. Med.* **2022**, *386*, 2436–2438. [CrossRef] [PubMed]
133. Truong, T.T.; Ryutov, A.; Pandey, U.; Yee, R.; Goldberg, L.; Bhojwani, D.; Aguayo-Hiraldo, P.; Pinsky, B.A.; Pekosz, A.; Shen, L.; et al. Increased viral variants in children and young adults with impaired humoral immunity and persistent SARS-CoV-2 infection: A consecutive case series. *eBioMedicine* **2021**, *67*, 103355. [CrossRef] [PubMed]
134. Kemp, S.A.; Collier, D.A.; Datir, R.P.; Ferreira, I.; Gayed, S.; Jahun, A.; Hosmillo, M.; Rees-Spear, C.; Mlcochova, P.; Lumb, I.U.; et al. SARS-CoV-2 evolution during treatment of chronic infection. *Nature* **2021**, *592*, 277–282. [PubMed]
135. Choi, B.; Choudhary, M.C.; Regan, J.; Sparks, J.A.; Padera, R.F.; Qiu, X.; Solomon, I.H.; Kuo, H.H.; Boucau, J.; Bowman, K.; et al. Persistence and Evolution of SARS-CoV-2 in an Immunocompromised Host. *N. Engl. J. Med.* **2020**, *383*, 2291–2293. [CrossRef]
136. Avanzato, V.A.; Matson, M.J.; Seifert, S.N.; Pryce, R.; Williamson, B.N.; Anzick, S.L.; Barbian, K.; Judson, S.D.; Fischer, E.R.; Martens, C.; et al. Case Study: Prolonged Infectious SARS-CoV-2 Shedding from an Asymptomatic Immunocompromised Individual with Cancer. *Cell* **2020**, *183*, 1901–1912.e9. [CrossRef]

137. Cele, S.; Karim, F.; Lustig, G.; San, J.E.; Hermanus, T.; Tegally, H.; Snyman, J.; Moyo-Gwete, T.; Wilkinson, E.; Bernstein, M.; et al. SARS-CoV-2 prolonged infection during advanced HIV disease evolves extensive immune escape. *Cell Host Microbe* **2022**, *30*, 154–162.e5. [CrossRef]
138. Tang, J.W.; Tambyah, P.A.; Hui, D.S. Emergence of a new SARS-CoV-2 variant in the UK. *J. Infect.* **2021**, *82*, e27–e28. [CrossRef]
139. Tao, K.; Tzou, P.L.; Nouhin, J.; Gupta, R.K.; de Oliveira, T.; Kosakovsky Pond, S.L.; Fera, D.; Shafer, R.W. The biological and clinical significance of emerging SARS-CoV-2 variants. *Nat. Rev. Genet.* **2021**, *22*, 757–773. [CrossRef]
140. Harari, S.; Tahor, M.; Rutsinsky, N.; Meijer, S.; Miller, D.; Henig, O.; Halutz, O.; Levytskyi, K.; Ben-Ami, R.; Adler, A.; et al. Drivers of adaptive evolution during chronic SARS-CoV-2 infections. *Nat. Med.* **2022**, *28*, 1501–1508. [CrossRef]
141. Pardi, N.; Parkhouse, K.; Kirkpatrick, E.; McMahon, M.; Zost, S.J.; Mui, B.L.; Tam, Y.K.; Karikó, K.; Barbosa, C.J.; Madden, T.D.; et al. Nucleoside-modified mRNA immunization elicits influenza virus hemagglutinin stalk-specific antibodies. *Nat. Commun.* **2018**, *9*, 3361. [CrossRef]
142. Krammer, F.; Palese, P. Influenza virus hemagglutinin stalk-based antibodies and vaccines. *Curr. Opin. Virol.* **2013**, *3*, 521–530. [CrossRef] [PubMed]
143. Manenti, A.; Maciola, A.K.; Trombetta, C.M.; Kistner, O.; Casa, E.; Hyseni, I.; Razzano, I.; Torelli, A.; Montomoli, E. Influenza Anti-Stalk Antibodies: Development of a New Method for the Evaluation of the Immune Responses to Universal Vaccine. *Vaccines* **2020**, *8*, 43. [CrossRef] [PubMed]
144. Guthmiller, J.J.; Han, J.; Utset, H.A.; Li, L.; Lan, L.Y.-L.; Henry, C.; Stamper, C.T.; McMahon, M.; O'Dell, G.; Fernández-Quintero, M.L.; et al. Broadly neutralizing antibodies target a haemagglutinin anchor epitope. *Nature* **2022**, *602*, 314–320. [CrossRef] [PubMed]
145. Cao, Y.; Wang, J.; Jian, F.; Xiao, T.; Song, W.; Yisimayi, A.; Huang, W.; Li, Q.; Wang, P.; An, R.; et al. Omicron escapes the majority of existing SARS-CoV-2 neutralizing antibodies. *Nature* **2022**, *602*, 657–663. [CrossRef] [PubMed]
146. Pinto, D.; Park, Y.-J.; Beltramello, M.; Walls, A.C.; Tortorici, M.A.; Bianchi, S.; Jaconi, S.; Culap, K.; Zatta, F.; De Marco, A.; et al. Cross-neutralization of SARS-CoV-2 by a human monoclonal SARS-CoV antibody. *Nature* **2020**, *583*, 290–295. [CrossRef] [PubMed]
147. Yuan, M.; Wu, N.C.; Zhu, X.; Lee, C.D.; So, R.T.Y.; Lv, H.; Mok, C.K.P.; Wilson, I.A. A highly conserved cryptic epitope in the receptor binding domains of SARS-CoV-2 and SARS-CoV. *Science* **2020**, *368*, 630–633. [CrossRef]
148. Grifoni, A.; Weiskopf, D.; Ramirez, S.I.; Mateus, J.; Dan, J.M.; Moderbacher, C.R.; Rawlings, S.A.; Sutherland, A.; Premkumar, L.; Jadi, R.S.; et al. Targets of T Cell Responses to SARS-CoV-2 Coronavirus in Humans with COVID-19 Disease and Unexposed Individuals. *Cell* **2020**, *181*, 1489–1501.e15. [CrossRef]
149. Le Bert, N.; Tan, A.T.; Kunasegaran, K.; Tham, C.Y.L.; Hafezi, M.; Chia, A.; Chng, M.H.Y.; Lin, M.; Tan, N.; Linster, M.; et al. SARS-CoV-2-specific T cell immunity in cases of COVID-19 and SARS, and uninfected controls. *Nature* **2020**, *584*, 457–462. [CrossRef]
150. Ng, O.W.; Chia, A.; Tan, A.T.; Jadi, R.S.; Leong, H.N.; Bertoletti, A.; Tan, Y.J. Memory T cell responses targeting the SARS coronavirus persist up to 11 years post-infection. *Vaccine* **2016**, *34*, 2008–2014. [CrossRef]

viruses

MDPI

Review

A Detailed Overview of SARS-CoV-2 Omicron: Its Sub-Variants, Mutations and Pathophysiology, Clinical Characteristics, Immunological Landscape, Immune Escape, and Therapies

Srijan Chatterjee [1,†], Manojit Bhattacharya [2,†], Sagnik Nag [3], Kuldeep Dhama [4] and Chiranjib Chakraborty [1,*,†]

1 Department of Biotechnology, School of Life Science and Biotechnology, Adamas University, Kolkata 700126, West Bengal, India
2 Department of Zoology, Fakir Mohan University, Vyasa Vihar, Balasore 756020, Odisha, India
3 Department of Biotechnology, School of Biosciences & Technology, Vellore Institute of Technology (VIT), Vellore 632014, Tamil Nadu, India
4 Division of Pathology, ICAR-Indian Veterinary Research Institute, Izatnagar, Bareilly 243122, Uttar Pradesh, India
* Correspondence: drchiranjib@yahoo.com; Tel.: +91-9871608125
† These authors contributed equally to this work.

Abstract: The COVID-19 pandemic has created significant concern for everyone. Recent data from many worldwide reports suggest that most infections are caused by the Omicron variant and its sub-lineages, dominating all the previously emerged variants. The numerous mutations in Omicron's viral genome and its sub-lineages attribute it a larger amount of viral fitness, owing to the alteration of the transmission and pathophysiology of the virus. With a rapid change to the viral structure, Omicron and its sub-variants, namely BA.1, BA.2, BA.3, BA.4, and BA.5, dominate the community with an ability to escape the neutralization efficiency induced by prior vaccination or infections. Similarly, several recombinant sub-variants of Omicron, namely XBB, XBD, and XBF, etc., have emerged, which a better understanding. This review mainly entails the changes to Omicron and its sub-lineages due to it having a higher number of mutations. The binding affinity, cellular entry, disease severity, infection rates, and most importantly, the immune evading potential of them are discussed in this review. A comparative analysis of the Delta variant and the other dominating variants that evolved before Omicron gives the readers an in-depth understanding of the landscape of Omicron's transmission and infection. Furthermore, this review discusses the range of neutralization abilities possessed by several approved antiviral therapeutic molecules and neutralizing antibodies which are functional against Omicron and its sub-variants. The rapid evolution of the sub-variants is causing infections, but the broader aspect of their transmission and neutralization has not been explored. Thus, the scientific community should adopt an elucidative approach to obtain a clear idea about the recently emerged sub-variants, including the recombinant variants, so that effective neutralization with vaccines and drugs can be achieved. This, in turn, will lead to a drop in the number of cases and, finally, an end to the pandemic.

Keywords: Omicron: sub-lineages; transmission and infection; disease severity

Citation: Chatterjee, S.; Bhattacharya, M.; Nag, S.; Dhama, K.; Chakraborty, C. A Detailed Overview of SARS-CoV-2 Omicron: Its Sub-Variants, Mutations and Pathophysiology, Clinical Characteristics, Immunological Landscape, Immune Escape, and Therapies. *Viruses* **2023**, *15*, 167. https://doi.org/10.3390/v15010167

Academic Editors: Ahmed El-Shamy and Mohamed Ibrahim

Received: 23 November 2022
Revised: 28 December 2022
Accepted: 4 January 2023
Published: 5 January 2023

1. Introduction

The SARS-CoV-2 virus is highly infectious, and it was the causative agent of the outbreak of the COVID disease in 2019. The WHO declared it to be a global pandemic [1,2]. More than 480 million cases have already been reported worldwide, with there having been over 6 million deaths since late 2019 [3]. Most of the infected people develop mild-to-moderate symptoms such as a cough, fever, the loss of smell and taste, a headache, sore throat, diarrhea, body aches, and tiredness. The virus kept evolving, and APOBEC-induced mutations contributed to SARS-CoV-2's evolution and fitness, and different variants were identified during the pandemic [4,5]. The variants were classified as variants under monitoring (VUMs), variants of concern (VOCs), and variants of interest (VOIs). These variants are Alpha (B.1.1.7), Beta (B.1.351), Gamma (P.1), Delta (B.1.617.2), and a novel or new variant, Omicron (B.1.1.529), which has a much faster infection rate than the other four variants do [6]. A new variant's threat depends on three main factors and their interactions. Those factors are its transmissibility, severity compared to other strain (fewer ICU hospitalizations), and immune evasion. The variants have evolved by multiple mutations in inconsistent combinations, mainly in the spike protein (S1 and S2 subunits) of the virus, which helps to initiate the coronavirus' life cycle. The important mutations that play a crucial role in the infectivity of the variants are Alpha, with an N501Y mutation in the RBD, Beta with N501Y, K417N, and E484K mutations in the RBD regions, Gamma with N501Y, K417T, and E484K mutations in the RBD regions, Delta with T478K, L452R mutations in the RBD regions, and Omicron with S371L, G339D, S375F, S373P, K417N, N440K, S477N, G446S, E484A, T478K, Q493K, Q498R, G496S, N501Y, and Y505H mutations in the RBD regions [7–10]. The symptoms of the Omicron infection are less dangerous than those of the other strains, but it is more transmissible and less susceptible to vaccines, even though the mortality rate is lower than those of other strains [11–13].

Omicron was first spotted in South Africa and Botswana in November 2021 [14] (Figure 1). More than 130 million cases including 500,000 deaths have been reported worldwide since Omicron was declared as a VOC by the WHO, leading to a 44% rise in the average number of COVID-19 cases. The basic reproduction number (R0) of the Omicron variant was reported as being 8.2, showing an increased rate of transmissibility compared to that of the Delta variant [15]. The R0 of the Delta variant was observed as being between 3.2 and 8 [16]. It is estimated that Omicron infects 3–6 times more people as compared to Delta during a given time frame [17]. During the period when Omicron predominated, the rate of weekly hospitalization per 100,000 adults peaked at 38.4 compared to that of the Delta variant, which was 15.5 during predominant period in the United States. The Omicron variant has generated a new wave, evidenced by high infection rates worldwide [18]. The new wave is called the Omicron wave. The peak of the Omicron wave is very high compared to the waves of the variants such as the Alpha and Beta, etc., (Figure 2). Due to mutations in the genomic region of the variants, several subtypes have emerged over time, which are named sub-variants. The noteworthy evolution of the genomic regions of the Omicron variants has resulted in the emergence of several sub-lineages or sub-variants, which are denoted as BA.1, BA.2, BA.3, BA.4, BA.5, and recombinant BA.1/BA.2 [19]. Several researchers have tried to understand the Omicron sub-variants in more detail [13,20]. The BA.1 sub-lineage was the most prevalent globally, but BA.2 progressively replaced BA.1 in numerous countries, and the transmissibility of BA.3 is very restricted, with it having the lowest number of cases [16] (Figure 1). The other two new lineages detected in South Africa during January and February 2022 were BA.4 and BA.5, respectively [21] (Figure 1). These two lineages were predominant in the 5th wave of COVID-19 pandemic that was initiated in South Africa, and it replaced BA.2, as more than 50% of the cases were due to the dominance of BA.4 (35%) and BA.5 (20%) [9]. Omicron has more mutations than any other variant does. Omicron's mutation helps it to bind more strongly with the ACE2 host cell receptors than the other reported variants can [22]. It also evades most of the virus-blocking antibodies or the 'neutralizing' antibodies (but not all of them) produced by vaccinated individuals or individuals infected with the other variants [23–25].

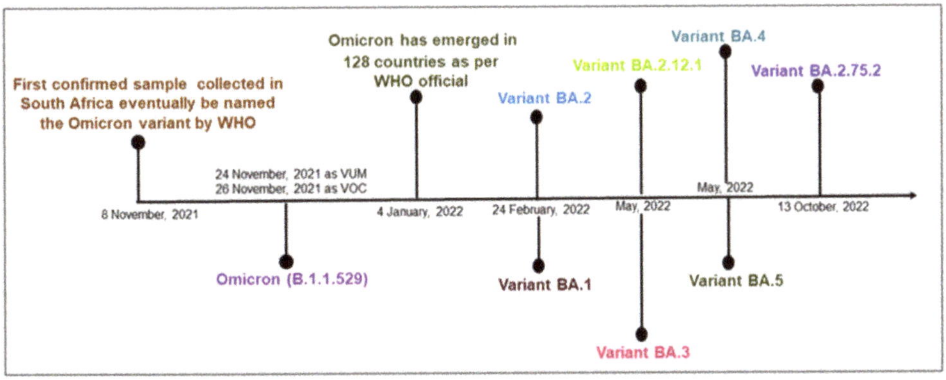

Figure 1. A timeline describes the origin of SARS-CoV-2 Omicron and different times of origin of Omicron's sub-variants.

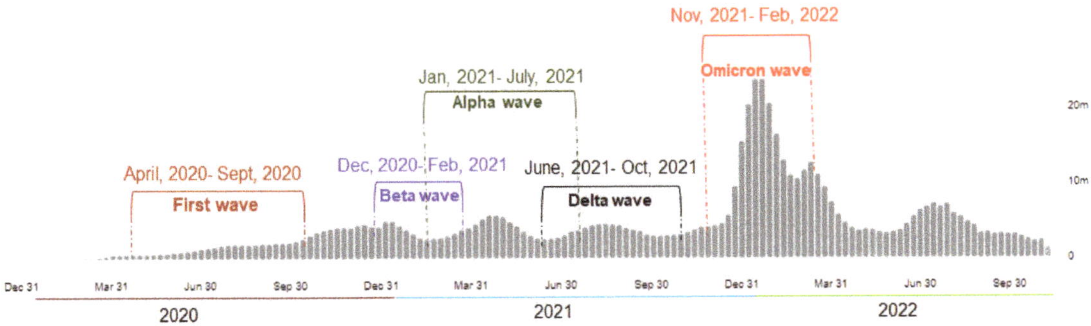

Figure 2. The new wave generated due to high infection worldwide due to Omicron's infection. The new wave is called the Omicron wave. The peak of the Omicron wave is very high compared to the other waves, such as the Alpha wave and the Beta wave, etc.

The main hindrance in the generation of antibodies and the development of suitable vaccines and therapeutic agents is due to the number of escape mutations generated in the genome of the SARS-CoV-2 virus and the sudden appearance of new strains. It is resistant to some existing vaccines and therapeutic agents. Studying the conformational dynamics of different antibody neutralization escape mutants is thus very important [26–29]. Additionally, understanding the antibodies targeted to different regions (S1/S2) of the spike protein which inhibit viral entry is essential for designing new antibodies. It can target the spike protein's local dynamics, decreasing the efficacy of viral inhibition by the antibodies. Generating synthetic vaccines depending on the conformational dynamics of the variants will also be economical and they will be easy to update as they contain parts that are easily replaceable to act against the new strains of the pathogens.

In the present article, we have enlisted and summarized the different mutations of the Omicron variant and its sub-variants, along with the pathophysiology, clinical characteristics, and associated disease severity. Subsequently, we have highlighted the infection, reinfection, and transmissibility of the Omicron variant and its sub-variants, including the specific immunological features inside the host cells. Furthermore, a particular emphasis is also placed on the proposed small molecules and antibody-based therapeutics against Omicron and its sub-variants. This evidenced-based review will support future researchers in formulating appropriate strategies to resist the infections caused by Omicron and its sub-variants in the future.

2. Sub-Variants of Omicron Variant

Omicron has many mutations in its viral genome. According to the reports published on April 2022 by the WHO, five sub-variants of the Omicron variant have been detected. They are BA.1, BA.2, BA.3, BA.4, and BA.5. These mutations have been prevalent worldwide in different quantities (Figure 3A) [30]. Kumar et al. applied an elucidated approach using certain computational tools to provide insight regarding the pathogenicity and infectivity of the S-glycoprotein of BA.1 and the corresponding sub-lineages, BA.1.1, BA.2 and BA.3. The BA.1 sub-lineage shares 39 substitutions in the genome, followed by 40 mutational changes residing in the genome of BA.1.1 [31]. On the contrary, the BA.2 and BA.3 variants also share 31 and 34 mutations in their genome. Significantly, 21 mutations are prevalent in all of the evolved sub-lineages of the Omicron variant. Furthermore, 11 usual mutational changes have been deciphered in the RBD of the Omicron variant and the evolved sub-variants. The T95I, V213R, Y505H, N856K, N786K, and N211I mutations residing in Omicron and its sub-variants are highly pathogenic (Figure 3B) [31]. Reports have highlighted that no substantial mutations have been observed in the BA.3 variant's spike glycoprotein. The mutations in the spike protein of BA.1 and BA.2 are only observable in the lately emerged BA.3 sub-lineage [32]. The data retrieved from the following website, https://outbreak.info/ (accessed on 6 August, 2022), provide the ratio, which shows the dominance of the Omicron sub-variants worldwide. The first sub-lineage, BA.1, accounts for 5% of the total number of cases in 161 countries, which is followed by the BA.1.1 variant, possessing 17% of the cases in the same countries. A steeper rise in the number of cases was prevalent for the BA.2 sub-variant, i.e., 9% of the reported cases across 163 different countries, with an extremely low prevalence of BA.3 cases until May 2022 [31]. The most alarming insight is the ability of these variants to escape the immune system and decrease the neutralization efficiency of vaccines. It was observed that the BA.1 sub-variant is more transmissible than the previously emerged Delta variant is, but the infected people rarely require hospital support. Owing to the presence of the H78Y mutation, the BA.2 sub-variant is more severe than the BA.1 sub-variant is [33]. The latest transmission rates account for the BA.3 sub-lineage because they lack six mutations in the genome, namely, L981F, G496S, ins214EPE, N856K, T547K, and S371L [32].

In early 2022, scientists also found two more sub-lineages of the Omicron variant from South African, namely, BA.4 and BA.5. After its report in South Africa, this sub-variant emerged in many areas across the globe. At the end of 2021, the BA.1 variant replaced the Delta one, and it was considered to be the leading causative agent of the fourth wave. Similarly, this BA.1 sub-variant was again replaced by BA.2, manifesting its dominance by April 2022 [33,34]. These two newly emerged sub-variants are the key factor responsible for the fifth wave of COVID-19. These variants are replacing all of the previously emerged sub-lineages of Omicron. The spike proteins of these recently evolved sub-lineages are somewhat similar. The BA.4 and BA.5 sub-variants possess certain extra mutations in the viral genome, with a synonymous substitution that is similar to the B.1.429 SARS-CoV-2 variant, which was also seen in BA.2 [35]. These two sub-lineages also show potent activities in evading the host's immune system. However, the appropriate information regarding the hospitalization rates of the victims of BA.4 or BA.5 has remained unexplored. One of the recently evolved sub-variants, BA.2.12.1, exhibits a key feature of invalidating the antibodies present in the host due to vaccination or prior infection with the Omicron variant [36]. In a nutshell, it can be concluded that the BA.4 and BA.5 sub-lineages, along with the BA.2.12.1 one, are more robust and can evade the host's humoral immunity [37].

Apart from these sub-lineages, several hybrid combinations of these sub-variants are also prevailing in the community, such as the XD, XE, and XF ones. The XE variant is a combination of the BA.1 and BA.2 sub-lineages, which has significantly worse effects, raising severe concerns for global health amidst the pandemic [38–41]. Similarly, XD is considered to be a recombinant form of the BA.1 and Delta variants, while the XF one is a product of the BA.1 sub-lineage and the Delta strains isolated from the United Kingdom.

The WHO has termed the XE variant as "stealth Omicron," possessing a ten times higher infectivity rate than the BA.2 sub-lineage does [41]. Scientists are worried about the severity of the infections caused by these recombinant variants. A closer look at the mutation profile of the XE variants revealed three new mutations, namely, C14599T, V1069I, and C3241T, which were not reported in the BA.1 and BA.2 sub-variants. The XD hybrid was first one to be reported from France. It contains a new mutation in the nsp2 gene, i.e., E172D, whereas the XF variant possesses a breakpoint at the end of the nsp3 gene, which is not common in the other sub-variants [40].

Recently, scientists noted that numerous other recombinant sub-variants have evolved during the post-Omicron period, such as the XBB, XBD, and XBF ones, etc., whose pathophysiology are yet to be studied [42].

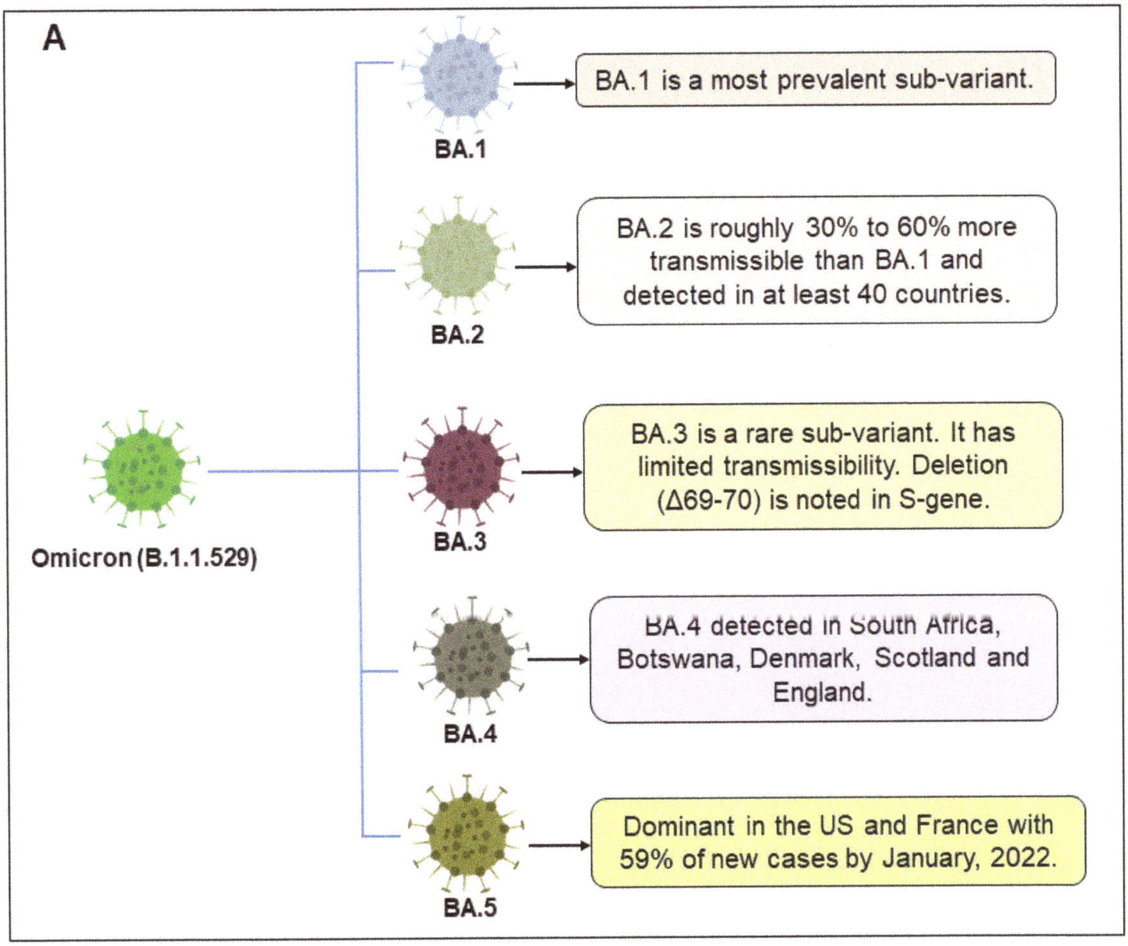

Figure 3. *Cont.*

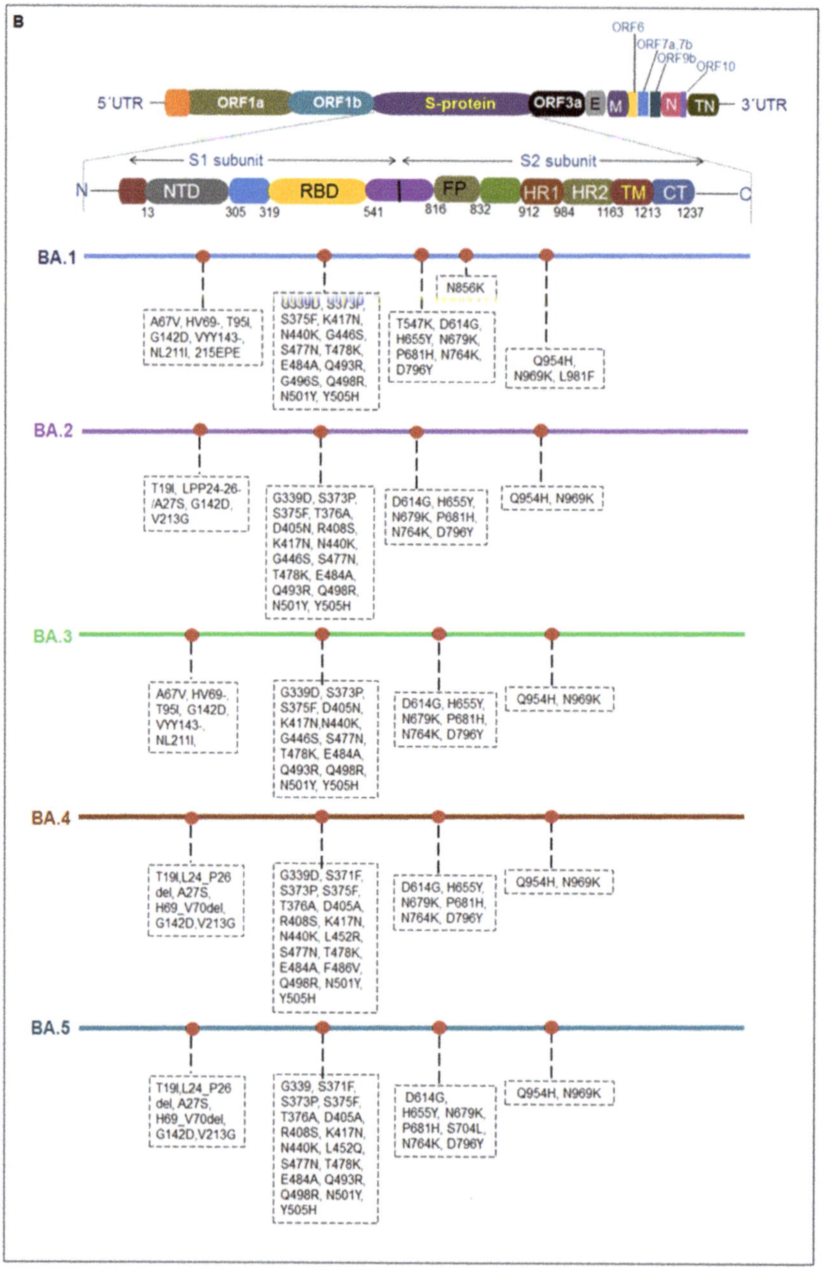

Figure 3. The figure illustrates different sub-variants of Omicron and their significant mutations in S-glycoprotein. (**A**) The figure describes different sub-variants of Omicron and their features. It describes the features of BA.1, BA.2, BA.3, BA.4, and BA.5. (**B**) The figure illustrates different mutations in S-glycoprotein of different sub-variants of Omicron. It describes the S-glycoprotein mutations of BA.1, BA.2, BA.3, BA.4, and BA.5.

3. Different Mutations and Pathophysiology Condition

Omicron has more than 50 known mutations, 32 of which are in the spike protein rather than the wild-type one [43]. The Delta strain, in comparison, has nine mutations in the spike protein itself and thirteen mutations in the added regions. Out of the fifty mutations, twenty-six of them are unique to Omicron, and it also has ten mutations that are unique to Delta and six mutations that are unique to the Beta strain [44]. The mutations that the Omicron lineage possesses are ORF1a-6 substitutions at K856R, A2710T, L2084I, P3395H, T3255I, and I3758V, two deletions at positions 2083 and 3674–3676, ORF1b-2 substitutions at P314L and I1566V, deletions at positions 27, 28, and 29, and a P10S substitution at ORF9b. The mutations in the structural proteins are an envelope (E)-T9I substitution, membrane (M)- D3G, Q19E, and A63T substitutions, a nucleocapsid (N)- a three residue deletion, and three residue substitutions. The significant spike (S) mutations are A67V, T95I, L212I, Y145D, G339D, S373P, S371L, K417N, S375F, N440K, G446S, S477N, E484A, T478K, Q493R, N856K, Q498R, G496S, N501Y, Y505H, T547K, Q954H, P681H, D614G, H655Y, N764K, N679K, N969K, and D796Y, etc., (Figure 4A). Some other mutations are an L981F substitution, H69/V70, G142/V143/Y144, and N211 deletions, and an insertion of amino acid EPE at position 214 [3,45].

The BA.2 lineage consists of 57 mutations, of which 31 are in the S protein, and the N-terminus is exclusively different from that of the BA.1 lineage, whereas 12 mutations are common in both the BA.1 and BA.2 lineages in the RBD region, which are G339D, K417N, S373P, S375F, T478K, N440K, S477N, E484A, Q498R, N501Y, Q493R, and Y505H (Figure 4B). G446S, S371L, and G496S are unique to the BA.1 lineage, and R346K is found in a member of the BA.1 lineage, namely, BA.1.1.

Omicron and its variants have several unique mutations in the RBD region. The RBD mutations might control the functionality of that specific RBD region. The BA.2 lineage has two unique mutations in RBD, R408S, and S371F, and T376A and D405N mutations are common with the BA.3 lineage. The newly evolved BA.4 and BA.5 lineages of Omicron are similar to the BA.2 lineage, except for the deletion of an amino acid at positions 69 and 70 and F486V and R493Q mutations in the RBD of the spike protein compared to the BA.1 lineage [9]. The mutation F486V in the spike protein is the leading cause of the infection. The BA.4 and BA.5 sub-lineages have substitution mutations in the RBD: L452R, F486V, and R493Q, compared to BA.2, which does not (Table 1).

The strength of the binding affinity of the RBD region of the Omicron variant to the receptor ACE2 is 1.5–2.8 times higher than that of the wild-type strain [46–49]. In comparison with the Delta variant, the Omicron RBD part has a weaker or a similar binding affinity to ACE2 [48–51]. However, the binding affinity of Omicron's RBD to ACE2 is weaker than that of the Alpha variant. The alpha variant has only one mutation (N501Y) in the RBD region [46,50]. The strength of the binding affinity of the Omicron variant's RBD to ACE2 is in between those of the RBDs of the wild type and the Delta variant of SARS-CoV-2. The mutations, namely, T478K, S477N, Q496S, Q493R, and Q498R, in addition to N501Y, are thought to potentiate the interaction between the Omicron variant and human ACE2 by forming new salt bridges or hydrogen bonds with the ACE2 receptor [46,48,52,53].

On the other hand, K417N and E484A can cause a loss of interaction between Omicron and the ACE2 receptor part, which the other mutations may have enhanced [46,53,54]. The N501Y mutation in Omicron was also seen in the Gamma, Alpha, and Beta variants, and it augments the binding strength of the spike protein with ACE2. The transmissibility increases further in the N501Y mutation with an added H69/V70 deletion [55,56]. The N679K and P681H mutations incorporate essential amino acids near the furin cleavage site. This facilitates spike protein cleavage in the S1 and S2 subunits, tighter binding, and enhanced virus infectivity [29]. This enhances fusion and virus infection [57]. However, the effects of most of the mutations in the Omicron variant are still unknown. Once all of the roles the mutations play have been identified, the generation of effective vaccines, and thus, the prevention of the disease becomes easier.

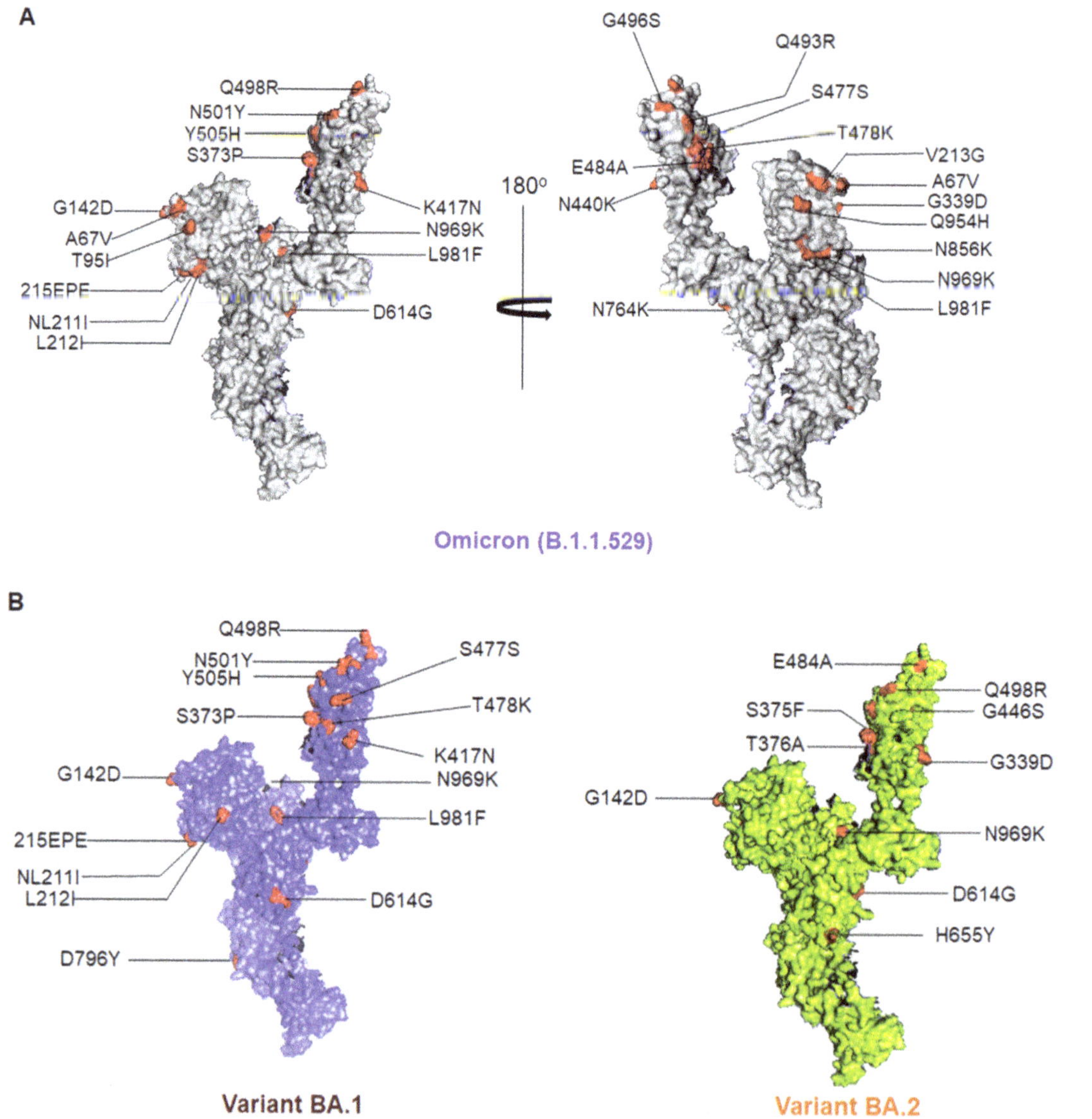

Figure 4. The figure shows different significant mutations in a 3D model of the S-glycoprotein of Omicron and its different sub-variants. (**A**) The figure shows different significant mutations in a 3D model of the S-glycoprotein of Omicron. (**B**) The figure shows different significant mutations in a 3D model of the S-glycoprotein of Omicron's sub-variants, BA.1 and BA.2.

Table 1. Omicron and its variants have several unique mutations in the RBD region. The RBD mutations might control the functionality of the RBD region.

Sl. No	Omicron Sub-Variant Name	Mutations in S Protein	
		RBD Region	Other Than RBD Region
1.	BA.1	G339D, S373P, S375F, K417N, N440K, G446S, S477N, T478K, E484A, Q493R, G496S, Q498R, N501Y, Y505H	A67V, HV69-, T95I, G142D, VYY143-, NL211I, 215EPE, T547K, D614G, H655Y, N679K, P681H, N764K, D796Y, N856K, Q954H, N969K, L981F

Table 1. *Cont.*

Sl. No	Omicron Sub-Variant Name	Mutations in S Protein	
		RBD Region	Other Than RBD Region
2.	BA.2	G339D, S373P, S375F, T376A, D405N, R408S, K417N, N440K, G446S, S477N, T478K, E484A, Q493R, Q498R, N501Y, Y505H	T19I, LPP24-26-/A27S, G142D, V213G, D614G, H655Y, N679K, P681H, N764K, D796Y, Q954H, N969K
3.	BA.3	G339D, S373P, S375F, D405N, K417N, N440K, G446S, S477N, T478K, E484A, Q493R, Q498R, N501Y, Y505H	A67V, HV69-, T95I, G142D, VYY143-, NL211I, D614G, H655Y, N679K, P681H, N764K, D796Y, Q954H, N969K
4.	BA.4	G339D, S371F, S373P, S375F, T376A, D405A, R408S, K417N, N440K, L452R, S477N, T478K, E484A, F486V, Q498R, N501Y, Y505H	T19I, L24_P26del, A27S, H69_V70del, G142D, V213G, D614G, H655Y, N679K, P681H, N764K, D796Y, Q954H, N969K
5.	BA.5	G339, S371F, S373P, S375F, T376A, D405A, R408S, K417N, N440K, L452Q, S477N, T478K, E484A, Q493R, Q498R, N501Y, Y505H	T19I, L24_P26del, A27S, G142D, V213G, D614G, H655Y, N679K, P681H, S704L, N764K, D796Y, Q954H, N969K

4. Omicron Variant-Associated Disease Intensity

Omicron shows a disease severity that is lower than that of the other variants, which may be because of its faster growth rate and transmissibility, detrimental changes in the epidemiology of the previous variants, the more virulent nature of the virus or the clinical presentation of the disease, the decreased effectiveness of the vaccines or other therapeutics, or the decreased effectiveness of the social or public health measures. Of all of the Omicron lineages, BA.4 and BA.5 are more transmissible than the others are, which could be because of the higher growth rate of BA.4 and BA.5 than the that of the other sub-lineages. BA.5, which is the most predominant one, was first identified in January, and it was detected by the WHO in April. They can readily evade the immune system, induced by vaccination or viral infections [58]. The original Omicron strain is less severe than the Delta strain is, but the BA.5 variant, along with the BA.4 strain, is most the contagious one, causing more than 50% of the cases due to this variant. All of the Omicron variants, in general, also have a much higher transmissibility rate than the Delta variant does [6]. Omicron is associated with milder symptoms, decreased hospitalization and mortality, and the generation of more asymptomatic carriers compared to infections with other variants [59–62]. By comparing the Omicron lineages with the other SARS-CoV-2 variants such as the Alpha or Delta ones, the data show that Delta is the most prevalent type in terms of severity. At the same time, Omicron is noted as the most transmissible variant [63]. The effective reproductive number of the Omicron variant (8.2) elicited a 3.8 times higher transmissibility rate than the Delta variant did [15]. The Omicron variant significantly multiplied the number of daily hospitalization cases by three the number of daily cases caused by the Delta variant. However, the number of daily ICU cases was lower in the case of the Omicron variant. The number of everyday hospitalizations during the peak of Omicron was around one time higher in the US and UK than it was during the peak of the Delta variant. This is true for both the minimum and the maximum number of cases. The maximum number of daily ICU cases was similar for the Delta and Omicron variants during the peak outbreak, and the number of daily ICU cases were reduced in every country. During the Delta outbreak, Brazil's maximum number of deaths was 1857.43 per million. In France, it was 11.86 per million, while in India, it was 3387.71 per million. The number in other countries of interest such as the UK it was 12 per million, and in the US, it was 432.29 per million. During the Omicron outbreak, the maximum numbers of deaths in Brazil, France, India, the UK, and the US are 831.14, 328.86, 1117.71, 86.86, and 2576.71 per million respectively. During the Omicron outbreak, Brazil and India had a lower number of daily deaths than the other countries did [15,64]. Similarly, the degree of the severity of the illness was much lower than it was during the Delta outbreak [65]. The vaccine's effectiveness also decreased much faster for the Omicron variant than it did for the pre-Omicron variants, and people infected with pre-Omicron variants have only 15% protection against the BA.4 and BA.5 variants.

5. Clinical Characteristics and Symptom Prevalence

Several scientists have studied the Omicron variant's disease intensity and found that it has an increased transmissibility rate and a higher growth rate. However, the Omicron's higher growth and transmissibility rates might be associated with the viral load. Studies have noted that the viral load is higher in the lungs during the infection of a wild strain of SARS-CoV-2. They also reported that the viral load is higher during an infection with Omicron in the upper airway, especially in the nose, windpipe, and throat, but not in the lower respiratory system [66]. The higher growth and augmented viral load may cause the virus particles to aggregate in the upper airway (Figure 5A).

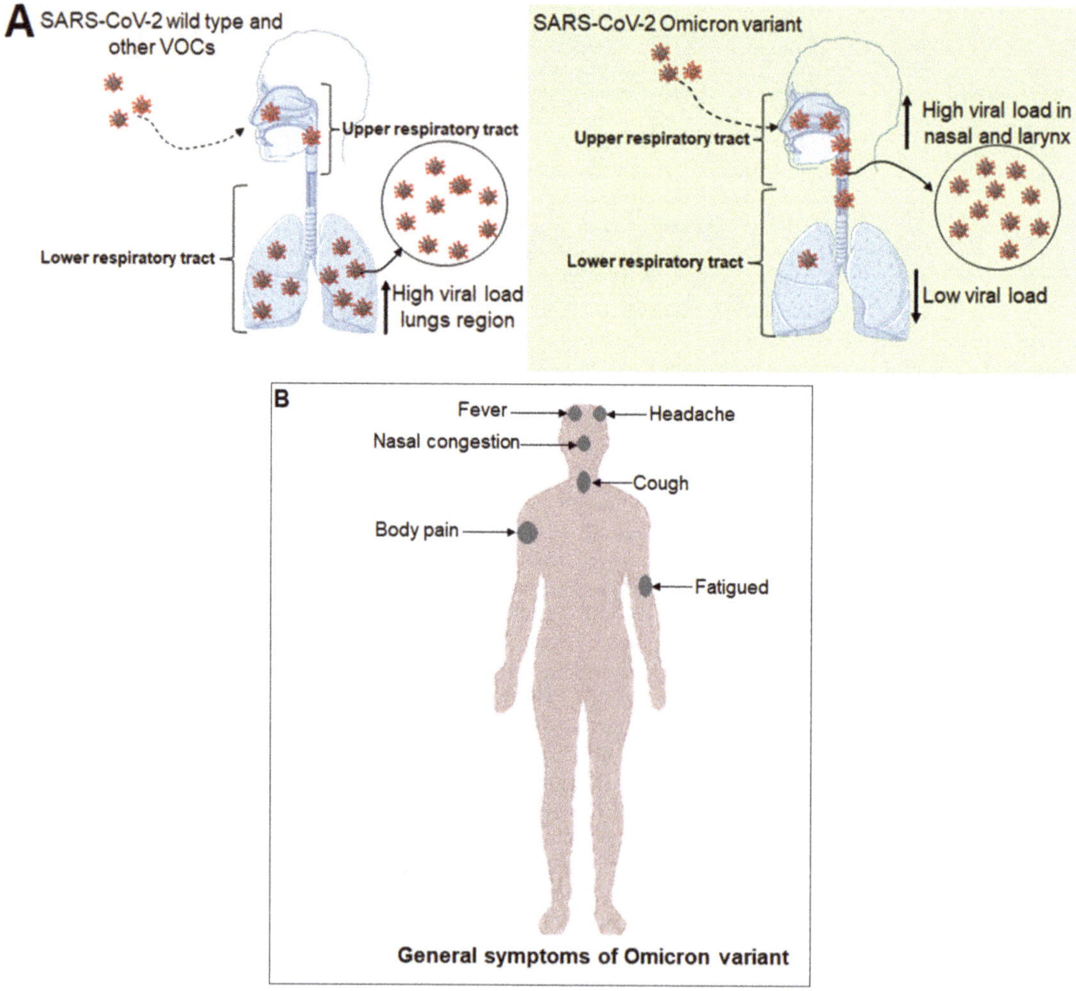

Figure 5. The figure shows viral load in the respiratory tract and the lungs during an infection with the wild strain of SARS-CoV-2 and Omicron. The figure also describes the common symptoms of Omicron-infected patients. (**A**) It shows high viral load in the respiratory tract during Omicron infection. It shows high viral load in lungs during an infection with the wild strain of SARS-CoV-2. (**B**) The figure depicting the common clinical symptoms of Omicron-infected patients.

Omicron is not worse than other coronavirus strains, and it is less severe (less ICU hospitalization) than the Delta variant. The number of individuals with oxygen supports was also smaller than it was during the previous waves due to other SARS-CoV-2 variants, specifically Delta [67]. The clinical characteristics of the SARS-CoV-2 Omicron variant are different from those of all of the previous variants [68–71]. The most common symptoms are a cough, runny nose, congestion, and fatigue (Figure 5B) [71]. The loss of smell and taste, fever, dizziness, headache, runny nose, hair loss, and blisters on the feet were more adequately prevented during the Delta outbreak than they were during the Omicron outbreak. A sore throat and a hoarse voice were more prevalent during the Omicron outbreak. Individuals infected with Omicron are less likely to show at least one of the three classic symptoms of COVID-19: the loss of smell, a fever, and a persistent cough, which associated with individuals infected with the Delta variant [71]. A study showed that acute symptoms prevailed for a more extended period in patients affected during the Delta outbreak (average of 8.89 days) than they did during Omicron outbreak (average of 6.87 days). It also showed that 1.9% of the vaccinated individuals were admitted to hospitals during the Omicron outbreak compared to 2.6% during the Delta outbreak [71–73]. A high number of asymptomatic carriers were identified during the outbreak of the Omicron variant, suggesting that it caused milder symptoms [59]. Respiratory distress is a common symptom in all age groups. Among the gastroenterological symptoms, vomiting is the most common one, and diarrhea and abdominal cramps are common in children aged 5–9 years who are infected with Omicron. Children in the age group of 9–11 show less severe symptoms than infants do, which is valid for both the Delta and the Omicron variants. There have been reports of seizure-associated infections caused by the variant [74]. In vivo studies showed that the Omicron variant did not cause a significant loss of body weight, the viral load was lower, and the amount of lung damage was significantly smaller, and the mortality rates were also lower compared to those of other variants [75–77]. Omicron tends to stay in the upper respiratory tract, such as in the nose, throat, and bronchi, rather than settling in the lungs [78,79]. However, in severe cases, pneumonia, respiratory failure, and death can also occur [80,81]. Bronchitis, hypertension, and diabetes are significant comorbidities in individuals infected with the Omicron variant. Another study showed that 36.1% of the Omicron-infected individuals did not show any antibody response, 62.7% of them produced IgG, and 1.2% of them produced IgG and IgM. Many of the Omicron-infected individuals showed abnormally high WBCs, lymphocytes, monocytes, and neutrophils levels, which can lead to monocytosis, neutrophilia, lymphocytopenia, and leukocytosis, while the RBCs and hemoglobin levels were in the normal range [82].

Regarding the transmissibility and effectiveness of the vaccines against the variants, Alpha is 50% more transmissible than the original Wuhan strain is, and it is associated with increased disease severity [83–85]. However, the vaccines and monoclonal antibodies remain effective against the variant. The Beta strain is again 50% more transmissible than the previous strains are, but it is not related a more severe disease. It has a reduced neutralization efficiency by the antibodies, and people who have been previously infected are at a greater risk of being reinfected. The Gamma variant is 1.7–2.4 times more transmissible than the non-VOCs are, and patients who have been previously infected with COVID-19 have 54–79% protection against the variant, and the existing vaccines work well against the variant [84]. The Delta variant is 40–60% more transmissible than the Alpha one is, and it is twice as transmissible as the Wuhan strain. Vaccines are less effective against the Delta variant [29,64,85,86]. The vaccines are least effective against the Omicron sub-lineages, especially the BA.4 and BA.5 ones. Another new sub-lineage of Omicron, BA.2.75, which was first found in India in June 2022 is spreading faster than the BA.5 variant did, and it also evades the protection by the immune system caused by a previous infection or antibody generation. However, there are no unique symptoms related to BA.2.75 infection, with a mild fever in most cases, and sometimes, the patients are even asymptomatic. The Omicron variant is the most transmissible one of all the other variants, but the severity of the disease is comparatively lower. As the Omicron variant has mutations that lead to higher transmissibility and better immune escape, the combined mutations are responsible for Omicron's dominance over the other variants.

The COVID-19 pandemic is a global challenge, and it is necessary to improve healthcare systems, especially the vaccination rates in developing countries. Omicron is highly transmissible, but it is less pathogenic than the other SARS-CoV-2 variants. Even double-dose-vaccinated people with comorbidity are not protected against Omicron [87]. However, immunity can prevent the severity of COVID-19. Increased immunity among the vaccinated population and them having been previously infected can reduce the severity of COVID-19, and SARS-CoV-2 can become endemic, similar to other seasonal viral infections. However, Omicron may still cause severe COVID-19 and death, especially in comorbid and unvaccinated individuals.

6. Infection, Reinfection, and Transmissibility

Mutations in viruses is are widespread phenomena. The SARS-CoV-2 virus is not exempt from this. The main question that is speculated by the entire scientific community is the plentiful number of mutations residing in the genome of the Omicron variant, which has significantly decreased the chances of the occurrence of primary infections, but it has resulted in a more significant increase in the chances of reinfecting individuals [88]. It uses the spike glycoprotein, which binds with the host ACE2 receptor and mediates the membrane fusion by utilizing furin and cathepsin L or TMPRSS2 [89]. Most importantly, this variant is more highly contagious than the previously evolved strains are [90]. Similar to the other mutated variants of SARS-CoV-2, Omicron too shares the same procedure of infecting the host cells. The infectivity rate of the Omicron variant is many folds greater than that of the Delta variant [16]. Before November 2021, the rate of reinfection worldwide was considerably low, around 2%, as implied by some international reports. After the emergence of the Omicron variant, the scenario changed. A deeper look at the reinfection rates of Omicron in a place in South Africa elucidated that this variant is more proficient at reinfecting individuals due to its capability of escaping the immune system [91–93]. According to the GISAID data, the Omicron variant consists of 11 mutations in the N-terminal domain with an insertion and six deletions. The ins214EPE and N211 mutations present in Omicron have not been reported in any other mutant variants that has evolved before this one [24]. Out of the five VOCs that have been declared to date, some of the mutations responsible for other viral fitness are D614G, T478K, N501Y, and K417N. Besides these mutations, Omicron possesses several more substitutions, which increases the infectivity rate of these variants by many times [94].

The transmission rate of Omicron is approximately 3.2 folds higher than that of the Delta variant, with a doubling time of ~3 days [95,96]. Among the evolving sub-lineages of the Omicron variant, the BA.2 one is found to be more transmissible than the BA.1 sub-lineage is among household contacts [97]. An incident reported in Norway details an alarming scenario about the transmission of the Omicron variant. Of 117 individuals who went to a party, 76% of them were Omicron victims. Out of all of them, 96% of the people who attended the party had been vaccinated. This alarming fact highlights the high transmission rates of this variant, even in the fully vaccinated subjects [98]. Notably, the elevated rates of Omicron transmission can also be due to its potent immune evading capacity, nullifying the vaccinated subjects' neutralization capabilities [99]. Apart from this, altered cellular tropism and different pathways of infecting host cells may contribute to the increased transmissibility of Omicron [100,101]. The infection landscape of the Omicron variant describes the silent transmission of the virus from one individual to another, as some victims of Omicron rarely show any symptoms [90]. Some of the ancillary factors responsible for Omicron transmission is the binding of the RBD with the hACE2. However, the exact facts about the viral loads after an infection with Omicron remain undiscovered [102,103].

7. Omicron Entry and Associated Immunological Features inside the Host Cells

The entry of the SARS-CoV-2 virus inside the host cells is mainly facilitated by the S-glycoprotein [104]. Recent investigations elaborated that this Omicron variant follows an altered cellular entry route. Instead of entering through the plasma membrane, the Omicron variant follows the endosomal entry pathway, which is enhanced by the cathepsins instead of TMPRSS2. Willett et al. also found that pseudotyped Omicron variant infection was more prominent in the cells with a lower expression of TMPRSS2 than it was in those with a high TMPRSS2 expression level. This, in turn, proves the affinity of Omicron's entry inside the cell through the endosomes [105]. Of several mutations, P681H, N679K, and H655Y reside in the region adjacent to the furin cleavage site. In the case of the previously emerged variants, Gamma and Alpha, the P681H mutation mediates the cleavage [106]. For Omicron, the scenario is slightly different. This variant's cleavage efficiency is lower than it is for the others, suggesting that the N679K and H655Y mutations impede the cleavage [107]. Unlike the other variants, Omicron possesses additional mutations in all of the structural proteins. The mutations in the N and S proteins escalate the cellular permeability of the Omicron variant.

Additionally, this mutation favors a more robust capsid assembly, which is almost three folds greater than that of the newly emerged Delta variant [108]. The Omicron variant also uses similar protein receptors as the other emerged variants do. Experimental evidence indicates that Omicron entry was predominant in cells with a higher number of ACE2 receptors [22]. Willet et al. elucidated that the changing of Omicron's preferred entry route indicates that it will have more replication fitness in the upper respiratory tract. Due to the enormous amount of alterations in the spike protein, the recently emerged sub-lineages of Omicron, especially the BA.1 and BA.2 ones, do not form syncytia, which are mainly formed during the initial stages of the processing of the spike protein at the boundary of the two subunits, namely, S1 and S2. These changes, along with the switched entry route, alter cellular tropism [109].

Kared et al. have mentioned the immunological events associated with the entry of Omicron variants inside the host cell, especially for vaccinated individuals. Omicron entry triggers the production of both the cytotoxic and follicular T helper cells, along with a massive surge of RBD and spike-related IgG+ B cells known as plasmablasts, along with some memory B cells [110]. The B cells follow the exact mechanism of neutralization as that which is seen in the wild-type SARS-CoV-2 variant. The B cells derived from the pool of live memory cells has a similar interaction pattern with the S-glycoprotein of Omicron as that of the wild-type variant [111].

8. Interaction of Host ACE2 and Capability of Binding with RBD

Compared to Delta, the Omicron variant exhibits an extreme affinity for the ACE2 receptor and accelerates the transmission rate of this variant by many folds. The mutations that result in the extremely high affinity of the Omicron RBD with the human ACE2 receptor are Q493R, T478K, S373P, N501Y, Q498R, S371L, and S375F (Table 1). Additionally, Omicron's S protein and RBD harbors some amino acids such as leucine and phenylalanine, which are naturally hydrophobic [112]. Some of the mutations in the Omicron variant even contribute to the formation of salt bridges or several hydrogen bonds, which contribute to the binding of the spike protein with hACE2. The polar contacts between Omicron and ACE2 can be significantly lost by K417N and E484A, negating some of the improved interactions created by other mutations [46,51,53].

A deeper look at the crystal structures of the RBD–ACE2 complex of Omicron indicates that the surface area that the Omicron variant can access for interaction with the host is much more prominent than it is for the Delta variant [51]. According to Jung et al., out of the 31 alterations in the spike protein of the Omicron variant in comparison to those in the wild-type variant from Wuhan, 12 changes are found in the S1 subunit of the spike protein, which reside very near the N-terminal region. Fifteen changes can be seen in the receptor-binding domain, with more than five mutations residing near the C terminal.

Moreover, the RBD, which forms a direct connection with hACE2 for binding, harbors ten significant mutations, thereby altering the affinity of the spike protein to bind with the host receptor [113]. Among all of the emerging variants, Omicron is highly transmissible. Computational studies regarding the RBD–hACE2 complex of Omicron evidence that it is incredibly stable due to the replacement of some uncharged amino acid residues with lysine and arginine [6]. T478K, Q498R, N440K, and Q493R are some of the mutations present in the RBD of Omicron's spike protein, where there are some replacements with positively charged residues, thereby improving the binding of RBD to the human ACE2 receptor. Owing to the growth of the side chain, the T478K mutation is situated very close to a solvent-prone area, permitting the interaction between ACE2 and Omicron's RBD. Furthermore, the Q493R mutation enables an advantageous interaction with certain damaging amino acids such as Glu35 and Asp38 in the ACE2 receptor. It also enables a powerful binding effect [6].

9. Phylogenomics and Distribution of Omicron and Its Sub-Variants

Several scientists have studied the phylogenomics of Omicron and its variants, and their studies have immense importance with respect to the evolution of the virus (Figure 6A). Recently, we have found the phylogenetics of Omicron and its sub-variants. Callebaut et al. described the phylogenetic properties of the BA.1 and BA.2 variants. Samples were collected from Omicron-infected patients [114]. Kandeel M, El-Deeb demonstrated the evolutionary relationships of the RBD of SARS-CoV-2 using a phylogenetic tree. The study placed the Omicron variant into a novel monophyletic class [115]. Additionally, it also described the rapid appearance of multiple sub-variants of Omicron and their divergence [116].

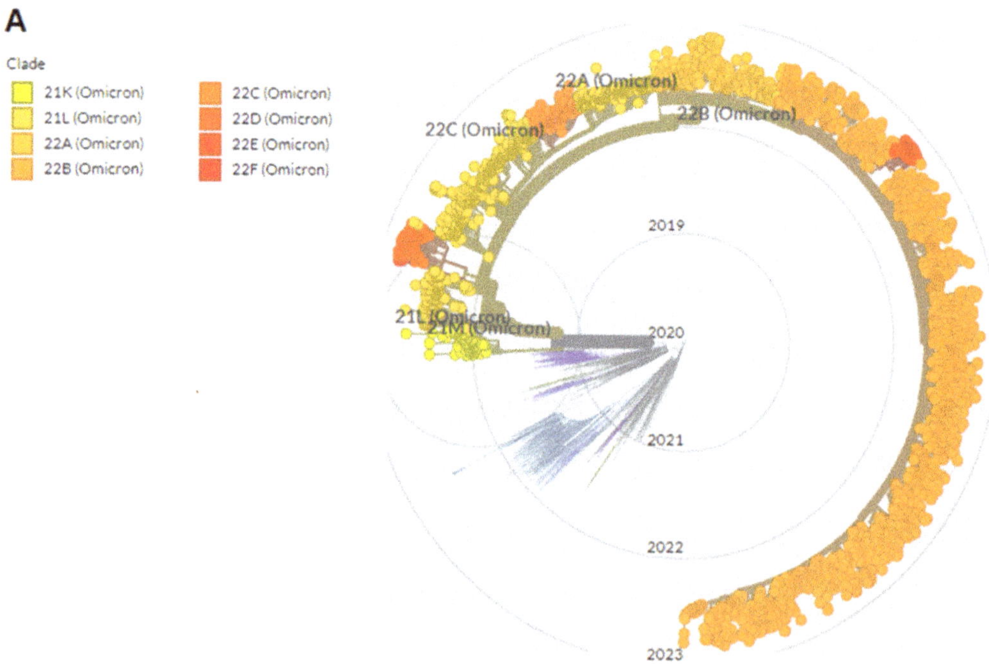

Figure 6. *Cont.*

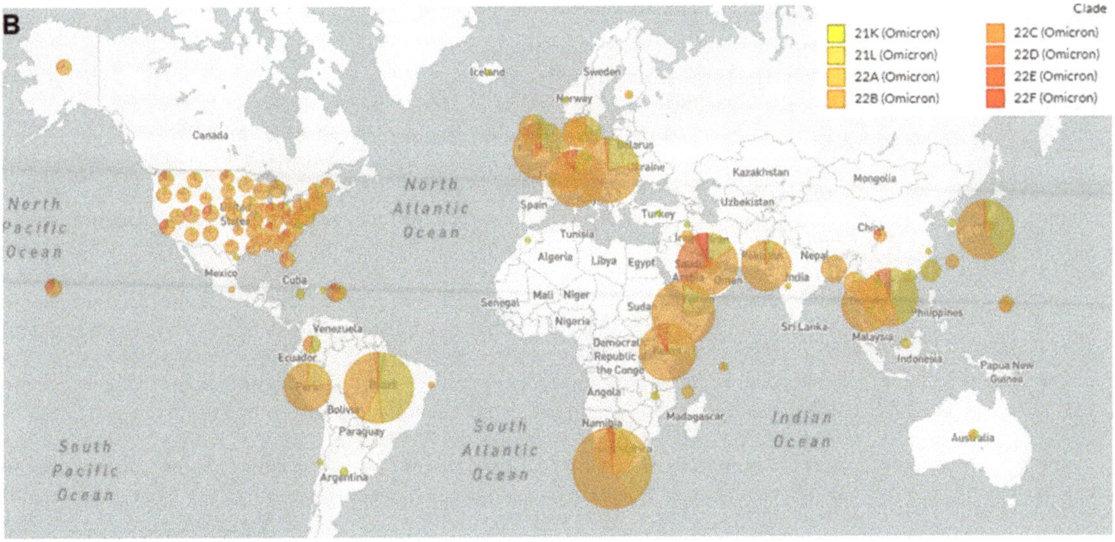

Figure 6. The figure shows the phylogenetic tree of the Omicron and its sub-variants. It also describes the distribution of Omicron and its sub-variants in the entire world. (**A**) The circular phylogenetic tree using the Omicron and its sub-variants (**B**) The distribution of Omicron and its sub-variants. These two figures (**A**,**B**) were developed using the next strain server.

It is also essential to understand the distribution of Omicron and its sub-variants. After the first identification of the Omicron variant in South Africa and Botswana, the variant spread throughout the globe, and several sub-variants, BA.1, BA.2, BA.3, BA.4, and BA.5, generated over time and were spotted throughout the world (Figure 6B). Recently, a new sub-variant, BA.2.75.2, was generated in India, which might be of global concern [20]. We need to obtain more detailed information about the distribution of several sub-variants of Omicron.

10. Immune Escape of Emerging Omicron Variant and Its Sub-Variant

In general, vaccine effectiveness against severe diseases is a matter of concern. The vaccine's effectiveness is not largely affected by the variants. This is because of the mutations of the variants which hinder the neutralization potency of any vaccine. New variants have developed as a result of certain mutations. Several mutations have been observed in the newly developed variants which alter the binding region of nAb, leading to antibody escape [117,118]. In the Omicron case, several mutations have been noted in the nAb binding region of the S protein, especially in RBD and NTD, which cause the nAb escape phenomenon [13,19,27–29,119,120]. Therefore, we can say that the Omicron variant possesses a partial vaccine escape ability.

Recent studies elucidate that the Omicron variant and the three sub-lineages, BA.1, BA.2, and BA.3, are very competent in escaping the immune system. The subjects who have taken one or two doses of the vaccine cannot protect against this variant significantly, thus, the neutralization efficiency of these vaccines is gradually decreasing. Most surprisingly, people who had received three shots of the vaccine only have partial protection from the infection of this variant. However, vaccine escape is a common phenomenon. Several researchers urge for the development of new vaccines against the Omicron variant. However, several researchers or pharmacological companies have developed new vaccines against the Omicron variant (Table 2). Similarly, a bivalent COVID-19 vaccine (ancestral and Omicron) can provide long-term protection. Recently, ModernaTX has developed an mRNA-based bivalent Omicron-containing vaccine. The study is now in Phase II and

Phase III. The study has evaluated the safety and immunogenicity of the mRNA vaccine boosters (bivalent Omicron-containing vaccine). Chalkias et al. have published data of a clinical trial and evaluated the immunogenicity, reactogenicity, and safety the bivalent Omicron-containing vaccine (mRNA-1273.214). In this study, they assessed three parameters of the bivalent vaccine on the 28th day after the booster dose. Here, the participants received either mRNA-1273 (n = 377) or 50 μg of mRNA-1273.214 (437 participants) as a second booster dose. The researchers found that mRNA-1273.214 (the bivalent Omicron-containing vaccine) elicited superior neutralizing antibody responses compared to the mRNA-1273 vaccine ones against the Omicron variant (ClinicalTrials.gov; Clinical trial: NCT04927065) [121]. Other than the bivalent vaccine of ModernaTX, Pfizer-BioNTech has also developed a bivalent COVID-19 vaccine. A recent clinical trial has been conducted to understand the vaccine's safety profile (ClinicalTrials.gov; NCT04977479). The first dosage of the mRNA vaccine produced a systemic allergic reactions in some individuals. The researchers want to study the safety profile of giving a second mRNA COVID-19 vaccine to individuals who had developed a systemic allergic reaction to their first dose. Similarly, these two bivalent vaccines' safety profiles have been assessed in kidney transplant recipients (ClinicalTrials.gov; NCT05518487). It is one of the most likely reasons responsible for the rapid spread of the Omicron variant in countries where people have natural immunity or a rapid vaccination rate [109]. The various mutations in the N-terminal and the RBD of the Omicron variant are the sites that are majorly targeted by the antibodies [122–125]. The high mutation rates in these positions are the key factors responsible for changing the antigenicity.

Moreover, this antigenic shift can even nullify the overall immunity in the host's system [126]. The conformation of the S protein and RBD is a significant factor dominating antibody recognition. The trimeric spike complex of the Omicron variant adopts a single "up" conformation, with the RBD keeping the other two in the "down" conformational state, which is a bit different from the previously emerged variants [46,127]. The stearic hindrances induced due to the mutations are solely responsible for altering the interactions in the antibody-binding sites. The unprecedented changes in the spike protein also interfere with the recognition of the antibodies [126,127]. Some of the mutations in the Omicron variant such as Q498R, S477N, Y505H, G496H, and Q493R, along with the other mutational changes prevalent in the VOCs such as T478K, N501Y, and E484K are majorly involved in altering its antigenicity, with this variant being more efficient at escaping the immune system [128]. Cui et al. have mentioned that the two main sites for neutralization, the RBD and the NTD, are heavily mutated in the Omicron variant, which causes severe changes in the conformation of several antigenic sites. The three minor deletions, four substitutions, and one insertion of a 3-residue-long amino acid in the N-terminal region have been the primary cause behind the immune escape strategy of the Omicron variant [46].

Furthermore, the recently reported sub-lineages of Omicron, BA.4, BA.5, and BA.2.12.1, have illustrated more robust strategies for escaping the immune system than the BA.1 and BA.2 sub-lineages have. The BA.1 variant can produce several copies of BA.1-specific antibodies that can be effective against BA.1 infection. However, the other sub-lineages, namely, the BA.4/BA.5 and the BA.2 variants, can invalidate the neutralization efficiency of these antibodies because of the presence of F486V and D405N mutations [37].

Table 2. Vaccines against Omicron in the clinical trial.

SL. NO.	Vaccine	Country of Origin	Company Name	Clinical Trial Number	Phase	Recruitment Status	Number of Partici-pants	Remark
1.	ABO1009-DP vaccine	China	Suzhou Abogen Biosciences Co., Ltd.	NCT05433194	Phase I	Not Yet recruiting	48	A clinical trial which aimed to monitor the safety and efficacy profile of this vaccine against Omicron in fully vaccinated subjects below 18 years
2.	Inactivated Omicron COVID-19 vaccine (Vero Cell) Inactivated	China	China National Biotec Group Company Limited	NCT05365724	Phase II	Recruiting	280	A non-randomized trial which aims to monitor the safety and efficacy profiles of the vaccine in non-vaccinated subjects below 18 years old
3.	mRNA-1273.214 Vaccine	Israel	Sheba Medical Center	NCT05383560	Phase II	Not Yet recruiting	150	A placebo controlled study aimed to evaluate the immunogenicity of Omicron-matched booster doses in adult subjects
4.	SCTV01E	China	Sinocelltech Ltd.	NCT05308576	Phase III	Not Yet recruiting	12,000	A randomized study which monitored the safety profile of SCTV01E in subjects aging 12 years or older
5.	BIBP Omicron Inactivated COVID-19 vaccine	Hong Kong	China National Biotec Group Company Limited	NCT05382871	Phase III	Recruiting	1800	A randomized study which monitors the safety and efficacy of this vaccine in subjects who previously received two or three doses of any vaccine
6.	mRNA-1273.214 (bivalent Omicron-containing vaccine)	United States	ModernaTX, Inc.	NCT04927065	Phase III	Active	5158	Immunogenicity and safety evaluation of bivalent mRNA vaccine boosters for SARS-CoV-2 variants
7.	Pfizer-BioNTech bivalent (Omicron-specific) vaccine	Australia	Murdoch Childrens Research Institute	NCT05543356	Phase III	Withdrawn	1143	Evaluation of bivalent Omicron-specific COVID-19 vaccine booster dose (Pfizer-BioNTech) in healthy adults
8.	Pfizer-BioNTech COVID-19 bivalent vaccine	United States	National Institute of Allergy and Infectious Diseases (NIAID)	NCT04977479	Phase II	Active	17	Safety analysis of the COVID-19 mRNA vaccine (2nd dose) to individuals who had a systemic allergic reaction to a prior dose
9.	Bivalent booster of mRNA based COVID-19 vaccine	United States	National Institute of Allergy and Infectious Diseases (NIAID)	NCT05518487	Phase II	Not Yet recruiting	80	Safety and immunogenicity study of single dose of bivalent (mRNA-based) vaccine to individuals (kidney transplant recipient) with a persistently low SARS CoV-2 antibody titer
10.	Bivalent mRNA COVID-19 vaccine	United States	National Institute of Allergy and Infectious Diseases (NIAID)	NCT05077254	Phase II	Recruiting	400	Evaluation of Ab response to an extra dose of bivalent (mRNA-based) COVID-19 vaccination in subject of immunosuppression reduction in organ (kidney and liver) transplant recipients

11. Antiviral Drugs and Antibody-Based Therapeutics against the Omicron and Its Sub-Variants

Several antiviral drugs and antibody-based therapeutics have been investigated and proposed over time against Omicron and its sub-variants. The investigated and proposed antiviral drugs and antibodies are discussed below.

11.1. *Efficacy of Antiviral Drugs*

Numerous antiviral options have been explored for emergency use in hospitalized and non-hospitalized patients to reduce the clinical severity in patients infected with the SARS-CoV-2 wild strain and other variants, including Omicron. A number of antivirals have been proposed against Omicron, such as Remdesivir, Molnupiravir, Camostat, and Ensovibep. These antivirals have been investigated over time to assess their antiviral activities against Omicron and its sub-variants [66]. Takashita et al. have recently evaluated the antiviral activity of three antiviral molecules, such as Remdesivir, Molnupiravir, and Lufotrelvir. In this study, the researchers used three chemicals, namely, 441524, EIDD-1931, and PF-00835231, as therapeutic molecules. The study indicated that these three compounds had efficacy against the Omicron variant. In this study, the researchers evaluated the drugs' susceptibility to GS-441524, EIDD-1931, and PF-00835231 using a 50% inhibitory concentration (IC50) value. The value for each of them was found to be 1.2, 0.8, and 0.7, respectively. However, the data are different due to the influence of different factors. Here, GS-441524 is an RdRp (RNA-dependent RNA polymerase) inhibitor, and the molecule is the active form of Remdesivir. Similarly, EIDD-1931 is also an RdRp inhibitor, and the molecule is an active form of Molnupiravir. At the same time, the study confirms that Omicron-infected patients can be treated with these drugs. PF-00835231 is a protease inhibitor, which is the active form of PF-07304814 [129]. PF-07304814 is known as Lufotrelvir, which was developed by Pfizer. Similar to remdesivir, this molecule can be administered by intravenous infusion.

Another oral protease inhibitor that has been found by researchers is Nirmatrelvir. Arbel et al. evaluated the activity of these molecules in 109,254 patients. During the study period, 4% of the total number of patients (3902) received Nirmatrelvir. The researchers found that the death and hospitalization rates were significantly lower among the Nirmatrelvir-receiving patients compared to that among the patients who did not receive any dose [130]. The USFDA approved the drug through EUA (an emergency use authorization) for treating of mild-to-moderately infected patients. It was approved for oral use in December 2021. Bojkova et al. have assessed the effects of some molecules, such as Remdesivir, Favipiravir, Ribavirin, EIDD-1931, PF-07321332, Camostat, Nafamostat, and Aprotinin, in Omicron-infected cell cultures. They found similar kinds of antiviral activity among the Delta and Omicron isolates [131]. Vangeel et al. performed an in vitro antiviral assay and reported that Remdesivir (parent nucleoside GS-441524), Molnupiravir (parent nucleoside EIDD-1931), and Nirmatrelvir showed antiviral activity against Omicron. These molecules have also shown antiviral activity against the wild strain of it and other VOCs [132].

The Nirmatrelvir–Ritonavir combination is now an essential antiviral option against the Omicron variant. Several scientists have explained the activity of the Nirmatrelvir–Ritonavir combination against Omicron. Recently, from a cohort study with COVID-19 patients (N = 41,255), Wong et al. stated that the molecules could be considered to be therapeutics for the early phase of the infection [133]. In another study, Wong found that the Nirmatrelvir–Ritonavir combination could have been a therapeutic agent during the early phase of the infection during Omicron BA.2's wave. The researchers concluded this from a cohort study in Hong Kong [134]. A recent article describes the recommended indications, antiviral activity, pharmacokinetics, mechanisms of action, clinical trial of drug interactions, and adverse affects of the antiviral molecules such as Molnupiravir and the Nirmatrelvir–Ritonavir combination (Paxlovid) against the Omicron variant [135]. Therefore, these two molecules (Molnupiravir and the Nirmatrelvir–Ritonavir combination) are significant additions to the early phase of the treatment of COVID-19, especially to Omicron and its sub-variants.

11.2. Efficiency Therapeutic Antibodies

Scientists are facing a real challenge to finding therapeutic antibodies against the Omicron variant because the therapeutic antibody escapes their neutralization efficacy due to certain properties of the Omicron variant [120,136]. Due to this, several scientists have tried to evaluate therapeutic antibodies against the Omicron variant over time and assess their efficacy of naturalization (Table 3). Recently Tao et al. published a meta-analysis and systematic review, where they found that several studies were involved in understanding the susceptibility of mAbs (monoclonal antibodies) against the Omicron variants [132]. Takashita et al. have assessed the antibodies against Omicron which are Bamlanivimab (LY-CoV555), Imdevimab (REGN10987), Casirivimab (REGN10933), Tixagevimab (COV2-2196), Cilgavimab (COV2-2130), and Sotrovimab precursors (S309). The researchers also evaluated a plethora of the combinations of monoclonal antibodies. Some of the combinations include Tixagevimab with Cilgavimab, Imdevimab with Casirivimab, and Etesevimab with Bamlanivimab. It was also found that these combinations of monoclonal antibodies could neutralize the wild strain as well as the Delta and Alpha variants. At the same time, the combined treatment of Bamlanivimab and Etesevimab highlighted a reduced neutralizing activity against the Gamma variant. Furthermore, these combinations have completely lost their neutralization efficacy against the Beta and Omicron variants [129].

Table 3. The efficiency of the antibodies effective against Omicron and its sub-variants.

Sl. No.	Therapeutic Antibodies	Neutralization Efficacy in Different Omicron Sub-Variants				
		BA.1	BA.2	BA.3	BA.4	BA.5
1.	Tixagevimab	Low	Low	Low	Low	Low
2.	Bamlanivimab	Low	Low	Low	Low	Low
3.	Imdevimab	Low	Moderate	Low	Moderate	Moderate
4.	Regdanvimab	Low	Low	-	-	-
5.	Sotrovimab	Moderate	Moderate	Moderate	Moderate	Moderate
6.	Casirivimab	Low	Low	Low	Low	Low
7.	Cilgavimab	Low	High	High	High	High
8.	Etesevimab	Low	Low	Low	Low	Low
9.	Bebtelovimab	High	High	High	High	High
10.	Bamlanivimab + Etesevimab	Low	Low	Low	Low	Low

Similarly, they also found that the Casirivimab and Imdevimab combination has shown activity against the Gamma and Beta variants. However, this combination failed to neutralize the Omicron one. However, it has been noted that the Cilgavimab–tixagevimab combination has shown significant neutralization potency against the Beta, Gamma, and Omicron ones [129]. Similarly, Tada et al. found from a study that Sotrovimab and Evusheld were partially effective against the Omicron pseudotype. On the other hand, Eli Lilly and Regeneron monoclonal antibodies were found to be ineffective against the Omicron pseudotype [137]. Several in silico studies have been performed to identify the therapeutic antibodies against Omicron. In this field, Shah and Woo have suggested that a cocktail of sotrovimab (GSK, S203 mAb) and Evusheld (AstraZeneca mAbs) could successfully neutralize the Omicron variant [138]. Researchers have also tried to understand the interaction between the neutralizing antibodies (nAB) and Omicron's spike protein. It might provide a deeper understanding of the specific interaction mechanisms possessed by these antibodies. A recent study informed us that ZCB11 is a promising antibody against the Omicron variant. Zhou et al. have elucidated the interaction between ZCB11 and the spike protein of the Omicron variant (PDB id: 7XH8). The study informed us that ZCB11 targets the viral RBD and neutralizes the spike protein of the SARS-CoV-2 variants such as Delta or Omicron [139] (Figure 7).

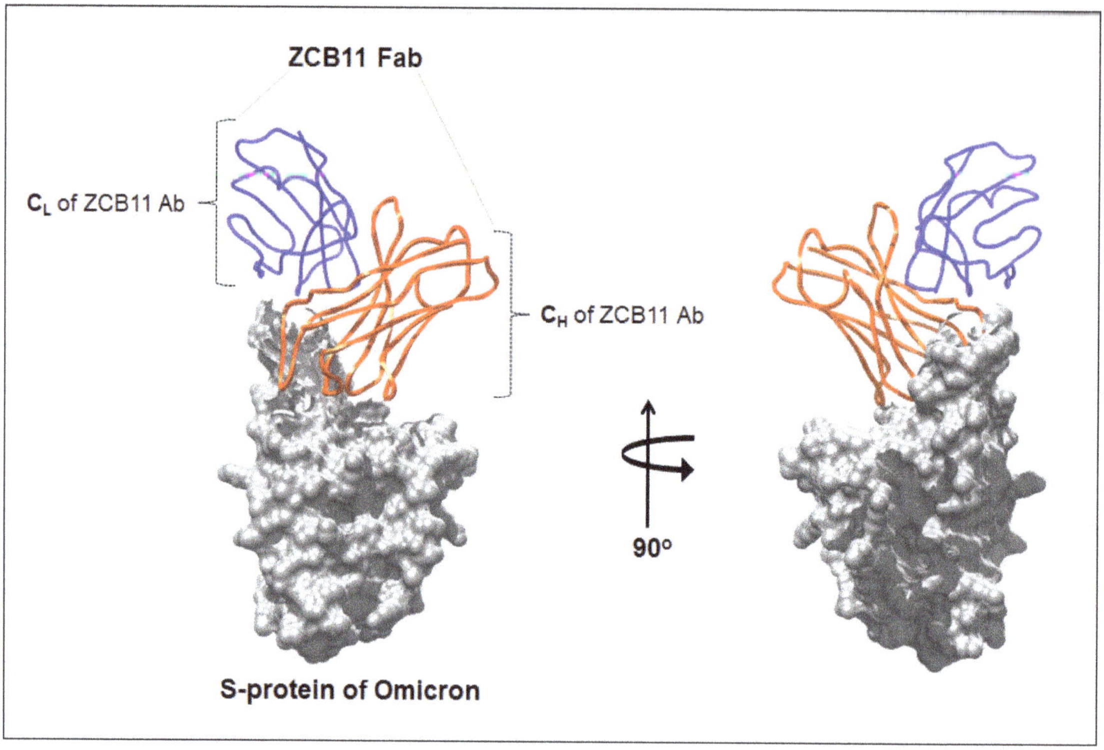

Figure 7. The figure shows the interaction structure of neutralizing antibodies (nAb) with the Omicron spike protein. It shows the interaction Fab fragment of ZCB11 against the SARS-CoV-2 Omicron spike. The structure was developed from a PDB file (PDB id: 7XH8).

12. Conclusions

Current data have informed us of the three significant properties of the Omicron variant. Firstly, the Omicron variant causes less severe infections. Secondly, the variant has a very high rate of transmissibility compared to that of other VOCs. Lastly, the Omicron variant has a high immune escape capacity and partial vaccine escape ability. Efforts are being made over time to develop the next-generation vaccine and mutation-proof vaccines [28,140,141]. At the same time, it has been observed that the Omicron variant and its sub-variant possess a very high number of mutations [27,29,120]. These mutations provide three significant properties to the Omicron variant.

Recently, it has been seen that hybrid immunity significantly provide more immune protective against SARS-CoV-2 and the other VOCs [141,142]. Therefore, we need to explore the possibility of hybrid immunity for protection against Omicron. At the same time, researchers have informed us that the Omicron variant might be a possible vaccine candidate. The viral strain can be used as a promising live-attenuated vaccine candidate. Therefore, a strategy has been proposed to find a possible solution to provide protective immunity against Omicron, which is known as the "virus against the virus" [143]. However, a bivalent Omicron-containing vaccine has recently been developed by ModernaTX, which can provide long-term protection against the Omicron variant [121]. Molnupiravir and the Nirmatrelvir–Ritonavir combination (Paxlovid) have been found to be effective therapeutic antiviral molecules against the Omicron variant and its sub-variants. However, further studies are needed on the Omicron variant to obtain a clear idea about its pathophysiology and the infection landscape, which will be beneficial for the development of suitable therapeutics.

Author Contributions: Conceptualization was conducted by C.C.; Writing—original draft preparation, investigation was performed by S.C. and M.B.; Data curation, validation and formal analysis conducted by S.C. and S.N.; Writing—review and editing was conducted by C.C.; Visualization and validation was conducted by K.D.; Supervision and Project administration was conducted by C.C. All authors have read and agreed to the published version of the manuscript.

Funding: This research received no funding.

Institutional Review Board Statement: Not applicable.

Informed Consent Statement: Not applicable.

Data Availability Statement: No new data were created or analyzed in this study. Data sharing is not applicable to this article.

Conflicts of Interest: The authors declare that they have no competing interests.

References

1. Zhu, N.; Zhang, D.; Wang, W.; Li, X.; Yang, B.; Song, J.; Zhao, X.; Huang, B.; Shi, W.; Lu, R. A novel coronavirus from patients with pneumonia in China, 2019. *N. Engl. J. Med.* **2020**, *382*, 727–733. [CrossRef] [PubMed]
2. Huang, C.; Wang, Y.; Li, X.; Ren, L.; Zhao, J.; Hu, Y.; Zhang, L.; Fan, G.; Xu, J.; Gu, X. Clinical features of patients infected with 2019 novel coronavirus in Wuhan, China. *Lancet* **2020**, *395*, 497–506. [CrossRef] [PubMed]
3. He, X.; Hong, W.; Pan, X.; Lu, G.; Wei, X. SARS-CoV-2 Omicron variant: Characteristics and prevention. *MedComm* **2021**, *2*, 838–845. [CrossRef] [PubMed]
4. Azgari, C.; Kilinc, Z.; Turhan, B.; Circi, D.; Adebali, O. The mutation profile of SARS-CoV-2 is primarily shaped by the host antiviral defense. *Viruses* **2021**, *13*, 394. [CrossRef] [PubMed]
5. Simmonds, P. Rampant C→ U hypermutation in the genomes of SARS-CoV-2 and other coronaviruses: Causes and consequences for their short-and long-term evolutionary trajectories. *Msphere* **2020**, *5*, e00408–e00420. [CrossRef]
6. Jung, C.; Kmiec, D.; Koepke, L.; Zech, F.; Jacob, T.; Sparrer, K.M.J.; Kirchhoff, F. Omicron: What Makes the Latest SARS-CoV-2 Variant of Concern So Concerning? *J. Virol.* **2022**, *96*, e0207721. [CrossRef]
7. Cherian, S.; Potdar, V.; Jadhav, S.; Yadav, P.; Gupta, N.; Das, M.; Rakshit, P.; Singh, S.; Abraham, P.; Panda, S. SARS-CoV-2 spike mutations, L452R, T478K, E484Q and P681R, in the second wave of COVID-19 in Maharashtra, India. *Microorganisms* **2021**, *9*, 1542. [CrossRef]
8. Davies, N.G.; Abbott, S.; Barnard, R.C.; Jarvis, C.I.; Kucharski, A.J.; Munday, J.D.; Pearson, C.A.; Russell, T.W.; Tully, D.C.; Washburne, A.D. Estimated transmissibility and impact of SARS-CoV-2 lineage B.1.1.7 in England. *Science* **2021**, *372*, eabg3055. [CrossRef]
9. Tegally, H.; Moir, M.; Everatt, J.; Giovanetti, M.; Scheepers, C.; Wilkinson, E.; Subramoney, K.; Moyo, S.; Amoako, D.G.; Althaus, C.L. Continued emergence and evolution of Omicron in South Africa: New BA.4 and BA.5 lineages. *medRxiv* **2022**. [CrossRef]
10. Volz, E.; Mishra, S.; Chand, M.; Barrett, J.C.; Johnson, R.; Geidelberg, L.; Hinsley, W.R.; Laydon, D.J.; Dabrera, G.; O'Toole, Á. Assessing transmissibility of SARS-CoV-2 lineage B.1.1.7 in England. *Nature* **2021**, *593*, 266–269. [CrossRef]
11. Callaway, E. What Omicron's BA.4 and BA.5 variants mean for the pandemic. *Nature* **2022**, *606*, 848–849. [CrossRef]
12. Brüssow, H. COVID-19: Omicron—The latest, the least virulent, but probably not the last variant of concern of SARS-CoV-2. *Microb. Biotechnol.* **2022**, *15*, 1927–1939. [CrossRef]
13. Dhama, K.; Nainu, F.; Frediansyah, A.; Yatoo, M.I.; Mohapatra, R.K.; Chakraborty, S.; Zhou, H.; Islam, M.R.; Mamada, S.S.; Kusuma, H.I.; et al. Global emerging Omicron variant of SARS-CoV-2: Impacts, challenges and strategies. *J. Infect. Public Health* **2022**, *16*, 4–14. [CrossRef]
14. Callaway, E. Heavily mutated Omicron variant puts scientists on alert. *Nature* **2021**, *600*, 21. [CrossRef]
15. Liu, Y.; Rocklöv, J. The effective reproductive number of the Omicron variant of SARS-CoV-2 is several times relative to Delta. *J. Travel Med.* **2022**, *29*, taac037. [CrossRef]
16. Fan, Y.; Li, X.; Zhang, L.; Wan, S.; Zhang, L.; Zhou, F. SARS-CoV-2 Omicron variant: Recent progress and future perspectives. *Signal Transduct. Target. Ther.* **2022**, *7*, 141. [CrossRef]
17. Callaway, E.; Ledford, H. How bad is Omicron? What scientists know so far. *Nature* **2021**, *600*, 197–199. [CrossRef]

18. Taylor, C.A.; Whitaker, M.; Anglin, O.; Milucky, J.; Patel, K.; Pham, H.; Chai, S.J.; Alden, N.B.; Yousey-Hindes, K.; Anderson, E.J.; et al. COVID-NET Surveillance Team. COVID-19-Associated Hospitalizations Among Adults During SARS-CoV-2 Delta and Omicron Variant Predominance, by Race/Ethnicity and Vaccination Status—COVID-NET, 14 States, July 2021-January 2022. *MMWR Morb. Mortal. Wkly. Rep.* **2022**, *71*, 466–473. [CrossRef]
19. Mohapatra, R.K.; Kandi, V.; Sarangi, A.K.; Verma, S.; Tuli, H.S.; Chakraborty, S.; Chakraborty, C.; Dhama, K. The recently emerged BA. 4 and BA. 5 lineages of Omicron and their global health concerns amid the ongoing wave of COVID-19 pandemic–Correspondence. *Int. J. Surg.* **2022**, *103*, 106698. [CrossRef]
20. Chakraborty, C.; Bhattacharya, M.; Sharma, A.R.; Dhama, K.; Lee, S.S. The rapid emergence of multiple sublineages of Omicron (B.1.1.529) variant: Dynamic profiling via molecular phylogenetics and mutational landscape studies. *J. Infect. Public Health* **2022**, *15*, 1234–1258. [CrossRef]
21. Callaway, E. Are COVID surges becoming more predictable? New Omicron variants offer a hint. *Nature* **2022**, *605*, 204–206. [CrossRef] [PubMed]
22. Cele, S.; Jackson, L.; Khoury, D.S.; Khan, K.; Moyo-Gwete, T.; Tegally, H.; San, J.E.; Cromer, D.; Scheepers, C.; Amoako, D.; et al. SARS-CoV-2 Omicron has extensive but incomplete escape of Pfizer BNT162b2 elicited neutralization and requires ACE2 for infection. *medRxiv* **2021**. [CrossRef]
23. Cao, Y.; Wang, J.; Jian, F.; Xiao, T.; Song, W.; Yisimayi, A.; Huang, W.; Li, Q.; Wang, P.; An, R. Omicron escapes the majority of existing SARS-CoV-2 neutralizing antibodies. *Nature* **2022**, *602*, 657–663. [CrossRef] [PubMed]
24. Hoffmann, M.; Kruger, N.; Schulz, S.; Cossmann, A.; Rocha, C.; Kempf, A.; Nehlmeier, I.; Graichen, L.; Moldenhauer, A.S.; Winkler, M.S.; et al. The Omicron variant is highly resistant against antibody-mediated neutralization: Implications for control of the COVID-19 pandemic. *Cell* **2022**, *185*, 447–456.e11. [CrossRef] [PubMed]
25. Wilhelm, A.; Widera, M.; Grikscheit, K.; Toptan, T.; Schenk, B.; Pallas, C.; Metzler, M.; Kohmer, N.; Hoehl, S.; Marschalek, R. Limited neutralisation of the SARS-CoV-2 Omicron subvariants BA. 1 and BA. 2 by convalescent and vaccine serum and monoclonal antibodies. *EBioMedicine* **2022**, *82*, 104158. [CrossRef]
26. Peter, A.S.; Grüner, E.; Socher, E.; Fraedrich, K.; Richel, E.; Mueller-Schmucker, S.; Cordsmeier, A.; Ensser, A.; Sticht, H.; Überla, K. Characterization of SARS-CoV-2 Escape Mutants to a Pair of Neutralizing Antibodies Targeting the RBD and the NTD. *Int. J. Mol. Sci.* **2022**, *23*, 8177. [CrossRef]
27. Chakraborty, C.; Bhattacharya, M.; Sharma, A.R.; Dhama, K.; Agoramoorthy, G. A comprehensive analysis of the mutational landscape of the newly emerging Omicron (B.1.1.529) variant and comparison of mutations with VOCs and VOIs. *GeroScience* **2022**, *44*, 2393–2425. [CrossRef]
28. Mohapatra, R.K.; Tiwari, R.; Sarangi, A.K.; Islam, M.R.; Chakraborty, C.; Dhama, K. Omicron (B.1.1.529) variant of SARS-CoV-2: Concerns, challenges, and recent updates. *J. Med. Virol.* **2022**, *94*, 2336–2342. [CrossRef]
29. Bhattacharya, M.; Sharma, A.R.; Dhama, K.; Agoramoorthy, G.; Chakraborty, C. Omicron variant (B.1.1.529) of SARS-CoV-2: Understanding mutations in the genome, S-glycoprotein, and antibody-binding regions. *GeroScience* **2022**, *44*, 619–637. [CrossRef]
30. Carrazco-Montalvo, A.; Herrera-Yela, A.; Alarcon-Vallejo, D.; Gutierrez-Pallo, D.; Armendariz-Castillo, I.; Andrade-Molina, D.; Munoz-Mawyin, K.; Fernandez-Cadena, J.C.; Morey-Leon, G.; Usfq Covid, C.; et al. Omicron Sub-Lineages (BA.1.1.529 + BA.*) Current Status in Ecuador. *Viruses* **2022**, *14*, 1177. [CrossRef]
31. Kumar, S.; Karuppanan, K.; Subramaniam, G. Omicron (BA.1) and sub-variants (BA.1.1, BA.2, and BA.3) of SARS-CoV-2 spike infectivity and pathogenicity: A comparative sequence and structural-based computational assessment. *J. Med. Virol.* **2022**, *94*, 4780–4791. [CrossRef]
32. Desingu, P.A.; Nagarajan, K.; Dhama, K. Emergence of Omicron third lineage BA.3 and its importance. *J. Med. Virol.* **2022**, *94*, 1808–1810. [CrossRef]
33. Mohapatra, R.K.; Kandi, V.; Verma, S.; Dhama, K. Challenges of the Omicron (B.1.1.529) Variant and Its Lineages: A Global Perspective. *Chembiochem* **2022**, *23*, e202200059. [CrossRef]
34. Khandia, R.; Singhal, S.; Alqahtani, T.; Kamal, M.A.; El-Shall, N.A.; Nainu, F.; Desingu, P.A.; Dhama, K. Emergence of SARS-CoV-2 Omicron (B.1.1.529) variant, salient features, high global health concerns and strategies to counter it amid ongoing COVID-19 pandemic. *Environ. Res.* **2022**, *209*, 112816. [CrossRef]
35. Tegally, H.; Moir, M.; Everatt, J.; Giovanetti, M.; Scheepers, C.; Wilkinson, E.; Subramoney, K.; Makatini, Z.; Moyo, S.; Amoako, D.G.; et al. Emergence of SARS-CoV-2 Omicron lineages BA.4 and BA.5 in South Africa. *Nat. Med.* **2022**. [CrossRef]
36. Rahman, S.; Hossain, M.J.; Nahar, Z.; Shahriar, M.; Bhuiyan, M.A.; Islam, M.R. Emerging SARS-CoV-2 Variants and Subvariants: Challenges and Opportunities in the Context of COVID-19 Pandemic. *Environ. Health Insights* **2022**, *16*, 11786302221129396. [CrossRef]
37. Cao, Y.; Yisimayi, A.; Jian, F.; Song, W.; Xiao, T.; Wang, L.; Du, S.; Wang, J.; Li, Q.; Chen, X.; et al. BA.2.12.1, BA.4 and BA.5 escape antibodies elicited by Omicron infection. *Nature* **2022**, *608*, 593–602. [CrossRef]
38. Basky, G.; Vogel, L. XE, XD & XF: What to know about the Omicron hybrid variants. *CMAJ* **2022**, *194*, E654–E655. [CrossRef]
39. Chakraborty, C.; Bhattacharya, M.; Sharma, A.R.; Dhama, K. Recombinant SARS-CoV-2 variants XD, XE, and XF: The emergence of recombinant variants requires an urgent call for research—Correspondence. *Int. J. Surg.* **2022**, *102*, 106670. [CrossRef]
40. Mohapatra, R.K.; Kandi, V.; Tuli, H.S.; Chakraborty, C.; Dhama, K. The recombinant variants of SARS-CoV-2: Concerns continues amid COVID-19 pandemic. *J. Med. Virol.* **2022**, *94*, 3506–3508. [CrossRef]

41. Rahimi, F.; Talebi Bezmin Abadi, A. Hybrid SARS-CoV-2 variants. *Int. J. Surg.* **2022**, *102*, 106656. [CrossRef] [PubMed]
42. Roemer, C.; Hisner, R.; Frohberg, N.; Sakaguchi, H.; Gueli, F.; Peacock, T.P. SARS-CoV-2 Evolution, Post-Omicron. virological.org. 911. Available online: https://virological.org/t/sars-cov-2-evolution-post-omicron/911 (accessed on 6 August 2022).
43. Ingraham, N.E.; Ingbar, D.H. The omicron variant of SARS-CoV-2: Understanding the known and living with unknowns. *Clin. Transl. Med.* **2021**, *11*, e685. [CrossRef] [PubMed]
44. Kupferschmidt, K. Where did 'weird' Omicron come from? *Science* **2021**, *374*, 1179. [CrossRef] [PubMed]
45. Karim, S.S.A.; Karim, Q.A. Omicron SARS-CoV-2 variant: A new chapter in the COVID-19 pandemic. *Lancet* **2021**, *398*, 2126–2128. [CrossRef] [PubMed]
46. Cui, Z.; Liu, P.; Wang, N.; Wang, L.; Fan, K.; Zhu, Q.; Wang, K.; Chen, R.; Feng, R.; Jia, Z.; et al. Structural and functional characterizations of infectivity and immune evasion of SARS-CoV-2 Omicron. *Cell* **2022**, *185*, 860–871.e13. [CrossRef] [PubMed]
47. Cameroni, E.; Bowen, J.E.; Rosen, L.E.; Saliba, C.; Zepeda, S.K.; Culap, K.; Pinto, D.; VanBlargan, L.A.; De Marco, A.; di Iulio, J. Broadly neutralizing antibodies overcome SARS-CoV-2 Omicron antigenic shift. *Nature* **2022**, *602*, 664–670. [CrossRef]
48. Mannar, D.; Saville, J.W.; Zhu, X.; Srivastava, S.S.; Berezuk, A.M.; Tuttle, K.S.; Marquez, A.C.; Sekirov, I.; Subramaniam, S. SARS-CoV-2 Omicron variant: Antibody evasion and cryo-EM structure of spike protein–ACE2 complex. *Science* **2022**, *375*, 760–764. [CrossRef]
49. Zhang, X.; Wu, S.; Wu, B.; Yang, Q.; Chen, A.; Li, Y.; Zhang, Y.; Pan, T.; Zhang, H.; He, X. SARS-CoV-2 Omicron strain exhibits potent capabilities for immune evasion and viral entrance. *Signal Transduct. Target. Ther.* **2021**, *6*, 430. [CrossRef]
50. Wu, L.; Zhou, L.; Mo, M.; Liu, T.; Wu, C.; Gong, C.; Lu, K.; Gong, L.; Zhu, W.; Xu, Z. SARS-CoV-2 Omicron RBD shows weaker binding affinity than the currently dominant Delta variant to human ACE2. *Signal Transduct. Target. Ther.* **2022**, *7*, 8. [CrossRef]
51. Han, P.; Li, L.; Liu, S.; Wang, Q.; Zhang, D.; Xu, Z.; Han, P.; Li, X.; Peng, Q.; Su, C. Receptor binding and complex structures of human ACE2 to spike RBD from omicron and delta SARS-CoV-2. *Cell* **2022**, *185*, 630–640.e10. [CrossRef]
52. Bhattacharya, M.; Chatterjee, S.; Lee, S.S.; Chakraborty, C. Therapeutic applications of nanobodies against SARS-CoV-2 and other viral infections: Current update. *Int. J. Biol. Macromol.* **2022**, *229*, 70–80. [CrossRef]
53. Yin, W.; Xu, Y.; Xu, P.; Cao, X.; Wu, C.; Gu, C.; He, X.; Wang, X.; Huang, S.; Yuan, Q.; et al. Structures of the Omicron spike trimer with ACE2 and an anti-Omicron antibody. *Science* **2022**, *375*, 1048–1053. [CrossRef]
54. Ledford, H. How severe are Omicron infections. *Nature* **2021**, *600*, 577–578. [CrossRef]
55. Yang, T.-J.; Yu, P.-Y.; Chang, Y.-C.; Liang, K.-H.; Tso, H.-C.; Ho, M.-R.; Chen, W.-Y.; Lin, H.-T.; Wu, H.-C.; Hsu, S.-T.D. Effect of SARS-CoV-2 B. 1.1. 7 mutations on spike protein structure and function. *Nat. Struct. Mol. Biol.* **2021**, *28*, 731–739. [CrossRef]
56. Leung, K.; Shum, M.H.; Leung, G.M.; Lam, T.T.; Wu, J.T. Early transmissibility assessment of the N501Y mutant strains of SARS-CoV-2 in the United Kingdom, October to November 2020. *Eurosurveillance* **2021**, *26*, 2002106. [CrossRef]
57. Zuckerman, N.S.; Fleishon, S.; Bucris, E.; Bar-Ilan, D.; Linial, M.; Bar-Or, I.; Indenbaum, V.; Weil, M.; Lustig, Y.; Mendelson, E. A unique SARS-CoV-2 spike protein P681H variant detected in Israel. *Vaccines* **2021**, *9*, 616. [CrossRef]
58. Tuekprakhon, A.; Nutalai, R.; Dijokaite-Guraliuc, A.; Zhou, D.; Ginn, H.M.; Selvaraj, M.; Liu, C.; Mentzer, A.J.; Supasa, P.; Duyvesteyn, H.M. Antibody escape of SARS-CoV-2 Omicron BA. 4 and BA. 5 from vaccine and BA. 1 serum. *Cell* **2022**, *185*, 2422–2433.e13. [CrossRef]
59. Garrett, N.; Tapley, A.; Andriesen, J.; Seocharan, I.; Fisher, L.H.; Bunts, L.; Espy, N.; Wallis, C.L.; Randhawa, A.K.; Ketter, N. High rate of asymptomatic carriage associated with variant strain omicron. *medRxiv* **2022**. [CrossRef]
60. Wolter, N.; Jassat, W.; Walaza, S.; Welch, R.; Moultrie, H.; Groome, M.; Amoako, D.G.; Everatt, J.; Bhiman, J.N.; Scheepers, C. Early assessment of the clinical severity of the SARS-CoV-2 omicron variant in South Africa: A data linkage study. *Lancet* **2022**, *399*, 437–446. [CrossRef]
61. Izcovich, A.; Ragusa, M.A.; Tortosa, F.; Lavena Marzio, M.A.; Agnoletti, C.; Bengolea, A.; Ceirano, A.; Espinosa, F.; Saavedra, E.; Sanguine, V. Prognostic factors for severity and mortality in patients infected with COVID-19: A systematic review. *PLoS ONE* **2020**, *15*, e0241955. [CrossRef]
62. Christensen, P.A.; Olsen, R.J.; Long, S.W.; Snehal, R.; Davis, J.J.; Saavedra, M.O.; Reppond, K.; Shyer, M.N.; Cambric, J.; Gadd, R. Signals of significantly increased vaccine breakthrough, decreased hospitalization rates, and less severe disease in patients with Coronavirus disease 2019 caused by the Omicron variant of severe acute respiratory syndrome Coronavirus 2 in Houston, Texas. *Am. J. Pathol.* **2022**, *192*, 642–652. [CrossRef] [PubMed]
63. Islam, F.; Dhawan, M.; Nafady, M.H.; Emran, T.B.; Mitra, S.; Choudhary, O.P.; Akter, A. Understanding the omicron variant (B.1.1.529) of SARS-CoV-2: Mutational impacts, concerns, and the possible solutions. *Ann. Med. Surg.* **2022**, *78*, 103737. [CrossRef]
64. Chakraborty, C.; Sharma, A.R.; Bhattacharya, M.; Agoramoorthy, G.; Lee, S.S. A Paradigm Shift in the Combination Changes of SARS-CoV-2 Variants and Increased Spread of Delta Variant (B.1.617.2) across the World. *Aging Dis.* **2022**, *13*, 927–942. [CrossRef]
65. Bhattacharya, M.; Chatterjee, S.; Sharma, A.R.; Lee, S.S.; Chakraborty, C. Delta variant (B.1.617.2) of SARS-CoV-2: Current understanding of infection, transmission, immune escape, and mutational landscape. *Folia Microbiol.* **2022**, 1–12. [CrossRef] [PubMed]
66. Duong, B.V.; Larpruenrudee, P.; Fang, T.; Hossain, S.I.; Saha, S.C.; Gu, Y.; Islam, M.S. Is the SARS CoV-2 Omicron Variant Deadlier and More Transmissible Than Delta Variant? *Int. J. Environ. Res. Public Health* **2022**, *19*, 4586. [CrossRef] [PubMed]

67. Zhou, H.; Mohlenberg, M.; Thakor, J.C.; Tuli, H.S.; Wang, P.; Assaraf, Y.G.; Dhama, K.; Jiang, S. Sensitivity to Vaccines, Therapeutic Antibodies, and Viral Entry Inhibitors and Advances To Counter the SARS-CoV-2 Omicron Variant. *Clin. Microbiol. Rev.* **2022**, *35*, e0001422. [CrossRef]
68. Kupferschmidt, K.; Vogel, G. How bad is Omicron? Some clues are emerging. *Science* **2021**, *374*, 1304–1305. [CrossRef]
69. Chen, N.; Zhou, M.; Dong, X.; Qu, J.; Gong, F.; Han, Y.; Qiu, Y.; Wang, J.; Liu, Y.; Wei, Y. Epidemiological and clinical characteristics of 99 cases of 2019 novel coronavirus pneumonia in Wuhan, China: A descriptive study. *Lancet* **2020**, *395*, 507–513. [CrossRef]
70. Guan, W.-J.; Ni, Z.-Y.; Hu, Y.; Liang, W.-H.; Ou, C.-Q.; He, J.-X.; Liu, L.; Shan, H.; Lei, C.; Hui, D.S. China medical treatment expert group for Covid-19. *Clin. Charact. Coronavirus Dis.* **2019**, *382*, 1708–1720.
71. Alemi, F.; Vang, J.; Wojtusiak, J.; Guralnik, E.; Peterson, R.; Roess, A.; Jain, P. Differential diagnosis of COVID-19 and influenza. *PLOS Glob. Public Health* **2022**, *2*, e0000221. [CrossRef]
72. Menni, C.; Valdes, A.M.; Polidori, L.; Antonelli, M.; Penamakuri, S.; Nogal, A.; Louca, P.; May, A.; Figueiredo, J.C.; Hu, C. Symptom prevalence, duration, and risk of hospital admission in individuals infected with SARS-CoV-2 during periods of omicron and delta variant dominance: A prospective observational study from the ZOE COVID Study. *Lancet* **2022**, *399*, 1618–1624. [CrossRef]
73. Iuliano, A.D. Trends in disease severity and health care utilization during the early Omicron variant period compared with previous SARS-CoV-2 high transmission periods—United States, December 2020–January 2022. *MMWR. Morb. Mortal. Wkly. Rep.* **2022**, *71*, 146–152. [CrossRef]
74. Lewnard, J.A.; Hong, V.X.; Patel, M.M.; Kahn, R.; Lipsitch, M.; Tartof, S.Y. Clinical outcomes associated with SARS-CoV-2 Omicron (B.1.1.529) variant and BA.1/BA.1.1 or BA.2 subvariant infection in southern California. *Nat. Med.* **2022**, *28*, 1933–1943. [CrossRef]
75. Ludvigsson, J.F. Convulsions in children with COVID-19 during the Omicron wave. *Acta Paediatr.* **2022**, *111*, 1023–1026. [CrossRef]
76. Suzuki, R.; Yamasoba, D.; Kimura, I.; Wang, L.; Kishimoto, M.; Ito, J.; Morioka, Y.; Nao, N.; Nasser, H.; Uriu, K. Attenuated fusogenicity and pathogenicity of SARS-CoV-2 Omicron variant. *Nature* **2022**, *603*, 700–705. [CrossRef]
77. Shuai, H.; Chan, J.F.-W.; Hu, B.; Chai, Y.; Yuen, T.T.-T.; Yin, F.; Huang, X.; Yoon, C.; Hu, J.-C.; Liu, H. Attenuated replication and pathogenicity of SARS-CoV-2 B.1.1.529 Omicron. *Nature* **2022**, *603*, 693–699. [CrossRef]
78. Bouzid, D.; Visseaux, B.; Kassasseya, C.; Daoud, A.; Fémy, F.; Hermand, C.; Truchot, J.; Beaune, S.; Javaud, N.; Peyrony, O.; et al. Comparison of Patients Infected With Delta Versus Omicron COVID-19 Variants Presenting to Paris Emergency Departments: A Retrospective Cohort Study. *Ann. Intern. Med.* **2022**, *175*, 831–837. [CrossRef]
79. Kneidinger, N.; Hecker, M.; Bessa, V.; Hettich, I.; Wald, A.; Wege, S.; Nolde, A.B.; Oldigs, M.; Syunyaeva, Z.; Wilkens, H.; et al. Outcome of lung transplant recipients infected with SARS-CoV-2/Omicron/B.1.1.529: A Nationwide German study. *Infection* **2022**, *9*, 1–9. [CrossRef]
80. Feikin, D.R.; Abu-Raddad, L.J.; Andrews, N.; Davies, M.A.; Higdon, M.M.; Orenstein, W.A.; Patel, M.K. Assessing vaccine effectiveness against severe COVID-19 disease caused by omicron variant. Report from a meeting of the World Health Organization. *Vaccine* **2022**, *40*, 3516–3527. [CrossRef]
81. Chen, X.; Yan, X.; Sun, K.; Zheng, N.; Sun, R.; Zhou, J.; Deng, X.; Zhuang, T.; Cai, J.; Zhang, J. Estimation of disease burden and clinical severity of COVID-19 caused by Omicron BA.2 in Shanghai, February-June 2022. *Emerg. Microbes Infect.* **2022**, *11*, 2800–2807. [CrossRef]
82. Halfmann, P.J.; Iida, S.; Iwatsuki-Horimoto, K.; Maemura, T.; Kiso, M.; Scheaffer, S.M.; Darling, T.L.; Joshi, A.; Loeber, S.; Singh, G. SARS-CoV-2 Omicron virus causes attenuated disease in mice and hamsters. *Nature* **2022**, *603*, 687–692. [CrossRef] [PubMed]
83. Akkız, H. The Biological Functions and Clinical Significance of SARS-CoV-2 Variants of Corcern. *Front. Med.* **2022**, *20*, 849217. [CrossRef] [PubMed]
84. Thye, A.Y.; Law, J.W.; Pusparajah, P.; Letchumanan, V.; Chan, K.G.; Lee, L.H. Emerging SARS-CoV-2 Variants of Concern (VOCs): An Impending Global Crisis. *Biomedicines* **2021**, *9*, 1303. [CrossRef]
85. Chakraborty, C.; Sharma, A.R.; Bhattacharya, M.; Mallik, B.; Nandi, S.S.; Lee, S.S. Comparative genomics, evolutionary epidemiology, and RBD-hACE2 receptor binding pattern in B.1.1.7 (Alpha) and B.1.617.2 (Delta) related to their pandemic response in UK and India. *Infect. Genet. Evol.* **2022**, *101*, 105282. [CrossRef] [PubMed]
86. Bian, L.; Gao, Q.; Gao, F.; Wang, Q.; He, Q.; Wu, X.; Mao, Q.; Xu, M.; Liang, Z. Impact of the Delta variant on vaccine efficacy and response strategies. *Expert Rev. Vaccines* **2021**, *10*, 1201–1209. [CrossRef]
87. Zhang, J.; Chen, N.; Zhao, D.; Zhang, J.; Hu, Z.; Tao, Z. Clinical Characteristics of COVID-19 Patients Infected by the Omicron Variant of SARS-CoV-2. *Front. Med.* **2022**, *9*, 912367. [CrossRef]
88. Andrews, N.; Stowe, J.; Kirsebom, F.; Toffa, S.; Rickeard, T.; Gallagher, E.; Gower, C.; Kall, M.; Groves, N.; O'Connell, A.-M. Covid-19 vaccine effectiveness against the Omicron (B.1.1.529) variant. *N. Engl. J. Med.* **2022**, *386*, 1532–1546. [CrossRef]
89. Kupferschmidt, K. Scientists see a 'really, really tough winter' with Omicron. *Science* **2021**, *374*, 1421–1422. [CrossRef]
90. Dong, Y.; Dai, T.; Liu, J.; Zhang, L.; Zhou, F. Coronavirus in Continuous Flux: From SARS-CoV to SARS-CoV-2. *Adv. Sci.* **2020**, *7*, 2001474. [CrossRef]
91. Dimeglio, C.; Migueres, M.; Mansuy, J.M.; Saivin, S.; Miedouge, M.; Chapuy-Regaud, S.; Izopet, J. Antibody titers and breakthrough infections with Omicron SARS-CoV-2. *J. Infect.* **2022**, *84*, e13–e15. [CrossRef]

92. Nguyen, N.N.; Houhamdi, L.; Hoang, V.T.; Delerce, J.; Delorme, L.; Colson, P.; Brouqui, P.; Fournier, P.E.; Raoult, D.; Gautret, P. SARS-CoV-2 reinfection and COVID-19 severity. *Emerg. Microbes Infect.* **2022**, *11*, 894–901. [CrossRef]
93. Sheehan, M.M.; Reddy, A.J.; Rothberg, M.B. Reinfection Rates Among Patients Who Previously Tested Positive for Coronavirus Disease 2019: A Retrospective Cohort Study. *Clin. Infect. Dis.* **2021**, *73*, 1882–1886. [CrossRef]
94. Pulliam, J.R.C.; van Schalkwyk, C.; Govender, N.; von Gottberg, A.; Cohen, C.; Groome, M.J.; Dushoff, J.; Mlisana, K.; Moultrie, H. Increased risk of SARS-CoV-2 reinfection associated with emergence of Omicron in South Africa. *Science* **2022**, *376*, eabn4947. [CrossRef]
95. Araf, Y.; Akter, F.; Tang, Y.D.; Fatemi, R.; Parvez, M.S.A.; Zheng, C.; Hossain, M.G. Omicron variant of SARS-CoV-2: Genomics, transmissibility, and responses to current COVID-19 vaccines. *J. Med. Virol.* **2022**, *94*, 1825–1832. [CrossRef]
96. Long, B.; Carius, B.M.; Chavez, S.; Liang, S.Y.; Brady, W.J.; Koyfman, A.; Gottlieb, M. Clinical update on COVID-19 for the emergency clinician: Presentation and evaluation. *Am. J. Emerg. Med.* **2022**, *54*, 46–57. [CrossRef]
97. UK Health Security Agency. *SARS-CoV-2 Variants of Concern and Variants under Investigation in England*; Technical Briefing 28; UK Health Security Agency: London, UK, 2021.
98. Mahase, E. Omicron sub-lineage BA.2 may have "substantial growth advantage," UKHSA reports. *BMJ* **2022**, *376*, o263. [CrossRef]
99. Brandal, L.T.; MacDonald, E.; Veneti, L.; Ravlo, T.; Lange, H.; Naseer, U.; Feruglio, S.; Bragstad, K.; Hungnes, O.; Odeskaug, L.E.; et al. Outbreak caused by the SARS-CoV-2 Omicron variant in Norway, November to December 2021. *Eurosurveillance* **2021**, *26*, 2101147. [CrossRef]
100. Chaguza, C.; Coppi, A.; Earnest, R.; Ferguson, D.; Kerantzas, N.; Warner, F.; Young, H.P.; Breban, M.I.; Billig, K.; Koch, R.T.; et al. Rapid emergence of SARS-CoV-2 Omicron variant is associated with an infection advantage over Delta in vaccinated persons. *Med* **2022**, *3*, 325–334.e4. [CrossRef]
101. Meng, B.; Abdullahi, A.; Ferreira, I.; Goonawardane, N.; Saito, A.; Kimura, I.; Yamasoba, D.; Gerber, P.P.; Fatihi, S.; Rathore, S.; et al. Altered TMPRSS2 usage by SARS-CoV-2 Omicron impacts infectivity and fusogenicity. *Nature* **2022**, *603*, 706–714. [CrossRef]
102. Hui, K.P.Y.; Ho, J.C.W.; Cheung, M.C.; Ng, K.C.; Ching, R.H.H.; Lai, K.L.; Kam, T.T.; Gu, H.; Sit, K.Y.; Hsin, M.K.Y.; et al. SARS-CoV-2 Omicron variant replication in human bronchus and lung ex vivo. *Nature* **2022**, *603*, 715–720. [CrossRef]
103. Riediker, M.; Briceno-Ayala, L.; Ichihara, G.; Albani, D.; Poffet, D.; Tsai, D.H.; Iff, S.; Monn, C. Higher viral load and infectivity increase risk of aerosol transmission for Delta and Omicron variants of SARS-CoV-2. *Swiss Med. Wkly.* **2022**, *152*, w30133. [CrossRef] [PubMed]
104. Migueres, M.; Dimeglio, C.; Tremeaux, P.; Abravanel, F.; Raymond, S.; Lhomme, S.; Mansuy, J.M.; Izopet, J. Influence of immune escape and nasopharyngeal virus load on the spread of SARS-CoV-2 Omicron variant. *J. Infect.* **2022**, *84*, e7–e9. [CrossRef] [PubMed]
105. Wrapp, D.; Wang, N.; Corbett, K.S.; Goldsmith, J.A.; Hsieh, C.L.; Abiona, O.; Graham, B.S.; McLellan, J.S. Cryo-EM structure of the 2019-nCoV spike in the prefusion conformation. *Science* **2020**, *367*, 1260–1263. [CrossRef] [PubMed]
106. Pia, L.; Rowland-Jones, S. Omicron entry route. *Nat. Reviews. Immunol.* **2022**, *22*, 144. [CrossRef]
107. Peacock, T.P.; Goldhill, D.H.; Zhou, J.; Baillon, L.; Frise, R.; Swann, O.C.; Kugathasan, R.; Penn, R.; Brown, J.C.; Sanchez-David, R.Y.; et al. The furin cleavage site in the SARS-CoV-2 spike protein is required for transmission in ferrets. *Nat. Microbiol.* **2021**, *6*, 899–909. [CrossRef]
108. Du, X.; Tang, H.; Gao, L.; Wu, Z.; Meng, F.; Yan, R.; Qiao, S.; An, J.; Wang, C.; Qin, F.X. Omicron adopts a different strategy from Delta and other variants to adapt to host. *Signal Transduct. Target. Ther.* **2022**, *7*, 45. [CrossRef]
109. Syed, A.M.; Ciling, A.; Khalid, M.M.; Sreekumar, B.; Chen, P.Y.; Kumar, G.R.; Silva, I.; Milbes, B.; Kojima, N.; Hess, V.; et al. Omicron mutations enhance infectivity and reduce antibody neutralization of SARS-CoV-2 virus-like particles. *medRxiv* **2022**. [CrossRef]
110. Willett, B.J.; Grove, J.; MacLean, O.A.; Wilkie, C.; De Lorenzo, G.; Furnon, W.; Cantoni, D.; Scott, S.; Logan, N.; Ashraf, S.; et al. SARS-CoV-2 Omicron is an immune escape variant with an altered cell entry pathway. *Nat. Microbiol.* **2022**, *7*, 1161–1179. [CrossRef]
111. Kared, H.; Wolf, A.S.; Alirezaylavasani, A.; Ravussin, A.; Solum, G.; Tran, T.T.; Lund-Johansen, F.; Vaage, J.T.; Nissen-Meyer, L.S.; Nygaard, U.C.; et al. Immune responses in Omicron SARS-CoV-2 breakthrough infection in vaccinated adults. *Nat. Commun.* **2022**, *13*, 4165. [CrossRef]
112. Goutam Mukherjee, A.; Ramesh Wanjari, U.; Murali, R.; Chaudhary, U.; Renu, K.; Madhyastha, H.; Iyer, M.; Vellingiri, B.; Valsala Gopalakrishnan, A. Omicron variant infection and the associated immunological scenario. *Immunobiology* **2022**, *227*, 152222. [CrossRef]
113. Kumar, S.; Thambiraja, T.S.; Karuppanan, K.; Subramaniam, G. Omicron and Delta variant of SARS-CoV-2: A comparative computational study of spike protein. *J. Med. Virol.* **2022**, *94*, 1641–1649. [CrossRef]
114. Da Costa, C.H.S.; de Freitas, C.A.B.; Alves, C.N.; Lameira, J. Assessment of mutations on RBD in the Spike protein of SARS-CoV-2 Alpha, Delta and Omicron variants. *Sci. Rep.* **2022**, *12*, 8540. [CrossRef]
115. Callebaut, K.; Stoefs, A.; Stylemans, D.; Soetens, O.; Crombe, F.; Vancutsem, E.; Imamura, H.; Wybo, I.; De Geyter, D.; Pierard, D.; et al. Healthcare-Associated SARS-CoV-2 Reinfection after 3 Months with a Phylogenetically Distinct Omicron Variant: A Case Report. *Viruses* **2022**, *14*, 1852. [CrossRef]

116. Kandeel, M.; El-Deeb, W. Omicron variant receptor-binding domain phylogenetics and molecular dynamics. *Comput. Biol. Med.* **2022**, *146*, 105633. [CrossRef]
117. Chakraborty, C.; Sharma, A.R.; Bhattacharya, M.; Lee, S.S. A Detailed Overview of Immune Escape, Antibody Escape, Partial Vaccine Escape of SARS-CoV-2 and Their Emerging Variants With Escape Mutations. *Front. Immunol.* **2022**, *13*, 801522. [CrossRef]
118. Chakraborty, C.; Bhattacharya, M.; Sharma, A.R. Emerging mutations in the SARS-CoV-2 variants and their role in antibody escape to small molecule-based therapeutic resistance. *Curr. Opin. Pharmacol.* **2022**, *62*, 64–73. [CrossRef]
119. Chakraborty, C.; Bhattacharya, M.; Sharma, A.R.; Mohapatra, R.K.; Chakraborty, S.; Pal, S.; Dhama, K. Immediate need for next-generation and mutation-proof vaccine to protect against current emerging Omicron sublineages and future SARS-CoV-2 variants: An urgent call for researchers and vaccine companies—Correspondence. *Int. J. Surg.* **2022**, *106*, 106903. [CrossRef]
120. Chakraborty, C.; Bhattacharya, M.; Sharma, A.R.; Mallik, B. Omicron (B.1.1.529)—A new heavily mutated variant: Mapped location and probable properties of its mutations with an emphasis on S-glycoprotein. *Int. J. Biol. Macromol.* **2022**, *219*, 980–997. [CrossRef]
121. Chalkias, S.; Harper, C.; Vrbicky, K.; Walsh, S.R.; Essink, B.; Brosz, A.; McGhee, N.; Tomassini, J.E.; Chen, X.; Chang, Y.; et al. A Bivalent Omicron-Containing Booster Vaccine against Covid-19. *N. Engl. J. Med.* **2022**, *387*, 1279–1291. [CrossRef]
122. Chakraborty, C.; Bhattacharya, M.; Dhama, K. Cases of BA.2.75 and recent BA.2.75.2 subvariant of Omicron are increasing in India: Is it alarming at the global level? *Ann. Med. Surg.* **2022**, *84*, 104963. [CrossRef]
123. Cerutti, G.; Guo, Y.; Zhou, T.; Gorman, J.; Lee, M.; Rapp, M.; Reddem, E.R.; Yu, J.; Bahna, F.; Bimela, J.; et al. Potent SARS-CoV-2 neutralizing antibodies directed against spike N-terminal domain target a single supersite. *Cell Host Microbe* **2021**, *29*, 819–833.e7. [CrossRef] [PubMed]
124. Dejnirattisai, W.; Zhou, D.; Ginn, H.M.; Duyvesteyn, H.M.E.; Supasa, P.; Case, J.B.; Zhao, Y.; Walter, T.S.; Mentzer, A.J.; Liu, C.; et al. The antigenic anatomy of SARS-CoV-2 receptor binding domain. *Cell* **2021**, *184*, 2183–2200.e22. [CrossRef] [PubMed]
125. McCallum, M.; De Marco, A.; Lempp, F.A.; Tortorici, M.A.; Pinto, D.; Walls, A.C.; Beltramello, M.; Chen, A.; Liu, Z.; Zatta, F.; et al. N-terminal domain antigenic mapping reveals a site of vulnerability for SARS-CoV-2. *Cell* **2021**, *184*, 2332–2347.e16. [CrossRef] [PubMed]
126. Zhang, L.; Cao, L.; Gao, X.S.; Zheng, B.Y.; Deng, Y.Q.; Li, J.X.; Feng, R.; Bian, Q.; Guo, X.L.; Wang, N.; et al. A proof of concept for neutralizing antibody-guided vaccine design against SARS-CoV-2. *Natl. Sci. Rev.* **2021**, *8*, nwab053. [CrossRef] [PubMed]
127. Sun, C.; Kang, Y.F.; Liu, Y.T.; Kong, X.W.; Xu, H.Q.; Xiong, D.; Xie, C.; Liu, Y.H.; Peng, S.; Feng, G.K.; et al. Parallel profiling of antigenicity alteration and immune escape of SARS-CoV-2 Omicron and other variants. *Signal Transduct. Target. Ther.* **2022**, *7*, 42. [CrossRef]
128. Cerutti, G.; Guo, Y.; Liu, L.; Zhang, Z.; Luo, Y.; Huang, Y.; Wang, H.H.; Ho, D.D.; Sheng, Z.; Shapiro, L. Cryo-EM structure of the SARS-CoV-2 Omicron spike. *Cell Rep.* **2022**, *38*, 110428. [CrossRef]
129. Takashita, E.; Kinoshita, N.; Yamayoshi, S.; Sakai-Tagawa, Y.; Fujisaki, S.; Ito, M.; Iwatsuki-Horimoto, K.; Chiba, S.; Halfmann, P.; Nagai, H.; et al. Efficacy of Antibodies and Antiviral Drugs against Covid-19 Omicron Variant. *N. Engl. J. Med.* **2022**, *386*, 995–998. [CrossRef]
130. Arbel, R.; Wolff Sagy, Y.; Hoshen, M.; Battat, E.; Lavie, G.; Sergienko, R.; Friger, M.; Waxman, J.G.; Dagan, N.; Balicer, R.; et al. Nirmatrelvir Use and Severe Covid-19 Outcomes during the Omicron Surge. *N. Engl. J. Med.* **2022**, *387*, 790–798. [CrossRef]
131. Bojkova, D.; Widera, M.; Ciesek, S.; Wass, M.N.; Michaelis, M.; Cinatl, J., Jr. Reduced interferon antagonism but similar drug sensitivity in Omicron variant compared to Delta variant of SARS-CoV-2 isolates. *Cell Res.* **2022**, *32*, 319–321. [CrossRef]
132. Vangeel, L.; Chiu, W.; De Jonghe, S.; Maes, P.; Slechten, B.; Raymenants, J.; Andre, E.; Leyssen, P.; Neyts, J.; Jochmans, D. Remdesivir, Molnupiravir and Nirmatrelvir remain active against SARS-CoV-2 Omicron and other variants of concern. *Antivir. Res.* **2022**, *198*, 105252. [CrossRef]
133. Wong, G.L.; Yip, T.C.; Lai, M.S.; Wong, V.W.; Hui, D.S.; Lui, G.C. Incidence of Viral Rebound After Treatment With Nirmatrelvir-Ritonavir and Molnupiravir. *JAMA Netw. Open* **2022**, *12*, e2245086. [CrossRef]
134. Wong, C.K.H.; Au, I.C.H.; Lau, K.T.K.; Lau, E.H.Y.; Cowling, B.J.; Leung, G.M. Real-world effectiveness of early molnupiravir or nirmatrelvir-ritonavir in hospitalised patients with COVID-19 without supplemental oxygen requirement on admission during Hong Kong's omicron BA.2 wave: A retrospective cohort study. *Lancet Infect. Dieases* **2022**, *12*, 1681–1693. [CrossRef]
135. Saravolatz, L.D.; Depcinski, S.; Sharma, M. Molnupiravir and Nirmatrelvir-Ritonavir: Oral COVID Antiviral Drugs. *Clin. Infect. Dis.* **2022**, ciac180. [CrossRef]
136. Bhattacharya, M.; Chatterjee, S.; Mallik, B.; Sharma, A.R.; Chakraborty, C. Therapeutic Role of Neutralizing Antibody for the Treatment against SARS-CoV-2 and Its Emerging Variants: A Clinical and Pre-Clinical Perspective. *Vaccines* **2022**, *10*, 1612. [CrossRef]
137. Tada, T.; Zhou, H.; Dcosta, B.M.; Samanovic, M.I.; Chivukula, V.; Herati, R.S.; Hubbard, S.R.; Mulligan, M.J.; Landau, N.R. Increased resistance of SARS-CoV-2 Omicron variant to neutralization by vaccine-elicited and therapeutic antibodies. *EBioMedicine* **2022**, *78*, 103944. [CrossRef]
138. Shah, M.; Woo, H.G. Omicron: A Heavily Mutated SARS-CoV-2 Variant Exhibits Stronger Binding to ACE2 and Potently Escapes Approved COVID-19 Therapeutic Antibodies. *Front. Immunol.* **2021**, *12*, 830527. [CrossRef]
139. Zhou, B.; Zhou, R.; Tang, B.; Chan, J.F.; Luo, M.; Peng, Q.; Yuan, S.; Liu, H.; Mok, B.W.; Chen, B.; et al. A broadly neutralizing antibody protects Syrian hamsters against SARS-CoV-2 Omicron challenge. *Nat. Commun.* **2022**, *13*, 3589. [CrossRef]

140. Fang, Z.; Monteiro, V.S.; Hahn, A.M.; Grubaugh, N.D.; Lucas, C.; Chen, S. Bivalent mRNA vaccine booster induces robust antibody immunity against Omicron lineages BA.2, BA.2.12.1, BA.2.75 and BA.5. *Cell Discov.* **2022**, *8*, 108. [CrossRef]
141. Bhattacharya, M.; Sharma, A.R.; Dhama, K.; Agoramoorthy, G.; Chakraborty, C. Hybrid immunity against COVID-19 in different countries with a special emphasis on the Indian scenario during the Omicron period. *Int. Immunopharmacol.* **2022**, *108*, 108766. [CrossRef]
142. Pilz, S.; Ioannidis, J.P.A. Does natural and hybrid immunity obviate the need for frequent vaccine boosters against SARS-CoV-2 in the endemic phase? *Eur. J. Clin. Investig.* **2022**, e13906. [CrossRef]
143. Nguyen, T.M.; Zhang, Y.; Pandolfi, P.P. Virus against virus: A potential treatment for 2019-nCov (SARS-CoV-2) and other RNA viruses. *Cell Res.* **2020**, *30*, 189–190. [CrossRef] [PubMed]

viruses

MDPI

Article

Humoral Immune Response Profile of COVID-19 Reveals Severity and Variant-Specific Epitopes: Lessons from SARS-CoV-2 Peptide Microarray

Arup Acharjee [1,†], Arka Ray [2], Akanksha Salkar [1], Surbhi Bihani [1], Chaitanya Tuckley [2], Jayanthi Shastri [3], Sachee Agrawal [3], Siddhartha Duttagupta [4] and Sanjeeva Srivastava [1,*]

1 Department of Biosciences and Bioengineering, Indian Institute of Technology Bombay, Mumbai 400076, India
2 Centre for Research in Nanotechnology and Science, Indian Institute of Technology Bombay, Mumbai 400076, India
3 Kasturba Hospital for Infectious Diseases, Mumbai 400011, India
4 Department of Electrical Engineering, Indian Institute of Technology Bombay, Mumbai 400076, India
* Correspondence: sanjeeva@iitb.ac.in
† Current address: Department of Zoology, Faculty of Science, University of Allahabad, Prayagraj 211002, India.

Abstract: The amaranthine scale of the COVID-19 pandemic and unpredictable disease severity is of grave concern. Serological diagnostic aids are an excellent choice for clinicians for rapid and easy prognosis of the disease. To this end, we studied the humoral immune response to SARS-CoV-2 infection to map immunogenic regions in the SARS-CoV-2 proteome at amino acid resolution using a high-density SARS-CoV-2 proteome peptide microarray. The microarray has 4932 overlapping peptides printed in duplicates spanning the entire SARS-CoV-2 proteome. We found 204 and 676 immunogenic peptides against IgA and IgG, corresponding to 137 and 412 IgA and IgG epitopes, respectively. Of these, 6 and 307 epitopes could discriminate between disease severity. The emergence of variants has added to the complexity of the disease. Using the mutation panel available, we could detect 5 and 10 immunogenic peptides against IgA and IgG with mutations belonging to SAR-CoV-2 variants. The study revealed severity-based epitopes that could be presented as potential prognostic serological markers. Further, the mutant epitope immunogenicity could indicate the putative use of these markers for diagnosing variants responsible for the infection.

Keywords: COVID-19; SARS-CoV-2; peptide microarray; humoral immunity; IgA; IgG; epitope mapping; SARS-CoV-2 variant

Citation: Acharjee, A.; Ray, A.; Salkar, A.; Bihani, S.; Tuckley, C.; Shastri, J.; Agrawal, S.; Duttagupta, S.; Srivastava, S. Humoral Immune Response Profile of COVID-19 Reveals Severity and Variant-Specific Epitopes: Lessons from SARS-CoV-2 Peptide Microarray. *Viruses* **2023**, *15*, 248. https://doi.org/10.3390/v15010248

Academic Editors: Ahmed El-Shamy and Mohamed Ibrahim

Received: 26 November 2022
Revised: 12 January 2023
Accepted: 14 January 2023
Published: 15 January 2023

1. Introduction

The COVID-19 pandemic has been an epoch-defining moment in human history. COVID-19 has left behind an inexplicable trail of death and disease [1]. Epidemiological studies indicate that the actual number of cases could be inordinately higher than those reported [2]. Dedicated instrumentation and the risk of handling live viruses add to the woes of using nucleic-acid-based technologies [3]. Serological techniques are more robust in this regard as they are less susceptible to false-positives [4,5]. Omics technologies and big data analytics often have led to a panel of biomarkers for the prediction of disease severity [6–11] from the early days of the pandemic. However, they are of little or no translational value, especially in the event of a colossal hospitalization burden, as is often the case in outbreaks. Though we are in the third year of the pandemic, we are yet to find a reasonably accessible mode of detecting disease severity. Increased susceptibility of a certain group, such as the elderly population, those with pre-existing comorbidities [12], and immunocompromised patients, remains the biggest challenge in controlling the pandemic. Moreover, the variant-driven resurgence in COVID-19 cases further complicates the scenario [13–15]. With NGS as the primary way of finding the causative strain of the virus [14,16,17], clinicians and frontline workers have little or no means of fighting this amaranthine scourge when fresh outbreaks occur due to the emergence of new variants. Serological tools to diagnose and prognose the disease development in patients can aid in this regard. Moreover, serological tools to detect variants can be a good alternative to NGS.

Before the development of peptide microarrays, antibody detection has been traditionally handled using tools, such as the Enzyme-Linked Immunosorbent Assay (ELISA), Western blot, radioimmunoassay (RIA), flow cytometry, and platforms such as Luminex. Advanced variants of ELISA, such as Chemiluminescent immunoassay (CLIA), have also come to the fore. A diagnostic ELISA technique was used to identify MERS-CoV viral antibodies in patients with Middle East Respiratory Syndrome [18]. Western blotting was developed to assess antibodies produced against SARS [19]. The amount of class-specific antibodies against the tick-borne encephalitis virus (TBEV) was evaluated using an RIA-based test [20]. The presence of dengue virus antibodies in human patients has been studied using flow cytometry [21]. To detect and quantify antibodies against multiple viruses such as bovine respiratory syncytial virus, parainfluenza 3 virus, bovine viral diarrhea virus, and bovine herpes virus, Anderson et al. designed and assessed a multiplex assay employing the Luminex platform [22]. Although these techniques have historically been employed in basic and clinical research, peptide microarrays have emerged as cutting-edge since they can identify novel epitopes at an amino acid resolution.

To this end, a thorough understanding of the host immune response to SARS-CoV-2 [23] and its variants will aid in developing diagnostic and prognostic tools as well as vaccines. A thorough characterization of humoral antibody responses to viral proteins requires tools that could unravel the epitopes available on the entire proteome of SARS-CoV-2. Protein microarray-based technologies have been used widely over the past two decades to investigate multiple pathogens [24,25]. It is a fantastic tool for researching the humoral immune response at the amino acid resolution. They have been frequently used to investigate the immunological response to infections and to help in pathogen identification and strain-typing [26]. Peptide microarray is a fast-expanding area that has the potential to be a robust diagnostic platform for a wide range of diseases [27]. Multiple studies in the recent past have used this technology to study the impact of COVID-19 [24,28–32]. While most of these studies gleaned new information on the immunological landscape of the infected populace and found discriminatory epitopes, none of these studies assessed the impact of mutations on immunological topography.

In the current pilot investigation, we studied a cohort of 14 patients using a high-density SARS-CoV-2 peptide microarray. We investigated the humoral immune response to SARS-CoV-2 infection to identify immunogenic areas in the SARS-CoV-2 proteome at a resolution of 2-amino acids. The SARS-CoV-2 proteome was reflected by 4932 peptides printed in duplicate on a chip. The microarray also housed the most relevant mutations from SARS-CoV-2 B.1.1.7 lineage and the Gamma SARS-CoV-2 variant (P.1) apart from other widely circulating SARS-CoV-2 mutations.

In addition to the humoral immune landscape of SARS-CoV-2, the study found severity-specific IgG and IgA epitopes. Additionally, the response to the mutations panel reflected on the effect of mutations on immunoreactivity and may point to their potential utility for identifying infection-causing variants.

2. Materials and Methods

2.1. Cohort Characteristics and Sample Details

This study was approved by the Institutional Review Board of the Kasturba Hospital for Infectious Diseases and IIT Bombay. Patient consent was waived as leftover samples from the routine analysis were used in the study. The pilot study included 14 patients who had a confirmed diagnosis of COVID-19 using a RT-PCR test. The patients with at least a seven-day interval between symptom onset and sample collection were included in the study. Stratification of patients based on the severity of the symptoms was done by the clinical team at Kasturba Hospital for Infectious Diseases. Parameters such as disease presentation, respiratory distress, SpO_2 levels, and oxygen supplementation status were used for classifying the patients into severe and non-severe COVID-19 groups (Table S1).

The leftover blood samples anticoagulated with EDTA after blood tests were centrifuged at 2000 rpm for 10 min to separate the plasma and aliquoted in sterile cryovials. The aliquots were then heat-treated at 56 °C for 30 min to inactivate the virus [33]. The heat-treated samples were then transported on ice to IIT Bombay. At IIT Bombay, the samples were divided into sub-aliquots and stored at −80 °C until further use to ensure minimum freeze–thaw cycles.

2.2. Peptide Microarray

Epitope-level antibody responses from the patient plasma were deciphered using SARS-CoV-2 whole proteome peptide microarray. The PEPperCHIP® SARS-CoV-2 Proteome Microarray (PEPperPRINT GmbH, Heidelberg, Germany) covers the entire proteome of SARS-CoV-2 isolate Wuhan-Hu-1 (GenBank ID: MN908947.3). The protein sequences of ORF1a/b, Spike, ORF3a, Envelope, Membrane glycoprotein, ORF6, ORF7a, ORF8, Nucleocapsid, and ORF10 were translated into 15 amino acid long peptides with an overlap of 13 amino acids. This resulted in 4883 individual peptides printed in duplicates. As internal controls, the PEPperCHIP® SARS-CoV-2 Microarray contained influenza hemagglutinin (HA) and polio control peptides (108 spots for each control peptide).

Individual patient plasma was thawed on ice and then diluted at 1:100 using staining buffer (PBS with 0.05% Tween20 and 10% Rockland blocking buffer MB 070, pH 7.4). One PEPperCHIP® Peptide Microarray slide was processed per patient plasma for the experiment. PEPperCHIP® Peptide Microarray slides were brought to room temperature, assembled onto the PEPperCHIP® incubation tray (PEPperPRINT GmbH, Germany), and equilibrated using the staining buffer for 15 min. The slides were then incubated with the diluted plasma samples overnight at 4 °C. All the incubations were done on an orbital shaker at 140 rpm unless otherwise stated. On the following day, slides were washed thrice for 1 min each using a washing buffer (PBS with 0.05% Tween20, pH 7.4) while incubating the slide at room temperature. The washing buffer was aspirated entirely after each wash using a micropipette. The slides were then incubated with a mixture of Cy5 conjugated rabbit anti-human IgA (antibodies.com, ABIN901561) and Cy3-conjugated donkey anti-human IgG (Rockland, 609-704-123) antibodies diluted 1:1000 in staining buffer for 45 min at room temperature in the dark. The slides were again washed, as explained earlier. The incubation tray was then disassembled, and the slides were dipped in dipping buffer (1 mM Tris buffer, pH 7.4) to remove PBS residues or dust. The slides were dried under the pressurized nitrogen stream from top to bottom. The slides were then scanned using a Molecular Devices, GenePix® 4000 B scanner, with 33% of the laser intensity for Cy3 and 100% for Cy5 signals.

Next, the slides were processed similarly to acquire signals for the influenza HA control spots for a quality check. The slides were re-assembled onto the incubation tray and equilibrated in staining buffer for 15 min at RT. This was followed by incubation with a PEPperCHIP® Cy5-conjugated anti-HA control antibody diluted 1:2000 in staining buffer for 45 min at RT in the dark. Further washing, drying, and scanning steps were performed, as mentioned earlier.

2.3. Data Acquisition

After scanning the chips, their images were acquired in GenePix®Pro 7 in .tiff format. The fluorescence or raw intensity and the background intensity of the individual spots were extracted using a PepSlide® Analyzer. An R-based script was used for statistical analysis to determine the immunogenic response against the peptides and analyze the severity-based discrimination of the SARS-CoV-2 patients. These steps were carried out on the datasets generated against IgG and IgA antibodies.

2.4. Data Analysis

The raw intensities of the spots were background-corrected using the "norm-exp" technique to adjust the spot intensities individually with respect to the background. An offset value of 50 was also added so that the weak signals from the features do not get suppressed [34]. The "limma" package of R programming language was used to implement it [35]. Subsequently, the intensities of the duplicate peptides were averaged. Further, to reduce the skewness of the dataset, a logarithmic transformation with a base value of 10 was applied to these values. The mean and standard deviation of intensities of all the peptides across all the samples were used to compute the z-score [32,36,37]. The peptides for which the z-scores exceeded the value of 3 [32] in at least one COVID-19 patient sample were classified as immunogenic. The z-scores were then used to generate heatmaps for visualizing the IgG and IgA response against the SARS-CoV-2 proteome. For determining peptides with discriminatory potential, the response of patients with severe and non-severe COVID-19 was compared using the Mann–Whitney U-test [32]. Peptides with a p-value less than 0.05 were considered to be statistically significant. R programming language was used to generate heatmaps, box plots, and dot plots for visualizing different comparisons made during analysis. The study design and the overall workflow is depicted in Figure 1.

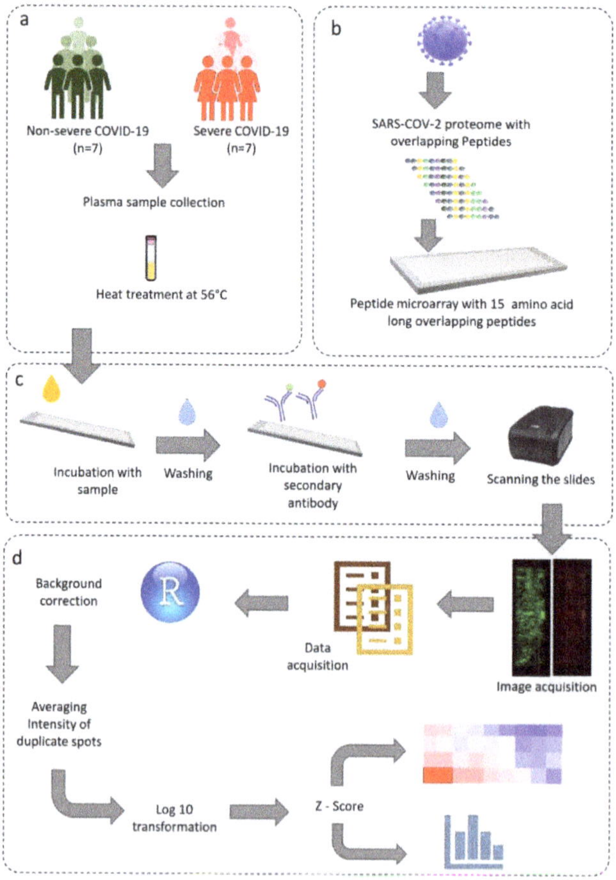

Figure 1. Schematic representation of the study design, microarray protocol, and data analysis. (**a**) Sample acquisition and heat inactivation of virus, (**b**) SARS-CoV-2 whole proteome microarray design, (**c**) microarray staining and image acquisition, (**d**) data analysis pipeline.

3. Results

We used the PEPperCHIP® Peptide Microarray slides to study the landscape of B-cell epitopes of IgA and IgG antibodies in COVID-19 patients during acute infection. In total, 14 patients were included in the study; 7 had non-severe COVID-19, and 7 had severe COVID-19 (Table S1). The plasma samples were used to detect the humoral immune response against COVID-19. The PEPperCHIP® Peptide Microarray had 15 amino acid-long 4932 peptides with 13 overlapping peptides, thus providing an epitope resolution as high as two amino acids. During data pre-processing, the data distribution of two samples, one severe (patient ID 69) and the other non-severe (patient ID 33), were skewed. Therefore, these two samples were removed from further data analysis.

3.1. Proteome-Wide Immunogenic Response for IgA and IgG

A total of 204 and 676 peptides were identified as immunogenic for IgA and IgG, respectively, based on z-scores. In addition, out of 49 peptides in the mutant panel, 5 and 10 peptides for IgA and IgG respectively, were found to be immunogenic. The peptides were considered immunogenic if the z-score was greater than 3 in any one of the patients.

IgA Response: Among the structural proteins, IgA immunoreactivity was found in 17, 5, and 1 peptides in spike, nucleocapsid, and membrane glycoprotein, respectively (Figure S1 and Table S2). No immunogenic response was observed in the envelope protein. Out of the 204 immunogenic peptides identified for IgA, 159 peptides were from the non-structural proteins encoded by ORF1a/b (Figure S1). 41 and 43 IgA-reactive peptides were identified in NSP3 (PLpro) and NSP12 (RdRp). In NSP3, VSELLTPLGIDLDEWSMATYYLFDE (aa81–aa105) epitope showed response for IgA. In NSP 12, we observed response from the N-terminal nidovirus RdRp-associated nucleotidyltransferase domain (RiRAN) and RdRp region. For accessory proteins, 9, 2, 3, and 8 immunogenic peptides were identified in ORF3a, ORF6, ORF7a, and ORF8, respectively (Figure S2a–k and Table S2). However, most of the consecutive regions exhibiting reactivity against IgA were identified in NSP 3, NSP 12, and spike protein (Table S4).

IgG Response: A total of 676 peptides were immunoreactive for IgG. There were 478 immunogenic peptides from the NSPs (Figure S1), of which 116 belonged to NSP 3 and 92 belonged to NSP 12. Similarly, 90, 31, 1, and 17 peptides were identified for spike, nucleocapsid, envelope, and membrane glycoprotein, respectively (Figure S2). Among accessory proteins, ORF3a, ORF6, ORF7a, and ORF8 had 27, 5, 12, and 14 immunoreactive peptides, respectively (Figure S3a–k and Table S3).

The top epitopes in the spike protein identified in at least one-fourth of the patients are "CEFQFCNDPFLGVYY" (aa131–aa145 located in the N-terminal domain), "VYYHKNNKSWMESEF" (aa143–aa157, N-terminal domain), "CLIGAEHVNNSYECD" (aa649–aa663, near furin cleavage site), "PSKPSKRSFIEDLLF" (aa809–aa823, near fusion peptide in S2 sub-unit), "ESLIDLQELGKYEQY" (aa1195–aa1209, HR2 region of the S2 sub-unit), "QELGKYEQYIKWPWY" (aa1201–aa1215, HR2 region of the S2 sub-unit) (Table S5). It is interesting to note that these highly immunogenic regions belong to regions other than the RBD. The immunoreactive peptides belonging to the RBD region were identified in a maximum of two patients only. Non-structural proteins 3 and 12 and accessory proteins ORF3a and ORF8 were also associated with strong immunoreactivity in at least one-fourth of the patients. Top immunogenic peptides identified in the nucleocapsid protein were located in the N-terminal domain. Strong immunoreactivity was not observed in envelope and membrane proteins in the majority of the patients. On the other hand we identified VSELLTPLGIDLDEWSMATYYLFDE (aa81–aa105), CSFYPPDEDEEEGDCEEEEFEPS (aa117–139), GDCEEEEFEPSTQYEYG (aa129–aa145), SAALQPEEEQEEDWLDDDS (aa161–aa179) and VLPNDDTLRVEAFEYYH (aa815–aa831) epitopes belonging to the NSP3 as highly reactive for IgG (Table S5).

The major regions in the SARS-CoV-2 proteome immunoreactive to IgA and IgG in at least one-fourth of the patients are shown in Figure 2.

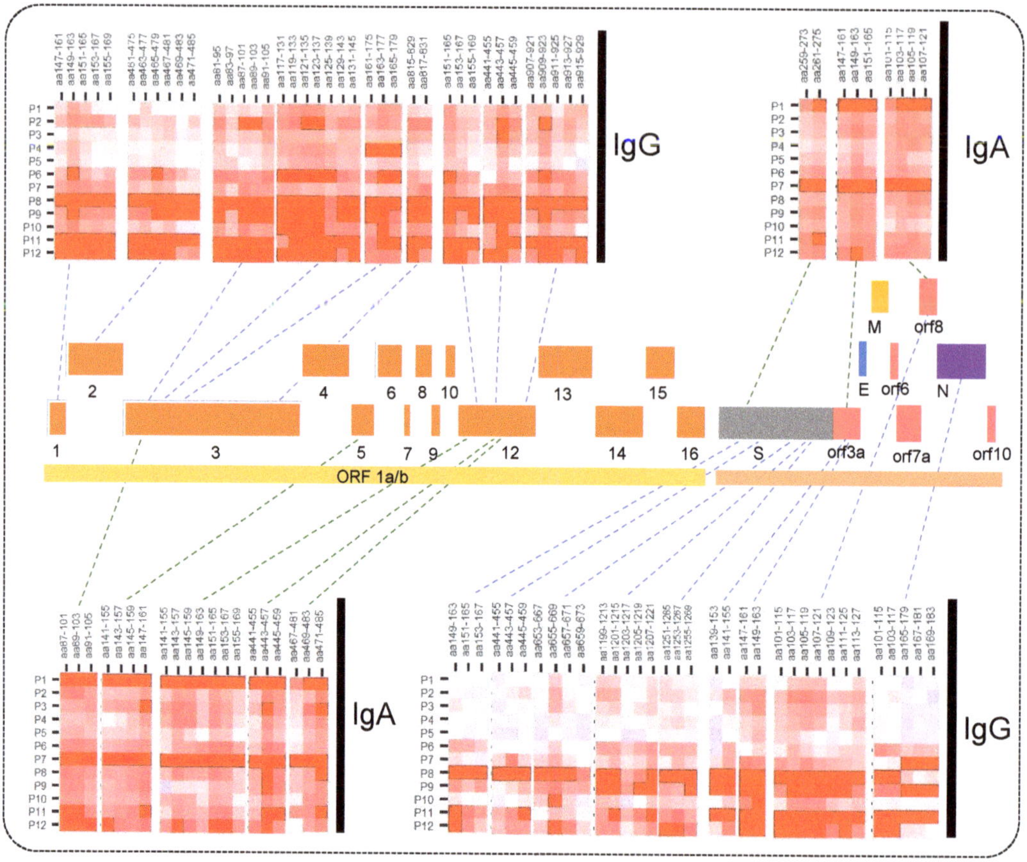

Figure 2. Heatmaps for IgA and IgG response showing major immunogenic regions identified in the SARS-CoV-2 whole proteome microarray. The printed proteome constitutes ORF1a/b polyprotein encoding 16 non-structural proteins (1–10 and 12–16), structural proteins (S, N, E, and M), and the accessory proteins (ORF3a, 6, 7a, 8, and 10).

3.2. Severity-Based Epitopes

We studied severity-based differences in the immune response. The significance of differential response was calculated using the Mann–Whitney U test. We observed that 6 and 319 peptides from ORF1a/b had a significant difference in IgA and IgG reactivity (Tables S6 and S7), respectively, in severe vs. non-severe patients. The response in severe and non-severe COVID-19 in IgA and IgG from some representative peptides has been illustrated in boxplots in Figure 3. Further, for the structural proteins such as spike, nucleocapsid, membrane, and envelope, we identified 57, 12, 8, and 1 peptide, respectively, with significantly differential responses. However, there was no significant difference in IgA response for these structural proteins. Further, we also identified 22, 2, 6, and 13 peptides eliciting severity-based significantly different IgG responses originating from accessory proteins like ORF3a, ORF6, ORF7a, and ORF8, respectively. Some of the discriminatory peptides showed responses both for IgG and IgA (Table 1). Of note, the discriminatory epitopes specific to the severe disease include "ANYFLCWHTNCYDYC" in ORF3a, "EILVTYNCCDDDYFN", "EVVDKYFDCYDGGCI", "VLTLDNQDLNGNWYD" and "YRNRDVDTDFVNEFY" in NSP12, "VSELLTPLGIDLDEWSMATYYLFDESGEF", "VLPNDDTLRVEAFEY", "CEEEEFEPSTQYEYG", "FYPPDEDEEEGDCEE", and

"FKWDLTAFGLVAEWF" in NSP3, and "CEFQFCNDPFLGVYY" in spike protein (Tables S8 and S9).

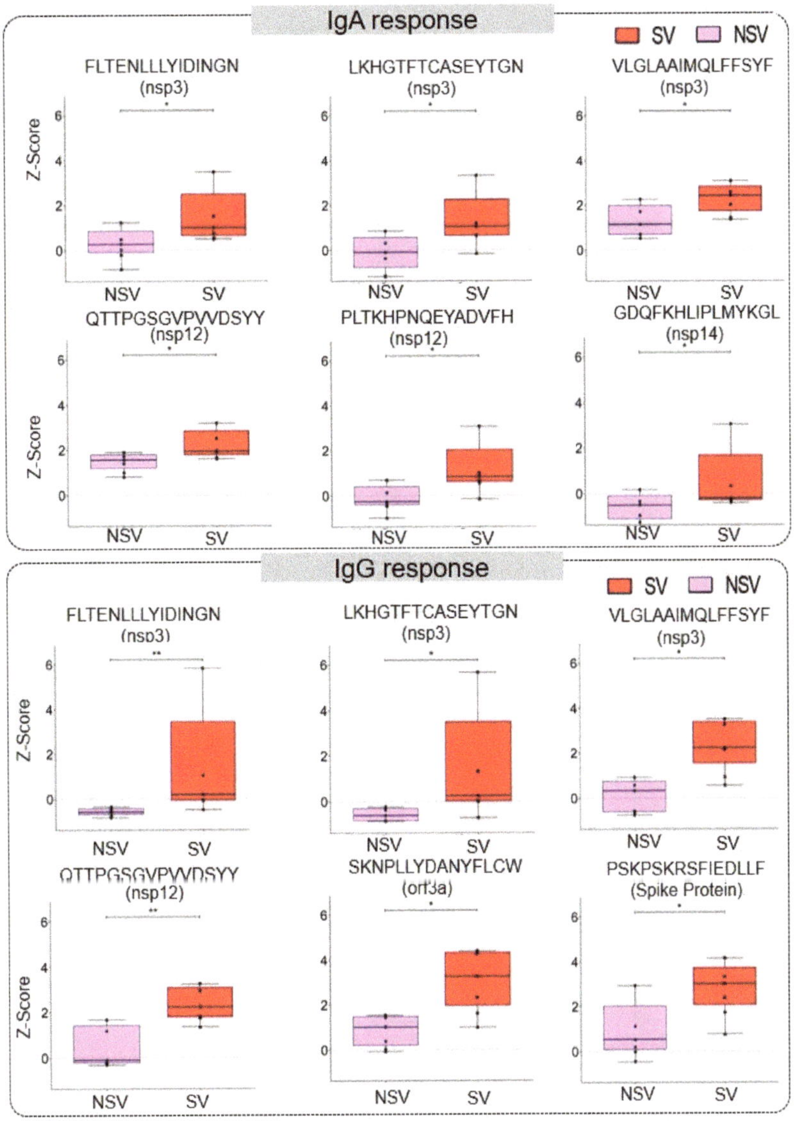

Figure 3. Severity-based discrimination in IgA and IgG response against SARS-CoV-2 peptides. The representative box plot showing IgA and IgG response against a particular peptide in severe (SV) and non-severe (NSV) COVID-19 patients. The significance was calculated using a Mann–Whitney U-Test with a p-value < 0.05 (the p-values between 0.05 and 0.01 are represented using '*' and those between 0.01 and 0.001 by '**').

Table 1. Viral protein-wise list of epitopes that seroconverted from IgA to IgG in patients. The IgA/IgG epitopes that could discriminate between severe and non-severe forms of the disease are indicated.

Protein Name	Seroconverted Epitope	Discrimination Status
NSP1	GEIPVAYRKVLLRKNGN	Non-significant $^{\alpha}$
NSP1	LKSFDLGDELGTDPYEDFQENWN	IgG $^{\beta}$
NSP2	GAYTRYVDNNFCGPDGYPLEC; NIVGDFKLNEEIAII; LDWLEEKFKEGVEFLRDGWEIVKFI	IgG
NSP3	QPVSELLTPLGIDLDEWSMATYYLFDESGEFKL; SAALQPEEEQEEDWLDDDSQQ	Non-significant
NSP3	MYCSFYPPDEDEEEGDCEEEEFEPSTQYEYGTEDDYQ; RTNVYLAVFDKNLYD; GIKIQEGVVDYGARFYFYT; FYVLPNDDTLRVEAFEYYH; TLRVEAFEYYHTTDPSFLGRY; IELKFNPPALQDAYY; AGEAANFCALILAYC; GVVCTEIDPKLDNYY; TFFPDLNGDVVAIDY; ITEEVGHTDLMAAYV; SYFAVHFISNSWLMWLI	IgG
NSP3	TLEETKFLTENLLLYIDINGN; LKHGTFTCASEYTGN; VLGLAAIMQLFFSYF	IgG/IgA $^{\lambda}$
NSP4	DTCFANKHADFDTWF; FATSACVLAAECTIF; EGSVRVVTTFDSEYCRH; VSFSTFEEAALCTFLLN	IgG
NSP5	QVTCGTTTLNGLWLDDVVYCPRH; LNGSCGSVGFNIDYDCVSFCYMHHMEL; LAWLYAAVINGDRWF; NGRTILGSALLEDEF; ILGSALLEDEFTPFDVVRQ	IgG
NSP6	LVQSTQWSLFFFLYE; MFLARGIVFMCVEYC; RGIVFMCVEYCPIFF; CLLNRYFRLTLGVYDYL; LTLGVYDYLVSTQEFRY	IgG
NSP8	NTCDGTTFTYASALWEI	IgG
NSP9	TTQTACTDDNALAYY	IgG
NSP9	VLGSLAATVRLQAGN	Non-significant
NSP12	TGTSTDVVYRAFDIYND; FQEKDEDDNLIDSYFVV;TKYTMADLVYALRHFDEGNCD; QTVKPGNFNKDFYDF; FFFAQDGNAAISDYDYYRYNL; ARLYYDSMSYEDQDALFAY; RLYECLYRNRDVDTDFVNEFYAY; HFSMMILSDDAVVCF	IgG
NSP12	NCDTLKEILVTYNCCDDDYFNKKDWYDFVEN; ADKYVRNLQHRLYECLY; FCSQHTMLVKQGDDYVYLPYP; MLTNDNTSRYWEPEFYEAMYTPHTVLQ	Non-significant
NSP12	QTTPGSGVPVVDSYY; PLTKHPNQEYADVFHLYLQYI	IgG/IgA
NSP13	ACIRRPFLCCKCCYD; DVTDVTQLYLGGMSYYC; FNAIATCDWTNAGDYIL; TQTVDSSQGSEYDYVIF	IgG

Table 1. *Cont.*

Protein Name	Seroconverted Epitope	Discrimination Status
NSP13	VNALPETTADIVVFDEISM; CPAEIVDTVSALVYD	Non-significant
NSP14	SDTYACWHHSIGFDYVYNPFMIDVQQWGF; WGFTGNLQSNHDLYC; HECFVKRVDWTIEYPII	IgG
NSP14	EYPIIGDELKINAAC; AQPCSDKAYKIEELFYS; IEELFYSYATHSDKFTD; ATHSDKFTDGVCLFWNC	Non-significant
NSP15	NLGVDIAANTVIWDY; TVFFDGRVDGQVDLFRN; QMEIDFLELAMDEFIERYK; LAMDEFIERYKLEGYAFEH; KLEGYAFEHIVYGDFSH; LAKRFKESPFELEDF; GSSKCVCSVIDLLLDDFVEIIKS; VKVTIDYTEISFMLW	IgG
NSP16	PTGTLLVDSDLNDFV; TEHSWNADLYKLMGHFAWW	Non-significant
NSP16	PIQLSSYSLFDMSKF	IgG
Spike	TQDLFLPFFSNVTWF; CEFQFCNDPFLGVYY; FRVYSSANNCTFEYV; KNLREFVFKNIDGYFKI; AGCLIGAEHVNNSYECDIP; DPLQPELDSFKEELDKYFK; LNESLIDLQELGKYEQY;DLQELGKYEQYIKWPWYIW; KGCCSCGSCCKFDEDDSEP	IgG
Spike	DTTDAVRDPQTLEILDI	Non-significant
Membrane Glycoprotein	LEQWNLVIGFLFLTW	IgG
Nucleocapsid	WFTALTQHGKEDLKF; IRGGDGKMKDLSPRWYFYY; GSSRGTSPARMAGNGGDAALALLLLDR	IgG
Orf3a	YSHLLLVAAGLEAPFLYLY; WKCRSKNPLLYDANYFLCW; YDANYFLCWHTNCYDYCIPYN; VKDCVVLHSYFTSDYYQLY	IgG
Orf6	IMRTFKVSIWNLDYI	IgG
Orf7a	ILFLALITLATCELYHYQECVRG	IgG
Orf8	KLGSLVVRCSFYEDFLEYHDVRVVLDF	IgG

α indicates seroconverted epitopes that do not discriminate severe and non-severe forms of the disease. β indicates IgG isotype of seroconverted epitopes that could discriminate severe and non-severe forms of the disease. λ indicates both IgG and IgA isotypes of seroconverted epitopes could discriminate severe and non-severe forms of the disease.

3.3. Response to Mutant Peptides

The PEPperCHIP® Peptide Microarray also had the peptides with the mutations from several SARS-CoV-2 variants printed on the slides (Table S10). In particular, the mutant peptides originated from the 501.V2, B1.1.7, and P.1 Manaus variants, as well as some other highly frequent mutations, were included. We observed strong IgG reactivity for 10 of the peptides harbouring mutation, of which 2 originated from ORF1a/b, 7 from the spike, and 1 from nucleocapsid protein. Seven mutant peptides from the spike proteins were associated with B1.1.7, 501.V2, and P.1. Manaus variants. In addition, 5 mutant peptides were reactive for IgA, of which 1 was from ORF1a/b, 3 from the spike, and 1 from nucleocapsid protein. The IgG and IgA response against mutant peptides has been illustrated using a heatmap (Figure 4a). We also looked at differences in the reactivity against the mutant peptide and the corresponding wildtype peptide to see the effect of mutations on the immune response. The patient-wise variation in immunoreactivity to the mutant peptides vs. the corresponding wildtype has been depicted in Figure 4b and Figure S3a–e and Figure S4a,b. We observed that the D138Y mutation in spike protein and the P80R mutation in the nucleocapsid protein increased the immunoreactivity of both IgA and IgG.

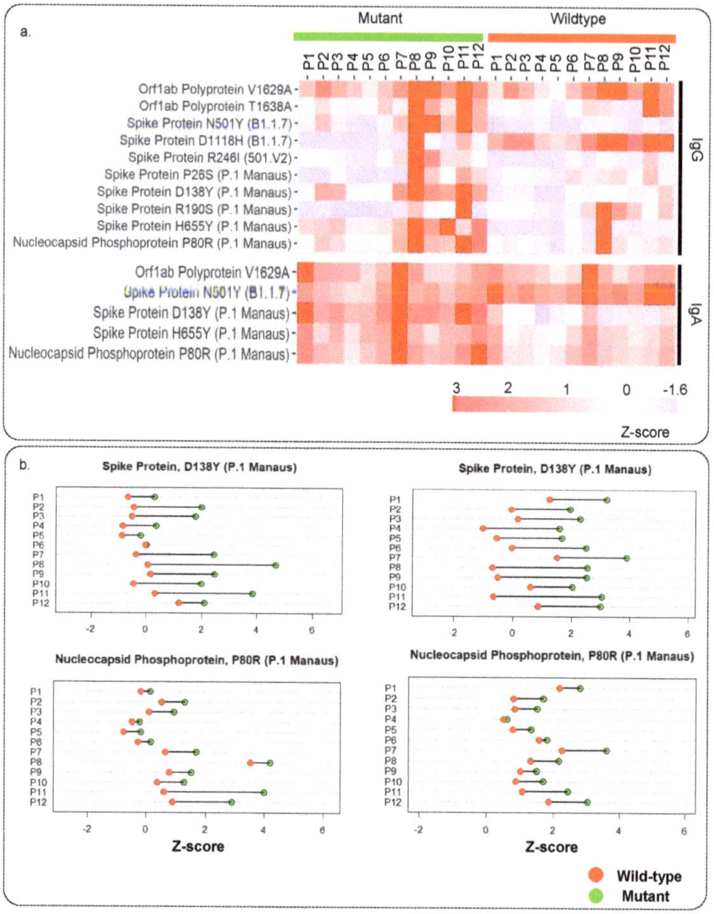

Figure 4. Immune response against mutant peptides for IgG and IgA: (**a**) Heatmap showing IgG and IgA response for peptides with mutations from different SARS-CoV-2 variants and their wildtype counterpart, (**b**) representative dot plots showing the patient-wise comparison of response against mutant and wildtype peptides.

4. Discussion

SARS-CoV-2 epitope-level proteome-wide analysis was done using peptide microarrays to study the antibody responses in COVID-19 patients. The high peptide-to-peptide overlap of the SARS-CoV-2 proteome array allowed a high-resolution epitope analysis giving a detailed picture of antibody binding patterns, contributing to better characterization of SARS-CoV-2-specific humoral immune responses. For this study, we chose IgA and IgG responses. A previous study [38] using Spike-RBD specific antibodies on patient samples indicated that during the initial days post-infection, that is, 4–10 day, around 88% of the patients were seropositive for IgA, and this response was most robust among IgA, IgM, and IgG. While on the other hand, both IgA and IgG responses were found to be most stable, and 100% of patients showed seroconversion beyond 15 days post-infection. IgM responses have been reported to be erratic [39], indicating that for SARS-CoV-2 it is not a reliable immunological indicator.

The top immunoreactive regions identified in the spike protein were located in the NTD, HR2 region, and near the furin cleavage site (Figure 5a). Some of these epitopes have also been reported by other groups to be immunogenic [28,31,32]. The receptor binding domain (RBD) region of the spike protein enables viral entry into the cell, therefore, antibodies against spike RBD have been forerunners in neutralising the virus. However, intensities from this region were observed in at most two patients. This could be due to the presence of more conformational than linear epitopes in this region. In addition, antibodies against the NTD, furin cleavage site, and HR2 have also been reported to inhibit viral entry [40–42]. We identified the "ESLIDLQELGKYEQY" (aa1195–aa1209) belonging to the conserved HR2 region of the S2 subunit to be highly immunogenic. The S2 subunit is more conserved among the beta-coronaviruses and is less susceptible to non-synonymous mutations [43]. Studies have reported cross-reactive and neutralising antibodies against specific regions in the S2 subunit [44–46]. Thus, immunoreactivity of this region can be crucial for protection against other beta-coronaviruses and this can be used to develop effective vaccines.

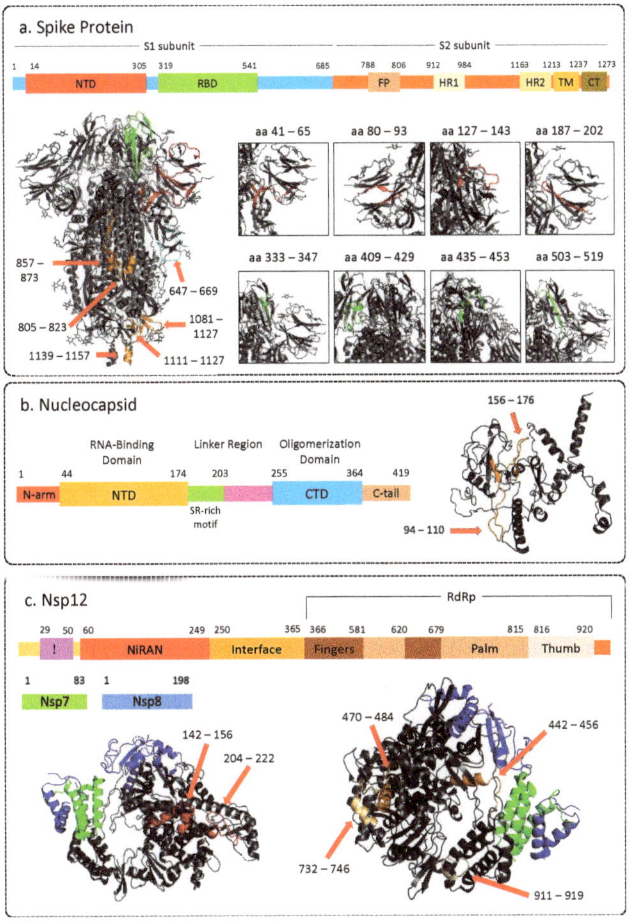

Figure 5. Representative images showing 3D structure of SARS-CoV-2 proteins highlighting immunogenic peptides, along with corresponding linear structure. (**a**) Spike protein (PDB: 6VXX); (**b**) Nucleocapsid (3D structure modeled by Zhang lab using I-Tasser). (**c**) Non-structural protein 12 (RNA-dependent RNA-polymerase) in complex with NSP7 and NSP8 (PDB: 6M71). NTD: N-terminal domain; RBD: Receptor binding domain; FP: Fusion; HR: Heptad repeat; TM: Transmembrane region; CT: Cytoplasmic tail; CTD: C-terminal domain. The 3-D structures of the proteins were visualised using PyMOL (version 2.5.2) [47].

Two regions in the N-terminal domain of the nucleocapsid phosphoprotein, that is, aa94–aa110 and aa156–aa176, were found to be highly immunoreactive (Figure 5b). Structure of the N-protein shows that part of both of these regions are on the surface and can be easily accessible to antibodies (Figure 5b). These regions have also been reported to be immunodominant by other studies [32,48] and thus can be considered as effective targets for vaccines and serological markers.

Apart from the structural proteins SARS-CoV-2 proteome also has 16 non-structural proteins, which regulate viral replication. Nsp 1 protein is crucial for hijacking the host translational machinery and is, therefore, a crucial virulence factor. Of the epitopes identified against IgG, LKSFDLGDELGTDPYEDFQENWN (aa147–aa169) epitope was identified from the C-terminal domain (CTD) of the Nsp1. A part of these epitopes was also found to be reactive for IgA. Even The region also has been reported to interact with host ribosomes to disarm the IFN-β or RIG-I governed immune responses [49]. This epitope was also reported to show differential responses in mild vs. severe patients in a previous study analyzing the epitope signatures of COVID-19 patients [31]. Therefore, targeting this epitope could be a potential approach for viral clearance.

Nsp2, the LDWLEEKFKEGVEFLRDGWEIVKFI (aa461–aa485) was identified to elicit an IgG response, whereas a part (aa465–aa481) of it was reactive for IgA. However, these epitopes originate from the residual region of the Nsp2 protein. The Nsp2 protein has an N-terminus and a residual domain. However, understanding the structure and function of disordered regions of NSP2 is in a nascent stage and remains elusive. Therefore, further studies targeting the disordered regions are required. It is the N-terminus domain that binds to the nucleic acids [50]. We also identified two immunogenic peptides from the N-terminus RGVYCCREHEHEIAW (aa59–aa73) and CCREHEHEIAWYTER (aa63–aa77) that elicited a response in at least three patients with significantly different severe vs. non-severe COVID-19. Schwarz et al. also observed a similar trend in comparing patients with severe symptoms to those with mild [31]. In another study by Heidepriem et al., the above-mentioned peptides showed a high IgA response in some of the patients during the early phase of infection. However, the IgG response remained low in comparison to IgA and IgM responses [51].

NSP 3, or papain-like protease, is critical for viral replication and suppression of host responses. It is responsible for hydrolyzing the polyprotein pp1a into Nsp 1, 2, and 3. It interferes with the immune response in the host, in particular, the interferon and NF-κB pathways [52]. NSP 3 is a multidomain protein and is divided into 10 domains. We observed reactivity against domains like ubiquitin-like domain 1, hypervariable region, macrodomain I, and macrodomain II. Interestingly, we found the highly reactive regions originating from ubiquitin-like domain 1, hypervariable region, and ubiquitin-like domain 2 regions. This hypervariable region is a Glu-rich region with probable interaction with other proteins like Nsp6, 8, and 9 and the NAB–βSM–TM1 of Nsp3 [53]. However, the exact role of this region remains unknown. Therefore, the potential role of HVR reactivity also remains elusive. Another highly reactive region was identified in ubiquitin-like 1 region of Nsp3; VSELLTPLGIDLDEWSMATYYLFDE (aa81–aa105). This region has been speculated to mimic the host ubiquitin enzymes and thus help in escaping the host degradation mechanism. Moreover, the regions have also been reported to interact with ssRNA, thereby indicating its role in viral RNA replication and processing [54]. This region also showed discriminatory potential based on disease severity. We also identified the reactive peptides from macrodomain I (mac1) and papain-like protease (PLpro) regions. Whereas the mac1 domain has the ADP-ribosyl hydrolyase activity by regulating the host-mediated antiviral adenosine diphosphate-ribosylation signaling. PLpro has proteolytic, deubiquitinating, and deISGylating activity. These immune reactive regions can be targeted for inhibiting the viral genome replication and transcription.

Nsp 12 protein consists of N-terminal nidovirus RdRp-associated nucleotidyltransferase domain (RiRAN), interface, and the RdRp domain. We identified epitopes originating from all three domains. LKEILVTYNCCDDDYFN (aa143–aa159) and NCCDDDYFNKKD-

WYDFVEN (aa151–169) were IgA and IgG-reactive epitopes belonging to the RiRAN domain. In addition, we also identified peptide DNQDLNGNWYDFGDF (aa209–aa223) to be reactive for IgG. This epitope was also reported as one of the top reactive epitopes in another study mapping the epitope response [30]. Further, we identified FFFAQDGNAAISDYDYYRY (aa441–aa459), QLLFVVEVVDKYFDCYDGGCI (aa469–aa489), LYYDSMSYEDQDALFAY (aa515–aa531) YRNRDVDTDFVNEFYAY (aa733–aa749) and TNDNTSRYWEPEFYEAMYTPH (aa909–929) epitopes originating from RdRp domain to be reactive against IgG with some peptides forming these epitopes also showing reactivity for IgA (Figure 5c). In addition, aa441–aa459, aa515–aa531, aa733–aa749, and aa911–aa929 were found to elicit a discriminant response based on severity. Since Nsp12 is the core of RTC, it has been reported to form complexes with other non-structural proteins during viral genome replication and transcription [52]. Therefore, understanding the immune response against Nsp12 becomes crucial for developing antiviral therapies.

Although many immunogenic peptides were identified in other non-structural proteins like Nsp4, 5, 6, 7, 8, 9, 10, 13, 14, 15, and 16, there were no immunodominant regions identified. Most of the reactivity was patient-specific and concentrated in patients 8 and 11 for IgG and patients 1 and 7 for IgA. Of particular interest was the peptides like SVGFNIDYDCVSFCY (aa147–aa161) from Nsp5 (main proteases) which elicited a response in both IgG and IgA and severity-based discrimination in terms of IgG response. Another noteworthy peptide was LGVYDYLVSTQEFRY (aa239–aa253) which showed reactivity for IgG. A previous study has also reported this peptide as one of the highly reactive peptides for IgG [30]. In addition, the discriminatory potential of this peptide has been observed [31]. However, the trend reported in the earlier study is not similar. Considering the current cohort size a definitive comparison is difficult. Nonetheless, despite the conserved sequences, these proteins can be explored further as potential therapeutic targets due to their role in the viral life cycle and host invasion or for developing prognostic assays. Further studies are required to study the protective role of antibodies against non-structural proteins.

ORF8 protein has been reported to be a secreted protein detectable in sera of COVID-19 patients and has been shown to elicit immune responses during the early stages of SARS-CoV-2 infection [32]. Hachim and colleagues have identified acute and convalescent antibody responses against ORF8 protein suggesting the possible use of ORF8 as a sensitive and specific method for detection of both early and late SARS-CoV-2 infection [55]. The protective role of anti-ORF8 antibodies, however, remains unclear. It has been reported that SARS-CoV-2 mediates the downregulation of MHC-I as a way of immune evasion [56]. Thus, it can be speculated that neutralization of ORF8 may salvage potential immune evasion by ORF8.

Immunoreactivity of IgGs and IgAs for several epitopes was higher in the severe cohort as compared to the non-severe cohort, as can be observed from Figure 3. Whereas, for some epitopes majorly from ORF1a/b, immunogenicity was higher in the non-severe group. The course of the disease and the outcomes thereof can be attributed to the intensity of immune response to various immunogenic regions in the SARS-CoV-2 proteome. A couple of studies have associated SARS-CoV-2 antibodies with disease severity and survivability [32,37,57]. However, the sample sizes were too small to comment on whether this difference in immunogenicity can be related to the severity and survivability of the patients.

The SARS-CoV-2 proteome microarray also had a panel of peptides with mutations associated with the common variants of concern. Immunoreactivity to several mutations, including those from the P.1 and B.1.1.7 lineages, was observed, indicating either infection with these variants or that these mutations do not affect the immunogenicity of the corresponding wildtype peptides. The information about which variant caused the infection in patients from whom the samples were collected is not known. However, the samples in the study were collected from March to April 2021 during which both P.1 (alpha) and B.1.1.7 (beta) variants were circulating and the B.1.615 (delta) variant was on the rise. The effect of mutations on the immunogenicity of epitopes can also be comprehended using the mutant panel. Both IgG and IgA response to the peptide "QSYGFQPTNGVGYQP" increased significantly with N501Y mutation of the B1.1.7 linage. The mutation N501Y affects the conformation of the receptor binding domain of the spike protein [58]. Thus, it can be concluded that this mutation increases the immunogenicity of the spike protein. Whereas some mutations decreased the immunogenicity of peptides for both IgG and IgA, such as K417T and D138Y, both of which are specific to P.1 lineage. K417T is another key mutation in the spike RBD and is involved in interaction with the ACE2 receptor, while D138Y lies in the highly immunogenic region of the spike NTD. Thus, it can be inferred that these mutations decrease the immunogenicity of the spike protein. Even so, despite mutations, by and large, most peptides remained immunogenic.

Finally, being a pilot study, we conducted the test on a small cohort of patients. Further work is currently being planned to study an extensive cohort of patients with a longitudinal follow-up. This may lead to a better understanding of disease progression and aggressiveness as well as the impact of variant-specific mutations on the immune signature of the patient.

5. Conclusions

In conclusion, the humoral response to SARS-CoV-2 proteins was studied, revealing several immunogenic regions and severity-based epitopes in the viral proteome at amino acid resolution. Immunogenic regions in the SARS-CoV-2 proteome can be associated with disease severity and survivability, and they have the potential to be used as serological markers for prognosis and disease stratification. In addition, the response to common mutations was also studied using the panel of mutant peptides. We found differences in immunoreactivity in response to some mutant peptides as compared to the corresponding wildtype peptides. This highlights the changes in the immunoreactivity due to mutations and mechanisms of immune escape by the variants.

Supplementary Materials: The following supporting information can be downloaded at: https://www.mdpi.com/article/10.3390/v15010248/s1, Figure S1: Heatmap showing IgG response for peptide against ORF1a/b non-structural protein NSP1, NSP2 (a), NSP3 (b); NSP4, NSP5 (c), NSP6, NSP7, NSP8, NSP9 (d), NSP10, NSP12 (e), NSP13 (f), NSP14, NSP15 (g), NSP14, and NSP15 (h), accessory proteins like ORF3a, ORF7a, ORF6, and ORF10 (i), and structural proteins envelope (h), membrane (h), spike (j) and nucleocapsid phosphoprotein (k). Figure S2: Heatmap showing IgA response for peptide against ORF1a/b non-structural protein NSP1, NSP2 (a), NSP3 (b); NSP4, NSP5 (c), NSP6, NSP7, NSP8, NSP9 (d), NSP10, NSP12 (e), NSP13 (f), NSP14, NSP15 (g), NSP14, and NSP15 (h), accessory proteins like ORF3a, ORF7a, ORF6, and ORF10 (i), and structural proteins envelope (h), membrane (h), spike (j) and nucleocapsid phosphoprotein (k). Figure S3: Representative dot plots showing patient-wise comparison of IgG response against mutant and wildtype peptides (a) i) Orf1a/b Polyprotein, T1638A ii) Spike Protein, D138Y (P.1 Manaus); (b) i) Spike Protein, D111H (B1.1.7) ii) Spike Protein, H655Y (P.1 Manaus); (c) i) Nucleocapsid Phophoprotein, P80R (P.1 Manaus) ii) Orf1a/b Polyprotein, V1629A; (d) i) Spike Protein, R190S (P.1 Manaus) ii) Spike Protein, R246I (501.V2); (e) i) Spike Protein, N501Y(B1.1.7) ii) Spike Protein, P26S (P.1 Manaus); Figure S4: Representative dot plots showing patient-wise comparison of IgA response against mutant and wildtype (a) i) Spike Protein, D138Y (P.1 Manaus), ii) Spike Protein, H655Y (P.1 Manaus); (b) i) Nucleocapsid Phophoprotein, P80R(P.1 Manaus), ii) Orf1a/b Polyprotein, V1629A, iii) Spike Protein, N501Y(B1.1.7). Table S1: Clinical Characteristics of patients. Table S2: IgA reactive peptides from the SARS-CoV-2 proteome. Table S3: IgG reactive peptides from SARS-CoV-2 proteome. Table S4: Epitopes identified based on IgA response against SARS-CoV-2 proteins. Table S5: Epitopes identified based on IgG response against SARS-CoV-2 proteins. Table S6: Severity-based discrimination in IgA response against SARS-CoV-2 peptides. Table S7: Severity-based discrimination in IgG response against SARS-CoV-2 peptides. Table S8: Discriminatory epitope for IgA. Table S9: Discriminatory epitopes for IgG. Table S10: List of peptides with mutations.

Author Contributions: Conceptualization, A.A. and S.S.; methodology, A.A., A.S. and S.B.; formal analysis, A.R. and C.T.; investigation, A.A., A.S. and S.B.; resources, A.S., J.S. and S.A.; data curation, A.R. and C.T.; writing—original draft preparation, A.S., S.B., A.R., A.A. and C.T.; writing—review and editing, S.S, S.D., A.S., S.B., A.R., A.A. and C.T.; visualization, A.A, A.R., S.B. and C.T.; supervision, S.S. and A.A.; project administration, A.A., A.S. and S.B.; funding acquisition, S.S., A.A., A.S., A.R. and S.B. have contributed equally to the manuscript. All authors have read and agreed to the published version of the manuscript.

Funding: The study was supported through the Ministry of Science and Technology, Department of Biotechnology, Government of India (BT/PR41020/COT/142/14/2020) and a special COVID seed grant (RD/0520-IRCCHC0-006) from IRCC, IIT Bombay, to S.S. The authors also thank MERCK-COE (DO/2021-MLSP) for their extended support.

Institutional Review Board Statement: The study was approved by the Institute Ethics Committee, IIT Bombay, and Kasturba Hospital for Infectious Diseases Institutional Review Board.

Informed Consent Statement: Patient consent was waived as leftover samples from routine blood test were used for the study.

Data Availability Statement: The microarray raw and processed data will be available upon request.

Acknowledgments: We would like to acknowledge active support from Ambarish Kunwar from the Department of Biosciences and Bioengineering to fabricate UV transport device for sample transport. We are grateful to Renuka Bankar and Viswanthram P. for their contribution in collecting the samples and clinical details for the project. The authors thank Kasturba Hospital for Infectious Diseases for sample collection and sharing sample-related information. S.B. is supported by CSIR fellowship for a AR and CT were supported by fellowship from DBT (BT/INF/22/SP23026/2017).

Conflicts of Interest: The authors declare no conflict of interest.

References

1. Dong, E.; Du, H.; Gardner, L. An Interactive Web-Based Dashboard to Track COVID-19 in Real Time. *Lancet Infect. Dis.* **2020**, *20*, 533–534. [CrossRef] [PubMed]
2. Bauchner, H.; Fontanarosa, P.B. Excess Deaths and the Great Pandemic of 2020. *JAMA* **2020**, *324*, 1504–1505. [CrossRef] [PubMed]
3. World Health Organization. *Laboratory Testing for Coronavirus Disease 2019 (COVID-19) in Suspected Human Cases: Interim Guidance*; World Health Organization: Geneva, Switzerland, 2020.
4. Wu, X.; Fu, B.; Chen, L.; Feng, Y. Serological Tests Facilitate Identification of Asymptomatic SARS-CoV-2 Infection in Wuhan, China. *J. Med. Virol.* **2020**, *92*, 1795–1796. [CrossRef] [PubMed]
5. Tahamtan, A.; Ardebili, A. Real-Time RT-PCR in COVID-19 Detection: Issues Affecting the Results. *Expert Rev. Mol. Diagn.* **2020**, *20*, 453–454. [CrossRef]
6. Aggarwal, S.; Acharjee, A.; Mukherjee, A.; Baker, M.S.; Srivastava, S. Role of Multiomics Data to Understand Host–Pathogen Interactions in COVID-19 Pathogenesis. *J. Proteome Res.* **2021**, *20*, 1107–1132. [CrossRef] [PubMed]
7. Ahern, D.J.; Ai, Z.; Ainsworth, M.; Allan, C.; Allcock, A.; Angus, B.; Ansari, M.A.; Arancibia-Cárcamo, C.V.; Aschenbrenner, D.; Attar, M.; et al. A Blood Atlas of COVID-19 Defines Hallmarks of Disease Severity and Specificity. *Cell* **2022**, *185*, 916–938.e58. [CrossRef]
8. Messner, C.B.; Demichev, V.; Wendisch, D.; Michalick, L.; White, M.; Freiwald, A.; Textoris-Taube, K.; Vernardis, S.I.; Egger, A.-S.; Kreidl, M. Ultra-High-Throughput Clinical Proteomics Reveals Classifiers of COVID-19 Infection. *Cell Syst.* **2020**, *11*, 11–24. [CrossRef] [PubMed]
9. Shen, B.; Yi, X.; Sun, Y.; Bi, X.; Du, J.; Zhang, C.; Quan, S.; Zhang, F.; Sun, R.; Qian, L.; et al. Proteomic and Metabolomic Characterization of COVID-19 Patient Sera. *Cell* **2020**, *182*, 59–72.e15. [CrossRef]
10. Suvarna, K.; Biswas, D.; Pai, M.G.; Acharjee, A.; Bankar, R.; P, V.; Salkar, A.; Verma, A.; Mukherjee, A.; Choudhury, M.; et al. Proteomics and Machine Learning Approaches Reveal a Set of Prognostic Markers for COVID-19 Severity with Drug Re-Purposing Potential. *Front. Physiol.* **2021**, *12*, 652799. [CrossRef]
11. Acharjee, A.; Stephen Kingsly, J.; Kamat, M.; Kurlawala, V.; Chakraborty, A.; Vyas, P.; Vaishnav, R.; Srivastava, S. Rise of the SARS-CoV-2 Variants: Can Proteomics Be the Silver Bullet? *Expert Rev. Proteomics* **2022**, *19*, 197–212. [CrossRef]
12. Yadav, R.; Acharjee, A.; Salkar, A.; Bankar, R.; Palanivel, V.; Agrawal, S.; Shastri, J.; Sabnis, S.V.; Srivastava, S. Mumbai Mayhem of COVID-19 Pandemic Reveals Important Factors That Influence Susceptibility to Infection. *EClinicalMedicine* **2021**, *35*, 100841. [CrossRef] [PubMed]
13. Baric, R.S. Emergence of a Highly Fit SARS-CoV-2 Variant. *N. Engl. J. Med.* **2020**, *383*, 2684–2686. [CrossRef] [PubMed]
14. Cyranoski, D. Alarming COVID Variants Show Vital Role of Genomic Surveillance. *Nature* **2021**, *589*, 337–338. [CrossRef] [PubMed]

15. Sabino, E.C.; Buss, L.F.; Carvalho, M.P.S.; Prete, C.A.; Crispim, M.A.E.; Fraiji, N.A.; Pereira, R.H.M.; Parag, K.V.; Peixoto, P.d.S.; Kraemer, M.U.G.; et al. Resurgence of COVID-19 in Manaus, Brazil, despite High Seroprevalence. *The Lancet* **2021**, *397*, 452–455. [CrossRef] [PubMed]
16. Choudhury, M.; Verma, A.; Halder, A.; Acharjee, A. NGS Technologies for Detection of SARS-CoV-2 Strains and Mutations. In *Multi-Pronged Omics Technologies to Understand COVID-19*; Srivastava, S., Ed.; CRC Press: Boca Raton, FL, USA, 2022; ISBN 978-1-00-322078-7.
17. John, G.; Sahajpal, N.S.; Mondal, A.K.; Ananth, S.; Williams, C.; Chaubey, A.; Rojiani, A.M.; Kolhe, R. Next-Generation Sequencing (NGS) in COVID-19: A Tool for SARS-CoV-2 Diagnosis, Monitoring New Strains and Phylodynamic Modeling in Molecular Epidemiology. *Curr. Issues Mol. Biol.* **2021**, *43*, 845–867. [CrossRef]
18. Lee, K.; Ko, H.L.; Lee, E.-Y.; Park, H.-J.; Kim, Y.S.; Kim, Y.-S.; Cho, N.-H.; Park, M.-S.; Lee, S.-M.; Kim, J.; et al. Development of a Diagnostic System for Detection of Specific Antibodies and Antigens against Middle East Respiratory Syndrome Coronavirus. *Microbiol. Immunol.* **2018**, *62*, 574–584. [CrossRef]
19. He, Q.; Chong, K.H.; Hee Chng, H.; Leung, B.; Ee Ling, A.; Wei, T.; Chan, S.-W.; Eong Ooi, E.; Kwang, J. Development of a Western Blot Assay for Detection of Antibodies against Coronavirus Causing Severe Acute Respiratory Syndrome. *Clin. Diagn. Lab. Immunol.* **2004**, *11*, 417–422. [CrossRef]
20. Frisch-Niggemeyer, W. A Solid-Phase Radioimmunoassay for Quantitative Measurement of Class-Specific Antibodies against Tick-Borne Encephalitis Virus. *J. Virol. Methods* **1982**, *3*, 319–328. [CrossRef]
21. de Alwis, R.; de Silva, A.M. Measuring Antibody Neutralization of Dengue Virus (DENV) Using a Flow Cytometry-Based Technique. In *Dengue: Methods and Protocols*; Padmanabhan, R., Vasudevan, S.G., Eds.; Methods in Molecular Biology; Springer: New York, NY, USA, 2014; pp. 27–39. ISBN 978-1-4939-0348-1.
22. Anderson, S.; Wakeley, P.; Wibberley, G.; Webster, K.; Sawyer, J. Development and Evaluation of a Luminex Multiplex Serology Assay to Detect Antibodies to Bovine Herpes Virus 1, Parainfluenza 3 Virus, Bovine Viral Diarrhoea Virus, and Bovine Respiratory Syncytial Virus, with Comparison to Existing ELISA Detection Methods. *J. Immunol. Methods* **2011**, *366*, 79–88. [CrossRef]
23. Bihani, S.; Aggarwal, S.; Acharjee, A. COVID-19 Pathogenesis and Host Immune Response. In *Multi-Pronged Omics Technologies to Understand COVID-19*; Srivastava, S., Ed.; CRC Press: Boca Raton, FL, USA, 2022; ISBN 978-1-00-322078-7.
24. Acharjee, A.; Barapanda, A.; Ren, J.; Yu, X. Protein Microarrays for COVID-19 Research: Biomarker Discovery, Humoral Response and Vaccine Targets. In *Multi-Pronged Omics Technologies to Understand COVID-19*; Srivastava, S., Ed.; CRC Press: Boca Raton, FL, USA, 2022; ISBN 978-1-00-322078-7.
25. Davies, D.H.; Liang, X.; Hernandez, J.E.; Randall, A.; Hirst, S.; Mu, Y.; Romero, K.M.; Nguyen, T.T.; Kalantari-Dehaghi, M.; Crotty, S.; et al. Profiling the Humoral Immune Response to Infection by Using Proteome Microarrays: High-Throughput Vaccine and Diagnostic Antigen Discovery. *Proc. Natl. Acad. Sci. USA* **2005**, *102*, 547–552. [CrossRef]
26. Natesan, M.; Ulrich, R.G. Protein Microarrays and Biomarkers of Infectious Disease. *Int. J. Mol. Sci.* **2010**, *11*, 5165–5183. [CrossRef]
27. Li, S.; Song, G.; Bai, Y.; Song, N.; Zhao, J.; Liu, J.; Hu, C. Applications of Protein Microarrays in Biomarker Discovery for Autoimmune Diseases. *Front. Immunol.* **2021**, *12*, 645632. [CrossRef] [PubMed]
28. Holenya, P.; Lange, P.J.; Reimer, U.; Woltersdorf, W.; Panterodt, T.; Glas, M.; Wasner, M.; Eckey, M.; Drosch, M.; Hollidt, J.-M.; et al. Peptide Microarray-Based Analysis of Antibody Responses to SARS-CoV-2 Identifies Unique Epitopes with Potential for Diagnostic Test Development. *Eur. J. Immunol.* **2021**, *51*, 1839–1849. [CrossRef] [PubMed]
29. Jiang, H.; Li, Y.; Zhang, H.; Wang, W.; Yang, X.; Qi, H.; Li, H.; Men, D.; Zhou, J.; Tao, S. SARS-CoV-2 Proteome Microarray for Global Profiling of COVID-19 Specific IgG and IgM Responses. *Nat. Commun.* **2020**, *11*, 1–11. [CrossRef]
30. Musicò, A.; Frigerio, R.; Mussida, A.; Barzon, L.; Sinigaglia, A.; Riccetti, S.; Gobbi, F.; Piubelli, C.; Bergamaschi, G.; Chiari, M.; et al. SARS-CoV-2 Epitope Mapping on Microarrays Highlights Strong Immune-Response to N Protein Region. *Vaccines* **2021**, *9*, 35. [CrossRef] [PubMed]
31. Schwarz, T.; Heiss, K.; Mahendran, Y.; Casilag, F.; Kurth, F.; Sander, L.E.; Wendtner, C.-M.; Hoechstetter, M.A.; Müller, M.A.; Sekul, R.; et al. SARS-CoV-2 Proteome-Wide Analysis Revealed Significant Epitope Signatures in COVID-19 Patients. *Front. Immunol.* **2021**, *12*, 629185. [CrossRef] [PubMed]
32. Wang, H.; Wu, X.; Zhang, X.; Hou, X.; Liang, T.; Wang, D.; Teng, F.; Dai, J.; Duan, H.; Guo, S.; et al. SARS-CoV-2 Proteome Microarray for Mapping COVID-19 Antibody Interactions at Amino Acid Resolution. *ACS Cent. Sci.* **2020**, *6*, 2238–2249. [CrossRef]
33. Darnell, M.E.R.; Subbarao, K.; Feinstone, S.M.; Taylor, D.R. Inactivation of the Coronavirus That Induces Severe Acute Respiratory Syndrome, SARS-CoV. *J. Virol. Methods* **2004**, *121*, 85–91. [CrossRef]
34. Ritchie, M.E.; Silver, J.; Oshlack, A.; Holmes, M.; Diyagama, D.; Holloway, A.; Smyth, G.K. A Comparison of Background Correction Methods for Two-Colour Microarrays. *Bioinformatics* **2007**, *23*, 2700–2707. [CrossRef]
35. Ritchie, M.E.; Phipson, B.; Wu, D.; Hu, Y.; Law, C.W.; Shi, W.; Smyth, G.K. Limma Powers Differential Expression Analyses for RNA-Sequencing and Microarray Studies. *Nucleic Acids Res.* **2015**, *43*, e47. [CrossRef]
36. Cheadle, C.; Vawter, M.P.; Freed, W.J.; Becker, K.G. Analysis of Microarray Data Using Z Score Transformation. *J. Mol. Diagn.* **2003**, *5*, 73–81. [CrossRef] [PubMed]

37. Cheng, L.; Zhang, X.; Chen, Y.; Wang, D.; Zhang, D.; Yan, S.; Wang, H.; Xiao, M.; Liang, T.; Li, H.; et al. Dynamic Landscape Mapping of Humoral Immunity to SARS-CoV-2 Identifies Non-Structural Protein Antibodies Associated with the Survival of Critical COVID-19 Patients. *Signal Transduct. Target. Ther.* **2021**, *6*, 1–14. [CrossRef] [PubMed]
38. Ma, H.; Zeng, W.; He, H.; Zhao, D.; Jiang, D.; Zhou, P.; Cheng, L.; Li, Y.; Ma, X.; Jin, T. Serum IgA, IgM, and IgG Responses in COVID-19. *Cell. Mol. Immunol.* **2020**, *17*, 773–775. [CrossRef] [PubMed]
39. Post, N.; Eddy, D.; Huntley, C.; van Schalkwyk, M.C.I.; Shrotri, M.; Leeman, D.; Rigby, S.; Williams, S.V.; Bermingham, W.H.; Kellam, P.; et al. Antibody Response to SARS-CoV-2 Infection in Humans: A Systematic Review. *PLoS ONE* **2020**, *15*, e0244126. [CrossRef]
40. Farrera-Soler, L.; Daguer, J.-P.; Barluenga, S.; Vadas, O.; Cohen, P.; Pagano, S.; Yerly, S.; Kaiser, L.; Vuilleumier, N.; Winssinger, N. Identification of Immunodominant Linear Epitopes from SARS-CoV-2 Patient Plasma. *PLoS ONE* **2020**, *15*, e0238089. [CrossRef]
41. Cerutti, G.; Guo, Y.; Zhou, T.; Gorman, J.; Lee, M.; Rapp, M.; Reddem, E.R.; Yu, J.; Bahna, F.; Bimela, J.; et al. Potent SARS-CoV-2 Neutralizing Antibodies Directed against Spike N-Terminal Domain Target a Single Supersite. *Cell Host Microbe* **2021**, *29*, 819–833.e7. [CrossRef]
42. Shah, P.; Canziani, G.A.; Carter, E.P.; Chaiken, I. The Case for S2: The Potential Benefits of the S2 Subunit of the SARS-CoV-2 Spike Protein as an Immunogen in Fighting the COVID-19 Pandemic. *Front. Immunol.* **2021**, *12*, 637651. [CrossRef]
43. Jaimes, J.A.; André, N.M.; Chappie, J.S.; Millet, J.K.; Whittaker, G.R. Phylogenetic Analysis and Structural Modeling of SARS-CoV-2 Spike Protein Reveals an Evolutionary Distinct and Proteolytically Sensitive Activation Loop. *J. Mol. Biol.* **2020**, *432*, 3309–3325. [CrossRef]
44. Ladner, J.T.; Henson, S.N.; Boyle, A.S.; Engelbrektson, A.L.; Fink, Z.W.; Rahee, F.; D'ambrozio, J.; Schaecher, K.E.; Stone, M.; Dong, W.; et al. Epitope-Resolved Profiling of the SARS-CoV-2 Antibody Response Identifies Cross-Reactivity with Endemic Human Coronaviruses. *Cell Rep. Med.* **2021**, *2*, 100189. [CrossRef]
45. Lai, S.-C.; Chong, P.C.-S.; Yeh, C.-T.; Liu, L.S.-J.; Jan, J.-T.; Chi, H.-Y.; Liu, H.-W.; Chen, A.; Wang, Y.-C. Characterization of Neutralizing Monoclonal Antibodies Recognizing a 15-Residues Epitope on the Spike Protein HR2 Region of Severe Acute Respiratory Syndrome Coronavirus (SARS-CoV). *J. Biomed. Sci.* **2005**, *12*, 711–727. [CrossRef]
46. Nguyen-Contant, P.; Embong, A.K.; Kanagaiah, P.; Chaves, F.A.; Yang, H.; Branche, A.R.; Topham, D.J.; Sangster, M.Y. S Protein-Reactive IgG and Memory B Cell Production after Human SARS-CoV-2 Infection Includes Broad Reactivity to the S2 Subunit. *mBio* **2020**, *11*, e01991-20. [CrossRef] [PubMed]
47. Schrödinger, LLC. The PyMOL Molecular Graphics System, Version 1.8. 2015. Available online: https://pymol.org/2/ (accessed on 25 November 2022).
48. Amrun, S.N.; Lee, C.Y.-P.; Lee, B.; Fong, S.-W.; Young, B.E.; Chee, R.S.-L.; Yeo, N.K.-W.; Torres-Ruesta, A.; Carissimo, G.; Poh, C.M.; et al. Linear B-Cell Epitopes in the Spike and Nucleocapsid Proteins as Markers of SARS-CoV-2 Exposure and Disease Severity. *eBioMedicine* **2020**, *58*, 102911. [CrossRef]
49. Thoms, M.; Buschauer, R.; Ameismeier, M.; Koepke, L.; Denk, T.; Hirschenberger, M.; Kratzat, H.; Hayn, M.; Mackens-Kiani, T.; Cheng, J.; et al. Structural Basis for Translational Shutdown and Immune Evasion by the Nsp1 Protein of SARS-CoV-2. *Science* **2020**, *369*, 1249–1255. [CrossRef]
50. Ma, J.; Chen, Y.; Wu, W.; Chen, Z. Structure and Function of N-Terminal Zinc Finger Domain of SARS-CoV-2 NSP2. *Virol. Sin.* **2021**, *36*, 1104–1112. [CrossRef] [PubMed]
51. Heidepriem, J.; Dahlke, C.; Kobbe, R.; Santer, R.; Koch, T.; Fathi, A.; Seco, B.M.S.; Ly, M.L.; Schmiedel, S.; Schwinge, D.; et al. Longitudinal Development of Antibody Responses in COVID-19 Patients of Different Severity with ELISA, Peptide, and Glycan Arrays: An Immunological Case Series. *Pathog. Basel Switz.* **2021**, *10*, 438. [CrossRef] [PubMed]
52. Yan, W.; Zheng, Y.; Zeng, X.; He, B.; Cheng, W. Structural Biology of SARS-CoV-2: Open the Door for Novel Therapies. *Signal Transduct. Target. Ther.* **2022**, *7*, 1–28. [CrossRef]
53. Lei, J.; Kusov, Y.; Hilgenfeld, R. Nsp3 of Coronaviruses: Structures and Functions of a Large Multi-Domain Protein. *Antiviral Res.* **2018**, *149*, 58–74. [CrossRef]
54. Serrano, P.; Johnson, M.A.; Almeida, M.S.; Horst, R.; Herrmann, T.; Joseph, J.S.; Neuman, B.W.; Subramanian, V.; Saikatendu, K.S.; Buchmeier, M.J.; et al. Nuclear Magnetic Resonance Structure of the N-Terminal Domain of Nonstructural Protein 3 from the Severe Acute Respiratory Syndrome Coronavirus. *J. Virol.* **2007**, *81*, 12049–12060. [CrossRef]
55. Hachim, A.; Kavian, N.; Cohen, C.A.; Chin, A.W.H.; Chu, D.K.W.; Mok, C.K.P.; Tsang, O.T.Y.; Yeung, Y.C.; Perera, R.A.P.M.; Poon, L.L.M.; et al. ORF8 and ORF3b Antibodies Are Accurate Serological Markers of Early and Late SARS-CoV-2 Infection. *Nat. Immunol.* **2020**, *21*, 1293–1301. [CrossRef]
56. Zhang, Y.; Chen, Y.; Li, Y.; Huang, F.; Luo, B.; Yuan, Y.; Xia, B.; Ma, X.; Yang, T.; Yu, F.; et al. The ORF8 Protein of SARS-CoV-2 Mediates Immune Evasion through down-Regulating MHC-I. *Proc. Natl. Acad. Sci. USA* **2021**, *118*, e2024202118. [CrossRef]

57. Shrock, E.; Fujimura, E.; Kula, T.; Timms, R.T.; Lee, I.-H.; Leng, Y.; Robinson, M.L.; Sie, B.M.; Li, M.Z.; Chen, Y.; et al. Viral Epitope Profiling of COVID-19 Patients Reveals Cross-Reactivity and Correlates of Severity. *Science* **2020**, *370*, eabd4250. [CrossRef] [PubMed]
58. Galloway, S.E. Emergence of SARS-CoV-2 B.1.1.7 Lineage—United States, December 29, 2020–January 12, 2021. *MMWR Morb. Mortal. Wkly. Rep.* **2021**, *70*, 95–99. [CrossRef] [PubMed]

viruses

MDPI

Review

Electrostatic Surface Potential as a Key Parameter in Virus Transmission and Evolution: How to Manage Future Virus Pandemics in the Post-COVID-19 Era

Jacques Fantini *, Fodil Azzaz, Henri Chahinian and Nouara Yahi

Department of Biology, Faculty of Medicine, University of Aix-Marseille, INSERM UMR_S 1072, 13015 Marseille, France
* Correspondence: jacques.fantini@univ-amu.fr

Abstract: Virus-cell interactions involve fundamental parameters that need to be considered in strategies implemented to control viral outbreaks. Among these, the surface electrostatic potential can give valuable information to deal with new epidemics. In this article, we describe the role of this key parameter in the hemagglutination of red blood cells and in the co-evolution of synaptic receptors and neurotransmitters. We then establish the functional link between lipid rafts and the electrostatic potential of viruses, with special emphasis on gangliosides, which are sialic-acid-containing, electronegatively charged plasma membrane components. We describe the common features of ganglioside binding domains, which include a wide variety of structures with little sequence homology but that possess key amino acids controlling ganglioside recognition. We analyze the role of the electrostatic potential in the transmission and intra-individual evolution of HIV-1 infections, including gatekeeper and co-receptor switch mechanisms. We show how to organize the epidemic surveillance of influenza viruses by focusing on mutations affecting the hemagglutinin surface potential. We demonstrate that the electrostatic surface potential, by modulating spike-ganglioside interactions, controls the hemagglutination properties of coronaviruses (SARS-CoV-1, MERS-CoV, and SARS-CoV-2) as well as the structural dynamics of SARS-CoV-2 evolution. We relate the broad-spectrum antiviral activity of repositioned molecules to their ability to disrupt virus-raft interactions, challenging the old concept that an antibiotic or anti-parasitic cannot also be an antiviral. We propose a new concept based on the analysis of the electrostatic surface potential to develop, in real time, therapeutic and vaccine strategies adapted to each new viral epidemic.

Keywords: pandemic; vaccine; antiviral; SARS-CoV-2; HIV-1; MERS-CoV; monkeypox virus; influenza virus; lipid raft; ganglioside; neutralization; electrostatic surface potential

Citation: Fantini, J.; Azzaz, F.; Chahinian, H.; Yahi, N. Electrostatic Surface Potential as a Key Parameter in Virus Transmission and Evolution: How to Manage Future Virus Pandemics in the Post-COVID-19 Era. *Viruses* **2023**, *15*, 284. https://doi.org/10.3390/v15020284

Academic Editors: Ahmed El-Shamy and Mohamed Ibrahim

Received: 26 December 2022
Revised: 14 January 2023
Accepted: 18 January 2023
Published: 19 January 2023

1. Introduction

Our experience in teaching biochemistry and molecular biology at university level has allowed us, over the years, to identify major concepts in biology that are insufficiently covered in biology courses [1]. Among these concepts we can cite the multiple functions of water molecules in biology [2], the temporal dimension of biological processes [3,4], quantum phenomena at work in biology [5], and the electrostatic surface potential of biomolecules [6]. This latter concept has taken on major importance over the past three years in explaining the structural dynamics of SARS-CoV-2 variants [7]. More generally, the electrostatic surface potential is a key element for understanding the evolution of viruses [8,9] and, more specifically, the evolution of virus-host relationships [10]. It therefore seemed important to us to devote this review article to defining the role of the electrostatic surface potential in the evolution of SARS-CoV-2 and other viruses, and to draw possible solutions from this in the face of future viral pandemics.

2. Definition of the Electrostatic Surface Potential

Electrostatic interactions play a central role in biology [11]. Intuitively, molecular interactions can be summed up in a double complementarity: geometric and electrostatic. An electronegative hollow is thus naturally adapted to be occupied by an electropositive protuberance [1]. This concept applies at several levels in biology, from molecular interactions to cell–cell associations, and vice versa to different types of repulsion. Although widely developed in the 20th century, it was not until 1982 with the advances in computer graphics that it was possible to visualize the electrostatic potential of biological macromolecules [6]. A universal color code was then adopted: red for electronegative zones, blue for electropositive zones, and white for neutral zones. In this princeps article, the authors represented for the first time the surface electrostatic potential of not only trypsin and an inhibitor attached to the enzyme, but also a DNA-protein complex. This visualization made obvious the notions of geometric and electrostatic complementarities, which represented a major advance for drug design.

A reflective exercise created for our students in the university's Evolutionary Biology course will allow us to illustrate the scope of this concept. Consider a series of peptide motifs of a virus protein whose amino acid sequence has gradually evolved over time (from t1 to t6):

t1: AEDEEDLDA
t2: AKDEEDLDA
t3: AKDERDLDA
t4: AKDERDLKA
t5: AKDRRDLKA
t6: AKDRRKLKA

Let us now look at Figure 1 in which the surface electrostatic potential of each of these patterns is represented in a random order. The question is simple: can we attribute to each peptide sequence its corresponding surface electrostatic potential?

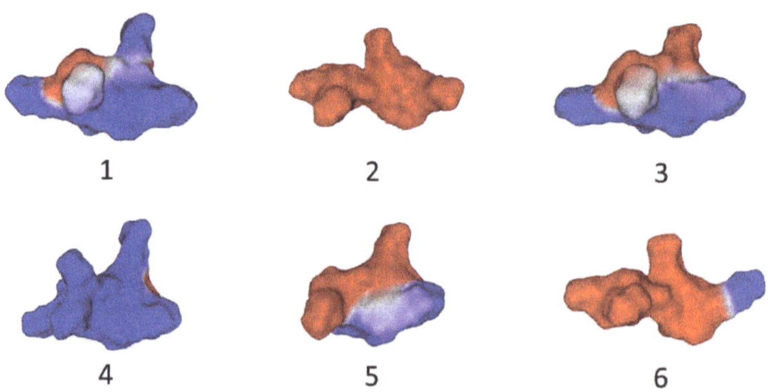

Figure 1. Electrostatic surface potential of an evolving peptide motif (see text for the amino acid sequences). Note that both the electrostatic surface potential and the shape of the motifs are affected by amino acid changes. The purpose of the exercise is to assign each peptide (represented by its electrostatic surface potential and identified by a number) to its corresponding amino acid sequence. Blue, positive; red, negative; white, neutral.

The answer is: yes, of course. First you must calculate the net charge of each peptide sequence at pH7: negatively charged side chains (D, aspartic acid; E, glutamic acid) are bold; positively charged side chains (K, lysine; R, arginine) are bold and underscored.

t1: A**EDEED**L**D**A: −6
t2: A**KDEED**L**D**A: −4
t3: A**KDERD**L**D**A: −2
t4: A**KDERD**L**K**A: 0

t5: A**KDRRD**L**K**A: +2
t6: A**KDRRK**L**K**A: +4

The relative position of negatively charged, positively charged, or neutral amino acids will help you assign each sequence its electrostatic surface potential (Figure 1):

t1: A**EDEED**L**D**A: Number 2 (all red)
t2: A**KDEED**L**D**A: Number 6 (mostly red + small lateral blue spot)
t3: A**KDERD**L**D**A: Number 5 (still mostly red, but blue zones become larger)
t4: A**KDERD**L**K**A: Number 3 (50% red, 50% blue)
t5: A**KDRRD**L**K**A: Number 1 (mostly blue + two small red spots)
t6: A**KDRRK**L**K**A: Number 4 (mostly blue + small lateral red spot)

We thus visualized the electrostatic logic of a biomolecule by converting the concept of surface electrostatic potential into a tricolor code. Schematically two types of analysis can visualize the electrostatic potential of a protein: (i) the electrostatic surface potential which superimposes the distribution of charges on the relief of the protein, and (ii) its spatial distribution (isopotential contours). Both types of representations are given in Figure 2. Using isopotential contours is especially useful to highlight slight differences on protein surface, e.g., when studying the evolution of mutants [12]. However, this representation gives a distorted picture of the protein structure, as can be seen in Figure 2. In the next part of this review, we will illustrate the impact of the electrostatic potential concept using different examples from biology.

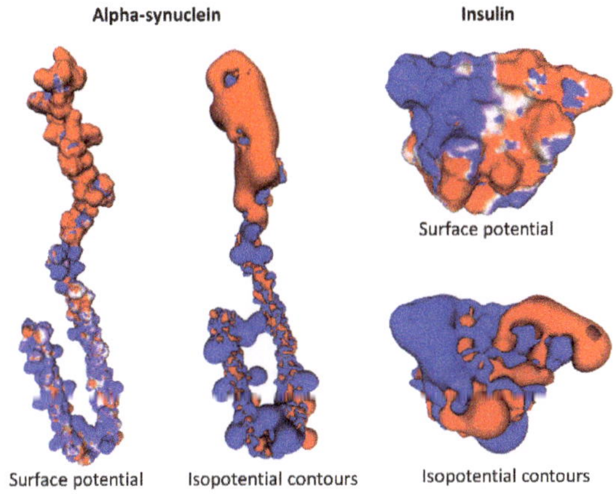

Figure 2. Surface electrostatic potential and isopotential contours. The models (pdb files 1XQ8 and 3I40 for alpha-synuclein and insulin, respectively) were generated by PBEQ-Solver in the biomolecular simulation program CHARMM. Two renditions are shown for each protein (electrostatic surface potential and isopotential contours). Blue, positive; red, negative; white, neutral.

3. Biological Significance of the Electrostatic Surface Potential

Technically, the electrostatic potential can be generated by solving the Poisson–Boltzmann equations, using the partial charges of all the atoms belonging to a given area of a molecule [13,14], or by using Coulomb's law [15]. In this review, we generally used Molegro Molecular Viewer (http://molexus.io/molegro-molecular-viewer/, accessed on 14 January 2023) to visualize the surface electrostatic potential. The electrostatic potential measured and illustrated by Molegro Molecular Viewer is the sum of the Coulomb potentials for each atom of the considered molecule, with a distance-dependent dielectric constant. Alternatively, the biomolecular simulation program CHARMM (http://www.charmm-gui.org, accessed on 14 January 2023) proposes the PBEQ-Solver module to solve the finite-difference Poisson–Boltzmann equation of submitted proteins. PBEQ-Solver gives the calculated electrostatic potential on the solvent-accessible surface as well as iso-electrostatic potential contours [16,17].

Biomolecules present immense diversity at the level of their surface potential, which is expressed not only by well-delimited zones presenting a positive or negative potential, but also by distribution gradients of charges visualized by the shades of the color code blue-red-white. However, there are major trends among biomolecules. Nucleic acids are negatively charged because of the phosphate groups of the 3′-5′-phosphodiester bonds [18]. Lipid rafts, which are rich in glycosphingolipids [19], are globally electronegative, especially when these glycosphingolipids are gangliosides [20]. Proteins are more ambivalent, as they can have anionic, cationic, and most often both amino acids in infinitely varied proportions. Unlike nucleic acids and lipid rafts, the electrostatic surface potential of proteins is therefore a characteristic property of each protein that needs to be studied independently for each protein. Moreover, as we have seen in Figure 1 with selected peptides, evolution by point mutations can have a very strong impact on the surface potential of a particular region of the protein, with potentially important functional consequences.

If we had to illustrate with a single example the importance of the electrostatic surface potential in biology, we would cite red blood cells. Despite their very large number in the blood, these cells do not aggregate under physiological conditions. On the contrary, two red blood cells repel each other if they get too close. The reason is that the plasma membrane of red blood cells contains negatively charged glycoproteins and glycolipids, which creates a repulsive electric potential (zeta) between cells and prevents their aggregation in the bloodstream [21]. Correspondingly, both neuraminidase (which removes negatively charged sialic acids from glycoproteins and gangliosides) and protease treatments of red blood cells reduce charge surface density and promote agglutination [22,23].

Another important area of biology in which the electrostatic surface potential plays a major role is the synapse [1]. Post-synaptic membranes are enriched in mono, di-, and tri-sialylated gangliosides [24,25] which confer a strong electronegative field [1,26]. This electrostatic shield repels glutamate away from the neuronal membrane, thus limiting the risk of excitotoxicity [1]. However, this mechanism implies that the binding site of glutamate and its agonists on their receptors is located outside the influence of the negative charges of gangliosides. This is particularly clear on metabotropic receptors, such as mGluR5, whose oversized Venus flytrap domain binds glutamate at a distance of about 80 Å from the membrane (Figure 3) [27]. Thus, we should consider not only the surface electrostatic potential of a protein, but also its dielectric constants which express the influence of the environment on protein–protein and protein–ligand interactions [28].

In the plasma membrane, gangliosides are not randomly distributed but concentrated in particular microdomains, referred to as lipid rafts [19]. Rafts are a privileged site of attack for many pathogens, especially viruses [29–32]. There are many explanations for this phenomenon, the first being topological. Rafts are relatively flat areas of the plasma membrane [33]. They therefore represent very accessible landing strips for pathogens (Figure 4A). The second reason is that many virus receptors and/or co-receptors are associated with lipid rafts. By directly targeting the rafts, viruses therefore have facilitated access to these receptors. This is the case for ACE2, the main SARS-CoV-2 receptor, and for CD4, the classical HIV-1 receptor. This situation moreover complicates the very notion of "virus receptor", since in certain cases raft gangliosides (or raft glycosphingolipids) can fulfill the virus receptor function. Indeed, gangliosides have been identified as bona fide receptors for various viruses, including influenza [34], Sendaï [35], SV40 [36], polyomaviruses [37], and rotavirus [38]. HIV-1 also uses ganglioside GM3 as a fusion cofactor [39–42] and galactosylceramide (GalCer) as an alternative receptor to infect CD4-negative neural [43] and intestinal epithelial cells [11].

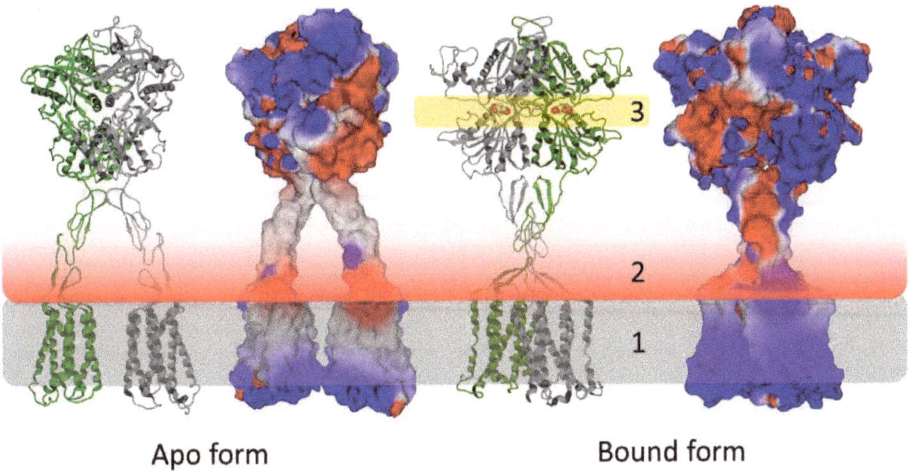

Figure 3. Post-synaptic membrane gangliosides (1, grey zone) generate a repulsive electrostatic field (2, red to white gradient) that determines the position of the agonist binding site (3, orange zone) in the extracellular Venus flytraps of the dimeric metabotropic glutamate receptor 5 (mGluR5). The structure of mGluR5 was retrieved from pdb 6N52.

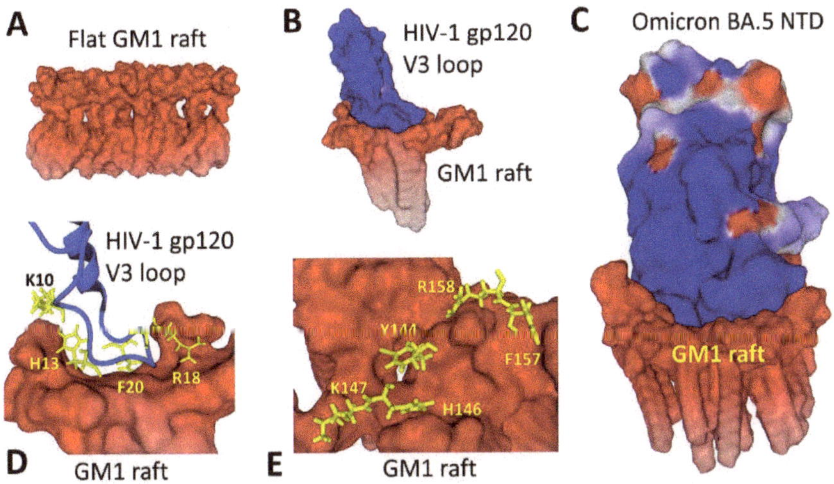

Figure 4. The electrostatic logic of virus-raft interactions. (**A**) The typical flat surface of a GM1 raft before virus binding. (**B**) A typical loop-shaped ganglioside binding domain (HIV-1 gp120 V3 loop). Note the raft curvature induced by the binding reaction. (**C**) Attachment of the N-terminal domain (NTD) of SARS-CoV-2 Omicron BA.5 variant to a GM1 raft. In this case, the membrane curvature induced by the binding reaction is particularly obvious. (**D**) Key amino acid residues controlling the binding of HIV-1 gp120 V3 loop to a GM1 raft. (**E**) Key amino acid residues controlling the binding of the BA.5 NTD to a GM1 raft. The structures were retrieved from pdb 1CE4 (V3 loop) and 7BNM (Omicron spike protein) and modeled with Hyperchem.

The structural basis of the interaction between gangliosides and these viruses is a ganglioside binding domain [19,26,45], which may be either a small loop [46] (Figure 4B), a large flat surface [47] (Figure 4C), or an annular binding domain [48]. Despite the lack of amino acid sequence homology, these domains display a combination of aromatic and cationic residues which are particularly adapted for optimal ganglioside binding [26,49].

For instance, the binding of HIV-1 gp120 to a cluster of three GM1 molecules involves a central aromatic residue (F20) surrounded by two cationic amino acids (K10, R18) and one histidine (H13) (Figure 4D). The binding of the N-terminal domain (NTD) of SARS-CoV-2 to a GM1 raft involves a similar panel of amino acids (Figure 4E): aromatic (Y144, F157), cationic (K147, R158), and one histidine (H146). A notable feature of this attachment of a virus protein to a raft is the deformation that this molecular interaction causes in the organization of gangliosides, inducing a local curvature of the raft, a phenomenon facilitated by cooperative interactions between gangliosides. This curvature allows the raft to form a kind of stabilization cocoon. The binding reaction is cooperative, starting with one ganglioside molecule and gradually reinforced by its neighbors in the raft. The kinetics of the reaction are controlled by attractive electrostatic forces between the electronegative (red) surface of the raft and the electropositive (blue) surface of the virus protein. Molecular dynamics simulations performed with the Hyperchem software [50] showed that the conformational rearrangements needed to fit the raft surface concerns chiefly the amino acid side chains of the virus protein that interact with gangliosides, rather than the secondary or tertiary structure of the protein. In other words, it is the lipid raft that adapts its shape to the viral protein surface, not the reverse. This phenomenon is well illustrated by comparing the raft surface before (Figure 4A) and after binding to the V3 loop of HIV-1 gp120 (Figure 4B) or to the SARS-CoV-2 Omicron Spike protein (Figure 4D).

4. Electrostatic Surface Potential in HIV-1 Evolution

The main HIV-1 receptor is the CD4 glycoprotein expressed by certain immune cells [51]. However, one of the most surprising characteristics of this retrovirus is the ability to use, in addition to CD4, a co-receptor necessary for the process of fusion between the virus envelope and the plasma membrane of the host cell [52]. When an individual is infected with HIV-1, it is usually a strain using the CCR5 co-receptor that is transmitted [53]. Then, as the virus evolves in the patient, a co-receptor switch occurs, allowing the virus to use another co-receptor, CXCR4 [54]. In the first case, we speak of R5 viruses; in the second case, of X4 viruses. The transition is marked by viruses that can use both types of co-receptors: these are the R5X4 viruses [55]. The co-receptor switch (from CCR5 to CXCR4) is associated with the emergence of more aggressive viruses, inducing a more rapid decline of $CD4^+$ lymphocytes [56,57]. The structural basis of this evolution is largely caused by an increase in the net charge of the V3 loop, due to an accumulation of mutations that dramatically affect its electrostatic surface potential [10]. This mechanism can be explained by considering the electrostatic surface potential of CCR5 and CXCR4 (Figure 5), which must be complementary to the V3 loop.

According to the analysis of Figure 5 performed with the ImageJ software [58], the isopotential contours of CXCR4 are 2.12 times more electronegative than CCR5 (and 1.65 times for the determinations based on the electrostatic surface potential). Correspondingly, the net charge of the V3 loop, and thus its electropositive surface potential, gradually increases as the virus evolves in an infected individual. When it reaches a threshold value (+4 or +5), the virus becomes able to use both CCR5 and CXCR4. Above this value, the virus definitely switches and uses CXCR4 [10]. Thus, the value of the net charge of the V3 loop makes it possible to predict the type of co-receptor used by each HIV-1 isolate. This type of analysis allows an understanding of this retrovirus' evolution since its emergence in the human species, giving us keys to anticipate its future evolution [59].

A representative example of V3 loop sequence evolution associated with the co-receptor switch is given in Figure 6. The V3 loop of a typical R5 isolate has a net positive charge of +3, which results from the compensation of 5 cationic and 2 acidic residues [60]. The electrostatic surface potential of this V3 loop is globally electropositive, but with a large central electronegative spot generated by aspartic acids D25 and D29. The evolution of this V3 loop led to a net charge of +6 due to the presence of a new cationic residue (R11) and the substitution of D25 and D29 by two amide residues (Q25 and N29) which are not electrically charged. These changes are associated with the R5 $\rightarrow$X4 co-receptor switch. It is obvious that the electrostatic surface potential of this X4 V3 loop, which is highly electropositive, is well adapted to interact with the large electronegative receptacle of CXCR4 (Figure 5).

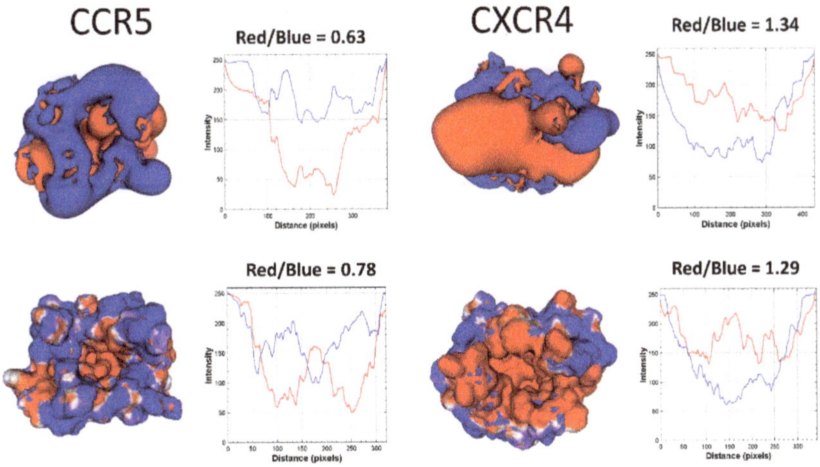

Figure 5. Electrostatic surface potential of CCR5 and CXR4 co-receptors. The structures were modeled from the coordinates of pdb 7F1T (CCR5) and 3OE0 (CXCR4). The upper panels show the isopotential contours and the lower panels the surface electrostatic potential as calculated by PBEQ-Solver. The determination of the blue (electropositive) and red (electronegative) areas performed with the ImageJ software showed that the isopotential contours give a clearer estimate of the relative charge distribution.

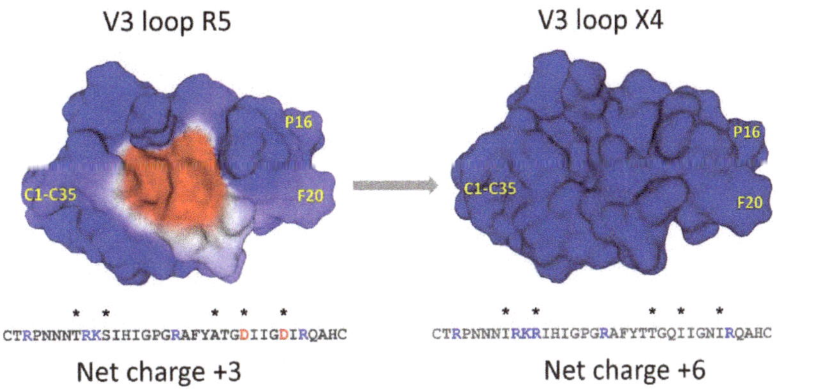

Figure 6. V3 loop evolution associated with the co-receptor switch. Amino acid changes are indicated by an asterisk (*): blue, amino acid with a positive charge; red, amino acid with a negative charge. The position of the disulfide bridge between cysteine C1 and C35 is indicated. The crown of the V3 loop is on the right side of the structure (GPGRAF motif). These structures were modeled from the coordinates of pdb 1CE4.

Aside from mutations that increase the electrostatic potential, the X4 V3 loop lacks the glycosylation site NNT which is changed to NNI (Figure 6). Glycans display an electronegative potential [61] which favors the use of CCR5 rather than CXCR4 [10,60]. Thus, the lack of a glycosylation site in the X4 V3 loop can be interpreted as the result of the selective pressure that allows the emergence of viruses with an increased electropositive surface potential.

Heterosexual transmission of HIV-1 is an imperfectly known process during which a single virus is selected and transmitted to the recipient. In the majority of cases, even if the sexual secretions contain a mixture of R5, R5X4, and X4 viruses, the transmitted founder virus is a R5 virus [62]. The mechanisms responsible for this selection are still the subject of debate. Several hypotheses have been put forward, some suggesting positive selection, others negative selection. Among the multiple barriers that could protect the recipient from X4 viruses, mucus is probably the more efficient, because it is not selective. Human cervical mucus is made of mucins, which are polyanionic glycoproteins [63]. Since the V3 loop of X4 viruses is more cationic than R5 viruses (Figure 6), this may result in the trapping of X4 strains to mucins, leaving the field open to the less electropositive R5 viruses [64]. In the same way, heparan sulfate proteoglycans that cover mucosal surfaces display an electronegative surface potential able to attract and inactivate the V3 loop of X4 viruses [65]. Sulfatides, which are negatively charged glycosphingolipids expressed by vaginal and intestinal epithelial cells, can also selectively inhibit the sexual transmission of highly cationic X4 viruses [66]. Vaginal epithelial cells are not infected by HIV-1 [67], but the specific sequestration of X4 strains by the genital epithelium could also contribute to the HIV-1 selection process [68]. Finally, the predominant transmission of R5 strains after sexual intercourse may also involve the preferential transmigration of R5 viruses associated with monocytes across the endocervical monolayer [69]. Taken together, these elements suggest that there is not a single "gatekeeper" [70] but rather multiple barriers that gradually select R5 over X4 HIV-1 strains after sexual intercourse [64]. Yet the problem is not easy to solve. Indeed, the selection of R5 viruses after direct intravenous contamination (e.g., transfusion with HIV-1 contaminated blood) suggests that post-mucosal gatekeeping mechanisms are also operative [64]. In this case, the infection of macrophages by R5 viruses might play a role, as these cells are less susceptible to cytotoxic lymphocytes [71]. What is clear is that R5 viruses systematically evolve towards X4 strains by increasing the electrostatic surface potential of the gp120 V3 loop by several mechanisms: (i) increase in the frequency of cationic amino acids, (ii) disappearance of electronegative amino acids, and (iii) suppression of the V3 loop glycosylation site (Figure 6) [12].

Differences in co-receptor usage have also been observed between genetic HIV-1 subtypes with a distinct geographical distribution [72]. Interestingly, López de Victoria et al. (2012) elegantly demonstrated that V3 loop subtypes with similar spatial distribution of electrostatic potential cluster together [12]. Thus, for X4 and R5 viruses, the electrostatic surface potential of the V3 loop is a fundamental property that can be used to characterize and classify HIV-1 subtypes. This notion of a reference threshold value to categorize variants, quasi-species, and/or subtypes of HIV-1 is in fact fairly standard for this retrovirus. Indeed, variations in the genomic sequence of HIV-1 subtypes can also be detected retrospectively in RT and/or protease sequence databases, when the divergence with a reference subtype B virus (HXB2) exceeds the cut-off value determined by the algorithm [73].

5. Electrostatic Surface Potential in Influenza Virus Evolution

If there is a virus for which the electrostatic surface potential should be studied with great interest, it is the influenza virus [9,74–76]. The basic reason for this is that this virus uses the sialic acid residues of glycoproteins and gangliosides to infect host cells and spread from animal species to humans, as well as from human to human [77–81]. The sialic acid binding site of influenza virus hemagglutinin displays the same pattern of cationic and aromatic residues as canonic ganglioside binding domains. This is perfectly illustrated by a ferret-transmissible H5 avian influenza virus (Figure 7) [82]. Indeed, the tip of this H5 hemagglutinin displays a high electropositive surface potential that fits with the electronegative potential of the sialic acid receptor. This adaptation renders the virus able to infect several animal species, representing a potential threat for humans. A totally distinct situation has been demonstrated for the bat influenza virus H17N10. In this case, electrostatic potential analyses revealed that its putative receptor-binding site is highly acidic, making it unfavorable to bind any negatively charged sialylated receptors [83]. This study highlights the power of the concept of surface electrostatic potential to predict the spillover of influenza viruses. Any mutation or genetic rearrangement that would render this domain electropositive would be considered a potential signal for future transmission to humans. Focusing on these hot spots would simplify the virological surveillance based on nucleotidic sequence studies. There are some advantages of this strategy. On the one hand, sequence homology is not always related to structural similarity, meaning we may need to consider structural homology instead [46,74]. On the other hand, a classification based on the electrostatic surface potential is immediately informative since it is directly related to virus-host interactions [7]. Furthermore, as developed for the monkeypox virus, there is a consistent overlapping between the cationic ganglioside binding motifs of virus glycoproteins and neutralizing epitopes [48]. In this respect, any increase in the receptor-binding affinity to gangliosides and related sialic acid receptors should alert us [81]. However, this requires complex physicochemical measurements and the real-time availability of recombinant hemagglutinin. Sequencing methods bypass these delicate and time-consuming steps. Identifying and periodically monitoring hot mutational spots in the genomic regions coding for ganglioside binding motifs will give valuable and timely information about the imminence of animal virus outbreaks, possible transmission to humans, and pandemic risks.

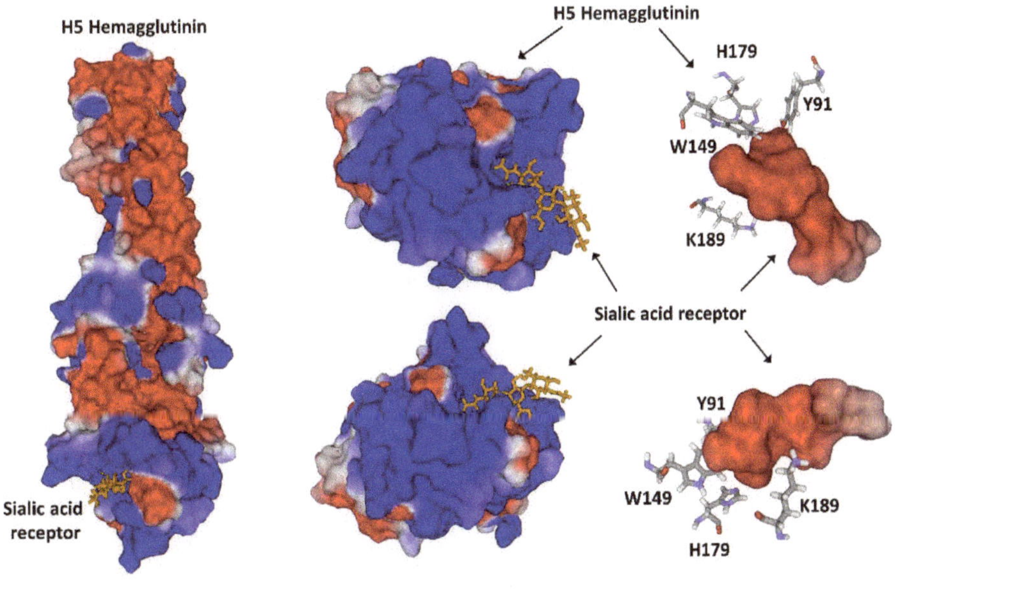

Figure 7. Sialic acid receptor binding site on influenza H5 hemagglutinin (retrieved from pdb 4BGY). The sialic acid receptor binding sites include cationic (K189), aromatic (Y91, W149), and one histidine residue (H179).

6. Electrostatic Surface Potential in Coronavirus Evolution

Three coronaviruses can trigger severe diseases in infected human individuals: SARS-CoV-1 [84], MERS-CoV [85], and SARS-CoV-2 [86]. The binding of these viruses to the host cell membrane is mediated by a spike protein arranged in a trimer configuration. Each monomer has a typical Y shape where the lateral branches of the letter correspond to the N-terminal domain (NTD) and the receptor binding domain (RBD). Sialic acids, gangliosides, and/or lipid rafts are involved in the entry of these viruses [47,87–91]. In most cases the NTD controls the initial interaction of the virus with lipid raft gangliosides, whereas the RBD is assigned to the recognition of a protein receptor, dipeptidyl peptidase 4 (DPP4, also known as CD26) for MERS-CoV [92], and ACE2 for both SARS-CoV-1 and SARS-CoV-2 [93].

If we compare the NTD of these three coronaviruses, we can see that this domain has evolved from an electronegative protuberance in SARS-CoV-1 to a curved electropositive domain in MERS-CoV, and finally to a flat and mostly electropositive surface for the initial SARS-CoV-2 strain (Figure 8). In parallel, the RBD was significantly rearranged in SARS-CoV-2 to acquire a curved and mostly electropositive surface that fits particularly well with the electronegative surface potential of ACE2 [7]. This evolution ensures both optimized access to lipid rafts through a kinetic effect and a slight increase in the affinity for ACE2, explaining why only SARS-CoV-2 has been pandemic. Moreover, the global electronegative potential of the NTD of SARS-CoV-1 may explain why this virus does not hemagglutinate red blood cells [89], in contrast with MERS-CoV [89] and SARS-CoV-2 [94]. In fact, the ability of a virus to hemagglutinate red blood cells requires the co-expression of a sialic acid recognition motif [95] and an electrostatic surface potential sufficiently positive to abolish the repulsion of these cells due to their zeta potential [94]. MERS-CoV and SARS-CoV-2 fulfill these criteria, whereas SARS-CoV-1 does not.

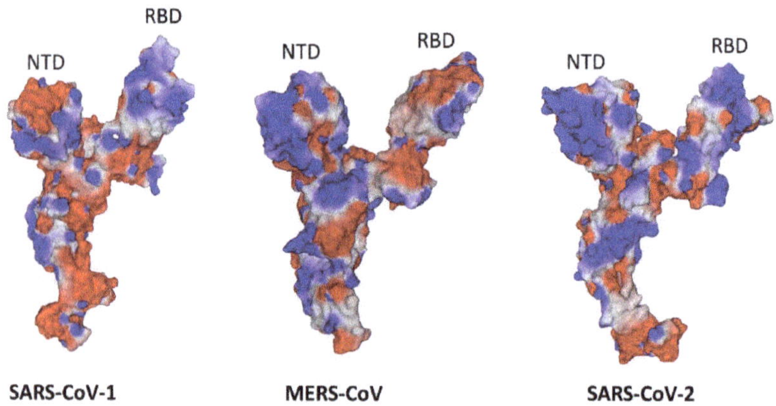

Figure 8. Comparison of the electrostatic surface potential of the three pathogenic coronaviruses for the human species. These structures were retrieved and modeled from pdb 5X5B (SARS-CoV-1), 5X59 (MERS-CoV), and 7BNM (SARS-CoV-2).

From the initial strain originating from Asia in 2019, SARS-CoV-2 variants emerged sequentially to rapidly reach a global distribution [96]. In these variants, mutations, deletions, and/or insertions have remodeled the NTD and the RBD according to a double selection pressure: (i) an immune escape progressively decreasing the effectiveness of neutralizing antibodies [97–99] and (ii) a faster access to lipid rafts determined by an increase in the electrostatic surface potential of the NTD, which tends to become increasingly electropositive [7]. In parallel, compensation mutations have appeared, allowing the RBD to retain its binding properties to the ACE2 receptor [100,101].

Analysis of the electrostatic surface potential of the spike trimers shows an overall increase of this potential towards strongly electropositive forms (Figure 9). These global changes mask the differences that may exist in the evolution of a particular domain such as the NTD [102]. Thus, the electrostatic potential of NTD steadily increased in the Alpha to Delta variant series [7]. On the other hand, it has decreased for Omicron, whereas the overall potential of the trimer is markedly increased compared to Delta [94,103], due to the very high electropositivity of the Omicron RBD [102]. Correspondingly, the Omicron spike trimer has a higher hemagglutination capacity compared to other variants, including Delta [94].

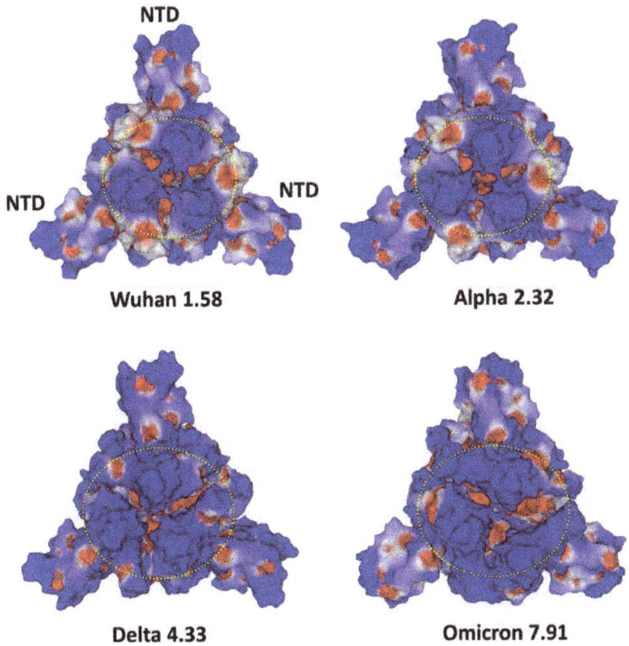

Figure 9. Comparison of the electrostatic surface potential of trimeric spikes of SARS-CoV-2 variants. NTD, N-terminal domain. The three receptor binding domains (RBD) are localized with a yellow dashed circle. The value of the electrostatic surface potential (positive) is indicated for each virus. Note that the NTD of Omicron is less electropositive than that of Delta. However, the surface of Omicron is globally more electropositive than Delta. These structures were retrieved and modeled from pdb file 7BNM.

This analysis demonstrates the usefulness of considering the surface electrostatic potential as a marker of the evolution of viruses, consistent with the notion that this parameter is one of the essential driving forces of variants [7]. In this respect, a clustering based on the spatial distribution of HIV-1 V3 loop subtypes electrostatic potential was successfully carried out by López de Victoria et al. (2012) [12]. It would also be interesting to compare these analyses with those obtained from antigenic maps of SARS-CoV-2 and influenza virus variants [104–107]. In this respect, the NTD antigenic mapping revealed a supersite of vulnerability for SARS-CoV-2 [107], which overlaps the ganglioside binding domain [47]. These findings strongly support the concept that ganglioside binding domains coincide with neutralizing epitopes. Thus, identifying these domains in a virus protein is a direct way to develop rapid vaccine formulations.

7. Clues for Managing Future Pandemics

How to manage a viral epidemic brutally striking the human species? Very recently we have developed a strategy that could be applied in the event of a new health crisis due to an infectious disease. We have illustrated this strategy using the example of the monkeypox virus [48]. This virus hit the headlines in the summer of 2022 with an unexpected outbreak outside its usual geographical area [108]. Our idea was to identify ganglioside binding motifs in proteins of this virus known to be the target of neutralizing antibodies.

For instance, the ganglioside binding domain located in the NTD of SARS-CoV-2 [47,109] overlaps with the neutralizing epitope of the 4A8 antibody [110].

A bibliographic search enabled us to identify the cell surface binding protein E8L. We generated a 3D structural model of this protein using data from the UniProt database (https://www.uniprot.org, accessed on 26 December 2022) and the Robetta server (https://robetta.

bakerlab.org, accessed on 26 December 2022). By a dedicated molecular modeling approach adapted to the topology of E8L [111], we have determined a possible mode of interaction of this protein, with a cluster of gangliosides mimicking a lipid raft (Figure 10A) [48]. The electrostatic surface potential at the level of the protein domain containing the ganglioside binding motif is strongly electropositive (Figure 10B), in agreement with all the viruses previously studied by our team [26]. This study allowed us to identify a new type of ganglioside binding domain, organized in an annular structure on the surface of the protein (Figure 10C). As expected, this motif contains the usual amino acids necessary for the recognition of gangliosides: cationic (arginine, lysine), aromatic (tyrosine), and histidine amino acids. Thus, apart from the novelty at the level of the annular organization of the motif, the ganglioside binding domain of the monkeypox virus fulfills the molecular criteria governing virus-ganglioside interactions [48].

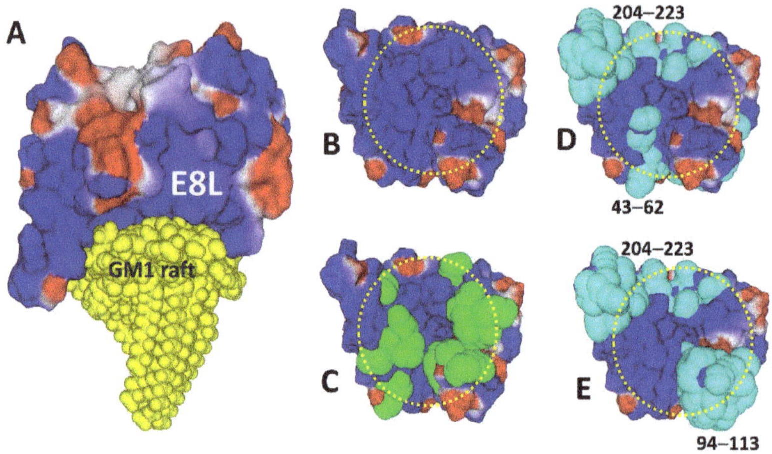

Figure 10. Ganglioside binding domain and B linear epitopes of the Monkeypox virus E8L protein. (**A**) Molecular model of E8L bound to a cluster of gangliosides GM1 in a typical raft organization. (**B**) Electrostatic surface potential of the E8L side facing the GM1 raft. (**C**) Annular distribution of amino acid residues (atomic green spheres) forming the ganglioside binding motif of E8L. (**D**) Epitopes 43–62 and 204–223 (in cyan) of E8L. (**E**) Epitopes 91–113 and 204–223 (in cyan) of E8L. In B–E the annular distribution of amino acids constituting the ganglioside binding domain is indicated by a yellow dashed circle.

The second step of our strategy consisted in identifying linear B epitopes [112] that could be easily incorporated into a vaccine formulation in the form of synthetic peptides. Finally, we selected, among all the potential epitopes, those which overlap with the ganglioside binding domain (Figure 10D,E). Given that this analysis is based on the 3D structure of the E8L protein, we were able to determine the most suitable formulation to promote synergies between epitopes and eliminate redundant epitopes. Indeed, neighboring, or even partially superimposed, domains in the 3D structure of the protein may in fact correspond to distant regions in the amino acid sequence. This is the case for epitopes 43–62 and 94–113 of E8L, which show some overlapping (Figure 10). In this case, the best antigenic formula would be to mix synthetic peptides 94–113 and 204–223. These peptides are well conserved among monkeypox virus strains and ideally localized in the structure of the E8L protein to efficiently trigger neutralizing antibodies against monkeypox virus [48].

Until effective vaccines are available, it is possible to use broad-spectrum antivirals to treat infected patients. Here again, the surface electrostatic potential of virus proteins explains the nonspecific antiviral effects of many compounds. Indeed, anionic polymers such as heparan sulfate [113] and glycosaminoglycans [114] are natural antivirals that can bind to the electropositive regions of viruses and prevent their initial adhesion to raft gangliosides. Low molecular weight anionic compounds such as suramin [115] or sulfatide [66], which bind to the V3 loop of HIV-1 gp120, also have potent antiviral activity. Cationic peptide dendrimers which bind to cell surface glycosphingolipids [116] block the infection of not only lymphocytes and macrophages but also CD4-negative cells by several HIV-1 and HIV-2 strains [117,118]. Conversely, synthetic analogues of glycosphingolipids, which interact with the cationic regions of viral proteins, have shown interesting anti-HIV activity [119–123].

More recently we unraveled a new antiviral mechanism for hydroxychloroquine, an antiparasitic drug used by some clinicians for treating SARS-CoV-2 infection [124]. We showed that hydroxychloroquine strongly interacts with raft gangliosides [125] and thus could provide protection against a broad range of viruses that use lipid rafts as the portal of entry [47]. Additionally, our modeling studies identified a potential synergy between hydroxychloroquine and azithromycin, a combination therapy also used for treating SARS-CoV-2 infections [124]. We showed that azithromycin binds to the conserved ganglioside binding domain located in the NTD of SARS-CoV-2 [126], confirming the synergy observed in vitro in infection studies [127]. The antiparasitic drug ivermectin also has broad antiviral properties [128]. In addition, this drug inhibits the hemagglutination of red blood cells induced by the spike trimers of SARS-CoV-2 variants, including Omicron [94].

Despite these important findings, the connection between lipid rafts, surface electrostatic potential, and antiviral activity has not been exploited enough by the medical community. It is a fact that we must now abandon the outdated dogma that an antibiotic cannot cure viral diseases [129]. At the molecular level, such a rigid classification is just nonsense. Azithromycin [130], ivermectin [131], hydroxychloroquine [132], suramin [133], and sulfolipids [134], to mention only a few, whatever they have been used or are used for by clinicians, are also antivirals. Their broad-spectrum antiviral activity has a common target—virus-raft interactions—as these drugs attach themselves either to rafts or to the ganglioside binding domains of viruses. Correspondingly, the antiviral properties of these drugs can be revealed with biological experiments using virus pseudotypes [135]. This assay focuses on viral entry mechanisms, excluding any other step in the replication cycle [136]. Virus pseudotypes have successfully demonstrated the antiviral effect of various antibiotics and antiparasitics, including atovaquone [137], carrimycin [138], azithromycin [139], hydroxychloroquine [140], and suramin [141], as well as glycodendrimers [142] and anionic drugs such as glycosphingolipid sulfatide [66]. By preventing the attachment of viruses to lipid rafts, these compounds could somehow mimic the selection barrier controlling transmission by the R5 strains of HIV-1, to the detriment of the X4 strains, which are blocked by various electronegative structures. The antiviral effect of all these molecules takes its logic from the natural history of virus-cell interactions, which are under the control of the surface electrostatic potential. It is time to incorporate this concept into our therapeutic arsenal in order to reposition old molecules [143–146] and/or for the design of new antivirals targeting virus-raft interactions [47].

Author Contributions: Conceptualization, J.F.; methodology, F.A., H.C., J.F. and N.Y.; formal analysis, F.A., H.C., J.F. and N.Y.; investigation, F.A., H.C., J.F. and N.Y.; writing—original draft preparation, J.F.; writing—review and editing, F.A., H.C., J.F. and N.Y.; supervision, J.F.; project administration, J.F.; funding acquisition, J.F. All authors have read and agreed to the published version of the manuscript.

Funding: F.A. is the recipient of a DGA (Direction Générale de l'Armement)/University of Aix-Marseille Ph.D. fellowship (grant number 2020007).

Institutional Review Board Statement: Not applicable.

Informed Consent Statement: Not applicable.

Data Availability Statement: Not applicable.

Acknowledgments: J.F. thanks the students of the Origin of Life and Molecular Evolution course (Physiology-Functional Genomics) at the University of Aix-Marseille (2022–2023). These students had to solve the exercise in ese students had to solve the exercise in Figure 1 of this article and their comments were particularly insightful.

Conflicts of Interest: The authors declare no conflict of interest.

References

1. Fantini, J.; Yahi, N. *Brain Lipids in Synaptic Function and Neurological Disease: Clues to Innovative Therapeutic Strategies for Brain Disorders*; Academic Press: Cambridge, MA, USA, 2015.
2. Chaplin, M. Do we underestimate the importance of water in cell biology? *Nat. Rev. Mol. Cell Biol.* **2006**, *7*, 861–866. [CrossRef] [PubMed]
3. Huang, S.; Wikswo, J. Dimensions of systems biology. *Rev. Physiol. Biochem. Pharmacol.* **2006**, *157*, 81–104. [PubMed]
4. Brodsky, W.Y. Protein synthesis rhythm. *J. Theor. Biol.* **1975**, *55*, 167–200. [CrossRef] [PubMed]
5. McFadden, J.; Al-Khalili, J. The origins of quantum biology. *Proc. R. Soc. A* **2018**, *474*, 20180674. [CrossRef] [PubMed]
6. Weiner, P.K.; Langridge, R.; Blaney, J.M.; Schaefer, R.; Kollman, P.A. Electrostatic potential molecular surfaces. *Proc. Natl. Acad. Sci. USA* **1982**, *79*, 3754–3758. [CrossRef] [PubMed]
7. Fantini, J.; Yahi, N.; Azzaz, F.; Chahinian, H. Structural dynamics of SARS-CoV-2 variants: A health monitoring strategy for anticipating Covid-19 outbreaks. *J. Infect.* **2021**, *83*, 197–206. [CrossRef] [PubMed]
8. Repits, J.; Sterjovski, J.; Badia-Martinez, D.; Mild, M.; Gray, L.; Churchill, M.J.; Purcell, D.F.; Karlsson, A.; Albert, J.; Fenyö, E.M. Primary HIV-1 R5 isolates from end-stage disease display enhanced viral fitness in parallel with increased gp120 net charge. *Virology* **2008**, *379*, 125–134. [CrossRef]
9. Heidari, A.; Righetto, I.; Filippini, F. Electrostatic variation of haemagglutinin as a hallmark of the evolution of avian influenza viruses. *Sci. Rep.* **2018**, *8*, 1929. [CrossRef]
10. Kalinina, O.V.; Pfeifer, N.; Lengauer, T. Modelling binding between CCR5 and CXCR4 receptors and their ligands suggests the surface electrostatic potential of the co-receptor to be a key player in the HIV-1 tropism. *Retrovirology* **2013**, *10*, 130. [CrossRef]
11. Honig, B.; Nicholls, A. Classical electrostatics in biology and chemistry. *Science* **1995**, *268*, 1144–1149. [CrossRef]
12. López de Victoria, A.; Kieslich, C.A.; Rizos, A.K.; Krambovitis, E.; Morikis, D. Clustering of HIV-1 subtypes based on gp120 V3 loop electrostatic properties. *BMC Biophys.* **2012**, *5*, 3. [CrossRef] [PubMed]
13. Fogolari, F.; Brigo, A.; Molinari, H. The Poisson–Boltzmann equation for biomolecular electrostatics: A tool for structural biology. *J. Mol. Recognit.* **2002**, *15*, 377–392. [CrossRef] [PubMed]
14. Sharp, K.A.; Honig, B. Calculating total electrostatic energies with the nonlinear Poisson-Boltzmann equation. *J. Phys. Chem.* **1990**, *94*, 7684–7692. [CrossRef]
15. Murray, J.S.; Politzer, P. The electrostatic potential: An overview. *Wiley Interdiscip. Rev. Comput. Mol. Sci.* **2011**, *1*, 153–163. [CrossRef]
16. Jo, S.; Vargyas, M.; Vasko-Szedlar, J.; Roux, B.; Im, W. PBEQ-Solver for online visualization of electrostatic potential of biomolecules. *Nucleic Acids Res.* **2008**, *36*, W270–W275. [CrossRef] [PubMed]
17. Im, W.; Beglov, D.; Roux, B. Continuum solvation model: Computation of electrostatic forces from numerical solutions to the Poisson-Boltzmann equation. *Comput. Phys. Commun.* **1998**, *111*, 59–75. [CrossRef]
18. Auffinger, P.; Hashem, Y. Nucleic acid solvation: From outside to insight. *Curr. Opin. Struct. Biol.* **2007**, *17*, 325–333. [CrossRef]
19. Fantini, J.; Garmy, N.; Mahfoud, R.; Yahi, N. Lipid rafts: Structure, function and role in HIV, Alzheimer's and prion diseases. *Expert Rev. Mol. Med.* **2002**, *4*, 1–22. [CrossRef]
20. Fantini, J.; Chahinian, H.; Yahi, N. Progress toward Alzheimer's disease treatment: Leveraging the Achilles' heel of Aβ oligomers? *Protein Sci. A Publ. Protein Soc.* **2020**, *29*, 1748–1759. [CrossRef]
21. Fontes, A.; Fernandes, H.; Barjas-Castro, M.; de Thomaz, A.; de Ysasa Pozzo, L.; Barbosa, L.; Cesar, C. *Red Blood Cell Membrane Viscoelasticity, Agglutination, and Zeta Potential Measurements with Double Optical Tweezers*; SPIE: Bellingham, WA, USA, 2006; Volume 6088.
22. Luner, S.J.; Sturgeon, P.; Szklarek, D.; McQuiston, D.T. Effects of Proteases and Neuraminidase on RBC Surface Charge and Agglutination: A Klinetic Study 1. *Vox Sang.* **1975**, *28*, 184–199. [CrossRef]
23. Van Oss, C.; Absolom, D. Hemagglutination and the closest distance of approach of normal, neuraminidase-and papain-treated erythrocytes. *Vox Sang.* **1984**, *47*, 250–256. [CrossRef] [PubMed]
24. Schnaar, R.L. Gangliosides of the Vertebrate Nervous System. *J. Mol. Biol.* **2016**, *428*, 3325–3336. [CrossRef] [PubMed]
25. Schnaar, R.L.; Gerardy-Schahn, R.; Hildebrandt, H. Sialic acids in the brain: Gangliosides and polysialic acid in nervous system development, stability, disease, and regeneration. *Physiol. Rev.* **2014**, *94*, 461–518. [CrossRef] [PubMed]
26. Azzaz, F.; Yahi, N.; Di Scala, C.; Chahinian, H.; Fantini, J. Ganglioside binding domains in proteins: Physiological and pathological mechanisms. *Adv. Protein Chem. Struct. Biol.* **2022**, *128*, 289–324. [CrossRef] [PubMed]
27. Koehl, A.; Hu, H.; Feng, D.; Sun, B.; Zhang, Y.; Robertson, M.J.; Chu, M.; Kobilka, T.S.; Laeremans, T.; Steyaert, J.; et al. Structural insights into the activation of metabotropic glutamate receptors. *Nature* **2019**, *566*, 79–84. [CrossRef]
28. Amin, M.; Küpper, J. Variations in proteins dielectric constants. *ChemistryOpen* **2020**, *9*, 691–694. [CrossRef]

29. Mañes, S.; del Real, G.; Lacalle, R.A.; Lucas, P.; Gómez-Moutón, C.; Sánchez-Palomino, S.; Delgado, R.; Alcamí, J.; Mira, E.; Martínez, A.C. Membrane raft microdomains mediate lateral assemblies required for HIV-1 infection. *EMBO Rep* **2000**, *1*, 190–196. [CrossRef]
30. Mañes, S.; del Real, G.; Martínez, A.C. Pathogens: Raft hijackers. *Nat. Rev. Immunol.* **2003**, *3*, 557–568. [CrossRef]
31. Ripa, I.; Andreu, S.; López-Guerrero, J.A.; Bello-Morales, R. Membrane rafts: Portals for viral entry. *Front. Microbiol.* **2021**, *12*, 631274. [CrossRef]
32. Omasta, B.; Tomaskova, J. Cellular Lipids—Hijacked Victims of Viruses. *Viruses* **2022**, *14*, 1896. [CrossRef]
33. Pike, L.J. Lipid rafts: Bringing order to chaos. *J. Lipid Res.* **2003**, *44*, 655–667. [CrossRef]
34. Suzuki, Y. Gangliosides as influenza virus receptors. Variation of influenza viruses and their recognition of the receptor sialo-sugar chains. Prog. Lipid Res. 1994, 33, 429–457. *Prog. Lipid Res.* **1994**, *33*, 429–457. [CrossRef]
35. Markwell, M.; Svennerholm, L.; Paulson, J.C. Specific gangliosides function as host cell receptors for Sendai virus. *Proc. Natl. Acad. Sci. USA* **1981**, *78*, 5406–5410. [CrossRef] [PubMed]
36. Campanero-Rhodes, M.A.; Smith, A.; Chai, W.; Sonnino, S.; Mauri, L.; Childs, R.A.; Zhang, Y.; Ewers, H.; Helenius, A.; Imberty, A. N-glycolyl GM1 ganglioside as a receptor for simian virus 40. *J. Virol.* **2007**, *81*, 12846–12858. [CrossRef]
37. Maginnis, M.S. Virus–receptor interactions: The key to cellular invasion. *J. Mol. Biol.* **2018**, *430*, 2590–2611. [CrossRef] [PubMed]
38. Rolsma, M.D.; Kuhlenschmidt, T.B.; Gelberg, H.B.; Kuhlenschmidt, M.S. Structure and function of a ganglioside receptor for porcine rotavirus. *J. Virol.* **1998**, *72*, 9079–9091. [CrossRef]
39. Hammache, D.; Piéroni, G.; Yahi, N.; Delézay, O.; Koch, N.; Lafont, H.; Tamalet, C.; Fantini, J. Specific interaction of HIV-1 and HIV-2 surface envelope glycoproteins with monolayers of galactosylceramide and ganglioside GM3. *J. Biol. Chem.* **1998**, *273*, 7967–7971. [CrossRef] [PubMed]
40. Hammache, D.; Yahi, N.; Maresca, M.; Piéroni, G.; Fantini, J. Human erythrocyte glycosphingolipids as alternative cofactors for human immunodeficiency virus type 1 (HIV-1) entry: Evidence for CD4-induced interactions between HIV-1 gp120 and reconstituted membrane microdomains of glycosphingolipids (Gb3 and GM3). *J. Virol.* **1999**, *73*, 5244–5248. [CrossRef] [PubMed]
41. Hammache, D.; Yahi, N.; Piéroni, G.; Ariasi, F.; Tamalet, C.; Fantini, J. Sequential interaction of CD4 and HIV-1 gp120 with a reconstituted membrane patch of ganglioside GM3: Implications for the role of glycolipids as potential HIV-1 fusion cofactors. *Biochem. Biophys. Res. Commun.* **1998**, *246*, 117–122. [CrossRef] [PubMed]
42. Hug, P.; Lin, H.M.; Korte, T.; Xiao, X.; Dimitrov, D.S.; Wang, J.M.; Puri, A.; Blumenthal, R. Glycosphingolipids promote entry of a broad range of human immunodeficiency virus type 1 isolates into cell lines expressing CD4, CXCR4, and/or CCR5. *J. Virol.* **2000**, *74*, 6377–6385. [CrossRef]
43. Harouse, J.M.; Bhat, S.; Spitalnik, S.L.; Laughlin, M.; Stefano, K.; Silberberg, D.H.; Gonzalez-Scarano, F. Inhibition of entry of HIV-1 in neural cell lines by antibodies against galactosyl ceramide. *Science* **1991**, *253*, 320–323. [CrossRef] [PubMed]
44. Yahi, N.; Baghdiguian, S.; Moreau, H.; Fantini, J. Galactosyl ceramide (or a closely related molecule) is the receptor for human immunodeficiency virus type 1 on human colon epithelial HT29 cells. *J. Virol.* **1992**, *66*, 4848–4854. [CrossRef] [PubMed]
45. Fantini, J.; Garmy, N.; Yahi, N. Prediction of glycolipid-binding domains from the amino acid sequence of lipid raft-associated proteins: Application to HpaA, a protein involved in the adhesion of Helicobacter pylori to gastrointestinal cells. *Biochemistry* **2006**, *45*, 10957–10962. [CrossRef] [PubMed]
46. Mahfoud, R.; Garmy, N.; Maresca, M.; Yahi, N.; Puigserver, A.; Fantini, J. Identification of a common sphingolipid-binding domain in Alzheimer, prion, and HIV-1 proteins. *J. Biol. Chem.* **2002**, *277*, 11292–11296. [CrossRef]
47. Fantini, J.; Chahinian, H.; Yahi, N. Leveraging coronavirus binding to gangliosides for innovative vaccine and therapeutic strategies against COVID-19. *Biochem. Biophys. Res. Commun.* **2021**, *538*, 132–136. [CrossRef]
48. Fantini, J.; Chahinian, H.; Yahi, N. A Vaccine Strategy Based on the Identification of an Annular Ganglioside Binding Motif in Monkeypox Virus Protein E8L. *Viruses* **2022**, *14*, 2531. [CrossRef]
49. Fantini, J. How sphingolipids bind and shape proteins: Molecular basis of lipid-protein interactions in lipid shells, rafts and related biomembrane domains. *Cell. Mol. Life Sci. CMLS* **2003**, *60*, 1027–1032. [CrossRef]
50. Froimowitz, M. HyperChem: A software package for computational chemistry and molecular modeling. *BioTechniques* **1993**, *14*, 1010–1013.
51. Bour, S.; Geleziunas, R.; Wainberg, M.A. The human immunodeficiency virus type 1 (HIV-1) CD4 receptor and its central role in promotion of HIV-1 infection. *Microbiol. Rev.* **1995**, *59*, 63–93. [CrossRef]
52. Moore, J.P.; Trkola, A.; Dragic, T. Co-receptors for HIV-1 entry. *Curr. Opin. Immunol.* **1997**, *9*, 551–562. [CrossRef]
53. Moore, J.P.; Kitchen, S.G.; Pugach, P.; Zack, J.A. The CCR5 and CXCR4 coreceptors—Central to understanding the transmission and pathogenesis of human immunodeficiency virus type 1 infection. *AIDS Res. Hum. Retrovir.* **2004**, *20*, 111–126. [CrossRef] [PubMed]
54. Regoes, R.R.; Bonhoeffer, S. The HIV coreceptor switch: A population dynamical perspective. *Trends Microbiol.* **2005**, *13*, 269–277. [CrossRef] [PubMed]
55. Briggs, D.R.; Tuttle, D.L.; Sleasman, J.W.; Goodenow, M.M. Envelope V3 amino acid sequence predicts HIV-1 phenotype (co-receptor usage and tropism for macrophages). *AIDS* **2000**, *14*, 2937–2939. [CrossRef] [PubMed]
56. Vicenzi, E.; Liò, P.; Poli, G. The puzzling role of CXCR4 in human immunodeficiency virus infection. *Theranostics* **2013**, *3*, 18. [CrossRef]
57. Philpott, S.M. HIV-1 coreceptor usage, transmission, and disease progression. *Curr. HIV Res.* **2003**, *1*, 217–227. [CrossRef]

58. Schindelin, J.; Rueden, C.T.; Hiner, M.C.; Eliceiri, K.W. The ImageJ ecosystem: An open platform for biomedical image analysis. *Mol. Reprod. Dev.* **2015**, *82*, 518–529. [CrossRef]
59. Tebit, D.M.; Arts, E.J. Tracking a century of global expansion and evolution of HIV to drive understanding and to combat disease. *Lancet Infect. Dis.* **2011**, *11*, 45–56. [CrossRef]
60. Pollakis, G.; Kang, S.; Kliphuis, A.; Chalaby, M.I.; Goudsmit, J.; Paxton, W.A. N-linked glycosylation of the HIV type-1 gp120 envelope glycoprotein as a major determinant of CCR5 and CXCR4 coreceptor utilization. *J. Biol. Chem.* **2001**, *276*, 13433–13441. [CrossRef]
61. Jo, S.; Song, K.C.; Desaire, H.; MacKerell, A.D., Jr.; Im, W. Glycan Reader: Automated sugar identification and simulation preparation for carbohydrates and glycoproteins. *J. Comput. Chem.* **2011**, *32*, 3135–3141. [CrossRef]
62. Nijmeijer, B.M.; Geijtenbeek, T.B. Negative and positive selection pressure during sexual transmission of transmitted founder HIV-1. *Front. Immunol.* **2019**, *10*, 1599. [CrossRef]
63. Wolf, D.P.; Sokoloski, J.E.; Litt, M. Composition and function of human cervical mucus. *Biochim. Et Biophys. Acta (BBA)-Gen. Subj.* **1980**, *630*, 545–558. [CrossRef]
64. Grivel, J.-C.; Shattock, R.J.; Margolis, L.B. Selective transmission of R5 HIV-1 variants: Where is the gatekeeper? *J. Transl. Med.* **2011**, *9*, S6. [CrossRef] [PubMed]
65. Moulard, M.; Lortat-Jacob, H.; Mondor, I.; Roca, G.; Wyatt, R.; Sodroski, J.; Zhao, L.; Olson, W.; Kwong, P.D.; Sattentau, Q.J. Selective interactions of polyanions with basic surfaces on human immunodeficiency virus type 1 gp120. *J. Virol.* **2000**, *74*, 1948–1960. [CrossRef]
66. Fantini, J.; Hammache, D.; Delézay, O.; Piéroni, G.; Tamalet, C.; Yahi, N. Sulfatide inhibits HIV-1 entry into CD4−/CXCR4+ cells. *Virology* **1998**, *246*, 211–220. [CrossRef] [PubMed]
67. Fantini, J.; Yahi, N.; Tourres, C.; Delezay, O.; Tamalet, C. HIV-1 transmission across the vaginal epithelium. *AIDS* **1997**, *11*, 1663–1664.
68. Berlier, W.; Bourlet, T.; Lawrence, P.; Hamzeh, H.; Lambert, C.; Genin, C.; Verrier, B.; Dieu-Nosjean, M.C.; Pozzetto, B.; Delézay, O. Selective sequestration of X4 isolates by human genital epithelial cells: Implication for virus tropism selection process during sexual transmission of HIV. *J. Med. Virol.* **2005**, *77*, 465–474. [CrossRef]
69. Lawrence, P.; Portran, D.; Terrasse, R.; Palle, S.; Olivier, T.; Fantini, J.; Bourlet, T.; Pozzetto, B.; Delezay, O. Selective transmigration of monocyte-associated HIV-1 across a human cervical monolayer and its modulation by seminal plasma. *AIDS* **2012**, *26*, 785–796. [CrossRef]
70. Margolis, L.; Shattock, R. Selective transmission of CCR5-utilizing HIV-1: The'gatekeeper'problem resolved? *Nat. Rev. Microbiol.* **2006**, *4*, 312–317. [CrossRef]
71. Schutten, M.; Van Baalen, C.; Guillon, C.; Huisman, R.; Boers, P.; Sintnicolaas, K.; Gruters, R.; Osterhaus, A.D. Macrophage tropism of human immunodeficiency virus type 1 facilitates in vivo escape from cytotoxic T-lymphocyte pressure. *J. Virol.* **2001**, *75*, 2706–2709. [CrossRef]
72. Tscherning, C.; Alaeus, A.; Fredriksson, R.; Björndal, Å.; Deng, H.; Littman, D.R.; Fenyö, E.M.; Albert, J. Differences in chemokine coreceptor usage between genetic subtypes of HIV-1. *Virology* **1998**, *241*, 181–188. [CrossRef]
73. Yahi, N.; Fantini, J.; Tourres, C.; Tivoli, N.; Koch, N.; Tamalet, C. Use of drug resistance sequence data for the systematic detection of non-B human immunodeficiency virus type 1 (HIV-1) subtypes: How to create a sentinel site for monitoring the genetic diversity of HIV-1 at a country scale. *J. Infect. Dis.* **2001**, *183*, 1311–1317. [CrossRef] [PubMed]
74. Righetto, I.; Milani, A.; Cattoli, G.; Filippini, F. Comparative structural analysis of haemagglutinin proteins from type A influenza viruses: Conserved and variable features. *BMC Bioinform.* **2014**, *15*, 363. [CrossRef] [PubMed]
75. Righetto, I.; Filippini, F. Normal modes analysis and surface electrostatics of haemagglutinin proteins as fingerprints for high pathogenic type A influenza viruses. *BMC Bioinform.* **2020**, *21*, 354. [CrossRef] [PubMed]
76. Jimenez-Alberto, A.; Alvarado-Facundo, E.; Ribas-Aparicio, R.M.; Castelán-Vega, J.A. Analysis of adaptation mutants in the hemagglutinin of the influenza A (H1N1) pdm09 virus. *PLoS ONE* **2013**, *8*, e70005. [CrossRef] [PubMed]
77. Weis, W.; Brown, J.; Cusack, S.; Paulson, J.; Skehel, J.; Wiley, D. Structure of the influenza virus haemagglutinin complexed with its receptor, sialic acid. *Nature* **1988**, *333*, 426–431. [CrossRef]
78. Kumlin, U.; Olofsson, S.; Dimock, K.; Arnberg, N. Sialic acid tissue distribution and influenza virus tropism. *Influenza Other Respir. Viruses* **2008**, *2*, 147–154. [CrossRef] [PubMed]
79. Franca, M.; Stallknecht, D.; Howerth, E. Expression and distribution of sialic acid influenza virus receptors in wild birds. *Avian Pathol.* **2013**, *42*, 60–71. [CrossRef]
80. Qi, L.; Kash, J.C.; Dugan, V.G.; Wang, R.; Jin, G.; Cunningham, R.E.; Taubenberger, J.K. Role of sialic acid binding specificity of the 1918 influenza virus hemagglutinin protein in virulence and pathogenesis for mice. *J. Virol.* **2009**, *83*, 3754–3761. [CrossRef]
81. Ayora-Talavera, G. Sialic acid receptors: Focus on their role in influenza infection. *J. Recept. Ligand Channel Res.* **2018**, *10*, 1–11. [CrossRef]
82. Xiong, X.; Coombs, P.J.; Martin, S.R.; Liu, J.; Xiao, H.; McCauley, J.W.; Locher, K.; Walker, P.A.; Collins, P.J.; Kawaoka, Y.; et al. Receptor binding by a ferret-transmissible H5 avian influenza virus. *Nature* **2013**, *497*, 392–396. [CrossRef]
83. Zhu, X.; Yu, W.; McBride, R.; Li, Y.; Chen, L.M.; Donis, R.O.; Tong, S.; Paulson, J.C.; Wilson, I.A. Hemagglutinin homologue from H17N10 bat influenza virus exhibits divergent receptor-binding and pH-dependent fusion activities. *Proc. Natl. Acad. Sci. USA* **2013**, *110*, 1458–1463. [CrossRef] [PubMed]

84. Weinstein, R.A. Planning for epidemics–the lessons of SARS. *N. Engl. J. Med.* **2004**, *350*, 2332–2334. [CrossRef] [PubMed]
85. Zaki, A.M.; van Boheemen, S.; Bestebroer, T.M.; Osterhaus, A.D.; Fouchier, R.A. Isolation of a novel coronavirus from a man with pneumonia in Saudi Arabia. *N. Engl. J. Med.* **2012**, *367*, 1814–1820. [CrossRef]
86. Zhou, P.; Yang, X.L.; Wang, X.G.; Hu, B.; Zhang, L.; Zhang, W.; Si, H.R.; Zhu, Y.; Li, B.; Huang, C.L.; et al. A pneumonia outbreak associated with a new coronavirus of probable bat origin. *Nature* **2020**, *579*, 270–273. [CrossRef]
87. Lu, Y.; Liu, D.X.; Tam, J.P. Lipid rafts are involved in SARS-CoV entry into Vero E6 cells. *Biochem. Biophys. Res. Commun.* **2008**, *369*, 344–349. [CrossRef]
88. Li, G.-M.; Li, Y.-G.; Yamate, M.; Li, S.-M.; Ikuta, K. Lipid rafts play an important role in the early stage of severe acute respiratory syndrome-coronavirus life cycle. *Microbes Infect.* **2007**, *9*, 96–102. [CrossRef] [PubMed]
89. Li, W.; Hulswit, R.J.; Widjaja, I.; Raj, V.S.; McBride, R.; Peng, W.; Widagdo, W.; Tortorici, M.A.; Van Dieren, B.; Lang, Y. Identification of sialic acid-binding function for the Middle East respiratory syndrome coronavirus spike glycoprotein. *Proc. Natl. Acad. Sci. USA* **2017**, *114*, E8508–E8517. [CrossRef]
90. Sun, X.-L. The role of cell surface sialic acids for SARS-CoV-2 infection. *Glycobiology* **2021**, *31*, 1245–1253. [CrossRef]
91. Pirone, L.; Del Gatto, A.; Di Gaetano, S.; Saviano, M.; Capasso, D.; Zaccaro, L.; Pedone, E. A multi-targeting approach to fight SARS-CoV-2 attachment. *Front. Mol. Biosci.* **2020**, *7*, 186. [CrossRef]
92. Raj, V.S.; Mou, H.; Smits, S.L.; Dekkers, D.H.; Müller, M.A.; Dijkman, R.; Muth, D.; Demmers, J.A.; Zaki, A.; Fouchier, R.A. Dipeptidyl peptidase 4 is a functional receptor for the emerging human coronavirus-EMC. *Nature* **2013**, *495*, 251–254. [CrossRef]
93. Hatmal, M.M.M.; Alshaer, W.; Al-Hatamleh, M.A.; Hatmal, M.; Smadi, O.; Taha, M.O.; Oweida, A.J.; Boer, J.C.; Mohamud, R.; Plebanski, M. Comprehensive structural and molecular comparison of spike proteins of SARS-CoV-2, SARS-CoV and MERS-CoV, and their interactions with ACE2. *Cells* **2020**, *9*, 2638. [CrossRef] [PubMed]
94. Boschi, C.; Scheim, D.E.; Bancod, A.; Militello, M.; Bideau, M.L.; Colson, P.; Fantini, J.; Scola, B.L. SARS-CoV-2 Spike Protein Induces Hemagglutination: Implications for COVID-19 Morbidities and Therapeutics and for Vaccine Adverse Effects. *Int. J. Mol. Sci.* **2022**, *23*, 15480. [CrossRef] [PubMed]
95. Stencel-Baerenwald, J.E.; Reiss, K.; Reiter, D.M.; Stehle, T.; Dermody, T.S. The sweet spot: Defining virus–sialic acid interactions. *Nat. Rev. Microbiol.* **2014**, *12*, 739–749. [CrossRef] [PubMed]
96. Guérin, P.; Yahi, N.; Azzaz, F.; Chahinian, H.; Sabatier, J.M.; Fantini, J. Structural Dynamics of the SARS-CoV-2 Spike Protein: A 2-Year Retrospective Analysis of SARS-CoV-2 Variants (from Alpha to Omicron) Reveals an Early Divergence between Conserved and Variable Epitopes. *Molecules* **2022**, *27*, 3851. [CrossRef] [PubMed]
97. Harvey, W.T.; Carabelli, A.M.; Jackson, B.; Gupta, R.K.; Thomson, E.C.; Harrison, E.M.; Ludden, C.; Reeve, R.; Rambaut, A.; Peacock, S.J. SARS-CoV-2 variants, spike mutations and immune escape. *Nat. Rev. Microbiol.* **2021**, *19*, 409–424. [CrossRef] [PubMed]
98. Lazarevic, I.; Pravica, V.; Miljanovic, D.; Cupic, M. Immune evasion of SARS-CoV-2 emerging variants: What have we learnt so far? *Viruses* **2021**, *13*, 1192. [CrossRef]
99. Hu, J.; Peng, P.; Cao, X.; Wu, K.; Chen, J.; Wang, K.; Tang, N.; Huang, A.-l. Increased immune escape of the new SARS-CoV-2 variant of concern Omicron. *Cell. Mol. Immunol.* **2022**, *19*, 293–295. [CrossRef]
100. Barton, M.I.; MacGowan, S.A.; Kutuzov, M.A.; Dushek, O.; Barton, G.J.; Van Der Merwe, P.A. Effects of common mutations in the SARS-CoV-2 Spike RBD and its ligand, the human ACE2 receptor on binding affinity and kinetics. *elife* **2021**, *10*, e70658. [CrossRef]
101. Moulana, A.; Dupic, T.; Phillips, A.M.; Chang, J.; Nieves, S.; Roffler, A.A.; Greaney, A.J.; Starr, T.N.; Bloom, J.D.; Desai, M.M. Compensatory epistasis maintains ACE2 affinity in SARS-CoV-2 Omicron BA.1. *Nat. Commun.* **2022**, *13*, 7011. [CrossRef]
102. Fantini, J.; Yahi, N.; Colson, P.; Chahinian, H.; La Scola, B.; Raoult, D. The puzzling mutational landscape of the SARS-2-variant Omicron. *J. Med. Virol.* **2022**, *94*, 2019–2025. [CrossRef]
103. Pascarella, S.; Ciccozzi, M.; Bianchi, M.; Benvenuto, D.; Cauda, R.; Cassone, A. The electrostatic potential of the Omicron variant spike is higher than in Delta and Delta-plus variants: A hint to higher transmissibility? *J. Med. Virol.* **2021**, *94*, 1277–1280. [CrossRef] [PubMed]
104. Mykytyn, A.Z.; Rissmann, M.; Kok, A.; Rosu, M.E.; Schipper, D.; Breugem, T.I.; van den Doel, P.B.; Chandler, F.; Bestebroer, T.; de Wit, M. Antigenic cartography of SARS-CoV-2 reveals that Omicron BA.1 and BA. 2 are antigenically distinct. *Sci. Immunol.* **2022**, *7*, eabq4450. [PubMed]
105. Smith, D.J.; Lapedes, A.S.; De Jong, J.C.; Bestebroer, T.M.; Rimmelzwaan, G.F.; Osterhaus, A.D.; Fouchier, R.A. Mapping the antigenic and genetic evolution of influenza virus. *Science* **2004**, *305*, 371–376. [CrossRef] [PubMed]
106. Liu, M.; Zhao, X.; Hua, S.; Du, X.; Peng, Y.; Li, X.; Lan, Y.; Wang, D.; Wu, A.; Shu, Y. Antigenic patterns and evolution of the human influenza A (H1N1) virus. *Sci. Rep.* **2015**, *5*, 14171. [CrossRef] [PubMed]
107. McCallum, M.; De Marco, A.; Lempp, F.A.; Tortorici, M.A.; Pinto, D.; Walls, A.C.; Beltramello, M.; Chen, A.; Liu, Z.; Zatta, F. N-terminal domain antigenic mapping reveals a site of vulnerability for SARS-CoV-2. *Cell* **2021**, *184*, 2332–2347.e16. [CrossRef]
108. Focosi, D.; Novazzi, F.; Baj, A.; Maggi, F. Monkeypox: An international epidemic. *Rev. Med. Virol.* **2022**, *32*, e2392. [CrossRef]
109. Das, T.; Mukhopadhyay, C. Identification of possible binding modes of SARS-CoV-2 spike N-terminal domain for ganglioside GM1. *Chem. Phys. Lett.* **2022**, *812*, 140260. [CrossRef]
110. Liu, L.; Wang, P.; Nair, M.S.; Yu, J.; Rapp, M.; Wang, Q.; Luo, Y.; Chan, J.F.W.; Sahi, V.; Figueroa, A.; et al. Potent neutralizing antibodies against multiple epitopes on SARS-CoV-2 spike. *Nature* **2020**, *584*, 450–456. [CrossRef]

111. Azzaz, F.; Yahi, N.; Chahinian, H.; Fantini, J. The Epigenetic Dimension of Protein Structure Is an Intrinsic Weakness of the AlphaFold Program. *Biomolecules* **2022**, *12*, 1527. [CrossRef]
112. Shantier, S.W.; Mustafa, M.I.; Abdelmoneim, A.H.; Fadl, H.A.; Elbager, S.G.; Makhawi, A.M. Novel multi epitope-based vaccine against monkeypox virus: Vaccinomic approach. *Sci. Rep.* **2022**, *12*, 15983. [CrossRef]
113. Baba, M.; Snoeck, R.; Pauwels, R.; De Clercq, E. Sulfated polysaccharides are potent and selective inhibitors of various enveloped viruses, including herpes simplex virus, cytomegalovirus, vesicular stomatitis virus, and human immunodeficiency virus. *Antimicrob. Agents Chemother.* **1988**, *32*, 1742–1745. [CrossRef]
114. Lin, L.-T.; Chen, T.-Y.; Lin, S.-C.; Chung, C.-Y.; Lin, T.-C.; Wang, G.-H.; Anderson, R.; Lin, C.-C.; Richardson, C.D. Broad-spectrum antiviral activity of chebulagic acid and punicalagin against viruses that use glycosaminoglycans for entry. *BMC Microbiol.* **2013**, *13*, 187. [CrossRef] [PubMed]
115. Yahi, N.; Sabatier, J.M.; Nickel, P.; Mabrouk, K.; Gonzalez-Scarano, F.; Fantini, J. Suramin inhibits binding of the V3 region of HIV-1 envelope glycoprotein gp120 to galactosylceramide, the receptor for HIV 1 gp120 on human colon epithelial cells. *J. Biol. Chem.* **1994**, *269*, 24349–24353. [CrossRef] [PubMed]
116. Delézay, O.; Hammache, D.; Fantini, J.; Yahi, N. SPC3, a V3 loop-derived synthetic peptide inhibitor of HIV-1 infection, binds to cell surface glycosphingolipids. *Biochemistry* **1996**, *35*, 15663–15671. [CrossRef]
117. Yahi, N.; Fantini, J.; Baghdiguian, S.; Mabrouk, K.; Tamalet, C.; Rochat, H.; Van Rietschoten, J.; Sabatier, J.-M. SPC3, a synthetic peptide derived from the V3 domain of human immunodeficiency virus type 1 (HIV-1) gp120, inhibits HIV-1 entry into CD4+ and CD4-cells by two distinct mechanisms. *Proc. Natl. Acad. Sci. USA* **1995**, *92*, 4867–4871. [CrossRef] [PubMed]
118. Yahi, N.; Sabatier, J.M.; Baghdiguian, S.; Gonzalez-Scarano, F.; Fantini, J. Synthetic multimeric peptides derived from the principal neutralization domain (V3 loop) of human immunodeficiency virus type 1 (HIV-1) gp120 bind to galactosylceramide and block HIV-1 infection in a human CD4-negative mucosal epithelial cell line. *J. Virol.* **1995**, *69*, 320–325. [CrossRef]
119. Faroux-Corlay, B.; Greiner, J.; Terreux, R.; Cabrol-Bass, D.; Aubertin, A.-M.; Vierling, P.; Fantini, J. Amphiphilic anionic analogues of galactosylceramide: Synthesis, anti-HIV-1 activity, and gp120 binding. *J. Med. Chem.* **2001**, *44*, 2188–2203. [CrossRef]
120. Faroux-Corlay, B.; Clary, L.; Gadras, C.; Hammache, D.; Greiner, J.; Santaella, C.; Aubertin, A.-M.; Vierling, P.; Fantini, J. Synthesis of single-and double-chain fluorocarbon and hydrocarbon galactosyl amphiphiles and their anti-HIV-1 activity. *Carbohydr. Res.* **2000**, *327*, 223–260. [CrossRef]
121. Kensinger, R.D.; Yowler, B.C.; Benesi, A.J.; Schengrund, C.-L. Synthesis of novel, multivalent glycodendrimers as ligands for HIV-1 gp120. *Bioconjugate Chem.* **2004**, *15*, 349–358. [CrossRef]
122. Garg, H.; Francella, N.; Tony, K.A.; Augustine, L.A.; Barchi Jr, J.J.; Fantini, J.; Puri, A.; Mootoo, D.R.; Blumenthal, R. Glycoside analogs of β-galactosylceramide, a novel class of small molecule antiviral agents that inhibit HIV-1 entry. *Antivir. Res.* **2008**, *80*, 54–61. [CrossRef]
123. Fantini, J.; Hammache, D.; Delézay, O.; Yahi, N.; André-Barrès, C.; Rico-Lattes, I.; Lattes, A. Synthetic soluble analogs of galactosylceramide (GalCer) bind to the V3 domain of HIV-1 gp120 and inhibit HIV-1-induced fusion and entry. *J. Biol. Chem.* **1997**, *272*, 7245–7252. [CrossRef] [PubMed]
124. Gautret, P.; Lagier, J.-C.; Parola, P.; Meddeb, L.; Mailhe, M.; Doudier, B.; Courjon, J.; Giordanengo, V.; Vieira, V.E.; Dupont, H.T. Hydroxychloroquine and azithromycin as a treatment of COVID-19: Results of an open-label non-randomized clinical trial. *Int. J. Antimicrob. Agents* **2020**, *56*, 105949. [CrossRef] [PubMed]
125. Fantini, J.; Di Scala, C.; Chahinian, H.; Yahi, N. Structural and molecular modelling studies reveal a new mechanism of action of chloroquine and hydroxychloroquine against SARS-CoV-2 infection. *Int. J. Antimicrob. Agents* **2020**, *55*, 105960. [CrossRef] [PubMed]
126. Fantini, J.; Chahinian, H.; Yahi, N. Synergistic antiviral effect of hydroxychloroquine and azithromycin in combination against SARS-CoV-2: What molecular dynamics studies of virus-host interactions reveal. *Int. J. Antimicrob. Agents* **2020**, *56*, 106020. [CrossRef]
127. Andreani, J.; Le Bideau, M.; Duflot, I.; Jardot, P.; Rolland, C.; Boxberger, M.; Wurtz, N.; Rolain, J.-M.; Colson, P.; La Scola, B. In vitro testing of combined hydroxychloroquine and azithromycin on SARS-CoV-2 shows synergistic effect. *Microb. Pathog.* **2020**, *145*, 104228. [CrossRef]
128. Jans, D.A.; Wagstaff, K.M. Ivermectin as a broad-spectrum host-directed antiviral: The real deal? *Cells* **2020**, *9*, 2100. [CrossRef]
129. Colson, P.; Raoult, D. Fighting viruses with antibiotics: An overlooked path. *Int. J. Antimicrob. Agents* **2016**, *48*, 349. [CrossRef]
130. Khoshnood, S.; Shirani, M.; Dalir, A.; Moradi, M.; Haddadi, M.H.; Sadeghifard, N.; Birjandi, F.S.; Yashmi, I.; Heidary, M. Antiviral effects of azithromycin: A narrative review. *Biomed. Pharmacother.* **2022**, *147*, 112682. [CrossRef]
131. Rizzo, E. Ivermectin, antiviral properties and COVID-19: A possible new mechanism of action. *Naunyn-Schmiedeberg's Arch. Pharmacol.* **2020**, *393*, 1153–1156. [CrossRef]
132. Rodrigo, C.; Fernando, S.D.; Rajapakse, S. Clinical evidence for repurposing chloroquine and hydroxychloroquine as antiviral agents: A systematic review. *Clin. Microbiol. Infect.* **2020**, *26*, 979–987. [CrossRef]
133. Tan, C.W.; Sam, I. C.; Chong, W.L.; Lee, V.S.; Chan, Y.F. Polysulfonate suramin inhibits Zika virus infection. *Antivir. Res.* **2017**, *143*, 186–194. [CrossRef]
134. Gustafson, K.R.; Cardellina, J.H.; Fuller, R.W.; Weislow, O.S.; Kiser, R.F.; Snader, K.M.; Patterson, G.M.; Boyd, M.R. AIDS-antiviral sulfolipids from cyanobacteria (blue-green algae). *JNCI J. Natl. Cancer Inst.* **1989**, *81*, 1254–1258. [CrossRef]

135. Xu, M.; Pradhan, M.; Gorshkov, K.; Petersen, J.D.; Shen, M.; Guo, H.; Zhu, W.; Klumpp-Thomas, C.; Michael, S.; Itkin, M. A high throughput screening assay for inhibitors of SARS-CoV-2 pseudotyped particle entry. *Slas Discov.* **2022**, *27*, 86–94. [CrossRef]
136. Steffen, I.; Simmons, G. Pseudotyping viral vectors with emerging virus envelope proteins. *Curr. Gene Ther.* **2016**, *16*, 47–55. [CrossRef] [PubMed]
137. Carter-Timofte, M.E.; Arulanandam, R.; Kurmasheva, N.; Fu, K.; Laroche, G.; Taha, Z.; van Der Horst, D.; Cassin, L.; van der Sluis, R.M.; Palermo, E. Antiviral potential of the antimicrobial drug atovaquone against SARS-CoV-2 and emerging variants of concern. *ACS Infect. Dis.* **2021**, *7*, 3034–3051. [CrossRef] [PubMed]
138. Yan, H.; Sun, J.; Wang, K.; Wang, H.; Wu, S.; Bao, L.; He, W.; Wang, D.; Zhu, A.; Zhang, T. Repurposing carrimycin as an antiviral agent against human coronaviruses, including the currently pandemic SARS-CoV-2. *Acta Pharm. Sin. B* **2021**, *11*, 2850–2858. [CrossRef] [PubMed]
139. Du, X.; Zuo, X.; Meng, F.; Han, C.; Ouyang, W.; Han, Y.; Gu, Y.; Zhao, X.; Xu, F.; Qin, F.X. Direct inhibitory effect on viral entry of influenza A and SARS-CoV-2 viruses by azithromycin. *Cell Prolif.* **2021**, *54*, e12953. [CrossRef] [PubMed]
140. Ou, T.; Mou, H.; Zhang, L.; Ojha, A.; Choe, H.; Farzan, M. Hydroxychloroquine-mediated inhibition of SARS-CoV-2 entry is attenuated by TMPRSS2. *PLoS Pathog.* **2021**, *17*, e1009212. [CrossRef]
141. Henß, L.; Beck, S.; Weidner, T.; Biedenkopf, N.; Sliva, K.; Weber, C.; Becker, S.; Schnierle, B.S. Suramin is a potent inhibitor of Chikungunya and Ebola virus cell entry. *Virol. J.* **2016**, *13*, 149. [CrossRef]
142. Rojo, J.; Delgado, R. Glycodendritic structures: Promising new antiviral drugs. *J. Antimicrob. Chemother.* **2004**, *54*, 579–581. [CrossRef]
143. Niaee, M.S.; Namdar, P.; Allami, A.; Zolghadr, L.; Javadi, A.; Karampour, A.; Varnaseri, M.; Bijani, B.; Cheraghi, F.; Naderi, Y. Ivermectin as an adjunct treatment for hospitalized adult COVID-19 patients: A randomized multi-center clinical trial. *Asian Pac. J. Trop. Med.* **2021**, *14*, 266.
144. Mahmud, R.; Rahman, M.M.; Alam, I.; Ahmed, K.G.U.; Kabir, A.H.; Sayeed, S.J.B.; Rassel, M.A.; Monayem, F.B.; Islam, M.S.; Islam, M.M. Ivermectin in combination with doxycycline for treating COVID-19 symptoms: A randomized trial. *J. Int. Med. Res.* **2021**, *49*, 03000605211013550. [CrossRef] [PubMed]
145. Seet, R.C.S.; Quek, A.M.L.; Ooi, D.S.Q.; Sengupta, S.; Lakshminarasappa, S.R.; Koo, C.Y.; So, J.B.Y.; Goh, B.C.; Loh, K.S.; Fisher, D.; et al. Positive impact of oral hydroxychloroquine and povidone-iodine throat spray for COVID-19 prophylaxis: An open-label randomized trial. *Int. J. Infect. Dis.* **2021**, *106*, 314–322. [CrossRef] [PubMed]
146. Li, X.; Wang, Y.; Agostinis, P.; Rabson, A.; Melino, G.; Carafoli, E.; Shi, Y.; Sun, E. Is hydroxychloroquine beneficial for COVID-19 patients? *Cell Death Dis.* **2020**, *11*, 512. [CrossRef]

viruses

MDPI

Article

Omicron Waves in Argentina: Dynamics of SARS-CoV-2 Lineages BA.1, BA.2 and the Emerging BA.2.12.1 and BA.4/BA.5

Carolina Torres [1,2,*], Mercedes Nabaes Jodar [2,3], Dolores Acuña [2,3], Romina Micaela Zambrana Montaño [1,2], Andrés Carlos Alberto Culasso [1,2], Ariel Fernando Amadio [2,4], Paula Aulicino [2,5], Santiago Ceballos [6], Marco Cacciabue [7], Humberto Debat [8], María José Dus Santos [9,10], María Florencia Eberhardt [2,4], Carlos Espul [11], Fabián Fay [12], María Ailén Fernández [13], Franco Fernández [8], Juan Manuel Fernandez Muñoz [14], Florencia Ferrini [15], Fernando Gallego [6], Adriana Angélica Giri [16], Agustina Cerri [16], Elisa Bolatti [16], María Ines Gismondi [7,17], Stephanie Goya [3], Iván Gramundi [6], José Matías Irazoqui [2,4], Guido Alberto König [7], Viviana Leiva [11], Horacio Lucero [18], Nathalie Marquez [8], Cristina Nardi [6], Belén Ortiz [11], Luis Pianciola [13], Carolina Beatriz Pintos [13], Andrea Fabiana Puebla [7], Carolina Victoria Rastellini [13], Alejandro Ezequiel Rojas [6], Javier Sfalcin [12], Ariel Suárez [19], Estefanía Tittarelli [19], Rosana Toro [20], Gabriela Vanina Villanova [2,21], María Cecilia Ziehm [13], María Carla Zimmermann [15], Sebastián Zunino [17,22], Proyecto PAIS Working Group [23,†], Laura Valinotto [2,3] and Mariana Viegas [2,3,*]

1. Instituto de Investigaciones en Bacteriología y Virología Molecular (IbaViM), Facultad de Farmacia y Bioquímica, Universidad de Buenos Aires, Ciudad Autónoma de Buenos Aires 1113, Argentina
2. Consejo Nacional de Investigaciones Científicas y Técnicas (CONICET), Ciudad Autónoma de Buenos Aires 1425, Argentina
3. Laboratorio de Virología Hospital de Niños Dr. Ricardo Gutiérrez, Ciudad Autónoma de Buenos Aires 1425, Argentina
4. Instituto de Investigación de la Cadena Láctea (IDICAL) INTA-CONICET, Rafaela 2300, Argentina
5. Laboratorio de Biología Celular y Retrovirus, Hospital de Pediatría "Prof. Juan P. Garrahan", Ciudad Autónoma de Buenos Aires 1245, Argentina
6. Hospital Regional Ushuaia; Universidad Nacional de Tierra del Fuego, Antártida e Islas del Atlántico Sur, UNTDF y CADIC-CONICET, Ushuaia 9410, Argentina
7. Instituto de Biotecnología/Instituto de Agrobiotecnología y Biología Molecular (INTA-CONICET), Hurlingham 1686, Argentina
8. Instituto de Patología Vegetal-Centro de Investigaciones Agropecuarias Instituto Nacional de Tecnología Agropecuaria (IPAVE-CIAP-INTA), Córdoba 5020, Argentina
9. Instituto de Virología e Innovaciones Tecnológicas (INTA-CONICET), Hurlingham 1686, Argentina
10. Laboratorio de Diagnóstico-UNIDAD COVID- Universidad Nacional de Hurlingham, Hurlingham 1688, Argentina
11. Laboratorio de Salud Pública, Godoy Cruz 5501, Argentina
12. CIBIC Laboratorio, Rosario 2000, Argentina
13. Laboratorio Central Neuquén, Ministerio de Salud de la Provincia del Neuquén, Neuquén 8302, Argentina
14. Instituto de Medicina y Biología Experimental de Cuyo (IMBECU—CONICET), Mendoza 5500, Argentina
15. Laboratorio de Medicina Genómica, Facultad de Medicina, Universidad Nacional del Nordeste, Corrientes 3400, Argentina
16. IBR-CONICET, Facultad de Ciencias Bioquímicas y Farmacéuticas, Universidad Nacional de Rosario, Rosario 2000, Argentina
17. Departamento de Ciencias Básicas, Universidad Nacional de Luján, Luján 6700, Argentina
18. Instituto de Medicina Regional, Resistencia 3500, Argentina
19. Departamento de Biología y Genética Molecular, IACA Laboratorios, Bahía Blanca 8000, Argentina
20. Laboratorio de salud pública, Facultad de Ciencias Exactas, UNLP, La Plata 1900, Argentina
21. Laboratorio de Biotecnología Acuática, Facultad de Ciencias Bioquímicas y Farmacéuticas, Universidad Nacional de Rosario, Centro Científico Tecnológico y Educativo Acuario del río Paraná, Rosario 2000, Argentina
22. Laboratorio de Virología Molecular, Hospital Blas L. Dubarry, Mercedes 6600, Argentina
23. PAIS Working Group, Ministerio de Ciencia Tecnología e Innovación, Ciudad Autónoma de Buenos Aires 1425, Argentina

* Correspondence: ctorres@ffyb.uba.ar (C.T.); viegasmariana@conicet.gov.ar (M.V.)

† PAIS Working Group at http://pais.qb.fcen.uba.ar/about.php.

Citation: Torres, C.; Nabaes Jodar, M.; Acuña, D.; Montaño, R.M.Z.; Culasso, A.C.A.; Amadio, A.F.; Aulicino, P.; Ceballos, S.; Cacciabue, M.; Debat, H.; et al. Omicron Waves in Argentina: Dynamics of SARS-CoV-2 Lineages BA.1, BA.2 and the Emerging BA.2.12.1 and BA.4/BA.5. *Viruses* **2023**, *15*, 312. https://doi.org/10.3390/v15020312

Academic Editors: Ahmed El-Shamy and Mohamed Ibrahim

Received: 30 December 2022
Revised: 14 January 2023
Accepted: 18 January 2023
Published: 22 January 2023

Abstract: The COVID-19 pandemic has lately been driven by Omicron. This work aimed to study the dynamics of SARS-CoV-2 Omicron lineages during the third and fourth waves of COVID-19 in Argentina. Molecular surveillance was performed on 3431 samples from Argentina, between EW44/2021 and EW31/2022. Sequencing, phylogenetic and phylodynamic analyses were performed. A differential dynamic between the Omicron waves was found. The third wave was associated with lineage BA.1, characterized by a high number of cases, very fast displacement of Delta, doubling times of 3.3 days and a low level of lineage diversity and clustering. In contrast, the fourth wave was longer but associated with a lower number of cases, initially caused by BA.2, and later by BA.4/BA.5, with doubling times of about 10 days. Several BA.2 and BA.4/BA.5 sublineages and introductions were detected, although very few clusters with a constrained geographical distribution were observed, suggesting limited transmission chains. The differential dynamic could be due to waning immunity and an increase in population gatherings in the BA.1 wave, and a boosted population (for vaccination or recent prior immunity for BA.1 infection) in the wave caused by BA2/BA.4/BA.5, which may have limited the establishment of the new lineages.

Keywords: SARS-CoV-2; Omicron; variants; evolution; South America; dynamics; BA.1; BA.2; BA.4; BA.5

1. Introduction

The COVID-19 pandemic has strongly impacted all populations worldwide and was lastly driven by the Omicron variant, causing new waves of infections in almost all regions of the world since the end of the year 2021. Molecular surveillance has been encouraged, especially since the emergence of SARS-CoV-2 variants, and has become an important tool to help prevent the COVID-19 burden when those results were used for public health purposes.

The diversity and evolution of SARS-CoV-2 are reflected by both variants and lineages. While variants of concern (VOCs), namely Alpha, Beta, Gamma, Delta and Omicron, have been defined by the World Health Organization to prioritize the monitoring of some groups of SARS-CoV-2 sequences [1], thousands of lineages have been defined under the Pango system to track the viral transmission and spread more in detail [2].

In particular, Omicron sequences have been classified into lineages BA.1 to BA.5, as well as several sublineages or derived lineages, which may present some biological or clinical differences. For instance, BA.2 cases showed lower or similar risks of death or hospital admission than BA.1 [3,4], and these lineages showed no differences in vaccine effectiveness or in the rate of immunity decline over time [5]. In addition, evasion of neutralizing antibodies was found for BA.2.12.1 and BA.4/BA.5, compared with BA.2, against vaccinated individuals or individuals with immunity elicited by BA.1 [6,7].

In contrast, South America suffered the first Omicron wave almost simultaneously with other regions in December 2021–March 2022, and since then, other Omicron waves have affected countries around the world with different impacts [8].

The number of SARS-CoV-2 genome sequences in databases increased during 2022, associated with the unprecedented number of Omicron infections and with the consolidation of massive sequencing capabilities worldwide. However, even though more than 14 million SARS-CoV-2 genomes have been uploaded to databases until the present, only about 2.5% belong to South American countries [9], and most of them have been analyzed only for lineage assignment.

Moreover, there is a lack of studies on the important evolutionary aspects of different Omicron lineages and their circulation in different geographical regions, which could help to understand the potential impact of its emerging lineages in different epidemiological contexts.

This work aimed to study the dynamics of SARS-CoV-2 lineages during the third and the fourth waves of COVID-19 in Argentina, driven by Omicron, and to analyze their evolutionary pattern and behavior in light of the local epidemiological scenario.

2. Materials and Methods

2.1. Samples

Molecular surveillance was performed on a total of 3431 samples from the capital city (City of Buenos Aires) and 14 provinces of the country, including the most populated districts, distributed as follows: City of Buenos Aires (n = 874), and provinces of Buenos Aires (n = 655), Chaco (n = 104), Corrientes (n = 103), Entre Ríos (n = 10), Jujuy (n = 26), La Pampa (n = 11), Mendoza (n = 95), Misiones (n = 76), Neuquén (n = 395), Río Negro (n = 1), Salta (n = 51), Santa Cruz (n = 3), Santa Fe (n = 918) and Tierra del Fuego (n = 109) (Figure 1).

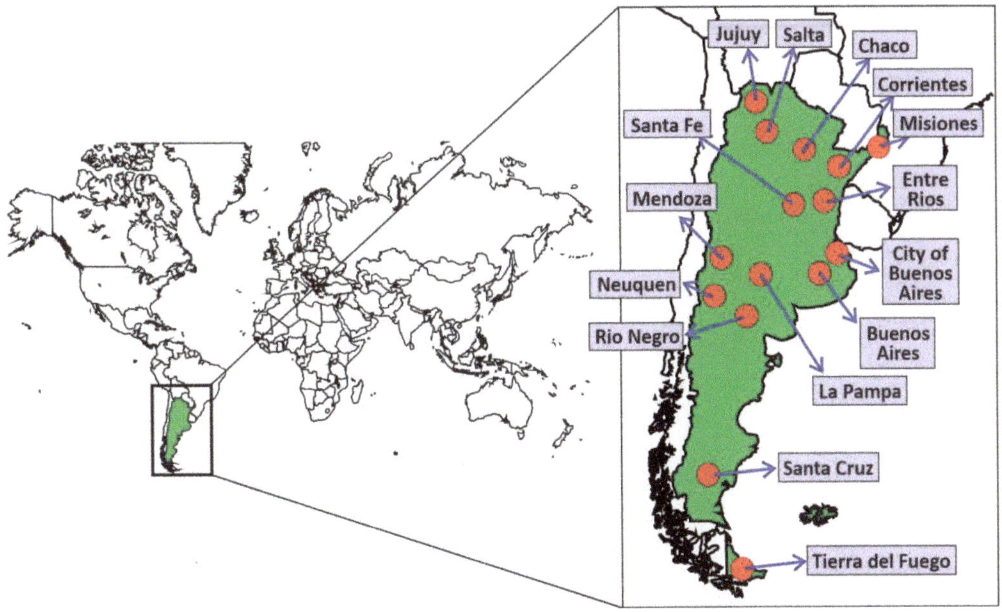

Figure 1. Sampling points for genomic surveillance of SARS-CoV-2 variants in Argentina.

Samples were collected between epidemiological weeks (EW) EW44/2021 and EW31/2022, covering the third and fourth waves of the COVID-19 pandemic in Argentina.

Surveillance was carried out in a fraction of 2.5–60% of the total positive cases weekly detected in different healthcare centers of the country, depending on the epidemiological situation at each moment and the sequencing capacity of the Proyecto PAIS sequencing nodes at each location [10]. Samples corresponded to randomly selected cases with no epidemiological link among them or with international travel.

2.2. Sequencing

The surveillance strategy was based on: i. Sanger sequencing of a 965 bp region of Spike spanning amino acids 428 to 750 (2074 samples), corresponding to the PCR29 fragment of the CDC Sanger sequencing protocol that uses primers CDC-29 Fw: W1_29F_22847: 5′-TTACAGGCTGCGTTATAGCTTGG-3′ and CDC-29 Rv: W1_29R_23812_5′-TGCTGCATTCAGTTGAATCACC-3′ [11], which allows for the identification of signature mutations associated with VOCs and VOIs, as previously described [12]; ii. Complete genome sequencing, for which the ARTIC protocol with the "midnight" primer set was used [13,14] with Oxford Nanopore or Illumina platforms (1357 samples). Nucleotide sequences generated for this study can be found in the GISAID database (https://www.gisaid.org/) under the GISAID Identifier: EPI_SET_230114um.

2.3. Statistical Analysis

The frequencies of variant detection and their 95% confidence intervals (CI95%) were estimated with the Wilson/Brown method, implemented in the Graph Pad Prism v.8.3 program (San Diego, CA, USA, www.graphpad.com).

2.4. Phylogenetic Analysis

Phylogenetic analysis was carried out for Omicron lineages BA.1, BA.2 and BA.4/BA.5 and their descendants to confirm the lineage assignment and to study their introduction and spread in Argentina.

Datasets included the Argentine sequences of each lineage and reference sequences for different SARS-CoV-2 lineages. For BA.2 and BA.4/BA.5, analyses also included the five most closely related sequences (with less than 10 SNPs), selected using the AudacityInstant application in the GISAID EpiCoV© database (https://www.gisaid.org) on 14 August 2022 [9].

Alignments were built using MAFFT v7.486 [15] and maximum likelihood trees were built using IQ-TREE v.2.1 [16], using the nucleotide evolutionary model according to the Bayesian Information Criterion estimated using ModelFinder [17]. The Shimodaira–Hasegawa-like approximate likelihood ratio test (SH-aLRT, 1000 replicates) [18] was used to evaluate the reliability of the groups. For BA.2 and BA4/BA5 phylogenies, the Ultrafast bootstrap Approximation (UFB, 1000 replicates) [14] was also used.

We gratefully acknowledge the authors from the originating laboratories responsible for obtaining the specimens and the submitting laboratories where genetic sequence data were generated and shared via the GISAID Initiative, on which part of this research is based (GISAID Identifiers: EPI_SET_221215va, EPI_SET_221215gu, EPI_SET_221215wf).

2.5. Phylodynamic Analysis

The doubling time, the exponential growth rate and the time to the most recent common ancestor (MRCA) were estimated from the genome data for lineages BA.1, BA.2 and BA.4/BA.5 using an exponential growth coalescent model and the uncorrelated lognormal molecular clock in the BEAST v.1.10.4 software package [19].

For these analyses, datasets included Argentine sequences only from the exponential period for each lineage (i.e., the beginning of the sustained detection in surveillance analyses until the peak of registered cases in each wave or the peak of lineage frequency (Table S1)). To reach the convergence of the analyses in affordable times, given the high number of sequences of lineage BA.1, a subsampling of the exponential period of the third wave was done (covering from EW50/2021 to EW02/2022), with a final dataset for lineage BA.1 that included 138 sequences (GISAID Identifier: EPI_SET_221215ph). The dataset for lineage BA.2 included sequences from EW11/2022 to EW20/2022 (n = 173 sequences) (GISAID Identifier: EPI_SET_221215am), whereas the dataset for lineages BA.4/BA.5 (that were analyzed together) included sequences from EW20/2022 to EW30/2022 (n = 102 sequences) (GISAID Identifier: EPI_SET_221229hr).

The temporal structure of the datasets was assessed through the Root-to-tip analysis with TempEst v1.5.3 [20], for which a positive correlation between genetic divergence and sampling time is expected in datasets suitable for a phylodynamic analysis with tip dating calibration. However, this exploratory analysis failed to confirm the temporal structure for datasets of lineages BA.1 and BA.4/BA.5. Thus, a mean rate of evolution of 1.2×10^{-3} substitutions/site/year (s/s/y) was used to calibrate all analyses, with a lognormal distribution on the mean rate prior, truncated between 0.92×10^{-3} and 1.49×10^{-3}, to cover the HPD95% intervals previously obtained for Omicron lineage BA.1 [21]. However, given that some temporal structure was observed for the dataset of lineage BA.2 (correlation coefficient ranged between 0.44–0.45 in the different functions of TempEst), another analysis using a tip-dating calibration was also carried out.

Analyses were run until convergence, assessed with an effective sample size higher than 200, and 10% of samples were discarded as burn-in. At least two runs were combined to summarize the posterior distribution of the parameters estimated.

2.6. Ethics Statement

The study was revised and approved by the Medical Ethics and Research Committees of "Ricardo Gutiérrez" Children's Hospital, Buenos Aires, Argentina (DI-2020-165-GCABA-HGNRG). Informed consent was not obtained because patient information was anonymized and de-identified before analysis.

3. Results

3.1. Molecular Surveillance of SARS-CoV-2 Variants in Argentina

The Omicron lineage BA.1 was first detected in local transmission cases in EW50/2021 (14 December 2021), reaching a frequency of 78.6% (CI95% = 72.6–83.6) two weeks later (EW52/2021), producing a third wave with a record number of cases in mid-January 2022 and fully displacing Delta by the end of January (EW04/2022) (Figure 2 and Table S1).

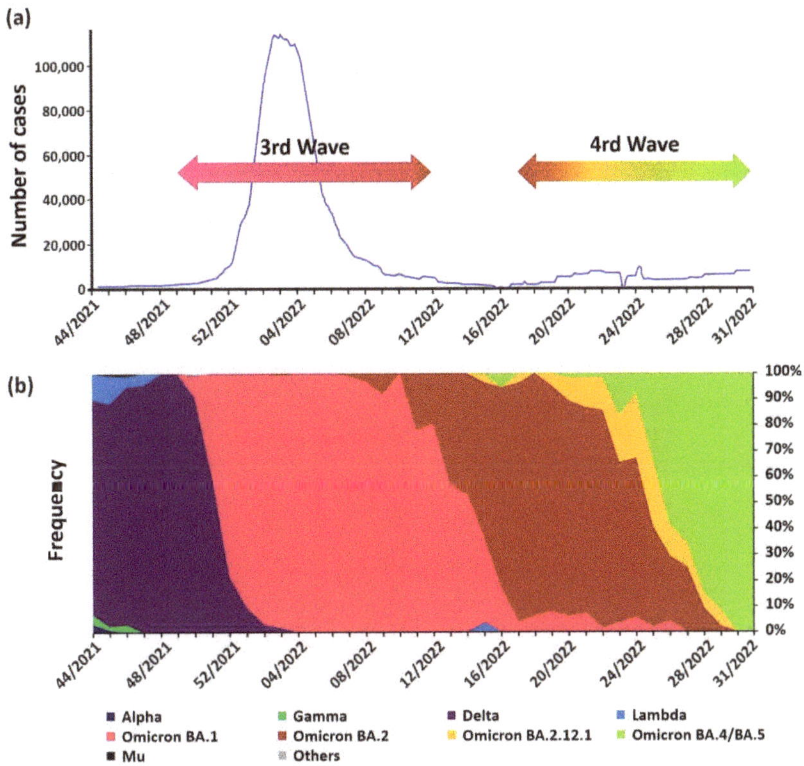

Figure 2. (**a**) Number of cases reported in Argentina from EW44/2021 to EW31/2022. Note that in EW15/2022 (between the third and the fourth waves), Argentina modified the COVID-19 massive testing criterion and since then, testing in public hospitals is only performed in populations under a higher risk for severe disease (>50 years or comorbidities). Data from https://ourworldindata.org/coronavirus#explore-the-global-situation, accessed on 1 December 2023. The arrows are colored according to the variant present at each moment of the wave. (**b**) Frequency of variant detection analyzed by epidemiological week, 2021–2022. Data from Spike and complete genome sequencing of samples from cases that did not present epidemiological link with travel (n = 3431).

Except for two sporadic cases in EW03/2022, the Omicron lineage BA.2 was mainly detected since EW07/2022 (13 to 19 February) and became predominant in EW15/2022 (10 to 16 April) (60.7%, CI95% = 42.4–76.4), when cases of BA.2.12.1 were also detected. Lineage BA.2 and its sublineages displaced BA.1 more slowly than when BA.1 displaced Delta, which happened in only three weeks, from EW50/2021 to EW52/2022 (Figure 2 and Table S1).

The lineages BA.4/BA.5 have been detected since EW16/2022 (17 to 23 April), boosting the fourth wave together with BA.2 and its sublineages, becoming predominant by the end of June (EW26/2022) with a frequency of 58.3% (CI95% = 38.8–75.5) and reaching 100% of the new cases by the end of July (EW30/2022) (Figure 2 and Table S1).

3.2. Evolutionary Analyses of Omicron BA.1 and its Sublineages

Lineage BA.1 and its sublineages were associated with the third wave of infections in Argentina, mainly driven by BA.1/BA.1.1 (91%); phylogenetic analysis did not separate BA.1.1 from BA.1, followed by BA.1.15 (5.9%) and other sublineages in minor proportions (Figure 3).

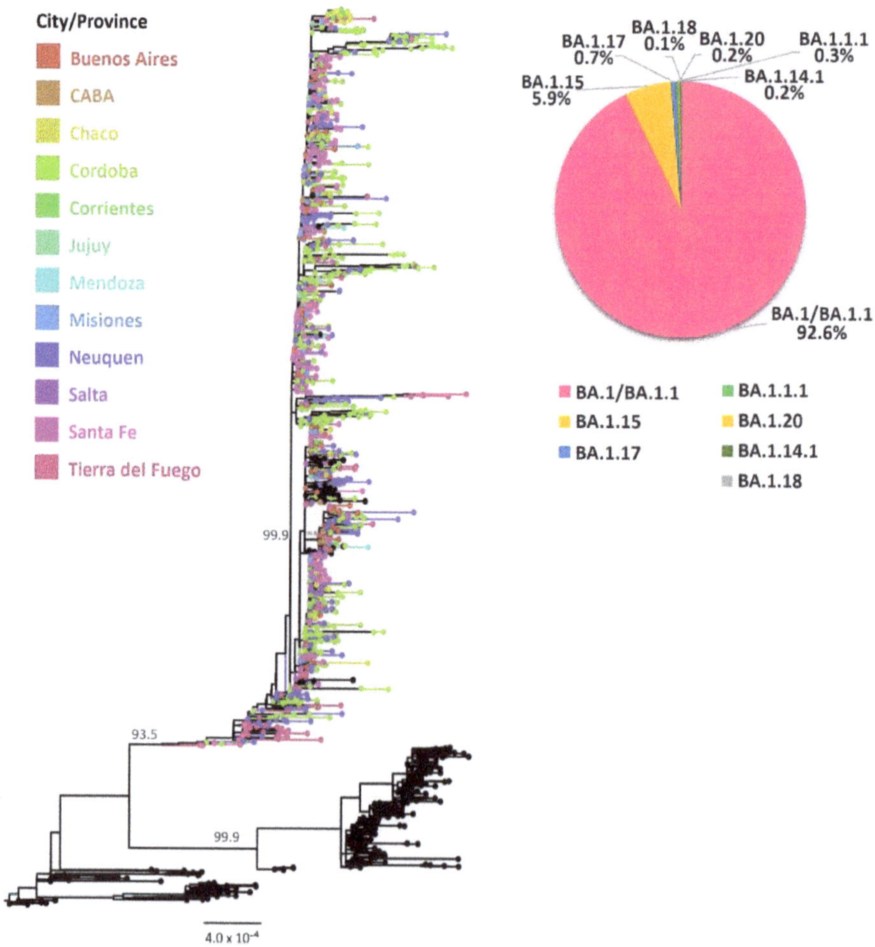

Figure 3. Phylogenetic tree of SARS-CoV-2 sequences focused on Omicron BA.1 and its sublineages. SH-aLRT support values are indicated at nodes for some groups. The branches and tips of the tree are colored according to the City or Provinces indicated in the legend. Black-colored tips represent sequences from other countries or sequences included as references for other lineages.

Phylodynamic analysis of the exponential phase of the BA.1 wave showed a doubling time of infections estimated in 3.3 days (HPD95% = 1.6–5.4), with an MRCA dated 12 November 2021 (HPD95% = 25 October–25 November) (Table 1).

Table 1. Population dynamic parameters estimated from Bayesian coalescent analyses [1].

Lineage	Rate of Evolution [s/s/y] (HPD95%)	Ancestral Date (HPD95%)	Doubling Time [days] (HPD95%)	Exponential Growth Rate [days^{-1}] (HPD95%)
BA.1	9.3×10^{-4} * (6.9×10^{-4}–1.2×10^{-3})	12 November 2021 (25 October–25 November)	3.3 (1.6–5.4)	0.213 (0.127–0.441)
BA.2	1.0×10^{-3} * (8.7×10^{-4}–1.1×10^{-3})	3 February 2022 (11 January–22 February)	9.9 (7.3–13.2)	0.070 (0.053–0.096)
	5.5×10^{-4} (3.1×10^{-4}–7.7×10^{-4})	24 November 2021 (23 August –26 January)	16.5 (9.5–26.4)	0.042 (0.026–0.073)
BA.4/BA.5	8.5×10^{-4} * (6.4×10^{-4}–1.0×10^{-3})	25 March 2022 (15 February–24 April)	10.8 (6.1–17.3)	0.064 (0.040–0.114)

[1] Median values are informed. * These analyses were performed by calibrating with a mean rate of evolution and the sampling times for datasets considering only the exponential phase of infections of each lineage (see Section 2.5).

3.3. *Evolutionary Analyses of Omicron BA.2 and its Sublineages*

Lineage BA.2 and its sublineages were associated with the early fourth wave of infections, mainly associated with lineages BA.2 (49.8%), BA.2.12.1 (12.4%), BA.2.3 (17.0%), BA.2.9 (10.4%), BA.2.72 (2.7%) and other sublineages in minor proportions (Figure 4). Notably, only one major monophyletic group from Argentinean sequences was observed including sequences from three provinces, whereas most of the other Argentine sequences were related to foreign sequences from several countries (Figure 4).

Phylodynamic analysis of the exponential phase of BA.2 circulation showed a doubling time of infections estimated in 9.9 days (HPD95% = 7.3–13.2) with an MRCA dated 3 February 2022 (HPD95% = 11 January–22 February) (Table 1).

In addition, when tip-dating calibration was used, the rate of evolution for the lineage BA.2 was estimated as 5.5×10^{-4} s/s/y (HPD95% = 3.1×10^{-4}–7.7×10^{-4}), with doubling times estimated in 16.5 days (HPD95%= 9.5–26.4), and the MRCA dated 24 November 2021 (HPD95% = 23 August–26 January) (Table 1).

3.4. *Evolutionary Analyses of Omicron BA.4/BA.5 and its Sublineages*

Lineages BA.4 and BA.5 were analyzed together for monitoring purposes and for phylodynamic estimations owing to their simultaneous introduction into the country, the inability of identifying them separately from the Spike sequencing strategy and their close phylogenetic relationship.

These lineages boosted the late fourth wave of infections into the country, mainly associated with lineages and sublineages BA.4 (18.4%) and BA.4.1 (13.6%), and with BA.5.1 (23.3%), BA.5.2 (4.9%), BA.5.2.1 (25.2%), BE.1 (2.9%) and other sublineages in minor proportions (Figure 5 and Table S1).

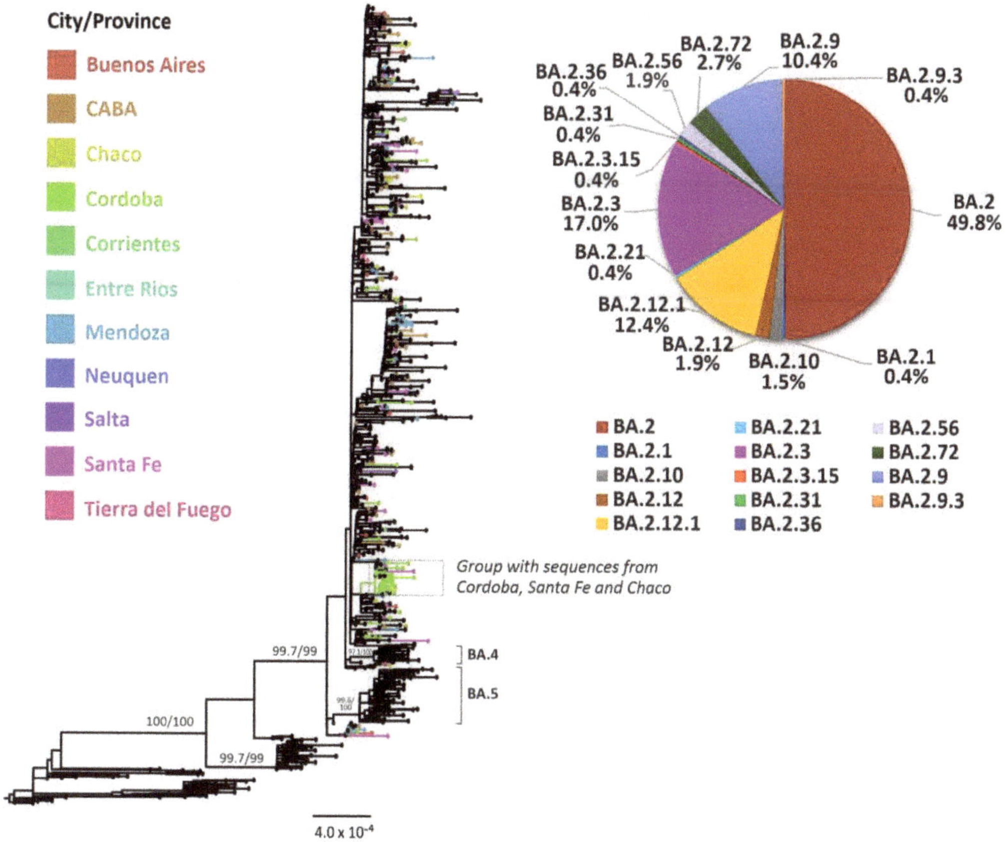

Figure 4. Phylogenetic tree of SARS-CoV-2 sequences, focused on Omicron BA.2 and its sublineages. SH-aLRT and UFB support values are indicated at nodes for some groups. The branches and tips of the tree are colored according to the City or Provinces indicated in the legend. Black-colored tips represent sequences from other countries or sequences included as references for other lineages.

Similar to what was observed for lineage BA.2, Argentine sequences from several sublineages were intermingled with sequences from numerous countries and only a few local clusters were observed, showing a circumscribed geographical distribution, limited to one or two provinces (Figure 5).

The phylodynamic analysis of the exponential phase of infections caused by BA.4/BA.5 showed a doubling time estimated in 10.8 days (HPD95%= 6.1–17.3), with the MRCA dated 25 March 2022 (HPD95%= 15 February–24 April) (Table 1).

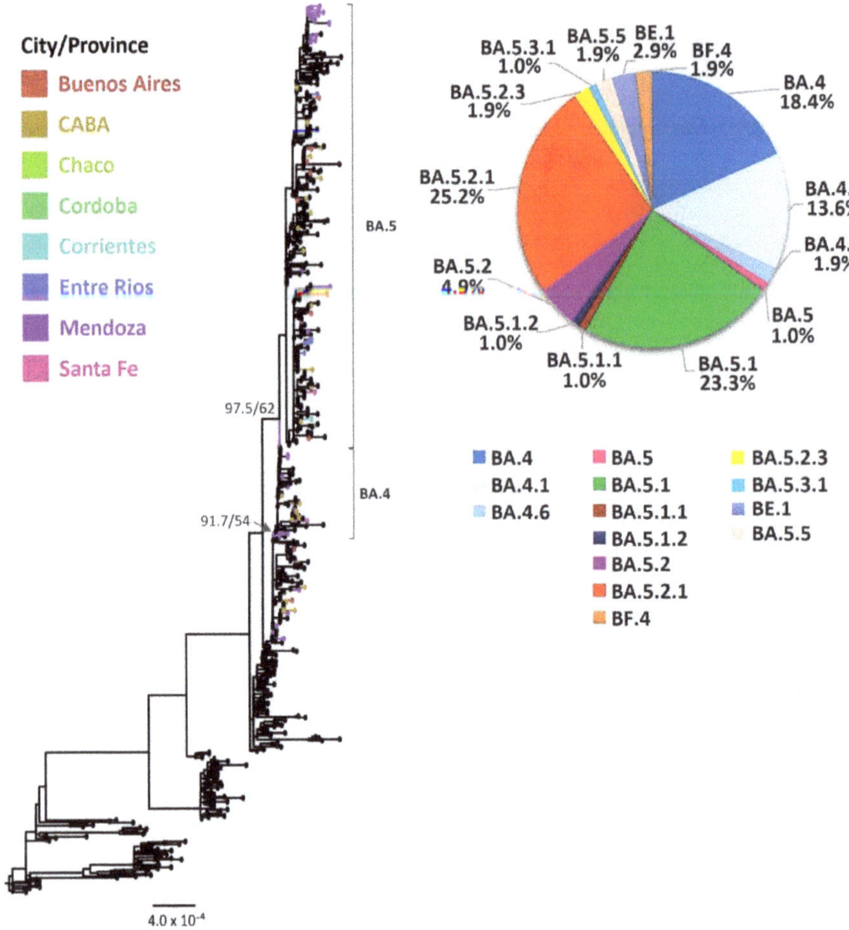

Figure 5. Phylogenetic tree of SARS-CoV-2 sequences, focused on Omicron BA.4/BA.5 and their sub-lineages. SH-aLRT and UFB support values are indicated at nodes for some groups. The branches and tips of the tree are colored according to the City or Provinces indicated in the legend. Black-colored tips represent sequences from other countries or sequences included as references for other lineages.

4. Discussion

The surveillance strategy implemented in Argentina allowed us to describe the introduction and establishment of the main SARS-CoV-2 Omicron lineages in the third and fourth waves of the COVID-19 pandemic in the country and analyze their different evolutionary dynamics.

Omicron waves worldwide were characterized by a very rapid expansion of several SARS-CoV-2 lineages that circulated almost without limits due to the absence of restrictions on international movement in the context of a massive application of vaccines, contrary to what happened during the dissemination of the previous circulating variants.

Before the emergence of Omicron, Delta was the dominant variant worldwide, whose introduction and establishment in Argentina were delayed in part due to an intense policy of international border controls implemented to reduce their impact while strengthening vaccination coverage in the population. For this and possibly other reasons, as a recent prior second wave, Delta was not associated with a COVID-19 wave in Argentina or other South American countries.

However, in December 2021–March 2022 (summer in the Southern hemisphere), Omicron BA.1 entered and caused the third wave of SARS-CoV-2 infections in Argentina. This wave was characterized by a very fast displacement of the Delta variant, showing a low level of clustering, with very similar viruses simultaneously infecting people worldwide. This was reflected in a relatively low number of lineages detected and in the intermingling of sequences from different Argentine provinces in the entire phylogeny. In other regions, the BA.1 epidemic wave has been traced back to a relatively small number of introductions with a later rapid expansion across the country, as was found in England [22].

By mid-April 2022, the COVID-19 testing criterion was modified, and tests were recommended and only enabled in public hospitals (at no cost to the patient) for prioritized groups, i.e., population with a higher risk for severe disease (>50 years or with comorbidities), which caused the registered number of cases to be incomparable to previous periods when testing was recommended and enabled in the public system for any symptomatic person. Considering that sequencing was mostly performed in samples from the public system, variant circulation proportions estimated before and after this change could have been affected by this distinct sampling, mainly oriented towards the prioritized population in the last period. As far as we know, there are no reports about a differential circulation of Omicron sublineages by age; however, due to the lack of age-stratified data, we were not able to perform a formal analysis on the influence of this factor.

Despite the change in the criterion and accessibility for COVID-19 testing, and consequently, in the case notification, the fourth wave was evidenced by an increase in the number of reported cases—although it was much lower than that observed in the third wave—and was initially associated with BA.2 and its sublineages.

Circulation of Omicron BA.2 in Argentina was characterized by the detection of several sublineages between EW03/2022 and EW29/2022 and very few clusters of Argentine sequences with a constrained distribution (one or few provinces), suggesting several introductions but limited transmission chains into the country.

Similarly, the sustained detection of Omicron BA.4/BA.5 was registered since EW20/2022, although, according to the results of phylogenetic analyses, these infections were mainly associated with several independent introductions and short transmission chains in the period analyzed in this work.

The evolutionary pattern of Omicron lineages detected in the third and fourth waves in Argentina reflects the distribution described in other countries from South America, the United States and Western Europe, as was observed in the pre-Omicron period [12]. However, this pattern differed from that of the first and second waves, when there was a high level of clustering and geographical structure within Argentina [12,23].

The introduction and establishment of SARS-CoV-2 lineages follow a complex pattern, balancing the intrinsic replication, immune evasion and transmission ability of viral variants, the waning immunity, and the policy of testing, reporting and case management of countries. Therefore, the epidemiological situation of countries could be determined by any of these factors, which could play a role in the variant dynamics.

In this work, the doubling time in the exponential period of the Omicron BA.1 wave in Argentina was estimated at 3.3 days, similar to what was observed in other regions [21,24], and was much faster than the estimations for BA.2 and BA.4/BA.5 in the country, which were estimated in about 10 days. In addition, ancestral times were also estimated from a subset of sequences that belongs to the period of the exponential growth of each lineage into the country. These ancestors were dated five (BA.1), six (BA.2) and eight (BA.4/BA.5) epidemiological weeks before their detection in local cases. These estimates along with the finding of multiple introductions suggest diversification outside Argentina and later entry into the country. For instance, for lineage BA.1, the ancestor for the Argentine sequences was dated to mid-November 2021, around one month after the time of the most recent common ancestor estimated from southern Africa sequences (mid-October 2021) [21].

The differences evidenced regarding the dynamics of Omicron waves could be due to several factors. The introduction of Omicron BA.1 into the country, displacing Delta,

occurred in a context of unrestricted international travel and during the beginning of the presential meeting and social gatherings, which were discouraged a few months before. Omicron BA.2 (and later, BA.4/BA.5) also entered the country without restrictions, but occurred when it was still suffering the BA.1 wave (or the BA.2 wave); therefore, the populational immunity elicited post-wave may have limited the establishment of the new lineages.

On the other hand, the vaccination campaign against COVID-19 in Argentina started in January 2021 with the Sputnik V (Gam-COVID-Vac), ChAdOx1 nCoV-19 (Oxford-AstraZeneca), and BBIBP-CorV (Sinopharm) vaccines, initially for the prioritized population [25]. By the end of October 2021, 77.1% of the Argentine population received at least one vaccine dose and 58.4% completed the vaccination protocol [26]; however, for many people it had already been more than five months since the last dose, and the booster application of the vaccine only began in mid-November 2021. Moreover, in 2021, Argentina suffered the second wave caused by Gamma and Lambda variants [12], with the maximum number of infections in May 2021. Therefore, the Omicron lineage BA.1 entered a population with waned immunity (for prior infection or vaccination), whereas lineages BA.2 and BA.4/BA.5 enter a boosted population.

Different countries showed a distinct relative speed of replacement of BA.1 by BA.2, and of BA.2 by BA.4/BA.5 [27]. These differences must be analyzed considering each epidemiological situation and integrating several sources of information (for instance, vaccination, time to the previous waves, testing and contact tracing policies, molecular surveillance capacities, social acceptance of non-pharmaceutical interventions, etc.). Since the combination of these factors is almost unique for each country, the need for a local or regional analysis is mandatory for a better understanding of the driving factors.

Finally, some new sublineages derived from BA.2, BA.4 and BA.5 have recently emerged, adding immune evasion properties to those already reported in both vaccinated and priorly infected individuals [28,29]. Some of these emerging lineages have already become the new players in the scenario of SARS-CoV-2 circulation worldwide, whose impact will have to be analyzed in the near future.

Supplementary Materials: The following supporting information can be downloaded at: https://www.mdpi.com/article/10.3390/v15020312/s1, Table S1: Frequency of SARS-CoV-2 variants in Argentina.

Author Contributions: Conceptualization, C.T. and M.V.; methodology, C.T., M.N.J., D.A., R.M.Z.M., A.C.A.C., A.F.A., P.A., S.C., M.C., H.D., M.J.D.S., M.F.E., C.E., F.F. (Fabián Fay), M.A.F., F.F. (Franco Fernández), J.M.F.M., F.F. (Florencia Ferrini), F.G., A.A.G., A.C., E.B., M.I.G., S.G., I.G., J.M.I., G.A.K., V.L., H.L., N.M., C.N., B.O., L.P., C.B.P., A.F.P., C.V.R., A.E.R., J.S., A.S., E.T., R.T., G.V.V., M.C.Z. (María Cecilia Ziehm), M.C.Z. (María Carla Zimmermann), S.Z., L.V. and PAIS working Group; formal analysis, C.T.; investigation, C.T., M.V.; resources, M.V.; writing—original draft preparation, C.T.; writing—review and editing, C.T. and M.V.; visualization, C.T.; supervision, C.T. and M.V.; project administration, M.V.; funding acquisition, M.V. All authors have read and agreed to the published version of the manuscript.

Funding: The sequencing strategy was supported by Proyecto IP COVID-19 N°08 (Ministerio de Ciencia, Tecnología e Innovación, Argentina) and Focem COF 03/11 Covid-19 (Fondo para la Convergencia Estructural del MERCOSUR).

Institutional Review Board Statement: The study was conducted following the Declaration of Helsinki and approved by the Medical Ethics and Research Committees of "Ricardo Gutiérrez" Children's Hospital, Buenos Aires, Argentina (DI-2020-165-GCABA-HGNRG).

Informed Consent Statement: Informed consent was waived due to patient information being anonymized and de-identified before analysis.

Data Availability Statement: Nucleotide sequences generated for this study can be found in the GISAID database (https://www.gisaid.org/), under the GISAID Identifier: EPI_SET_230114um.

Conflicts of Interest: The authors declare no conflict of interest. The funders had no role in the design of the study; in the collection, analyses, or interpretation of data; in the writing of the manuscript; or in the decision to publish the results.

References

1. World Health Organization Tracking SARS-CoV-2 Variants. Available online: https://www.who.int/en/activities/tracking-SARS-CoV-2-variants/ (accessed on 1 December 2022).
2. Rambaut, A.; Holmes, E.C.; O'Toole, Á.; Hill, V.; McCrone, J.T.; Ruis, C.; du Plessis, L.; Pybus, O.G. A Dynamic Nomenclature Proposal for SARS-CoV-2 Lineages to Assist Genomic Epidemiology. *Nat. Microbiol.* **2020**, *5*, 1403–1407. [CrossRef] [PubMed]
3. Webster, H.H.; Nyberg, T.; Sinnathamby, M.A.; Aziz, N.A.; Ferguson, N.; Seghezzo, G.; Blomquist, P.B.; Bridgen, J.; Chand, M.; Groves, N.; et al. Hospitalisation and Mortality Risk of SARS-COV-2 Variant Omicron Sub-Lineage BA.2 Compared to BA.1 in England. *Nat. Commun.* **2022**, *13*, 6053. [CrossRef] [PubMed]
4. Strasser, Z.H.; Greifer, N.; Hadavand, A.; Murphy, S.N.; Estiri, H. Estimates of SARS-CoV-2 Omicron BA.2 Subvariant Severity in New England. *JAMA Netw. Open* **2022**, *5*, e2238354. [CrossRef] [PubMed]
5. Kirsebom, F.C.M.; Andrews, N.; Stowe, J.; Toffa, S.; Sachdeva, R.; Gallagher, E.; Groves, N.; O'Connell, A.M.; Chand, M.; Ramsay, M.; et al. COVID-19 Vaccine Effectiveness against the Omicron (BA.2) Variant in England. *Lancet Infect. Dis.* **2022**, *22*, 931–933. [CrossRef] [PubMed]
6. Khan, K.; Karim, F.; Ganga, Y.; Bernstein, M.; Jule, Z.; Reedoy, K.; Cele, S.; Lustig, G.; Amoako, D.; Wolter, N.; et al. Omicron BA.4/BA.5 Escape Neutralizing Immunity Elicited by BA.1 Infection. *Nat. Commun.* **2022**, *13*, 4686. [CrossRef]
7. Cao, Y.; Yisimayi, A.; Jian, F.; Song, W.; Xiao, T.; Wang, L.; Du, S.; Wang, J.; Li, Q.; Chen, X.; et al. BA.2.12.1, BA.4 and BA.5 Escape Antibodies Elicited by Omicron Infection. *Nature* **2022**, *608*, 593–602. [CrossRef]
8. World Health Organization. COVID-19 Weekly Epidemiological Update. Edition 125 Published 11 January 2023. Available online: https://www.who.int/publications/m/item/weekly-epidemiological-update-on-covid-19---11-january-2023 (accessed on 1 December 2022).
9. Khare, S.; Gurry, C.; Freitas, L.; Schultz, M.B.; Bach, G.; Diallo, A.; Akite, N.; Ho, J.; Lee, R.T.C.; Yeo, W.; et al. GISAID's Role in Pandemic Response. *China CDC Wkly.* **2021**, *3*, 1049–1051. [CrossRef]
10. Proyecto PAIS. Proyecto Argentino Interinstitucional de Genómica de SARS-CoV-2. 2022. Available online: http://pais.qb.fcen.uba.ar/about.php (accessed on 1 December 2022).
11. Paden, C.; Tao, Y.; Queen, K.; Zhang, J.; Li, Y.; Uehara, A.; Tong, S. Rapid, Sensitive, Full-Genome Sequencing of Severe Acute Respiratory Syndrome Coronavirus 2. *Emerg. Infect. Dis. J.* **2020**, *26*, 2401. [CrossRef]
12. Torres, C.; Mojsiejczuk, L.; Acuña, D.; Alexay, S.; Amadio, A.; Aulicino, P.; Debat, H.; Fay, F.; Fernández, F.; Giri, A.A.; et al. Cost-Effective Method to Perform SARS-CoV-2 Variant Surveillance: Detection of Alpha, Gamma, Lambda, Delta, Epsilon, and Zeta in Argentina. *Front. Med.* **2021**, *8*, 755463. [CrossRef]
13. Quick, J. NCoV-2019 Sequencing Protocol v3 (LoCost) V.3. Available online: https://www.protocols.io/view/ncov-2019-sequencing-protocol-v3-locost-bh42j8ye (accessed on 1 December 2022).
14. Freed, N.; Silander, O. SARS-CoV2 Genome Sequencing Protocol (1200 bp Amplicon "Midnight" Primer Set, Using Nanopore Rapid Kit) V.6. Available online: https://www.protocols.io/view/sars-cov2-genome-sequencing-protocol-1200bp-amplic-rm7vz8q64vx1/v6 (accessed on 1 December 2022).
15. Katoh, K.; Standley, D.M. MAFFT Multiple Sequence Alignment Software Version 7: Improvements in Performance and Usability. *Mol. Biol. Evol.* **2013**, *30*, 772–780. [CrossRef]
16. Minh, B.Q.; Schmidt, H.A.; Chernomor, O.; Schrempf, D.; Woodhams, M.D.; von Haeseler, A.; Lanfear, R. IQ-TREE 2: New Models and Efficient Methods for Phylogenetic Inference in the Genomic Era. *Mol. Biol. Evol.* **2020**, *37*, 1530–1534. [CrossRef] [PubMed]
17. Kalyaanamoorthy, S.; Minh, B.Q.; Wong, T.K.F.; Von Haeseler, A.; Jermiin, L.S. ModelFinder: Fast Model Selection for Accurate Phylogenetic Estimates. *Nat. Methods* **2017**, *14*, 587–589. [CrossRef] [PubMed]
18. Guindon, S.; Dufayard, J.F.; Lefort, V.; Anisimova, M.; Hordijk, W.; Gascuel, O. New Algorithms and Methods to Estimate Maximum-Likelihood Phylogenies: Assessing the Performance of PhyML 3.0. *Syst. Biol.* **2010**, *59*, 307–321. [CrossRef]
19. Drummond, A.J.; Suchard, M.A.; Xie, D.; Rambaut, A. Bayesian Phylogenetics with BEAUti and the BEAST 1.7. *Mol. Biol. Evol.* **2012**, *29*, 1969–1973. [CrossRef] [PubMed]
20. Rambaut, A.; Lam, T.T.; Carvalho, L.M.; Pybus, O.G. Exploring the Temporal Structure of Heterochronous Sequences Using TempEst (Formerly Path-O-Gen). *Virus Evol.* **2016**, *2*, vew007. [CrossRef] [PubMed]
21. Viana, R.; Moyo, S.; Amoako, D.G.; Tegally, H.; Scheepers, C.; Althaus, C.L.; Anyaneji, U.J.; Bester, P.A.; Boni, M.F.; Chand, M.; et al. Rapid Epidemic Expansion of the SARS-CoV-2 Omicron Variant in Southern Africa. *Nature* **2022**, *603*, 679–686. [CrossRef] [PubMed]
22. Tsui, J.L.-H.; Lambert, B.; Bajaj, S.; McCrone, J.T.; Inward, R.P.D.; Bosetti, P.; Hill, V.; Pena, R.E.; Zarebski, A.E.; Peacock, T.P.; et al. Genomic Assessment of Invasion Dynamics of SARS-CoV-2 Omicron BA.1. *medRxiv* **2023**. [CrossRef]
23. Zambrana Montaño, R.; Culasso, A.C.A.; Fernández, F.; Marquez, N.; Debat, H.; Salmerón, M.; Zamora, A.M.; Ruíz de Huidobro, G.; Costas, D.; Alabarse, G.; et al. Evolution of SARS-CoV-2 during the First Year of the COVID-19 Pandemic in Northwestern Argentina. *Virus Res.* **2023**, *323*, 198936. [CrossRef] [PubMed]

24. Grabowski, F.; Kochańczyk, M.; Lipniacki, T. The Spread of SARS-CoV-2 Variant Omicron with a Doubling Time of 2.0–3.3 Days Can Be Explained by Immune Evasion. *Viruses* **2022**, *14*, 294. [CrossRef]
25. Rearte, A.; Castelli, J.M.; Rearte, R.; Fuentes, N.; Pennini, V.; Pesce, M.; Barbeira, P.B.; Iummato, L.E.; Laurora, M.; Bartolomeu, M.L.; et al. Effectiveness of RAd26-RAd5, ChAdOx1 NCoV-19, and BBIBP-CorV Vaccines for Risk of Infection with SARS-CoV-2 and Death Due to COVID-19 in People Older than 60 Years in Argentina: A Test-Negative, Case-Control, and Retrospective Longitudinal Study. *Lancet* **2022**, *399*, 1254–1264. [CrossRef]
26. Our World in Data. Vaccination in Argentina. Available online: https://ourworldindata.org/covid-vaccinations (accessed on 1 December 2022).
27. UK Health Security Agency. *SARS-CoV-2 Variants of Concern and Variants under Investigation in England: Technical Briefing 42*; UK Health Security Agency: London, UK, 2022.
28. Akerman, A.; Milogiannakis, V.; Jean, T.; Esneu, C.; Silva, M.R.; Ison, T.; Fitcher, C.; Lopez, J.A.; Chandra, D.; Naing, Z.; et al. Emergence and Antibody Evasion of BQ and BA.2.75 SARS-CoV-2 Sublineages in the Face of Maturing Antibody Breadth at the Population Level. *medRxiv* **2022**. [CrossRef]
29. Uraki, R.; Ito, M.; Furusawa, Y.; Yamayoshi, S.; Iwatsuki-Horimoto, K.; Adachi, E.; Saito, M.; Koga, M.; Tsutsumi, T.; Yamamoto, S.; et al. Humoral Immune Evasion of the Omicron Subvariants BQ.1.1 and XBB. *Lancet. Infect. Dis.* **2022**, *23*, 2022–2024. [CrossRef] [PubMed]

viruses

MDPI

Article

The Outcome of BNT162b2, ChAdOx1-Sand mRNA-1273 Vaccines and Two Boosters: A Prospective Longitudinal Real-World Study

Sanaa M. Kamal *, Moheyeldeen Mohamed Naghib, Moataz Daadour, Mansour N. Alsuliman, Ziad G. Alanazi, Abdulaziz Abdullah Basalem, Abdulaziz M. Alaskar and Khaled Saed

Department of Internal Medicine, Prince Sattam Bin Abdulaziz University College of Medicine, Al-Kharj 16278, Saudi Arabia

* Correspondence: s.kamal@psau.edu.sa

Abstract: To date, the effectiveness of COVID-19 vaccines and booster doses has yet to be evaluated in longitudinal head-to-head studies. This single-center longitudinal study assessed the effectiveness of ChAdOx1 nCoV-19, BNT162b2, and mRNA-1273 vaccines and assessed two BNT162b2 boosters in 1550 participants, of whom 26% had comorbidities. In addition, the SARS-CoV-2 antibody dynamics was monitored. A group of 1500 unvaccinated subjects was included as the controls. The study's endpoint was the development of virologically-proven COVID-19 cases after vaccine completion, while the secondary endpoint was hospitalizations due to severe COVID-19. Overall, 23 (4.6%), 16 (3%), and 18 (3.8%) participants vaccinated with ChAdOx1 nCoV-19, BNT162b2, and mRNA-1273, respectively, developed COVID-19 after vaccine completion, with an effectiveness of 89%, 92%, and 90%. Ten COVID-19 cases were reported in participants with comorbidities, three of whom were hospitalized. No hospitalizations occurred after boosters. SARS-CoV-2 antibody levels peaked 2–4 weeks after the second vaccine dose but declined after a mean of 28.50 ± 3.48 weeks. Booster doses significantly enhanced antibody responses. Antibody titers ≤ 154 U/mL were associated with a higher risk of COVID-19 emergence. Thus, COVID-19 vaccines effectively reduced COVID-19 and prevented severe disease. The vaccine-induced SARS-CoV-2 antibody responses declined after 28–32 weeks. Booster doses induced significant maintained responses. SARS-CoV-2 antibody levels may help determine the timing and need for vaccine booster doses.

Keywords: COVID-19; vaccines; booster vaccines doses; effectiveness; COVID-19-related disease; SARS CoV-2 antibody

Citation: Kamal, S.M.; Naghib, M.M.; Daadour, M.; Alsuliman, M.N.; Alanazi, Z.G.; Basalem, A.A.; Alaskar, A.M.; Saed, K. The Outcome of BNT162b2, ChAdOx1-Sand mRNA-1273 Vaccines and Two Boosters: A Prospective Longitudinal Real-World Study. *Viruses* **2023**, *15*, 326. https://doi.org/10.3390/v15020326

Academic Editor: Ahmed El-Shamy

Received: 6 December 2022
Revised: 6 January 2023
Accepted: 10 January 2023
Published: 24 January 2023

1. Introduction

COVID-19, the infection caused by SARS-CoV-2, which belongs to the *Coronaviridae* family, has resulted in devastating global public health and economic crises [1]. As of 15 November 2022, over 636 million confirmed cases and over 6.6 million deaths have been reported globally [2]. The high global transmissibility and *pandemicity* of SARS-CoV-2 are multifactorial. The virus transmission occurs along a spectrum that includes droplet transmission, contact with virus-contaminated surfaces, or possible airborne spread [3–5]. Besides symptomatic patients, asymptomatic, pauci-symptomatic, or pre-symptomatic people can also spread infection [5,6]. As SARS-CoV-2 evolves, new variants of concern (VOCs) emerge, such as 1.351 (Beta), B.1.617.2 (Delta), and B.1.1.529 (Omicron). There have been concerns that such variants could increase COVID-19 by escaping from immunity generated through previous infection or vaccination or by inducing more severe disease [7–10].

The genome of SARS-CoV-2 consists of 30,000 nucleotides which encode four structural proteins. The nucleocapsid (N) protein is responsible for packaging the RNA genome, and the membrane (M) protein is accountable for shaping the virions. In addition, the spike (S) and envelope (E) proteins are critical for virions assembly and release, as well as several

non-structural proteins (NSP) [11–14]. SARS-CoV-2 can evade innate recognition, signaling, IFN induction, and IFN-stimulated genes (ISGs) through the viral proteins that block these pathways [15,16]. Infection by SARS-CoV-2 evokes humoral and cellular immune responses. $CD4^+$ and $CD8^+$ T cells elicited by SARS-CoV-2 infection target various antigens, including structural and non-structural proteins, and are significantly associated with milder disease [17]. Most infected individuals with mild-to-moderate COVID-19 develop neutralizing antibody responses against the viral spike protein, which persist for several months after infection [18,19].

To date, no antivirals directed against SARS-CoV-2 have demonstrated promising efficacy in treating COVID-19. Antiviral therapy started at the early stages of COVID-19 mainly aims to avoid severe infection complications in high-risk patients [20,21]. Thus, mass COVID-19 vaccination campaigns are critical in controlling and slowing the pandemic. At the time of writing the research, more than 100 vaccines have been developed, and 26 vaccines have been evaluated in phase III clinical trials, according to the World Health Organization (WHO) [22]. Within less than 12 months after the beginning of the pandemic, several research teams rose to the challenge and developed vaccines that protect from SARS-CoV-2. The vaccines evoke immune responses, ideally neutralizing antibodies (NAbs) against the SARS-CoV-2 spike protein. Most of the available vaccines are either mRNA vaccines BNT162b2 (Pfizer/BioNTech) and mRNA-1273 (Moderna) that use genetically engineered RNA to generate a protein that safely prompts an immune response or adenoviral-vectored vaccines such as ChAdOx1 nCoV-19 (Oxford/AstraZeneca), Gam-COVID-Vac (Gamaleya), or Ad26.COV2.S (Johnson and Johnson) which are replication-deficient chimpanzee adenovirus (ChAd) vectored vaccines encoding the SARS-CoV-2 spike protein [23]. As of 15 November 2022, approximately 70% of the world's population has received at least one dose of a COVID-19 vaccine. To date, 12.9 billion doses have been administered globally [2]. However, the current challenge is making these vaccines available to people worldwide and achieving equity in COVID-19 vaccination [24].

The Kingdom of Saudi Arabia (KSA) has reported 825 million COVID-19 cases and 9435 deaths since the pandemic's start [25]. COVID-19 mass vaccinations in KSA began in mid-*December 2020*. The BNT162b2 mRNA *was the first vaccine approved by* the Saudi Ministry of Health (MOH) and the Saudi Food and Drug Authority (SFDA), followed by ChAdOx1 nCoV-19 and mRNA-1273. To date, 68 million vaccine doses have been administrated, more than 75% of the population is fully vaccinated, and more than 60% have received one or two booster doses [24]. Besides mass vaccination, the Kingdom imposed strict prophylactic measures, including masking and physical distancing.

To date, very few head-to-head studies compared the outcome of different COVID-19 vaccines, their impact on COVID-19 cases, and their progression to a severe disease requiring hospitalization or resulting in deaths. Therefore, the current study was designed to assess the effectiveness of three vaccines (BNT162b2, mRNA-1273, and ChAdOx1- nCoV-19) against concerned outcomes (RT-PCR confirmed COVI-19 cases after vaccination and boosters and COVID-19-related hospitalizations) in vaccinated versus non-vaccinated persons in real-world settings. In addition, this longitudinal study also evaluated the anti-SARS-CoV-2 antibody levels after the different COVID-19 vaccines to investigate if the antibody titers can prioritize persons for vaccine boosters.

2. Materials and Methods

2.1. Study Design, Study Participants, and Follow-up

This prospective, longitudinal, single-center, real-world study was conducted at Prince Sattam Bin Abdul Aziz University (PSAU) Hospital and vaccination center, Saudi Arabia, to compare the effectiveness of the three vaccines: ChAdOx1-S vaccine and the mRNA vaccines, BNT162b2 and mRNA-1273. Vaccine effectiveness measures the proportionate reduction in RT-PCR-proven COVID-19 among the vaccinated group after complete vaccination (two doses) with or without booster doses. The study also monitored the dynamics of SARS-CoV-2 antibodies after the various vaccines and booster doses. Men and women

(18 years and above) eligible for any studied vaccines were enrolled and followed between March 2021 and September 2022.

In the initial phases of this real-life study, not all COVID-19 vaccines were available. Due to the unavailability of an alternative vaccine, the vaccination of individuals with an allergy to a component of the available COVID-19 vaccine or individuals with severe allergic reactions or anaphylaxis after a previous dose of the available COVID-19 vaccine was postponed until other vaccines were available. Patients with medical conditions that warrant delaying the vaccination were temporarily exempted until resolution, invited to join the study, and followed until they could join. If they agreed and provided informed consent, they were enrolled and followed as a control group.

Vaccinated participants and non-vaccinated control subjects were followed at specific time intervals until the end of the study. Participants (vaccinated) and non-vaccinated (controls) were advised to report to the hospital to get tested for COVID-19 using RT-PCR when they developed any symptoms, even mild, or were in contact with COVID-19-positive patients. Participants who tested positive were considered ***"cases"***. In addition, COVID-19-related hospitalizations for participants were monitored and reported.

Enrolled subjects were followed at specific time points, namely two weeks after the first dose, then two, four, six, and eight months after the second vaccine dose and first booster dose, respectively. After the second booster, one or two follow-up visits were scheduled. Follow-up visits included physical examination, the COVID-19 antigen test, and SARS-CoV-2 antibody testing.

All vaccinated participants (active groups) and temporarily unvaccinated individuals (control group) provided written informed consent before vaccination, enrollment in the study, and study-related investigations. The study was conducted according to the standards of medical research, including the International Conference on Harmonization (ICH), the Council for International Organizations of Medical Sciences (CIOMS), and the Declaration of Helsinki. Prince Sattam Bin Abdulaziz University Review Board approved and monitored the study (*Approval number: REC-HSD-103-2021*).

2.2. Inclusion, Exclusion Criteria, and Control Group

All adults aged 18 years or older were eligible for inclusion if they consented to enrollment in the study and agreed to attend the follow-up visits and perform the necessary laboratory investigations. Individuals with a proven allergy to any of the components of a COVID-19 vaccine were not vaccinated with the vaccine that contains that component.

Individuals with anaphylaxis to a previous dose of one of the COVID-19 vaccines were not provided the second dose of the same vaccine. Persons with a history of severe recurrent thrombosis, thrombosis with thrombocytopenia (TTS), or heparin-induced thrombocytopenia, previous episodes of capillary leak syndrome, cerebral venous sinus thrombosis, antiphospholipid syndrome with thrombosis not enrolled in the ChAdOx1 nCoV-19 vaccine study arm, and were considered for one of the mRNA vaccines.

Vaccination was postponed, and individuals were designated temporary exemption according to the criteria of the Public Health Authorities in the following conditions:

- Inflammatory cardiac illness within the past three months; myocarditis or pericarditis; acute rheumatic fever or acute rheumatic heart disease (i.e., with active myocardial inflammation); or acute decompensated heart failure.
- Vaccination was temporarily deferred up to 4 months after a confirmed infection with COVID-19.
- Until recovery from a severe medical condition (such as major surgery or hospital admission for a serious illness).
    - Until recovery from a multisystem inflammatory syndrome in adults.
    - The serious adverse event caused by a previous dose of a COVID-19 vaccine, with no acceptable alternative vaccine available.
- Treatment with a monoclonal antibody for COVID-19.

- o Initially, data were limited about the vaccine's safety in immunocompromised patients (HIV, patients on immunosuppressant drugs, and immunocompromising diseases). The treating physician and the patient discussed the available safety data before administering the vaccine to evaluate risks and benefits.
- o In the early phase of the study, data about the vaccine's safety in pregnant women were limited. Thus, for pregnant women, particularly those with high-risk pregnancies, vaccination was postponed until delivery.

Control group: Subjects with temporary vaccine exemption were invited to join the study as a control group. If they provided informed consent, they were followed until the cause of exemption was resolved and they received the vaccine.

2.3. Assessment of Vaccine Effectiveness and Measurement of the Study Outcomes

This study's primary endpoint was to compare the effectiveness of each vaccine in preventing RT-PCR-confirmed COVID-19 cases and COVID-19-related hospitalizations *versus* those not vaccinated. The secondary endpoint was to compare and monitor the SARS-Cov-2 antibody responses to the different vaccines. During the study, vaccinated or unvaccinated participants developing fever or respiratory symptoms were tested immediately by COVID-19 RT-PCR. Positive cases were followed through home isolation or hospitalization and beyond.

2.4. COVID-19 Vaccines Registration, Documentation, and Management

Saudi Arabia has developed a unique and advanced COVID-19 reporting and vaccine registration system through a smartphone or web-based applications: Sehaty, Anat, and Tawakalna (https://ta.sdaia.gov.sa/ accessed on 25 November 2022), respectively. The web-based platforms are considered central repository systems that include COVID-19 PCR results in COVID-19 vaccine data, including the type of delivered vaccine, date, the timing of vaccines, location of vaccination center, and recorded adverse events. In addition, each vaccinated individual was provided with an electronic health passport with a barcode that verifies a person's immunization status.

2.5. Types, Preparation, Storage, and Administration of Vaccines

- ***ChAdOx1 nCoV-19* COVID-19** *vaccine***:** The multi-dose vial is stored in cold chain conditions of 2 °C to 8 °C for six months. Once opened, the vial can be used after 6 h at room temperature or 48 h in the refrigerator (2 °C to 8 °C). Vial content does not need to be mixed with a diluent.
- ***BNT162b2 COVID-19*** *vaccine* (Pfizer-Biontech, Inc., NY, USA) is provided as multi-dose vials. The vaccine is stored frozen and is thawed before dilution. Vials are thawed either in the refrigerator (2 °C and eight °C (35.6 °F and 46 °F)) or at room temperature (up to 25 °C). The BNT162b2 vaccine vial was removed from the refrigerator and kept at room temperature. It should be diluted within 2 h. The BNT162b2 vaccine is administered intramuscularly in the deltoid muscle after cutting the vial. The diluent is 1.8 mL of 0.9% sodium chloride (normal saline, preservative-free). After dilution, the vial contains five 0.3 mL doses of the vaccine that can be used. The interval between the two doses is 21–28 days
- *mRNA-1273* **COVID-19** *vaccine***:** The multi-dose vial is stored frozen and must be thawed before use either in the refrigerator (2 °C and 8 °C) or at room temperature (8 °C and 25 °C). The vial does not need to be mixed with a diluent. The Moderna vaccine is administered intramuscularly in the deltoid muscle. The interval between the two doses is 21–28 days.

2.6. Adverse Events (AEs) Recording and Reporting

All adverse events were reported by the study participants and recorded. Participants with AEs were provided with the appropriate management and advice. Severe and unusual adverse events were registered in a special Ministry of Health platform for the recording of AEs.

2.7. Measurement of Anti–SARS-CoV-2 Antibody Levels

Serum was collected using standard sampling tubes or tubes containing separating gel. Neutralizing antibody IgG against the SARS-CoV-2 spike protein subunit (S1) was measured using enzyme-linked immunosorbent assay (ELISA) kits according to the manufacturer's instructions (Elecsys® Anti-SARS-CoV-2 assay, Roche Diagnostics, NJ, USA). The assay uses a recombinant protein representing the nucleocapsid (N) antigen in a double-antigen sandwich assay format, which favors the detection of high-affinity antibodies against SARS-CoV-2. In addition, Elecsys® Anti-SARS-CoV-2 detects antibody titers, which have been shown to correlate positively with neutralizing antibodies in neutralization assays.

2.8. Statistical Analysis

We used descriptive statistics to describe the demographic and clinical data of the participants receiving the BNT162b2, mRNA-1273, or ChAdOx1-nCoV-19 vaccines. As appropriate, a comparison of means between vaccinated participants and controls was carried out using the Analysis of Variance (ANOVA) or Kruskal–Wallis test. A Mann–Whitney test was used for pairwise comparisons as a post hoc or follow-up analysis if the test showed differences between the groups. The tested vaccine's effectiveness (VE) (percentage reduction in COVID-19 incidence in a vaccinated group compared to a non-vaccinated group) was assessed by assessment of parameters. The relative risk (RR) is the risk of the outcome in the treated group (Y) compared to the risk in the control group. The RR, its standard error, and 95% confidence interval are calculated according to Altman, 1991. Absolute risk reduction (ARR) is the difference in risk between the control group (X) and the treatment group (Y). Relative risk reduction (RRR) is the percent reduction in risk in the treated group (Y) compared to the control group (X). The effectiveness = 1-RR × 100%. The number needed to treat (NNT) is the number of patients that need to be treated to prevent one additional bad outcome. The same approach assessed the rate of COVID-19-related hospitalizations [25]. Receiver operating curve (ROC) analysis was used to evaluate the anti-SARS-CoV-2 antibody test diagnostic ability to discriminate COVID-19 positive and negative subjects and the optimal cutoff values. Statistical analysis was performed using SPSS 20 statistical software.(IBM SPSS Statistics for Windows, Version 23.0. Armonk, NY, USA; IBM. Corp)

3. Results

3.1. Demographics and Characteristics of Enrolled Subjects

This study was conducted at the Prince Sattam Bin Abdulaziz University Hospital and Vaccination Center, Kharj, Kingdom of Saudi Arabia, between March 2021 and September 2022. Initially, 4145 vaccinees were screened for eligibility for the study. Eligible individuals who provided informed consent (n = 1500) were enrolled and prospectively followed until the end of the study (Figure 1). Participants received one of the COVID-19 vaccines: ChAdOx1 nCoV-19 (Cambridge, UK) (Group A; n = 503), BNT162b2 (BioNTech-Pfizer, Mainz, Germany) (Group B; n = 521), or mRNA-1273, (Moderna, Cambridge, MA, USA) (Group C; n = 476). The study cohort completed the two vaccination doses and received the BNT162b2 vaccine's first and second booster doses. The participants' follow-ups ranged between 12 to 18 months after vaccination.

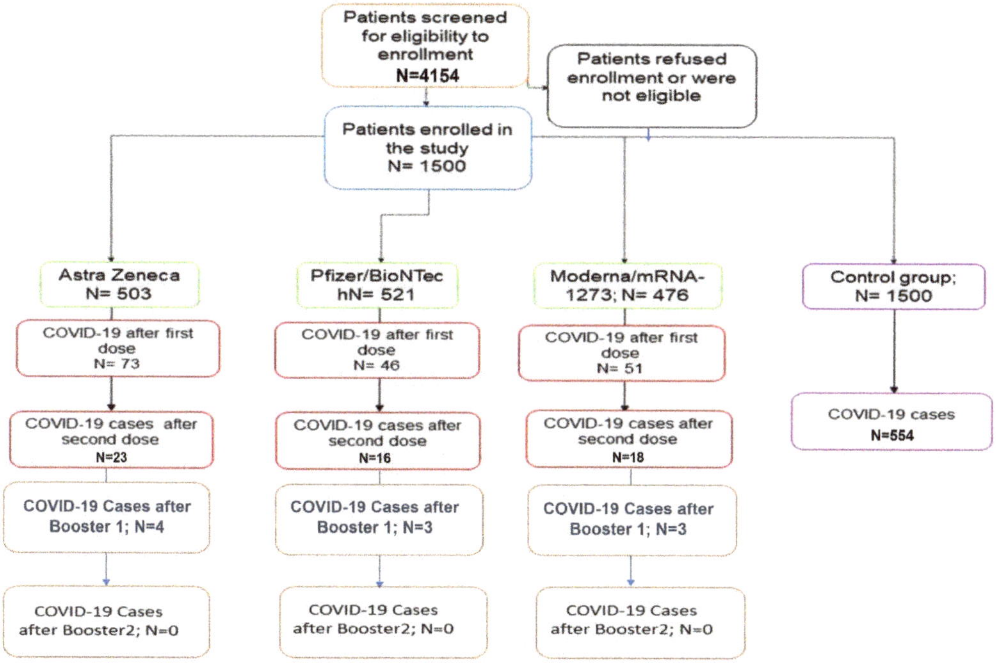

Figure 1. The flow of participants through the study. The figure shows the flow of participants through the study. Initially, 4154 subjects were screened. A total of 3535 were eligible; however, only 1500 subjects agreed to join the study, provided informed consent, and were enrolled. Subjects with temporary vaccine exemption were those recovering from natural COVID-19, those treated with immunosuppressive therapy, and pregnant women.

The control group included 1500 individuals who were either ineligible due to a documented allergy to components of both vaccines (95; 6.3%) or developed myocarditis or pericarditis after a dose of **mRNA vaccine (9; 0.6%) or** were temporarily exempted from vaccination as a control group. The causes of temporary exemption were previously confirmed COVID-19 (683; 45.53%), severe allergic reactions, or adverse events caused by a previous dose of a COVID-19 vaccine (ChAdOx1- nCoV-19) (187; 12.47%) or contraindication to a vaccine with no available alternative vaccine (174; 11.6%), severe medical conditions or major surgeries (153; 10.2%), high-risk pregnancy (275; 18.33%), and 172 (11.46%) due to other causes such as patients receiving immunotherapy or monoclonal antibody for COVID-19. The control subjects with temporary exemptions were followed until the cause of postponing vaccination was resolved (3–9 months). By the end of the study, that caused 1089 subjects (78%) to be vaccinated.

The active vaccinated cohort included 767 (51.13%) Saudis and 733 (48.87%) non-Saudis ($p = 0.02$) (Table 1). There were no significant differences in participants' ages or gender in the three vaccinated groups and the control subjects. Among the vaccinated participants, 1416 (94.4%) did not experience previous natural COVID-19 infection and were vaccine naïve. Eighty-four (5.7%) participants reported previous attacks of COVID-19 4–6 months before enrollment in the study. Among the enrolled subjects, 1114 (74.3%) vaccinated participants reported no diseases or health disorders, while 386 (25.7%) participants had one or more comorbidities. The chronic health disorders included diabetes, hypertension, dyslipidemia, cardiac disease, chronic liver disease, chronic renal disease, hemoglobinopathies, neurologic or psychiatric disorders, autoimmune disorders, and malignancies (Table 1). Among the patients with hemoglobinopathies, 88% and 12% had sickle cell anemia and thalassemia, respectively. The autoimmune diseases included systemic

lupus erythematosus (SLE), rheumatoid arthritis, psoriasis, and Hashimoto's thyroiditis. The participants with malignancies included patients with breast cancer, lymphoma, and prostate cancer who completed their therapeutic approaches. The control group had more comorbidities since patients with advanced chronic diseases were initially temporarily exempted from vaccination.

3.2. *The Outcome of Vaccination in the Three Groups*

The study endpoint was the occurrence of new COVID-19 cases after vaccine completion. The study demonstrated that 23, 16, and 18 participants developed COVID-19 RT-PCR-proven COVID-19 after taking the second dose of ChAdOx1 nCoV-19, *BNT162b2*, and *mRNA-1273* vaccines, respectively;($p = 0.369$), *versus* 554 non-vaccinated subjects (control group) ($p < 0.0001$). The effectiveness of ChAdOx1-nCoV-19, BNT162b2, and mRNA-1273 vaccines was 89%, 92% (95% CI of 95.03% to 98.14%), and 90% (95% CI of 80.8% to 95.5%.), respectively. A significant difference was observed only between BNT162b2 and ChAdOx1 nCoV-19 ($p = 0.0026$) (Table 2). Similarly, a significant difference in absolute risk reduction (ARR) was detected between COVID-19-infected participants within the vaccinated and unvaccinated groups. Only three vaccinated participants, one patient with SLE, one with colorectal cancer, and an obese individual, developed severe COVID-19-related disease and were hospitalized; however, no mortalities were reported in vaccinated participants. In contrast, 103 (18.6%) unvaccinated control subjects were hospitalized with COVID-19-related complications with three mortalities ($p < 0.0001$).

Although the endpoint was the development of COVID-19 cases after vaccine completion, the study also monitored the COVID-19 instances after the first vaccine dose. As shown in Table 1, there was a significant difference in the COVID-19 cases that developed after the first doses of ChAdOx1 nCoV-19, BNT162b2, and mRNA-1273, with the highest issues developing after ChAdOx1 nCoV-19.

Table 3 summarizes the emergence of COVID-19 cases after vaccine completion and booster doses versus unvaccinated control subjects. COVID-19 infections were significantly higher in unvaccinated individuals, with the highest COVID-19 incidence in patients with autoimmune diseases, malignancies, people with diabetes, particularly those with uncontrolled diabetes, and patients with hemoglobinopathies (Table 3). Among the participants who developed COVID-19 after completing the second dose of the vaccines, 73.7% had comorbidities. The reported COVID-19 incidents were significantly lower in vaccinated *versus* unvaccinated participants with comorbidities. The vaccine effectiveness in patients with diabetes mellitus, obesity, autoimmunity, and malignancies ranged between 80% and 86%, which is lower than VE in participants without chronic diseases and those with other comorbidities. None of those who received booster vaccines developed a severe illness that required hospitalization. Figure 2 is a Kaplan–Meier graph that demonstrates the time to event (the occurrence of COVID-19 cases in the three study groups and control subjects).

Table 1. Demographics baseline clinical features enrolled subjects according to COVID-19 vaccine type.

Parameters	Group A (n = 503) ChAdOx1 nCoV-19 Vaccine	Group B (n = 521) BNT162b2 Vaccine	Group C (N = 476) mRNA-1273 Vaccine	*p* Value between the Three Vaccinated Groups	Control Group (n = 1500)	*p* Value between Vaccinated and Non-Vaccinated
Age (years) Mean ± SD	46.31 ± 14.99	44.63 ± 15.53	43.79 ± 15.62	0.08	45.09 ± 16.19	0.06
Males; n (%)	295 (58.65)	301 (57.7	258 (54.2)	0.36	846 (56.4)	0.45
Nationality						
Saudi; n (%)	256 (50.89)	267(51.25	244 (51.26)	0.89	743 (49.53)	0.38(1.976% to 5.170%)
Non-Saudi; n (%)	247 (49.15)	254 (48.7	232 (48.75)		757 (50.47)	
	Patients with Comorbidities: Total 386 (25.7%)			*p* value between the three vaccinated groups; *p* (95% CI)		*p* value between vaccinated and non-vaccinated; *p* (95% CI)
Diabetes mellitus; n (%)	67 (13.32)	65 (12.48	58 (12.18)	0.54 (−2.344% to 4.878%	149 (9.30)	0.41 (−1.363% to 3.364%
Hypertension; n (%)	51 (10.14)	46 (8.83	40 (8.40)	0.33 (−1.479% to 4.887%	96 (6.4)	0.08; (−0.209% to 4.114%)
Dyslipidemia; n (%)	50 (9.94)	49 (9.4)	43 (9.03)	0.64 (−2.292% to 4.097%0	83 (5.53)	0.06; (−0.063% to 4.328%)
Overweight/Obesity; n (%)	49 (9.74)	58 (11.13	55 (11.55)	0.35 (−1.934% to 4.607%)	265 (17.7)	$p < 0.0001$ (17.303% to 22.946%)
Cardiac disease; n (%)	21 (4.17)	18 (3.45	19 (3.99)	0.62 (−1.447% to 2.865%)	46 (3.07)	0.002 **; (0.9527% to 4.1357%)
Chronic liver disease; n (%)	12 (2.34)	9 (1.73)	13 (2.73)	0.89 (−1.391% to 1.992%)	34 (2.30)	<0.0001 **; (1.747% to 4.149%)
Chronic renal disease; n (%)	9 (1.79)	8 (1.54)	6 (1.26)	0.562 (−0.873% to 2.064%	59 (3.90)	<0.0001 **; 2.4695% to 5.0889%
Hemoglobinopathies; n (%)	17 (3.38)	20 (3.84	18 (3.78)	0.72 (−1.841% to 2.229%	65 (4.33)	0.241; (−0.569% to 2.243%)
Neurologic disorders; n (%)	8 (1.59)	10 (1.92	7 (1.47)	0.88 (−1.534% to 1.386%	57 (3.80)	<0.0001 **; (2.152% to 4.779%)
Autoimmune disorders; n (%)	16 (3.18)	21 (4.03	19 (3.99)	0.42 (−1.362% to 2.686%)	68 (4.53)	0.473; (−0.929% to 1.997%)
Malignancy (patients with breast cancer, lymphoma, and prostate cancer); n (%)	6 (1.19)	11 (2.11	8 (1.68)	0.32 (−0.841% to 1.93%	103 (6.87)	<0.0001 **; (3.789% to 6.692%)

Among the enrolled subjects, 1114 (74.27%) vaccinated participants reported no diseases or health disorders; 386 (25.73%) participants had one or more comorbidities, for example, diabetes and hypertension, liver or renal disease, or overweight. Significant. ** Highly significant.

Table 2. RT-PCR confirmed COVID-19 cases and COVID-19-related hospitalizations after vaccines and booster doses.

COVID-19 Cases and Hospitalizations after Vaccines and Booster Doses	Group A (n = 503) ChAdOx1 nCoV-19Vaccine	Group B (n = 521) BNT162b2 Vaccine	Group C (N = 476) mRNA-1273	p Value and 95% CI between the Three Vaccines	COVID-19 Cases and Hospitalizations in the Control Group throughout the Study (n = 1500)	p Value between Vaccinated and Non-Vaccinated Control Subjects
Participants with confirmed RT-PCR COVID-19 (Cases) after the first vaccine dose; (n,%)	113 (22.46%)	66 (12.67%)	67(14.08%)	<0.0001 **		
Participants with confirmed RT-PCR COVID-19 (cases) after the second vaccine dose (n,%)	23 (4.6)	16 (3.07)	18 (3. 80)	0.37 (−1.118% to 3.5%)		
Relative risk (RR) (95% CI)	0.124 (0.083 to 0.186)	0.083 (0.051 to 0.135)	0.10 (0.065 to 0.162)			
Vaccine effectiveness (VE); Relative risk reduction (RRR)	89% 0.89	92% 0.92	90% 0.90	ChAdOx1 nCoV-19 vs. BNT162b2: 0.025 ChAdOx1 nCoV-19 vs. mRNA-1273: 0.298 BNT162b2 vs. mRNA-1273: 0.244	564 (38%)	<0.0001 **
Absolute risk reduction % (ARR)	37.36%;	38.75%;	38.15%;	0.68 (−4.14% to 6.22%)		
Number needed to vaccinate (NNV) (benefit)	3.09	2.95	3.02			
95% CI	2.724 (Benefit) to 3.570 (Benefit)	2.625 (Benefit) to 3.375 (Benefit)	2.660 (Benefit) to 3.483 (Benefit)			
COVID-19 (cases after the first booster dose; (n,%)	4 (0.79)	3 (0.58)	3 (0.63)	0.67 (−0.664% to 1.461%)		
Relative risk after the first booster (95% CI)	0.024 (0.009 to 0.057)	0.014; (0.0044 to 0.0425)	0.015; (0.0049 to 0.047)	0.19; (−0.444% to 2.784%)		
Relative risk reduction (RRR)	0.98	0.99	0.99	0.51; −(0.844% to 2.158%)		
Absolute risk reduction (ARR)	41.10%	41.40%	41.30%	0.93; −(5.046% to 5.474%)		
Participants with confirmed RT-PCR COVID-19 (Cases) after the second booster dose; (n,%)	0	0	0			
Hospitalizations after the vaccine's first dose	12 (2.38)	5 (0.95)	7 (1.47)	0.09 (−0.255% to 3.206%)		
Hospitalizations after the vaccine's second dose	2 (3.9)	1 (0.19)	0	0.05 *	103/554 (18.6%)	<0.0001
Hospitalizations after vaccine boosters 1 or 2	0	0	0			

The relative risk (RR), its standard error, 95% confidence interval, relative risk, absolute risk reduction, and number needed to vaccinate (NNV) is calculated according to Altman, 1991.
* Significant. ** Highly significant.

Table 3. Vaccine effectiveness in vaccinated participants with comorbidities and COVID-19 incidence in unvaccinated control subjects.

Vaccinated Participants with Comorbidities: (n = 386; 25.73% of the Study Cohort)	COVID-19 Cases after Vaccine Second Dose in Vaccinated Participants with Comorbidities; n (%)	COVID-19 cases after Booster 1	COVID-19-Related Hospitalizations after Vaccines Second Dose or Booster 1	Relative Risk (RR.) (95% CI)	Vaccines Effectiveness in Participants with Comorbidities	COVID-19 Cases in Unvaccinated Controls	*p* Value between COVID-19 Cases in Vaccinated and Unvaccinated Participants with Comorbidities
Diabetes mellitus; n = 190	11/190 (5.8%)	2 (1.05)	1 (0.5%)	0.137 (0.075 to 0.25)	86.3%	63/149 (35.2%)	0.0001
Hypertension; n = 137	2 (1.5%)	0	0	0.04 (0.009 0.163)	96%	35/96 (21.7%)	0.0001
Dyslipidemia; n = 142	1 (0.7%)	0	0	0.019 (0.0054 to 0.251	98%	31/83 (17.9%)	<0.0001
Overweight/Obesity; n = 162 BMI 30–40 BMI > 40	3 (1.85) 17 (10.49)	0 2 (1.23%)	0 0	0.285 (0.186 to 0.439)	82%	166/265 (27.1%)	<0.0001
Cardiac disease; n = 58	0	0	0	0.017 (0.003 to 0.649)	98.3%	23/46 (24%)	0.004**
Chronic liver disease; N = 34	0	0	0	0.04 (0.0052 to 1.218)	98%	12/34 (18.75%)	0.02*
Chronic renal disease; n = 23	1 (4.3%)	1 (4.3%)	0	0.066 (0.009 to 0.451)	93.4%	39/59 (48.1%)	0.005**
Hemoglobinopathies; n = 55	2 (3.6%)	0	0	0.074 (0.019 to 0.294)	92.6%	32/65 (32.3%)	0.002**
Neurologic/psychiatric disorders; n = 25	0	0	0	0.057 (0.004 to 0.912)	94.3%	19/57 (24. 5%)	0.04*
Autoimmune disorders; n = 56	10 (17.9%)	3 (5.4%)	1 (1.79%)	0.229 (0.129 to 0.408)	80%	53/68 (78%)	<0.0001**
Malignancy; n = 25	4 (16%)	2 (8%)	1 (4%)	0.204 (0.082 to 0.502)	80%	81/103 (71.8%)	0.0006**

Several vaccinated and unvaccinated participants had more than one comorbidity. The relative risk (RR), standard error, and 95% confidence interval are calculated (Altman, 1991). * Significant. ** Highly significant.

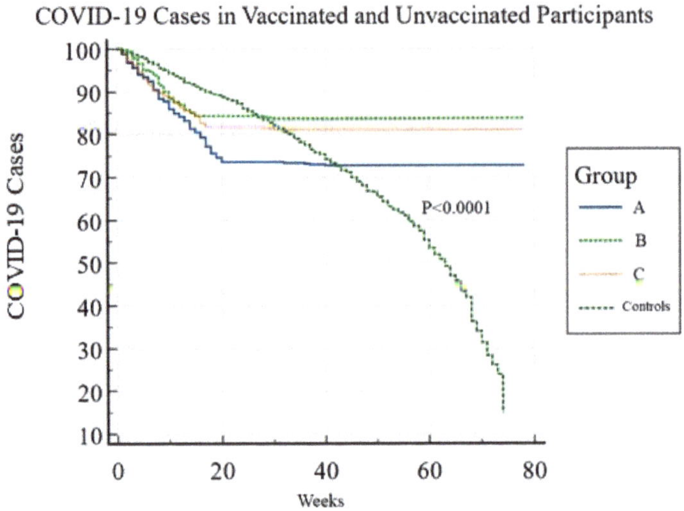

Figure 2. Time to COVID-19 infection in the three study groups and control subjects. Kaplan–Meier survival curves show the time to endpoint (occurrence of COVID-19) in vaccinated and unvaccinated participants throughout the study. In vaccinated participants, more COVID-19 cases occurred after the first dose, given that the post-vaccine immune responses were not wholly developed. After the second vaccine dose, 23 (4.6%), 16 (3.1%), and 18 (3.8%) (p = 0.3.69) cases occurred, with 82% of confirmed COVID-19 infections occurring within 5–18 days after the second dose. Seven (0.47%) COVID-19 cases occurred after the first booster dose. No cases were reported after the second booster. A reduction in the rate of confirmed COVID-19 infection was observed over time as compared with the unvaccinated control subjects. The hazard ratios with a 95% confidence interval for the vaccinated groups (Groups A, B, and C) and the unvaccinated controls are shown in the KM hazards table. The hazard ratio measures the magnitude of the difference between the two curves in the Kaplan–Meier plot. In contrast, the *p-value* measures the statistical significance of this difference.

3.3. Vaccine Adverse Events

Several adverse events were reported after receiving the different COVID-19 vaccines; however, most were not serious (Table 4). The most frequent adverse events included pain at the site of injection (89.5%), fatigue (39.6%), headache (25%), low-grade fever (17.8%), and muscle pain (11.3%). Less frequent adverse events included chills (5.8%), abdominal pain (4%), allergic skin reaction (6.2%), joint pains (3.3%), diarrhea, or vomiting (2.1%). In addition, about 2% of those receiving the vaccines reported palpitations (2%), numbness (1.8%), sleep disorder (1.7%), and anxiety/restlessness (0.5%). Fever > 38 extending beyond three days, headache, and abdominal and joint pain were significantly higher among participants receiving ChAdOx1 nCoV-19 (Table 4). Ten participants (0.66%) with comorbidities (diabetes mellitus, hypertension, obesity, sickle cell disease, cancer) developed serious adverse events, including severe allergic reactions/anaphylaxis, Guillain–Barré syndrome, thrombosis without or with thrombocytopenia, seizures, and myocarditis. Two previously controlled sickle cell disease (SCD) participants developed vaso-occlusive crisis (VOC), and a patient with diabetes and valvular heart disease developed deep venous thrombosis shortly after the ChAdOx1 nCov-19 vaccine. One participant vaccinated with the BNT162b2 vaccine developed Guillain–Barré syndrome one day after vaccination. In addition, one participant developed optic neuritis one day after vaccination, and she responded to corticosteroid therapy.

Table 4. Adverse events after the different COVID-19 vaccines.

Parameters	Group A (n = 503) ChAdOx1 nCoV-19 Vaccine	Group B (n = 521) BNT162b2 Vaccine	Group C (n = 476) mRNA-1273 Vaccine	p Value; (95% CI)
Pain at the injection site	526 (89.31)	375 (89.07)	308 (90.59)	0.880; (−3.413% to 4.525%)
Fever > 38.5 °C	137 (23.26)	46 (10.93)	57 (16.76)	<0.0001 **; (5.0826% to 13.5058%)
Fatigue > 2 days	294 (49.92)	113 (26.84)	127 (37.35)	$p < 0.0001$ **; (12.2457% to 22.6466%)
Headache > 2 days	174 (29.54)	72 (17.10)	69 (20.29)	$p < 0.0001$ **; (6.8979% to 16.1204%)
Chills	35 (5.94)	19 (4.51)	24 (7.06)	0.975; (−2.474% to 2.690%)
Muscle pain	78 (13.24)	43 (10.21)	31 (9.12)	0.084; (−0.406% to 6.534)
Skin reaction at the injection site; n (%)	45 (7.64)	18 (4.28)	21 (6.1)	0.11; (−0.493% to 4.959%)
Vomiting or diarrhea	13 (2.21)	7 (1.66)	8 (2.35)	0.895 (−1.407% to 1.762%)
Abdominal pain	31 (5.26)	10 (2.31)	13 (3.82)	0.003 **; (0.969% to 5.280%)
Thrombosis	11	1	1	
Numbness	17 (2.89)	4 (0.95)	3 (0.88)	0.005 **; (0.595% to 3.723%)
Joint pains	29 (4.92)	9 (2.14)	7 (2.06)	<0.0001 **; (2.339% to 6.181%)
Sleep disorders	18 (3.06)	3 (0.71)	2 (0.59)	0.457; (−1.091% to 2.591%)
Anxiety/restlessness	5 (0.85)	1 (0.24)	1 (0.29)	0.1020; (−0.194% to 1.765%)
Palpitations; n (%)	17 (2.89)	6 (1.43)	4 (1.18)	0.026 *; (0.1823% to 3.4698%)
Thrombosis	3 (1.8)	0	0	0.013 *; (0.32% to 3.19%)
Myocarditis	1	1	0	1.0000 (−1.4% to 1.4%)
Visual disturbances	1 (0.34)	0	0 (0.00)	1.0000; (−1.1129% to 0.855%)
Anaphylaxis	0	1	1	1.0000 (−1.4% to 1.5%)
Guillain–Barré syndrome	0	1	0	1.0000 (−1.4% to 1.4%)
Seizures	0	0	1	

* Significant. ** Highly significant.

3.4. SARS-CoV-2 Antibody Dynamics

The study demonstrated that the SARS-CoV-2 antibodies were slightly higher in participants vaccinated with either BNT162b2 or mRNA-1273 than ChAdOx1 nCoV-19; however, the difference was insignificant (p = 0.716) *(data are not shown)*. As shown in Figure 2, SARS-CoV-2 antibodies rose gradually after the first vaccine dose and peaked two to four weeks after the second dose. Although the levels of antibodies remained elevated until six months in the three groups, a gradual decline was detected between 24 to 32 weeks (mean 28.50 $\pm$ 3.48 weeks) post-vaccination (Figure 3a). The booster doses resulted in the antibody titers with levels that exceeded those after the second vaccine dose. The ROC curve analysis showed that antibody titers $\leq$ 154 U/mL were associated with an increased risk of acquiring COVID-19 with a sensitivity and specificity of 98% and 99%, respectively (Figure 3b). Patients with comorbidities tended to have slightly lower humoral responses than healthy participants. Furthermore, antibody titers declined earlier (between 16–28 weeks) in vaccinated participants with comorbidities compared to those without.

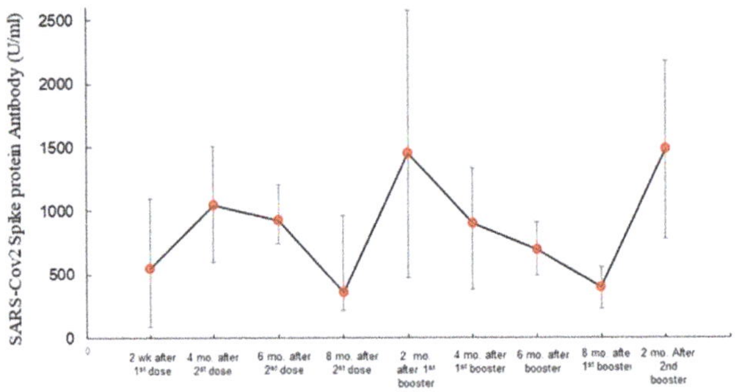

(a)

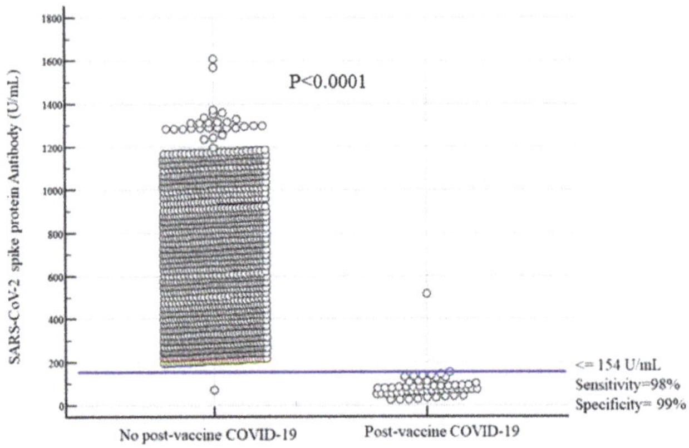

(b)

Figure 3. (**a**) SARS-CoV-2 antibody dynamics in vaccinated subjects. (**b**) SARS-CoV-2 antibodies titers in participants without and with post-vaccine COVID-19. (**a**) The figure demonstrates the kinetics of antibody titers against SARS-CoV-2 spike protein during follow-up. A gradual rise in the antibody titers started two weeks after the first vaccine doses. The antibody levels peaked after the second vaccine dose. Antibody levels started to decline 24–32 weeks after the second vaccine SARS-CoV-2 responses. (**b**) The figure shows a significant difference in the SARS-CoV-2 antibody levels in participants who did not develop post-vaccine COVID-19 compared to those who developed COVID-19 after vaccination. A cutoff of 154 U/mL was associated with an increased risk of developing COVID-19.

4. Discussion

This prospective, longitudinal, test-negative, real-world study compared the effectiveness, outcomes, and safety of three COVID-19 vaccines: ChAdOx1 nCoV-19, BNT162b2, and mRNA-1273, and evaluated the role of the vaccine boosters in a broad, representative patient population at different ages and ethnicities in diverse clinical settings, where many patients have multiple comorbidities. The study also measured the anti-SARS-CoV-2 antibodies' levels at other time points after vaccination. Overall, the study provides real-world evidence that the three tested vaccines conferred significant protection against COVID-19 infections and reduced severe outcomes that require hospitalization and COVID-19-related mortalities. The three vaccines were also well tolerated. However, the anti-SARS-CoV-2 antibody titers and, consequently, COVID-19 vaccine effectiveness waned with time but were regained with the booster doses.

The current study demonstrated that the effectiveness of ChAdOx1- nCoV-19, BNT162b2, and mRNA-1273 was 89%, 92%, and 90%, respectively, where the effectiveness of BNT162b2 was significantly higher than ChAdOx1 nCoV-19; however, no significant difference was detected between ChAdOx1 nCoV-19 and mRNA-1273 vaccines. The effectiveness of the vaccines in the current study was comparable with the efficacy or effectiveness rates reported by previous studies [26–29]. Although the study's endpoint was the development of COVID-19 cases following the second vaccine dose, we also monitored patients after the first vaccine dose to investigate the magnitude of protection conferred by such a dose. Although the immune response to the vaccines has not been fully developed, our findings showed that vaccinated participants, irrespective of the type, were at lower risk of developing COVID-19 and disease-related hospitalizations compared to unvaccinated individuals in similar prophylactic settings. However, the COVID-positive cases were significantly higher after the first dose of ChAdOx1 nCoV-19 than those detected after BNT162b2 or mRNA-1273 vaccination. This finding may be attributed to the long interval between the first and second ChAdOx1 nCoV-19 and the changes in the adopted prophylactic measures. Given the first dose's beneficial effects, distributing the vaccines among those eligible as quickly as possible, after providing for the most vulnerable broad mass vaccination campaigns may be beneficial for controlling the viral spread and severe COVID-19-related morbidities and mortalities in resource-limited countries with irregular vaccine supplies [30].

The current study demonstrated a significant reduction in post-vaccination COVID-19 incidents and severe disease after the second dose of the three vaccines. Thus, the present study results lend further credence to the findings of several studies [28,31–33] that showed the critical role of vaccine completion in reducing infection and COVID-19-related severe disease. The confirmed COVID-19 infections detected after vaccine completion were 57/1500 (3%), and none of the cases was associated with severe symptoms in this study cohort. Our findings are comparable to those of a study that showed 39 of 1497 (2.8%) fully vaccinated healthcare workers developed post-vaccination SARS-CoV-2 breakthrough incidents [34]. Vaccine breakthrough infections may occur following several vaccines that may result from viral factors, including the emergence of variants, transmissibility rates, immune evasion, or host determinants such as age, comorbidities, and immune status may result in post-vaccination confirmed infections [35]. This study did not assess the prevailing SARS-CoV-2 variants; however, the impact of comorbidities was investigated.

Few countries provided COVID-19 vaccine boosters. In KSA, two booster doses were administered, the first 6–8 months after vaccine completion and the second eight months after the first booster dose. In KSA, initially, priority was given first to elderly persons and high-risk individuals then boosters were offered on a larger scale. After the first vaccine booster, only 10 minimally symptomatic vaccine breakthroughs were detected. No confirmed COVID-19 infections occurred after the second vaccine booster. The current study also demonstrated that the anti-SARS-CoV-2 titers decline 24–28 weeks after vaccine completion per other reports. However, administering the first and second COVID-19 vaccine boosters was associated with a significant increase in anti-SARS-CoV-2 levels and a subsequent sharp reduction in breakthrough infections. Overall, this study showed that

booster doses have significantly reduced breakthrough vaccine infections and the risk of hospitalization or death in participants without and with comorbidities.

This study showed that some comorbidities influenced the vaccines' effectiveness and odds of acquiring COVID-19 after vaccination. Most post-vaccination symptomatic COVID-19 cases and related hospitalizations were observed in patients with comorbidities. Patients with diabetes mellitus, autoimmune diseases, malignancies, and participants with BMI > 40 showed lower VE and were more susceptible to infections and severe COVID-19 than healthy participants. Our findings and others [36,37] demonstrated that unvaccinated obese participants (BMI > 40) showed high susceptibility to acquiring and developing severe COVID-19 and increased risk of hospitalization. Furthermore, obese participants, particularly those with severe obesity (BMI > 40 kg/m^2), had lower VE than participants with reasonable BMI. Obesity is prevalent in KSA, as a national survey conducted between 1995 and 2000 found that the overall prevalence of obesity among Saudi adults was 36% [38]. Thus, obese patients are considered a high-risk group that must be prioritized for vaccination. In agreement with previous studies, this study showed increased incidence and severity of COVID-19 in patients with diabetes. Diabetes is a public health problem in Saudi Arabia; the World Health Organization (WHO) reported that Saudi Arabia ranks second highest in the Middle East and is seventh in the world for the rate of diabetes [39]. This study showed that COVID-19 infection occurred in 35% of diabetic unvaccinated control diabetic subjects. Some patients, particularly those with uncontrolled diabetes, developed a severe disease that needed hospitalization despite implementing strict prophylactic measures. Other studies reported similar observations [40–45]. Diabetes is a chronic disease associated with several metabolic, vascular, and immunologic abnormalities that can affect our response to pathogens. Hyperglycemia and insulin resistance induce oxidative stress and enhance the production of glycosylation end products (AGEs) and proinflammatory cytokines, which mediate tissue inflammation [46]. The inflammatory process may lead to higher infection susceptibility [46]. Unvaccinated cancer patients and those with autoimmune diseases were found in the current study to be highly susceptible to COVID-19 infection with severe symptoms. One study [47] enrolled 23,266 participants with cancer and 1,784,293 without cancer. A total of 10,404 positive COVID-19 cases were reported. Compared with participants without cancer, those living with cancer had a 60% increased risk of acquiring COVID-19. Cancer patients on current chemotherapy or immunotherapy treatment had a 2.2-fold increased risk of a positive test [47]. The present study showed reasonable vaccine response in patients with comorbidities, including diabetes, obesity, cancer, and autoimmune disease. After the second booster vaccine dose, no vaccine breakthrough was reported in this cohort. The findings of this study and another study [48] stress the benefit of booster vaccine doses in the high-risk population.

The COVID-19 vaccines' effectiveness demonstrated in this study was coupled with a low risk of adverse events. The adverse events of the three types of vaccines assessed in the current study are comparable to those reported in several studies [28–33]. Serious vaccine adverse events, severe allergic reaction/anaphylaxis, Guillain–Barré syndrome, thrombosis, thrombocytopenia, seizures, and myocarditis observed in this study were rare as they occurred in 10 (0.66%) participants, all of whom had comorbidities. Thrombosis without or with thrombocytopenia has been reported after COVID-19 vaccines [49,50]. The severe allergic reactions that occurred shortly after vaccination were promptly managed in the hospital. Two pre-vaccine controlled sickle cell disease (SCD) study participants developed vaso-occlusive crisis (VOC) and thrombosis shortly after AstraZeneca, an observation previously reported by some studies [51,52]. SCD patients vaccinated with BNT162b2, or mRNA-1273 did not suffer post-vaccine VOC. A patient with diabetes and valvular heart disease developed deep venous thrombosis following ChAdOx1 nCov-19 vaccine. One patient vaccinated with ChAdOx1 nCov-19 vaccine developed optic neuritis four days after receiving the vaccine and responded to steroid therapy. Recently, multiple reports were published on the development of optic neuropathy following COVID-19 vaccination [53]. Thus, high-risk patients with hypercoagulable states should be thoroughly investigated

before vaccination, and careful selection of the suitable COVID-19 vaccine is recommended. Our study demonstrated myocarditis in two patients following BNT162b2 and ChAdOx1 nCov-19 vaccines, a post-vaccination observation reported by several studies [54–56]. Post-SARS-CoV-2 vaccination, Guillain–Barre syndrome occurred in the current study in a diabetic participant vaccinated with the BNT162b2r vaccine; however, the patient responded to steroids [57–59]. Collectively, our findings and the results of several studies demonstrated the safety of the COVID-19 vaccines and the rare occurrence of serious adverse events. Thus, the benefits of vaccination against SARS-CoV-2 are substantial and outweigh the associated risks.

In agreement with several studies [60–62], SARS-CoV-2 antibodies rose gradually after the first vaccine dose. However, significantly high titers were achieved after the second dose of ChAdOx1 nCoV-19, BNT162b2, and mRNA-1273 vaccines. A gradual decline in antibody titers over time was detected in our cohort between 24 and 32 weeks, an observation reported by several groups [60–63]. The booster doses evoked high SARS-CoV-2 antibody levels that exceeded those after the second vaccine. Although participants with comorbidities had slightly lower humoral immune responses and showed earlier antibody titer decay, they responded well to the booster vaccines as breakthrough infections were detected after the second booster. In this study cohort, participants with antibody titers $\leq$154 U/mL were at increased risk of acquiring COVID-19. Establishing a cutoff of SARS-CoV-2 that may predict an increased risk of post-vaccination breakthrough infections may help prioritize persons for COVID-19 boosters, individualizing the timing of boosters and conferring better protection, particularly with patients with comorbidities. Further prospective extensive studies are needed to investigate if monitoring the dynamics of SARS-CoV-2 antibodies SARS-CoV-2 antibodies may play a role in prioritizing the use of a limited supply of vaccines.

This study has several limitations. First, the study has not addressed the impact of vaccines on the various SARS-CoV-2 variants that emerged since the introduction of the COVID-19 vaccines. Diagnosis of post-vaccine COVID-19 was based on participants' self-reporting of symptoms. Thus, there may be a probability of missing asymptomatic cases. Finally, the study has not assessed vaccine efficacy and safety in individuals younger than 18 and children younger than 12.

In conclusion, the three evaluated COVID-19 vaccines were tolerable, safe, and effective in reducing COVID-19 and preventing severe disease. Furthermore, COVID-19 vaccination showed effectiveness and safety in individuals with comorbidities and those without any underlying medical disorders. In addition, the vaccine-induced SARS-CoV-2 antibody responses gradually decline. Booster vaccine doses induced significant maintained responses. SARS-CoV-2 antibody levels may help determine the timing and need for vaccine booster doses, particularly in patients with comorbidities.

Author Contributions: Conceptualization, S.M.K.; methodology, S.M.K., M.M.N., M.N.A., Z.G.A., A.A.B. and K.S.; validation, S.M.K., M.D. and Z.G.A.; formal analysis, S.M.K., M.M.N. and A.M.A.; investigation, S.M.K., M.M.N., M.N.A., A.A.B. and K.S.; resources, S.M.K., M.M.N., M.D., M.N.A., Z.G.A., A.M.A. and K.S.; data curation, S.M.K., M.D. and M.N.A.; writing—original draft preparation, S.M.K., M.M.N., M.D., Z.G.A. and A.M.A.; writ-ing—review and editing, S.M.K.; visualization, Z.G.A. All authors have read and agreed to the published version of the manuscript.

Funding: This research received no external funding. The authors thank Prince Sattam University College of Medicine, Prince Sattam University Hospital, and Prince Sattam University Research Unit for continuous support through the study.

Institutional Review Board Statement: Prince Sattam Bin Abdulaziz University Review Board approved and monitored the study (Approval number: REC-HSD-103-2021).

Informed Consent Statement: No patients were identified in the study.

Data Availability Statement: Not application.

Conflicts of Interest: The authors declare no conflict of interest.

References

1. World Health Organisation (WHO). Coronavirus Disease (COVID-19) (Internet). Available online: https://www.who.int/health-topics/coronavirus#tab=tab_1 (accessed on 16 November 2022).
2. Johns Hopkins Corona Resource Center (Internet). Available online: https://coronavirus.jhu.edu/map.html (accessed on 20 November 2022).
3. Li, Q.; Guan, X.; Wu, P.; Wang, X.; Zhou, L.; Tong, Y.; Ren, R.; Leung, K.S.M.; Feng, Z.; Xiang, N.; et al. Early Transmission Dynamics in Wuhan, China, of Novel Coronavirus-Infected Pneumonia. *N. Engl. J. Med.* **2020**, *382*, 1199–1207. [CrossRef] [PubMed]
4. van Doremalen, N.; Bushmaker, T.; Morris, D.H.; Holbrook, M.G.; Gamble, A.; Williamson, B.N.; Munster, V.J.; Tamin, A.; Lloyd-Smith, J.O.; de Wit, T.; et al. Aerosol and Surface Stability of SARS-CoV-2 Compared with SARS-CoV-1. *N. Engl. J. Med.* **2020**, *382*, 1564–1567. [CrossRef]
5. Bak, A.; Mugglestone, M.A.; Ratnaraja, N.V.; Wilson, J.A.; Loveday, H.P.; Rivett, L.; Wilson, A.; Islam, J.; Loveday, H.P.; Price, J.; et al. SARS-CoV-2 routes of transmission and recommendations for preventing acquisition: Joint British Infection Association (BIA), Healthcare Infection Society (HIS), Infection Prevention Society (IPS) and Royal College of Pathologists (RCPath) guidance. *J. Hosp. Infect.* **2021**, *114*, 79–103. [CrossRef] [PubMed]
6. Jones, T.C.; Biele, G.; Mühlemann, B.; Veith, T.; Schneider, J.; Beheim-Schwarzbach, J.; Drosten, C.; Corman, V.M.; Edelmann, A.; Stein, A.; et al. Estimating infectiousness throughout SARS-CoV-2 infection course. *Science* **2021**, *373*, eabi5273. [CrossRef]
7. Walensky, R.P.; Walke, H.T.; Fauci, A.S. SARS-CoV-2 Variants of Concern in the United States-Challenges and Opportunities. *JAMA* **2021**, *325*, 1037–1038. [CrossRef]
8. Gu, H.; Krishnan, P.; Ng, D.Y.M.; Chang, L.D.J.; Liu, G.Y.Z.; Cheng, S.S.M.; Hui, M.M.Y.; Fan, M.C.Y.; Wan, J.H.L.; Lau, L.H.K.; et al. Probable Transmission of SARS-CoV-2 Omicron Variant in Quarantine Hotel, Hong Kong, China, November 2021. *Emerg. Infect. Dis.* **2022**, *28*, 460–462. [CrossRef] [PubMed]
9. Zhang, W.; Davis, B.D.; Chen, S.S.; Sincuir Martinez, J.M.; Plummer, J.T.; Vail, E. Emergence of a Novel SARS-CoV-2 Variant in Southern California. *JAMA* **2021**, *325*, 1324–1326. [CrossRef] [PubMed]
10. Boehm, E.; Kronig, I.; Neher, R.A.; Eckerle, I.; Vetter, P.; Kaiser, L. Geneva Centre for Emerging Viral Diseases. Novel SARS-CoV-2 variants: The pandemics within the pandemic. *Clin. Microbiol. Infect.* **2021**, *27*, 1109–1117. [CrossRef]
11. V'kovski, P.; Kratzel, A.; Steiner, S.; Stalder, H.; Thiel, V. Coronavirus biology and replication: Implications for SARS-CoV-2. *Nat. Rev. Microbiol* **2021**, *19*, 155–170. [CrossRef]
12. Hillen, H.S.; Kokic, G.; Farnung, L.; Dienemann, C.; Tegunov, D.; Cramer, P. Structure of replicating SARS-CoV-2 polymerase. *Nature* **2020**, *584*, 154–156. [CrossRef]
13. Troyano-Hernáez, P.; Reinosa, R.; Holguín, Á. Evolution of SARS-CoV-2 Envelope, Membrane, Nucleocapsid, and Spike Structural Proteins from the Beginning of the Pandemic to September 2020: A Global and Regional Approach by Epidemiological Week. *Viruses* **2021**, *13*, 243. [CrossRef] [PubMed]
14. Ke, Z.; Oton, J.; Qu, K.; Cortese, M.; Zila, V.; McKeane, L.; Briggs, A. Structures and distributions of SARS-CoV-2 spike proteins on intact virions. *Nature* **2020**, *588*, 498–502. [CrossRef] [PubMed]
15. Merad, M.; Blish, C.A.; Sallusto, F.; Iwasaki, A. The immunology and immunopathology of COVID-19. *Science* **2022**, *375*, 1122–1127. [CrossRef]
16. Kasuga, Y.; Zhu, B.; Jang, K.J.; Yoo, J.S. Innate immune sensing of coronavirus and viral evasion strategies. *Exp. Mol. Med.* **2021**, *53*, 723–736. [CrossRef] [PubMed]
17. Rydyznski Moderbacher, C.; Ramirez, S.I.; Dan, J.M.; Grifoni, A.; Hastie, K.M.; Weiskopf, D.S.; Crotty, S. Antigen-Specific Adaptive Immunity to SARS-CoV-2 in Acute COVID-19 and Associations with Age and Disease Severity. *Cell* **2020**, *183*, 996–1012.e19. [CrossRef]
18. Choi, A.; Koch, M.; Wu, K.; Dixon, G.; Oestreicher, J.; Legault, H.; Stewart-Jones, G.B.E.; Colpitts, T.; Pajon, R.; Bennett, H.; et al. Serum Neutralizing Activity of mRNA-1273 against SARS-CoV-2 Variants. *J. Virol.* **2021**, *95*, e0131321. [CrossRef] [PubMed]
19. Long, Q.X.; Liu, B.Z.; Deng, H.J.; Wu, G.C.; Deng, K.; Chen, Y.K. Antibody responses to SARS-CoV-2 in patients with COVID-19. *Nat. Med.* **2020**, *26*, 845–848. [CrossRef] [PubMed]
20. Wong, C.K.H.; Au, I.C.H.; Lau, K.T.K.; Lau, E.H.Y.; Cowling, B.J.; Leung, G.M. Real-world effectiveness of molnupiravir and nirmatrelvir plus ritonavir against mortality, hospitalization, and in-hospital outcomes among community-dwelling, ambulatory patients with confirmed SARS-CoV-2 infection during the omicron wave in Hong Kong: An observational study. *Lancet* **2022**, *400*, 1213–1222.
21. Gottlieb, R.L.; Nirula, A.; Chen, P.; Boscia, J.; Heller, B.; Morris, J. Effect of Bamlanivimab as Monotherapy or in Combination With Etesevimab on Viral Load in Patients With Mild to Moderate COVID-19: A Randomized Clinical Trial. *JAMA* **2021**, *325*, 632–644. [CrossRef]
22. World Health Organisation (WHO). COVID-19 Vaccines (Internet). Available online: https://www.who.int/emergencies/diseases/novel-coronavirus-2019/covid-19-vaccines (accessed on 20 October 2022).
23. Creech, C.B.; Walker, S.C.; Samuels, R.J. SARS-CoV-2 Vaccines. *JAMA* **2021**, *325*, 1318–1320. [CrossRef] [PubMed]
24. Grumbach, K.; Carson, M.; Harris, O.O. Achieving Racial and Ethnic Equity in COVID-19 Vaccination: From Individual Readiness to Health System Readiness. *JAMA Health Forum.* **2021**, *2*, e211724. [CrossRef]
25. Saudi Ministry of Health (Internet). Available online: https://covid19.moh.gov.sa/ (accessed on 20 November 2022).

26. Falsey, A.R.; Sobieszczyk, M.E.; Hirsch, I.; Sproule, S.; Robb, M.L.; Corey, L.; Neuzil, K.M.; Hahn, W.; Hunt, J.; Mulligan, M.J.; et al. Phase 3 Safety and Efficacy of AZD1222 (ChAdOx1 nCoV-19) Covid-19 Vaccine. *N. Engl. J. Med.* **2021**, *385*, 2348–2360. [CrossRef] [PubMed]
27. Martínez-Baz, I.; Trobajo-Sanmartín, C.; Miqueleiz, A.; Guevara, M.; Fernández-Huerta, M.; Burgui, C.; Casado, I.; Portillo, M.E.; Navascués, A.; Ezpeleta, C.; et al. Product-specific COVID-19 vaccine effectiveness against secondary infection in close contacts, Navara, August 2021. *Eurosurveillance* **2021**, *26*, 2100894. [CrossRef] [PubMed]
28. El Sahly, H.M.; Baden, L.R.; Essink, B.; Doblecki-Lewis, S.; Martin, J.M.; Anderson, E.J. Efficacy of the mRNA-1273 SARS-CoV-2 vaccine at completion of blinded phase. *N. Engl. J. Med.* **2021**, *385*, 1774–1785. [CrossRef] [PubMed]
29. Skowronski, D.M.; De Serres, G. Safety and Efficacy of the BNT16. mRNA Covid-19 Vaccine. *N. Engl. J. Med.* **2021**, *384*, 1576–1578.
30. Pasquale, S.; Gregorio, G.L.; Caterina, A.; Francesco, A.; Beatrice, P.M.; Vincenzo, P.; Caterina, P.M. COVID-19 in Low- and Middle-Income Countries (LMICs): A Narrative Review from Prevention to Vaccination Strategy. *Vaccines* **2021**, *9*, 1477. [CrossRef]
31. Voysey, M.; Clemens, S.A.C.; Madhi, S.A.; Weckx, L.Y.; Folegatti, P.M.; Aley, P. Oxford COVID Vaccine Trial Group. Safety and Efficacy of the ChAdOx1 nCoV-19 vaccine (AZD1222) against SARS-CoV-2: An interim analysis of four randomised controlled trials in Brazil, South Africa, and the UK. *Lancet* **2021**, *397*, 99–111. [CrossRef]
32. Dagan, N.; Barda, N.; Kepten, E.; Miron, O.; Perchik, S.; Katz, M.A.; Hernán, M.A.; Lipsitch, M.; Reis, B.; Balicer, R.D. BNT162b2 mRNA Covid-19 Vaccine in a Nationwide Mass Vaccination Setting. *N. Engl. J. Med. Mass. Med. Soc.* **2021**, *384*, 1412–1423. [CrossRef] [PubMed]
33. Baden, L.R.; El Sahly, H.M.; Essink, B.; Kotloff, K.; Frey, S.; Novak, R. Efficacy and Safety of the mRNA-1273 SARS-CoV-2 Vaccine. *N. Engl. J. Med. Mass. Med. Soc.* **2021**, *384*, 403–416. [CrossRef]
34. Bergwerk, M.; Gonen, T.; Lustig, Y.; Amit, S.; Lipsitch, M.; Cohen, C.; Mandelboim, M.; Levin, E.G.; Rubin, C.; Indenbaum, V.; et al. COVID-19 Breakthrough Infections in Vaccinated Health Care Workers. *N. Engl. J. Med.* **2021**, *385*, 1474–1484. [CrossRef]
35. Lipsitch, M.; Krammer, F.; Regev-Yochay, G.; Lustig, Y.; Balicer, R.D. SARS-CoV-2 breakthrough infections in vaccinated individuals: Measurement, causes and impact. *Nat. Rev. Immunol.* **2022**, *22*, 57–65. [CrossRef] [PubMed]
36. Sattar, N.; McInnes, I.B.; McMurray, J.J.V. Obesity a risk factor for severe COVID-19 infection: Multiple potential mechanisms. *Circulation* **2020**, *142*, 4–6. [CrossRef] [PubMed]
37. Simonnet, A.; Chetboun, M.; Poissy, J.; Raverdy, V.; Noulette, J.; Duhamel, A. High prevalence of obesity in severe acute respiratory syndrome coronavirus-2 (SARS-CoV-2) requiring invasive mechanical ventilation. *Obesity* **2020**, *28*, 1195–1199. [CrossRef] [PubMed]
38. El Bcheraoui, C.; Afshin, A.; Charara, R.; Khalil, I.; Moradi-Lakeh, M.; Kassebaum, N.J.; Collison, M.; Daoud, F. GBD 2015 Eastern Mediterranean Region Obesity Collaborators. The burden of obesity in the Eastern Mediterranean Region: Findings from the Global Burden of Disease 2015 study. *Int. J. Public Health* **2018**, *63*, 165–176.
39. Singh, A.S.; Gupta, R.; Ghosh, A.; Misra, A. Diabetes in COVID-19: Prevalence, pathophysiology, prognosis and practical considerations. *Diabetes Metab. Syndr.* **2020**, *14*, 303–310. [CrossRef]
40. Al Dawish, M.A.; Robert, A.A.; Braham, R.; Al Hayek, A.A.; Al Saeed, A.; Ahmed, R.A.; Al Sabaan, F.S. Diabetes Mellitus in Saudi Arabia. *Curr. Diabetes Rev.* **2016**, *12*, 359–368. [CrossRef]
41. Shang, J.; Wang, Q.; Zhang, H.; Wang, X.; Wan, J.; Yan, Y.; Gao, Y.; Cheng, J.; Li, Z.; Lin, J. The Relationship Between Diabetes Mellitus and COVID-19 Prognosis: A Retrospective Cohort Study in Wuhan, China. *Am. J. Med.* **2021**, *134*, e6–e14. [CrossRef]
42. Robert, A.A.; Al Saeed, A.; Al Dawish, M.A. COVID-19 among people with diabetes mellitus in Saudi Arabia: Current situation and new perspectives. *Diabetes Metab. Syndr. Clin. Res. Rev.* **2021**, *15*, 102231. [CrossRef]
43. Alguwaihes, A.M.; Al-Sofiani, M.E.; Megdad, M.; Albader, S.S.; Alsari, M.H.; Alelayan, A.; Saad, H. Diabetes and COVID-19 among hospitalized patients in Saudi Arabia: A single-center retrospective study. *Cardiovasc. Diabetol.* **2020**, *19*, 20. [CrossRef]
44. Saha, A.; Ahsan, M.M.; Quader, T.-U.; Naher, S.; Akter, F.; Mehedi, H.H.; Chowdhury, A.A.U.; Karim, H.; Rahman, T.; Parvin, A. Clinical characteristics and outcomes of COVID-19 infected diabetic patients admitted in ICUs of the southern region of Bangladesh. *Diabetes Metab. Syndr. Clin. Res. Rev.* **2021**, *15*, 229–235. [CrossRef]
45. Aihong, W.; Weibo, Z.; Zhangrong, X.; Jianwen, G. Timely blood glucose management for the outbreak of 2019 novel coronavirus disease (COVID-19) is urgently needed. *Diabetes Res. Clin. Pract.* **2020**, *162*, 108118.
46. Moutschen, M.P.; Scheen, A.J.; Lefebvre, P.J. Impaired immune responses in diabetes mellitus: Analysis of the factors and mechanisms involved. Relevance to the increased susceptibility of diabetic patients to specific infections. *Diabete Metab.* **1992**, *18*, 187–201.
47. Lee, K.A.; Ma, W.; Sikavi, D.R.; Drew, D.A.; Nguyen, L.H.; Bowyer, R.C.E.; Cardoso, M.J.; Fall, T.; Freidin, M.B.; Gomez, M.; et al. Cancer and Risk of COVID-19 Through a General Community Survey. *Oncologist* **2021**, *26*, e182–e185. [CrossRef] [PubMed]
48. Fabiani, M.; Puopolo, M.; Filia, A.; Sacco, C.; Mateo-Urdiales, A.; Alegiani, S.; Pezzotti, P. Effectiveness of an mRNA vaccine booster dose against SARS-CoV-2 infection and severe COVID-19 in persons aged ≥60 years and other high-risk groups during predominant circulation of the delta variant in Italy, July 19 to December 12 2021. *Expert Rev. Vaccines* **2022**, *21*, 1–8. [CrossRef]
49. de Oliveira, P.M.N.; Mendes-De-Almeida, D.P.; Porto, V.B.G.; Cordeiro, C.C.; Teixeira, G.V.; Pedro, R.S.; Takey, P.R.G.; Lignani, L.K.; Xavier, J.R.; da Gama, V.C.D.; et al. Vaccine-induced immune thrombotic thrombocytopenia after COVID-19 Vaccination: Description of a series of 39 cases in Brazil. *Vaccine* **2022**, *40*, 4788–4795. [CrossRef] [PubMed]
50. Greinacher, A.; Thiele, T.; Warkentin, T.E. Thrombotic thrombocytopenia after ChAdOx1 nCov-19 vaccination. *N. Engl. J. Med.* **2021**, *9*, 33835769. [CrossRef]

51. Alkindi, S.; Elsadek, R.; Al-Madhani, A.; Al-Musalhi, M.; Al-Khadouri, G.; Al Rawahi, B.; Al-Ruqeishi, S.; Al-Yazeedi, J.; Wali, Y.; Al Shamakhi, S.; et al. Impact of COVID-19 on vasooclusive crisis in patients with sickle cell anaemia. *Int. J. Infect. Dis.* **2021**, *106*, 128–133. [CrossRef]
52. Alkindi, S.; Elsadek, R.A.; Pathare, A.V. Safety Warning for ChAdOx1 nCov-19 Vaccine in Patients with Sickle Cell Disease. *Mediterr. J. Hematol. Infect. Dis.* **2021**, *13*, e2021059. [CrossRef]
53. Elnahry, A.G.; Al-Nawaflh, M.Y.; Gamal Eldin, A.A.; Solyman, O.; Sallam, A.B.; Phillips, P.H.; Elhusseiny, A.M. COVID-19 Vaccine-Associated Optic Neuropathy: A Systematic Review of 45 Patients. *Vaccines* **2022**, *10*, 1758. [CrossRef]
54. Diaz, G.A.; Parsons, G.T.; Gering, S.K.; Meier, A.R.; Hutchinson, I.V.; Robicsek, A. Myocarditis and Pericarditis After Vaccination for COVID-19. *JAMA* **2021**, *326*, 1210–1212. [CrossRef]
55. Montgomery, J.; Ryan, M.; Engler, R.; Hoffman, D.; McClenathan, B.; Collins, L.; Loran, D.; Hrncir, D.; Herring, K.; Platzer, M.; et al. Myocarditis Following Immunization With mRNA COVID-19 Vaccines in Members of the US Military. *JAMA Cardiol.* **2021**, *6*, 1202. [CrossRef]
56. Kim, H.W.; Jenista, E.R.; Wendell, D.C.; Azevedo, C.F.; Campbell, M.J.; Darty, S.N.; Parker, M.A.; Kim, R.J. Patients With Acute Myocarditis Following mRNA COVID-19 Vaccination. *JAMA Cardiol.* **2021**, *6*, 1196. [CrossRef]
57. Allen, C.M.; Ramsamy, S.; Tarr, A.W.; Tighe, P.J.; Irving, W.L.; Tanasescu, R.; Evans, J.R. Guillain-Barré Syndrome Variant Occurring after SARS-CoV-2 Vaccination. *Ann. Neurol.* **2021**, *90*, 315–318. [CrossRef] [PubMed]
58. Finsterer, J. Exacerbating Guillain-Barré Syndrome Eight Days after Vector-Based COVID-19 Vaccination. *Case Rep. Infect. Dis.* **2021**, *2021*, 3619131. [CrossRef] [PubMed]
59. Maramattom, B.V.; Krishnan, P.; Paul, R.; Padmanabhan, S.; Nampoothiri, D.S.C.V.; Syed, A.A.; Mangat, H.S. Guillain-Barré Syndrome following ChAdOx1-S / nCoV -19 Vaccine. *Ann. Neurol.* **2021**, *90*, 312–314. [CrossRef] [PubMed]
60. Hatzakis, A.; Karabinis, A.; Roussos, S.; Pantazis, N.; Degiannis, D.; Chaidaroglou, A.; Petsios, K.; Pavlopoulou, I.; Tsiodras, S.; Paraskevis, D.; et al. Modelling SARS-CoV-2 Binding Antibody Waning 8 Months after BNT162b2 vaccination. *Vaccines* **2022**, *10*, 285. [CrossRef]
61. Levin, E.G.; Lustig, Y.; Cohen, C.; Fluss, R.; Indenbaum, V.; Amit, S.; Doolman, R.; Asraf, K.; Mendelson, E.; Ziv, A.; et al. Waning Immune Humoral Response to BNT162b2 Covid-19 Vaccine over 6 Months. *N. Engl. J. Med.* **2021**, *385*, e84. [CrossRef]
62. Israel, A.; Shenhar, Y.; Green, I.; Merzon, E.; Golan-Cohen, A.; Schäffer, A.A.; Ruppin, E.; Vinker, S.; Magen, E. Large-Scale Study of Antibody Titer Decay following BNT162b2 mRNA Vaccine or SARS-CoV-2 Infection. *Vaccines* **2021**, *10*, 64. [CrossRef]
63. Doria-Rose, N.; Suthar, M.S.; Makowski, M.; O'Connell, S.; McDermott, A.B.; Flach, B.; Ledgerwood, J.E.; Mascola, J.R.; Graham, B.S.; Lin, B.C.; et al. Antibody Persistence through 6 Months after the Second Dose of mRNA-1273 Vaccine for Covid-19. *N. Engl. J. Med.* **2021**, *384*, 2259–2261. [CrossRef]

viruses

MDPI

Article

Combined Use of RT-qPCR and NGS for Identification and Surveillance of SARS-CoV-2 Variants of Concern in Residual Clinical Laboratory Samples in Miami-Dade County, Florida

Yamina L. Carattini [1], Anthony Griswold [2], Sion Williams [3], Ranjini Valiathan [1], Yi Zhou [1,3], Bhavarth Shukla [4], Lilian M. Abbo [4], Katiuska Parra [5], Merce Jorda [1,3], Stephen D. Nimer [3,4], Corneliu Sologon [3], Hilma R. Gallegos [3], Roy E. Weiss [4], Tanira Ferreira [4], Abdul Memon [6], Peter G. Paige [6,†], Emmanuel Thomas [1] and David M. Andrews [1,*]

1 Department of Pathology and Laboratory Medicine, University of Miami Miller School of Medicine, Miami, FL 33136, USA
2 John P. Hussman Institute for Human Genomics, University of Miami Miller School of Medicine, Miami, FL 33136, USA
3 Sylvester Comprehensive Cancer Center, University of Miami Miller School of Medicine, Miami, FL 33136, USA
4 Department of Medicine, University of Miami Miller School of Medicine, Miami, FL 33136, USA
5 Laboratory Services, Pathology Department, Jackson Memorial Hospital, Miami, FL 33136, USA
6 Jackson Health System, Miami, FL 33136, USA
* Correspondence: dandrews@med.miami.edu
† Current address: Albany Medical Center, Albany, NY 12208, USA.

Abstract: Over the course of the COVID-19 pandemic, SARS-CoV-2 variants of concern (VOCs) with increased transmissibility and immune escape capabilities, such as Delta and Omicron, have triggered waves of new COVID-19 infections worldwide, and Omicron subvariants continue to represent a global health concern. Tracking the prevalence and dynamics of VOCs has clinical and epidemiological significance and is essential for modeling the progression and evolution of the COVID-19 pandemic. Next generation sequencing (NGS) is recognized as the gold standard for genomic characterization of SARS-CoV-2 variants, but it is labor and cost intensive and not amenable to rapid lineage identification. Here we describe a two-pronged approach for rapid, cost-effective surveillance of SARS-CoV-2 VOCs by combining reverse-transcriptase quantitative polymerase chain reaction (RT-qPCR) and periodic NGS with the ARTIC sequencing method. Variant surveillance by RT-qPCR included the commercially available TaqPath COVID-19 Combo Kit to track S-gene target failure (SGTF) associated with the spike protein deletion H69-V70, as well as two internally designed and validated RT-qPCR assays targeting two N-terminal-domain (NTD) spike gene deletions, NTD156-7 and NTD25-7. The NTD156-7 RT-qPCR assay facilitated tracking of the Delta variant, while the NTD25-7 RT-qPCR assay was used for tracking Omicron variants, including the BA.2, BA.4, and BA.5 lineages. In silico validation of the NTD156-7 and NTD25-7 primers and probes compared with publicly available SARS-CoV-2 genome databases showed low variability in regions corresponding to oligonucleotide binding sites. Similarly, in vitro validation with NGS-confirmed samples showed excellent correlation. RT-qPCR assays allow for near-real-time monitoring of circulating and emerging variants allowing for ongoing surveillance of variant dynamics in a local population. By performing periodic sequencing of variant surveillance by RT-qPCR methods, we were able to provide ongoing validation of the results obtained by RT-qPCR screening. Rapid SARS-CoV-2 variant identification and surveillance by this combined approach served to inform clinical decisions in a timely manner and permitted better utilization of sequencing resources.

Keywords: SARS-CoV-2; COVID-19; delta; VOC; RT-qPCR; surveillance

Citation: Carattini, Y.L.; Griswold, A.; Williams, S.; Valiathan, R.; Zhou, Y.; Shukla, B.; Abbo, L.M.; Parra, K.; Jorda, M.; Nimer, S.D.; et al. Combined Use of RT-qPCR and NGS for Identification and Surveillance of SARS-CoV-2 Variants of Concern in Residual Clinical Laboratory Samples in Miami-Dade County, Florida. *Viruses* **2023**, *15*, 593. https://doi.org/10.3390/v15030593

Academic Editors: Ahmed El-Shamy and Mohamed Ibrahim

Received: 8 December 2022
Revised: 14 February 2023
Accepted: 17 February 2023
Published: 21 February 2023

1. Introduction

Throughout the COVID-19 pandemic, SARS-CoV-2 has continuously acquired mutations throughout its genome, resulting in the emergence of over a thousand genetic variations of the original Wuhan strain, categorized into lineages and larger phylogenetic groups, or clades [1]. A subset of these variants have been designated as variants of concern (VOC) by public health organizations including the WHO [2], the Centers for Disease Control and Prevention (CDC) [3], and the European Centre for Disease Prevention and Control (ECDC) [4] based on clinical and epidemiological data and on the presence of one or more mutations of interest, predominantly in the spike (S) glycoprotein (e.g., D614G, N501Y, N417K, E484K/Q, L452R, T478K, and P681R). Such mutations have been associated with increased transmissibility and/or virulence, enhanced ability to evade protective immunity from natural infection or vaccination, and/or reduced efficacy of therapeutic monoclonal antibodies [5–9]. Genetic mutations of interest may confer additional functional properties to the virus and have emerged through parallel evolutionary selection in several different lineages without defining a particular lineage or strain. In March 2022, the WHO and the CDC recognized two VOCs, Delta (B.1.617.2, including AY lineages) and Omicron (B.1.1.529, including BA. lineages) [3,10]

The Delta variant emerged in India in late 2020 and was classified as a VOC by the WHO in May 2021. With increased transmissibility and enhanced immune escape capabilities compared to previous variants [11–14], Delta caused waves of new COVID-19 infections around the world and rapidly outcompeted other lineages and VOCs including Alpha (B.1.1.7) and Gamma (P.1), achieving dominance in most countries by the fall of 2021 [15]. Soon after the culmination of the Delta-driven wave of COVID-19 infections, the Omicron variant emerged in South Africa with an unprecedented and alarming number of mutations in the S glycoprotein and was subsequently designated a VOC by the WHO on 26 November 2021 [16]. Of approximately 33 S gene mutations in present in the Omicron variant, 15 are in the receptor-binding domain (RBD), a region which represents the primary target of monoclonal antibody therapies and most SARS-CoV-2 vaccines [17]. More transmissible and better at immune evasion than previous variants [18–21], Omicron rapidly became the dominant strain throughout the world and drove the global number of new COVID-19 infections from 269 million in mid-December 2021 to more than 430 million by late-February 2022. The number of global COVID-19-related deaths rose from 5.3 million to over 6 million in the same period [22]. Iketani et al. showed disparate efficacy of monoclonal antibodies against different Omicron lineages (BA.1 vs. BA.2) [23], which rapidly changed the therapeutic landscape. Additionally, through its increased transmissibility and immune evasion properties, Omicron is characterized by high susceptibility to breakthrough infection, even in fully vaccinated individuals [24]. As Omicron continues its genetic evolution, ongoing surveillance of locally circulating lineages is necessary in order to monitor for the emergence of novel strains designated as Omicron subvariants under monitoring by the WHO. As of 13 September 2022, new COVID-19 cases in the United States (US) are predominantly of the Omicron BA.5 sublineage (~89%), with BA.4 comprising ~10%, and BA.2.12.2 representing <1% [25].

Although NGS is recognized as the gold standard for SARS-CoV-2 variant identification, it is cost- and labor-intensive, limiting its widespread use [26]. In addition, the turnaround time for results can take as much as 1 week to be finalized given the complex workflows. Furthermore, widely used multiplex-polymerase-chain-reaction (PCR)-based sequencing approaches (i.e., ARTIC SARS-CoV-2 sequencing method) harbor inherent technical limitations compared to more robust sequencing approaches. One limitation of the ARTIC method is high sequencing failure rates for samples with PCR Ct values >32 [27]. Another limitation associated with this method is primer-variant overlap with heavily mutated lineages (e.g., Omicron), which can result in PCR amplification failure and a subsequent decrease in coverage [28–31]. In contrast, PCR-based approaches represent a simpler approach to track specific VOCs [32]. Molecular techniques involving PCR have played an important role in COVID-19 diagnostic testing and are widely used for viral nucleic acid detection due to their high sensitivity. Importantly, PCR-based tests are also more amenable to providing semi-quantitative results pertaining to viral load [33].

In addition to their utility in COVID-19 diagnostic testing, PCR-based methods provide a feasible alternative to NGS for SARS-CoV-2 variant detection and surveillance. A commercially available Reverse Transcriptase quantitative PCR (RT-qPCR) COVID-19 diagnostic test (TaqPath COVID-19 Combo Kit, Thermo Fisher, Inc., Waltham, MA) became a useful tool for tracking the Alpha (B.1.1.7) variant in the first half of 2021, after discovery that a 6-base-pair (bp) deletion at nucleotide positions 21,765–21,770 in the N-terminal domain of the S gene, corresponding to amino acids (aa's) Histidine-69 and Valine-70 (H69-V70) of the S glycoprotein, results in failed or delayed amplification (defined here by a mean S-gene target cycle threshold (Ct) delay = 6.3 compared to the N-gene target Ct), referred to as S-gene target failure (SGTF) or S-gene target late amplification (SGTL), while amplification of the other two targets in the assay (N-gene and ORF1ab-gene) are retained [34,35]. More recently, the prevalence of VOC Omicron lineages BA.1, BA.4, and BA.5, which also harbor the H69-V70 S deletion, have been successfully tracked by this method. We found excellent correlation between SGTF and SGTL with the TaqPath assay and NGS-confirmed Alpha and Omicron lineages (Ct <30), supporting the continued use of the TaqPath assay for tracking the prevalence of SARS-CoV-2 lineages carrying the H69-V70 deletion. In contrast to the Omicron BA.1, BA.4, and BA.5 lineages, the BA.2 lineages lack the H69-V70 S deletion that results in SGTF/SGTL and cannot be detected with the TaqPath assay. Other investigators have proposed RT-qPCR strategies targeting lineage-defining mutations as a reliable complement to NGS for detection and surveillance of SARS-CoV-2 variants [36–40]. Here we utilized a combined approach involving the TaqPath assay, targeted RT-qPCR, and NGS to follow changes in the SARS-CoV-2 variant pool.

2. Materials and Methods

2.1. Study Design

Deidentified, residual SARS-CoV-2-positive nasal and/or nasopharyngeal swab (NPS) samples (in transport media/solution) were used in this study. SARS-CoV-2 samples previously identified as positive by PCR were retrieved randomly from routine COVID-19 PCR testing performed in local clinical laboratories. From January 2021 to April 2022, we obtained 6482 residual COVID-19-positve samples for SARS-CoV-2 variant screening by PCR. Of these, 1668 samples were submitted for genomic characterization by NGS. Residual patient samples were obtained from two health systems in Miami-Dade County, Florida: University of Miami Health System (UHealth) and Jackson Health System (JHS), a large public county health system distributed throughout Miami-Dade County. Residual samples were also obtained from routine University of Miami COVID-19 surveillance screening and testing programs involving students and staff. All samples used in this study were permanently de-identified and unlinked from the patient or student/staff records and were used for surveillance purposes only. Consequently, study samples were considered by the Institutional Review Board (IRB) to represent non-human studies research material and did not require IRB approval.

2.2. NGS/ARTIC SARS-CoV-2 Sequencing Method

Sample sequencing was performed at the University of Miami Miller School of Medicine, Sylvester Comprehensive Cancer Center Onco-genomics Shared Resource (OGSR) facility using the NEBNext ARTIC SARS-CoV-2 FS Library Prep Kit (Illumina, San Diego, CA, USA) (NEB #E7658), according to manufacturer's instructions. Briefly, following reverse transcription of purified viral RNA, multiplexed PCR amplification of cDNA with a primer scheme produces multiple amplicons (~98) spanning the viral genome. Preparation of final PCR products for sequencing follows with barcoded adaptor ligation and addition of index primers in accordance with Illumina library preparation instructions. Sequencing was performed on the NextSeq 500 (Illumina, San Diego, CA, USA).

2.3. Bioinformatics

Bioinformatics analysis of raw FASTQ files was performed using the resources of the Institute for Data Science and Computing and performed by personnel at the John P. Hussman Institute for Human Genomics, both at the University of Miami Miller School of Medicine. The protocol used was based on a pipeline instituted at the Utah Department of Health and made publicly available via the Centers for Disease Control and Prevention GitHub page [41]. Secondary analysis of the resulting consensus FASTA file was performed in a two-part approach. First, the FASTA file was analyzed using Phylogenetic Assignment of Named Global Outbreak LINeages software (Pangolin https://github.com/cov-lineages/pangolin). Prior to each analysis, the version of Pangolin and its associated software and databases were updated to the most current version to account for software evolution and recently emerging annotations. The second part utilized the Nextclade web-based software (all versions: 10.5281/zenodo.5726680) [42]. Nextclade performs alignment of the FASTA file against the SARS-CoV-2 reference and quality control, calls variants, and assigns clades for each sample. A visual output of coverage and variant calls are provided as well as a tabular format for cataloging lineage information.

2.4. RT-qPCR Target Selection and Primer/Probe Design

To develop a targeted RT-qPCR assay that could distinguish Delta from other circulating variants, we searched SARS-CoV-2 Delta variant sequences retrieved from the GISAID database and subsequently selected a 6 bp deletion located in the NTD of the S gene as the target for the fluorescence resonance energy transfer (FRET) probe (qPCR 5-prime nuclease probe). The target mutation spans three codons and results in a 6 bp deletion of glutamate (E) at position 156 and phenylalanine (F) at position 157, also resulting in the non-synonymous substitution of arginine (R) for glycine (G) at position 158. We downloaded SARS-CoV-2 reference sequences from GISAID (hCoV-19/Wuhan/WIV04/2019) and NCBI (NC_045512.2) (https://www.gisaid.org/) to select the location and obtain the sequence for the primers and probe. We then modified the reference sequence for probe binding to the E156-F157 six-base-pair NTD deletion at nucleotide positions 22,029–22,034 and selected the location of the primers to ensure an amplicon size < 100 bp. Oligos were obtained from Integrated DNA Technologies (IDT, Carolville, IA) and were optimized in silico using the IDT OligoAnalyzer tool for determining melting temperatures (Tm) and potential for formation of secondary structures (hairpins or primer-dimer). The 24-base FRET probe was labeled at the 5′-end with the reporter molecule 6-carboxyfluorescein (FAM) and with a double quencher, ZEN (internal), and Iowa Black (3IABkFQ) at the 3′-end. The amplicon size for the NTD156-7 RT-qPCR assay is 96 bp. Using the same approach, we designed a mutation-specific RT-qPCR assay to detect and monitor the local prevalence of the Omicron BA.2, which subsequently served for surveillance of the BA.4 and BA.5 lineages as well. In this case, the NTD 9 bp deletion selected as the target spans four codons resulting in a non-synonymous substitution of leucine (L) with serine (S) at position 24, and deletion of two proline (P) and one alanine (A) residues at positions 25–27 of the S glycoprotein. The 22 base FRET probe was labeled at the 5′-end with the reporter molecule Cyanine 5 (Cy5) and with TAO and Iowa Black (3IAbRQSp) as internal and 3′-end quenchers, respectively. The amplicon size for the NTD25-7 assay is 88 bp.

2.5. In Silico Validation of Targeted RT-qPCR Assays

To determine the rate of occurrence of mutations in the selected primers and probe sequences, we utilized the PrimerChecker function provided by the GISAID EpiCov resource (https://www.gisaid.org/). In short, the PrimerChecker performs a search using basic local alignment search tool (BLAST) parameters for short sequence matches to search input primer sequences against high quality (<1% N and <0.05% non-synonymous mutations) genome sequences deposited in GISAID. We analyzed sequences deposited from 23 February 2021 to 22 August 2021 (1,305,468 sequences) for the NTD156-7 primers and probe and from 19 December 2021 to 19 March 2022 (220,000 sequences) for the NTD25-7 primers and probe. These searches provided a list of sequences with one or more mutations in the binding regions for primers and FRET probes and captured the cumulative mutation rate during the period analyzed.

2.6. In Vitro Validation of Targeted RT-qPCR Assays

To validate the primers and FRET probe for the Delta-specific RT-qPCR assay (NTD156-7) in vitro, we obtained extracted RNA from 50 COVID-19-positive samples of variants previously identified by NGS. RNA extractions were performed on a Chemagic 360 instrument using the Chemagic Viral DNA/RNA 300 Kit H96 (PerkinElmer, Downers Grove, ILfollowing manufacturer's instructions. Of the 50 samples, n = 24 were identified as Delta variant (B.1.617.2 (9), AY.2 (1), AY.3 (3), AY.3.1 (2), AY.4 (1), AY.5 (1), AY.12 (1), AY.24 (1), AY.25 (5)), and n = 26 were non-Delta variants (B.1 (1), B.1.1.7 (Alpha) (6), B.1.427 (2), B.1.429 (2), B.1.526 (3), B.1.621 (Mu) (2), B.1.621.1 (Mu) (2), B.1.623 (1), B.1.628 (1), C.37 (Lambda) (3), P.1 (Gamma) (3)). For a 1 × 20 μL qPCR reaction, we used 5 μL of TaqPath 1-Step Multiplex Master Mix, no ROX (Thermo Fisher, Waltham, MA), 0.8 μL each forward and reverse primer from a 10 μM working concentration (final concentration of 400 nM each primer), 0.4 μL of FRET probe from a 10 μM working concentration (final concentration of 200 nM), 8 μL of molecular grade water, and 5 μL of purified sample RNA. RT-qPCR was performed on a QuantStudio 7 instrument. The assay was performed with the following PCR cycling conditions: 25 °C for 2 min, 53 °C for 30 min (reverse transcription, RT), 95 °C for 3 min, and 45 cycles of 95 °C for 10 s and 60 °C for 30 s.

For the Omicron-specific NTD25-7 assay, we used the same reaction and cycling conditions allowing for an eventual transition to a multiplexed assay containing both NTD156-7 and NTD25-7 RT-qPCR assays. As other SARS-CoV-2 variants have been displaced by the emergence of dominant strains, such as Delta and Omicron, we included two Delta lineages (AY.3 and AY.47) which were circulating within our study population at the time, albeit in low numbers (~2% of sequenced samples), and Omicron lineages BA.1, BA.1.1, and BA.2 for the in vitro validation of the Omicron-specific RT-qPCR assay. We followed the same protocol as described above for 56 COVID-19-positive samples previously identified by NGS as AY.3 (n = 2), AY.47 (n= 1), BA.1 (n = 9), BA.1.1 (n = 17), and BA.2 (n = 27) lineages to assess the performance of the NTD25-7 assay. The TaqPath COVID-19 Combo Kit provides negative, positive, and internal controls to monitor the reliability of the results for the entire batch of specimens from RNA extraction to PCR amplification. According to manufacturer's instructions, an internal control Ct < 37 is considered positive, and N, ORF1ab, and S Ct >37 is considered negative for SARS-CoV-2. In our study, the Ct value cutoff was <35, and samples lacking amplification (Ct = 0) for at least one of the three targets (except for the S gene target in the case of SGTF) were considered "not evaluable".

3. Results

3.1. ARTIC SARS-CoV-2 Sequencing Method Performance Is Directly Related to Sample Ct Values

In our study, we found a direct relationship between sample Ct values obtained from the TaqPath assay and sequencing failure rate with the ARTIC method. We performed a retrospective analysis of 1650 samples submitted for NGS and used the TaqPath N-gene target as the reference Ct to determine the relationship between Ct value and sequencing failure rate. The rationale for selecting the N-gene as the reference Ct for this analysis was based on the observation that the Ct values for the ORF1ab and S gene targets (TaqPath) generally trended higher compared to the N gene target, suggesting that the N gene target is more analytically sensitive. Furthermore, the N gene is highly conserved and has a lower mutation rate than the S gene [43,44]. From this retrospective analysis, we found 126/246 of our sequenced samples with N-gene target Ct values ranging from 30 to 35 (TaqPath COVID-19 Combo Kit RT-qPCR assay) failed, representing a sequencing failure rate of 51%. This failure rate was significantly decreased in samples with Ct values < 30, where only 54/1404 (4%) resulted in sequencing failure (Figure 1). To better understand the impact of Ct value on NGS performance with the ARTIC method, we grouped the 246 samples that failed NGS into Ct value ranges of 30.1–30.9, 31.0–31.9, 32.0–32.9, 33.0–33.9, and 34.0–34.9 and found sequencing failure increased in a Ct-dependent manner (correlation $R^2 = 0.999$), 28%, 40%, 56%, 69%, and 83%, respectively (Figure 2).

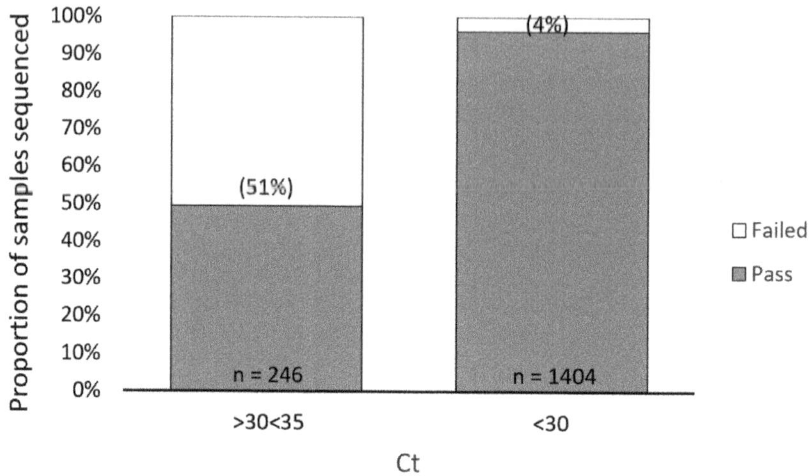

Figure 1. Impact of Ct Value on ARTIC SARS-CoV-2 Sequencing Performance. A retrospective analysis showing 126/246 of sequenced samples with N-gene target Ct values ranging from 30 to 35 (TaqPath COVID-19 Combo Kit RT-qPCR assay) failed, representing a sequencing failure rate of 51%. This failure rate was significantly decreased in samples with Ct values < 30, where only 54/1404 (4%) resulted in sequencing failure.

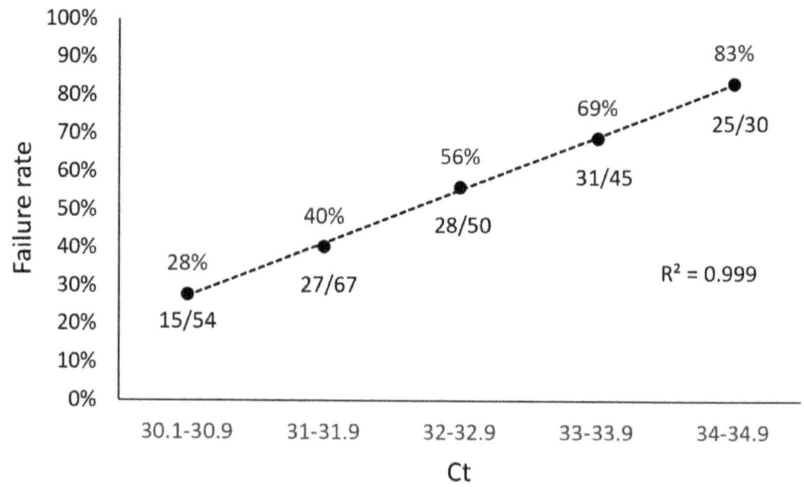

Figure 2. Correlation Between ARTIC SARS-CoV-2 Sequencing Failure Rate and Sample Ct. To determine the impact of Ct value on NGS performance with the ARTIC SARS-CoV-2 sequencing method, 246 sample that failed NGS were grouped into Ct value ranges of 30.1–30.9, 31.0–31.9, 32.0–32.9, 33.0–33.9, and 34.0–34.9. Sequencing failure rate increased in a Ct-dependent manner (correlation $R^2 = 0.999$), ranging from 28% (15/54 samples, Ct 30.1–30.9) to 83% (25/30 samples, Ct 34–34.9).

3.2. Targeted RT-qPCR for Detection and Surveillance of VOCs Delta and Omicron

The NTD156-7 assay was designed to target a highly conserved 6 bp deletion located in the N-terminal-domain (NTD) of the S gene at nucleotide positions 22,029–22,034 corresponding to aa glutamate (E) at position 156 and phenylalanine (F) at position 157 (E156-F157) of the Delta variant, including the AY lineages (Figure 3a). Similarly, the NTD25-7 assay targets a 9 bp deletion in the NTD of the S gene at nucleotide positions 21,633–21,641, which spans four codons resulting in a non-synonymous substitution of leucine (L) with serine (S) at position 24, and deletion of two proline (P) and one alanine (A) residues at positions 25–27 (P25-P26-A27) (Figure 3b).

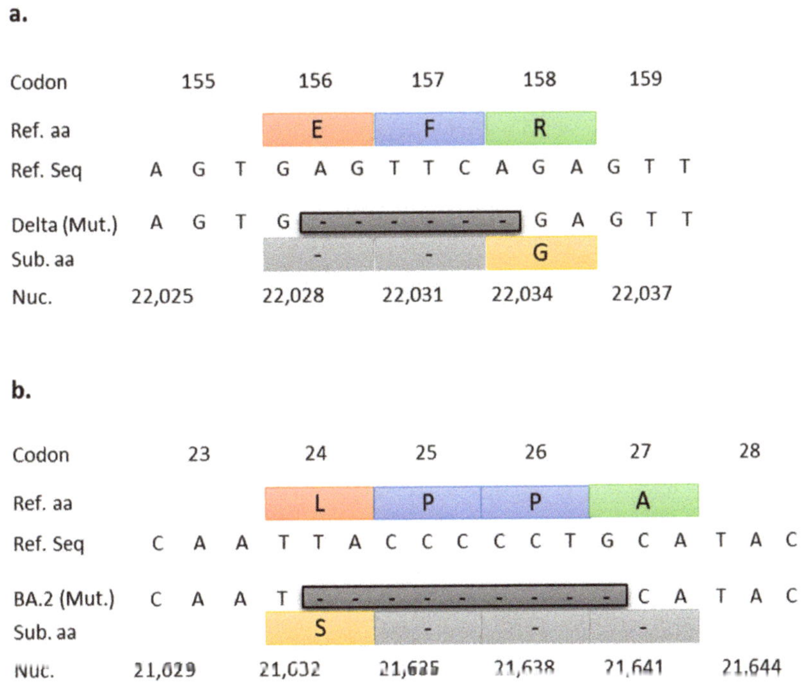

Figure 3. RT-qPCR target deletions for VOCs Delta and Omicron. (**a**) Alignment of Delta E156-F157 deletion with Wuhan reference sequence. This 6 bp deletion (Delta (Mut.)) spans three codons and results in the deletion of amino acids glutamate (E) at position 156 and phenylalanine (F) at position 157, and a non-synonymous substitution at position 158 (arginine (R) for glycine (G)). (**b**) Alignment of Omicron BA.2 lineage P25-P26-A27 deletion (also present in BA.4 and BA.5 lineages) with Wuhan reference sequence. The 9 bp deletion spans four codons resulting in a non-synonymous substitution of leucine (L) with serine (S) at position 24, and deletion of two proline (P) and one alanine (A) residues at positions 25–27. Ref. aa (reference amino acids), Ref. seq (reference sequence), Mut. (Mutation), Sub aa (substitution amino acids), Nuc. (nucleotides).

3.3. In Silico Validation Results

Primers and probes for the two targeted RT-qPCR assays (Table 1) were initially validated in silico to ensure a high level of assay performance with minimal requirement for in vitro optimization. The overall rate of mutation for the NTD156-7primers and FRET probe was 1.9%, based on a search of GISAID SARS-CoV-2 sequences deposited over the 6-month search period, 23 February 2021 to 22 August 2021 (1,305,468 sequences) (Table 2). A non-synonymous mutation (glycine to aspartic acid) in the fourth base from the 5-prime end of the forward primer at nucleotide position 21,987 in the NTD of the S gene (associated with the amino acid substitution G142D and seen predominantly in the Delta

Plus sub-lineages AY.1 and AY.2) had the highest mutation frequency (1.34%) [44]. This mutation at position 21,987 was not expected to have a significant impact on performance of the forward primer since the remaining 15 bases from the 3-prime end contain very low overall mutation frequencies. To determine the prevalence of this mutation in our study population, we performed an internal review of all Delta variant sequences from our surveillance population and found the G21987 base to be mutated in only 8/395 samples (2%). We tested the 8 samples containing the A21987 genotype (G142D substitution), and the NTD156-7 RT-qPCR assay detected 8/8 samples (CT values < 20). Sequence variability for the 24 -nucleotide FRET probe target revealed very low mutational frequency over the 6-month period analyzed. A nucleotide BLAST (performed 6 September 2021) of the probe, querying Betacoronavirus sequences deposited in the NCBI database, revealed 100% (24/24) concordance with 4997/5000 Delta variant targets, or a variation frequency of 0.06%. In silico validation of the NTD25-7 (Omicron BA.2, BA.4, and BA.5 lineages) primers and FRET probe showed an overall rate of mutation of 3.7%, based on 220,732 GISAID SARS-CoV-2 sequences deposited from 19 December 2021 to 19 March 2022 (Table 3). The forward primer was the oligonucleotide with the highest variability (2.37%); however, the rate of mutation was low within the last five bases at the 3′-end (0.29%). The reverse primer and FRET probe had overall rates of mutation of 1.27% and 0.07%, respectively, with low 3′-end mutation rates of 0.23% (reverse primer) and 0.01% (probe).

Table 1. Primers and probes for targeted RT-qPCR assays.

	Primer/Probe	Sequence 5′-3′
Delta (NTD156-7)	Forward Primer	TGGGTGTTTATTACCACAA
	Reverse Primer	GGCTGAGAGACATATTCAAA
	Probe	FAM-ATGGAAAGT/ZEN/GGAGTTTATTCTAGT-3IABkFQ
Omicron BA.2, BA.4, BA.5 (NTD25-7)	Forward Primer	TTTATTGCCACTAGTCTCTAGTCAG
	Reverse Primer	GGTAATAAACACCACGTGTGAAAG
	Probe	Cy5-AGAACTCAA/TAO/TCATACACTAATT-3IAbRQSp

Table 2. In silico Validation of NTD156-7 (Delta) Primers and Probe.

	NTD156-7 Forward Primer	Mutated Sequences		NTD156-7 Reverse Primer	Mutated Sequences		NTD156-7 Probe	Mutated Sequences	
	Ref Base	#	%	Ref Base	#	%	Ref Base	#	%
5′	T	48	0.004%	G	40	0.003%	A	196	0.015%
	G	3	0.000%	G	722	0.055%	T	187	0.014%
	G	4	0.000%	C	638	0.049%	G	1011	0.077%
	G	17,428	1.335%	T	68	0.005%	G	16	0.001%
	T	0	0.000%	G	346	0.027%	A	35	0.003%
	G	2	0.000%	A	512	0.039%	A	10	0.001%
	T	28	0.002%	G	77	0.006%	A	31	0.002%
	T	30	0.002%	A	455	0.035%	G	126	0.010%
	T	15	0.001%	G	45	0.003%	T	251	0.019%
	A	152	0.012%	A	19	0.001%	G	274	0.021%
	T	76	0.006%	C	142	0.011%	G	57	0.004%
	T	15	0.001%	A	142	0.011%	A	60	0.005%
	A	0	0.000%	T	4	0.000%	G	22	0.002%
	C	3	0.000%	A	156	0.012%	T	26	0.002%
	C	3	0.000%	T	36	0.003%	T	35	0.003%

Table 2. *Cont.*

	NTD156-7 Forward Primer	Mutated Sequences		NTD156-7 Reverse Primer	Mutated Sequences		NTD156-7 Probe	Mutated Sequences	
	Ref Base	**#**	**%**	**Ref Base**	**#**	**%**	**Ref Base**	**#**	**%**
	A	2	0.000%	T	7	0.001%	T	34	0.003%
	C	3	0.000%	C	12	0.001%	A	3	0.000%
	A	2	0.000%	A	10	0.001%	T	60	0.005%
	A	29	0.002%	A	181	0.014%	T	12	0.001%
				A	447	0.034%	C	54	0.004%
							T	18	0.001%
							A	20	0.002%
							G	149	0.011%
3′							T	29	0.002%

NTD156-7 Oligos	Sequence 5′-3′	% Sequences with any mutation	% Sequences with mutation in last 5 bases
Forward Primer	TGGGTGTTTATTACCACAA	1.367	0.003
Reverse Primer	GGCTGAGAGACATATTCAAA	0.311	0.050
Probe	ATGGAAAGTGGAGTTTATTCTAGT	0.210	0.020

Table 3. In silico Validation of NTD25-7 (Omicron) Primers and Probe.

	NTD25-7 Forward Primer	Mutated Sequences		NTD25-7 Reverse Primer	Mutated Sequences		NTD25-7 Probe	Mutated Sequences	
	Ref Base	**#**	**%**	**Ref Base**	**#**	**%**	**Ref Base**	**#**	**%**
5′	T	18	0.008%	G	24	0.011%	A	0	0.000%
	T	12	0.005%	G	385	0.174%	G	15	0.007%
	T	6	0.003%	T	19	0.009%	A	0	0.000%
	A	32	0.014%	A	12	0.005%	A	0	0.000%
	T	27	0.012%	A	81	0.037%	C	3	0.001%
	T	4	0.002%	T	284	0.129%	T	34	0.015%
	G	69	0.031%	A	281	0.127%	C	1	0.000%
	C	18	0.008%	A	9	0.004%	A	0	0.000%
	C	31	0.014%	A	6	0.003%	A	1	0.000%
	A	28	0.013%	C	926	0.420%	T	45	0.020%
	C	48	0.022%	A	8	0.004%	C	2	0.001%
	T	3	0.001%	C	8	0.004%	A	1	0.000%
	A	8	0.004%	C	10	0.005%	T	3	0.001%
	G	31	0.014%	A	9	0.004%	A	4	0.002%
	T	18	0.008%	C	14	0.006%	C	14	0.006%
	C	3672	1.664%	G	5	0.002%	A	0	0.000%
	T	17	0.008%	T	15	0.007%	C	8	0.004%
	C	480	0.217%	G	134	0.061%	T	12	0.005%
	T	38	0.017%	T	66	0.030%	A	0	0.000%
	A	31	0.014%	G	337	0.153%	A	1	0.000%
	G	242	0.110%	A	135	0.061%	T	12	0.005%
	T	33	0.015%	A	9	0.004%	T	0	0.000%
	C	30	0.014%	A	13	0.006%			
	A	52	0.024%	G	2	0.001%			

Table 3. *Cont.*

	NTD25-7 Forward Primer	Mutated Sequences		NTD25-7 Reverse Primer	Mutated Sequences		NTD25-7 Probe	Mutated Sequences	
	Ref Base	**#**	**%**	**Ref Base**	**#**	**%**	**Ref Base**	**#**	**%**
3′	G	285	0.129%						

NTD25-7 Oligos	Sequence 5′-3′	% Sequences with any mutation	% Sequences with mutation in last 5 bases
Forward Primer	TTTATTGCCACTAGTCTCTAGTCAG	2.371	0.291
Reverse Primer	GGTAATAAACACCACGTGTGAAAG	1.265	0.225
Probe	AGAACTCAATCATACACTAATT	0.071	0.011

The probe for the Delta assay (NTD156-7) was labeled at the 5′-end with the reporter molecule 6-carboxyfluorescein (FAM) and with a double quencher, ZEN (internal), and Iowa Black (3IABkFQ) at the 3′-end. The Omicron-specific assay (NTD25-7) probe is labeled at the 5′-end with the reported Cy5 and with quenchers TAO (internal) and Iowa Black at the 3′-end.

The overall rate of mutation for the NTD156-7 primers and FRET probe was 1.9%, based on a search of GISAID SARS-CoV-2 sequences deposited over the 6-month search period 23 February 2021 to 22 August 2021 (1,305,468 sequences). Reference bases are displayed vertically in a 5′ (top) to 3′ (bottom) orientation.

The overall rate of mutation for the NTD25-7 primers and FRET probe was 3.7%, based on 220,732 SARS-CoV-2 sequences deposited in GISAID from 19 December 2021 to 19 March 2022. Reference bases are displayed vertically in a 5′ (top) to 3′ (bottom) orientation.

3.4. In Vitro Validation Results

The performance of the NTD156-7assay was tested on 50 samples identified previously by NGS as Delta variants as well as an assortment of samples identified previously by NGS as non-Delta variants (see methods). The NTD156-7 RT-qPCR assay correctly identified 24/24 Delta variant samples with Ct values ranging from 20.19 to 38.6 (mean Ct = 28.43) representing a positive percent agreement (PPA) of 100%. The assay showed no amplification for 26/26 non-Delta variant samples of different lineages, representing a negative percent agreement (NPA) of 100% (Table 4). The NTD25-7 RT-qPCR assay performed equally well in vitro. Of the 53 BA.2 lineages (BA.2 (51), BA.2.3 (2)) previously identified by NGS, 53 showed amplification by our targeted RT-qPCR assay (mean CT = 27.2 and 20.1, respectively) (PPA = 100%), and 29/29 non-BA.2 lineage samples by NGS showed no amplification (NPA = 100%) (Table 5). Beyond the initial in vitro validation of the two targeted RT-qPCR assays, we evaluated assay performance on an ongoing basis. Both assays continued to demonstrate 100% PPA/NPA, as confirmed by randomly selected samples for NGS characterization.

Of the diverse variants previously identified by NGS, 50 samples were selected for the initial validation. Delta variant samples (n = 24) include B.1.617.2 (9), AY.2 (1), AY.3 (3), AY.3.1 (2), AY.4 (1), AY.5 (1), AY.12 (1), AY.24 (1), and AY.25 (5). Non-Delta samples (n = 26) comprise B.1 (1), B.1.1.7 (Alpha) (6), B.1.427 (2), B.1.429 (2), B.1.526 (3), B.1.621 (Mu) (2), B.1.621.1 (Mu) (2), B.1.623 (1), B.1.628 (1), C.37 (Lambda) (3), and P.1 (Gamma) (3). The NTD156-7 RT-qPCR assay correctly identified 24/24 Delta variant samples (positive percent agreement (PPA) = 100%) and showed np amplification (na) for 26/26 non-Delta variant samples of different lineages (negative percent agreement (NPA) = 100%). Observed CT values for Delta variant samples ranged from 20.19 to 38.6.

Table 4. In vitro Validation of NTD156-7 Assay.

Variant (NGS)	*n*	NTD156-7 Result		Mean Ct	Percent Agreement
		Positive	Negative		
B.1.617.2 (Delta)	9	9	0	28.60	100%
AY.2 (Delta)	1	1	0	25.20	100%
AY.3 (Delta)	3	3	0	28.90	100%
AY.3.1 (Delta)	2	2	0	31.05	100%
AY.4 (Delta)	1	1	0	26.94	100%
AY.5 (Delta)	1	1	0	37.73	100%
AY.12 (Delta)	1	1	0	26.77	100%
AY.24 (Delta)	1	1	0	20.19	100%
AY.25 (Delta)	5	5	0	27.86	100%
B.1	1	0	1	na	100%
B.1.1.7 (Alpha)	6	0	6	na	100%
B.1.427 (Epsilon)	2	0	2	na	100%
B.1.429 (Epsilon)	2	0	2	na	100%
B.1.526 (Iota)	3	0	3	na	100%
B.1.621 (Mu)	2	0	2	na	100%
B.1.621.1 (Mu)	2	0	2	na	100%
B.1.623	1	0	1	na	100%
B.1.628	1	0	1	na	100%
C.37 (Lambda)	3	0	3	na	100%
P.1 (Gamma)	3	0	3	na	100%
Total	50	24	26		

Table 5. In vitro Validation of the NTD25-7 Assay.

Variant (NGS)	*n*	NTD25-7 Results		Mean Ct	Percent Agreement
		Positive	Negative		
BA.2 (Omicron)	51	51	0	27.2	100%
BA.2.3	2	2	0	20.1	100%
BA.1 (Omicron)	9	0	9	na	100%
BA.1.1 (Omicron)	17	0	17	na	100%
AY.3 (Delta)	2	0	2	na	100%
AY.47 (Delta)	1	0	1	na	100%
Total	82	53	29		

The BA.2-specific RT-qPCR assay correctly identified 53/53 BA.2 (n = 51) and BA.2.3 (n = 2) variants (PPA = 100%) and showed no amplification (na) for 29/29 (NPA = 100%) non-BA.2 lineages (BA.1 (9), BA.1.1 (17), AY.3 (2), and AY.47 (1)).

3.5. SARS-CoV-2 VOC Surveillance Workflow

The current workflow for SARS-CoV-2 variant identification and surveillance (Figure 4) can effectively detect Omicron BA.1, BA.2, and BA.4/BA.5 lineages and the Delta variant by RT-qPCR methods. The initial RT-qPCR assay (TaqPath) serves as a quality assurance step to confirm positivity and assess sample integrity. In addition, this step allows us to identify samples with N gene target Ct values < 30 as a criterion for NGS. In the process of variant identification, the first RT-qPCR assay (TaqPath) serves to identify SGTF/SGTL samples associated with the H69-V70 deletion present in the Omicron BA.1, BA.4, and BA.5 lineages. For this assay, our N gene target Ct cutoff is 35, with similar amplification

(<2 Ct difference) for the ORF1ab and S gene targets in samples showing S gene target amplification (SGTA) and the same rule applies for the ORF1ab gene target in SGTF/SGTL samples. Next, samples (SGTF/SGTL and SGTA) were screened with the NTD25-7 RT-qPCR assay. The Ct value cutoff for this and the NTD156-7 assay was <38, as we observed variant confirmation by NGS within this Ct range. Samples positive with the NTD25-7 assay that also exhibit SGTF/SGTL are presumed BA.4 or BA.5 lineages, with lineage confirmation established by NGS. The SGTF/SGTL samples found to be negative with the NTD25-7 assay are presumed BA.1 lineage. Samples exhibiting SGTA and positive with the NTD25-7 assay were presumed BA.2 lineage. Non-BA.2 SGTA samples were subsequently screened for the Delta variant by targeted RT-qPCR (NTD156-7). If negative, these samples were sequenced for lineage identification. Figure 5 provides a snapshot of how we integrated NGS into variant prevalence monitoring by PCR for the first 13 weeks of 2022, when Omicron BA.1 was outcompeted and eventually replaced by the BA.2 lineage. Over this period, we performed 976 RT-qPCR reactions, typically with one or two runs per week and selected a subset of samples (n = 169) for periodic NGS confirmation. The results from NGS analysis were in complete concordance with our RT-qPCR conclusions and provided ongoing validation of our targeted RT-qPCR methods. For the five sequencing timepoints illustrated in Figure 5, concordance between NGS and RT-qPCR results was 100%.

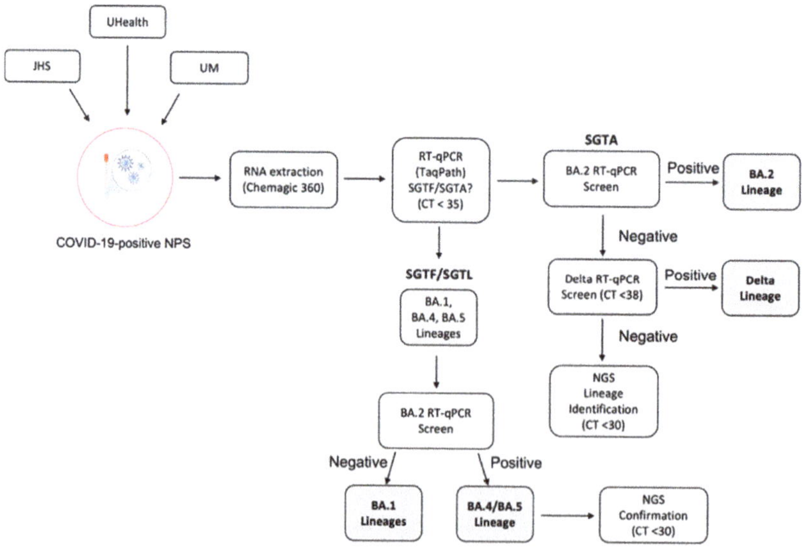

Figure 4. SARS-CoV-2 VOC Surveillance Workflow. In the workflow depicted, nasopharyngeal swab (NPS) samples were obtained from Jackson Health System (JHS), University of Miami Health System (UHealth), and University of Miami (UM) for variant identification. The first RT-qPCR assay (TaqPath) serves to identify SGTF/SGTL associated with the H69-V70 deletion present in the Omicron BA.1, BA.4, and BA.5 lineages. For this assay, our N gene target Ct cutoff was 35, with similar amplification (<2 Ct difference) for the ORF1ab and S gene targets in samples showing S gene target amplification (SGTA) and the same rule applied for the ORF1ab gene target in SGTF/SGTL samples. Next, samples (SGTF/SGTL and SGTA) were screened with the NTD25-7 RT-qPCR assay, for this and the Delta assay we were able to increase the Ct cutoff to 38 as we found reliable and reproducible results within this Ct range. Samples positive with the NTD25-7 assay, which also exhibit SGTF/SGTL, were presumed BA.4 or BA.5 lineages and were subsequently sequenced for lineage confirmation. SGTF/SGTL samples negative with the NTD25-7 assay are presumed BA.1 lineage. SGTA samples positive with the NTD25-7 assay were presumed BA.2 lineage. BA.2-negative, SGTA samples were subsequently screened for the Delta variant by targeted RT-qPCR (NTD156-7). If negative for Delta, these samples were sequenced for lineage identification.

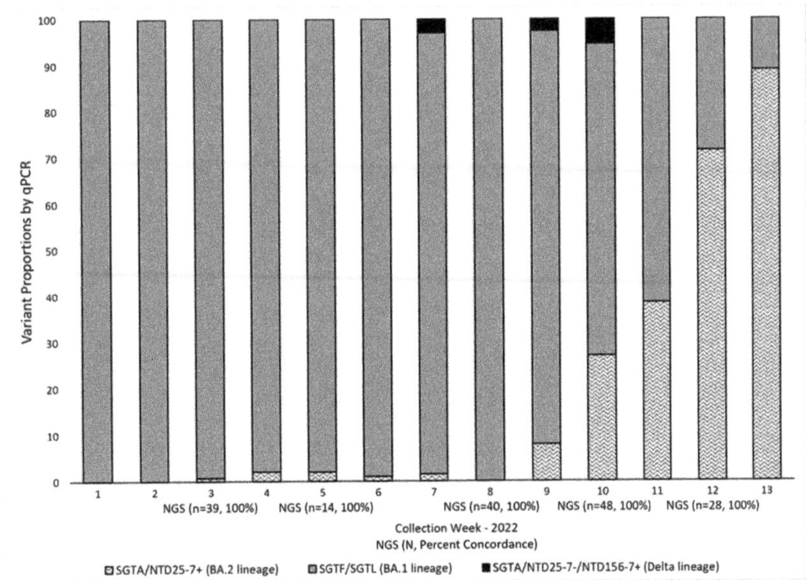

Figure 5. SARS-COV-2 Variant Proportions by RT-qPCR and NGS. Variant proportions were determined by RT-qPCR and NGS in residual COVID-19-positive samples collected from week 1 (1/3–1/9) to week 13 (3/23–4/1), 2022. Variant surveillance by RT-qPCR (n = 976) was performed weekly using the TaqPath assay in combination with internally designed and validated targeted RT-qPCR assays (NTD156-7 and NTD25-7). Subsets of these samples (n = 169) (Ct < 30) were sequenced on weeks 3 (n = 39), 5 (n = 14), 8 (n = 40), 10 (n = 48), and 12 (n = 28). Periodic confirmation by NGS of results obtained by RT-qPCR screenings provided confidence in the ongoing variant surveillance by targeted RT-qPCR methods. For the five sequencing timepoints illustrated, concordance between NGS and RT-qPCR results was 100%. SGTA (S-gene target amplification detected with the TaqPath assay); NTD25-7 (RT-qPCR assay targeting the N-terminal-domain deletion corresponding to amino acids 25–27 of the spike (S) protein in Omicron BA.2 lineages); SGTF/SGTL (S-gene target failure/S-gene target late amplification (TaqPath) corresponding to NTD deletion 69–70 of the S protein in Omicron BA.1 lineages); NTD156-7 (RT-qPCR assay targeting the NTD deletion corresponding to amino acids 156–157 of the S protein in the Delta lineage).

4. Discussion

The targeted RT-qPCR assays presented here for detection and surveillance of VOCs Delta and Omicron in combination with the TaqPath assay for identification of Omicron lineages harboring the H69-V70 deletion, resulting in the characteristic SGTF/SGTL signature, provide a simple and cost-effective alternative for near real-time variant identification, which has important clinical implications. While NGS remains the gold standard for genomic characterization of SARS-CoV-2 variants, the widely used ARTIC SARS-CoV-2 sequencing method has inherent limitations including high failure rates for samples with Ct values > 30, which can dampen variant surveillance efforts by excluding samples with lower viral loads from genomic characterization. Another limitation associated with this sequencing approach is primer/variant overlap in lineages with heavily mutated S genes, as is the case with the Omicron lineages. Primer/variant overlap results in amplification failure and subsequent overall decrease in NGS coverage. In addition to these limitations, NGS is cost- and labor-intensive, not readily available in many regions, and is associated with slower workflows compared to qPCR-based methods. It is important to note that the PCR methods proposed here do not exclude NGS as we recognize its tremendous value in providing ongoing confirmation that our qPCR assays are accurately reflecting local variant prevalence. Given the dynamic nature of SARS-CoV-2 and its ongoing genetic evolution, it is important to perform periodic method quality checks by randomly selecting samples screened by qPCR to be confirmed by sequencing. Sequencing methods are able to detect changes in areas that are distant and distinct genomic locations from the primer/probe binding sites. This underscores the importance of periodic random sequencing of samples. This combined approach significantly reduces the cost of variant surveillance while providing ongoing validation of targeted qPCR methods.

From January 2021 to February 2022, we performed RT-qPCR-based variant screening of over 5000 COVID-19 positive samples, of which 1668 were genetically characterized by NGS. This two-pronged approach allowed for cost-effective utilization of sequencing resources and continuous tracking of the prevalence and dynamics of locally circulating SARS-COV-2 variants. Ultimately, our efforts helped inform local public and private health systems, and University of Miami leaders' decisions in their efforts to ameliorate the spread of COVID-19 in the Miami-Dade County community and allocate resources including staffing in preparation for hospitalization surges. Variant monitoring also allowed our antimicrobial stewardship and infection prevention programs to target appropriate isolation policies and therapeutics. As some monoclonals became ineffective, we were able to pivot our hospital and outpatient formularies to provide appropriate therapies based on risk factors and therapeutic effectiveness. In addition to monitoring variant prevalence and dynamics by qPCR and NGS, we also followed national and Miami-Dade County variant prevalence data available from the CDC COVID Data Tracker. The VOC prevalence observed in our cohort mirrored closely the CDC Variant Proportions reports. Our combined qPCR and NGS approach provided information that was directly relevant to our surveillance population and was available in near real time. Epidemiologically, continuous monitoring of the prevalence and dynamics of circulating variants can better inform local health leaders' decisions to adopt preventive measures to control community transmission of COVID-19 infections.

Additionally, by incorporating tracking of variant dynamics by RT-qPCR-based methods, near real-time changes in variant prevalence can be determined, with subsequent confirmation by NGS. The continued genetic evolution of SARS-CoV-2 and the rapid emergence and rise to dominance of new VOCs with different mutational landscapes underscores the value of an equally dynamic surveillance approach. For this reason, revision of variant tracking workflows is necessary to account for these genetic variations. New targeted qPCR assays may need to be developed and implemented as part of the workflow as novel variant-defining mutations emerge. Surveillance accuracy will be enhanced by incorporating several targeted qPCR assays in communities where multiple dominant variants circulate. Despite these limitations, in addition to providing a rapid, cost-effective alternative for tracking the prevalence of SARS-CoV-2 variants, RT-qPCR-based surveillance strategies permit better utilization of NGS resources by avoiding sequencing redundancy (and associated investment of resources) in an environment where a dominant variant prevails in the population. In addition, PCR approaches provide a feasible option for variant surveillance in communities where NGS is not readily available. Importantly, rapid near-real-time emergence of known variants not yet circulating can be accomplished more easily with PCR approaches in conjunction with sequencing support. Confidence in calling variant progression by PCR is made possible by regular NGS analysis of subsets of samples. Future studies should address the balance between qPCR surveillance methods and parallel NGS analysis for optimal accuracy of variant calling. Therefore, RT-qPCR serves as complementary methods to allow active, accurate identification and surveillance of changes in variant prevalence over time that can impact public health and improve direct patient care. The ability to detect changes in VOC prevalence in our surveillance population reinforced messaging to the community regarding the importance of vaccination uptake and vigilance. VOC prevalence (especially BA.2) informed strategic decisions regarding anti-SARS-CoV-2 monoclonal antibody therapies (e.g., sotrovimab) to include in hospital formularies. By having active VOC prevalence available, health system leaders were able to make critical decisions with greater confidence and in a timelier manner.

Author Contributions: Conceptualization, D.M.A. and E.T.; methodology, Y.L.C., A.G., S.W., R.V., K.P., C.S. and H.R.G.; software, A.G.; validation, Y.L.C. and A.G.; formal analysis, Y.L.C. and A.G.; investigation, Y.L.C. and R.V.; resources, S.W., Y.Z., B.S., L.M.A., M.J., S.D.N., R.E.W., T.F. and P.G.P.; data curation, Y.L.C.; writing—original draft preparation, Y.L.C.; writing—review and editing, Y.L.C., E.T., D.M.A. and R.V.; visualization, Y.L.C.; supervision, D.M.A. and E.T.; project administration, D.M.A. and A.M.; funding acquisition, M.J. and S.D.N. All authors have read and agreed to the published version of the manuscript.

Funding: This research received no external funding.

Institutional Review Board Statement: Samples used in this study were permanently de-identified and unlinked from subjects and were used for surveillance purposes only. Consequently, study samples were considered by the Institutional Review Board (IRB) to represent non-human studies research material and did not require IRB approval.

Informed Consent Statement: Not applicable.

Acknowledgments: The Miller School of Medicine Department of Pathology and Laboratory Medicine is gratefully acknowledged for support as well as the Jackson Health System laboratory services team for sample acquisition. The Oncogenomics Sequencing Resource team is gratefully acknowledged for engagement throughout.

Conflicts of Interest: All authors declare no conflict of interest.

References

1. O'Toole, Á.; Scher, E.; Underwood, A.; Jackson, B.; Hill, V.; McCrone, J.T.; Colquhoun, R.; Ruis, C.; Abu-Dahab, K.; Taylor, B.; et al. Assignment of epidemiological lineages in an emerging pandemic using the pangolin tool. *Virus Evol.* **2021**, *7*, veab064. [CrossRef] [PubMed]
2. WHO. World Health Organization (WHO) Tracking SARS-CoV-2 Variants. Available online: https://www.who.int/en/activities/tracking-SARS-CoV-2-variants/ (accessed on 26 February 2021).
3. CDC. SARS-CoV-2 Variant Classifications and Definitions. Available online: https://www.cdc.gov/coronavirus/2019-ncov/variants/variant-classifications.html (accessed on 26 February 2021).
4. ECDC. SARS-CoV-2 Variants of Concern as of 24 February 2022. Available online: https://www.ecdc.europa.eu/en/covid-19/variants-concern (accessed on 26 February 2022).
5. Plante, J.A.; Liu, Y.; Liu, J.; Xia, H.; Johnson, B.A.; Lokugamage, K.G.; Zhang, X.; Muruato, A.E.; Zou, J.; Fontes-Garfias, C.R.; et al. Spike mutation D614G alters SARS-CoV-2 fitness. *Nature* **2021**, *592*, 116–121. [CrossRef] [PubMed]
6. Volz, E.; Hill, V.; McCrone, J.T.; Price, A.; Jorgensen, D.; O'Toole, Á.; Southgate, J.; Johnson, R.; Jackson, B.; Nascimento, F.F.; et al. Evaluating the Effects of SARS-CoV-2 Spike Mutation D614G on Transmissibility and Pathogenicity. *Cell* **2021**, *184*, 64–75.e11. [CrossRef] [PubMed]
7. Greaney, A.J.; Loes, A.N.; Crawford, K.H.D.; Starr, T.N.; Malone, K.D.; Chu, H.Y.; Bloom, J.D. Comprehensive mapping of mutations in the SARS-CoV-2 receptor-binding domain that affect recognition by polyclonal human plasma antibodies. *Cell Host Microbe* **2021**, *29*, 463–476. [CrossRef] [PubMed]
8. Cherian, S.; Potdar, V.; Jadhav, S.; Yadav, P.; Gupta, N.; Das, M.; Rakshit, P.; Singh, S.; Abraham, P.; Panda, S.; et al. SARS-CoV-2 Spike Mutations, L452R, T478K, E484Q and P681R, in the Second Wave of COVID-19 in Maharashtra, India. *Microorganisms* **2021**, *9*, 1542. [CrossRef] [PubMed]
9. Deng, X.; Garcia-Knight, M.A.; Khalid, M.M.; Servellita, V.; Wang, C.; Morris, M.K.; Sotomayor-González, A.; Glasner, D.R.; Reyes, K.R.; Gliwa, A.S.; et al. Transmission, infectivity, and neutralization of a spike L452R SARS-CoV-2 variant. *Cell* **2021**, *184*, 3426–3437. [CrossRef] [PubMed]
10. WHO. Tracking SARS-CoV-2 Variants. Available online: https://www.who.int/activities/tracking-SARS-CoV-2-variants (accessed on 22 August 2021).
11. Liu, Y.; Rocklöv, J. The reproductive number of the Delta variant of SARS-CoV-2 is far higher compared to the ancestral SARS-CoV-2 virus. *J. Travel Med.* **2021**, *28*, taab124. [CrossRef]
12. Liu, Y.; Liu, J.; Johnson, B.A.; Xia, H.; Ku, Z.; Schindewolf, C.; Widen, S.G.; An, Z.; Weaver, S.C.; Menachery, V.D.; et al. Delta spike P681R mutation enhances SARS-CoV-2 fitness over Alpha variant. *bioRxiv* **2021**. [CrossRef]
13. Lopez Bernal, J.; Andrews, N.; Gower, C.; Gallagher, E.; Simmons, R.; Thelwall, S.; Stowe, J.; Tessier, E.; Groves, N.; Dabrera, G.; et al. Effectiveness of Covid-19 Vaccines against the B.1.617.2 (Delta) Variant. *N. Engl. J. Med.* **2021**, *385*, 585–594. [CrossRef]
14. Planas, D.; Veyer, D.; Baidaliuk, A.; Staropoli, I.; Guivel-Benhassine, F.; Rajah, M.M.; Planchais, C.; Porrot, F.; Robillard, N.; Puech, J.; et al. Reduced sensitivity of SARS-CoV-2 variant Delta to antibody neutralization. *Nature* **2021**, *596*, 276–280. [CrossRef]

15. Bolze, A.; Cirulli, E.T.; Luo, S.; White, S.; Wyman, D.; Rossi, A.D.; Machado, H.; Cassens, T.; Jacobs, S.; Schiabor Barrett, K.M.; et al. SARS-CoV-2 variant Delta rapidly displaced variant Alpha in the United States and led to higher viral loads. *medRxiv* **2021**. [CrossRef]
16. Viana, R.; Moyo, S.; Amoako, D.G.; Tegally, H.; Scheepers, C.; Althaus, C.L.; Anyaneji, U.J.; Bester, P.A.; Boni, M.F.; Chand, M.; et al. Rapid epidemic expansion of the SARS-CoV-2 Omicron variant in southern Africa. *Nature* **2022**, *603*, 679–686. [CrossRef]
17. Liu, L.; Iketani, S.; Guo, Y.; Chan, J.F.; Wang, M.; Luo, Y.; Chu, H.; Huang, Y.; Nair, M.S.; Yu, J.; et al. Striking antibody evasion manifested by the Omicron variant of SARS-CoV-2. *Nature* **2022**, *602*, 676–681. [CrossRef] [PubMed]
18. Liu, Y.; Rocklöv, J. The effective reproduction number for the omicron SARS-CoV-2 variant of concern is several times higher than Delta. *J. Travel Med.* **2022**, *29*, taac037. [CrossRef] [PubMed]
19. Aggarwal, A.; Stella, A.O.; Walker, G.; Akerman, A.; Milogiannakis, V.; Brilot, F.; Amatayakul-Chantler, S.; Roth, N.; Coppola, G.; Schofield, P.; et al. SARS-CoV-2 Omicron: Evasion of potent humoral responses and resistance to clinical immunotherapeutics relative to viral variants of concern. *medRxiv* **2021**. [CrossRef]
20. Cao, Y.; Wang, J.; Jian, F.; Xiao, T.; Song, W.; Yisimayi, A.; Huang, W.; Li, Q.; Wang, P.; An, R.; et al. Omicron escapes the majority of existing SARS-CoV-2 neutralizing antibodies. *Nature* **2022**, *602*, 657–663. [CrossRef]
21. Planas, D.; Saunders, N.; Maes, P.; Guivel-Benhassine, F.; Planchais, C.; Buchrieser, J.; Bolland, W.H.; Porrot, F.; Staropoli, I.; Lemoine, F.; et al. Considerable escape of SARS-CoV-2 Omicron to antibody neutralization. *Nature* **2022**, *602*, 671–675. [CrossRef]
22. Iketani, S.; Liu, L.; Guo, Y.; Chan, J.F.; Huang, Y.; Wang, M.; Luo, Y.; Yu, J.; Chu, H.; Chik, K.K.; et al. Antibody evasion properties of SARS-CoV-2 Omicron sublineages. *Nature* **2022**, *604*, 553–556. [CrossRef] [PubMed]
23. Kuhlmann, C.; Mayer, C.K.; Claassen, M.; Maponga, T.; Burgers, W.A.; Keeton, R.; Riou, C.; Sutherland, A.D.; Suliman, T.; Shaw, M.L.; et al. Breakthrough infections with SARS-CoV-2 omicron despite mRNA vaccine booster dose. *Lancet* **2022**, *399*, 625–626. [CrossRef] [PubMed]
24. CDC. COVID Data Tracker. Available online: https://covid.cdc.gov/covid-data-tracker/#variant-proportions (accessed on 22 August 2021).
25. Nasir, J.A.; Kozak, R.A.; Aftanas, P.; Raphenya, A.R.; Smith, K.M.; Maguire, F.; Maan, H.; Alruwaili, M.; Banerjee, A.; Mbareche, H.; et al. A Comparison of Whole Genome Sequencing of SARS-CoV-2 Using Amplicon-Based Sequencing, Random Hexamers, and Bait Capture. *Viruses* **2020**, *12*, 895. [CrossRef]
26. Baker, D.J.; Aydin, A.; Le-Viet, T.; Kay, G.L.; Rudder, S.; de Oliveira Martins, L.; Tedim, A.P.; Kolyva, A.; Diaz, M.; Alikhan, N.F.; et al. CoronaHiT: High-throughput sequencing of SARS-CoV-2 genomes. *Genome Med.* **2021**, *13*, 21. [CrossRef]
27. Tyson, J.R.; James, P.; Stoddart, D.; Sparks, N.; Wickenhagen, A.; Hall, G.; Choi, J.H.; Lapointe, H.; Kamelian, K.; Smith, A.D.; et al. Improvements to the ARTIC multiplex PCR method for SARS-CoV-2 genome sequencing using nanopore. *bioRxiv* **2020**. [CrossRef]
28. GISAID. SARS-CoV-2 Lineage Variant Summary: Primer Monitor Tool. Available online: https://primer-monitor.neb.com/lineages (accessed on 3 June 2021).
29. Elbe, S.; Buckland-Merrett, G. Data, disease and diplomacy: GISAID's innovative contribution to global health. *Glob. Chall.* **2017**, *1*, 33–46. [CrossRef]
30. Shu, Y.; McCauley, J. GISAID: Global initiative on sharing all influenza data—From vision to reality. *Euro Surveill.* **2017**, *22*, 30494. [CrossRef]
31. Thomas, E.; Delabat, S.; Carattini, Y.L.; Andrews, D.M. SARS-CoV-2 and Variant Diagnostic Testing Approaches in the United States. *Viruses* **2021**, *13*, 2492. [CrossRef] [PubMed]
32. Thomas, E.; Delabat, S.; Andrews, D.M. Diagnostic Testing for SARS-CoV-2 Infection. *Curr. Hepatol. Rep.* **2021**, *20*, 166–174. [CrossRef] [PubMed]
33. Bal, A.; Destras, G.; Gaymard, A.; Stefic, K.; Marlet, J.; Eymieux, S.; Regue, H.; Semanas, Q.; d'Aubarede, C.; Billaud, G.; et al. Two-step strategy for the identification of SARS-CoV-2 variant of concern 202012/01 and other variants with spike deletion H69-V70, France, August to December 2020. *Euro Surveill.* **2021**, *26*, 2100008. [CrossRef]
34. Borges, V.; Sousa, C.; Menezes, L.; Gonçalves, A.M.; Picão, M.; Almeida, J.P.; Vieita, M.; Santos, R.; Silva, A.R.; Costa, M.; et al. Tracking SARS-CoV-2 lineage B.1.1.7 dissemination: Insights from nationwide spike gene target failure (SGTF) and spike gene late detection (SGTL) data, Portugal, week 49 2020 to week 3 2021. *Euro Surveill.* **2021**, *26*, 2100130. [CrossRef] [PubMed]
35. Vogels, C.B.F.; Breban, M.I.; Ott, I.M.; Alpert, T.; Petrone, M.E.; Watkins, A.E.; Kalinich, C.C.; Earnest, R.; Rothman, J.E.; Goes de Jesus, J.; et al. Multiplex qPCR discriminates variants of concern to enhance global surveillance of SARS-CoV-2. *PLoS Biol.* **2021**, *19*, e3001236. [CrossRef]
36. Bedotto, M.; Fournier, P.E.; Houhamdi, L.; Levasseur, A.; Delerce, J.; Pinault, L.; Padane, A.; Chamieh, A.; Tissot-Dupont, H.; Brouqui, P.; et al. Implementation of an in-house real-time reverse transcription-PCR assay for the rapid detection of the SARS-CoV-2 Marseille-4 variant. *J. Clin. Virol.* **2021**, *139*, 104814. [CrossRef]
37. Peterson, S.W.; Lidder, R.; Daigle, J.; Wonitowy, Q.; Dueck, C.; Nagasawa, A.; Mulvey, M.R.; Mangat, C.S. RT-qPCR detection of SARS-CoV-2 mutations S 69-70 del, S N501Y and N D3L associated with variants of concern in Canadian wastewater samples. *Sci. Total Environ.* **2022**, *810*, 151283. [CrossRef]
38. Gand, M.; Vanneste, K.; Thomas, I.; Van Gucht, S.; Capron, A.; Herman, P.; Roosens, N.H.C.; De Keersmaecker, S.C.J. Deepening of In Silico Evaluation of SARS-CoV-2 Detection RT-qPCR Assays in the Context of New Variants. *Genes* **2021**, *12*, 565. [CrossRef]

39. Yaniv, K.; Ozer, E.; Shagan, M.; Lakkakula, S.; Plotkin, N.; Bhandarkar, N.S.; Kushmaro, A. Direct RT-qPCR assay for SARS-CoV-2 variants of concern (Alpha, B.1.1.7 and Beta, B.1.351) detection and quantification in wastewater. *Environ. Res.* **2021**, *201*, 111653. [CrossRef] [PubMed]
40. Utah DoH ARTIC/Illumina Bioinformatic Workflow. Available online: https://github.com/CDCgov/SARS-CoV-2_Sequencing/tree/master/protocols/BFX-UT_ARTIC_Illumina (accessed on 29 January 2021).
41. Aksamentov, I.; Roemer, C.; Hodcroft, E.B.; Neher, R.A. Nextclade: Clade assignment, mutation calling and quality control for viral genomes. *J. Open Source Softw.* **2021**, *6*, 3773. [CrossRef]
42. Dutta, N.K.; Mazumdar, K.; Gordy, J.T. The Nucleocapsid Protein of SARS-CoV-2: A Target for Vaccine Development. *J. Virol.* **2020**, *94*, e00647-20. [CrossRef]
43. Rahman, M.S.; Islam, M.R.; Alam, A.S.M.R.; Islam, I.; Hoque, M.N.; Akter, S.; Rahaman, M.M.; Sultana, M.; Hossain, M.A. Evolutionary dynamics of SARS-CoV-2 nucleocapsid protein and its consequences. *J. Med. Virol.* **2021**, *93*, 2177–2195. [CrossRef]
44. Kannan, S.R.; Spratt, A.N.; Cohen, A.R.; Naqvi, S.H.; Chand, H.S.; Quinn, T.P.; Lorson, C.L.; Byrareddy, S.N.; Singh, K. Evolutionary analysis of the Delta and Delta Plus variants of the SARS-CoV-2 viruses. *J. Autoimmun.* **2021**, *124*, 102715. [CrossRef] [PubMed]

MDPI
St. Alban-Anlage 66
4052 Basel
Switzerland
www.mdpi.com

Viruses Editorial Office
E-mail: viruses@mdpi.com
www.mdpi.com/journal/viruses

www.ingramcontent.com/pod-product-compliance
Lightning Source LLC
Chambersburg PA
CBHW061559220926
51850CB00005B/375